# THE BIG BOOK OF ASTRONAUTS & COSMONAUTS

Graeme K. Tyree

ISBN 978-0-473-71311-9. Paperback. First edition, June 2024.
ISBN 978-0-473-71313-3. Apple Books.
ISBN 978-0-473-71312-6. Epub.

"We are all of us in the gutter,
but some of us are looking at the stars"

Oscar Wilde (1854-1900)

Images, previous page: Yuri A. Gagarin *(top left)*; Alan B. Shepard, Jr. *(top right)*, Valentina V. Tereshkova *(bottom left)*, and Dr. Sally K. Ride *(bottom right)*.

# Contents

# Preface

An astronaut (***astron***, from the Ancient Greek, meaning 'star', and ***nautes***, meaning 'sailor') is a person trained, equipped, and deployed by a human spaceflight program to serve as a commander or crew member aboard a spacecraft. Although generally reserved for professional space travelers, the term is sometimes applied to anyone who travels into space; including scientists, politicians, journalists, and "space participant" tourists, who, beginning in 2009, have flown to the International Space Station. "Astronaut" technically applies to all human space travelers regardless of nationality. However, astronauts fielded by Russia or the former Soviet Union are typically known as cosmonauts, an Anglicization of kosmonavt (from the Russian *kosmos*, meaning 'space'). On May 30, 2020, Doug Hurley and Robert "Bob" Behnken became the first astronauts to launch on a private crewed spacecraft, the ***SpaceX*** Crew Dragon.

Recent developments in crewed spaceflight made by China have led to the rise of the term taikonaut (from the Mandarin ***taikong***, meaning 'space'), although its use is somewhat informal. In China, the People's Liberation Army Astronaut Corps and their foreign counterparts are all officially called *hangtian yuan*, meaning *"navigating celestial-heaven personnel"* or literally "heaven-sailing staff". Until 2002, astronauts were sponsored and trained exclusively by governments, either by the military or civilian space agencies. With the suborbital flight of the ***Virgin Galactic*** privately funded *SpaceShipOne* in 2004, a new category of astronaut was created: the commercial astronaut. One who flies to the International Space Station (ISS) as part of a "privately funded, dedicated commercial spaceflight on a commercial launch vehicle dedicated to the mission ... to conduct approved commercial and marketing activities on the ISS (or in a commercial segment attached to the Station)" is considered by NASA to be a *private astronaut*.

The Federation Aeronautique Internationale (FAI) Sporting Code for astronautics recognizes only spaceflights that exceed the Karman line, at an altitude of 100 kilometres (62 miles). In the United States, professional, military, and commercial astronauts who have traveled above an altitude of eighty kilometres (fifty miles) are awarded astronaut wings. On July 20, 2021, in response to the private suborbital flights of Jeff Bezos's ***Blue Origin*** and Richard Branson's ***Virgin Galactic***, the FAA issued an order redefining the eligibility criteria to be an astronaut. The new criteria states that one must have *"demonstrated activities during flight that were essential to public safety, or contributed to human space flight safety"*. This new definition excludes the activities of the Bezos and Branson space tourism flights.

Most astronauts are very highly qualified people. While the original "Mercury Seven" NASA astronauts were chosen from a group of Air Force test pilots, astronauts of the 1980s often were B.Sc. qualified in aeronautics and astronautics, or as electrical or mechanical engineers, while in the 21st century they are more likely to have a Ph.D. in physics. Graduating from the USAF Test Pilot School is still a feature of many astronaut Resumes. Many have a distinguished previous career in the military, and all require a high level of mental and physical fitness, initiative and resilience.

In President John F. Kennedy's historic speech at Rice University, Texas, on September 12, 1962, he said, *"The exploration of space will go ahead, whether we join in it or not, and it is one of the great adventures of all time, and no nation which expects to be the leader of other nations can expect to stay behind in the race for space. ... We mean to be a part of it - we mean to lead it. ... In short, our leadership in science and in industry, our hopes for peace and security, our obligations to ourselves as well as others, all require us to make this effort, to solve these mysteries, to solve them for the good of all men, and to become the world's leading space-faring nation. ... We set sail on this new sea because there is new knowledge to be gained, and new rights to be won, and they must be won and used for the progress of all people. ... There is no strife, no national conflict in outer space as yet. ... Its hazards are hostile to us all. Its conquest deserves the best for all mankind, and its opportunity for peaceful co-operation may never come again. ... We choose to go to the Moon.* ***We choose to go to the Moon in this decade and do the other things, not because they are easy, but because they are hard****".* Unquote.

The golden age of crewed human spaceflight may now be over. In 2024, it is far safer and more economical to send autonomous robots to explore asteroids and other planets. All they need is a power source and a propulsion system. While there are immediate plans for humans to return to the Moon, and possibly land on Mars by 2030, major technological challenges remain to be overcome for long-duration missions to succeed in deep space. This work you hold is to detail the background, training, and flight experience of the now over six hundred men and women who have put their lives on the line as mankind ventured into the heavens. To make it personal, the author has included material on their post-mission life, and honors received. The thirty years of Space Shuttle missions showed that the survivability rate of flying to space is 97% at best. Work on the International Space Station has contributed to better understanding of the physiological challenges of living in microgravity. It certainly takes a lot of the right stuff to launch on top of a barely controled chemical explosion, and later re-enter the Earth's atmosphere traveling at Mach 32!

The space programs of the Soviet Union and of the United States of America were sharply contrasted in their first decades. While the Soviets were keen to demonstrate the technical capabilities of their interplanetary probes and make propaganda mileage with their pioneering human spaceflights, their hardware was fully automated and missions were controlled from the ground. American astronauts were fond of the derisory phrase *"spam in a can"*. Astronauts of America flying the Mercury and Gemini series were insistent that their test pilot skills should be put to good use, and so NASA reluctantly agreed to manual controls being fitted, allowing the spacecraft to change trajectory and make re-entry more likely to succeed if automated systems should fail. At first, the Russians had the more powerful rocket boosters, and could lift larger payloads into Earth orbit. The United States had a series of embarrasing rocket failures, and national pride was dented. On May 25, 1961, newly-elected President John F. Kennedy set the ambitious goal of making a successful Moon landing mission *"before the end of the decade, ... of landing a man on the Moon and returning him safely to Earth".*

The Space Race was underway, with the unspoken agenda to secure military dominance with rockets capable of landing intercontinental ballistic missiles on the ideological foes. The Cold War drove intense competition between the two nations, and not until the joint Apollo-Soyuz Test Project of mid-1975 did international relations begin to ease. The International Space Station has fostered collaboration between the major spacefaring nations for over two decades since 2000, but America is very wary of China's military ambition. A worrying trend is for the future militarization of space. In May 2022, Russian and China announced they have joined forces to plan for the ***International Lunar Research Station*** to be built at the Moon's south pole. China stated it intends to land taikonauts on the Moon before 2030. India, now the globe's most populous nation, has a burgeoning space program. Its *Chandrayaan-3* mission successfully touched down at the lunar south pole, and deployed the *Vikram* lander on August 27, 2023. This book is about the brave space travelers who made history as mankind first dared to explore the Universe.

Larry Niven, one of science fiction's foremost authors wrote, *"The dinosaurs became extinct because they didn't have a space program. And if we become extinct because we don't have a space program, it'll serve us right!"*. Many astronauts find that their space missions make them feel more connected to their earthly home. They go to space to support the international mission of science and exploration, yet while being there discover their connection to home, and their view of Earth as their home deepened and expanded. Nicole Stott wrote, *"When I was farthest from the planet, I found myself feeling more connected than I had ever felt before when I'd had my feet on the ground"*.

The number of people who have been to outer space depends upon the definition of the boundary of space. According to the **Federation Aeronautique Internationale** (FAI), this is at an altitude of one hundred kilometres (62 miles) above sea level; while the United States considers anyone who has flown above eighty kilometres altitude (50 miles) to be an astronaut. The brief sub-orbital "space tourist" flights of **Virgin Galactic** (beginning with the *Galactic 02* flight on June 09, 2021, reaching an altitude of 88 km) and **Blue Origin**'s flight 'NS-16' (reaching 104 km altitude) generally fall in between. With ticket prices from USD 250,000 to 350,000, these commercial operators provide their passengers with six minutes' experience in a weightless environment. Launched from a carrier aircraft at 50,000 feet, the **Virgin Galactic** *SpaceShipTwo* spaceplane separates, fires its' rocket engine, and the entire flight lasts about 90 minutes to touchdown.

The **Blue Origin** capsule is boosted by a reusable *New Shepard* rocket which lands vertically seven minutes after liftoff. On September 16, 2021, **SpaceX**'s 'Inspiration 4' mission using a *Dragon* capsule reached 1,368 km into the cosmos during a three-day orbital trip. Based on these criteria for the boundary of space, the number of space travelers varies: by the FAI definition, 643 persons had reached the altitude of space as of November 07, 2023. By the US definition, 676 persons had journeyed to space by the same date. By the stricter definition of **reaching Earth orbi**t, as of June 2023, 606 people had been to space. Of these, twenty-four men from the USA have gone beyond low-Earth orbit and traveled around or walked on the Moon.

These were all NASA astronauts on Apollo missions from late December 1968 to early December 1972, until Yusaku Maezawa, a Japanese billionaire was scheduled to fly around the Moon with eight other passengers on a **SpaceX *Starship*** mission in December 2023. Unfortunately, there have been three test flight booster explosions up to mid-January 2024! On April 20, 2023, the *Starship* rocket had a "rapid unscheduled disassembly", spinning out of control four minutes after liftoff, following a failed stage separation; on November 18, 2023, contact with the *Starship* upper stage was lost near the end of the burn, just before reaching orbit; on January 14, 2024, the *Starship* IFT-2 launch saw the *Super Heavy* booster (with 16.7 million pounds of thrust) explode eight minutes after launch, and *Ship 25* was lost.

The NASA *Orion* four-person capsule is in development, and due to circumnavigate the Moon in September 2025. The crew for the *Artemis II* mission are slated to be astronauts Commander Reid Wiseman, Victor J. Glover, Christina H. Koch (who holds the record for the longest single spaceflight by a woman, at 328 days), and Jeremy Hansen [a Canadian, of the European Space Agency (ESA)]. This mission will make history: with the first female lunar astronaut (Koch), and first person of color (Glover) to travel to the Moon.

PART ONE

# Operation Paperclip

This was a covert plan to bring Nazi scientists to America, originally to access nerve gas and V-2 missile technology so the United States could run a bombing campaign in Japan to end World War II. By mid-1945, sixty-seven Japanese cites had been firebombed, causing over 1,000,000 deaths, but the nation did not surrender until atomic weapons were dropped on Hiroshima and Nagasaki in early August 1945. **Operation Paperclip** was initiated to collect scientists and engineers who had been at the center of Nazi Germany's advanced military technologies, assessed to be twenty-five years ahead of US technology. In August 1945, one hundred and twenty-seven missile specialists led by Dr. Wernher von Braun signed work contracts with the United States Army Ordnance Corps. Most of them had worked under von Braun on the V-2 missile development at Peenemunde, on the Baltic coast. These specialists were sent to Fort Bliss, Texas, USA, joining the US Army's newly formed Research and Development Division, Sub-office (Rocket).

The German rocket scientists were imported to give the United States the strategic benefit of advanced designs and materials for missiles. For the next five years, Dr. von Braun and his team of scientists and engineers were primarily engaged in adapting and improving the V-2 missile for US applications. Testing was conducted at the nearby White Sands Proving Grounds, New Mexico. Von Braun was allowed to use a WAC Corporal rocket as a second stage for a V-2. The combination, called Bumper, reached a record-breaking 393 km (244 miles) altitude on February 24, 1949. Other talented German scientists had their documents processed and their war records either erased or **adjusted to make them acceptable for immigration**. The program later became a "denial program" where everything previously considered morally unacceptable was done to keep these men from working for the communist Soviet Union.

At the close of World War II in Europe, Supreme Headquarters Allied Expeditionary Forces (SHAEF, active from 1943-45) had set up internment centers where more than 1,500 scientists were being held separate from other German prisoners of war. The US Army had about 500 scientists in custody in Garmisch-Partenkirchen, in the Bavarian Alps, including the von Braun and Dornberger group; there were 444 persons of interest detained in Heidenheim, north of Munich; 200 were in Zell am See, in Austria; and 30 were being kept at Chateau du Grande Chesnay, France. The US Army Air Forces had about 150 Luftwaffe engineers and technicians in Bad Kissengen, Germany, most of whom had been rounded up by Colonel Donald Putt. General Dwight D. Eisenhower sought clarification on policy regarding the future of these men. He wrote, *"Restraint and control of future German scientific and technical investigations are clearly indicated, but this headquarters is without guidance on the matter and is in no position to formulate long-term policy"*. The US War Department let him know his query was considered a "matter of urgency".

Major General Kenneth B. Wolfe of the United States Army Air Forces took matters into his own hands. When he saw the vast numbers of V-weapons left behind, he became convinced that a US program to exploit Nazi science had to happen immediately. Upon his return to Washington, D.C., he said America needed to hire the German scientists and engineers and put them to work in America. *"If steps to this end are taken, the double purpose of preventing Germany's resurgence as a war power and and* ***advancing our***

***own industrial future*** *may be served"*. US Army Ordnance believed that the V-2 rocket could help win the Pacific war. With the help of Karl Fleischer, more than fourteen tons of crated V-2 plans and blueprints were recovered in the Nordhausen area of Germany from a sealed mine shaft.

John Dolibois was a young US Army officer and fluent German speaker who worked for Military Intelligence (G-2). He had interrogated the major war criminals of the Nazi party at the "Ashcan" internment facility in Luxembourg, Europe. Albert Einstein was the most esteemed figure to publicly denounce **Operation Paperclip**. One of Germany's most famous pre-war scientists, Einstein left Germany just months after Hitler took power, declaring that science and justice were now in the hands of *"a raw and rabid mob of Nazi militia"*. In an impassioned letter to President Truman, written on behalf of his colleagues, Einstein said, *"**We hold these individuals to be potentially dangerous.** ... Their former eminence as Nazi Party members and supporters raises the issue of their fitness to become American citizens and hold key positions in American industrial, scientific and educational institutions"*. Unquote. Einstein believed that anyone who served Hitler was unfit for United States citizenship.

When Friedrich Hoffmann arrived at the US's Edgewood Arsenal in February 1947, the facility had been producing war gases for thirty years (since World War One, in 1917). Colonel Loucks had been ordered by the US Chemical Corps to determine tabun gas on an industrial scale under a classified program code-named AI.13. US Army intelligence believed the Soviet's chemical weapons program to be considerably ahead, and the Chemical Corps was under pressure to catch up. Through a parallel exploitation program called Operation Osoaviakhim, the Soviets had captured the entire IG Farben laboratory at Dyhernfurth (and some of the team of Farben chemists, led by Dr. Von Bock) and re-assembled the laboratory at Beketovka, outside Stalingrad, under the code-name Chemical Works No. 91. They also began producing Sarin gas. Dr. von Braun had admitted to ***New Yorker*** writer Daniel Lang that what he cared about most was seeing *"how the golden cow could be milked most successfully [in America]"*, and the newly created Central Intelligence Agency (CIA) saw the **Operation Paperclip** scientists in similar quid pro quo terms. There was advantage to be had in using men who had everything to lose and were, at the same time, uniquely focused on personal gain. It was in the CIA that **Operation Paperclip** found its strongest supporting partner.

By the Spring of 1948, half of the 1,000 German scientists bound for America had arrived, but not one of them had a Visa. On May 11, 1948, military intelligence chief General Stephen J. Chamberlin took matters into his own hands. He went to meet Federal Bureau of Investigation (FBI) director J. Edgar Hoover to enlist his help. Cold War paranoia was on the rise, and both men were staunch anti-Communists. Chamberlin claimed the success of **Operation Paperclip** was *"essential to national security. The FBI now had the Communists to fear, not the Nazis"*. In June 1950, North Korean forces moved across the thirty eighth parallel, and the Korean War began. The idea that Communists were about to invade Western Europe panicked the US Pentagon. On July 14, the commander at Wright-Patterson Air Force Base sent an urgent memo to Operation Paperclip's in-house champion, Colonel Donald Putt: *"Due to the threat of impending hostilities in Europe and the possibility that forces of the USSR may rapidly overun the continent, this command is concerned with with the problem of the immediate implementation of an evacuation program for German and Austrian scientists"*.

Were these scientists to *"fall into enemy hands ... they would constitute a threat to our national security"*. US Air Force intelligence recommended a *"mass procurement effort"* in Germany. Dr. Karl Nordstrom was assigned to be the German liaison to a new Joint

Intelligence Objectives Agency (JIOA) program named **Accelerated Paperclip** to move *"especially dangerous top level scientists"* out of Germany in a *"modified Denial Program"* that needed them to be kept away from the Soviets at all costs. The US Joint Chiefs of Staff approved a USD one million dollar budget to help entice these *"especially dangerous top level scientists"* to come to America; the equivalent of about USD ten million in 2013. With the new policy in place, Class I offenders could now be put on a JIOA list. This list included Dr. Schreiber, still serving as physician at Camp King Post (a Cold War "Black site" using extreme interrogation techniques and behaviour modification programs as part of CIA operations) and Dr. Kurt Blome, former deputy Surgeon General of the Third Reich and Hitler's biological weapons expert, along with Dr. Otto Ambros, the war criminal convicted at the Nuremburg trial of slavery and mass murder.

In the winter of 1951, Otto Ambros was still incarcerated at Landsberg Prison. The Peck Panel had finished its review process and recommended *"substantial reductions of sentences"*. John J. McCloy's office drastically reduced the sentences of sixty-four out of seventy-four remaining war criminals, which meant that one-third of the inmates tried at Nuremburg were then freed. The political mood had now changed: Lt. Col. Harry R. Smith wrote, *"Based on available records,* ***Subjects were not war criminals*** *but undoubtedly were ardent Nazis"*. He also wrote that Dr. Blome was not likely to become a security threat. *"It is the opinion of the United States High Commissioner for Germany that they are not likely to become security threats to the United States"*. His report was signed and dated September 27, 1951.

**Major General Dr. Walter Schreiber** was the Surgeon General of the Third Reich. Five of his physician-colleagues who testified at the Nuremburg doctor's trial revealed that Schreiber had overseen a host of "racial hygiene"-related Reich medical programs in which large numbers of non-consenting humans were sacrificed in the name of scientific research. Included were details of experiments with yellow fever epidemic jaundice, sulfanilamide, euthanasia by phenol experiments, and the notorious typhus vaccination program "with its 90% death rate". *"The most sinister crime in which Schreiber was involved was the introduction of intravenous lethal phenol injections as a quick and convenient means of executing troublemakers"*, explained Dr. Leopold Alexander, a war crimes investigator and expert consultant at the Nuremburg doctor's trial. Paperclip contracts secured by Dr. Schreiber were: US Army, Camp King, Germany; and US Air Force, Texas, USA. **Dr. Kurt Blome** was Hitler's biological weapons maker and the deputy Surgeon General of the Third Reich. He had nearly completed a bubonic plague weapon when the Red Army captured his research institute in Poland.

**Erich Traub** was a virologist, microbiologist, and doctor of veterinary medicine. He weaponized Rinderpest (cattle plague) at the request of Heinrich Himmler. During WWII, Traub traveled to Turkey to acquire a black-market sample of the Rinderpest virus. **Major General Walter Dornberger** was in charge of V-weapons development for the Reich. Arrested by the British and held for nearly two years on war crimes charges, Dornberger was released into US custody with the warning that he was a ***"menace of the first order"***. He secured a Paperclip contract with US Army Air Forces. In 1957, Dornberger attempted to sell Bell Aircraft's BoMi (bomber missile) to the Pentagon. This was a rocket-powered manned spacecraft designed for nuclear combat in space! When an audience of Air Force officials shouted "abusive and insulting remarks" at him, he is said to have turned on his audience and insisted that BoMi would receive a lot more respect if he had had a chance to fly it against the United States during a war. There was reportedly a "deafening silence" in the room. Based on an insider's spying tip, a 1958 FBI investigation honed in on Dornberger's cunning and duplicitous nature: *"It is believed that Subject could carry on satisfactorily in the role of a double [Soviet] agent"*.

**Arthur Rudolph** specialized in V-weapons assembly and served as Operations Director at the slave labor facility in Nordhausen, Germany. In America he would become known as the "Father of the Saturn Rocket". In 1985, he told journalist John Huber, *"I read **Mein Kampf** and agree with lots of things in it. Hitler's first six years, until the war started, were really marvelous"*. In 1983, the US Department of Justice began its investigation of Rudolph on war crimes. He was told to prepare to stand trial, or to renounce his US citizenship and leave the country. Rudolph chose to leave America! **Georg Rickhey** oversaw tunnel operations for Hitler's Fuhrerbunker headquarters in Berlin. On the V-2 program he was General Manager of the slave labor facility and appeared as a defendant in the Nordhausen war crimes trial. He secured a Paperclip contract with US Army Air Forces.

**Kurt Debus** was a V-weapons engineer who oversaw mobile rocket launches as well as those at Peenemunde. An ardent Nazi, he wore the SS uniform to work. According to his US government security report, Debus turned a colleague over to the Gestapo for making anti-Nazi remarks. In America, Debus became the first Director of NASA's **John F. Kennedy Space Center** in Florida. Paperclip contract: US Army, Texas. **Walther Riedel** was an engineer with the V-weapons design bureau, and part of the von Braun rocket team. His Army interrogator classified him as an *"ardent Nazi"*, but after Riedel threatened his handler that he would go work for the Russians, he was hired and brought to America. Paperclip contract: US Army, Texas. **Otto Ambros** was Hitler's most valuable chemist, co-discoverer of Sarin poison gas, and Chief of the Reich's Committee-C for chemical warfare. The US Army coveted his knowledge. Tried at Nuremburg, Ambros was convicted of mass murder and slavery, then was later granted clemency by High Commissioner John J. McCloy. Paperclip contract: US Department of Energy.

**Friedrich "Fritz" Hoffman** was a chemist and philosopher. When captured by Allied forces he carried a paper signed by a US diplomat stating he was anti-Nazi. In America, Hoffman synthesized Nazi nerve gas stockpiles and worked in the CIA's assassination-by-poison program. Paperclip contract: US Army Chemical Corps, Maryland. **Jurgen von Klenck** was a chemist, SS officer, and Deputy Chief of the Committee-C for chemical warfare. Surprised by how much information he provided, his interrogators concluded *"that a lesser secret has been admitted to deflect the investigation from a more important secret"*. Paperclip contract: US Army, Heidelberg, USA. **Dr. Hubertus Strughold** was in charge of the Aviation Medical Research Institute of the Reich Air Ministry in Berlin. Despite being sought for war crimes, he was hired by the US Army Air Forces and became America's "Father of Space Medicine". He went to great lengths to whitewash a dubious past, *"Only the janitor and the man who took care of the animals"* were members of the Nazi party, he told a journalist in 1961, referring to his Institute, which was filled with hardcore Nazis. Paperclip contracts: US Army Air Forces, Heidelberg; US Air Force, Texas.

**Dr. Theodore Benzinger** directed the Experimental Station of the Air Force Research Center, Rechlin, under Hermann Goring and was an officer with the SA (Storm Troopers). While working for the US Army in Heidelberg, Benzinger was arrested, imprisoned at Nuremburg, and listed as one of the defendants in the doctors' trial. Shortly thereafter, he was mysteriously released. Paperclip contracts: US Army Air Forces, Heidelberg, USA; Naval Medical Research Institute, Maryland, USA. **Dr. Konrad Schafer** was a physiologist and chemist who developed a wartime process to separate salt from seawater in sea emergencies. Medical experiments at the Dachau concentration camp were based on the Schafer process. He was tried at Nuremburg and acquitted. Paperclip contracts: US Army Air Forces, Heidelberg, USA; US Air Force, Texas.

**Dr. Herman Becker-Freyseng** was an aviation physiologist who worked under Dr. Strughold in Berlin and oversaw medical experiments on prisoners at Dachau. He was tried and convicted at Nuremburg, then contributed to Strughold's US Army work from his prison cell. Paperclip contract: US Army Air Forces, Heidelberg. **Dr. Siegfried Ruff** directed the Aero Medical Division of the German Experimental Station for Aviation Medicine in Berlin and was a close colleague and co-author of Dr. Strughold. At Dachau, Dr. Ruff supervised medical murder experiments. Paperclip contract: US Army Air Forces, Heidelberg, USA.

**Emil Salmon**, aircraft engineer and SS officer, was implicated in the burning down of a Synagogue during World War II. *"This Command is cognizant of Mr. Salmon's Nazi activities and certain allegations made by some of his associates in Europe,"* wrote the US Army Air Forces, but they found Salmon's expertise *"difficult, if not impossible, to duplicate"*. Paperclip contract: US Army Air Forces, Ohio, USA. **Dr. Wilhelm Beiglbock** oversaw the salt water experiments at Dachau and brutally removed a piece of prisoner Karl Hollenrainer's liver without anesthesia. During the trial, Hollenrainer, one of the only experiment survivors, tried to stab Beiglbock with a dagger.

**Dr. Wernher M.M. von Braun** was a German-American aerospace engineer and space architect. He was a member of the Nazi Party and *Allgemeine SS*, and the leading figure in the development of rocket technology in Nazi Germany. By 1950, the US Army had decided that it required a much larger facility, so the Fort Bliss German-led rocket team moved from Texas, USA, to the Redstone Arsenal, near Huntsville, Alabama, to begin work on the US Army's Jupiter ballistic missile. They had launched sixty-four of the V-2 missiles from White Sands, New Mexico. In the Summer of 1952, Von Braun published a piece in *American Magazine* entitled "Why I Chose America". He claimed that during World War II he opposed Nazism and was never in a position to do anything but follow orders. Von Braun was photographed with his team at the Redstone Arsenal in 1958. In the group were Ernst Stuhlinger, Helmut Hoelzer, Karl Heimburg, E.D. Geissler, E.W. Neubert, Walter Haeussermann, W.A. Mrazek, Hans Hueter, Eberhard Rees, Kurt Debus and Hans Maus. Von Braun became the biggest celebrity of the **Operation Paperclip** scientists.

**Harry Armstrong** set up the US Army Air Forces Aero Medical Center in Germany and hired fifty-eight Nazi doctors to continue work they had been doing for the Third Reich. The Center violated the Potsdam Accord and was shut down after two years. Thirty-four Nazi doctors followed Armstrong to the USAF School of Aviation Medicine in Texas, USA. **Charles E. Loucks** oversaw the Paperclip scientists working on chemical weapons at Edgewood Arsenal. After being transferred to US Occupied Germany, Loucks created an off-book working group on Sarin gas production and invited Hitler's former chemists and Himmler's right-hand man to weekly roundtable discussions at his home. Col. **Donald L. Putt**, accomplished test pilot and engineer, was one of the first wartime officers to arrive at Herman Goring's secret aeronautical research center at Volkenrode. Amazed by what he saw, Putt recruited dozens of Nazi scientists and engineers for **Operation Paperclip** and oversaw their work at Wright Field, USA.

The Nuremburg trial had twenty-three former Nazi defendants. No one outside a small group of American medical military elite had any idea that four of the accused had already worked for the US Army under **Operation Paperclip**. This fact would remain secret for forty years. Neither Siegfried Ruff, Konrad Schafer, Hermann Becker-Freyseng, nor Oskar Schroder mentioned this damning fact during the trial. On August 06, 1947, the verdicts were read. Seven defendants were given death sentences, and nine were given prison sentences of between ten years and life, including Wilhelm Beiglbock (15 years), Hermann Becker-Freyseng (20 years), and Oskar Schroder (life). Seven Nazi doctors were acquitted, including Karl Blome, Konrad Schafer, Siegfried Ruff and Georg Welz. Of Blome's acquittal, the Nuremburg judges noted,*"It may well be that defendant Blome was preparing to experiment on human beings in connection with bacteriological warfare, but the record fails to disclose that fact, or that he ever actually conducted the experiments"*.

# NASA organization facilities:

## John F. Kennedy Space Center

The **John F. Kennedy Space Centre** (KSC) is one of the National Aeronautics and Space Administration's (NASA) ten field centers. It is located on Merritt Island in Florida, USA, and has been NASA's primary launch center of human spaceflight, research and technology since December 1968. KSC launches manned space vehicles, unmanned planetary spacecraft, scientific, meteorological and communications satellites, and served for thirty years as the launch site for the reusable Space Shuttle. It is also used for International Space Station payload integration and launching.

On July 29, 1958, President Dwight D. Eisenhower signed **Public Law 85-568** that established the **National Aeronautics and Space Administration** (NASA). On October 01, the new independent agency of the US federal government began operation with the mission to perform civilian research related to space flight and aeronautics. On September 01, 1961, NASA requested appropriations for initial land purchases on Merritt Island to support the Apollo Lunar Landing Program. In November 1963, President Lyndon B. Johnson designated the facilities of the Launch Operations Center and Station No. 1 of the Atlantic Missile Range as the **John F. Kennedy Space Center**. Plans were made for Launch Complex 39 facilities which include the Launch Control Center, Pads A and B as well as the huge hangar called the Vehicle Assembly Building (VAB). In February 1964, construction began on the Central Instrumentation Facility. In recent years, the Cape Canaveral Space Force Station (CCSFS) and KSC have been jointly referred to as the Cape Canaveral Spaceport. This name reflects a growing partnership between KSC and the Forty Fifth Space Wing, and a vision for the future of Florida's Space Coast.

## Marshall Space Flight Center

The **George C. Marshall Space Flight Center** (MSFC) is located in Redstone Arsenal, Alabama, USA. It is the US government's civilian rocketry and spacecraft propulsion research center. On July 01, 1960, the Marshall Space Flight Center was created out of the old Redstone Arsenal, and placed under the jurisdiction of NASA, with Dr. Wernher von Braun appointed as the first NASA Director. About USD 100 million worth of buildings and equipment, 1,840 acres of land and 4,670 civilian employees were transferred from AOMC/ABMA to the new MSFC. The initial technical leaders of the MSFC were all (except for Heinz-Hermann Koelle) former colleagues of Dr. Wernher von Braun, starting back at Peenemunde, Germany, before WWII.

As the largest NASA center, MSFC's first mission was developing the Saturn launch vehicle for the Apollo program. Marshall has been the lead center for the Space Shuttle main propulsion and external tank (ET); payloads and related crew training; International Space Station (ISS) design and assembly; computers, networks, and information management; and the Space Launch System (SLS). The Marshall Center contains the Huntsville Operations Support Center (HOSC), also known as the International Space Station Payload Operations Center. This facility supports the ISS launch, payload, and experiment activities at the Kennedy Space Center (KSC). The HOSC also monitors rocket launches from Cape Canaveral Space Force Station when a Marshall Center payload is on board.

The initial main project at MSFC was the final preparation of a Redstone rocket for Project Mercury to lift a space capsule carrying the first American into space. During the 1960s, MSFC activities were largely devoted to the Apollo Program, with the Saturn family of launch vehicles. Additionally, the prime contractors for launch vehicles and major related items (including North American Aviation, Chrysler, Boeing, Douglas Aircraft, Rocketdyne and IBM) collectively had a similar number of employees working in MSFC facilities. Later MSFC major roles included Skylab, the Space Shuttle, Spacelab and other experimental activities which made use of the Shuttle's cargo bay. Between 2004 and early 2010, the **Constellation Program** was a major NASA activity. MSFC was responsible for the propulsion of the proposed *Ares 1* and *Ares V* heavy-lift vehicles. Starting in 2006, the MSFC Exploration Launch Projects Office began work on the Ares Projects. On October 28, 2009, an *Ares I-X* test rocket lifted off from the newly-modified Launch Complex 39B at KSC for a two-minute powered flight, traveling 240 km downrange.

# Johnson Space Center

The **Lyndon B. Johnson Space Center** (JSC) is located in Houston, Texas, USA. Originally named the Manned Spaceflight Center, it is where human spaceflight training, research, and flight control are conducted. JSC had its origins in NASA's Space Task Group (STG). Starting on November 05, 1958, Langley Research Center engineers under Robert R. Gilruth directed Project Mercury and follow-on American crewed space programs. The STG originally reported to the Goddard Space Flight Center organization in Maryland, with a total staff of forty-five, including thirty-seven engineers and eight secretaries and human "computers". In 1959, the center added thirty-two Canadian engineers. In 1961, the US Congress held hearings and passed a USD 1.7 billion NASA Appropriations Bill which included $60 million for the new crewed spaceflight laboratory. The land for the new facility was 400 hectares situated in an undeveloped area forty kilometres southeast of Houston, adjacent to Clear Lake near Galveston Bay. More land was acquired, and the 1,620-acre (6.6 square km) facility was officially opened for business in September 1963.

As plans for Project Gemini began in 1961, it became clear that the Mercury Control Center located at the Cape Canaveral Air Force Station launch center would become inadequate to control missions with maneuverable spacecraft such as *Gemini* and *Apollo*. Construction for the new Mission Control Center began in 1963. The new facility had two Mission operations Control Rooms, allowing training and preparation for a later mission to be carried out while a live mission was in progress. It was brought online for testing purposes during the uncrewed *Gemini 2* flight in January 1965. Since *Gemini 4*, JSC has been the primary flight control center for all subsequent United States crewed space missions. On April 14, 2011, NASA named the Mission Control Center the **Christopher C. Kraft Jr. Mission Control Center,** (MCC-H) which directed all Space Shuttle missions, and currently directs American activities aboard the International Space Station. The Apollo Mission Control Center, a National Historic Landmark, is in Building 30.

JSC handles most of the planning and training of the US Astronaut Corps and houses training facilities such as the Sonny Carter Training Facility and the Neutral Buoyancy Laboratory, a critical component in training astronauts for spacewalks. Building 31-N houses the Lunar Sample Laboratory Facility, which stores, analyzes, and processes most of the samples returned from the Moon during the Apollo program. The Center is also responsible for direction of operations at White Sands Test Facility in New Mexico, which served as a backup Space Shuttle landing site. The visitor center has been the adjacent Space Center Houston since 1994. About 3,200 civil servants, including 110 astronauts, are employed at JSC. Astronaut candidates maintain their flying proficiency by flying fifteen hours per month in NASA's fleet of T-38 jets based at nearby Ellington Field. The bulk of the JSC workforce consists of over 11,000 contractors. As of October 2014, **Stinger Gaffarian Technologies** took over United Space Alliance's primary contract.

Johnson Space Center leads NASA's human spaceflight-related scientific and medical research programs. Technologies developed for spaceflight are now in use in many areas of medicine, energy, transportation, agriculture, communications and electronics. The Astromaterials Research and Exploration Science (ARES) office performs the physical science research at the center. ARES directs and manages all functions and activities of the ARES scientists who perform basic research in Earth, planetary, and space sciences. ARES scientists and engineers provide support to the human and robotic spaceflight programs. The several responsibilities of ARES include interaction with the Office of Safety and Mission Assurance and the Human Space Flight Programs. JSC put in a bid to display one of the retired Space Shuttle orbiters, but was not selected.

# Preparing to go into space

In the 1960s, preparing for a space mission was an intricate and meticulous process. The six steps involved were:

## 1. Astronaut selection and training.

NASA carefully chose candidates based on their qualifications, physical fitness and mental resilience. These individuals underwent rigorous medical examinations and interviews. Astronauts participated in extensive training programs covering various aspects: learning about the spacecraft's systems, controls, and emergency procedures; practicing maneuvers in simulators to simulate space conditions; maintaining peak physical condition through exercise and fitness routines; training for emergency landings and survival in different environments; learning how to perform extravehicular activities (spacewalks); understanding orbital dynamics; learning Russian language (for joint missions with Soviet cosmonauts); familiarizing themselves with scientific experiments; and coping with stress, isolation, and confinement.

## 2. Mission planning and design.

Defining the purpose of the mission (e.g., scientific research, technology testing, or spacewalks); determining the spacecraft's trajectory and orbital parameters; choosing experiments and equipment to carry; creating detailed mission timelines.

## 3. Spacecraft development and testing.

Building and testing the spacecraft (e.g., Mercury, Gemini, or Apollo); ensuring all systems worked seamlessly together; conducting ground tests and unmanned flights.

## 4. Launch preparation.

Selecting the crew for the specific mission; intensive simulations of launch, orbit and re-entry; customizing spacesuits for each astronaut; preparing the rocket (e.g., *Saturn V* for Apollo missions); final checks, fueling, and countdown procedures.

## 5. Mission execution.

Astronauts boarded atop the rocket; achieving the desired orbit; conducting experiments, spacewalks, and observations; safely returning to Earth.

## 6. Post-Mission activities.

Reviewing the mission's successes and challenges; analyzing scientific data collected during the mission; sharing mission achievements with the public.

Overall, the entire process - from astronaut selection to mission execution - could take several years. The dedication, training, and teamwork of astronauts, engineers, and mission controllers were crucial in achieving successful space missions.

# The Space Shuttle Era

The Space Shuttle was a partially reusable low Earth orbital spacecraft system operated by the United States' **National Aeronautics and Space Administration** (NASA) as part of the Space Shuttle program. Its official name was the Space Transportation System (STS), taken from a 1969 plan for a system of reusable spacecraft. The first (STS-1) of four orbital test flights occurred in 1981, leading to operational flights beginning with STS-5 in 1982. Five complete Space Shuttle orbiter vehicles were built. The first orbiter, *Enterprise*, was built in 1976 and used in Approach and Landing Tests (ALT), but had no orbital capability. Four fully operational orbiters were initially built: *Columbia, Challenger, Discovery,* and *Atlantis*.

From April 12, 1981 to July 08, 2011, there were one hundred and thirty-five Space Shuttle missions. Two ended in disaster, with the loss of fourteen crew members lives: seven aboard *Challenger* in 1986, and seven aboard *Columbia* in 2003. The shuttles were built by **Rockwell International** at their Palmdale, California, assembly plant. The Shuttles were 37.2 metres long, 17.2 metres high, with a wingspan of 23.7 metres. Their weight, with three Shuttle Main Engines was 68,635 kg (about 151,314 lb). The shuttle mass at launch was 110,000 kg (about 242,509 lb or 112.73 tons). The underside of the shuttle was protected by special ceramic tiles to counter the heat of re-entry, and after reconditioning the craft was ready for its next space mission. To move payloads from the Shuttle payload bay, the Shuttle had its own fifteen-metre-long robotic arm, called *Canadarm*. This arm had six moveable joints and was completely hollow inside. In the weightless environment of space, the *Canardarm* could lift loads of over 265,000 kg (about 584,224 lb) and was remotely controlled from inside the Orbiter by an astronaut using a joystick.

The final Shuttle spaceplane version ended up much bigger than first planned; able to carry the United States military payloads that helped to offset the cost of the project. Originally budgeted at USD 200,000,000, over the Shuttle's lifetime the average cost per launch was USD 1,350 million. Operational missions launched numerous satellites, interplanetary probes, and the Hubble Space Telescope, conducted science experiments in orbit, participated in the *Shuttle-Mir* program with Russia, and facilitated the construction and servicing of the International Space Station. The fleet's total mission time was 1,323 days. The Shuttle *Atlantis* was nearly 3.5 tons lighter than the *Columbia* version, so it could carry heavier payloads to orbit. A fifth operational orbiter, *Endeavour*, was built in 1991 to replace *Challenger*. The three surviving operational vehicles were retired from service after July 21, 2011. The United States then relied on the Russian *Soyuz* spacecraft to transport astronauts to the International Space Station (ISS) from the final Shuttle flight of *Atlantis* until the launch of the SpaceX *Crew Dragon* 'Demo-2' mission to the ISS in May 2020.

Space Shuttle components included the Orbiter Vehicle (OV) with three clustered Rocketdyne RS-25 main engines (SSMEs) supplied with fuel from the large External Tank and providing thrust during launch; a pair of recoverable solid rocket boosters (SRBs); and the expendable external tank (ET) containing liquid hydrogen and liquid oxygen (LOX) that separated when empty to fall down and burn up in Earth's atmosphere. The Space Shuttle was launched vertically, like a conventional rocket, with the two SRBs operating in parallel with the orbiter's three main engines, which were fueled from the ET. The detachable SRBs were jettisoned before the vehicle reached orbit and were parachuted down to be recovered for re-use, while the main engines continued to operate. The ET was jettisoned after main engine cutoff and just before orbital insertion, which used the orbiter's two Orbital Maneuvering System (OMS) engines.

The first orbiter to be built, *Enterprise* was launched from a modified Boeing 747 carrier aircraft, and was used for atmospheric test flights. Plans to upgrade it to orbital capacity were ultimately canceled. The shuttle *Endeavour* was built from spare parts to replace *Challenger* which was destroyed 73 seconds after launch in an explosion on January 28, 1986. The Shuttles remained in use for twice the intended lifespan. In 135 missions, they hauled freight and hosted landmark in-orbit construction and repairs. During that period the fleet logged 1,322 days, 19 hours, 21 minutes and 23 seconds of flight time (about three and a half years). **The longest orbital flight of any Space Shuttle was STS-80 at 17 days, 15 hours and 20 minutes**. At 365 days in total flight time, Shuttle *Discovery* accumulated the most days in space. The Shuttles docked with the Russian space station *Mir* nine times and visited the ISS 37 times. The highest altitude (apogee) achieved by any Shuttle was 621 km (386 miles) when deploying the Hubble Space Telescope.

Initially the Space Shuttle launches were given sequential numbers indicating order of launch, such as STS-7. Subsequent to the *Apollo 13* mishap, and due to Administrator of NASA James M. Beggs' superstition about the number 13 (triskaidekaphobia) and consequent unwillingness to number a forthcoming flight as STS-13, beginning in 1984 each mission was assigned a code, such as STS-41-B, with the first digit indicating the federal fiscal year offset into the program (so 41-B was scheduled for FY 1984, and the third flight in FY 1985 would have been named 51-C). The second digit indicated the launch site: 1 was Kennedy Space Center, and 2 was Space Launch Complex 6 at Vandenberg Air Force Base, although that site was never used. These codes were assigned when the launches were initially scheduled and were not changed as missions were delayed or rescheduled.

The nose section of the Space Shuttle orbiter contained as many as seven crew, and the pilot could view through the front and side windows. The mid-deck, under the flight deck, had more seats, equipment lockers, sleeping facilities, and a toilet compartment. A basic kitchen included hot and cold-water dispensers, an oven, serving trays for food, and a water heater; but no freezer or refrigerator. Some meals were ready-to-eat, while others needed water added. The shuttle payload bay held the cargo e.g., satellites, or components for the International Space Station, while the Shuttle engines were in the orbiter's tail. A Space Shuttle could be used to retrieve artificial satellites that needed repair, service them, and return them to orbit.

The Space Shuttles were all launched from the **Kennedy Space Center** (KSC) in Florida, USA. Launch Control oversaw the Space Shuttle all through launch preparation, until the message "tower clear" was announced. Then, Mission Control at the **Johnson Space Centre** in Houston, Texas, took over and remained in charge until the Shuttle had safely landed back on Earth. If everything was working normally, the Shuttle used its own four linked computers to fly into and out of orbit. For both launch and re-entry, astronauts wore pressure suits to protect against fire, and in case the spacecraft pressurization systems failed. They wore different suits for extravehicular activity (EVA), to work outside the spacecraft or space station. Crews on the Shuttle carried a change of clothes for every day of the mission. On the ISS there is no washing machine and little storage space, so astronauts change their shirts and pants about once every three or four days. Dirty clothing is placed with other waste in an unpiloted *Progress* supply vehicle that has carried supplies to the ISS, and this is sent on a trajectory to burn up in the Earth's atmosphere.

Astronauts aboard the space shuttles "bathed' by rubbing wet cloths over their body, using one for washing and one for rinsing. They washed their hair with rinse-less shampoo, and used toothpaste that could be swallowed or suctioned out. Urine was recycled into drinking water. The space toilets worked like vacuum cleaners, with fans that sucked air and waste into a storage tank. In 2008, astronauts began recycling all the water aboard the ISS.

To train for EVA, astronauts used the Neutral Buoyancy Laboratory, one of the world's largest indoor pools. This simulated the microgravity conditions that crews experienced in space. They also used a reduced-gravity simulator, to learn to walk and ride on the Moon or Mars.

Space Shuttles usually stayed in orbit on missions of eight to seventeen days duration. The Space Shuttle de-orbit routine was very precise. Astronauts donned pressure suits, chilled the cabin temperature to make them less likely to become airsick, strapped themselves into their seats, and started "fluid loading". They each drank at least one litre of salty water; to increase blood volume and help counteract orthostatic intolerance. This is the inability to remain standing upright when the body cannot maintain adequate arterial pressure and experiences decreased blood flow to the brain when upright.

One hour later, while flying over a spot on the opposite side of the planet from the landing zone, some tiny thrusters were fired to flip the shuttle around, into a tail-first position for the de-orbit burn. It then flew belly-up to the Sun, to warm up the tires. The orbital maneuvering engines, the orbiter's two most powerful thrusters, are fired to initiate de-orbit; slowing the shuttle enough to start the descent into the Earth's atmosphere. This maneuver was precisely choreographed to get the shuttle to the correct angle and speed. There is only one chance to get it right: if it goes too fast or too steep, the shuttle will burn up or break apart; going too shallow would cause the shuttle to "skip" off the atmosphere and miss the landing site altogether; going too slow would land the craft short of the runway in a crash. A few minutes after the de-orbit burn was completed, the shuttle flipped around to have the nose facing forward, to minimize the heat during descent.

Around thirty minutes later, at an altitude of 400,000 feet (80 miles) and about 5,000 miles from the landing site; and thirty-one minutes away from landing, the craft was traveling at the top speed of Mach 25. The exterior of the shuttle became surrounded by a hot, glowing orange plasma that reaches up to 3,000 degrees Fahrenheit. As it entered Earth's atmosphere, the Shuttle's protective tiled bottom glowed orange-red. Continuing to descend, the shuttle began to fly like a glider, and the astronauts began to feel the force of Earth's gravity again. About ten minutes before landing, the force peaks at 1.5 G's. Around 83,000 feet altitude, having slowed to about Mach 2.5 (1,700 mph), the shuttle would be sixty miles from the landing runway and 5.5 minutes to touchdown. The shuttle slowed to sub-sonic velocity when about twenty-five miles from the landing site. Throughout the final approach, the shuttle descended at more than 10,000 feet per minute (about 20 times faster and seven times steeper than a commercial airliner).

The commander took over flying the craft manually; circling and continued to steeply descend over the runway until it was lined up for the final approach. About seven miles out, the shuttle had slowed to about 400 mph, and was about 1.5 minutes from touchdown. At 2,000 feet, the commander raised the nose sharply and slowed the rate of descent. At 300 feet, the landing gear was put down, and the shuttle crossed the end of the runway, having slowed to a landing speed of around 354 km/hr (about 220 miles/hr). To slow down, it braked and released a twelve-metre drag chute from its tail, bringing it to a standstill. The final part of the descent from 40,000 feet altitude to touchdown took just three and a half minutes. The journey from space to Earth took only one hour. A 150-point checklist was followed through, then the crew disembarked. After their time in space, some crew experienced "Earth sickness" and vomited, while others needed assistance to walk again in Earth's gravity. If the landing occurred at Edwards Air Force Base (AFB), California, the orbiter was flown back to KSC atop the Shuttle Carrier Aircraft (SCA), a specially modified Boeing 747 airplane designed to carry the Shuttle above it.

# 1981

Launched on April 12, 1981, the United States flew its first space-worthy Space Shuttle, *Columbia* on maiden mission STS-1. With an empty payload, and a minimum flight crew of veteran Commander John W. Young and pilot Robert "Bob" L. Crippen, the mission spent two days and six hours in thirty-seven orbits at an altitude of 306 km (190 miles). The mission objectives were to demonstrate safe launch into orbit and safe return of the orbiter and crew, and to verify the combined performance of the entire Shuttle vehicle. Systems tested included the flight control system, the data processing system, and thermal protection system. STS-1 arrived safely at Edwards Air Force Base, California for the first-ever wheeled spacecraft landing on April 14, 1981.

The Space Shuttle *Columbia* mission STS-2 launched on November 12, 1981. The crew were Joe H. Engle and Richard H. Truly, who landed two days, six hours, 13 minutes and 12 seconds later. Engle and Truly had served as one of the two shuttle crews during the Approach and Landing Tests (ALT) program in 1977. STS-2 marked the first time that a crewed, reusable orbital vehicle returned to space. On board was the Canadarm robotic arm, which performed so well that in 1982 NASA extended an invitation to fly a Canadian in space. This was the beginning of the Canadian astronaut program.

Among other things, mission STS-2 tested the Shuttle Imaging Radar (SIR) as part of the Office of Space and Terrestrial Applications (OSTA-1) payload. One of the feats accomplished was various tests on the Orbital Maneuvering System (OMS), including starting and restarting the engines while in orbit, and various adjustments to the Shuttle orbit. During the mission, United States President Ronald Reagan called the crew of STS-2 from Mission Control Center in Houston, Texas. The launch mass was 104,647 kg, the landing mass was 92,650 kg and the payload mass was 8,517 kg (18,777 lb). Their flight traveled 1.73 million km (1.07 million miles).

# 1982

The Space Shuttle *Columbia* on mission STS-3 launched on schedule on March 22, 1982, and landed eight days, four minutes and 46 seconds later. Prior to the launch, *Columbia* had spent only seventy days (about two and a half months) in the Orbiter Processing Facility – a record checkout time. Crewed by Commander Jack R. Lousma (on his second and last spaceflight) and C. Gordon Fullerton (on his first spaceflight), the mission involved extensive orbital endurance testing of *Columbia* itself, as well as numerous science experiments. Primary objectives included testing the Remote Manipulator System (Canadarm) and carrying out extensive thermal testing of *Columbia* by exposing its tail, nose and top to the sun for varying periods of time. Rolling the orbiter was necessary to balance temperatures, and to avoid slight warping of the cargo bay doors. Their 130-orbit flight traveled 5.3 million km (3.3 million miles). The launch mass was 106,782 kg, the landing mass was 93,924 kg and the payload mass was 10,301 kg (22,710 lb).

Several minor problems were experienced during the flight. The orbiter's onboard toilet malfunctioned; both crew members experienced some space sickness; and on March 26 three communications links were lost. STS-3 was the first shuttle launch with an unpainted external tank, and the only mission to land at the Northrop Strip, White Sands Space Harbor near Almogordo, New Mexico. The landing was dramatic, with the landing gear deploying at an altitude of just forty-six metres at a speed of 509 km/hr (about 316 miles/hr) and *locking in place just five seconds before touchdown!* The original planned landing site at Edwards Air Force Base was subject to flooding, and the crew had previously trained to land at White Sands. The unexpectedly difficult landing and post-flight conditions damaged the space shuttle, requiring extensive repair at Kennedy Space Center. Gypsum dust continued to be found in the spacecraft for the rest of its career.

In the early days of the Space Shuttle program, some of the missions were run jointly by NASA and the United States military. This was in part because the National Reconnaissance Office (NRO) had successfully requested the shuttle's payload bay – the part of the shuttle that carried satellites into space – be enlarged to accommodate large military satellites. NRO also wanted polar shuttle missions, since they make it possible to see the Earth's entire surface below. Between 1982 and 1992, NASA launched eleven shuttle flights with classified payloads, honoring a deal with the NRO that dated back to 1969. The missions STS-41C, 41-G, 61-A, 31, 37, 44, 46, 48, 52, 53, 60, 63 and STS-93 were classified (secret) Department of Defense payload missions. This list does not include satellites deployed from the space shuttle and retrieved during the same mission, tiny satellites for universities, etc., as "Getaway Specials" (GAS) or other add-on payloads. The astronauts encountered considerable challenges in keeping information secret since the space shuttles did not have secure information channels, and their movements and training operations had to be somehow kept separate in a normally open public agency.

The shuttle *Columbia* on her fourth mission, STS-4 launched on June 27, 1982. This was the final shuttle test flight. The two crew were Commander Thomas K. "Ken" Mattingly (on his second spaceflight) and Henry Hartsfield on his first spaceflight. Their 113-orbit flight traveled 4.7 million km (2.9 million miles) in seven days, one hour, nine minutes and 31 seconds; landing on American Independence Day, July 04. US President Ronald Reagan was in California to watch the landing and welcome the crew home. The launch mass was 109,616 kg, landing mass was 94,774 kg and the payload mass was 11,109 kg (24,491 lb).

The STS-4 classified payload was known as Cryogenic Infrared Radiance Instrument for Shuttle (CIRRIS), which was supposed to test infrared sensors for a future surveillance satellite called *Teal Ruby*, according to *America Space*. The lens cap on CIRRIS failed to open, and so the experiment failed. *Teal Ruby* was first shifted onto STS-39, and finally canceled after the disastrous *Challenger* incident. By the time mission STS-39 lifted off in April 1991, it carried not *Teal Ruby* but an updated version of CIRRIS. Apparently, by the time it would have been ready to launch, the *Teal Ruby* technology – considered "advanced" in the late 1970s – would be virtually obsolete, because sensor technology was advancing rapidly.

The *Columbia* STS-5 mission launched on November 11, and it was the first commercial shuttle mission to deploy multiple comsats. This was the first officially "operational" mission of the shuttle program. The crew included Commander Vance D. Brand (on his second spaceflight), Pilot Robert F. Overmyer, and Mission Specialists Joseph P. Allen (both on their first spaceflights) and William B. Lenoir (on his only spaceflight). The two satellites were *SBS-3*, owned by Satellite Business Systems; and *Anik-C3*, owned by Telesat Canada; both were **Hughes Aircraft Company**-built HS-376-series satellites. In addition, STS-5 carried a West German-sponsored microgravity "Getaway Special" (GAS) experiment canister in the payload bay. The first EVA of the shuttle program was canceled due to problems with both spacesuits. Their eighty-one-orbit mission lasted five days, two hours, 14 minutes and 26 seconds; traveling 3,397,082 km (about 2,110,849 miles). The launch mass was 112,088 kg, the landing mass was 91,841 kg and the payload mass was 14,551 kg (32,079 lb).

# 1983

Launched on April 04, the shuttle *Challenger* made her first flight, mission STS-6. It deployed a *Tracking and Data Relay Satellite* (TDRS-A). A crew member made the first EVA from a Space Shuttle, with a duration of four hours and seventeen minutes. The crew of four were Commander Paul J. Weitz (on his second and last spaceflight), Pilot Karol J. Bobko, and Mission Specialists F. Story Musgrave and Donald H. Peterson (on his only spaceflight). Bobko and Musgrave were on their first spaceflights, and they landed at Edwards Air Force Base on April 09. Their eighty-one-orbit mission duration was five days, 23 minutes and

42 seconds; traveling 3,370,437 km (2,094,292 miles). The launch mass was 116,457 kg, the landing mass was 86,330 kg and the payload mass was a very heavy 21,305 kg (46,969 lb).

Paul Weitz had previously served as Pilot on *Skylab-2*, the first crewed mission, where he lived and worked in the *Skylab* station for nearly a month from May to June 1973. After *Skylab*, Weitz became the Deputy Chief of the Astronaut Office under Chief Astronaut John Young. Karol Bobko originally became an astronaut for the Air Force's Manned Orbiting Laboratory (MOL) program but later joined NASA in 1969 after cancellation of the MOL program. Prior to STS-6 he participated in the Skylab Medical Experiment Altitude Test (SMEAT) and worked as a member of the support crew for the *Apollo-Soyuz Test Project*. Donald Peterson was also a transfer from the MOL program and was a member of the support crew for *Apollo 16*. Story Musgrave joined NASA in 1967 as part of the second scientist-astronaut group and was the backup Science Pilot for the first *Skylab* mission. He also participated in the design of equipment that he and Peterson used during their EVA on the STS-6 mission.

Europe's first contribution to space station technology was called *Spacelab*. It was a small, reusable station with a variety of interchangeable components, with the major one being a crewed laboratory that could fit into the Space Shuttle's cargo bay and be returned to Earth. However, the habitable module did not have to be flown to conduct a 'Spacelab'-type mission. There was a variety of five external pallets built by **British Aerospace** (designed to carry out experiments in zero gravity) and other hardware supporting space research. The habitable 'Spacelab' laboratory module comprised a cylindrical environment in the rear of the Space Shuttle orbiter payload bay, connected to the orbiter crew compartment by a tunnel. The laboratory module consisted at minimum of a core segment, which could be used alone in a "short module" configuration. The "long module" configuration included an additional experiment segment. Two laboratory modules were built, identified as LM1 and LM2. LM1 is now on display at the Steven F. Udvar-Hazy Center at the **Smithsonian Air and Space Museum**, behind the Space Shuttle *Discovery*.

The 'Spacelab 1' mission had experiments in the fields of space plasma physics, solar physics, atmospheric physics, astronomy, and Earth observation. There were twenty-two 'Spacelab' shuttle missions between November 1983 and April 1998; sixteen with full 'Spacelab' modules, and six were 'Spacelab' pallet-only missions. In addition to the laboratory module, there were the external pallets for experiments in vacuum, and a pressurised *Igloo* (a cylinder connected to the Space Shuttle orbiter crew area) containing the subsystems needed for the pallet-only flight configuration operation.

The 'Spacelab Pallet' was a U-shaped platform for mounting instrumentation, large instruments, experiments requiring exposure to space, and instruments requiring a large field of view, such as telescopes. The pallet could be used in single configuration, or stacked end-to-end in double or triple configurations. Up to five pallets could be configured in the shuttle cargo bay by using a double pallet plus a triple pallet configuration. 'Spacelab' could carry a crew of six or seven, was seven metres long, 4.12 metres in diameter, and weighed 12,923 kg (about 28,490 lb) at launch. A 'Spacelab Pallet' was recommissioned in 2000 for flight on STS-99. The 'Spacelab Pallet-Deployable 1' (SLP-D1) with the Canadian-made robotic *Dextre* (Special Purpose Dexterous Manipulator) was launched on STS-123.

Launched on June 18, the shuttle *Challenger* on mission STS-7 carried Sally K. Ride (1951-2012), the first American woman in space. She and four other astronauts made a six-day flight. The crew were Commander Robert "Bob" L. Crippen (on his second spaceflight), Pilot Frederick Hauck, and Mission Specialists John M. Fabian, Sally K. Ride, and Norman E. Thagard (all on their first spaceflight). There were multiple comsat deployments, and the first deployment and retrieval of a shuttle pallet satellite. Sally was the first known LGBT astronaut, the youngest American astronaut to fly in space, the first woman to use the robotic arm in space (deploying two communications satellites), the first woman to serve as a NASA spacecraft ground controller, and was co-founder of **Sally Ride Science**, a company that promotes learning and careers in science, technology, engineering and math (STEM) for students, especially girls.

With the help of a payload assist module, Canada's *Anik C2* satellite was deployed on June 23. The STS-7 mission returned to Earth at Edwards Air Force Base on June 24, was refurbished, mated with its twin Solid Rocket Boosters and External Tank, and the stack was rolled out to Launch Pad 39A on August 02, a then-record turnaround time. Mission STS-7 made 97 orbits in six days, two hours, 23 minutes and 59 seconds; traveling 3.57 million km (2.220 million miles). The launch mass was 113,025 kg, the landing mass was 92,550 kg and the payload mass was 16,839 kg (37,124 lb).

Launched on August 30 aboard shuttle *Challenger* for STS-8, Guion "Guy" S. Bluford, Jr., became the first African American to go into space. This was only the second night launch in the American human space flight program. Bluford deployed the *Insat-1B* satellite for India, the mission's primary payload. The STS-8 crew were Commander Richard H. Truly (on his second and last spaceflight), Pilot Daniel C. Brandenstein, and Mission Specialists Dale A. Gardner, Guion "Guy" S. Bluford, Jr., and Dr. William E. Thornton, a medical doctor who at age 54 was the oldest person in space up to that time. Brandenstein, Gardner, Bluford and Thornton were all on their first spaceflights. Richard Truly had flown as the pilot of STS-2 in 1981, and for two of the Approach and Landing Tests aboard the prototype shuttle *Enterprise* in 1977.

The Indian National Satellite System (INSAT) is a series of multipurpose geostationary satellites to satisfy telecommunications, broadcasting, meteorology, and search and rescue operations. *INSAT* was the largest domestic communication system in the Indo-Pacific region. *KALPANA-1* was an exclusive meteorological satellite. The satellites incorporate transponders for receiving distress alert signals, as ISRO is a member of the Cospas-Sarsat program. Some of the satellites had the Very High Resolution Radiometer (VHRR), CCD cameras for meteorological imaging. There were five satellites in the *INSAT-2* series.

On September 05, the STS-8 crew made the first night landing of the shuttle program at Edwards Air Force Base (AFB) in California. Their 98-orbit mission lasted six days, one hour, eight minutes and 43 seconds; traveling 4,046,660 km (2,514,480 miles). The launch mass was 110,108 kg, the landing mass was 92,508 kg and the payload mass was 12,011 kg (26,480 lb). Components for the United States 'Spacelab', a reusable laboratory, traveled in the cargo bay of NASA's space shuttles on various missions from 1983 to 1997. Experiments in astronomy, biology and medicine were carried out on 'Spacelab'.

Space Shuttle *Columbia* for STS-9 launched on November 28, 1983, carrying 'Spacelab 1'. The crew were Commander John W. Young (on his sixth and last spaceflight), Pilot Brewster H. Shaw, Mission Specialists Dr. Owen K. Garriott (his second and last spaceflight), Robert A. Parker, and Payload Specialists Ulf Merbold, who became the first ESA astronaut on a shuttle mission, and Byron K. Lichtenberg. Astronauts Shaw, Parker, Merbold and Lichtenberg were all on their first spaceflights. Their 167-orbit mission lasted ten days, seven hours, 47 minutes and 24 seconds; traveling 6,913,504 km (4,295,852 miles). The launch mass was 112,918 kg, the landing mass was 99,800 kg and the payload mass was 15,068 kg (32,219 lb). After this mission, *Columbia* was taken out of service for renovations and did not fly again until mission STS-61-C in early January 1986.

## 1984

Launched on February 03, the shuttle *Challenger* on mission STS-41-B (NASA's tenth Space Shuttle mission) had five crew: Commander Vance D. Brand (on his third spaceflight), Pilot Robert L. Gibson, and Mission Specialists Bruce McCandless II, Robert L. Stewart, and Ronald E. McNair (on his only spaceflight). Gibson, McCandless, Stewart and McNair were on their first spaceflights. Two comsats were deployed, and Bruce McCandless made the first untethered spacewalk with the *Manned Maneuvering Unit*. This mission was a dry run for the Solar Maximum Mission rescue. Their 128-orbit flight duration was seven days, 23 hours, 15 minutes and 55 seconds; traveling 5,329,150 km (3,311,380 miles). The launch mass was 113,603 kg, the landing mass was 91,280 kg and the payload mass was 12,815 kg (28,252 lb).

Launched on April 06, shuttle *Challenger* on her fifth mission, STS-41-C (formerly STS-13) marked the first direct ascent trajectory for a space shuttle mission. The crew were Commander Robert "Bob" L. Crippen, Pilot Francis R. "Dick" Scobee, and Mission Specialists Terry Hart, James van Hoften, and George D. "Pinky" Nelson. The crew captured and repaired the malfunctioning *Solar Maximum Mission* satellite (aka *Solar Max*), the first on-orbit repair and deployed the Long Duration Exposure Facility (LDEF) experimental apparatus. The LDEF is a platform the size of a school bus and contained 57 materials experiments. Their 108-orbit mission lasted for six days, 23 hours, 40 minutes and seven seconds; traveling 4.62 million km (2.87 million miles). The launch mass was 115,328 kg, the landing mass was 89,346 kg and the payload mass was 15,345 kg (33,830 lb). The landing on April 13 took place on Runway 17 at Edwards Air Force Base, instead of at Kennedy Space Center, Florida, as had been planned.

Launched on August 30, *Discovery* on her first flight, mission STS-41-D (formerly STS-14) deployed three commercial comsats, and tested the OAST-1 Solar Array (a prototype extendable solar array that would eventually form the basis of the main solar arrays on the ISS). The mission was delayed more than two months from its original planned launch date, having experienced the Space Shuttle program's first launch abort at T-6 seconds on June 26, 1984. The crew were Commander Henry W. Hartsfield, Jr., (on his second spaceflight), Pilot Michael L. Coats, Mission Specialists Richard M. "Mike" Mullane, Steven A. Hawley, Judith A. Resnik (on her first mission; she was to die during the launch explosion of *Challenger* in January 1986) and Payload Specialist Charles D. Walker. Astronauts Coats, Mullane, Hawley and Walker were all on their first spaceflights. Their ninety-seven-orbit mission lasted six days, 56 minutes and four seconds; traveling 4.01 million km (2.49 million miles). The shuttle launch mass was 119,511 kg (263,477 lb), the landing mass was 91,418 kg (201,542 lb), and the payload mass was 18,681 kg (41,185 lb).

Sally Ride had her second mission STS-41-G aboard the space shuttle *Challenger* from October 05 to 13 1984. This was the thirteenth flight of NASA's Space Shuttle program, and the sixth flight of the shuttle *Challenger*, which made the second shuttle landing at Kennedy Space Center. The mission deployed a classified radar imaging satellite. The crew were Commander Robert L. Crippen on his fourth and last spaceflight (the first American to complete two space missions in the same calendar year), Pilot Jon A. McBride, three Mission Specialists; David C. Leestma, Sally K. Ride, and Kathryn D. Sullivan; and two Payload Specialists: Paul D. Scully-Power (the first Australian-born astronaut) and Marc Garneau (of CSA), the first Canadian in space. McBride and Scully-Power were on their only spaceflights, while Sullivan, Leestma and Garneau were on their first spaceflights.

This was the first space mission to include seven crew members, the first flight of a Canadian Payload Specialist (Marc Garneau), the first to include two women, the first spacewalk by an American woman (Kathryn D. Sullivan), the first crewman to fly a fourth shuttle mission (Commander Robert L. Crippen), the first demonstration of a satellite refueling technique, and the first flight with a re-entry profile crossing the eastern United States. The craft entered the atmosphere over the Atlantic Ocean and flew over several states before landing in Florida. Sally Ride spent more than of 343 hours (about two weeks) in space, then served on the **Rogers Commission**, which investigated the causes and then recommended remedies for the accident which destroyed the shuttle *Challenger* in January 1986.

Nine hours after liftoff, the 2,307 kg (about 5,086 lb) *Earth Radiation Budget Satellite* (ERBS) was deployed by the Canadarm robotic arm, and its on-board thrusters boosted it into orbit 560 km (about 348 miles) above the Earth. Kathryn Sullivan and David Leestma performed a three-hour EVA on October 11, demonstrating the Orbital Refueling System (ORS) and proving the feasibility of refueling satellites in orbit. Another major mission activity was the operation of the Shuttle Imaging Radar-B (SIR-B), part of the OSTA-3 experiment in the payload bay, which also included the Large Format Camera (LFC) to photograph the Earth, another camera called MAPS which measured air pollution, and a feature identification and location experiment called FILE, which consisted of two TV cameras and two 70-mm film size (about 2.76 inch) still cameras.

Payload Specialist Paul Scully-Power, an employee of the United States **Naval Research Laboratory** (NRL), performed a series of oceanography observations during the mission. Marc Garneau conducted a series of experiments sponsored by the Canadian government, called CANEX, which were related to medical, atmospheric, climatic, materials, and robotic science. Several "Getaway Special" (GAS) canisters, covering a wide variety of materials testing and physics experiments were also flown. Their mission lasted eight days, five hours, 23 minutes and 33 seconds. *Challenger* traveled 5,293,847 km (about 3,289,444 miles) and completed 133 orbits. The launch mass was 110,120 kg, the landing mass was 91,746 kg and the payload mass was just 8,753 kg (18,900 lb).

The STS-41-G mission was later described in detail in the book *Oceans to Orbit: The Story of Australia's First Man in Space, Paul Scully-Power*, by space historian Colin Burgess. Almost forty years after the flight, one of the OMS engines installed on *Challenger* for STS-41-G was reused and repurposed as the main engine for the *Orion* capsule on the *Artemis I* test flight around the Moon in 2022. In 1989, Marc Garneau retired from the military and became deputy Director of the Canadian astronaut program.

Launched from Kennedy Space Center on November 08, the shuttle *Discovery* on STS-51-A (its' second flight) deployed two **Hughes Aircraft Company**-built communications satellites (the Canadian *Anik D2*, one of the biggest communications satellites of that era and *Syncom IV-1*, (aka *Leasat-1*), and retrieved two stranded satellites from orbit: *Palapa B2* and *Westar 6*, which was later sold to Hong Kong based **AsiaSat**, refurbished and reflown. The latter had been placed into improper orbits due to the malfunctioning of their kick motors. The crew were Commander Frederick H. Hauck (on his second spaceflight, Pilot David M. Walker (on his first spaceflight), and Mission Specialists Joseph P. Allen (on his second and last spaceflight), Anna Lee Fisher (on her only spaceflight) and Dale A. Gardner (on his second and last spaceflight). Their 127-orbit flight lasted for seven days, 23 hours, 44 minutes and 56 seconds; traveling 5,293,786 km (3,289,406 miles), then landing on Runway 15 at KSC, Florida. This was the third shuttle landing at Kennedy Space Center, and the fifth and last shuttle mission of 1984. The launch mass was 119,442 kg, the landing mass was 94,123 kg and the payload mass was 17,375 kg (38,305 lb).

## 1985

Launched on January 24, very little is known about mission STS-51C aboard *Discovery's* classified US Department of Defense payload besides this terse line on the NASA website: "The US Air Force Inertial Upper Stage (IUS) booster was deployed and met the mission objectives". The crew were Commander Thomas K. Mattingly II (on his third and last spaceflight), Pilot Mr. Loren J. Shriver, 'Mission Specialist 1' Ellison S. Onizuka (on his only spaceflight), 'Mission Specialist 2' James F. Buchli, and 'Payload Specialist 1' Gary E. Payton, MSE on his only spaceflight. Shriver and Buchli were on their first spaceflights. Multiple sources suggest that the satellite deployed was called *Magnum/ORION ELINT*, a signals intelligence program about which little is known. Before launch, no pre-flight commentary was available until nine minutes before liftoff – a first for secrecy in the space shuttle program. The mission lasted for three days, one hour, 33 minutes and 23 seconds; traveling 2.01 million km (1.25 million miles). The launch mass was 113,802 and the other weight details were classified.

Launched on April 12, *Discovery* on her fourth flight (mission STS-51-D) deployed multiple comsats, including Canada's *Anik C1* and made the first impromptu EVA to fix *Syncom F3* (aka *Leasat 3*). It was the first spaceflight of a sitting politician, Jack Garn (Republican, UT). The crew were Commander Karol J. Bobco (on his second spaceflight), Pilot Donald E. Williams, Mission Specialists Margaret " Rhea" Seddon, S. David Griggs (on his only spaceflight) Jeffrey A. Hoffman, Payload Specialists Charles D. Walker (on his second spaceflight) and Jake Garn (United States Senator, Utah) on his only spaceflight. Jake Garn was the first sitting member of the United States Congress to fly in space. The 110-orbit mission lasted six days, 23 hours, 55 minutes and 23 seconds; traveling 4,650,658 km (about 2,889,785 miles).

The launch mass was 113,802 kg, the landing mass was 89,818 kg and the payload mass was 13,039 kg (28,746 lb). *Discovery* suffered extensive brake damage and a ruptured tire during landing. This forced all subsequent shuttle landings to be done at Edwards Air Force Base, Califiornia, until the development and implementation of nose wheel steering made safer landings at Kennedy Space Center more feasible.

Launched on April 29, after a two minute and 18 second delay due to a launch processing failure, the shuttle *Challenger* mission STS-51-B carried 'Spacelab 3', the first module in a fully operational configuration. The crew were Commander Robert F. Overmyer (on his second and last spaceflight), Pilot Frederick D. Gregory (on his first spaceflight), Mission Specialists Don L. Lind (on his only spaceflight), Norman E. Thagard (on his second spaceflight), William E. Thornton (on his second and last spaceflight), and Payload Specialists Lodewijk van den Berg of **EG & G Energy Management, Inc.** (on his only spaceflight) and Taylor G. Wang of the Jet Propulsion Laboratory (JPL) on his only spaceflight. The average age of 48.6 years was the oldest for an American space mission, and the crew conducted experiments in microgravity. Their 111-orbit mission lasted seven days, eight minutes and 46 seconds; traveling 4,651,621 km (about 2,890,383 miles). The launch mass was 111,980 kg, the landing mass was 96,373 kg and the payload mass was 15,610 kg (34,410 lb).

The gravity gradient attitude of the orbiter proved quite stable, allowing the delicate experiments in materials processing and fluid mechanics to proceed normally. The crew operated around the clock in two teams covering twelve-hour shifts; being supported by a temporary Payload Operations Control Center, located at the Johnson Space Center. Two "Getaway Special" (GAS) experiments required that they be deployed from their canisters, a first for the program. These were *NUSAT* (Northern Utah Satellite), which deployed successfully; and *Global Low Orbiting Message Relay* satellite (*GLOMR*), which did not deploy and was returned to Earth.

While participating in the investigation into the destruction of the shuttle *Challenger* during mission STS-51-I, Robert Overmyer discovered that a problem with the shuttle's O-rings, similar to that which led to the disaster, had emerged during the launch of STS-51-B. **Morton-Thiokol** engineers told astronaut Don Lind after the mission that "***you came within three-tenths of one second of dying***". It was the problem with the O-rings on the left Solid Rocket Motor (SRM) on this launch that prompted Roger Biosjoly to write a memo to Bob Lund about the potential for the O-rings to cause catastrophic failure. From National Archives, identifier 596263: *"It is my honest and very real fear that if we do not take immediate action to dedicate a team to solve the problem with the field joint having the number one priority, then* ***we stand in jeopardy of losing a flight*** *along with all the launch pad facilities".* Unquote.

Launched on June 17, the space shuttle *Discovery* on her fifth flight (mission STS-51-G) deployed multiple comsats, and carried the first member of royalty, Saudi, Muslim and Arab in space: Sultan bin Salman Al Saud. The crew were Commander Daniel C. Brandenstein (on his second spaceflight), Pilot John O. Creighton, Mission Specialists John M. Fabian, Steven R. Nagel, Mrs. Shannon Wells Lucid, and Payload Specialists Patrick Baudy of CNES (on his only spaceflight), and Prince Sultan bin Salman Al Saud (of RSAF) on his only spaceflight. Their 112-orbit mission lasted seven days, one hour, 38 minutes and 52 seconds; traveling 4,693,051 km (about 2,916,127 miles). The launch mass was 116,357 kg, the landing mass was 92,610 kg and the payload mass was 17,280 kg (38,100 lb).

Launched on July 29, the space shuttle *Challenger* on mission STS-51-F carried as primary payload the 'Spacelab 2' laboratory module. A faulty temperature sensor in Engine 1 Center SSME engine incorrectly indicated that fuel turbine discharge temperature exceeded the limit. Therefore, one main engine was shut down at T + 345 seconds (almost six minutes after launch), and the shuttle had to perform an "Abort to Orbit" (ATO) emergency procedure, resulting in a slightly lower orbit than planned. The crew were Commander C. Gordon Fullerton (on his second and last spaceflight), Pilot Roy D. Bridges, Jr., (on his only spaceflight), Mission Specialists Karl G. Henize (on his only spaceflight), F. Story Musgrave (on his second spaceflight), Anthony W. England (on his only spaceflight), and Payload Specialists Loren W. Acton (on her only spaceflight) and John-David F. Bartoe (on his only spaceflight).

All mission objectives were achieved, but five of the crew never flew again, probably something to do with that emergency procedure! A helium-cooled infrared telescope (IRT) was also flown on this mission, and while it did have some problems, it observed 60% of the galactic plane in infrared light. Their 127-orbit mission lasted seven days, 22 hours, 45 minutes and 26 seconds; traveling 5,284,350 km (3,282,540 miles). The launch mass was 114,693 kg, landing mass was 98,309 kg and the payload mass was 16,309 kg (35,955 lb).

Launched on August 27, the shuttle *Discovery* on her sixth flight (mission STS-51-I) deployed three comsats, and rescued *Syncom F3* (aka *Leasat 3*). The crew were Commander Joe H. Engle (on his second and last spaceflight), Pilot Richard O. Covey, and Mission Specialists John M. "Mike" Lounge, James D.A. van Hoften (on his second and last spaceflight), and William F. Fisher on his only spaceflight. Covey and Lounge were on their first spaceflights. Their 112-orbit mission lasted for seven days, two hours, 17 minutes and 42 seconds; traveling 4,698,602 km (about 2,919,576 miles). The launch mass was 118,981 kg, the landing mass was 89,210 kg and the payload mass was 17,540 kg (38,670 lb).

Launched from October 03-07, the maiden flight of shuttle *Atlantis*, designation Orbiter Vehicle 104 (OV-104) on mission STS-51-J was the second classified Department of Defense mission, which deployed a DoD satellite. The system was intended to support secure data and voice transmissions for military users from across the globe. The crew were Commander Karol J. Bobko (on his third and last spaceflight), Pilot Ronald J. Grabe, Mission Specialists David C. Hilmers, Robert L. Stewart (on his second and last spaceflight), and Payload Specialist William A. Pailes of MSE on his only spaceflight. Grabe and Hilmers were on their first spaceflights. Their 64-orbit mission lasted four days, one hour, 44 minutes and 38 seconds; traveling 2,707,948 km (about 1,682,641 miles). The landing mass was 86,400 kg and the payload mass was a hefty 19,968 kg (44,022 lb).

The shuttle *Challenger* launched on October 30 for mission STS-61A, carrying the 'D-1 Spacelab' mission - the first with German mission management, and controlled from the German Space Operations Center. West Germany was responsible for the scientific research carried out during the seven-day mission. The Spacelab mission was conducted in a long module and featured seventy-five numbered experiments. The crew were divided into two shifts, enabling twenty-four-hour laboratory work to be performed. STS-61A with its crew of eight set a record as the highest number of people in space on a single spacecraft.

The crew were Commander Henry W. Hartsfield, Jr., (on his third and last spaceflight, Pilot Steven R. Nagel (on his second spaceflight), RMS Specialist Bonnie J. Dunbar (on her first spaceflight), Flight Engineer James F. Buchli (on his second spaceflight), (Science) Mission Specialist Guion "Guy" S. Bluford, Ernst Messerschmid, Reinhard Furrer, and Wubbo Ockels who became the first Dutch citizen in space; while Guion Bluford became the first African American to fly in space twice. Guy Bluford would later fly on STS-39 in 1991 and on STS-53 in 1992. Furrer, Messerschmid and Ockels each made only one spaceflight.

The primary task of mission STS-61A was to conduct a series of experiments in 'Spacelab D-1', almost all related to functions in microgravity. Other assignments were to deploy the *Global Low Orbiting Message Relay Satellite* (GLOMR) out of a "Getaway Special" (GAS) canister in the cargo bay, and to operate five materials processing experiments, which were mounted in the orbiter's payload bay on a separate device called the German Unique Support Structure. Their 112-orbit mission lasted for seven days, 44 minutes and 51 seconds; traveling 4,682,148 km (2,909,353 miles), and landing at Edwards Air Force Base on November 06. The launch mass was 110,568 kg, the landing mass was 97,144 kg and the payload mass was 14,451 kg (31,859 lb).

The shuttle *Atlantis* launched on her second flight (mission STS-61-B) on November 26, 1985, only fifty days (seven weeks) after its previous mission and she was used for ten flights from 1988 to 1992. The crew were Commander Brewster H. Shaw, Jr. (on his second spaceflight), Pilot Bryan D. O'Connor, Mission Specialists Jerry L. Ross, Mary L. Cleave, Sherwood C. Spring (on his only spaceflight), and Payload Specialists Charles D. Walker (on his third and last spaceflight), and Rodolfo Neri Vela (the first Mexican in space) on his only spaceflight. O'Connor, Ross and Cleave were on their first spaceflights. The mission had two primary objectives: to deploy three communications satellites, and to test techniques of constructing structures in orbit. The objective of the EASE/ACCESS experiment was to grow single crystals that were larger and purer than any that could be grown on Earth. Their 109-orbit STS-61-B mission lasted for six days, 21 hours, four minutes and 49 seconds; with crew members making five spacewalks. They traveled 3,970,181 km (2,466,956 miles). The launch mass was 118,664 kg, the landing mass was 93,316 kg and the payload mass was a massive 21,791 kg (48,041 lb), the second-heaviest ever carried on a shuttle.

## 1986

Launched on January 12, the space shuttle *Columbia* on mission STS-61-C deployed a comsat, and flew United States Representative Bill Nelson. It was the twenty-fourth space shuttle mission, and the seventh flight of shuttle *Columbia*. The crew were Commander Robert L. Gibson (on his second space flight), Pilot Charles F. Bolden, Jr., George D. "Pinky" Nelson (on his second spaceflight), Steven A. Hawley (on his second spaceflight), Franklin Chang-Diaz, and Payload Specialists Bill Nelson (a US Democratic Representative from Florida) and Robert C. Cenker of RCA. Bolden and Chang-Diaz were on their first spaceflights. Their ninety-eight-orbit mission lasted six days, two hours, three minutes and 51 seconds; traveling 4,069,481 km (2,528,658 miles). The launch mass was 116,121 kg, the landing mass was 95,325 kg and the payload mass was a moderate 14,724 kg (32,461 lb).

On January 28, 1986, the United States space shuttle *Challenger*, on mission STS-51-L exploded seventy-three seconds after liftoff when traveling at Mach 1.92. A rubber O-ring on a Solid-fuel Rocket Booster (SRB) tank failed after fifty-eight seconds into a launch held in sub-freezing temperatures. Both primary and secondary redundant O-ring seals were breached, and hot pressurised gas from within the right-side Solid Rocket Booster leaked through the joint and burned through the aft attachment strut connecting it to the external propellant tank (ET), then into the tank itself, igniting the fuel into a fireball. The spacecraft disintegrated fourteen kilometres (about 8.7 miles) above the Atlantic Ocean, and all seven astronauts were killed when their capsule hit the ocean at over 200 mph. The Space Shuttle design was fatally flawed in that the orbiter did not have an escape system for the astronauts.

It was the tenth flight for the orbiter *Challenger*, and the twenty-fifth flight of the space shuttle fleet. Engineers from Morton-Thiokol (makers of the SRBs) were concerned over the effect the record-cold temperatures would have on the seal provided by the SRB O-rings for the launch. During a conference call on the evening of January 27, engineers argued that they did not have enough data to determine whether the O-rings would seal at temperatures below 12 degrees Celsius. An overnight measurement taken by the Kennedy Space Center Ice Team recorded the left SRB was minus four degrees Celsius, and the right SRB was minus 13 degrees Celsius. These measurements were recorded for engineering data and not reported, because the temperature of the SRBs was not part of the Launch Commit Criteria. The craft launched at an air temperature of two degrees Celsius; hot gas and soot were able to travel past the O-rings and severely erode them (as had been seen on earlier shuttle missions).

The STS-51-L crew was scheduled to deploy a communications satellite and study Halley's Comet while they were in orbit, in addition to taking schoolteacher S. Christa McAuliffe into space. There was higher than usual media interest and coverage of the mission; the launch and subsequent disaster were seen live in many schools across the United States. The seven crew were Commander Francis R. "Dick" Scobee; Pilot Michael J. Smith; Mission Specialists Judith A. Resnik (the second female NASA astronaut, on her second mission); Ronald McNair and Ellison S. Onizuka (the first Asian-American astronaut); Christa McAuliffe; and Hughes Aircraft Company engineer Gregory Jarvis. At the time of cabin separation, the crew experienced between 12 and 20 G-s. Within two seconds that dropped to below 4-G, and within ten seconds the cabin was in free fall. The pilot, Michael Smith, had made electrical switch changes, attempting to restore power to the cockpit after the crew cabin detached from the rest of the orbiter. The crew possibly (but not certainly) lost consciousness in the few seconds following the Orbiter breakup, due to in-flight loss of crew module cabin pressure. Debris rained into the Atlantic Ocean for more than an hour after the explosion.

A three-month search and recovery operation followed. Several crew members are thought to have survived the initial breakup of the spacecraft, but the orbiter design had no escape system. The shuttle orbiter could not survive high-speed separation from the External Tank (ET) until the last seconds of the booster's two-minute burn. The disaster resulted in a thirty-two-month hiatus in the US Space Shuttle program. The Rogers Commission criticised NASA's organizational culture and decision-making processes that contributed to the accident. Test data since 1977 had revealed a potentially catastrophic flaw in the SRB's O-rings. A memo sent in January 1978 from the chief of the Solid Rocket Motor branch at Marshall SFC to his superior specifically pointed out problems with the O-ring and stated that proper sealing of the joint maintained by O-ring pressure was *"mandatory to prevent hot gas leaks and resulting catastrophic failure"*. Despite numerous objections, the design was not changed. NASA minimized the issue, while **Morton-Thiokol**, the manufacturers, stated, *"The condition is not desirable but is acceptable"*.

NASA managers had also disregarded engineer's warnings about the dangers of launching in cold temperatures and did not report these technical concerns to their superiors. What emerged was an appalling pattern of assumptions that the space vehicle could survive minor mishaps and be pushed even further. As a result of the disaster, NASA established the '**Office of Safety, Reliability, and Quality Assurance**', and arranged for deployment of commercial satellites from expendable launch vehicles rather than from a crewed orbiter. Since the *Challenger* disaster, all astronauts aboard the Space Shuttle fleet wore pressure suits during the launch and re-entry phases.

The Space Shuttle *Discovery* mission STS-62A was canceled after the *Challenger* explosion. The mission was supposed to be the first one using the Air Force pad in Vandenburg, California, and its main mission was to put the *Teal Ruby* into orbit, according to NASASpaceflight.com Also scheduled as payload on STS-62A was the *COsmic Background Explorer* (*COBE*), to take precise measurements of the diffuse radiation between one micrometre and one centimetre over the whole celestial sphere. On November 18, 1986, NASA's *COBE* became the first orbital cosmic microwave observatory, and operated for four years from 1989-93, providing key evidence for the expanding Universe "Big Bang" theory. It was placed into orbit aboard a *Delta 5920-8* launch vehicle.

# 1988

Launched on September 29, shuttle *Discovery* on her seventh flight (mission STS-26) deployed a *Tracking and Data Relay Satellite* (TDRS-C), and eleven scheduled mid-deck scientific and technological experiments were carried out. Learning from the January 1986 *Challenger* disaster, the crew wore pressure suits for both launch and landing. The redesigned post-*Challenger* Solid Rocket Boosters showed no signs of leakage or overheating at any of the joints. The crew were Commander Frederick H. "Rick" Hauck (on his third and last spaceflight), Pilot Richard O. Covey (on his second spaceflight), and Mission Specialists John M. "Mike" Lounge (on his second spaceflight), David C. Hilmers (on his second spaceflight) and George D. "Pinky" Nelson (on his third and last spaceflight). Their 64-orbit mission lasted for four days, one hour and eleven seconds; traveling just 2,703,000 km (about 1,679,566 miles). The launch mass was 115,487 kg, the landing mass was 88,078 kg and the payload mass was a very heavy 21,082 kg (46,478 lb). Richard Covey was the NASA CAPCOM operator during the STS-51-L launch who uttered the words, *"Challenger, go at throttle up"*, shortly after which the orbiter disintegrated.

Launched on December 02, the shuttle *Atlantis* on her third flight (mission STS-27) probably released a classified *Lacrosse surveillance satellite* called *ONYX*, which had radar on board capable of observing targets on the ground (through any kind of weather or cloud cover). According to *Air & Space Magazine*, one of the satellite's antenna dishes did not open, and the crew possibly did a spacewalk to fix that problem. During the launch, a piece of insulation shed from the right SRB and struck the underside of *Atlantis*, severely damaging over seven hundred tiles and removing one altogether. The crew were instructed to use the remote manipulator system to survey the condition of the underside of the right wing, and they found substantial damage. They thought that they would die during re-entry. The survival of the crew is attributed to a steel L-band antenna plate which was positioned directly under the missing tile.

Heavy damage to the thermal protection system resulted in extreme heat damage to the right wing upon re-entry. A similar situation would eventually lead to the loss of the shuttle *Columbia* in 2003, albeit on the more critical reinforced carbon-carbon. The STS-27 crew were Commander Robert L. Gibson (on his third spaceflight), Pilot Guy S. Gardner, and Mission Specialists Richard M. "Mike" Mullane (on his second spaceflight), Jerry L. Ross (on his second spaceflight) and William M. Shepherd. Gardner and Shepherd were on their first spaceflight. Their sixty-eight-orbit mission lasted four days, nine hours, five minutes and 37 seconds; traveling 2,916,252 km (1,812,075 miles). The landing mass was 86,616 kg and the payload mass was 14,500 kg (32,000 lb).

## 1989

Launched on March 13, the shuttle *Discovery* on her eighth flight (mission STS-29R) deployed the third and final component of the *Tracking and Data Relay Satellite System* (TDRS-D) and carried out a Space Station Heat Pipe Advanced Radiator Element I experiment.
The launch was originally scheduled for February 18, but was postponed allowing for the replacement of faulty liquid oxygen turbopumps on the three main engines. There were two further delays, including a day to replace a faulty fuel preburner oxidizer valve (FPOV). The three on-orbit satellites were stationed over the equator at about 35,900 km (22,308 miles) above the Earth. Two of them were positioned 130 degrees apart, while the third was located between them as an on-orbit spare. The crew were Commander Michael L. Coats (on his second spaceflight), Pilot John E. Blaha, and Mission Specialists Robert C. Springer, James F. Buchli (on his third spaceflight) and James P. Bagian. Astronauts Blaha, Springer and Bagian were on their first spaceflights.

On Flight Day 1, one of the three cryogenic hydrogen tanks which supplied the shuttle's electricity-generating fuel cells developed erratic pressure fluctuations. It was deactivated while engineers studied the problem, and the crew was told to conserve electrical power. The tank was reactivated two days later and operated successfully thereafter. The eighty-orbit mission lasted four days, 23 hours, 38 minutes and 50 seconds; traveling 3.2 million km (about 1,988,388 miles). The launch mass was 116,281 kg, the landing mass was 88,353 kg and the payload mass was 17,280 kg (38,100 lb).

Launched on May 04, the shuttle *Atlantis* on her fourth flight (mission STS-30R) deployed the US *Magellan* Venus probe; the first shuttle to launch an interplanetary probe, and the first American interplanetary mission in eleven years. *Magellan* traveled to Venus to map its surface using a large dish that bounced radar radio waves through Venus's thick, poisonous atmosphere and back to the probe. *Magellan* showed there was crater damage on Venus's surface, meaning some objects had been able to pass through its atmosphere. The STS-30R crew were Commander David M. Walker (on his second spaceflight), Pilot Ronald J. Grabe (on his second spaceflight), and Mission Specialists Mark C. Lee (on his first spaceflight), Norman E. Thagard (on his third spaceflight) and Mary L. Cleave (on her second and last spaceflight). Their 65-orbit mission lasted four days, 56 minutes and 27 seconds; traveling 2,377,800 km (1,477,500 miles). The launch mass was 118,441 kg, the landing mass was 87,296 kg and the payload mass was an extremely heavy 20,833 kg (45,929 lb).

The launch of STS-30R was originally scheduled for April 28, 1989, the first day of a thirty-one-day launch period when Earth and Venus were properly aligned. However, the liftoff was scrubbed at T-3 seconds because of a problem with the liquid hydrogen recirculation pump on the Space Shuttle Main Engine (SSME) No. 1, and a vapor leak in the liquid hydrogen recirculation line between the orbiter and the External Tank. On the rescheduled liftoff date of May 04, the launch was again delayed until the final five minutes of the launch window due to cloud cover and excessive crosswinds at KSC's Shuttle Landing Site Facility (SLF). Good landing conditions were required at the SLF in case of a Return to Launch Site (RTLS) abort occurring early in the flight. A major glitch during the flight occurred on May 07, when one of the four general-purpose computers programmed to operate the orbiter failed. The crew replaced it, part of a redundant set, with a backup one. It was the first time a control computer had been replaced while in orbit.

Launched on August 08, the shuttle *Columbia* STS-28R mission hauled the first classified *Satellite Data System-2* spacecraft into orbit. SDS was supposed to relay imagery from other military satellites. The crew were Commander Brewster H. Shaw (on his third and last spaceflight), Pilot Richard N. Richards, and Mission Specialists James C. Adamson, David

C. Leestma (on his second spaceflight) and Mark N. Brown. Richards, Adamson and Brown were on their first spaceflights. Their eighty-one-orbit mission lasted five days, one hour and eight seconds; traveling 3.4 million km (2.1 million miles). The landing mass was 90,86 kg and the payload mass was a hefty 19,600 kg (43,000 lb).

Launched on October 18, 1989, aboard the shuttle *Atlantis* on mission STS-34, the *Galileo* spacecraft was deployed, and from December 07, 1995, became the first craft to orbit Jupiter. The crew of STS-34 were Commander Donald E. Williams (on his second and last spaceflight), Pilot Michael J. McCulley (on his only spaceflight), and Mission Specialists Mrs. Shannon Lucid (on her second spaceflight), Franklin Chang-Diaz (a native of Costa Rica, on his second spaceflight) and Ellen S. Baker, on her first spaceflight. Several anomalies occurred during the flight, but none had a major impact on the 79-orbit mission, which lasted four days, 23 hours, 39 minutes and 21 seconds; traveling 2.9 million km (1.8 million miles). The launch mass was 116,831 kg, the landing mass was 88,881 kg and the payload mass was a record-setting 22,064 kg (48,643 lb), the most ever carried on a Shuttle.

Demonstrators had protested at the launch because the flight carried a nuclear device to power the *Galileo* spacecraft. Chang-Diaz said they almost aborted the flight in orbit three times because of malfunctions, but continued because the alternative was to land with the nuclear device at an airport in Senegal, which could have caused an "international incident". Another tricky part of the mission was when he had only a tight six-second opportunity to deploy the *Galileo* spacecraft. Operating for eight years, *Galileo* sent an entry probe into Jupiter's atmosphere, and made flybys and took images of asteroids Gaspra, Ida and their moons. It discovered evidence of salt water below the surfaces of the moons Europa, Ganymede and Callisto. *Galileo* also discovered evidence of subsurface oceans on the moon Europa and observed that the moon may hold ice or liquid water. *Galileo* ended its mission by plunging into Jupiter in 2003.

Launched on November 22 from the shuttle *Discovery*, the STS-33 mission was the fifth classified Department of Defense (DoD) mission, with an unconfirmed satellite payload, apparently named *Magnum*. The launch set for November 20 was rescheduled to allow changeout of suspect integrated electronic assemblies on the twin solid rocket boosters. The crew were Commander Frederick D. Gregory (on his second spaceflight), Pilot John E. Blaha (on his second spaceflight), and Mission Specialists F. Story Musgrave (on his third spaceflight), Manley L. "Sonny" Carter, Jr., (on his only spaceflight), and Kathryn C. Thornton on her first spaceflight. Their seventy-nine-orbit mission lasted five days, six minutes and 48 seconds; traveling 3.379 million km (2.1 million miles). The landing mass was 88,125 kg and the payload mass was a very heavy 21,000 kg (46,000 lb).

## 1990

Launched on January 09, the shuttle *Columbia* on mission STS-32 deployed the *Syncom IV-F5* military communications satellite (aka *Leasat 5*) and retrieved the Long Duration Exposure Facility (LDEF) on January 12. The LDEF retrieval had been delayed 54 months (four and a half years) by scheduling changes and the *Challenger* disaster of January 28, 1986. The retrieval was filmed with an IMAX camera and appeared in the IMAX film ***Destiny in Space*** in 1994. Earth observation footage from the camera also appeared in the 1990 film ***Blue Planet***. The STS-32 crew were Commander Daniel Brandenstein (on his third spaceflight), Pilot Jim Wetherbee, and Mission Specialists Bonnie J. Dunbar (on her second spaceflight), Marsha S. Ivins, and G. David Low. Wetherbee, Ivins and Low were on their first spaceflights.

STS-32 made the third shuttle night landing; after a mission lasting ten days, 21 hours and 36 seconds: traveling 7,258,096 km (4,509,972 miles). STS-32 set a new shuttle flight duration record of ten days, 21 hours, zero minutes and 36 seconds; traveling 7,258,096 km (4,509,972 miles). The launch mass was 116,117 kg, the landing mass was 103,571 kg and the payload mass was 12,014 kg (26,486 lb). *Columbia* landed on January 20 at Runway 22 of Edwards Air Force Base, California. The roll-out landing distance was 3,271 metres (about the length of thirty city blocks), and the roll-out time was sixty-two seconds. The orbiter was returned to KSC on January 26, 1990.

Launch Complex 39A was extensively modified in preparation for the STS-32 launch. NASA made improvements to the crew emergency egress system and the shuttle payload room, increased anti-freeze protection for the water systems, installed debris traps used during propellant loading, and added more weather protection features and an umbilical to provide power, instrumentation and controls to the heaters for the solid rocket booster field joints. MLP-3, the oldest of the three Apollo-era launch structures, also underwent extensive remodeling for use with the space shuttle. Modifications included the removal of the umbilical tower, reconfiguring three exhaust holes, and amendments to the electrical and mechanical ground support systems.

Launched on February 28, the shuttle *Atlantis* on her sixth flight (mission STS-36) had much speculation about what classified US Department of Defense payload was aboard, but nothing has been officially confirmed. The crew were Commander John O. Creighton (on his second spaceflight), Pilot John Caspar, and Mission Specialists Pierre J. Thuot, David C. Hilmers (on his third spaceflight) and Richard M. "Mike" Mullane (on his third and last spaceflight). Casper and Thuot were on their first spaceflights. The shuttle's ground track took it as high as 62 degrees latitude, which was a record for the shuttle program. The payload mass was 19,600 kg (43,200 lb), and the shuttle landing mass was 84,900 kg (187,200 lb). Their seventy-two-orbit mission lasted four days, ten hours and 18 minutes; traveling 2,957,913 km (1,837,962 miles). The launch mass was classified, the landing mass was 84,900 kg and the payload mass was a heavy 19,600 kg (43,200 lb).

Launched on April 24, the thirty-fifth space shuttle flight, with shuttle *Discovery* on her tenth flight (mission STS-31) carried the **Hubble Space Telescope**, the first large optical space telescope, into a low-Earth orbit at 380-statute miles. STS-31 also carried a classified payload; part of the total payload of 11,878 kg (26,187 lb). The crew were Commander Loren Schriver (on his second spaceflight), Pilot Charles F. Bolden, Jr., (on his second spaceflight), and Mission Specialists Bruce McCandless, II (on his second and last spaceflight), Steven A. Hawley (on his third spaceflight) and Kathryn D. Sullivan (on her second spaceflight). Landing occurred during the eightieth orbit, with the mission lasting five days, one hour, 16 minutes and six seconds; traveling 3,328,466 km (about 2,068,213 miles). The launch mass was 117,586 kg, and the landing mass was 85,947 kg.

The Hubble Space Telescope (HST) is equipped with a main 2.4 metre primary mirror, a smaller secondary mirror, and scientific instruments that study the Universe in infrared, visible or ultraviolet light (which cannot penetrate the Earth's atmosphere). These light sources are collected by Hubble's main mirror, reflected onto its smaller mirror, and then sent to its scientific instruments to be analysed. The information is then beamed to Earth. Using its high-gain antenna, Hubble records and beams back about 120 gigabytes of information every week. Among the telescope's images was the ***Hubble Deep Field***, a photograph showing around 1,500 galaxies! Hubble's camera can either take wide-field images of a stretch of space, or a high-resolution image of a single object such as a planet. Because it is in space, Hubble's camera can take an image with a resolution ten times greater than any telescope on Earth. It can also detect objects that are fifty times fainter. Hubble is the size of a bus, and cost over USD 1.4 billion, but initially the telescope's main mirror was not correctly aligned; being out of position by 1.3 mm (about 0.05 inches).

A 1993 shuttle repair mission fixed that problem. Hubble was later serviced five times: replacing cameras, instruments, solar panels, and gyroscopes.

Launched on October 06, the shuttle *Discovery* on her eleventh flight (mission STS-41) carried the joint ESA and NASA *Ulysses* probe as part of the International Solar Polar Mission (ISPM). *Ulysses* was the first human-made object to travel in space outside the ecliptic plane of solar orbit. For the first time, a Canadian scientific instrument was launched beyond Earth's orbit. The High Flux Telescope (HFT), designed by the National Research Council's **Herzberg Institute of Astrophysics**, was one of the many instruments carried aboard *Ulysses*. The STS-41 crew were Commander Richard N. Richards (on his second spaceflight), Pilot Robert D. Cabana, and Mission Specialists Bruce E. Melnick, William M. Shepherd (on his second spaceflight), and Thomas D. Akers. Cabana, Melnik and Akers were on their first spaceflights. Their sixty-six-orbit mission lasted four days, two hours, ten minutes and four seconds; traveling 2,747,866 km (about 1,707,445 miles). The launch mass was 117,749 kg (259,592 lb), the payload mass was 15,362 kg (33,867 lb), and the landing mass was 89,278 kg (196,868 lb).

At launch, the Jupiter orbiter and probe together had a mass of 2,562 kg (5,648 lb) and stood 6.15 metres tall. The STS-34 mission total payload was 22,064 kg (48,643 lb). *Galileo* used gravity assist while en-route to Jupiter, passing by at a minimum distance of 16,000 km (about 9,942 miles). The mission was planned to last eight years and one month, with seven years and nine months in Jupiter orbit, but in fact it lasted for thirteen years, eleven months and three days.

Launched on November 15, the shuttle *Atlantis* on her seventh flight (mission STS-38) was also the seventh classified Department of Defense mission. It was originally scheduled for July 09, 1990, however, a liquid hydrogen leak found on shuttle *Columbia* during the STS-35 countdown prompted three precautionary tanking tests on *Atlantis* at the pad [on June 29, July 13, and on July 25, 1990]. Tests confirmed the leak on the external tank side of the ET/orbiter quick disconnect umbilical. After some unrelated minor damage was repaired, the fourth mini-tanking test was performed on October 24, with no excessive hydrogen or oxygen leakage detected. The launch was set for November 09, but deferred to November 15 due to payload problems. The crew were Commander Richard O. Covey (on his third spaceflight), Pilot Frank L. Culbertson, Jr., and Mission Specialists Carl J. Meade, Robert C. Springer (on his second and last spaceflight) and Charles D. Gemar. Culbertson, Meade and Gemar were on their first spaceflights. The payload was deployed into orbit, and the thirty-seventh shuttle mission, which made seventy-nine orbits lasted for four days, 21 hours, 54 minutes and 31 seconds; traveling 3,291,199 km (2,045,056 miles). The landing mass was 86,677 kg (191,090 lb).

Launched on December 02, the shuttle *Columbia* on mission STS-35 made use of the ASTRO-1 Spacelab observatory, consisting of four telescopes. The crew were Commander Vance D. Brand (on his fourth and last spaceflight), Pilot Guy S. Gardner (on his second and last spaceflight), Mission Specialists Jeffrey A. Hoffman (on his second spaceflight), John M. "Mike" Lounge (on his third and last spaceflight), Robert A. Parker (on his second and last spaceflight), and Payload Specialists Samuel T. Durrance and Ronald A. Parise (both on their first spaceflight). Their 144-orbit mission lasted for eight days, twenty-three hours, five minutes and eight seconds; traveling 6,000,658 km (about 3,728,636 miles). The launch mass was a very heavy 121,344 kg, the landing mass was 102,208 kg and the payload mass was 12,095 kg (26,665 lb).

# 1991

Launched on April 05, the *Atlantis* STS-37 mission deployed the primary payload, the Compton Gamma Ray Observatory (CGRO) on flight day three. The crew were Commander Steven R. Nagel (on his third spaceflight), Pilot Kenneth D. Cameron, and Mission Specialists Jerome "Jay" Apt III, Jerry L. Ross (on his third spaceflight) and Linda M. Godwin. Cameron, Apt and Godwin were on their first spaceflights. The *CGRO Observatory* high-gain antenna failed to deploy on command (after six attempts), so an unscheduled spacewalk by astronauts Jay Apt and Jerry Ross freed and manually deployed it. *CGRO* was the second of NASA's four Great Observatories and was launched on a two-year mission to search for high-energy celestial gamma ray emissions, which cannot penetrate Earth's atmosphere. At around 16,000 kg (about 35,274 lb), *CGRO* was the heaviest satellite to be deployed by a space shuttle to date. It was also designed to be the first satellite that could be refueled in orbit by Shuttle crews. The mission lasted five days, 23 hours, 32 minutes and 44 seconds; traveling 4,002,559 km (2,487,075 miles). The launch mass was 116,040 kg, the landing mass was 86,227 kg and the payload mass was 17,204 kg (37,928 lb).

Launched on April 28, the shuttle *Discovery* on her twelfth flight (mission STS-39) had a classified payload consisting of a Multi-Purpose Release Canister (MPRC). The other unclassified payload was known as Air Force Program-675 (AFP-675), which was an updated version of CIRRIS. The crew were Commander Michael Coats (on his third and last spaceflight), Pilot L. Blaine Hammond, and Mission Specialists Gregory J. Harbaugh, Donald R. McMonagle, Guion "Guy" Bluford (on his third spaceflight) Charles L. Veach and Richard J. Hieb. Hammond, Harbaugh, McMonagle, Veach and Hieb were on their first spaceflights. Their 134-orbit mission lasted for eight days, seven hours, 22 minutes and 23 seconds; traveling 5,584,423 km (3,470,000 miles). The launch mass was 112,207 kg, the landing mass was 102,755 kg and the payload mass was a very light 5,663 kg (12,485 lb).

Launched on June 05, the shuttle *Columbia* on her eleventh flight (mission STS-40) carried 'Spacelab' Module LM1 for Spacelab Life Sciences. The crew (the first including three women) were Commander Bryan D. O'Connor (on his second and last spaceflight), Pilot Sidney M. Gutierrez (on his first spaceflight), Mission Specialists James P. Bagian (on his second and last spaceflight), Tamara G. Jernigan (on her first spaceflight), Margaret "Rhea" Seddon (on her second spaceflight), and Payload Specialists F. Drew Gaffney (on his only spaceflight) and Millie Hughes-Fulford (on her only spaceflight). Their 146-orbit mission lasted nine days, two hours, 14 minutes and 20 seconds; traveling 6,083,223 km (about 3,779,939.5 miles). The launch mass was 114,290 kg, the landing mass was 102,283 kg and the payload mass was 12,374 kg (27,280 lb).

Launched on August 02, the shuttle *Atlantis* on her ninth flight (mission STS-43) deployed a *Tracking and Data Relay Satellite* (TDRS-E). The flight also tested an advanced heatpipe radiator for potential use on the then-future ISS and conducted a variety of medical and science investigations. The crew were Commander John E. Blaha (on his third spaceflight), Pilot Michael A. Baker (on his first spaceflight), and Mission Specialists Mrs. Shannon Lucid (on her third spaceflight), G. David Low (on his second spaceflight) and James C. Adamson (on his second and last spaceflight). Their 142-orbit mission lasted eight days, 21 hours, 21 minutes and 25 seconds; traveling 5,955,217 km (3,700,400 miles). The launch mass was 117,650 kg, the landing mass was 89,239 kg and the payload mass was a very heavy 21,067 kg (46,445 lb), the third-heaviest ever carried on a Shuttle.

Launched on September 12, 1991, the shuttle *Discovery* on mission STS-48 carried the classified primary payload of the *Upper Atmosphere Research Satellite* (UARS), the first satellite of NASA's Mission to Planet Earth program. Secondary payloads included:

Ascent Particle Monitor (APM), Middeck O-Gravity Dynamics Experiment (MODE), Shuttle Activation Monitor (SAM), Cosmic Ray Effects and Activation Monitor (CREAM), Physiological and Anatomical Rodent Experiment (PARE), Protein Crystal Growth II-2 (PCG II-2), Investigations into Polymer Membrane Processing (IPMP), and Air Force Maui Optical Site (AMOS) experiment. The crew were Commander John O. Creighton (on his third and last spaceflight), Pilot Kenneth S. Reightler, Jr. (on his first spaceflight), and Mission Specialists Charles D. Gemar (on his second spaceflight), James F. Buchli (on his fourth and last spaceflight) and Mark N. Brown (on his second and last spaceflight). During orbit on September 15, video showed a flash of light and several objects that appeared to be flying in a controlled fashion near STS-48. NASA explained the objects as ice particles reacting to engine jets. Their eighty-one-orbit flight lasted five days, eight hours, 27 minutes and 38 seconds; traveling 3,530,369 km (2,193,670 miles) and landing at Edwards AFB on September 18. The payload mass was 7,865 kg (17,339 lb).

Launched on November 24, the shuttle *Atlantis* on mission STS-44 deployed the Department of Defense's satellite/inertial upper stage (IUS) into a 195 nautical mile Earth orbit at an inclination of 28.45 degrees. The payload mass of 20,240 kg (44,620 lb) was for the Defense Support Program. The crew were Commander Frederick D. Gregory (on his third and last spaceflight), Pilot Terence T. Henricks, Mission Specialists James S. Voss, F. Story Musgrave, Mario Runco, Jr., and Payload Specialist Thomas J. Hennen (on his only spaceflight). Henricks, Voss and Runco were on their first spaceflights.

The crew performed all operations necessary to support the requirements of several secondary payloads/ experiments, including: Terra Scout, Military Man in Space (M88-1), Air Force Maui Optical System Calibration Test (AMOS), Cosmic Radiation Effects and Activation Monitor (CREAM), Shuttle Activation Monitor (SAM), Radiation Monitoring Equipment-3 (RME-3), Visual Function Tester-1 (VFT-1), and the Interim Operational Contamination Monitor (IOCM). Their classified 110-orbit mission lasted six days, 22 hours, 50 minutes and 44 seconds; traveling 4,651,112 km (2,890,067 miles). It was the last space shuttle mission to land on a dry lakebed runway. The launch mass was 117,766 kg, the landing mass was 87,919 kg and the payload mass was a hefty 20,240 kg (44,620 lb).

## 1992

Launched on January 22 after a one-hour delay due to weather constraints, the shuttle *Discovery* on mission STS-42 carried Module LM2 for 'Spacelab'. The main goal of this mission was to study the effects of microgravity on a variety of organisms. The seven crew were Commander Ronald L. Grabe (on his third spaceflight), Pilot Stephen S. Oswald, Mission Specialists Norman E. Thagard (on his fourth spaceflight), William F. Readdy, and David C. Hilmers (on his fourth and last space flight), plus Payload Specialists Dr. Roberta Bondar (of CSA), and Ulf Merbold of ESA (on his second spaceflight). Oswald, Readdy and Bondar were on their first spaceflight. Their 129-orbit mission lasted eight days, one hour, 14 minutes and 44 seconds; traveling 4,701,140 km (2,921,150 miles). Dr. Roberta Bondar became the second Canadian, and the first Canadian woman in space. She participated in the first International Microgravity Laboratory Mission. The launch mass was 110,400 kg, the landing mass was 98,924 kg and the payload mass was 13,066 kg (28,806 lb).

Launched on March 24, the shuttle *Atlantis* on her eleventh flight (mission STS-45) (aka mission ATLAS-1) carried the *Igloo* (two Pallets; F004 and F005). This mission had a non-deployable payload of twelve scientific instruments. STS-45 carried the first Atmospheric Laboratory for Applications and Science (ATLAS-1) experiments, placed on 'Spacelab' pallets mounted in the orbiter's payload bay. These studied atmospheric chemistry, solar radiation, space plasma physics and ultraviolet astronomy. The mission was extended by a day to continue science experiments. The crew were Commander Charles F. Bolden, Jr., (on his third spaceflight), Pilot Brian Duffy, Mission Specialists C. Michael Foale (of the UK), David C. Leestma (on his third and last spaceflight), and Kathryn D. Sullivan (on her third and last spaceflight), plus Payload Specialists Dirk Frimout (the first Belgian citizen in space, on his only spaceflight) and Byron K. Lichtenberg (on his second and last spaceflight). Duffy and Foale were on their first spaceflights. Their 143-orbit mission lasted eight days, 22 hours, nine minutes and 28 seconds; traveling 5,211,340 km (3,238,180 miles). The launch mass was 105,982 kg, the landing mass was 93,009 kg and the payload mass was only 9,947 kg (21,929 lb).

Launched on May 07, the maiden flight of the shuttle *Endeavour* on mission STS-49 retrieved *Intelsat VI* (aka *Intelsat 603*) which failed to leave low-Earth orbit two years before, attached it to a new upper stage, and relaunched it to its intended geosynchronous orbit. The first and only three-person EVA, a record-breaking eight hours and twenty-nine minutes was accomplished. The crew were Commander Daniel Brandenstein (on his fourth and last spaceflight), Pilot Kevin P. Chilton (on his first spaceflight), and five Mission Specialists: Richard J. Hieb (on his second spaceflight), Bruce E. Melnik (on his second and last spaceflight), Pierre J. Thuot (on his second spaceflight), Kathryn C. Thornton (on her second spaceflight) and Thomas D. Akers (on his second spaceflight). Their 141-orbit mission lasted eight days, 21 hours, 17 minutes and 38 seconds; traveling 5,948,166 km (3,696,019 miles). The space shuttle made its first landing with a drag parachute. The launch mass was 116,390 kg, landing mass was 91,279 kg and the payload mass was 16,984 kg (37,443 lb).

Launched on June 25, the shuttle *Columbia* on her twelfth mission (STS-50) carried Module LM1 on a 'spacelab' mission. The US Microgravity Laboratory-1 (USML-1) conducted experiments in material science, fluid physics and biotechnology. It was the first flight of a Space Shuttle with the Extended Duration Orbiter (EDO) hardware, allowing longer flight durations. Secondary experiments were Investigations into Polymer Membrane Processing (IPMP); Shuttle Amateur Radio Experiment (SAREX-II); and Ultraviolet Plume Instrument (UVPI). *Columbia* landed at KSC for the first time ever, due to bad weather at Edwards Air Force Base caused by the remnants of Hurricane Darby. The crew were Commander Richard N. Richards (on his third spaceflight), Pilot Ken Bowersox (on his first spaceflight), Mission Specialists Bonnie J. Dunbar (on her third spaceflight), Ellen S. Baker (on her second spaceflight), and Carl J. Meade (on his second spaceflight), plus Payload Specialists Lawrence J. DeLucas and Eugene H. Trinh. Bowersox and Trinh were on their first spaceflights. Their 221-orbit mission lasted thirteen days, 19 hours, 30 minutes and four seconds; traveling 9.2 million km (5.7 million miles). The launch mass was 116,693 kg, the landing mass was 103,814 kg, and the payload mass was 12,101 kg (26,678 lb).

Launched on July 31, the shuttle *Atlantis* on mission STS-46 (mission TSS-1) carried one Pallet (F003) and landed on August 08. The crew were Commander Loren Shriver (on his third and last spaceflight), Pilot Andrew M. Allen, and five Mission Specialists: Claude Nicollier (Swiss-born, of ESA) the first non-American Mission Specialist, Marsha S. Ivins (on her second spaceflight), Jeffrey A. Hoffman (on his third spaceflight), Franklin Chang-Diaz (on his third spaceflight), and Franco Malerba (of ASI) on his only spaceflight. Allen and Nicollier were on their first spaceflights. Their primary objective was to deploy the ESA's *European Retrievable Carrier* (EURECA) and the joint NASA/ASI (Italian Space Agency) *Tethered Satellite System* (TSS-1). TSS-1 had problems with a jammed tether line, failing to reach the planned twenty km (about 12.43 mile) separation from the orbiter, so the satellite was stowed for return to Earth. [It was reflown in 1996 on STS-75, with astronauts Allen, Hoffman, Nicollier and Chang-Diaz also flying again on that mission]. The payload mass was 12,164 kg (26,817 lb). With a flight duration of seven days, 23 hours, 15 minutes and three seconds, mission STS-46 completed 127 orbits; traveling 5,344,643 km (3,321,007 miles). The launch mass was 116,134 kg, the landing mass was 94,676 kg and the payload mass was 12,164 kg (26,817 lb).

Launched on September 12, the shuttle *Endeavour* on her second mission, STS-47 (J) (aka Mission Spacelab-J), [the fiftieth NASA Space Shuttle mission, which was funded by Japan], carried Module LM2, a collaborative laboratory inside the shuttle's payload bay. Mamoru Mohri was the first professional Japanese astronaut aboard the space shuttle; Mark C. Lee and Ms N. Jan Davis were the first married couple to fly to space together (contrary to NASA policy, but they did not disclose their relationship until the flight was imminent); and Mae C. Jemison became the first African American woman in space. The crew were Commander Robert L. Gibson (on his fourth spaceflight), Pilot Curtis Brown, Mission Specialists Mark C. Lee (on his second spaceflight), Jerome "Jay" Apt III (on his second spaceflight), Ms N. Jan Davis, and Mae Jemison (on her only spaceflight), plus Japanese 'Payload Specialist' Mamoru Mohri of NASDA. Brown, Davis and Mohri were all on their first spaceflights. Their 126-orbit mission lasted seven days, 22 hours, 30 minutes and 24 seconds; traveling 5,265,523 km (3,271,844 miles). The launch mass was 117,335 kg, the landing mass was 99,450 kg and the payload mass was 12,485 kg (27,525 lb).

Launched on October 22, the shuttle *Columbia* STS-52 mission had the primary objectives of deploying the *Laser Geodynamics Satellite 2* (*LAGEOS-2*) and operation of the US Microgravity Payload-1 (USMP-1). *LAGEOS-2*, a joint effort between NASA and the ASI, was boosted into an initial elliptical orbit by ASI's Italian Research Interim Stage (IRIS), and the craft's apogee kick motor later circularised its orbit at its operational altitude of 5,900 km (3,700 miles). There were numerous secondary payloads. The crew were Commander Jim Wetherbee (on his second spaceflight), Pilot Michael A. Baker (on his second spaceflight), Charles L. Veach (on his second and last flight), William M. Shepherd (on his third spaceflight), Tamara E. Jernigan (on her second spaceflight), and Steve MacLean (of CSA), the third Canadian astronaut in space. Their 159-orbit mission lasted nine days, 20 hours, 56 minutes and 13 seconds; traveling 6,645,026 km (4,129,028 miles). The launch mass was 113,460 kg, the landing mass was 97,574 kg and the payload mass was a small 8,078 kg (17,809 lb).

The shuttle *Discovery* launched on December 02 for mission STS-53, but details of the 'USA-89', or DoD-1 payload for the tenth United States Department of Defense mission remains partially classified. The launch was delayed for one hour and twenty-five minutes because of ice buildup on the External Tank. The *ODERACS* payload was unable to be deployed because of payload equipment malfunction. The crew were Commander David M. Walker (on his third spaceflight) Pilot Robert D. Cabana (on his second spaceflight), and Mission Specialists Guion "Guy" S. Bluford (on his fourth and last spaceflight), Michael R. Clifford (on his first spaceflight) and James S. Voss (on his second spaceflight). A classified primary payload 'USA-89' (1992-086B) was carried for the US Department of Defense. The 116-orbit mission lasted seven days, seven hours, 19 minutes and 17 seconds; traveling 4,883,843 km (3,034,679 miles). The launch mass was 1110,655 kg, the landing mass was 87,565 kg and the payload mass was 11,860 kg (26,150 lb).

The DoD satellite was the third launch of a *Satellite Data System-2* (SDS 2-3) military communications satellite (after *USA-40* on mission STS-28, and *USA-67* deployed on the STS-38 mission). Secondary payloads were: Orbital Debris Radar Calibration Spheres (ODERACS) satellites and the combined Shuttle Glow Experiment/Cryogenic Heat Pipe Experiment (GCP), while the mid-deck experiments included Battlefield Laser Acquisition Sensor Test (BLAST), Cloud Logic to Optimize Use of Defense Systems (CLOUDS), Fluid Acquisition and Resupply Experiment (FARE), 'Hand-held, Earth-oriented, Real-time, Cooperative, User-friendly, Location-targeting and Environmental System' (HERCULES), Microcapsules in Space (MIS-I), Radiation Monitoring Equipment (RME-III), Space Tissue Loss (STL), and Visual Function Tester (VFT-2).

# 1993

Launched on January 13, the shuttle *Endeavour* on her third flight (mission STS-54) deployed a *Tracking and Data Relay Satellite* (TDRS-F). The crew were Commander John Casper (on his second spaceflight), Pilot Donald R. McMonagle (on his second spaceflight) and three Mission Specialists: Mario Runco, Jr., (on his second spaceflight), Gregory J. Harbaugh (on his second spaceflight) and Susan J. Helms on her first spaceflight. Their ninety-six-orbit mission lasted five days, 23 hours, 38 minutes and 19 seconds; traveling 4.0 million km (2.5 million miles). The payload mass was 18,559 kg (40,916 lb).

Launched on April 08, the shuttle *Discovery* STS-56 (mission ATLAS-2) carried *Igloo* (one Pallet F008) and the ATLAS-2 science platform. The crew were Commander Kenneth D. Cameron (on his second spaceflight), Pilot Stephen S. Oswald (on his second spaceflight), and three Mission Specialists: C. Michael Foale (on his second spaceflight), Kenneth Cockrell and Ellen L. Ochoa, both on their first spaceflight. The primary payload was the Atmospheric Laboratory for Applications and Science-2 (ATLAS-2), designed to collect data on the relationship between the Sun's energy output and Earth's middle atmosphere and how these factors affect the ozone layer. The mission also deployed the *Shuttle Point Autonomous Research Tool for Astronomy-201* (*SPARTAN-201*), a free-flying science instrument platform designed to study velocity and acceleration of the solar wind and observe the Sun's corona. The crew made numerous radio contacts to schools around the world using the Shuttle Amateur Radio Experiment (SAREX II), including brief radio contact with the *Mir* space station. Their 148-orbit mission lasted nine days, six hours, eight minutes and 24 seconds; traveling 6,202,407 km (3,853,997 miles). The landing mass was 93,683 kg, and the payload mass was only 7,026 kg (15,490 lb).

Launched on April 26, the shuttle *Columbia* on her 14th flight (mission STS-55, or Deutschland 2) (mission Spacelab D2) carried Module LM1. The mission was funded by Germany. The crew were Commander Steven R. Nagel (on his fourth and last spaceflight), Pilot Terence T. Henricks (on his second spaceflight), Mission Specialists Jerry L. Ross (on his fourth spaceflight), Charles J. Precourt, Bernard A. Harris, Jr. (both on their first spaceflight), and Payload Specialists Ulrich Walter of Germany's DLR (on his only spaceflight) and German astronaut Hans Schlegel (of DLR) on his first spaceflight. Their 160-orbit mission lasted nine days, 23 hours, 39 minutes and 59 seconds; traveling 6,701,603 km (4,164,183 miles). The landing mass was 103,191 kg, and the payload mass was 11,539 kg (25,439 lb).

Launched on June 21, shuttle *Endeavour* on mission STS-57 carried Commander Ronald J. Grabe (on his fourth and last spaceflight), Pilot Brian Duffy (on his second spaceflight), Payload Commander G. David Low (on his third and last spaceflight), and Mission Specialists Nancy Jane Sherlock (later Currie-Gregg), Janice E. Voss, and Peter J.K. Wisoff, all on their first spaceflight. The flight was originally scheduled for mid-May, 1993 but was moved to June to allow lift-off and landing in daylight. The mission was extended by two days, owing to poor weather at KSC. Their 155-orbit mission lasted nine days, 23 hours, 44 minutes and 54 seconds; traveling 6,608,628 km (4,106,411 miles). The landing mass was 101,657 kg and the payload mass was 13,074 kg (28,823 lb).

STS-57 was the first flight of the commercial 'Spacehab' pressurized laboratory, which effectively doubled the pressurized workspace available for astronauts to tend experiments. The crew captured the *European Retrievable Carrier* (*EURECA*) orbiting science satellite deployed in August 1992 during STS-46. On June 25, Low and Wisoff spent the first portion of a five-hour, fifty-minute spacewalk manually folding the craft's antennas for stowage. The remainder of the EVA was spent testing space station assembly techniques and procedures necessary for upcoming Hubble Space Telescope repair missions.

Launched on September 12, the shuttle *Discovery* on her sixth flight (mission STS-51), the twentieth of NASA's Space Shuttle program deployed the *Advanced Communications Technology Satellite* (ACTS), and deployed and retrieved the *Orbiting Retrievable Far and Extreme Ultraviolet Spectrometer* (ORFEUS), with its IMAX camera. The crew were Commander Frank L. Culbertson, Jr., (on his second spaceflight), Pilot William F. Readdy (on his second spaceflight), and three Mission Specialists: James H. Newman, Daniel W. Bursch and Carl E. Walz, all of whom were on their first spaceflight. Their 157-orbit mission lasted nine days, 20 hours, 11 minutes and six seconds; traveling 6,608,628 km (4,106,411 miles). The landing mass was 92,371 kg, and the payload mass was 18,947 kg (41,771 lb).

Mission STS-51 was notable for having been scrubbed three times on the launchpad, each time after the crew were aboard. On July 17, there was a flaw in the pyrotechnic initiator that triggers the release of the SRBs from the mobile launcher platform. On July 24, problems arose with a hydraulic power unit in one of the SRBs. Because of the Perseids meteor shower, the next launch window did not open until the second week of August 1993. On August 12, the count reached the T-3 second mark, at which point the Space Shuttle Main Engine (SSMEs) had ignited. A shutdown was then triggered by faulty fuel flow sensors in one of the SSMEs. The successful launch was on September 12, 1993.

Launched on October 18, the shuttle *Columbia* STS-58 carried module LM2 on a 'Spacelab' flight. The mission was primarily devoted to experiments concerning the physiological effects in space. This was the first in-flight use of the Portable In-flight Landing Operations Trainer (PILOT) simulation software. The crew were Commander John E. Blaha (on his fourth spaceflight), Pilot Richard A. Searfoss, Mission Specialists Margaret "Rhea" Seddon (Payload Commander, on her third and last spaceflight), William S. McArthur, David A. Wolf, and Mrs. Shannon Lucid (on her fourth spaceflight), plus Payload Specialist Martin J. Fettman on his only spaceflight. Their 225-orbit mission lasted fourteen days, zero hours, 12 minutes and 32 seconds; traveling 9.4 million km (5.8 million miles). The landing mass was 103,146 kg, and the payload mass was 11,803 kg (26,021 lb).

Launched on December 02, the shuttle *Endeavour* on her fifth flight (mission STS-61) carried one Pallet (F009), and then repaired the three-year-old orbiting Hubble Space Telescope. Hubble was berthed on December 04 until December 09, for five days and one hour. A finely-ground corrective lens package (COSTAR) was installed to correct the optics, enabling the telescope to better capture and send images in detail never seen before. The Hubble main camera was replaced; plus, there were instrument upgrades and new solar arrays. The crew were Commander Richard O. Covey (on his fourth and last spaceflight), Pilot Ken Bowersox (on his second spaceflight), and Mission Specialists Kathryn C. Thornton (on her third spaceflight), Claude Nicollier of ESA (on his second spaceflight), Jeffrey A. Hoffman (on his fourth spaceflight), F. Story Musgrave (on his fifth spaceflight) and Thomas D. Akers (on his third spaceflight).

The crew of STS-61 made a record five EVAs, which were shared between two pairs of different astronauts, with Musgrave and Hoffman making three EVA. Their durations lasted from 6.5 to almost eight hours. During the mission, Jeffrey Hoffman spun a dreidel for the Jewish holiday of Hanukah to a live audience watching via satellite feed. Their 163-orbit mission lasted for ten days, 19 hours, 58 minutes and 33 seconds; traveling 7,135,464 km (4,433,772 miles). The launch mass was 113,541 kg, the landing mass was 94,972 kg, and the payload mass was just 8,011 kg (17,661 lb).

Space Shuttles flew four subsequent missions to the Hubble Telescope, the last one in 2009. The *Space Shuttle-Mir* Program started in 1993, with the first crewed flight in 1994 and the last in 1998. The International Space Station program also started in 1993, with the first crewed mission in 1998, and activities are ongoing; but will perhaps be ending by 2028-30.

# 1994

Launched on February 03, the shuttle *Discovery* STS-60 mission was the first of the *Shuttle-Mir* Program, which carried Sergei K. Krikalev, the first Russian cosmonaut to fly aboard a Space Shuttle. The mission carried the *Wake Shield Facility* experiment and a 'SPACEHAB' module, (developed by SPACEHAB Inc.) into orbit. Astronauts carried out a live bi-directional audio and downlink link-up with cosmonauts abord the Russian space station *Mir*. The *Discovery* crew were Commander Charles F. Bolden, Jr., (on his fourth and final spaceflight), Pilot Kenneth S. Reightler, Jr. (on his second and last spaceflight), Ms N. Jan Davis (on her second spaceflight), Ronald M. Sega (on his first spaceflight), Franklin R. Chang-Diaz (on his fourth spaceflight), and 'Mission Specialist 4' Sergei K. Krikalev of Roscosmos (RKA, the Russian Federal Space Agency) on his third spaceflight. Their mission duration was eight days, seven hours, nine minutes and 22 seconds; traveling 5,535,667 km (3,439,704 miles). The landing mass was 97,448 kg; payload mass was 10,231 kg (22,555 lb).

Launched on March 04, the shuttle *Columbia* on mission STS-62 carried the primary payloads of the 'USMP-02' microgravity experiments package and the 'OAST-2' engineering and technology payload, both in the orbiter's cargo bay. The mission also featured several biomedical experiments focusing on the effects of long-duration spaceflight. The landing was chronicled by the 1994 ***Discovery Channel*** special about the Space Shuttle program and served as the show's opening. A C.F. Martin backpack guitar was also flown aboard *Columbia* during this mission. The crew were Commander John H. Caspar (on his third spaceflight), Pilot Andrew M. Allen (on his second spaceflight), and three Mission Specialists: Pierre J. Thuot (on his third and last spaceflight), Charles D. Gemar (on his third and last spaceflight) and Marsha S. Ivins (on her third spaceflight). Their 224-orbit mission lasted for thirteen days, 23 hours, 16 minutes and 41 seconds; traveling 9,366,617 km (5,820,146 miles). The landing mass was 102,861 kg and the payload mass was 8,759 kg (19,310 lb).

Launched on April 09, the shuttle *Endeavour* STS-59 carried one Pallet (F006). The crew were Commander Sidney M. Gutierrez (on his second and last spaceflight), Pilot Kevin P. Chilton (on his second spaceflight), and four Mission Specialists: Jerome "Jay" Apt III (on his third spaceflight), Michael R. Clifford (on his second spaceflight), Linda M. Godwin (on her second spaceflight) and Thomas D. Jones on his first spaceflight. Their 183-orbit mission lasted eleven days, five hours, forty-nine minutes and 30 seconds; traveling 7,571,762 km (4,704,875 miles). The payload mass was 12,490 kg (27,540 lb).

Scheduled for July 15, STS-62-A was a planned NASA Space Shuttle mission to deliver the *Teal Ruby* reconnaissance satellite payload into polar orbit, but it was not launched. The satellite was designed to detect heat and give early warning of enemy aircraft crossing the polar region towards the United States during the Cold War. High-performance CCD-multiplexer arrays were to transduce infrared radiation into electric signals, detecting sources in the 2-to-16-micron band. The Cold War shaped international relations, influenced military strategies, and led to significant advancements in science and technology.

Launched on July 08, the shuttle *Columbia* STS-65 carried Module LM1 (payload mass 10,811 kg) for a 'Spacelab' mission. The crew were Commander Robert D. Cabana (on his third spaceflight; he would later lead the Kennedy Space Center), Pilot James D. Halsell, Mission Specialists Richard J. Hieb (on his third and last spaceflight), Carl E. Walz (on his second spaceflight), Leroy Chiao, Donald A. Thomas and Payload Specialist Chiaki Mukai (of NASDA) on his first spaceflight. Halsell, Chiao, Thomas and Mukai were all on their first spaceflights. Their 235-orbit mission lasted fourteen days, 17 hours and 55 minutes; traveling 9,886,200 km (6,143,000 miles). The payload mass was 10,811 kg (23,834 lb). For the STS-65 mission, CSA astronaut Marc Garneau became the first non-American CapCom (Capsule Communicator) at Mission Control in Houston, Texas. On September 02, Chris A. Hadfield of CSA completed Mission Specialist training and was assigned to Shuttle mission STS-74.

Launched on September 09, the shuttle *Discovery* on STS-64 carried one Pallet (F007) and multiple science experiments. The crew were Commander Richard N. Richards (on his fourth and last spaceflight), Pilot L. Blaine Hammond, Jr. (on his second and last spaceflight), and four Mission Specialists: Jerry M. Linenger (on his first spaceflight), Susan J. Helms (on her second spaceflight), Carl J. Meade (on his third and last spaceflight) and Mark C. Lee (on his third spaceflight). Their 176-orbit mission lasted ten days, 22 hours, 49 minutes and 57 seconds; traveling 7,242,048 km (4.5 million miles). The payload mass was 9,260 kg.

Launched on September 30, the shuttle *Endeavour* on mission STS-68 carried one Pallet (F006), and experiments aboard the Space Radar Laboratory-2. The crew were Commander Michael A. Baker (on his third spaceflight), Pilot Terrence W. Wilcutt, and four Mission Specialists: Steven L. Smith, Daniel W. Bursch (on his second spaceflight), Peter J.K. Wisoff (on his second spaceflight) and Thomas D. Jones (on his second spaceflight). Wilcutt and Smith were on their first spaceflight. Their 182-orbit mission lasted eleven days, five hours, 46 minutes and eight seconds; traveling 7,569,092 km (4,703,216 miles).

Launched on November 03, the shuttle *Atlantis* on STS-66 (mission ATLAS-3) carried the Igloo Pallet (F008) and an ATLAS-3 science platform. The crew were Commander Donald R. McMonagle (on his third and last spaceflight), Pilot Curtis L. Brown, Jr., (on his second spaceflight), and four Mission Specialists: Ellen L. Ochoa (on her second spaceflight), Joseph R. Tanner, Jean-Francois Clervoy (of CNES), and Scott E. Parazynski, all on their first spaceflight. Their 174-orbit mission lasted ten days, 22 hours, 34 minutes and two seconds; traveling 7,330,226 km (4,554,791 miles). The payload mass was 10,544 kg (23,246 lb).

# 1995

Launched on February 03, the shuttle *Discovery* on mission STS-63 was the second of the US/Russian *Shuttle-Mir* Program. The diverse crew included Commander James D. Wetherbee (on his third spaceflight); two women (Eileen Collins, the first woman to pilot a space shuttle, and Mission Specialist Janice E. Voss on her second spaceflight); C. Michael Foale (on his third spaceflight); an African American, Bernard A. Harris, Jr., (on his second and last spaceflight); and a cosmonaut, Vladimir G. Titov (on his third spaceflight). The mission included the first EVAs for both a UK-born astronaut (Foale), and a United States astronaut of African heritage, Bernard Harris. The primary objectives were to perform a rendezvous and fly around the Russian space station *Mir*, accomplish the 'Spacehab-3' experiments in a pressurised module, and deploy and retrieve the *Shuttle Pointed Autonomous Research Tool for Astronomy-204* (SPARTAN-204) payload. Known as the "Near Mir" mission, the crew completed 129 orbits in eight days, six hours, 28 minutes and 15 seconds; traveling 4,816,454 km (2,992,806 miles). The payload mass was only 8,641 kg (19,050 lb).

Launched on March 02, the shuttle *Endeavour* on mission STS-67 deployed the *ASTRO-2* satellite. Along with the Igloo, *Endeavour* carried two Pallets (F002 and F010) + IPS + EDO. The crew were Commander Stephen S. Oswald (on his third and last spaceflight), Pilot William G. Gregory (on his only spaceflight), Mission Specialists John M. Grunsfeld, Wendy B. Lawrence, and Tamara E. Jernigan (on her third spaceflight), plus Payload Specialists Samuel T. Durrance his (on his second and last spaceflight) and Ronald A. Parise (on his second and last spaceflight). Grunsfeld and Lawrence were on their first spaceflights. Their 262-orbit mission lasted sixteen days, fifteen hours, eight minutes and 48 seconds; traveling 11.1 million km (6.9 million miles). The payload mass was 13,116 kg (28,916 lb).

On June 27, the shuttle *Atlantis* on mission STS-71 carrying Module LM2 linked up with Russia's space station, *Mir* in an historic first docking of these spacecraft from the two once-competitive nations. STS-71 was the one hundredth US human space launch conducted from Cape Canaveral and had the first on-orbit changeout of shuttle crew. It delivered supplies, and a relief crew of cosmonauts Anatoly Solovyev and Nikolai Budarin to *Mir*. While docked to *Mir* for five days, the two craft became the largest spacecraft in orbit at the time. Various on-orbit joint life science investigations were done in 'Spacelab', along with the Shuttle Amateur Radio Experiment-II (SAREX-II) experiment. The shuttle crew were Commander Robert L. Gibson (on his fifth and last spaceflight), Pilot Charles J. Precourt (on his second spaceflight), and five Mission Specialists: Ellen S. Baker (on her third and last spaceflight), Gregory J. Harbaugh (on his third spaceflight), Bonnie J. Dunbar (on her fourth spaceflight), Anatoly Solovyev of RKA (on his fourth spaceflight) and Nikolai Budarin (of RKA), the latter two joining mission 'EO-19'.

Returning to Earth from mission 'EO-18' on *Mir* were Gennady Strekalov (on his fifth and last spaceflight), Vladimir Dezhurov (on his first spaceflight) and Norman E. Thagard (on his fifth and last spaceflight). Their 153-orbit STS-71 mission lasted nine days, 19 hours, 23 minutes and nine seconds; traveling 6,598,310 km (about 4.1 million miles). The payload mass was 12,191 kg (26,877 lb). The shuttle *Atlantis* landed on July 07 at Kennedy Space Center.

Launched on July 13, the shuttle *Discovery* on her twenty-first flight (mission STS-70) deployed the *Tracking and Data Relay Satellite* (TDRS-G). This was the first shuttle mission controlled from the new Mission Control Center room at the **Johnson Space Center** in Houston, Texas. STS-70 was also the first flight of the new Block 1 orbiter main engine, designed to improve both engine performance and safety. Engine number 2036 featured the new high-pressure liquid oxygen turbopump, a two-duct powerhead, baffleless main injector, single-coil heat exchanger and start sequence modifications. *Discovery* launched only six days after the landing of sister ship *Atlantis*, marking the fastest turnaround between flights in the space shuttle program. The crew were Commander Terence T. Henricks (on his third spaceflight), Pilot Kevin J. Kregel, and three Mission Specialists: Donald A. Thomas (on his second spaceflight), Nancy J. Currie (on her second spaceflight) and Mary E. Weber. Kregel and Weber were on their first spaceflights. Their 143-orbit mission lasted eight days, 22 hours, 20 minutes and five seconds; traveling 6.0 million km (3.7 million miles). The payload mass was a heavy 20,159 kg (44,443 lb).

STS-70 had originally moved ahead of STS-71 because of a delay in the launch of the Russian *Spektr* laboratory module to the *Mir* station. However, on May 31 shuttle managers assessed damage to the External Tank of STS-70 caused by nesting flicker woodpeckers, who had made about seventy-one holes in the ET's thermal protection foam insulation. On June 02, NASA managers decided to delay the launch of *Discovery* to make repairs to the holes, which ranged in size from 0.5 to four inches in diameter.

Launched on September 07, the shuttle *Endeavour* on mission STS-69 carried the *Wake Shield Facility* (WSF) and deployed and retrieved the *SPARTAN 201* astronomy satellite. The purpose of the WSF was to grow thin films in a near-perfect vacuum created by the wake of the satellite as it moved through space. The crew were Commander David M. Walker (on his fourth and last spaceflight), Pilot Kenneth Cockrell (on his second spaceflight), and three Mission Specialists: James S. Voss (on his third spaceflight), James H. Newman (on his second spaceflight) and Michael L. Gernhardt (on his first spaceflight). This was the 100th successful crewed NASA spaceflight. Crew members performed a six-hour EVA to test assembly techniques for the ISS, and tested thermal improvements made to spacesuits used during space walks. Their 171-orbit mission lasted ten days, 20 hours, 29 minutes and 56 seconds; traveling 7.2 million km (4.5 million miles). The payload mass was 11,499 kg (25,351 lb).

Launched on October 20, the shuttle *Columbia* on mission STS-73 carried Module LM1 for a 'Spacelab' mission. The crew were Kenneth D. Bowersox (on his third spaceflight), Pilot Kent V. Rominger, Mission Specialists Catherine G. Coleman, Michael Lopez-Alegria of Spain, and Kathryn C. Thornton (on her fourth and last spaceflight), plus Payload Specialists Fred W. Leslie and Albert Sacco, Jr., both on their only spaceflights. Rominger, Coleman and Lopez-Alegria were all on their first spaceflights. The USML-2 payload built on the USML-1 mission, which provided insights into theoretical models of fluid physics, the role of gravity in combustion and flame spreading, and how gravity affects the formation of semiconductor crystals. Data collected from several protein crystals grown on USML-1 enabled scientists to determine the molecular structures of those proteins. Their 255-orbit mission lasted fifteen days, 21 hours, 53 minutes and 16 seconds; traveling 10.6 million km (6.6 million miles). The payload mass was a medium 15,250 kg (33,620 lb).

Launched on November 12, the shuttle *Atlantis* on mission STS-74 made the fourth docking of a Space Shuttle with the *Mir* space station, delivering the docking module and the IMAX cargo bay camera. It was the second in a series of seven straight missions to the station flown by *Atlantis*. The shuttle delivered a pair of solar arrays, along with the fifteen-foot-long Russian-built Mir Docking Module to allow docking with the station by the shuttle without having to move *Mir*'s *Kristall* module. Aboard *Mir* were cosmonauts Yuri Gidzenko and Sergei Avdeyev, carrying out the 'Mir EO-20' expedition. During the three-day docking, the Russian, Canadian and American crew transferred supplies, water and equipment, and upgraded the station with new equipment.

The STS-74 crew were Commander Kenneth D. Cameron (on his third and last spaceflight), Pilot James L. Halsell, and three Mission Specialists: Chris A. Hadfield (of CSA), the fourth Canadian in space, and the first Canadian aboard *Mir*; Jerry L. Ross (on his fifth spaceflight); and William S. McArthur, Jr., on his second spaceflight. Their 128-orbit mission lasted eight days, four hours, 31 minutes and 42 seconds, traveling 5,500,000 kilometres (3.4 million miles). The launch mass was 112,358 kg, the landing mass was 92,701 kg, and the payload mass was 6,134 kg (13,523 lb), a rather light cargo.

## 1996

Launched on January 11, the shuttle *Endeavour* on mission STS-72 retrieved Japan's microgravity research *Space Flyer Unit* after ten months in orbit, and there were two EVAs. The crew were Commander Brian Duffy (on his third spaceflight), Pilot Brent W. Jett, and Mission Specialists Leroy Chiao (on his second spaceflight), Winston E. Scott, Koichi Wakata (of JAXA), and Daniel T. Barry, all of whom were on their first spaceflight. Their 142-orbit mission lasted eight days, 22 hours, one minute and 47 seconds; traveling 6.0 million km (3.7 million miles). The launch mass was 112,182 kg, the landing mass 98,549 kg, and the payload mass was just 6,510 kg (14,350 lb).

Launched on February 22, the shuttle *Columbia* on mission STS-75 (TSS-1R/USMP-3) carried one Pallet (F003) and two MPESS + EDO. There was a tethered satellite reflight; lost due to a broken tether. The crew were Commander Andrew M. Allen (on his third and last spaceflight), Pilot Scott J. Horowitz, and Mission Specialists Jeffrey A. Hoffman (on his fifth and last spaceflight), Maurizio Cheli (of ESA) on his only spaceflight, Claude Nicollier of ESA (on his third spaceflight), Franklin R. Chang-Diaz (Payload Commander, and on his fifth spaceflight), and Umberto Guidoni (of ASI) on his first spaceflight. Crew members Allen, Hoffman, Nicollier and Chang-Diaz had previously been members of the STS-46 crew, which had flown the TSS-1 experiment in 1992. Their 252-orbit STS-75 mission duration was fifteen days, 17 hours, 40 minutes and 22 seconds; traveling 10.5 million km (6.5 million miles). The payload mass was 10,592 kg (23,351 lb).

Launched on March 22, the shuttle *Atlantis* on her sixteenth mission (STS-76) made a *Shuttle-Mir* docking. The crew were Commander Kevin P. Chilton (on his third and last spaceflight), Pilot Richard A. Searfoss (on his second spaceflight), and Mission Specialists Ronald M. Sega (on his second and last spaceflight), Michael R. Clifford (on his third and last spaceflight), Linda M. Godwin (on her third spaceflight, and Mrs. Shannon Lucid (on her fifth and last record-making spaceflight). Mrs. Lucid replaced Norman E. Thagard aboard *Mir* for mission 'EO-21'. The STS-76 145-orbit mission lasted nine days, five hours, sixteen minutes and 48 seconds; traveling 6.1 million kn (3.8 million miles). The launch mass was 111,740 kg, the landing mass 95,396 kg, and the payload mass was just 6,753 kg (14,888 lb).

Launched on May 19, the space shuttle *Endeavour* on mission STS-77 carried the 'SPACEHAB' single module for microgravity research, and deployed and retrieved the *SPARTAN 207*/ IAE Inflatable Antenna Experiment. *Endeavour* also made rendezvous with a test satellite. The 'Spacehab' module carried nearly 1,400 kg (about 3086 lb) of experiments and support equipment for twelve commercial space product development payloads in the areas of biotechnology, electronic materials, polymers and agriculture; as well as several experiments for other NASA payload organizations. The Commercial Float Zone Facility (CFZF) was developed through international collaboration between the United States, Canada and Germany. It heated various samples of electronic and semiconductor material throught the float-zone technique.

A suite of four technology experiments known as the Technology Experiments for Advancing Missions in Space (TEAMS) also flew in the shuttle's payload bay. The Canadian "Getaway Special" (GAS) GAScan experiments were performed in a can stored in the cargo bay with little intervention required by astronauts. The STS-77 crew were Commander John H. Caspar (on his fourth and last spaceflight), Pilot Curtis L. Brown, Jr., (on his third spaceflight), and Mission Specialists Andrew S. Thomas (on his first spaceflight), Daniel W. Bursch (on his third spaceflight), Mario Runco, Jr., (on his third and last spaceflight) and Marc Garneau of CSA, the first Canadian to fly in space twice. Their 161-orbit mission lasted for ten days, forty minutes and ten seconds; traveling 6.6 million km (4.1 million miles). The launch mass was 115,456 kg, the landing mass was 92,701 kg and the payload mass was 12,233 kg (26,969 lb).

Launched on June 20, the shuttle *Columbia* on mission STS-78/Life and Microgravity Sciences (LMS) carried Module LM2 on a 'Spacelab' mission. This was the fifth dedicated Life and Microgravity Spacelab mission, flown partly in preparation for the ISS project. The crew were Commander Terence T. Henricks (on his fourth and last spaceflight), Pilot Kevin J. Kregel (on his second spaceflight), Mission Specialists Richard M. Linnehan, Susan J. Helms (on her third spaceflight), Charles E. Brady, Jr., and Payload Specialists Jean-Jacques Favier (of CNES) and Robert Thirsk (of CSA), the fifth Canadian in space. Linnehan and Thirsk were on their first spaceflights. Their 271-orbit mission lasted sixteen days, 21 hours, 48 minutes and 30 seconds; traveling 11.0 million km (6.8 million miles). The payload mass was 9,649 kg (21,272 lb).

Launched on September 16, the shuttle *Atlantis* on her seventeenth mission (STS-79) made a *Shuttle-Mir* docking. The crew were Commander William F. Readdy (on his third and last spaceflight), Pilot Terrence W. Wilcutt (on his second spaceflight), and Mission Specialists Jerome "Jay" Apt III (on his fourth and last spaceflight), Thomas D. Akers (on his fourth and last spaceflight), Carl E. Walz (on his third spaceflight) and John E. Blaha (on his fifth and last spaceflight, and who became a member of mission 'EO-22' for about four months). Three experiments were transferred: the Biotechnology System (BTS) for study of cartilage development; the Material in Devices as Superconductors (MIDAS) experiment to measure electrical properties of high-temperature semiconductor materials; and the Commercial Generic Bioprocessing Apparatus (CGBA), containing several smaller experiments, including self-contained aquatic systems.

The mission returned Mrs. Shannon Lucid after 188 days (over six months) in space; the first American crewmember exchange aboard Space Station *Mir*. Shannon Lucid's long-duration spaceflight set a new American record, as well as worldwide spaceflight record for a woman astronaut. During her stay on *Mir*, Mrs Lucid conducted research in the fields of Earth sciences, fundamental biology, human life sciences, microgravity research and space sciences. Blood serum and saliva samples were collected. *Atlantis* spent four days, 22 hours, 18 minutes and 16 seconds docked to space station *Mir*. Their 160-orbit mission lasted ten days, three hours, 19 minutes and 28 seconds; traveling 6.3 million km (3.9 million miles). No launch mass data was available.

Launched on November 19, the shuttle *Columbia* on mission STS-80 carried the *Wake Shield Facility* (WSF) for the third time, and the *Orbiting and Retrievable Far and Extreme Ultraviolet Spectrometer* Shuttle Pallet (ORFEUS). The flight was originally scheduled for October 31, but was delayed for several reasons. Likewise, the landing was pushed back two days after bad weather on Earth. The crew were Commander Kenneth D. Cockrell (on his third spaceflight), Pilot Kent V. Rominger (on his second spaceflight), and Mission Specialists F. Story Musgrave (on his sixth and last spaceflight), Thomas D. Jones (on his third spaceflight) and Tamara E. Jernigan (on her fourth spaceflight). Although two EVAs were planned, both were canceled after problems with the airlock prevented astronauts Jones and Jernigan from exiting the orbiter. Their 279-orbit mission lasted seventeen days, 15 hours and 53 minutes; traveling 11 million km (6.8 million miles). It was the longest Space Shuttle mission ever flown. The payload mass was 13,006 kg (28,673 lb).

# 1997

Launched on January 12, 1997, the shuttle *Atlantis* on mission STS-81 (NASA's 100th manned space mission), carried a docking pod to attach to *Mir*'s *Kristall* module. This was the fifth of nine planned missions to *Mir* and the second one involving an exchange of US astronauts. John Blaha who had been on *Mir* for 118 days (about four months) since September 19, 1996 (on his fifth and last spaceflight) was replaced by Jerry Linenger, who then spent over four months on *Mir*, returning on STS-84 to complete his second and last spaceflight. Linenger performed a Canadian experiment to measure astronaut's sleep patterns and processed 102 samples of another Canadian experiment, the QUELD-2 furnace (Queens University Experiment in Liquid-metal Diffusion). The crew of STS-81 were Commander Michael A. Baker (on his fourth and last spaceflight), Pilot Brent W. Jeff, Jr., (on his second spaceflight), Peter J.K. Wisoff (on his third spaceflight), John M. Grunsfeld (on his second spaceflight), Marsha S. Ivins (on her fourth spaceflight), and Jerry M. Linenger (on his second and last spaceflight). John E. Blaha was returned from mission 'EO-22', ending his fifth and last spaceflight.

Mission STS-81 carried the 'Spacehab' double module, providing additional mid-deck locker space for secondary experiments. The mission included several experiments in the fields of advanced technology, Earth sciences, fundamental biology, human life sciences, microgravity, and space sciences. STS-81 involved the transfer of 2,710 kg (about 5,974 lb) of logistics to and from *Mir*, the largest transfer of items to date. This included 635 kg (1,400 lb) of water, and 516.1 kg (1,138 lb) of United States science equipment, plus 1,000.7 kg (2,206 lb) of Russian logistics. The Shuttle spent four days, 22 hours and 21 minutes docked to *Mir*. Their 160-orbit mission lasted ten days, four hours, 56 minutes and 30 seconds; traveling 6.1 million km (about 3,790,364 miles).

Launched on February 21, the shuttle *Discovery* on her twenty-second flight (mission STS-82) (HST SM 02) carried one Pallet (F009), and then serviced the Hubble Space Telescope for the second time. The crew were Commander Kenneth D. Bowersox (on his fourth spaceflight), Pilot Scott J. Horowitz (his second spaceflight), and Mission Specialists Joseph R. Tanner (on his second spaceflight), Steven A. Hawley (on his fourth spaceflight), Gregory J. Harbaugh (on his fourth and last spaceflight), Mark C. Lee (on his fourth and last spaceflight) and Steven L. Smith on his second spaceflight. Their 149-orbit mission lasted nine days, 23 hours, 38 minutes and nine seconds; traveling 6.1 million km (3.8 million miles). The launch mass was 116,884 kg (about 257,685 lb), and the landing mass was 83,122 kg.

Launched on April 04, the shuttle *Columbia* on mission STS-83 carried Module LM1 on a 'Spacelab' mission, which was truncated due to a fuel cell problem, when Fuel Cell 2 passed permissible voltage thresholds. Prior to launch, and continuing through the early part of the mission, flight controllers on the ground were monitoring an anomaly within the electrical power generating fuel cells. It appeared that oxygen and hydrogen might be starting to uncontrollably mix, which could lead to detonation (a similar scenario that caused the explosion on *Apollo 13*). Despite troubleshooting, the anomaly persisted and appeared to worsen. With only two of three fuel cells working, that invoked another Flight Rule which required the mission to be terminated early. Loss of a second cell would require severe and dangerous 'power-downs', although the Shuttle operates normally on two fuel cells. Payload Specialists described the mission as *"an exercise in crisis management. The main bus alarm was going off continually"*.

The crew on STS-83 were Commander James D. Halsell (on his third spaceflight), Pilot Susan L. Still, Mission Specialists Janice E. Voss (on her third spaceflight), Michael L. Gernhardt (on his second spaceflight), Donald A. Thomas (on his third spaceflight), and Payload Specialists Roger Crouch and Greg Linteris (both on their first spaceflight). The mission was planned to be fifteen days and six hours, but was shortened to three days, 23 hours, 13 minutes and 38 seconds; traveling 2.4 million km (1.5 million miles). The launch mass was 117,546 kg, the landing mass was 106,725 kg, and the payload mass was 11,377 kg (25,082 lb). The same crew flew the reflight (which was designated STS-94) three months later in July 1997.

Colonel Chris A. Hadfield was CAPCOM during mission STS-83, and cited NASA's decision to terminate the mission as a positive example of applying the Flight Rules body of knowledge to ensure astronaut safety: *"The beauty of* ***Flight Rules*** *is that they create certainty when we have to make tough calls ... In real time, the temptation to take a chance is always higher. However, the flight rules were unequivocal: the Shuttle had to return to Earth"*. Chris Hadfield became an international celebrity after recording David Bowie's *Space Oddity* song while in orbit aboard the ISS in 2013: in less than a day he quickly gained one million hits on ***YouTube*** and had over ten million views in the first three days online.

Launched on May 15, shuttle *Atlantis* on mission STS-84 carried Commander Charles Precourt (on his third spaceflight), Pilot Eileen M. Collins (on her second spaceflight), and five Mission Specialists: Jean-Francois Clervoy (on his second spaceflight), Carlos I. Noriega, Edward T. Lu, (both on their first spaceflight), Yelena K. Kondakova of RKA (on her second and last spaceflight) and C. Michael Foale on his fourth spaceflight, joining 'EO-23'. Jerry M. Linenger was returned from mission 'EO-23', ending his second and last spaceflight. Their 144-orbit mission lasted nine days, four hours, 19 minutes and 55 seconds; traveling six million km (3.7 million miles). The landing mass was 100,285 kg (221,091 lb).

Launched on July 01, the shuttle *Columbia* on mission STS-94 carried Module LM1 for 'Spacelab'. The primary payload was the Microgravity Science Laboratory (MSL); a collection of experiments housed inside a European Spacelab Long Module (LM). It built upon the cooperative and scientific foundation of eight previous International Microgravity Laboratory missions. The crew were the same as for mission STS-83, with this mission lasting fifteen days, 16 hours, 45 minutes and 29 seconds. Landing occurred during orbit two hundred and fifty-one. The craft traveled 10.0 million km (6.2 million miles). The landing mass was 117,802 kg, and the payload mass was 10,169 kg (22,419 lb).

Launched from August 07-19, the shuttle *Discovery* on mission STS-85 both deployed and retrieved the *Cryogenic Infrared Spectrometers and Telescopes for the Atmosphere-Shuttle Pallet Satellite-2* (CRISTA-SPAS), designed to study Earth's middle atmosphere. The crew were Commander Curtis L. Brown, Jr., (on his fourth spaceflight), Pilot Kent V. Rominger (on his third spaceflight), Mission Specialists N. Jan Davis (on her third and last spaceflight), Robert L. Curbeam, Jr. (on his third spaceflight), Stephen K. Robinson (on his first spaceflight), and Payload Specialist Bjarni V. Tryggvason (of CSA) on his only spaceflight. Tryggvason became the sixth Canadian astronaut in space, and deployed the CRISTA-SPAS pallet. As a Payload Specialist, he tested the next-generation Microgravity Isolation Mount (MIM), a unique Canadian device that he co-designed. Their 185-orbit mission lasted eleven days, 20 hours, 28 minutes and seven seconds; traveling 7.6 million km (4.7 million miles). The payload mass was 9,191 kg (20,263 lb).

Launched on September 25, the shuttle *Atlantis* on mission STS-86 made a *Shuttle-Mir* docking. The crew were Commander James D. Wetherbee (on his fourth spaceflight), Pilot Michael J. Bloomfield (on his first spaceflight), and Mission Specialists Vladimir G. Titov of RKA (on his fourth and last spaceflight), Scott E. Parazynski (on his second spaceflight), Jean-Loup Cretien of CNES (on his third and last spaceflight), Wendy B. Lawrence (on her second spaceflight) and David A. Wolf, who joined 'Mission EO-24' on his second spaceflight. Astronaut C. Michael Foale returned from *Mir* to Earth, ending his fourth spaceflight. David Wolf was originally scheduled to fly on the STS-89 mission to *Mir* and join the 'Mir EO-24' crew. In four months, he performed the CAPE experiments; some 800 space protein crystal growth (PCG) experiments with samples from fifteen universities and twelve secondary schools. Their 170-orbit mission STS-86 lasted for ten days, 19 hours, 22 minutes and 12 seconds; traveling 7.0 million km (4.3 million miles). The landing mass was 114,185 kg and the payload mass was a modest 8,375 kg (18,464 lb).

Launched on November 19, the shuttle *Columbia* on her twenty-fourth mission (STS-87) made microgravity experiments, two EVAs, deployed the *SPARTAN-201* experiment, and was the first spaceflight by a woman of Indian origin; Kalpana Chawla. The crew were Commander Kevin J. Kregel (on his third spaceflight), Pilot Steven W. Lindsey (on his first space flight), Mission Specialists Kalpana "KC" Chawla (on her first spaceflight), Winston E. Scott (on his second and last spaceflight), Dr. Takao Doi of JAXA (on his first spaceflight), and Payload Specialist Leonid Kadeniuk (of NSAU) on his only spaceflight. Their mission lasted fifteen days, 16 hours, 35 minutes and one second; traveling 10.5 million km (6.5 million miles). The landing mass was 102,717 kg, and the payload mass was a very light 4,451 kg (9,813 lb).

# 1998

Launched on January 22, the shuttle *Endeavour* on mission STS-89 made the eighth *Shuttle-Mir* docking. The crew were Commander Terrence W. Wilcutt (on his third spaceflight), Pilot Joe F. Edwards, Jr., Mission Specialists James F. Reilly II, Michael P. Anderson, Bonnie J. Dunbar (on her fifth and last spaceflight, setting a joint US-women's record), Salizhan Sharipov (of Roscosmos/RKA), and Andrew S.W. Thomas, who joined mission 'EO 24' on his second spaceflight. Astronaut Thomas replaced David A. Wolf aboard *Mir*. Wolf had spent 119 days (about four months) aboard the *Mir* station, and a total of 128 days (about four months) in space. Mission STS-89 lasted eight days, 19 hours, 48 minutes and four seconds; traveling 5.8 million km (3.6 million miles). The landing mass was 114,131 kg, and the payload mass was only 7,748 kg (17,081 lb).

Launched on April 17, the shuttle *Columbia* STS-90 (mission Neurolab) carried Module LM2 on the last of the ESA's 'Spacelab' missions. The crew were Commander Richard A. Searfoss (on his third and last spaceflight), Pilot Scott D. Altman, Mission Specialists Richard E. Linnehan (on his second spaceflight), Kathryn P. Hire, Daffyd "Dave" Williams of CSA, the seventh Canadian in space, and Payload Specialists Jay C. Buckey (on his only spaceflight), and James A. Pawelczyk (on his only spaceflight). Altman and Hire were on their first spaceflights. Their mission lasted fifteen days, 21 hours, 50 minutes and 58 seconds; traveling 10.0 million km (6.2 million miles). The landing mass was 105,462 kg, and the payload mass was 10,788 kg (23,783 lb).

Launched on June 02, the shuttle *Discovery* on mission STS-91 made the ninth and last *Shuttle-Mir* docking (at an altitude of 208 miles). This mission was the first to use the super lightweight external tank (SLWT) which was the same size, at forty-seven metres long and 8.2 metres in diameter, but 7,500 pounds (3,400 kg) lighter. The tank was made of an aluminum lithium alloy, and its structural design had also been improved, making it 30% stronger and 5% less dense. The walls of the redesigned hydrogen tank were machined in an orthogonal waffle-like pattern, providing more strength and stability than the previous design. These improvements would later provide additional payload capacity for missions to the International Space Station.

The crew for STS-91 were Commander Charles J. Precourt (on his fourth and last spaceflight), Pilot Dominic L. Pudwill Gorie, and Mission Specialists Franklin R. Chang-Diaz (on his sixth spaceflight), Wendy B. Lawrence (on her third spaceflight), Janet L. Kavandi, and Valery Ryumin of RKA (on his fourth and last spaceflight). Pudwill Gorie and Kavandi were on their first spaceflights. Astronaut Andrew S.W. Thomas was returned from mission 'EO-25' on *Mir*. The mission STS-91 duration was nine days, 19 hours, 54 minutes and two seconds. The landing mass was 117,861 kg, and the payload mass was 16,537 kg (36,458 lb).

Launched on October 29, the shuttle *Discovery* on her twenty-fifth mission (STS-95) carried a variety of experiments in the pressurised 'Spacehab' module, and had operations with the Hubble Space Telescope Orbiting Systems Test (HOST) and the International Extreme Ultraviolet Hitchhiker payloads carried in the payload bay. Aboard was 77-year-old Senator John Glenn, who returned to space after thirty-six years as the oldest astronaut in history, and he was the oldest person to reach Earth orbit. This record stood for twenty-three years, until 82-year-old Ms Wally Funk flew on a suborbital flight on *Blue Origin NS-16*, launching on July 20, 2021; which record was in turn broken by William Shatner (again, flying only sub-orbitally) at age ninety years on October 13, 2021. The mission's objectives involved investigating life sciences experiments, using the 'Spacehab' module to perform these experiments on Senator Glenn. The *SPARTAN 201* spacecraft was released, flying free from the Shuttle, studying the acceleration of the solar wind that originates in the Sun's solar corona, then was retrieved.

The STS-95 crew were Commander Curtis L. Brown, Jr., (on his fifth spaceflight), Pilot Steven W. Lindsey (on his second spaceflight), Mission Specialists Stephen K. Robinson (on his second spaceflight), Scott E. Parazynski (on his third spaceflight), Pedro Duque of ESA (the first Spaniard in space), and Payload Specialists Chiaki Mukai of NASDA and John H. Glenn (US Senator; Democrat-Ohio) on his second and last spaceflight. Their mission lasted eight days and 21 hours; traveling 5.8 million km (3.6 million miles). The landing mass was 103,321 kg and the payload mass was 11,130 kg (24,540 lb). Bill Clinton became the second sitting United States President to witness a crewed space launch, joined by his wife Hillary on the roof of the Launch Control Center, and the only one to witness a Space Shuttle launch. President Richard Nixon had witnessed the launch of *Apollo 12* on November 14, 1969.

Launched on December 04, the shuttle *Endeavour* on mission STS-88/2A was the first Shuttle mission to the International Space Station (ISS). This mission carried the *Unity* node, the first US-built section of the ISS, into orbit. The upgraded Canadarm grappled the *Unity* module to the Functional Cargo Block *Zarya* module, connecting the first two components of the ISS. During this process, the use of the Orbiter Space Vision System (OSVS), an offshoot of the Canadian Space Vision System was essential, since the *Unity* node blocked the view of the approaching Russian module. The STS-88 crew were Commander Robert D. Cabana (on his fourth and last spaceflight), Pilot Frederick W. Sturckow, and Mission Specialists Jerry L. Ross (on his sixth spaceflight), Nancy J. Currie (on her third spaceflight), James H. Newman (on his third spaceflight) and Sergei K. Krikalev (of RKA) on his fourth spaceflight. Their 186-orbit mission lasted for eleven days, 19 hours, 18 minutes and 47 seconds; traveling 7.6 million km (4.7 million miles). The launch mass was 119,715 kg, and the landing mass was 90,853 kg (200,296 lb).

More than fifteen nations are partners in the International Space Station project. The ISS's two Orbital Segments are made up of 16 modules: five from Russia, eight from America, two from Japan, and one from Europe. Launched in 2000, the original *Zvezda* module contains the ISS's living quarters. At around fifty square metres (about twice the area of a parking space) in size, the module contains sleeping compartments for two crew members at a time, a kitchen galley with a refrigerator, an exercise bike and treadmill, and a toilet. This module also contains an oxygen recycling system that takes wastewater molecules and recycles them into oxygen (for breathing) and hydrogen (waste) gas. To provide electricity while orbiting 386 km (about 240 miles) above Earth, the ISS uses a series of solar panels, which are organised into wing-like arrays. If placed together, these panels would cover over 2,500 square metres (the area of a large mansion). The solar arrays angle themselves to track the Sun, and so gain the greatest possible exposure to its rays.

# 1999

On March 12, a 17.6-metre-long robotic arm, a key element of Canada's Mobile Servicing System (MSS) for the ISS was officially transferred to CSA. The USD 1.4 billion Canadarm-2 was shipped to Florida to the Space Station Processing Facility on May 16, to undergo integration testing with other elements of the International Space Station. The SSMMS was scheduled for launch in 2001, on mission STS-100/6A.

Launched on May 27, the shuttle *Discovery* on mission STS-96/2A.1 was the first shuttle to dock at and supply the ISS. On this first logistics mission, it carried the 'Spacehab' module, filled with cargo for outfitting the ISS. The basketball-shaped *STARSHINE* educational satellite was launched from the cargo bay by Julie Payette. The crew were Commander Kent V. Rominger (on his fourth spaceflight), Pilot Rick D. Husband, EVA Specialists Daniel T. Barry (on his second spaceflight) and Tamara E. Jernigan (on her fifth and last spaceflight), Flight Engineer Ellen L. Ochoa (on her third spaceflight), Julie Payette (of CSA) [the eighth Canadian in space] and Mission Specialist Valeri I. Tokarev (of Roscosmos/RKA), on his first spaceflight. The flight lasted nine days, 19 hours, 13 minutes and 57 seconds; traveling 6.0 million km (3.7 million miles). The launch mass was 118,857 kg, the landing mass was 100,230 kg and the payload mass was a modest 9,097 kg (20,056 lb).

Launched on July 23, the shuttle *Columbia* on her 26th mission, STS-93 had problems during the main engine ignition sequence, and the redundant set of engine primary digital control units (DCUs) saved the *Columbia* and her crew from potential catastrophe, as a shutdown of two engines at that early point in the flight would have resulted in a very risky contingency abort. Five seconds after liftoff an electrical short was caused by poorly routed wiring, which had rubbed on an exposed screw head. This issue led to a program-wide inspection of the wiring in all Space Shuttle orbiters.

The mission's primary objective was to deploy the *Chandra X-ray Observatory* (CXO) with its Inertial Upper Stage booster. *Chandra* was designed to observe X-rays from high-energy regions of the Universe, such as hot gas in the remnants of exploded stars. CXO is a 'Flagship-class' space telescope sensitive to X-ray sources 100 times fainter than any previous X-ray telescope, enabled by the high angular resolution of its mirrors. Since the Earth's atmosphere absorbs most X-rays, they are not detectable from Earth-based telescopes; therefore space-based telescopes are required. *Chandra* is an Earth satellite in a 64-hour orbit, and as of mid-2023 its mission is ongoing. *Chandra* is one of the "Great Observatories", along with the Hubble Space Telescope, Compton Gamma Ray Observatory (1991-2000), and the Spitzer Space Telescope (2003-2020). Its mission is like that of ESA's *XMM-Newton* spacecraft (also launched in 1999) but the two telescopes have different design foci. *Chandra* has a much higher angular resolution, and *XMM-Newton* has higher spectroscopy throughput.

The STS-93 mission carried twelve other payloads. Eileen Collins became the first woman (while on her third spaceflight) to command a Space Shuttle. Other crew members were Pilot Jeffrey S. Ashby, and Mission Specialists Michel Tognini of France's CNES (on his second and last spaceflight), Steven A. Hawley (on his fifth and last spaceflight), and Catherine G. Coleman on her second spaceflight. Their mission lasted four days, 22 hours, 49 minutes and 34 seconds; traveling 2.89 million km (1.796 million miles). The launch mass was 122,534 kg, landing mass was 99,781 kg, and the payload mass was a very massive 22,780 kg (50,222 lb).

Launched on December 19 from KSC, the shuttle *Discovery* on mission STS-103 (HST SM 03A) carried one Pallet (F009). This mission serviced the Hubble Space Telescope and was the only Space Shuttle mission to orbit through the Christmas period. Hubble was given a major increase in computer and data storage capability, new battery components, an upgraded guidance system and new gyroscopes. A 1970s-style mechanical tape recorder was replaced by a second solid state data recorder, which, at twelve gigabytes capacity could hold about ten times as much data as the old unit. The EVA crew replaced the telescope's outer insulation that had degraded. This helped protect Hubble from the severe and rapid temperature changes it experiences during each ninety-minute orbit as it moves from sunlight to darkness.

The STS-103 crew were Commander Curtis L. Brown, Jr., (on his sixth and final spaceflight), Pilot Scott J. Kelly (on his first spaceflight), and Mission Specialists John M. Grunsfeld (on his third spaceflight), Jean-Francois Clervoy of ESA (on his third and last spaceflight) [who operated the Canadarm for the 54th time during a space mission], C. Michael Foale (on his fifth spaceflight), Steven L. Smith (on his third spaceflight), and Claude Nicollier of ESA (on his fourth and last spaceflight). Their mission lasted for seven days, 23 hours, 11 minutes and 34 seconds; traveling 5.23 million km (3.25 million miles). The launch mass was 112,493 kg and the landing mass was 95,768 kg (211,132 lb).

## 2000

Launched on February 11, the shuttle *Endeavour* on mission STS-99, the first Space Shuttle mission of the 2000s, carried one Pallet (F006) and conducted the Shuttle Radar Topography Mission (mission SRTM). The SRTM was an international project spearheaded by the National Imagery and Mapping Agency (now the National Geospatial-Intelligence agency), an agency within the US Department of Defense, and NASA, with participation of the German Aerospace Center, DLR. Its objective was to obtain the most complete high-resolution digital topographic database of the Earth. The radar system gathered about eight terabytes of data to produce high-quality 3D images of the Earth's surface. The crew were Commander Kevin R. Kregel (on his fourth and last spaceflight), Pilot Dominic L. Pudwill

Gorie (on his second spaceflight), and Mission Specialists Gerhard P.J. Thiele (of ESA, on his only spaceflight), Janet L. Kavandi (on her second spaceflight), Janice E. Voss [(1956-2012) on her fifth and last spaceflight; jointly holding *the number of missions record for American women*, along with Mrs. Shannon Lucid, Bonnie Dunbar, Tamara Jernigan, Marsha Ivins and Susan Helms], and Mamoru Mohri of NASDA, Japan, on his second and last spaceflight.

SRTM used C-band and X-band interferometric synthetic aperture radar to acquire data between 60 degrees north and 56 degrees south; with 16 metre absolute vertical height accuracy and 20 metre absolute horizontal circular accuracy. Besides contributing to the production of better maps, these measurements could lead to improved water drainage modeling, more realistic flight simulators, better locations for cell phone towers, and enhanced navigation safety. The mission lasted eleven days, five hours, 39 minutes and 41 seconds; traveling 6.540 million km (about 4,063,768 miles). The launch mass was 116,376 kg, the landing mass was 102,363 kg and the payload mass was 13,154 kg (29,000 lb).

Launched on May 19, the shuttle *Atlantis* on mission STS-101 supplied the ISS. This mission was delayed three times in April, due to high winds. The crew were Commander James D. Halsell (on his fifth and last spaceflight), Pilot Scott J. Horowitz (on his third spaceflight), and Mission Specialists Mary E. Weber (on her second and last spaceflight), Jeffrey N. Williams, James S. Voss (on his fourth spaceflight), Susan J. Helms (on her fourth spaceflight) and Yury V. Usachov (of RKA) on his third spaceflight. Their 155-orbit mission lasted nine days, 21 hours, ten minutes and ten seconds; traveling 6,598,310 km (about 4.1 million miles). The landing mass was 100,369 kg, and the payload mass was the lightest at 1,801 kg (3,971 lb).

Launched on September 08, the space shuttle *Atlantis* on mission STS-106 again supplied the ISS. The crew were Commander Terrence W. Wilcutt (on his fourth and last spaceflight), Pilot Scott D. Altman (on his second spaceflight), and Mission Specialists Dr. Edward "Ed" T. Lu (on his second spaceflight), Richard A. Mastracchio, Daniel C. Burbank (both on their first spaceflight), Yuri Malenchenko of RKA, a Colonel in the Russian Air Force (on his second spaceflight) and Boris Morukov (of RKA) on his only spaceflight. The goal of the flight was to prepare the *Zvezda* module for the arrival of the first resident crew. Dr. Ed Lu and Col. Yuri Malenchenko made one EVA of six hours and fourteen minutes during the mission. They routed and connected nine power, data and communications cables between the *Zvezda* module and the other Russian-built module, *Zarya*, as well as installing the six-foot-long magnetometer. Their 185-orbit mission lasted eleven days, nineteen hours, twelve minutes and 15 seconds; traveling 7.9 million km (about 4,908,833 miles). The launch mass was 115,259 kg, the landing mass 100,369 kg, and the payload mass was 10,219 kg (22,529 lb).

Launched on October 11, the shuttle *Discovery* on the one hundredth Space Shuttle mission STS-92 (for ISS assembly 3A) carried one Pallet (F005) with a Z1 truss, Control Moment Gyros, Pressurized Mating Adapter-3 (PMA-3), and two DDCU (heat pipes). The crew were Commander Brian Duffy (on his fourth and last spaceflight), Pilot Pamela A. Melroy (on her first spaceflight), and Mission Specialists Koichi Wakata of JAXA (on his second spaceflight), William S. McArthur (on his third spaceflight), Peter J.K. Wisoff (on his fourth and last spaceflight), Michael E. Lopez-Alegria (on his second spaceflight) and Leroy Chiao on his third spaceflight.

The Z1 truss was the first exterior framework installed on the ISS and allowed the first US solar arrays to be temporarily installed on the *Unity* module for early power during flight 4A. The Ku-band communication system supported early science capabilities and US television on flight 6A. PMA-3 provided a shuttle docking port for solar array installation on Flight 4A and for *Destiny* lab installation on Flight 5A. During the fourth EVA, astronauts Peter Wisoff and Michael Lopez-Alegria tested the *SAFER* jet backpack, flying up to fifty feet while remaining tethered to the spacecraft. Their mission lasted twelve days, 21 hours, 43 minutes and 47 seconds; spending six days, 21 hours and 23 minutes docked to the ISS. The launch mass was 115,127 kg, the landing mass was 92,741 kg and the payload mass was a modest 9,513 kg (20,973 lb).

Launched on November 30, the shuttle *Endeavour* on mission STS-97 was an ISS assembly flight 4A, carrying P6 solar arrays and radiators. The crew were Commander Brent W. Jett (on his third spaceflight), Pilot Michael J. Bloomfield (on his second spaceflight), and Mission Specialists Joseph R. Tanner (on his third spaceflight), Marc Garneau of CSA (on his third and last spaceflight), and Carlos I. Noriega on his second and last spaceflight. The primary mission objective was to deliver and connect the first huge set of US-provided solar arrays and the P6 truss to the ISS.

The shuttle spent six days, 23 hours and 13 minutes docked to the ISS. During three EVA the astronauts prepared a docking port for the arrival of the *Destiny* Laboratory Module, installed Floating Potential Probes to measure the electrical potential surrounding the station, installed a camera cable outside the *Unity* module, and transferred supplies, equipment and refuse between the craft. On Flight Day 3, Commander Jett linked *Endeavour* to the ISS while 370 km (230 miles) above northeast Kazakhstan. Mission STS-97 was the fifteenth flight of *Endeavour* and the 101st Space Shuttle mission. It lasted ten days, 19 hours, 58 minutes and 20 seconds; traveling 7.203 million km (4,476,000 miles). The shuttle launch mass was 120,742 kg, the landing mass was a very light 89,758 kg and the payload mass was just 7,906 kg (17,430 lb).

## 2001

Launched on February 07, the shuttle *Atlantis* on mission STS-98 carried Commander Kenneth D. Cockrell (on his fourth spaceflight), Pilot Mark L. Polansky (on his first spaceflight), and Mission Specialists Robert L. Curbeam (on his second spaceflight), Marsha S. Ivins (on her fifth and last record-equaling spaceflight) and Thomas D. Jones, on his fourth and last spaceflight. This mission occurred while the first ISS crew was aboard the new space station. The shuttle crew delivered the *Destiny* Laboratory Module; manufactured by **Boeing** at the Michoud Assembly Facility and the Marshall Space Flight Center in 1997. This module is 8.5 metres (28 feet) long and 4.3 metres (14 feet) wide. The shuttle spent six days, 21 hours and 14 minutes docked to the ISS while the laboratory was attached, and three EVAs were conducted to complete its assembly. This mission also saw the one hundredth spacewalk in US history. Their 171-orbit mission duration was twelve days, 21 hours, and 21 minutes exactly; traveling 8.5 million km (5.3 million miles), before landing at Edwards AFB Runway 22 on February 20. The launch mass was 115,529 kg, the landing mass was 90,225 kg and the payload mass was 14,55 kg (32,000 lb).

Launched on March 08 aboard the shuttle *Discovery*, during mission STS-102 (Space Station Assembly Flight ISS-5A) James Voss and Susan J. Helms made the longest spacewalk in history on March 11; of eight hours and fifty-six minutes. The mission of twelve days, 19 hours, 51 minutes and 57 seconds was to supply the ISS, and for crew rotation. It made first use of the Multi Purpose Logistics Module *Leonardo* to bring supplies to the ISS. The steel modules were equipped with up to sixteen International Standard Payload Racks for installation in the US Lab. The mission traveled 8.5 million km (5.3 million miles). The launch mass was a relatively light 99,503 kg, the landing mass was 90,043 kg and the payload mass was very light at 5,760 kg (12,700 lb).

The STS-102 crew were Commander James D. Wetherbee (on his fifth spaceflight), Pilot James M. Kelly (on his first spaceflight), and Mission Specialists Andrew S.W. Thomas (on his third spaceflight) and Paul W. Richards on his only spaceflight. Joining 'Expedition 2' were

Yury V. Usachov of RKA (on his fourth and last spaceflight), James S. Voss (on his fifth and last spaceflight) and Susan J. Helms (on her fifth and last spaceflight). Returning to Earth on *Discovery* from 'Expedition 1' were ISS Commander William M. Shepherd (on his fourth and last spaceflight), ISS *Soyuz* Commander Yuri P. Gidzenko of RKA (on his second spaceflight) and ISS Flight Engineer Sergei K. Krikalev (of Roscosmos/RKA) on his fifth spaceflight.

Launched on April 19, the shuttle *Endeavour* on mission STS-100 (ISS assembly 6A) carried one Pallet (F004), with *Canadarm2*, a robotic arm. The crew were Commander Kent V. Rominger (on his fifth and last spaceflight), Pilot Jeffrey S. Ashby (on his second spaceflight), Chris A. Hadfield of CSA (on his second spaceflight), and Mission Specialists John L. Phillips, Scott E. Parazynski (on his fourth spaceflight), Umberto Guidoni of ESA (on his second and last spaceflight), and Yury Lonchakov of RKA (on his first spaceflight). This mission saw the first spacewalk by a Canadian, Col. Chris A. Hadfield as EV1 (lead spacewalker). Chris Hadfield experienced severe eye irritation during the first seven-hour ten-minute spacewalk, due to the anti-fog solution used to polish his spacesuit visor; temporarily blinding him and forcing him to vent oxygen into space.

The highest priority of this mission was the installation, activation and checkout of the *Canadarm2*, which went into operation on April 28. Designed and built by Canada, the *Canadarm2* is a seventeen-metre-long robotic arm that was used to assemble the ISS in space. A permanent fixture, it is used to move equipment, undock cargo and perform "cosmic catches": the capturing of unmanned spacecraft that travel up from Earth bringing all the ISS's supplies, including food and water. The robotic arm was critical to continue assembly of the ISS.

While docked, *Endeavour* boosted the altitude of the ISS, and later performed a fly-around survey of the complex, including recording views of the station with an IMAX cargo bay camera. Another mission objective was to berth the *Raffaello* logistics module to the ISS, activate it, transfer cargo, and reberth *Raffaello* in the shuttle's payload bay. *Raffaello* was the second of three Italian Space Agency-developed Multi-Purpose Logistics Module, manufactured out of stainless steel at the **Cannes Mandelieu Space Center**. The *Leonardo* module was launched and returned on the previous shuttle flight, STS-102, in March. The shuttle spent eight days, three hours and 35 minutes docked to the ISS. This mission lasted eleven days, 21 hours, 31 minutes and 14 seconds. There were two EVAs, with a total duration of fourteen hours and fifty minutes. The launch mass was 103,506 kg, the landing mass 99,742 kg, and the payload mass was a very light 4,899 kg (10,800 lb).

Launched on July 12, the shuttle *Atlantis* on mission STS-104 (ISS assembly 7A) carried two Pallets (F002 and F010). The primary objectives were to install the *Quest* Joint Airlock, and help perform maintenance on the ISS. The crew were Commander Steven W. Lindsey (on his third spaceflight), Pilot Charles O. Hobaugh (on his first spaceflight), and Mission Specialists Michael L. Gernhardt (on his fourth and last spaceflight), Janet L. Kavandi (on her third and last spaceflight) and James F. Reilly (on his second spaceflight). The Joint Airlock is a pressurised flight element consisting of two cylindrical chambers attached end-to-end by a connecting bulkhead and hatch. Once installed and activated, the ISS airlock became the primary path for ISS spacewalk entry and departure for United States spacesuits, which are known as *Extra-Vehicular Mobility Units*, or EMU's. In addition, the Joint Airlock is designed to support the Russian *Orlan* spacesuit for EVA activity. Their 200-orbit mission lasted twelve days, 18 hours and 36 minutes; traveling 8.5 million km (5.3 million miles). The launch mass was 117,129 kg, the landing mass was 94,009 kg and the payload mass was 8,241 kg (18,168 lb).

Launched on August 10, the shuttle *Discovery* on mission STS-105 was for ISS supply and crew rotation. This mission was *Discovery*'s final mission until STS-114, because she was grounded for a refit, which included an update of the flight deck to the glass cockpit layout (which was already installed on *Atlantis* and *Columbia*). The STS-105 crew were Commander Scott J. Horowitz (on his fourth and last spaceflight), Pilot Frederick J. Sturckow (on his second spaceflight), and Mission Specialists Patrick G. Forrester, Daniel T. Barry (on his third and last spaceflight), Frank L. Culbertson, Jr., (on his third and last spaceflight, becoming 'Expedition 3' ISS Commander), Mikhail Tyurin of RKA (becoming ISS Flight Engineer), and Vladimir N. Dezhurov of RKA (on his second and last spaceflight, becoming ISS *Soyuz* Commander).

Three ISS crew were returned to Earth in crew rotation: Yury V. Usachev (of RKA) on 'Expedition 2' (his fourth and last spaceflight, being both ISS and *Soyuz* Commander), James S. Voss (ending his fifth and last spaceflight), and Susan J. Helms, 'ISS Flight Engineer 2', ending her record-equaling fifth and last spaceflight. This mission lasted eleven days, 21 hours, 13 minutes and 52 seconds. The launch mass was 116,914 kg, the landing mass was 100,824 kg, and the payload mass was 9,072 kg (20,000 lb). The shuttle was docked with the ISS for seven days, 20 hours and nine minutes.

Launched on December 05, the shuttle *Endeavour* on mission STS-108 was for ISS supply and crew rotation. It was the first shuttle flight to the ISS since the installation of the Russian airlock called *Pirs*. Aboard were two Canadian science experiments: the innovative Extra-Vehicular Activity Radiation Monitors (EVARM), measuring the amount of radiation received during spacewalks; and the Hoffman-Reflex experiment (H-Reflex), measuring how human reflexes are affected by microgravity.

The STS-108 crew were Commander Dominic L. Pudwill Gorie (on his third spaceflight), Pilot Mark E. Kelly, and Mission Specialists Linda M. Godwin (on her fourth and last spaceflight), Daniel M. Tani, Yuri I. Onufrienko of RKA (on his second and last spaceflight), Carl E. Walz (on his fourth and last spaceflight), and Daniel W. Bursch (on his fourth and last spaceflight). The 'Expedition 3' crew returned to Earth were Frank L. Culbertson, Jr., (ISS Commander, on his third and last spaceflight), Mikhail Tyurin of RKA (ISS Flight Engineer), and Vladimir N. Dezhurov of RKA (ISS *Soyuz* Commander, ending his second and last spaceflight). Their mission lasted eleven days, 19 hours, 36 minutes and 45 seconds; traveling 7.7 million km (4.8 million miles). The payload mass was a very light 4,082 kg (8,999 lb).

# 2002

Launched twenty-five minutes before sunrise on March 01, the shuttle *Columbia* on her twenty-seventh flight (mission STS-109) (HST SM 03B) [before her fatal mission STS-107] carried one Pallet (F009). *Columbia* carried out the fourth servicing of the Hubble Space Telescope; delivering the Advanced Camera for Surveys (ACS), new rigid solar arrays (SA3), a new Power Control Unit (PCU), a Reaction Wheel Assembly and an experimental cryocooler for the Near Infrared Camera and Multi-Object Spectrometer (NICMOS) unit. Astronauts performed a total of five spacewalks in five consecutive days to service and upgrade the Hubble Space Telescope. *Columbia* also boosted Hubble to a higher orbit.

The STS-109 crew were Commander Scott D. Altman (on his third spaceflight), Pilot Duane G. Carey (on his only spaceflight), and Mission Specialists John M. Grunsfeld (on his fourth spaceflight), Nancy J. Currie (on her fourth and last spaceflight), Richard M. Linnehan (on his third spaceflight), James H. Newman (on his fourth and last spaceflight) and Michael J. Massimino on his first spaceflight. Their 105-orbit mission lasted ten days, 22 hours, 11 minutes and nine seconds; traveling 6.3 million km (3.9 million miles). The launch mass was 116,987 kg and the landing mass was 100,564 kg (221,706 lb).

Launched on April 08, the shuttle *Atlantis* on mission STS-110 was for ISS assembly. Flight 8A carried the S0 truss, which forms the backbone of the truss structure on the ISS. The crew were

Commander Michael J. Bloomfield (on his third and last spaceflight), Pilot Stephen N. Frick, and Mission Specialists Rex J. Walheim, Ellen L. Ochoa (on her fourth and last spaceflight), Lee M.E. Morin (on his only flight), Jerry L. Ross (on his record-setting seventh and last spaceflight) and Steven L. Smith (on his fourth and last spaceflight). The crew made four EVAs, and the 171-orbit flight duration was ten days, 19 hours and 44 minutes; traveling 7,240,000 km (4.5 million miles).

Launched on June 05, the shuttle *Endeavour* on mission STS-111 was for ISS assembly and crew rotation. Canada's newest contribution – the Mobile Base System was carried up and installed by astronauts Chang-Diaz and Perrin from June 09. The crew were Commander Kenneth D. Cockrell (on his fifth and last spaceflight), Pilot Paul S. Lockhart, and Mission Specialists Phillipe Perrin of CNES (on his only spaceflight), Franklin Chang-Diaz (on his joint record-setting seventh and last spaceflight), Valery G. Korzun of Roscosmos/RKA (on his second and last spaceflight), Dr. Peggy A. Whitson, and Sergei Y. Treshchov of Roscosmos/RKA (on his only spaceflight).

The 'Expedition 4' crew returned to Earth were: Yuri I. Onefrienko (both ISS and *Soyuz* Commander, ending his second and last spaceflight), ISS Flight Engineer Carl E. Walz (ending his fourth and last spaceflight) and ISS Flight Engineer Daniel W. Bursch (ending his fourth and last spaceflight). *Endeavour* spent seven days, 22 hours and seven minutes docked to the ISS. Their mission lasted thirteen days, 20 hours, 35 minutes and 56 seconds; traveling 9.3 million km (5.8 million miles). The launch mass was 116,523 kg, the landing mass was 99,385 kg, and the payload mass was a medium 12,058 kg (26,583 lb). STS-111 was the last flight of a CNES astronaut, the French agency having disbanded its astronaut group and transferred them to the ESA.

Launched on October 07, the shuttle *Atlantis* on mission STS-112 was ISS assembly Flight 9A, carrying the 27,717-pound S1 truss. **The Boeing Company** and **Lockheed Martin** started manufacturing the truss in May 1998, and it was completed in March 1999. The shuttle crew were Commander Jeffrey S. Ashby (on his third and last spaceflight), Pilot Pamela A. Melroy (on her second spaceflight), and Mission Specialists David A. Wolf (on his third spaceflight), Piers Sellers, Sandra H. Magnus and Fyodor N. Yurchikhin of RKA (all were on their first spaceflight). Their 170-orbit mission lasted ten days, 19 hours, 58 minutes and 44 seconds; traveling 7.2 million km (4.5 million miles). The launch mass was 116,538 kg, the landing mass was 91,390 kg, and the payload mass was 12,572 kg (27,717 lb).

Arguably the most significant event from the launch of STS-112 was the ET bipod ramp shedding a chunk of foam, estimated to be four-by-five-by twelve inches, that broke away and hit the lower left SRB-ET Attach Ring near the Integrated Electronics Assembly box, causing a dent about four inches wide and three inches deep into the solid metal. Prior to approval for the next mission, the situation was analyzed, and NASA decided to press ahead under the justification that *"The ET is safe to fly with no new concerns (and no added risk) of further foam strikes"*. **This fateful decision set the stage for the STS-107 tragedy just two launches later.** The CAIB report did not highlight the significance of video data from this being the first flight with the "rocketcam" video camera mounted to the upper part of the *Atlantis* external tank. The video was near perfect until the two SRBs were jettisoned.

Launched on November 23, the shuttle *Endeavour* on mission STS-113 was for crew rotation and ISS assembly Flight 11A, carrying the P1 truss, which extended the backbone of the ISS. The crew were Commander James D. Wetherbee (on his sixth and last spaceflight), Pilot Paul S. Lockhart (on his second and last spaceflight), and Mission Specialists Michael Lopez-Alegria (on his third spaceflight), John B. Herrington (on his only spaceflight), Kenneth D. Bowersox (on his fifth and last spaceflight, becoming ISS Commander for 'Expedition 6'), Nikolai M. Budarin (on his third and last spaceflight, becoming an ISS Flight Engineer and *Soyuz* Commander, and Donald R. Pettit (on his first spaceflight, becoming an ISS Flight Engineer).

'Expedition 5' crew returned to Earth were: Valery G. Korzun (both ISS and *Soyuz* Commander, ending his second and last spaceflight), ISS Flight Engineer Dr. Peggy A. Whitson, and ISS Flight Engineer Sergei Y. Treshchov of RKA (ending his only spaceflight). Mission STS-113 lasted for thirteen days, 18 hours, 48 minutes and 38 seconds; traveling 9.0 million km (5.6 million miles). The launch mass was 116,460 kg, the landing mass was 91,498 kg, and the payload mass was 12,477 kg (27,507 lb). Because *Endeavour* entered its Orbiter Major Modification period after the *Columbia* shuttle disaster, this was the last space shuttle mission to fly with an analogue-style cockpit.

# 2003

Launched on January 16, the space shuttle *Columbia* on her twenty-eighth flight (mission STS-107), the 113th flight of the Space Shuttle program, broke apart on February 01 as it re-entered Earth's atmosphere, and all the seven crew members died. The crew were Commander Rick D. Husband (on his second spaceflight), Pilot William "Willie" C. McCool (on his first spaceflight), Mission Specialist Kalpana "KC" Chawla (on her second spaceflight), David M. Brown (on his first spaceflight), Laurel B. Clark (on her first spaceflight), Payload Specialist Michael P. Anderson on his second spaceflight and Israel's first astronaut, Ilan Ramon. The crew had conducted eighty separate experiments, comprised of hundreds of samples and test points. The seven astronauts had worked around the clock in two alternating shifts. Their mission lasted fifteen days, 22 hours, 20 minutes and 32 seconds.

An official enquiry followed, held by the **Columbia Accident Investigation Board.** Video footage of the launch revealed a chunk of foam insulation torn loose by vibration from the left fuel tank strut striking the leading edge of a wing at liftoff and damaging the thermal protection system. This enabled hot plasma gases to cause a catastrophic failure during re-entry, when *Columbia* tumbled over and was broken apart by aerodynamic forces, depressurising the crew cabin. The entire United States Space Shuttle fleet was grounded for over two years during 2003-04 due to safety concerns. Careful inspections of all the Space Shuttle heat shield tiles were then made when subsequent missions docked at the ISS, and special epoxy-based repair kits were developed.

# 2004

There were no United States Space Shuttle flights for the next thirty months; during the period from January 16, 2003, to July 26, 2005.

# 2005

Launched on July 26, the space shuttle *Discovery* on mission STS-114 was a post-*Columbia* flight safety and evaluation test. It included ISS supply and repair and carried the Multi-Purpose Logistics Module (MPLM) *Raffaello*. The crew were Commander Eileen Collins (on her fourth and last spaceflight), Pilot James M. Kelly (on his second and last spaceflight), and Mission Specialists Soichi Noguchi (of JAXA), Stephen K. Robinson (on his third spaceflight), Andrew S.W. Thomas (on his fourth and last spaceflight), Wendy B. Lawrence (on her fourth and last spaceflight) and Charles J. Camarda (on his only spaceflight). Their 219-orbit mission lasted thirteen days, 21 hours, 33 minutes and 48 seconds; traveling 9.3 million km (5.8 million miles). The launch mass was 121,483 kg, and the landing mass was 102,913 kg. *Discovery* was docked to the PMA-2 (*Destiny* forward) port for eight days, 19 hours and 54 minutes. Returned to Earth were 'Expedition 6' ISS Commander Ken Bowersox, ISS Flight Engineer Nikolai Budarin, and ISS Flight Engineer Donald Pettit.

# 2006

Launched on July 04 (American Independence Day), the shuttle *Discovery* on mission STS-121 was ISS supply Flight ULF1.1, and was also for crew rotation. It carried the MPLM *Leonardo*, and the mission tested new safety and repair techniques introduced following the *Columbia* re-entry disaster of February 01, 2003. The crew were Commander Steven Lindsey (on his fourth spaceflight), Pilot Mark C. Kelly, and Mission Specialists Michael E. Fossum, Lisa Nowak (on her only spaceflight), Stephanie Wilson (on her first spaceflight), Piers Sellers (on his second spaceflight), and Thomas Reiter (of ESA, Germany) who joined 'Expedition 13' as ISS Flight Engineer, for his second and last spaceflight. The launch mass was 121,092 kg (266,962 lb). Their 202-orbit mission lasted twelve days, 18 hours, 37 minutes and 54 seconds; traveling 8.5 million km (5.3 million miles). *Discovery* was docked to the ISS PMA-2 (*Destiny* forward) port for eight days, 19 hours and 16 minutes.

Launched on September 09, the shuttle *Atlantis* on mission STS-115 (aka ISS assembly Flight 12A), carrying the second port-side P3/P4 truss and a pair of solar arrays (2A and 4A) with batteries, which doubled the power capacity of the ISS. There were three EVAs, during which the crew connected the systems on the installed trusses, prepared them for deployment, and did other maintenance work. The crew were Commander Brent W. Jett, Jr., (on his fourth and last spaceflight), Pilot Christopher Ferguson, and Mission Specialists Steven MacLean of CSA (on his second and last spaceflight, and first EVA), Daniel C. Burbank (on his second spaceflight), Joseph R. Tanner (on his fourth and last spaceflight) and Heidemarie M. Stefanyshyn-Piper (on her first spaceflight). MacLean operated the *Canadarm2*, and its extension boom to help inspect the surface and tiles of the orbiter for signs of damage that may have occurred during launch. *Atlantis* was docked to the ISS for six days, two hours and four minutes. The 188-orbit mission lasted eleven days, 19 hours, six minutes and 35 seconds; traveling 7.840 million km (4,870,000 miles).

After a previous launch attempt on December 07 had been canceled due to cloud cover, space shuttle *Discovery* was launched on December 09 for mission STS-116 (aka ISS assembly Flight 12A.1), carrying the P5 truss and 'SPACEHAB-SM'. Also providing for crew rotation, it was the final shuttle launch from Launch Complex 39B. A major mission goal was rewiring the ISS power system. The crew were Commander Mark L. Polansky (on his second spaceflight), Pilot William "Bill" Oefelein (on his only spaceflight), and Mission Specialists Nicholas Patrick, Robert L. Curbeam, Jr., (on his fourth and last spaceflight), Mr. Christer Fuglesang of ESA (the first spaceflight of a Scandinavian astronaut), Joan E. Higginbotham (on her only spaceflight) and Sunita "Suni" Williams, on her first spaceflight; joining ISS 'Expedition 14' as Flight Engineer.

Robert Curbeam made a record four spacewalks, the most so far during a single spaceflight. ISS Flight Engineer Thomas Reiter was returned to Earth from the ISS 'Expedition 14', ending his second and last spaceflight. STS-116 saw *Discovery* docked to the ISS for seven days, 23 hours and 58 minutes. The launch mass was 120,413 kg, and the landing mass was 102,220 kg. Mission duration was twelve days, 20 hours, 44 minutes and 16 seconds; traveling 8.5 million km (5.3 million miles).

# 2007

Launched on June 08, the shuttle *Atlantis* on mission STS-117 was ISS assembly Flight 13A, carrying the S3/S4 second and third starboard truss segments and another set of solar arrays. The crew were Commander Frederick W. Sturckow (on his third spaceflight), Pilot Lee J. Archambault, and Mission Specialists Patrick G. Forrester (on his second spaceflight), Steven R. Swanson, John D. Olivas (both on their first spaceflight), James F. Reilly (on his third and last spaceflight), and Clayton Anderson, who became an ISS Flight Engineer on

'Expedition 15'. Providing for crew rotation (bringing up Clayton Anderson, and returning with 'Expedition 14' ISS Flight Engineer Sunita "Suni" Williams), the 219-orbit mission lasted thirteen days, twenty hours, twelve minutes and 44 seconds. The shuttle traveled 9.334 million km (5.8 million miles), and the landing mass was 90,492 kg (199,501 lb).

Launched on August 08, the shuttle *Endeavour* on mission STS-118 was ISS assembly Flight 13A.1. The 22nd Shuttle flight to the ISS, it was the first for *Endeavour* since mission STS-113 in November 2002, and carried the S5 starboard truss and Spacehab-SM and 'ESP-3'. The crew were Commander Scott J. Kelly (on his second spaceflight), Pilot Charles O. Hobaugh (on his second spaceflight), and Mission Specialists Tracy Caldwell-Dyson, Richard Mastracchio, (on his second spaceflight), Daffyd "Dave" Williams of CSA (on his second and last spaceflight), Barbara Morgan and Alvin Drew. Their 201-orbit flight lasted twelve days, 17 hours and 55 minutes; traveling 8,489,253 km (about 5,274,977 miles). Dave Williams performed three EVAs.

Launched on October 23, the shuttle *Discovery* on mission STS-120 was ISS assembly Flight 10A, carrying the United States *Harmony* module, and providing for crew rotation. It reconfigured a portion of the ISS in preparation for future assembly missions. The crew were Commander Pamela Melroy (on her third and last spaceflight), the second woman after Eileen Collins to command a Space Shuttle mission, Pilot George D. Zamka, and Mission Specialists Douglas H. Wheelock, Stephanie Wilson (on her second spaceflight), Scott E. Parazynski (on his fifth and last spaceflight), Paolo A. Nespoli (of the ESA), and Daniel M. Tani (on his second and last spaceflight). Daniel Tani became an ISS Flight Engineer, while Clayton Anderson returned from the ISS. Their 238-orbit mission lasted fifteen days, two hours and 24 minutes; traveling 10.06 million km (6.25 million miles).

# 2008

Launched on February 07 (the third attempt), the shuttle *Atlantis* on mission STS-122 (ISS assembly Flight 1E), carried the European Space Agency's *Columbus* science laboratory, and provided for crew rotation. The crew were Commander Stephen Frick (on his second and last spaceflight), Pilot Alan G. Poindexter, and Mission Specialists Leland D. Melvin, Rex J. Walheim (on his second spaceflight), Hans Schlegel of ESA (on his second and last spaceflight), Stanley G. Love (on his only spaceflight), and Leopold Eyharts of ESA, who joined 'Expedition 16' as ISS Flight Engineer for his second and last spaceflight. Returned to Earth was ISS Flight Engineer Daniel M. Tani, ending his second and last spaceflight. The STS-122 mission of 202-orbits lasted for twelve days, 18 hours, 21 minutes and 50 seconds; traveling 8.5 million km (5.3 million miles). The launch mass was 121,264 kg and the landing mass was 93,536 kg (206,212 lb).

Launched on March 11, the shuttle *Endeavour* on mission STS-123 was ISS assembly Flight 1J/A, carrying Experiment Logistics Module-Pressurized Section (ELM PS) and SPDM on one Pallet (F004). This was the twenty-fifth space shuttle mission to visit the ISS, and delivered the first module of the Japanese laboratory, Japanese Experiment Module, and the Canadian robotic *Dextre*, the Special Purpose Dexterous Manipulator (SPDM) to the station, which completed the Mobile Servicing System. Like *Canadarm2* and the Mobile Base System, *Dextre* can be controlled from a workstation inside the ISS or by controllers on the ground in Mission Control centers in Houston, Texas and at CSA headquarters in Longeuil, Quebec. The crew were Commander Dominic L. Pudwill Gorie (on his fourth and last spaceflight), Pilot Gregory H. Johnson, and Mission Specialists Robert "Bob" L. Behnken, Michael Foreman (both on their first spaceflight), Richard M. Linnehan (on his fourth and last spaceflight), Dr. Takao Doi and Garrett Reisman, who joined 'Expedition 16' as ISS Flight Engineer. Returned to Earth was Leopold Eyharts (of ESA), an ISS Flight Engineer ending his second and last spaceflight. Their 250-orbit mission lasted fifteen days, 18 hours, 12 minutes and 27 seconds; traveling 10,585,900 km (about 6,577,773 miles). The launch mass was 122,364 kg, and the landing mass was 94,158 kg (207,582 lb).

Launched on May 31, the shuttle *Discovery* on mission STS-124 (ISS assembly flight ISS-1J), carried the Japanese modules *Kibo*, the largest module of the ISS, i.e., the Japanese Experiment Module pressurised section, and RMS. *Kibo* was berthed to the *Harmony* module, and the pressurized section of the JEM Experiment Logistics Module was moved from *Harmony* to the JEM-PM. The Japanese Remote Manipulator System, a robotic arm, was also delivered and attached to *Kibo*. The entire *Kibo* laboratory was brought up over three missions. All the modules were manufactured at the **Tsukuba Space Center** from stainless steel and titanium. The launch mass was 122,072 kg (269,123 lb) and the payload mass was 17,530 kg (about twice the weight of a school bus).

The STS-124 crew were Commander Mark E. Kelly (on his third spaceflight), Pilot Kenneth T. Ham, and Mission Specialists Karen L. Nyberg, Ronald "Ron" J. Garan, Jr., Michael E. Fossum (on his second spaceflight), Akihiko Hoshide of JAXA (on his first spaceflight) and Gregory E. Chamitoff, joining 'Expedition 17' as ISS Flight Engineer, on his first spaceflight. Returned to Earth was Garret E. Reisman, an ISS Flight Engineer, ending his first spaceflight. Chamitoff brought the first bagels into space: three bags of 18 sesame seed Montreal-style. Flying with the crew was an action figure of Buzz Lightyear, a fictional character in the ***Toy Story*** franchise. Their 217-orbit mission lasted thirteen days, 18 hours, 13 minutes and seven seconds; traveling 9,230,623 km (5,735,643 miles).

Launched on November 14, the shuttle *Endeavour* on mission STS-126 was ISS assembly Flight ULF2, carrying MPLM *Leonardo*. *Endeavour* delivered equipment and supplies, serviced the Solar Alpha Rotary Joints (SARJ), and repaired the problem in the starboard SARJ that had limited its use since mission STS-120. The crew were Commander Christopher J. Ferguson (on his second spaceflight), Pilot Eric A. Boe, and Mission Specialists Donald R. Pettit (on his second spaceflight), Stephen G. Bowen, Heidemarie M. Stefanyshyn-Piper (on her second and last spaceflight), Robert S. Kimbrough, and Sandra H. Magnus, who on her second spaceflight became an ISS Flight Engineer, while Gregory E. Chamitoff was returned to Earth in crew rotation. The crew performed four EVAs. STS-126 flew the first bovine embryos as well as porcine embryonic cells, for an experiment to evaluate effects of the space environment on embryonic development. Their 251-orbit flight lasted fifteen days, 20 hours, 30 minutes and 30 seconds; a new shuttle record. The craft traveled 10,645,986 km (about 6,615,109 miles).

# 2009

Launched on March 15, the space shuttle *Discovery* on mission STS-119 (HST SM 04) was ISS assembly flight 15A. It carried one Pallet (F009), the S6 fourth starboard truss, and the fourth set of solar arrays. The crew were Commander Lee Archambault (on his second and last spaceflight), Pilot Dominic A. Antonelli, and Mission Specialists Joseph M. Acaba, Steven Swanson (on his second spaceflight), Richard R. Arnold, John L. Phillips (on his third and last spaceflight) and Koichi Wakata (of JAXA), who became an ISS Flight Engineer, while Sandra Magnus was returned to Earth from the ISS. This mission delivered the final set of solar arrays and batteries needed to complete the station's complement of electricity-generating panels, and through them supported the station's expanded crew of six in 2009. There were four EVAs, assembling the fourth starboard Integrated Truss Segment. The shuttle was docked to the ISS for seven days, 22 hours and 33 minutes. Their 202-orbit mission lasted twelve days, 19 hours and 29 minutes. The launch mass was 120,859 kg, and the landing mass was 91,166 kg (200,986 lb).

Launched on May 11, the shuttle *Atlantis* on her thirtieth flight (mission STS-125) was the fifth and last Hubble Space Telescope servicing mission, and the final non-ISS flight. STS-125 upgraded Hubble to its technologically most advanced state since its launch nineteen years before and made it more powerful. The upgrades made it possible for Hubble to see deeper into the universe and further into the past, closer to the time of the "Big Bang". *Atlantis* was berthed with Hubble from May 13 to May 19; five and three-quarter days. The crew were Commander Scott Altman (on his fourth and last spaceflight), Pilot Gregory C. Johnson (on his only spaceflight), and Mission Specialists Michael T. Good, Megan McArthur (both on their first spaceflight), John M. Grunsfeld (on his fifth and last spaceflight), Michael "Mike" J. Massimino (on his second and last spaceflight) and Andrew "Drew" J. Feustel on his first spaceflight. Their 197-orbit mission duration was twelve days, 21 hours, 37 minutes and nine seconds; traveling 8.5 million km (about 5,281,655 million miles). The payload was 13,104 kg (28,889 lb).

The first launch attempt for mission STS-127 on June 13 was scrubbed due to a gaseous hydrogen leak observed during tanking. The second launch opportunity on June 17 was also scrubbed due to hydrogen leak issues seen from the Ground Umbilical Carrier Plate. Because of lightning strikes near the launch pad during the evening of July 10, NASA scrubbed the launch for the third time, and rescheduled it for July 12. Due to a Return to Launch Site (RTLS) weather violation, NASA scrubbed the launch for the fourth time on the evening of July 12. STS-127's fifth launch attempt on July 13 was scrubbed due to anvil clouds and lightning within ten nautical miles (19 km) of the launch site.

Shuttle *Endeavor* eventually launched on July 15 on her twenty-third flight (mission STS-127); ISS assembly Flight 2J/A, and pieces of foam were observed falling off the External Tank during the ascent, the same occurrence that led to the loss of *Columbia* in 2003. However, *Endeavour* received only minor scuffs to its heat shield, and the damage was not enough to cause concern over re-entry. The crew were Commander Mark L. Polansky (on his third and last spaceflight), Pilot Douglas "Doug" G. Hurley, and Mission Specialists Christopher J. Cassidy (the five hundredth person to fly in space), both on their first spaceflight, Julie Payette of CSA (on her second and last spaceflight), Thomas H. Marshburn (on his first spaceflight), David A. Wolf (on his fourth and last spaceflight) and Timothy Kopra, joining 'Expedition 20' on his first spaceflight. Koichi Wakata (of JAXA) was returned to Earth from the ISS, ending his third spaceflight.

Mission STS-127 marked the first time that two Canadian astronauts (Robert "Bob" Thirsk and Julie Payette) were in space at the same time. *Endeavour* carried the final two components of the Japanese Experiment Modules Exposed Facility (EF) and the 'Exposed Section' of the Experiment Logistics Module (ELM ES). When *Endeavour* docked with the ISS on this mission, it set a record for the most humans (thirteen) in space at the same time in the same vehicle. Five EVAs were performed, totaling thirty hours and thirty minutes. *Endeavour* spent ten days, 23 hours and 41 minutes docked to the ISS. Their 248-orbit mission lasted fifteen days, sixteen hours, 44 minutes and 58 seconds; traveling 10,537,748 km (6.55 million miles).

Launched on August 28, the shuttle *Discovery* on mission STS-128 was ISS assembly Flight 17A, carrying the Multi-Purpose Logistics Module (MPLM) *Leonardo* to the ISS. This was a large pressurised container for taking materials to and from the station. The mission's first launch attempt was delayed due to weather concerns, including multiple weather violations in NASA's launch rules. The second attempt on August 26 was called off due to an anomaly in one of the orbiter's fuel valves. The crew were Commander Frederick W. Sturckow (on his fourth and last NASA spaceflight), Pilot Kevin A. Ford, and Mission Specialists Patrick G. Forrester (on his third and last spaceflight), two Hispanic Americans: Jose M. Hernandez of Stockton, California, and John D. "Danny" Olivas of El Paso, Texas (on his second and last spaceflight). Mission Specialists were Mr. Christer Fuglesang of ESA (a Scandinavian on his second and last spaceflight) and Nicole Stott, who joined 'Expedition 20' on the ISS; while Timothy Kopra returned to Earth, ending his first spaceflight.

Three EVAs were carried out, which removed and replaced a materials processing experiment outside ESA's *Columbus* module and returned an empty ammonia tank assembly. Nicole Stott was originally scheduled to return aboard *Soyuz TMA-15*, but a change in the flight plan was made due to the possible flight delays in future shuttle missions, which could have extended Canadian astronaut Robert Thirsk's mission beyond the six-month duration preferred for ISS crew members. On September 17, the CSA's *Canadarm2* successfully captured the *HTV* free-flying Japanese vehicle – a first for the robotic arm on the ISS. CSA astronaut R. "Bob" Thirsk took over the controls of the arm and docked the spacecraft to the ISS to finalize the rendezvous. After providing ISS crew rotation, the spacecraft then landed at Edwards AFB. Their 219-orbit mission lasted thirteen days, 20 hours and 54 minutes; traveling 9,262,217 km (about 5,755,275 miles).

Launched on November 16, the shuttle *Atlantis* on mission STS-129 was ISS assembly Flight ULF3, bringing two large ExPRESS Logistics Carriers (ELCs) 1 and 2, and two spare gyroscopes. The payload bay also carried two nitrogen tank assemblies, two pump modules, an ammonia tank assembly, a spare latching and end effector for the station's robotic arm, a spare trailing umbilical system for the Mobile Transporter, and a high-pressure gas tank. This thirty-first shuttle mission to the ISS was the final Space Shuttle crew rotation flight to or from the ISS. The crew were Commander Charles O. Hobaugh (on his third and last spaceflight), Pilot Barry E. Wilmore (on his first spaceflight), and Mission Specialists Leland D. Melvin (on his second and last spaceflight), Randolf Bresnik, Michael Foreman (on his second and last spaceflight) and Robert Satcher (on his only spaceflight). Nicole Stott was returned from 'Expedition 21' on the ISS, ending her first spaceflight. *Atlantis* was docked to the ISS for six days, 17 hours and two minutes. The mission duration was ten days, 19 hours, 16 minutes and 13 seconds; traveling 7,226,177 km (4,490,138 miles). The launch mass was 120,848 kg, and the dry landing mass was 93,063 kg.

## 2010

Launched on February 08, the shuttle *Endeavour* on mission STS-130 was ISS assembly Flight 20A, carrying the *Tranquility* module Node 3 and the *Cupola* (a robotic control station with six windows around its sides and another in the center, providing a 360-degree view around the ISS. The crew were Commander George D. Zamka (on his second and last spaceflight), Pilot Terry W. Virts, Jr. (on his first spaceflight), and Mission Specialists Kathryn P. Hire (on her second and last spaceflight), Stephen K. Robinson (on his fourth and last spaceflight), Nicholas Patrick (on his second and last spaceflight) and Robert "Bob" L. Behnken (on his second spaceflight). Their 217-orbit mission lasted thirteen days, 18 hours, six minutes and 24 seconds; traveling 9.25 million km (about 5,747,684 million miles). The launch mass was 121,230 kg, the landing mass was 91,033 kg, and the payload mass was 19,000 kg (42,000 lb).

Launched on April 05, the shuttle *Discovery* on her thirty-eighth flight (mission STS-131) was ISS assembly Flight 19A, carrying the Utility and Logistics Flight 4: Multi-Purpose Logistics Module (MPLM) *Leonardo*. The crew were Commander Alan Poindexter (on his second and last spaceflight), Pilot James Dutton (on his only spaceflight), and Mission Specialists Richard Mastracchio (on his third spaceflight), Dorothy M. Metcalf-Lindenburger (on her only spaceflight), Stephanie Wilson (on her third spaceflight), Naoko Yamazaki of JAXA (on her only spaceflight) and Clayton Anderson (on his second and last spaceflight). This was the last night launch of the Shuttle Program, and the longest flight of *Discovery*.

It was the second flight of the *TriDAR*, a 3-D dual-sensing laser camera, intended for potential use as an autonomous rendezvous and docking sensor. *TriDAR* provides guidance information that can be used to guide a vehicle during rendezvous and docking operations in space. Working in real-time, *TriDAR* does not rely on any reference markers, such as reflectors positioned on the target spacecraft. The 238-orbit mission lasted 15 days, two hours, 47 minutes and 11 seconds; traveling 10,029,810 km (about 6,232,235 miles). The launch mass was 121,047 kg, and the landing mass was 102,039 kg (224,957 lb).

Launched on May 14, the shuttle *Atlantis* on mission STS-132 was ISS assembly flight ULF4. It delivered an Integrated Cargo Carrier-Vertical Light Deployable (ICC-VLD), and a Russian-built *Rassvet* Mini Research Module-1. Three EVAs were conducted while *Atlantis* was docked to the orbiting laboratory. The crew were Commander Kenneth "Hock" Ham (on his second and last spaceflight), Pilot Dominic A. "Tony" Antonelli (on his second and last spaceflight), and Mission Specialists Garrett Reisman (on his second and last spaceflight), Michael T. Good (on his second and last spaceflight), Stephen G. Bowen (on his second spaceflight) and Piers Sellers (on his third and last spaceflight). Their 186-orbit mission lasted eleven days, 18 hours and 29 minutes; traveling 7,853,563 km (about 4,879,978 miles).

## 2011

Launched on February 24, the shuttle *Discovery* on her thirty-ninth and final flight (mission STS-133) was ISS assembly Flight ULF5, carrying the Permanent Multipurpose Module *Leonardo* (named after the famed Italian Renaissance inventor, Leonardo da Vinci, and which was left permanently docked to a port of the ISS); a humanoid robot called *Robonaut*; and ExPRESS Logistics Carrier 4 (ELC), a steel platform designed to support external payloads mounted to the ISS starboard and port trusses (with either deep space or Earthward views). The ELC carried several Orbital Replacement Units (ORUs). Among these were a Heat Rejection System Radiator (HRSR), and Flight Support Equipment (FSE0), which takes up one whole side of the ELC. The other primary ORUs were the ExPRESS Pallet Controller Avionics 4 (ExPCA #4). The HRSR launching on ELC4 was a spare, for one of the six radiators that are part of the Space Station's external active thermal control system.

The STS-133 crew were Commander Steven Lindsey (on his fifth and last spaceflight), Pilot Eric Boe (on his second spaceflight), and Mission Specialists Nicole Stott (on her second and last spaceflight), Alvin Drew, the last African American to fly on the Shuttle (on his second and last spaceflight), Michael Barratt (on his second spaceflight) and Stephen G. Bowen (on his third spaceflight). All but Steven Lindsey were part of NASA's Astronaut Group 18, all being selected in the year 2000. Having flown on mission STS-132, Bowen became the first and only NASA astronaut to be launched on two consecutive missions; until Douglas "Doug" G. Hurley launched aboard '*Crew Dragon* Demo-2' in May 2020, after having previously launched on mission STS-135. *Discovery* was docked to the ISS for eight days, 16 hours and 46 minutes.

The STS-133 mission carried the signatures of more than 500,000 students who had participated in the 2010 **Student Signatures in Space** program (SSiS), which was jointly sponsored by NASA and Lockheed Martin. SSiS has been active since 1997. In that time, nearly seven million student signatures from 6,552 schools were flown on ten Space Shuttle missions. Mission STS-133 lasted twelve days, nineteen hours, four minutes and fifty seconds; traveling 8,536,190 km (about 5,304,143 miles). The orbiter launch mass was 121,840 kg, with a dry mass of 92,867 kg. With the External Tank and two SRBs, the stack weighed 2,052,610 kg (2,052 tons, or 4,525,220 lb).

The Space Shuttle had been scheduled to be retired from service after STS-133, but controversy over the cancellation of several ISS components, most notable the Alpha Magnetic Spectrometer, to meet deadlines for the retirement of the Space Shuttle, caused the United States government to consider ordering an additional mission *"to deliver science experiments to the station"*. United States President George W. Bush had previously opposed any additional Shuttle missions, as they could delay the transition to the ambitious

**Project Constellation** (later canceled). In the spring of 2009, the Obama Administration included funds for the STS-134 mission in its proposed 2010 NASA budget.

Launched on May 16, the shuttle *Endeavour* on its twenty-fifth and final flight, mission STS-134 (aka ISS assembly Flight ULF6), carried ELC 3, and the Alpha Magnetic Spectrometer-2 (AMS-2), attached to the ISS S3 truss segment. The AMS-2 is a particle physics detector which contains a large permanent magnet. It was designed to search for antimatter and investigate the origin of dark matter. The ExPRESS Logistics Carrier 3 (ELC3) carried several Orbital Replacement Units (ORU) that were too large or too heavy for other spacecraft to carry to the ISS. These ORU included a High Pressure Gas Tank (HPGT), an Ammonia Tank Assembly (ATA), the S-band Antenna Sub-System Assembly #2 & 3 (SASA), a Special Purpose Dextrous Manipulator Arm (SPDM) with Orbital Replacement Unit change-out mechanism, a Space Test Program Houston 3 Department of Defense payload, and a spare ELC pallet controller avionics box.

Mission STS-134 delivered the Materials on International Space Station Experiment 8 (MISSE-8) experiments and returned to Earth the completed MISSE-7 experiments (which went up on mission STS-129 in 2009). *Endeavour* brought specialised nutrition bars, called "STEM Bars" to the ISS. These were created by high school students and sisters Mikayla and Shannon Diesch. The nutrition bars were certified for spaceflight by meeting a specific NASA-developed nutritional profile and had to pass strict microbial testing. The STEM Bars were flown to support the work of the Battle Creek, Michigan-based sisters to raise awareness of the importance of STEM education (the technical disciplines of Science, Technology, Engineering and Mathematics) among their peers, an outreach effort which they started after they won the 2010 Conrad Foundation 'Spirit of Innovation' Awards.

The first launch attempt for mission STS-134 on April 29 was scrubbed by launch managers due to problems with two heaters on one of the orbiter's Auxiliary Power Units (APU). The crew were Commander Mark E. Kelly (on his fourth and last spaceflight), Pilot Gregory H. Johnson (on his second and last spaceflight), and Mission Specialists Michael Fincke (on his third spaceflight), Roberto Vittori of ESA (on his third and last spaceflight), Andrew "Drew" J. Feustel (on his second spaceflight) and Gregory E. Chamitoff (on his second and last spaceflight). If STS-134 had launched during 'Expedition 26', then Mark E. Kelly and 'Expedition 26' Commander Scott J. Kelly would have become the first siblings (and twins) to fly in space at the same time.

On Flight Day 6, members of *Endeavour's* crew performed a focused inspection of an area of thermal protection tiles on the bottom of the orbiter. These tiles were damaged during launch, and detailed data provided by the Orbiter Boom Sensor System was needed to make sure the orbiter could re-enter Earth's atmosphere safely. The inspection process took two hours to complete and resulted in the Thermal Protection System (TPS) being cleared for re-entry. The joint crew spoke with Pope Benedict XVI, answering his several questions. This marked the first time a Pope had spoken to astronauts while they were in space.

On Flight Day 7, the second EVA was conducted by Drew Feustel and Mike Fincke. It lasted eight hours and seven minutes; the sixth-longest in the history of spaceflight at that time, and the second-longest EVA from the ISS. The astronauts hooked up a jumper to transfer 2.3 kg of ammonia to the Port 6 Photovoltaic Thermal Control System (PVTCS), lubricated the Solar Alpha Rotary Joint (SARJ) and one of the "hands" on robotic *Dextre*, and installed a stowage beam on the Starboard 1 (S1) truss. The spacewalk was the seventh for Michael Fincke and the fifth for Andrew Feustel. Mike Fincke had conducted his previous six spacewalks in Russian *Orlan* suits.

On Flight Day 8, Commander Mark E. Kelly and Mike Fincke conducted an in-flight interview with four hundred students from Mesa Verde Elementary School in Tucson, Arizona. Later in the crew day, Mission Specialist Robert Vittori and 'Expedition 27' Flight Engineer Paulo Nespoli answered questions from Italian President Giorgio Napolitano. *Endeavour* was docked to the ISS for 11 days, 17 hours, and 41 minutes. Their STS-134 mission lasted for fifteen days, 17 hours, 38 minutes and 51 seconds; traveling 10,477,185 km (about 6,510,221 miles). The launch mass was 121,830 kg, the landing mass was 92,240 kg, and the payload mass was 15,770 kg (34,760 lb).

Launched on July 08, 2011, the shuttle *Atlantis* on her thirty-third and final mission (STS-135) made the final Space Shuttle flight, which was extended by one day. It was the one hundred and sixty-sixth NASA crewed space flight, the one hundredth day-time Space Shuttle launch, and used hardware originally processed for the STS-335 contingency mission. The four crew were Commander Christopher Ferguson (on his third and last NASA spaceflight), Pilot Douglas G. Hurley (on his second spaceflight), and Mission Specialists Sandra Magnus (on her third and last spaceflight) and Rex J. Walheim (on his third and last spaceflight). The primary payload was the Multi-Purpose Logistics Module (MPLM) *Raffaello* full of supplies, and a Lightweight Multi-Purpose carrier (LMC). This flight with *Raffaello* was the only time that *Atlantis* carried an MPLM. The shuttle was docked to the ISS for eight days, 15 hours and 21 minutes. Mission STS-135 lasted for twelve days, 18 hours, 28 minutes and 50 seconds; traveling 8,505,161 km (about 5,284,862 miles).

Ending the Space Shuttle program, *Atlantis*, the last of the world's heaviest space-plane gliders came in hard and fast for the final time on July 21, landing at the John F. Kennedy Space Centre, Florida. Since then, the United States had to rely upon Russian *Soyuz* spacecraft to carry United States astronauts to the ISS, paying over USD 40 million per astronaut launched. Russian and Japanese cargo vessels also bring supplies to the ISS. The United States' **Commercial Crew** Program started in 2011, with the first crewed mission to the ISS in May 2020; it is ongoing.

Mission STS-135 marked the one hundred thirty-third Space Shuttle landing, the seventy-eighth landing at Kennedy Space Center, the twenty-sixth night landing, and the twentieth shuttle night landing at Kennedy Space Center. By the end of that mission, *Atlantis* had orbited the Earth a total of 4,848 times, traveling nearly 203 million km (about 126,138,352 miles), which is more than 525 times the distance from Earth to the Moon. The craft was built by **Rockwell International** at their Palmdale, California, assembly plant. Its weight, with three shuttle main engines was 68,635 kg (about 151,314 lb), the length was 37.2 metres, the height was 17.2 metres, and the wingspan was 23.7 metres. *Atlantis* was nearly 3.5 tons lighter than the shuttle *Columbia* version.

Over the entire Space Shuttle program, one initial emergency flight abort (RTLS) sub-orbital test mission was canceled due to high risk. Many other planned missions were canceled due to the late development of the Shuttle, and because of the *Challenger* and *Columbia* disasters. Four missions were cut short by a day or more while in orbit: STS-2 (equipment failure); STS-35 (weather); STS-44 (equipment failure); and STS-83 (equipment failure, but later relaunched as STS-94).

Astronaut Mark Kelly said during his 2006 flight on Space Shuttle *Discovery*, about the risks of flying the shuttle: *"The Space Shuttle's a very complex machine. It's got a lot of moving parts that move and operate at pretty much the limit of what we've been able to engineer. Spaceflight is risky. I think with regards to the tank, we've reduced some of the risks there. We've changed the design a little bit and we've made some pretty big strides in trying to get foam not to shed from the tank anymore. So there is some risk reduction there, and, I guess, overall, the risk is probably a little less. But this is a risky business, but it's got a big reward. Everybody on board Discovery and the Space Station here thinks it's worthwhile".* Unquote.

# The International Space Station

The motto of the International Space Station is ***"Off the Earth, for the Earth"***.

There is a growing consensus that a mega-asteroid impact at the end of the Cretaceous geologic period, sixty-five million years ago caused the mass extinction of dinosaurs on Earth. Science-fiction writer Larry Niven wrote: "*The dinosaurs became extinct because they didn't have a space program. And if we [humans] become extinct because we don't have a space program, it'll serve us right!*" On November 20, 1998, the **International Space Station**, a venture involving the USA (NASA), Russia (Roscosmos/RKA), Europe (ESA), Japan (JAXA) and Canada (CSA) became the first multinational space station. It is the largest artificial object built in space to date. The **International Space Station** (ISS) is seen as an asset to help understand and mitigate several human health risks anticipated on long-duration missions, as well as test and mature the technologies and spacecraft systems necessary for humans to operate safely and productively in deep space.

Anatomical changes associated with long-duration space flight are certainly negative: the immune system weakens, the heart shrinks (because it doesn't have to strain against gravity), eyesight tends to degrade, the spine lengthens as little sacs of fluid between the vertebrae expand, and bone mass decreases as the body sheds calcium. As Kevin Fong explained in an article for ***Wired*** magazine: "*Deprived of gravitational load, bones fall prey to a kind of spaceflight-induced osteoporosis. And because 99% of our body's calcium is stored in the skeleton, as it wastes away, that calcium finds its way into the bloodstream, causing yet more problems from constipation to renal stones to psychotic depression*", he said. To preserve their health, astronauts aboard the ISS commit to exercising for at least two hours each day; they regularly take part in medical monitoring of their physical functions.

More than fifteen nations are partners in the ISS project. The ISS's two Orbital Segments are made up of 16 modules: five from Russia, eight from America, two from Japan, and one from Europe. Launched in 2000, the *Zvezda* module contains the ISS's living quarters. At around fifty square metres (about twice the area of a parking space) in size, the module contains sleeping compartments for two crew members at a time, a kitchen galley with a refrigerator, an exercise bike and treadmill, and a toilet. This module also contains an oxygen recycling system that takes wastewater molecules and recycles them into oxygen (for breathing) and hydrogen (waste) gas. To provide electricity while orbiting 386 km (about 240 miles) above Earth, the ISS uses a series of solar panels, which are organised into wing-like arrays. If placed together, these panels would cover over 2,500 square metres (the area of a large mansion). The solar arrays angle themselves to track the Sun, and so gain the greatest possible exposure to its rays.

The **International Space Station** (ISS) is now 72.8 metres long (the size of a football field), 108.5 metres in diameter, it orbits the Earth fifteen times each day, and has been continuously occupied since April 2018. The usual crew is of six persons, with up to thirteen aboard at changeover of crews. Up to August 2023, the ISS program had cost USD 150 billion. It has several laboratory modules, including the European Space Agency's *Columbus* module, which was launched aboard Space Shuttle *Atlantis* in 2008. The loss of muscle mass and the weakening of bones are two of the negative effects of living in space, so astronauts must do a range of cardiovascular and strength exercises for at least 2.5 hours every day to maintain strong bodies. The ISS is funded as far ahead as 2028, or perhaps 2033, when it will be de-orbited.

On November 20, 1998, the Russians launched the *Zarya* aka "Dawn" module, also known as Functional Cargo Block or FGB, to begin construction of the **International Space Station**. *Zarya* provided the ISS with electrical power during its early assembly, and it was built with three docking points. With the launch and assembly in orbit of other modules with more specialised functionality, as of August 2021, *Zarya* was primarily used for storage, both inside the pressurised section and in the externally mounted fuel tanks. The *Zarya* is a descendant of the TKS spacecraft designed for the Soviet *Salyut* space station program. The name *"Zarya"* was given to the FGB because it signified *the dawn of a new era* of international cooperation in space. Although it was built by a Russian company, *Zarya* is owned by the United States.

Launched on December 04, 1998, the shuttle *Endeavour* on mission STS-88/2A was the first Shuttle mission to the ISS. This mission carried the *Unity* node, the first US-built section of the ISS, into orbit. The upgraded Canadarm grappled the *Unity* module to the Functional Cargo Block *Zarya* module, connecting the first two components of the ISS. During this process, the use of the Orbiter Space Vision System (OSVS), an offshoot of the Canadian Space Vision System was essential, since the *Unity* node blocked the view of the approaching Russian module.

PART TWO

# The European Space Agency

The European Space Agency (ESA) is a twenty-two member intergovernmental body devoted to space exploration. The ESA had ten founding member states: Belgium, Denmark, France, West Germany, Italy, the Netherlands, Spain, Sweden, Switzerland and the United Kingdom. Founded in 1975, with its headquarters in Paris and a staff of around 2,200 people globally as of 2022, the annual budget in 2024 was 7.8 billion Euro. The Treaty establishing the ESA reads: ***The purpose of the agency shall be to provide for and to promote, for exclusively peaceful purposes, cooperation among European States in space research and technology and their space applications, with a view to their being used for scientific purposes and for operational space applications systems***.

Although the ESA had relied on cooperation with NASA in previous decades, especially the 1990s, changed circumstances (such as tough legal restrictions on information sharing by the United States military) led to decisions to rely more on itself and on cooperation with Russia. The ESA's space flight program includes human spaceflight (mainly through participation in the International Space Station program); the launch and operation of crewless exploration missions to other planets (such as Mars) and the Moon; Earth observation, science and telecommunication; designing launch vehicles; and maintaining a major spaceport, the Guiana Space Center at Kourou (French Guiana), France.

The main European launch vehicle *Ariane 6* will be operated through **Arianespace** with ESA sharing in the costs of launching and further developing this space vehicle. The agency is also working with NASA to manufacture the *Orion* spacecraft service module that flies on the Space Launch System. On January 21, 2019, ArianeGroup and Arianespace announced a one-year contract with the ESA to study and prepare for a mission to mine the Moon for lunar regolith. July 01, 2023 saw the launch of the *Euclid* spacecraft, developed jointly with the **Euclid Consortium**. After ten years of planning and building, *Euclid* is designed to better understand dark energy and dark matter by accurately measuring the accelerating expansion of the Universe. At present, dark energy is understood as a repulsive force (perhaps the opposite of gravity, which "sucks").

# Astronauts of the European Space Agency, China, Japan and others

## ACTIVE ASTRONAUTS:

As of 2023, there were six active members of the European Astronaut Corps. Five of these were selected in 2009, and one in 2015. Timothy Peake, a member of the 2009 group, retired in 2023.

### Christoforetti, Samantha. Soyuz TMA-15M/ Expedition 42/43, SpaceX Crew 4/Expn 67/68.

Samantha Christoforetti was born in April 1977 in Milan, Italy. The first Italian woman in space, she is an accomplished astronaut who has spent time aboard the International Space Station (ISS) and contributed to scientific research in space. At age eighteen years, she participated in an AFS exchange program to the United States and attended Space Camp. She graduated from the Technical University of Munich with a degree in Mechanical Engineering, then studied at the Ecole nationale superieure de l'aeronautique et de l'espace in Toulouse, France and the Mendeleev Russian University of Chemistry and Technology in Moscow, Russia. She also graduated from the Accademia Aeronautica in Pozzouli, Italy, with a degree in Aeronautics Sciences; becoming one of the first women to be a lieutenant and fighter pilot in the Italian Air Force. She is a former Italian Air Force pilot and engineer, with over 500 hours flying time in six types of military aircraft.

In 2009, Christoforetti was selected by the European Space Agency (ESA) from a pool of 8,000 applicants. Her basic astronaut training included general survival skills, scuba diving, lessons in orbital mechanics, and piloting. She completed the Euro-NATO Joint Jet Pilot program, learned about the ISS systems, trained for space walks, and traveled to Russia to learn how to pilot the *Soyuz* spacecraft. On November 23, 2014, she launched aboard *Soyuz TMA-15M*, docking with the ISS about six hours later to become Flight Engineer for ISS 'Expedition 42/43'. She was the primary operator during the undocking of the *European ATV-5* spacecraft. In February 2015, she began the outreach program **Mission V: Train Like an Astronaut** where students 8-12 years old were challenged to get fit during a nine-week program while Christoforetti trained in space. In April 2015, the SpaceX *Dragon* capsule delivered the ISSpresso, the first zero-G espresso machine which she installed. On May 03, 2015, Christoforetti brewed the first cup of espresso in space. A one-month delay after the failure of two Russian rockets extended her stay in space past the European astronaut and female astronaut endurance records.

At age forty-five years, Samantha Christoforetti launched atop a *Falcon-9* booster on April 27, 2022, with SpaceX 'Crew-4' aboard Crew Dragon *Freedom*, to reach the ISS about seventeen hours later for 'Expedition 67/68', becoming Commander of Expedition 68 on

September 28, 2022. The mission lasted 170 days, 13 hours and three minutes; returning to Earth on October 14, 2022, aboard Crew Dragon *Endurance*. Christoforetti was the first person to make a ***TikTok*** video on board the ISS, and some of her videos have been viewed millions of times. She has logged a total of 370 days, five hours and 45 minutes in space; including one EVA of seven hours and five minutes. She held the record for the longest single spaceflight by a woman (199 days and 16 hours), until it was broken by Dr. Peggy Whitson on June 05, 2017, and later on December 28, 2019, by Christina H. Koch. Christoforetti still holds the record for the longest uninterrupted spaceflight of a European astronaut.

From June 10-22, 2019, Christoforetti commanded the NEEMO mission on the sea floor, to test technologies and objectives for deep space missions and lunar exploration. She is married to a French engineer, Lionel Ferra, and they have a daughter and son. She speaks Italian, English, German, French, Russian and Chinese. Samantha Christoforetti appeared in the final of the Eurovision Song Contest 2022, which was held in Turin, Italy. She made a pre-recorded message from the ISS. She has received the award Commander of the Order of Merit of the Italian Republic (2013); and Knight Grand Cross of the Order of Merit of the Italian Republic (2015). On July 16, 2015, she was awarded the Order of Merit of the Italian Republic by Italian President Mattarella, who said, *"She has been followed with affection and love by all Italians"*. The 2016 IMAX documentary film ***A Beautiful Planet*** shows scenes of Earth which feature Christoforetti and other ISS astronauts.

## Gerst, Alexander. Soyuz TMA-13M/Expedition 40/41, Soyuz MS-09/ Expedition 56/57.

Dr. Alexander Gerst was born in May 1976 in Kunzelsau, Baden-Wurttemburg, Germany. He is a veteran astronaut and geophysicist who has completed two missions on the ISS, conducting experiments and maintaining the station. He received a degree in Geophysics (with distinction) from the University of Karlsruhe (now Karlsruhe Institute of Technology), Germany, then from 1998 to 2003 he participated in several international scientific collaborations and field experiments, including in some remote locations such as Antarctica. From 2001-03, Gerst studied for a Master's degree at Victoria University of Wellington in New Zealand. While researching his thesis, he developed volcano monitoring techniques that might improve forecasts of volcanic eruptions. From 2004-09, he worked as a researcher at the Institute of Geophysics at the University of Hamburg, and received his Doctorate in Natural Sciences from there in 2010; with a dissertation on geophysics and volcanic eruption dynamics at Mount Erebus volcano in Antarctica. In 2007, Gerst received the Bernd Rendel award for outstanding research from the German Research Foundation (DFG). He has published several papers, including one in *Nature*.

Gerst was selected by the ESA in 2009, and launched on *Soyuz TMA-13M* for 'Expedition 40/41' from May 28 to November 10, 2014. His six-month mission was named *Blue Dot*, reminiscent of Carl Sagan's description of Earth as a Pale Blue Dot, as seen on a photograph taken by NASA's *Voyager* mission. The ISS mission included experiments in physical science, biology, human physiology, radiation research, and technology demonstrations. Educational outreach included educational videos as part of the "flying classroom" and a set of experiments performed in microgravity. Launched on June 06, 2018, aboard *Soyuz MS-09* for 'Expedition 56/57', he became Commander of the ISS for 'Expedition 57', the youngest to hold this role at age 42 years as of October 2018. Alarmingly, a hole of about 2-mm diameter was found in the shell of the *Soyuz*, and sealed the same day with several layers of epoxy resin. The pressure inside the ISS soon stabilized. On July 20, 2018, he took part in a live concert by the German electronic music group Kraftwerk in the German city of Stuttgart. He addressed the audience from the ISS and played a synthesizer during the performance of the song *Spacelab*. After 197 days in space, Gerst returned to Earth on December 20, 2018. After the end of his second mission, and before being surpassed by Luca Parmitano in 2020 with 367 days, he held the record for the most time in space (362 days) of any active ESA astronaut.

Alexander Gerst logged a total of 362 days, one hour and 50 minutes in space; including one EVA of six hours and 13 minutes. He is single, and a licensed radio amateur (callsign KF5ONO); he has participated in several ARISS (Amateur Radio on the ISS) educational contacts. Some of his background and appearance is incorporated into the character of Alex Vogel, a German astronaut and part of the crew of the *Hermes* vessel in the 2015 science-fiction film ***The Martian***.

## Maurer, Matthias J.. SpaceX Crew 3/Expedition 66/67.

Dr. Matthias Josef Maurer was born in March 1970 in St. Wendel, Saarland, West Germany. He is a newer addition to the active astronaut roster. In 2008, he applied to the ESA, along with almost 8,500 other candidates, and was one of ten to pass the selection procedure. In 2010, he started working for the ESA as as a crew support engineer and for EuroCom for the *Columbus* flight control team at the European Astronaut Center (EAC) in Cologne, Germany. Before he formally joined the European Astronaut Corps in July 2015, he took a lead role in projects to prepare for future spaceflight and lunar operations with new international partners, and to extend the Agency's human exploration expertise beyond the International Space Station. In March 2018, Maurer gained certification in the American *EMU* spacesuit, and graduated from basic astronaut training. On May 12, 2020, he arrived at the Johnson Space Center in Houston, Texas, for training (amidst the Covid-19 pandemic). He launched on November 10, 2021, with SpaceX Crew-3, for 'Expedition 66/67'. He landed in the Gulf of Mexico on May 06, 2022, after 176 days, two hours and 39 minutes in space; including one EVA of six hours and 54 minutes. Matthias Maurer was awarded the NASA Distinguished Public Service Medal; and the Space Exploration Medal.

## Mogensen, Andreas E.. Soyuz TMA-18M, SpaceX Crew 7/Expedition 69/70.

Dr. Andreas "Andy" Enevold Mogensen was born in November 1976 in Copenhagen, Denmark. He is known for being the first Danish astronaut to fly in space, participating in the ESA's *Iriss* mission. In 1995, he received an International Baccalaureate from Copenhagen International School, then went to the Imperial College, London, where he obtained a Master of Engineering degree in Aeronautical Engineering in 1999. From 2001-03, he worked as an engineer in the R & D department of **Vestas Wind Systems** in Denmark. He received a Doctorate in Aerospace Engineering from the University of Texas at Austin in 2007. Next, he worked in Germany for **HE Space Operations** associated with the *SWARM* mission, launched in November 2013, then as a contractor for **EADS Astruim** in Portugal at the Institute for Systems and Robotics. When selected by the ESA in May 2009, he was in an aerospace engineering position at Surrey Space Center, University of Surrey, UK. On June 10, 2014, NASA announced that Mogensen would serve as an aquanaut aboard the *Aquarius* underwater laboratory during the NEEMO 19 undersea exploration mission, which ran for seven days from September 07, 2014.

Launched on September 02, 2015, aboard *Soyuz TMA-18M* to the ISS, he returned to Earth after ten days aboard *Soyuz TMA-16M*. Because of the short mission, the crew worked up to 9.5 hour days, instead of the eight-hour workdays that are usual on the ISS. His tasks included remote control of a robot on Earth, and filming Red Sprites and Blue Jets lightnings above thunderclouds, directed from Earth. Mogensen also tried a new kind of Skinsuit to alleviate back-pain astronauts feel due to lengthening of their spine, and used augmented reality goggles during his maintenance tasks. He reurned to Earth on September 12, 2015, just over three hours after departing the ISS. He was awarded the Danish Royal Medal of Recompense for his efforts. In March 2022, Mogensen was selected as pilot of SpaceX Crew-7, for 'Expedition 69/70'. He returned to space in August 2023, aboard SpaceX *Crew Dragon* as the first non-American to serve as a pilot. He became Commander of 'Expedition 70', with the European segment of the mission called *Huginn*. Total time logged: as of March 2024, he is currently in space.

## Parmitano, Luca. Soyuz TMA-09M/Expedition 50/51, Soyuz MS-13/ Expedition 60/61.

Col. Luca Parmitano was born in September 1976 in Paterno, Sicily, Italy. Parmitano has been part of two ISS expeditions and performed spacewalks to repair equipment. In 1995, he graduated from the Liceo Scientifico Statale "Galileo Galilei" in Catania, Italy. In 1999, he completed a Master's degree in Political Sciences at the University of Naples Federico II, Italy, with a thesis on International Law. In 2000, he graduated with the Sparviero IV academic course from the Italian Accademia Aeronautica, Pozzuoli, Italy. Parmitano completed basic training with the US Air Force at the Euro-NATO Joint Jet Pilot Training Program at Shepard Air Force Base, Texas, in 2001. In 2003, he qualified as Electronic Warfare Officer at the Reparto Supporto Tecnico Operativo Guerra Electronica (ReSTOGE) in Pratica di Mare, Italy. In 2005, he completed the Tactical Leadership Programme (TLP) in Florennes, Belgium. In July 2009, Parmitano received a Master's degree in experimental Flight Test Engineering at the Institut Superieur de l'Aeronautique et de l'Espace (ISAE), in Toulouse, France. Luca Parmitano has logged over 2,000 hours flying time, and is qualified on over twenty military aircraft types.

Selected in the May 2009 ESA Group, Parmitano in 2014 became a "cavenaut" serving in ESA CAVES training alongside Scott Tingle, Alexander Misurkin, Sergey Kud-Sverchkov and Matthias Maurer. He became an aquanaut in July 2015 when he served as Commander of the NEEMO 20 crew. In 2016, he took part in the first ESA PANGEA mission, alongside Matthias Maurer and Pedro Duque. In May 2017, he took part in the CAVES eXpedition One, carried out in the extreme environments of some hot caves in Sicily. Launched on May 28, 2013, aboard *Soyuz TMA-09M*, he arrived at the ISS on May 29 for 'Expedition 50/51'. On July 09, he became the first Italian to make an EVA; part of preparations for the new Russian multipurpose module planned to replace the *Pirs* docking compartment by the end of 2013. His second EVA on July 16 was terminated after one hour and 32 minutes when the helmet of his Extravehicular Mobility Unit (EMU) suit began filling with water, due to contamination clogging one of the suit's filters. He returned to Earth aboard *Soyuz TMA-09M* on November 11, 2013.

Luca Parmitano launched on July 20, 2019, as Flight Engineer aboard *Soyuz MS-13* to serve for 'Expedition 60', and became Commander for 'Expedition 61'. On August 13, he became the first "space DJ" when he played a set of electronic music from the ISS for a music festival in Ibiza. On November 15, he made the first of four spacewalks to repair the Alpha Magnetic Spectrometer (AMS); the first time astronauts had cut and reconnected cooling lines while in orbit. The EVAs were described as the *"most challenging since Hubble repairs"*. 'Expedition 61' ended with the *Soyuz MS-13* landing on February 06, 2020. Parmitano has logged a total of 366 days, 23 hours and one minute in space. Asteroid 37627 Lucaparmitano is named after him. He is divorced, and has two daughters with his ex-wife.

## Pesquet, Thomas G.. Soyuz MS-03/Expedition 50, SpaceX Crew 2/ Expedition 65/66.

Thomas Gautier Pesquet was born in February 1978 in Rouen, Normandy, France. He is a French aerospace engineer, pilot, actor and writer, and ESA astronaut who has lived and worked on the ISS, conducting experiments and outreach activities. In 1996, he graduated from the Lycee Pierre Corneille in Rouen. In 2001, he received a Master's degree from the Ecole Nationale Superieure de l'Aeronautique et de l'Espace in Toulouse, France, majoring in Space Systems and space vehicular mechanics. His final year (before graduation) was spent at the Ecole Polytechnique de Montreal, Canada, as an exchange student on the Aeronautics and Space Master course. Pesquet graduated from the Air France flight school in 2006, and this led to an Airline Transport Pilot Licence-Instrument Rating (ATPL-IR). He speaks French, English, Spanish, Chinese, German, and Russian, and is a member of the French Aeronautics and Astronautics Association (3AF), and of the American Institute of Aeronautics and Astronautics (AIAA).

From October 2001, Pesquet worked in Spain as a spacecraft dynamics engineer on remote sensing missions. From 2002-04, he worked as a research engineer at the French Space Agency (CNES). From late 2002, he was a CNES representative at CCSDS, the Consultative Committee for Space Data Syatems. A private pilot, he trained with Air France, becoming a commercial pilot, and began flying the *Airbus A320* in 2006. Pesquet has logged over 2,000 hours on various commercial airliners, and has qualified as a type-rating flight instructor on the A320. In 2018, he gained his *Airbus A310* type rating and is qualified as a Novespace Zero-G aircraft pilot. Selected by ESA in May 2009, he completed his basic training in November 2010. In 2014, he was chosen for a six-month mission to the ISS. He was also backup to Andreas Morgensen who flew to the ISS in September 2015.

Pesquet launched on November 17, 2016, as a Flight Engineer aboard *Soyuz MS-03*, docking on November 19 for ISS 'Expedition 50'. His arrival marked the beginning of the European *Proxima* mission, named for Proxima Centauri (continuing the French astronaut's tradition of naming their missions after stars and constellations), which included fifty experiments for ESA and CNES. After 196 days, 17 hours and 49 minutes in space, 'Expedition 51' ended on June 02, 2017. On July 28, 2020, Pesquet was assigned to SpaceX Crew-2 as a Mission Specialist. Crew Dragon *Endeavour* was launched on April 23, 2021, with SpaceX Commercial Crew-2, joining ISS 'Expedition 65'. On October 04, 2021, Pesquet became ISS Commander for 'Expedition 66'; the first French astronaut to command a space vehicle. In the next months, he performed four EVAs. After 199 days (six months) in space, Crew-2 landed in the Gulf of Mexico on November 09, 2021. Thomas Pesquet has logged a total of 396 days, 11 hours and 34 minutes in space; including six EVAs totaling 39 hours and 54 minutes (a European astronaut record).

At the closing ceremony of the 2020 Summer Olympics, during the portion introducing Paris, France, as the host of the 2024 games, Pesquet performed the closing bars of the *La Marsellaise* on a saxophone, while aboard the ISS. On April 12, 2021, he was nominated Goodwill Ambassador for the Food and Agriculture Organization of the United Nations. Thomas Pesquet wrote two books: a comic about his life *Dans la combi de Thomas Pesquet* (In the suit of Thomas Pesquet), and an autobiography, *Ma vie sans gravity* (My life without gravity). He also portrayed himeself in the movie ***Hawa***.

# CANADA

In chronological order of missions,

## Garneau, J-P. Marc. STS-41-G, STS-77, STS-97.

Joseph Jean-Pierre "Marc" Garneau (PC CC CD) was born in February 1949 in Quebec City, Quebec, Canada. He received a Bachelor of Science degree in Engineering Physics from the Royal Military College of Canada in 1970; and a Ph.D. in Electrical Engineering from the Imperial College of Science and Technology in London, England, in 1973. He served as a Canadian naval officer from 1974-89, beginning as a naval combat systems engineer aboard *HMCS Algonquin*. He attended the Canadian Forces Command and Staff College from 1982-83. In January 1986, he was promoted to Captain. Selected in 1983 NRC Group, and 1992 NASA Group 14, he left the Navy in 1989 to devote himself to his astronaut career as deputy director of the Canadian astronaut program.

Garneau flew as 'Payload Specialist 2' on STS-41-G aboard *Challenger* (October 05-13, 1984), the third shuttle mission to carry an IMAX camera to document the flight. He operated several Canadian experiments. He was 'Mission Specialist 4' on STS-77 aboard *Endeavour* (May 19-29, 1996). The defense and aerospace technology company **L'Garde** was responsible for the design and manufacture of the Antenna in the *SPARTAN-207*/ IAE Inflatable Antenna Experiment, a key component of the STS-77 mission. A suite of four technology experiments (TEAMS) also flew in the Space Shuttle's payload bay. Marc Garneau was 'Mission Specialist 2' on STS-97 aboard *Endeavour* (December 01-11, 2000) to the ISS, when crew installed the first set of solar arrays to the station, prepared a docking port for the arrival of the *Destiny* Laboratory Module, and delivered supplies. He logged a total of 29 days, two hours and one minute (over 677 hours) in space. In February 2001, Garneau became executive vice president of the CSA, and in November 2001 became the CSA's President.

He resigned as President of the CSA to run for the Liberal Party of Canada in the 2006 federal election, but lost by over 9,000 votes. In the general election of October 2008, he won the Montreal riding of Westmount-Ville-Marie by over 9,000 votes. As a member of the Liberal Party, he was in office from October 14, 2008, to March 08, 2023, and served as a Cabinet Minister of Transport from November 04, 2015, to January 12, 2021; then became Minister of Foreign Affairs. He gave his farewell speech in the House of Commons on March 08, 2023, and retired from politics the same day. Garneau was appointed an Officer of the Order of Canada (1984); given the Canadian Force's Decoration (C.D.) for twelve years' honorable service with the Canadian Forces; and was inducted into the International Space Hall of Fame in 1992. He received the One Hundred Twenty-fifth Anniversary of the Confederation of Canada Medal (1993); the Queen Elizabeth II Golden Jubilee Medal for Canada (2002); Companion of the Order of Canada (2003); and Queen Elizabeth II Diamond Jubilee Medal for Canada (2012). He has ten honorary degrees, is married, and has three children.

## Bondar, Dr. Roberta. STS-42.

Dr. Roberta Bondar was born in December 1945 in Sault Ste. Marie, Ontario, Canada. She received a Bachelor of Science degree from the University of Guelph; a Master of Science degree from the University of western Ontario; a Ph.D. From the Univerity of Toronto; and an M.D. from McMaster University. She is a neurologist and researcher of the nervous system.

She was one of the six original Canadian astronauts selected in 1983 NRC Group, and began astronaut training at NASA in February 1984. In 1992, she became the first Canadian woman and the second Canadian astronaut to fly in space. She flew as a Payload Specialist on STS-42 aboard *Discovery* (January 22-30, 1992), during which she performed over forty experiments in the 'Spacelab'. She logged eight days, one hour and 14 minutes in space.

After returning from space, Dr. Bondar left the CSA in September 1992 and continued her research. For the next ten years, she led a reserch team at NASA studying information from dozens of space missions to analyze the human body's mechanisms for recovering from exposure to space flight. Her research considerdd the linkage to Parkinson's disease in addition to other neurological effects. She also developed a new career as a nature photographer. From 2003-09, she was Chancellor of Trent University, has received 28 honorary degrees, and has published four books. Dr. Bondar has received many honors, including appointment as a Companion of the Order of Canada; the Order of Ontario; the Loyola Medal (2009); and was inducted into the Canadian Medical Hall of Fame; and the International Women's Forum Hall of Fame.

## MacLean, Steven G.. STS-52, STS-115.

Dr. Steven Glenwood MacLean (FRCGS) was born in December 1954 in Ottawa, Ontario, Canada. From 1974-76, he worked in sports administration and public relations at York University, Toronto, Canada, and taught part-time there from 1980-83. He received a Bachelor of Science degree in Physics in 1977; a Master of Science; and a Ph.D. in Physics from York University in 1983. In 1977, he received the President's Award at York University; and was a recipient of a Natural Sciences and Engineering Research Council postgraduate scholarship in 1980, two Ontario graduate scholarships (1981 and 1982), and a Natural Sciences and Engineering Research Council postdoctoral fellowship in 1983. He is a laser physicist, and his research has included work on electro-optics, laser-induced fluorescence of particles and crystals and multi-photon laser spectroscopy.

MacLean was selected in NASA Group 16 (1983), a class of 44 astronauts (nicknamed The Sardines) recruited for the anticipated needs of the International Space Station crew and the ongoing Shuttle program, with five members from Europe, two from Canada and two from Japan. He flew as 'Payload Specialist 1' on STS-52 aboard *Columbia* (October 22-November 01, 1992) with the primary payloads being Italy's *Laser Geodynamics Satellite* (*LAGEOS-II*) (destined for a high orbit to conduct precise measurements of tectonic-plate shifting within Earth's surface), and operation of the US Microgravity Payload-1 (USMP-1). Some of the ashes of ***Star Trek*** creator Gene Roddenbury were carried aboard the orbiter. MacLean was 'Mission Specialist 1' on STS-115 aboard *Atlantis* (September 09-21, 2006), which delivered the second port-side truss segment (ITS P3/P4), a pair of solar arrays (2A and 4A) and batteries to the ISS. MacLean became the first Canadian to operate Canadarm2 and its Mobile Base in space, and was the second Canadian to perform a spacewalk. He logged a total of 21 days, 16 hours and two minutes in space.

From 1993-94, Dr. MacLean was the chief science advisor to the ISS. In 1994, Steven MacLean received an honorary Doctorate from the College Militaire Royal de Saint-Jean in Quebec, Canada. He is an honorary fellow of Norman Bethune College at York University, an honorary fellow of the Royal Canadian Geographical Society; and president of the Board of Directors for the Mont Megantic Observatory project. In 2013, he returned to physics research as an associate member of the Institute for Quantum Computing (IQC) at the University of Waterloo. Steven MacLean is married, with three children.

## Hadfield, Chris A.. STS-74, STS-100, Soyuz TMA-07M/Expedition 35.

Chris Austin Hadfield was born in August 1959 in Sarnia, Ontario, Canada. He is a retired fighter pilot and musician. He wanted to be an astronaut from an early age. Chris received a glider pilot license at age fifteen years, then began to fly airplanes. Prior to his career as an astronaut, he served in the Canadian Armed Forces for twenty-five years as an Air Command fighter pilot. In 1982, he graduated with a Bachelor of Science degree in Mechanical Engineering from the Royal Military College in Kingston, Ontario. In 1992, Hadfield obtained a Master's degree in Aviation Systems at the University of Tennessee Space Institute. He learned to fly more than seventy types of aircraft while in the military. Hadfield flew for the North American Aerospace Defense Command (NORAD), and eventually became a test pilot, flying experimental planes. Hadfield was accepted into the Canadian astronaut program in 1992.

Chris Hadfield flew on two Space Shuttle missions: STS-74 in November 1995, and visited the space station *Mir*; and mission STS-100 in April 2001 to the ISS. He then served as Chief of Robotics for the NASA Astronaut Office and Chief of International Space Station Operations. Hadfield was the first Canadian to perform an extravehicular activity in space. From 1996 to 2000, he was chief astronaut at the CSA. From 2001 to 2003 Hadfield was director of NASA operations at the **Yuri Gagarin Cosmonaut Training Centre** in Star City, Russia. He was chief of robotics from 2003-06 and operations chief of the ISS from 2006-08.

Beginning his third spaceflight on *Soyuz TMA-07M* in December 2012, Hadfield became the commander of the ISS as part of 'Expedition 35', responsible for a crew of five astronauts. During the five-month mission he helped to run dozens of scientific experiments dealing with the impact of low gravity on the human body. He chronicled life aboard the station by taking pictures and posting them on various social media platforms including ***Twitter***. Hadfield was a guest on television news and talk shows. He gained popularity by making a music video of himself playing *Space Oddity* on the ISS's guitar. Posted on ***YouTube***, the video quickly received millions of views. When he launched into space, he had about 20,000 followers on ***Twitter***; when he returned to Earth, he had more than a million!

In 2014, Chris Hadfield was named an Officer of the Order of Canada. Hadfield has published five books including his autobiography the NYT-bestseller *An Astronaut's Guide to Life on Earth.* His picture book for children is titled *The Darkest Dark* (2016), and his thrillers set during the Cold War are *The Apollo Murders* (2021) and *The Defector* (2023).

## Thirsk, Robert B.. STS-78, Soyuz TMA-15/Expedition 21.

Dr. Robert "Bob" Brent Thirsk (OC OBC) was born in August 1953 in New Westminster, British Columbia, Canada. He received a Bachelor of Science degree in Mechanical Engineering from the University of Calgary in 1976; a Master of Science degree in Mechanical Engineering from the Massachusetts Institute of Technology (MIT) in 1978; and an M.D. from McGill University in 1982. As a Sloan Fellow, he completed an M.B.A. at the MIT Sloan School of Management in 1998. Dr. Thirsk was selected in December 1983 by the National Research Council of Canada to join the Canadian astronaut program. He served as backup to Marc Garneau for STS-41-G of October 1984. He led an international research team investigating the effect of weightlessness on the heart and blood vessels. The team designed and tested an experimental "anti-gravity suit" that may help astronauts withstand the effects of extended spaceflight on the cardiovascular system. He served as Chief Astronaut of the CSA in 1993-94.

Bob Thirsk flew as a Payload Specialist on STS-78 aboard *Columbia* (June 20-July 07, 1996), the Life and Microgravity 'Spacelab' mission. He wrote two columns for the ***Calgary Sun*** newspaper: the first occasion in which an astronaut wrote and filed a story to a newspaper – and had it published – while in orbit. In October 2004, he served as the commander of the

NEEMO 7 mission, living and working underwater for eleven days. He trained at the Gagarin Academy and became certified as a Flight engineer for the *Soyuz* spacecraft. Launched on May 27, 2009, aboard *Soyuz TMA-15* for ISS 'Expedition 21', he said, *"It will be the supreme thrill of my life. Throughout the mission I will examine the long-term effects of zero gravity as both a test subject and as a physician. My findings will undoutedly contribute to the future understanding of space station living".* He returned to Earth aboard *Soyuz TMA-15* in November 2009. Robert Thirsk holds the Canadian record for the most time spent in space: 204 days and 18 hours.

While aboard the ISS on July 08, 2009, he received an honorary Doctorate from the University of Calgary; the first person to receive a university dregree while in space. Dr. Thirsk is a member of Professional Engineers Ontario; the College of Family Physicians of Canada; the Canadian Aeronautics and Space Institute; the Aerospace Medical Association; the College of Physicians and Surgeons of Ontario; and the International Space University. He was named to the Order of British Columbia in 2012, and became an Officer of the Order of Canada in 2013. Dr. Thirsk has received the University of Calgary Distinguished Alumni Award, and the Association of Professional Engineers, Geologists and Geophysicists of Alberta Gold Medal. He is married, with three children.

## Tryggvason, Bjami V.. STS-85.

Bjami Valdimar Tryggvason (1945-2022) was born in September 1945 in Reykjavik, Iceland. He was one of the six Canadian astronauts selected in December 1983, and was the first Icelandic astronaut. Tryggvason was the principal investigator in the development of the Large Motion Isolation Mount; for the Microgravity vibration Isolation Mount (MIM), operated on space station *Mir* from April 1996 until January 1998; and for the MIM-2, which flew on STS-85. He flew as a Payload Specialist on STS-85 aboard *Discovery* (August 07-19, 1997), a twelve-day mission to study changes in the Earth's atmosphere. In August 1998, he became part of NASA Astronaut Group 17, with two years of physical and academic training related to future missions. He retired from the CSA in June 2008, and returned to teaching at the University of Western Ontario.

Bjami Tryggvason was a member of the Canadian Aeronautics and Space Institute, was awarded an honorary D.Sc. Degree from the University of Western Ontario in 1998, an honorary Dr. Techn. degree from the University of Iceland in 2000, and an honorary Doctor of Engineering degree from the University of Victoria in 2005. He received the Knight's Cross of the Icelandic Order of the Falcon in 2000; and the Innovators Award of the CSA in 2004. He had two children, and died on April 05, 2022, aged 76 years.

## Payette, Julie. STS-96, STS-127.

Julie Payette (CC CMM CQ CD FCAE) was born in October 1963 in Montreal, Canada. She received a Bachelor of Engineering degree from McGill University in 1986; and a Master of Applied Science degree in Computer Engineering from the University of Toronto in 1990. From 1986-88, she worked as a systems engineer for **IBM Canada**'s Science Engineering division. In early 1991, she joined the communications and science department of the IBM Zurich Research Laboratory in Switzerland. She joined the astronaut corps of the Canadian Space Agency in the June 1992 CSA-2 Group. In preparation for a space assignment, she obtained her commercial pilot license and logged 120 hours as a research operator on reduced gravity aircraft. She has logged over 1,300 hours flying time, including 600 hours on jet aircraft. Payette flew as a Mission Specialist on STS-96 aboard *Discovery* (May 27-June 06, 1999), an ISS logistics and supply mission; and her main responsibility was to operate the Canadarm robotic arm from the space station.

Her second flight was on STS-127 aboard *Endeavour* (July 15-31, 2009), which delivered and installed the third and last piece of the Japanese Exposed Facility to *Kibo* aboard the ISS, and the Japanese Experiment Logistics Module – Exposed Section (JEM ELM-ES). She logged a total of 25 days, 11 hours and 58 minutes in space, then served as the Chief Astronaut of the CSA from 2000-07. She also took on the role of CapCom at NASA Mission Control Center in Houston, Texas. In July 2013, Payette became the Chief Operating Officer for the **Montreal Science Center**. On July 13, 2017, Prime Minister Justin Trudeau announced that Queen Elizabeth II had approved Payette's appointment, and she served as the twenty-ninth **Governor-General of Canada** from October 02, 2017-21. She resigned in January 2021, following a workplace review of the environment at Rideau Hall. Payette is fluent in French and English, and conversational in Italian, Spanish, German and Russian. She has served on several boards of directors, and is now divorced, with one child.

## Williams, Daffyd R.. STS-90, STS-118.

Dr. Daffyd "Dave" Rhys Williams (OC, Oont, FCFPC, FRCPC, FRCP, FRCGS) was born in May 1954 in Saskatoon, Saskatchewan, Canada. He is a physician, public speaker, author and retired CSA astronaut. He earned a Bachelor of Science degree in Biology from McGill University in 1976; a Master of Science in Physiology, a Doctor of Medicine and Master of Surgery from McGill University in 1983. He received postgraduate training in advanced invertebrate physiology at the Friday Harbour Laboratories of the University of Washington. In 1988, he became an emergency physician at Sunnybrook Health Science Center as well as a lecturer with the Department of Surgery at the University of Toronto. From 1989-90, Williams served as an emergency physician with the Emergency Associates of Kitchener-Waterloo and as medical director of the Westmount Urgent Care Clinic. He remains active in life science and space medicine research.

Selected by the CSA in June 1992, he completed basic training and in May 1993 was appointed manager of the Missions and Space Medicine Group within the astronaut program. In January 1995, he was selected to join the 1995 international class of NASA Mission Specialist candidates, completing training and evaluation in May 1996. He flew the STS-90 Neurolab mission aboard *Columbia* (April 17-May 03, 1998), when the crew served as both experiment subjects and operators for twenty-six individual life science experiments. He served as an aquanaut on the first NEEMO crew aboard the *Aquarius* underwater laboratory in October 2001. In April 2006, he commanded the NEEMO 9 eighteen-day mission. Williams was a Mission Specialist on STS-118 aboard *Endeavour* (August 08-21, 2007), an assembly mission to the ISS, and completed three EVAs with a total time of 17 hours and 47 minutes. He logged a total of 28 days, 15 hours and 47 minutes in space. In April 2008, Dr. Williams was recruited by McMaster University as a physician-scientist where he was the director for the new McMaster Center for Medical Robotics at St. Joseph's Healthcare Hamilton. On May 18, 2011, he became the new president and CEO of Southlake Regional Health Center, continuing in that role until October 2017 when he left to work as a healthcare and aerospace consultant.

Dr. Williams is a member of eight medical associations and societies. He has received twenty-three honors, including the FAI Medaille De la Vaulx Award; the V.M. Komarov Diploma; the NASA Exceptional Service Medal; the Queen Elizabeth II Diamond Jubilee Medal; Officer of the Order of Canada (2013); the Order of Ontario (2015); the Award of Excellence from the College of Family Physicians of Canada (2015); and five Honorary Doctorates. From 2016-18, he published five space-themed books, including *Defying Limits: Lessons From the Edge of the Universe,* (2018); *Mighty Mission Machines: From Rockets to Rovers* (2018); and *To Burp or Not to Burp: A Guide to Your Body in Space,* (2016). He is married, with three children.

## Laliberte, Guy. Soyuz TMA-16/14.

Guy LaLiberte was born in September 1959 in Quebec City, Quebec, Canada. He is a billionaire businessman, entrepreneur and poker player. Along with Gilles Ste-Croix, he was the co-founder of Cirque de Soleil in 1984. Selected in June 2009, and billed as the world's first "space clown", he became the seventh private citizen to pay his way to space as a "space participant". He flew up on September 30 aboard *Soyuz TMA-16*, and returned to Earth on October 11, 2009, aboard *Soyuz TMA-14*. For his ten-day, 21-hour and 17 minute trip to the ISS, he paid a reported USD 35 million to Russia, through the US firm **Space Adventures**. He devoted his mission to raising awareness about his charity, the **One Drop Foundation**, which is dedicated to water conservation. Laliberte published a book of gorgeous photographs of the Earth from space. He belongs to the Rhinocerous political party, has an ex de-facto spouse, and five children.

## Saint-Jacques, David. Soyuz MS-11/Expedition 58/59.

Dr. David Saint-Jacques (OC OQ FRCGS) was born in January 1979 in Quebec City, Quebec, Canada. He is an astrophysicist, engineer and a physician. He holds a Bachelor of Engineering degree in Engineering Physics from the Ecole Polytechnique de Montreal (1993); and a Doctor of Philosophy in Astrophysics from the Cavendish Astrophysics Group at Corpus Christi College, Cambridge, UK, in 1998. He also holds a Doctor of medicine from Universite Laval in 2005, and completed a Family medicine residency at McGill University (2007). From 2007, he was a medical doctor and the Co-chief of Medicine at Inuulitsivik Health Center in Puvirnituq, Quebec. Selected in May 2009 CSA Group as one of two from a pool of 5,351 candidates, he moved to Houston, Texas, to become one of fourteen members of the 2009 NASA Group 20 astronaut class. He served as an aquanaut aboard the *Aquarius* underwater laboratory, offshore at Key Largo, Florida, from October 20-26, 2011. In 2013, he served as a cavenaut into the ESA CAVES training in Sardinia, Italy.

Saint-Jacques flew as 'Flight Engineer 1' on *Soyuz MS-11* (December 03, 2018 to June 25, 2019) for ISS 'Expedition 58/59'. He made an EVA of seven hours on April 08, 2019, and landed in June 2019 some 148 km southeast of Dzhezkazgan, Kazakhstan. This was the one hundred and fortieth flight of a *Soyuz* spacecraft. He logged 203 days, 15 hours and 16 minutes in space. Saint-Jacques is fluent in French and English, and has a basic understanding of Russian, Spanish and Japanese. He has received five awards, including Fellow of the Royal Canadian Geographic Society; Officer of the National Order of Quebec (2020); and Officer of the Order of Canada (2022). He is married to Dr. Veronique Morin and has three children.

## Pathy, Mark. Axiom Mission 1.

See later, Recent Commercial Missions, page 101.

# FRANCE

In chronological order of missions,

## Cretien, Jean-Loup J.M.. Soyuz T-6, Soyuz TM-7/TM6, STS-86.

General de Brigade Jean-Loup Jacques Marie Cretien was born in August 1938 in La Rochelle, France. He was educated at l'Ecole communale a Ploujean, and the Lycee de Morlaix. He then studied at l'Ecole de l'Air, graduating in 1961 with a Master's degree in Aeronautical Engineering. He earned his fighter pilot wings in 1962 and joined the Fifth Fighter Wing in Orange, France. After seven years, he became a test pilot at the Istres-Le Tube Air Base, where he supervised the flight test program for the Mirage F-1 fighter. Selected in CNES-1 in June 1980 as a Research Cosmonaut, he was the first French citizen and first western European in space. He flew on two Franco-Soviet space missions and one NASA Space Shuttle mission. Cretien flew on *Soyuz T-6*, launched on June 24-July 02, 1982, to the *Salyut 7* space station. He landed sixty-five km (forty miles) northeast of Arkalyk, Kazakhstan. He flew on the *Soyuz TM-7/Mir Aragatz* mission (November 26-December 21, 1988), returning to Earth aboard *Soyuz TM-6*.

Cretien's third spaceflight was as a Mission Specialist on STS-86 aboard *Atlantis* (September 26-October 06, 1997). He logged a total of 43 days, 11 hours and 17 minutes in space, and received seven awards, including Hero of the Soviet Union (1982); the Order of Lenin; the Order of the Red Banner of Labour; Commandeur de la Legion d'Honneur; Chevalier de l'Ordre National du Merite; and is Holder of the Aeronautics Medal. Cretien is fluent in French, English and Russian. He is a member of the board of the Academie de l'Air et de l'Espace; the Musee de l'Air et de l'Espace; the American Institite of Aeronautics and Astronautics, the International Academy of Astronautics; and the Association of Space Explorers. He was married and later divorced, with five children (one deceased).

## Baudry, Patrick. STS-51-G.

Lt. Col. Patrick Baudry was born in March 1946 in Douala, Cameroon. His parents were part of the French resistance in World War II, and the family received the Resistance Medal for their courage and effort. He was educated at Pyrtanee National Militaire (1967), and received a Master of Science degree in Aeronautical Engineering from the French Air Force Academy, Ecole de l'Air in 1969. He completed flight training at Salon-de-Provence and Tours, then served as a fighter pilot, completing numerous operational missions in several African countries. In 1978, he entered the Empire Test Pilot School, and was awarded the Patuxent Trophy. He logged over 4,000 hours flying time in over 100 aircraft types. Selected in 1980 CNES Group, he flew as a Payload Specialist on STS-51-G aboard *Discovery* (June 17-24, 1985), becoming the second French citizen in space. He conducted experiments in physiology, biology, materials processing in space, and astronomy. He logged seven days, one hour and 38 minutes (over 169 hours) in space, then worked as a test pilot for Airbus. Baudry belongs to the Association of European Astronauts; is a correspondent of the Air and Space Academy, and has received five honors, including Chevalier of the Legion of Honor; Chevalier of the National Merit Order; the French Astronautics Medal; the Soviet Order of Friendship of Peoples; and the Soviet Order of Gagarine. He is married, with three children.

## Tognini, Michel. Soyuz TM-15/14, STS-93.

Brig. Gen. Michel Ange-Charles Tognini was born in September 1949 in Vincennes, France. He received an advanced degree in Mathematics from EPA Grenoble (military school) in 1970; and an Engineering degree from Ecole de l'Air in 1973. After graduating from the French

Air Force Academy, he served as an operational fighter pilot at the Twelvth Escadre de Chasse. He earned his test pilot diploma at the Empire Test Pilot School in the UK in 1982, then joined the Cazaux Flight Test Center in France, where he tested a wide range of French flight hardware. He has logged 4,300 flying hours in eighty aircraft types. Selected by CNES in 1985, he was assigned as a backup crew member for the *Soyuz TM-7* mission.

Tognini flew on the *Antares* mission aboard *Soyuz TM-15* (July 27-August 10, 1992) to space station *Mir*, where he carried out a program of Soviet-French experiments, and returned aboard *Soyuz TM-14*. He flew on STS-93 aboard *Columbia* (July 22-27, 1999), when his primary task was to assist in the development of the Chandra X-ray observatory. From January 01, 2005, he served as head of the European Astronaut Center of the ESA, until retiring on November 01, 2011. He was awarded Commander of "l'Ordre de la Legion d'Honneur"; Chevalier de "l'Ordre National de Merite"; the NASA Spaceflight Medal; the V.N. Komarov Cosmonaut Diploma (USSR); the Hawker Hunter Trophy; and the Patuxent Shield Trophy. He is married, with four children.

## Haignere, Jean-Pierre. Soyuz TM-17/16, Soyuz TM-29.

Jean-Pierre Haignere was born in May 1948 in Paris, France. He joined the French Air Force, and trained as a test pilot serving as Flight Commander of the 13e Colmar Fighter wing, and as a test pilot involved with development of the Mirage 2000N. Selected by CNES in September 1985, in 1993 he visited the Russian orbital space station *Mir* for three weeks as part of the eleventh visiting mission *Altair*. He launched aboard *Soyuz TM-17* on July 01, and returned to Earth aboard *Soyuz TM-16* on July 22, 1993. In 1999, he flew as Flight Engineer aboard *Soyuz TM-29* on February 20, undertook the *Mir Perseus* (Mir EO-27) mission of 186 days, and made an EVA. He captured the first picture of the shadow of a solar eclise from space. He returned to Earth on August 28, 1999, and retired in May 2003. He logged a total of 209 days, 12 hours and 24 minutes in space. In a unique honor, the asteroid 135268 is named after Haignere and his wife, Claudie. He is involved with the European space tourism initiative, the Astronaut Club Europeen (ACE), which he co-founded. He married, with three children.

## Clervoy, Jean-Francois. STS-66, STS-84, STS-103.

Brig. Gen. Jean-Francois Clervoy was born in November 1958 in Longeville-les-Metz, Lorraine, France. From an early age, he took an interest in space exploration and dreamt of becoming an astronaut. He said, *"When I was in second grade, my teacher used to tell us that when we would be grown up, we would be able to fly in space, the same way he was able to buy a ticket to go to the United States"*. He received a Baccalaureat from College Militaire de Saint-Cyr-l'Ecole in 1976, and graduated from Ecole Polytechnique, Paris, in 1981. He graduated from Ecole Nationale Superieure de l'Aeronautique et de l'Espace, Toulouse, in 1983, and completed Flight Test Engineer training at Ecole du personnel navigant d'essais et de reception, Istres, in 1987. In 1983, Clervoy joined the French Space Agency (CNES), where he worked on autopilot systems for projects including Earth observation satellite SPOT, the optical inter-satellite space link STAR, and the comet probe VEGA. In 1992, he joined the astronaut corps of the ESA.

Selected in CNES-2/ESA-2 in September 1985, he started intensive Russian language training. In 1991, he trained in Star City, Moscow, on the *Soyuz* and *Mir* systems. He joined the astronaut corps of the ESA in Cologne in 1992. Clervoy flew as a Mission Specialist on three Space Shuttle missions, and was known as Billy Bob. STS-66 aboard *Atlantis* (November 03-14, 1994) was the ATLAS-3 mission, part of an ongoing program to determine the Earth's energy balance and atmospheric changes over an eleven-year solar cycle. STS-84 aboard *Atlantis* (May 15-24, 1997) was NASA's sixth Shuttle mission to dock with *Mir*. He managed over twenty experiments, the operation of the docking system, the double module

'SPACEHAB', and the transfer of four tons of equipment. Mission STS-103 aboard *Discovery* (December 19-28, 1999) had the primary objective to repair the Hubble Space Telescope. He was the Flight Engineer for ascent, space rendezvous and entry; and used the robotic arm to capture and deploy the telescope.

Jean-Francois Clervoy logged a total of 28 days, three hours and four minutes (28.13 days) in space, and retired on December 01, 2018. From 2001-08, he was assigned Senior Advisor Astronaut of the Automated Transfer Vehicle ESA project in Les Mureax, France. In 2008, he was appointed a member of the selection board for the new ESA astronaut class. Clervoy has received six honors, including Officer of the Legion of Honor; Knight of the National Order of Merit; two NASA Exceptional Service Medals; and the K.M. Komarov and S.P. Koroliev Awards from the FAI. He is married, with two children.

## Favier, Jean-Jacques. STS-78.

Dr. Jean-Jacques Favier (1949-2023) was born in April 1949 in Kehl, Germany. He received an Engineering degree from the INPG-ENSEEG (Grenoble Institute of Technology) in 1971; a Ph.D. in Engineering from the Ecoles des mines de Paris (1977); and a Ph.D. In Metallurgy and Physics from the University of Grenoble in 1977. He was a research engineer, Commissariat a l'Energie Atomique (CEA), 1976-79. Favier flew on STS-78 aboard *Columbia* (June 20-July 07, 1996), a 16-day, 21-hour Life and Microgravity 'Spacelab' mission. It included studies sponsored by ten nations and five space agencies, and served as a model for future studies on board the ISS. He received twelve honors, including the French Legion of Honor; The Gold Medal of the City of Grenoble; the Societe Francaise de Metallurgie; Member of the Space Science Committee of the European Science Foundation (ESF); several Patents on crystall growth processes, furnaces and in situ diagnosis; and has published over 130 research articles in refereed scientific journals and books. Favier died on March 19, 2023, in Albi, aged seventy-three years. He was married, with four children.

## Haignere, Claudie Andre-Deshays. Soyuz TM-24/23, Soyuz TM-33/32.

Dr. Claudie Andre-Deshays Haignere was born in May 1957 in Le Creusot, Burgundy, France. She studied medicine at the Faculte de Medecine (Paris-Cochin) and Faculte des Sciences (Paris-VII). She went on to obtain certificates in biology and sports medicine (1981), aviation medicine and space medicine (1982), and rheumatology (1984). In 1986, she received a Diploma in the biomechanics and physiology of movement, and received her Doctorates in Rheumatology (1984) and Neuroscience (1992). Selected out of 10,000 candidates, she was selected along with six men by France's space center in CNES-2/EAC as a Research Cosmonaut in September 1985 and became the first French woman in space. She served as a backup crew member for the 1993 *Mir Altair* mission, where her future husband, Jean-Pierre Haignere, participated.

From August 17 to September 02, 1996, she flew on the French-Russian *Cassiopee* mission, (*Soyuz TM-24/23*) and spent 15 days, 18 hours and 23 minutes in space; visiting space station *Mir*, and conducting experiments in physiology, developmental biology, fluid physics and technology. In 1999, she became the first woman qualified to command a *Soyuz* capsule during re-entry. In 2001, Haignere became the first European woman to visit the ISS as the Flight Engineer on *Soyuz TM-33*. The mission lasted from October 21-31, and she returned to Earth aboard *Soyuz TM-32*. She logged a total of 25 days, 14 hours and 22 minutes in space, retired from the ESA in June 2002, and transitioned into French politics.

From 2002-04, she served as the minister delegate for Research and New Technologies, then became minister delegate for European Affairs from 2004-05. Claudie Haignere received many special honors for her spaceflight career, including the Chevalier of the Legion d'Honneur; Chevalier of the Ordre National du Merite; the Order of Friendship (Russia); the Medal "For Merit in Space Exploration" (Russia); and the Medal for Personal Valour. She is a member of the International Academy of Astronautics (IAA); and of the Academie de l'Air et de l'Espace (ANAE). She is married, with one child.

## Eyharts, Leopold P.P.. Soyuz TM-27/26, STS-122/Expedition 16/STS-123.

Brig. Gen. Leopold "Leo" Paul Pierre Eyharts was born in April 1957 in Biarritz, Bay of Biscay, Basque Country, France. He joined the French Air Force Academy of Salon-de-Provence and graduated as an Aeronautical Engineer in 1979. Eyharts logged 3,800 hours flying time in over fifty aircraft types, and completed twenty-one parachute jumps. Selected in July 1990 CNES-3/EAV as a Research Cosmonaut, he flew on *Soyuz TM-27* (January 29-February 19, 1998) to the *Mir* space station, returning to Earth on *Soyuz TM-26.* Eyharts flew on STS-122 aboard *Atlantis* to the ISS (February 07-March 27, 2008), joining 'Expedition 16' and participating in the installation and configuration of the *Columbus* European Laboratory module; returning to Earth on STS-123 aboard *Endeavour.* He logged a total of 68 days, 21 hours and 28 minutes in space, and retired on May 01, 2017. Eyharts has received seven awards, including Chevalier of the Legion d'Honneur; Chevalier of the Ordre National du Merite; the Medaille d'Outre-Mer; Cavalier of the Order of Courage (Russia); and the Medal "For Merits in Development of Space", (April 2011). He is married, with one child.

## Perrin, Philippe. STS-111.

Col. Philippe Perrin was born in January 1963 in Meknes, Morocco. He entered the Ecole Polytechnique in 1982, graduating in 1985. In 1993, he received his Test Pilot license from the EPNER; and his Air Line Pilot Certificate in 1995. He has logged over 3,000 flying hours in over thirty aircraft types, with twenty-six combat missions. Selected in July 1990 CNES-3/EAC Group, Perrin was temporarily detached to the French Space Agency (CNES) and sent to Star City, Russia, for two months of training. In August 1996, he reported to JSC for two years' training. He served as a Mission Specialist on STS-111 aboard *Endeavour* (June 05-19, 2002), when he made three EVAs attaching new hardware to the ISS, and replacing a wrist joint on the robotic Canadarm2. He logged 13 days, 20 hours and 35 minutes (over 332 hours) in space. He did not fly in space with ESA, and retired from the European Astronaut Corps in May 2004. Perrin is currently working as a test pilot for **Airbus**. He received the French Overseas Medal (Gulf War, 1991); two French National Defense Medals; and was made an Officier, Legion d'Honneur. He is married, with three children.

## Pesquet, Thomas. Soyuz MS-03, SpaceX Crew 2.

Refer above, active ESA astronauts, pages 72, 104.

# GERMANY

In chronological order of missions,

## Jahn, Sigmund W.P.. Soyuz 31/29.

See later: Part Four, Intercosmos cosmonauts, pages 377-8.

## Merbold, Ulf D.. STS-9, STS-42, Soyuz TM-20/19.

Dr. Ulf Dietrich Merbold was born in June 1941 in Greiz, in the Vogtland area of Thuringia, Germany. From age nineteen years, he studied physics at the University of Stuttgart, graduating with a Diploma in 1968, then went on to gain his Doctorate (focusing on the effect of radiation on iron). Selected in May 1978, he was the first ESA astronaut to fly in space. He flew as Payload Specialist on STS-9 aboard *Columbia* (November 28-December 08, 1983), the United States 'Spacelab-1' mission. He was the first non-American to fly on a NASA spacecraft.

He was a Payload Specialist on STS-42 aboard *Discovery* (January 22-30, 1992), the International Microgravity Laboratory-1 (IML-1) Spacelab mission. In November 1992, ESA decided to start cooperating with Russia on human spaceflight, aiming to gain experience on long-duration spaceflights, which were not possible with NASA at the time. On October 04, 1994, he embarked on *Euromir '94*, launched aboard *Soyuz TM-20* and spending 31 days, 12 hours and 35 minutes on the *Mir* space station mission. He served as backup astronaut for the German 'Spacelab D-1' mission, and as a science coordinator for the second German 'Spacelab D-2' mission in 1993. He logged a total of 49 days, 21 hours and 36 minutes in space. Merbold worked on the *Columbus* program, and served as the head of the German Aerospace Center (DLR's) astronaut office, until retiring in 2004. He is married, with two children.

## Messerschmid, Ernst W.. STS-61-A.

Dr. Ernst Willi Messerschmid was born in May 1945 in Reutlingen, Germany. He studied at the Technisches Gymnasium in Stuttgart until 1965, then served two years in the military, followed by studies in physics at the University of Tubingen and Bonn. He earned his Diploma degree in 1972; and a Doctorate in 1976. While studying, he worked as a visiting scientist at CERN in Geneva, focusing on proton beams in accelerators and plasmas. From 1975-76, he worked at the University of Freiburg and the Brookhaven National Laboratory, New York, USA. In 1977, he joined DESY in Hamburg to work on the beam optics of the PETRA storage ring. From 1978-82, he worked at the DFVLR (the precursor of the DLR) in the Institute of Communications Technology in Oberpfaffenhofen on space-borne communications.

Selected on December 19, 1982, as one of the science astronauts for the first German 'Spacelab' mission D-1, he flew as a Payload Specialist on STS-61-A aboard *Challenger* (October 30-November 06, 1985), logging seven days, zero hours and 44 minutes (over 168 hours) in space, then retired from the space program. Messerschmid then became a professor at the Institut fur Raumfartsysteme at the University of Stuttgart. From 2000-04, he served as the head of the European Astronaut Center in Cologne, Germany. In January 2005, he returned to the University of Stuttgart, teaching subjects related to Astronautics and Space Stations. Messerschmid is married.

## Furrer, Reinhard. STS-61-A.

Prof. Dr. Reinhard Furrer (1940-95) was born in November 1940 in Worgl, Ostmark (now Austria). He joined the University of Kiel to study Physics, and later transferred to the Free University of Berlin, where he received a Diploma in 1969 and a Doctorate in Physics in 1972. During his time as a student in Berlin, he was involved in the building of the 145-metre-long "Tunnel 57" below the Berlin Wall, which was the escape route of 57 people from East Berlin to the West. From the files of the GDR State Security Service, it is known that he was a close personal friend of the former head of the "Action Group against Inhumanity". A permanent exhibition, dedicated to Furrer (code name Kurt Jacob), was opened in the Checkpoint Charlie Museum. In 1974, he became assistant professor in Stuttgart, and in 1979 qualified for full professorship. He spent time during 1980-81 at the University of Chicago and during 1981 at the Argonne National Laboratory in Chicago, USA.

Selected as one of two chosen in December 1982 German Group DLR-1, he flew on STS-61-A aboard *Challenger* (October 30-November 06, 1985), logging seven days, zero hours and 44 minutes in space. Furrer retired from the DLR on November 06, 1985, and became a professor in 1987, as well as the Director of the Institute of Space Sciences at the Free University of Berlin. Furrer was an avid pilot, earning his license in 1974, and doing several long-distance flights in one engine planes. He died as a passenger in a plane crash after a flight show on the Johannisthal Air Field while performing a roll at low altitude on September 09, 1995, in Berlin, Germany. Furrer was unmarried, with no children.

## Flade, Klaus-Dietrich. Soyuz TM-14/13.

Klaus-Dietrich Flade was born in August 1952 in Budesheim, West Germany. He is a German Air Force pilot and former astronaut with the German Aerospace Center (DFVLR). Initially trained as an airplane engineer, he later pursued aerospace engineering at Bundeswehr University Munich from 1976-80. From 1998-99, Flade trained as a test pilot, and in October 1990 was selected as part of the German astronaut team for the *Euromir '92* flight. After two years of training, he flew as a Research Cosmonaut on *Soyuz TM-14* in March 1992, logging seven days, 21 hours and 57 minutes aboard the *Mir* space station; returning to Earth aboard *Soyuz TM-13*. He then returned to the German Air Force, and served as a test pilot at **Airbus Industrie** in Toulouse, France.

## Walter, H. Ulrich. STS-55.

Dr. Hans Ulrich Walter was born in February 1954 in Iserlohn, West Germany. He studied physics at the University of Cologne, was awarded a Diploma in 1980; and a Doctorate in 1985, both in the field of Solid-state Physics. After two post-doctoral positions at the Argonne National laboratory, Chicago, Illinois, and the University of California at Berkeley, he was selected in the 1987 German Group. After basic training from 1988-90, he flew as Payload Specialist on STS-55 aboard *Columbia* (April 26-May 06, 1993), 'Spacelab D-2, logging nine days, 23 hours and 39 minutes in space. Walter then worked another four years at DLR, managing a space imaging database project; then worked at **IBM Germany**.

In 2003, he became full professor at the Technische Universitat Munchen, holding the chair of the Institute of Astronautics at the University's faculty of mechanical engineering. He serves on the advisory board of Deutches Museum, on the advisory council of Giordano Bruno Foundation, and is president of the Hermann Oberth Space Travel Museum in Feucht. Walter received the Order of Merit of the Federal Republic of Germany; the Wernher von Braun Medal; an honorary Doctorate from the Kyiv Polytechnic Institute (2012); and was made Honorary Professor, National Pedagogical Dragomanov University, Ukraine. He is married, with two children.

## Schlegel, Hans W.. STS-55, STS-122.

Hans Wilhelm Schlegel was born in August 1951 in Uberlingen, Germany. As an international exchange student, he graduated from Lewis Central High School in Council Bluffs, Iowa, USA. He later attended Hansa Gymnasium in Cologne, Germany, graduuating in 1970. He pursued physics studies at RWTH Aachen University from 1972-79, graduating with a Diploma in Physics. He conducted research in semiconductor physics , then, selected in August 1987 he became an astronaut. Basic Astronaut Training at the DLR was from 1988-90. In 1993, he flew as a DLR Payload specialist on STS-55 aboard *Columbia* (April 26-May 06, 1993), carrying the German-sponsored 'Spacelab D-2' research module. From 1995-97, he trained as the backup crew member for the German-Russian *Mir '97* mission. In 1998, he became a member of the European Astronaut Corps. Launched on STS-122 aboard *Atlantis* (February 07-20, 2008), the mission carried the *Columbus* Laboratory to the ISS. He performed an EVA, replacing an empty nitrogen tank on the P1 truss. Schlegel logged a total of 22 days, 18 hours and one minute in space. He received the Federal Service Cross, First Class; the Medal of Friendship of Russia; and the NASA Exceptional Service Medal. Schlegel is married to astronaut candidate Heike Walpot, with seven children.

## Ewald, Reinhold. Soyuz TM-25/24.

Dr. Reinhold Eward was born in December 1956 in Monchengladbach, West Germany. He earned a Diploma in Experimental Physics from the University of Cologne in 1983; and a Ph.D. (with a minor in Human Physiology) in 1986. From 1983-87, he worked as a research scientist at the University of Cologne, focusing on the structure and dynamics of interstellar molecular clouds, which are believed to be the birthplaces of new stars. Ewald joined the DLR in 1987, and managed various projects related to extraterrestrial science, including the SOFIA airborne stratospheric observatory. In 1990, he became part of the German national astronaut team and trained for the German-Russian *Mir '92* mission. As the backup astronaut, he played a crucial role as the Crew Interface Coordinator at the Russian mission control center. Ewald supported the 'Spacelab-D2' mission on STS-55, and participated in the German-Russian *Mir '97* mission. He flew to space station *Mir* aboard *Soyuz TM-25* in February 1997, spending twenty days in space. In February 1999, he joined the European Astronaut Corps at the European Astronaut Center (EAC) in Cologne, Germany.

He received Russia's Order of Friendship in 1992; Russia's Medal for Personal Courage in 1997; the German Federal Cross of Merit (First Class) in 1997; and the Russian Medal for Achievements in Spaceflight (2011). From 2006-11, he headed the Flight Operations Division within ESA's Operations department at the Columbus Control Center near Munich. He lectured on spaceflight operations at the Technical University of Munich and the International Space University in Strasbourg, France. In September 2015, he was appointed as Professor for Astronautics and Space Stations at the Institute of Space Systems at the University of Stuttgart. Ewald is married, with three children.

## Thiele, Gerhard P.J.. STS-99.

Dr. Gerhard Paul Julius Thiele was born in September 1953 in Heidenheim an der Brenz, West Germany. He attended the Friedrich-Schiller Gymnasium in Ludwigsburg, then volunteered for the German Navy, serving as an Operations/Weapons Officer aboard fast patrol boats. In 1976, he began studying physics at the University of Munich and later at the University of Heidelburg. His doctoral research, completed in 1985, focused on Environmental Science. From 1987-87, he was a postdoctoral student at Princeton University. Selected for the German astronaut team in August 1987, he underwent basic training at the DLR (German

Aerospace Center). He launched aboard *Endeavour* (February 11-22, 2000) as a Mission Specialist on the STS-99 Shuttle Radar Topography Mission, logging 11 days, five hours and 38 minutes in space. Thiele trained in Russia as the backup for Andre Kuipers on the *Soyuz TMA-4* mission in 2003-04. He retired from the ESA astronaut corps in October 2005, and since April 2010 he has been a Resident Fellow with the European Space Policy Institute in Vienna, Austria. From July 01, 2013, Thiele assumed the role of head of ESA's Human Spaceflight and Operations Strategic Planning and Outreach Office (HSO-K). He is married, with four children.

## Reiter, Thomas A.. Soyuz TM-22, STS-121/Expedition 13/STS-116.

Brig. Gen. Thomas Arthur Reiter was born in May 1958 in Frankfurt, West Germany. He received a Diploma in Aerospace Engineering from the Bundeswehr University, Munich, in 1982; and an Honorary Doctorate from there in 2010. He was onboard Engineer for the Euromir '95/*Soyuz TM-22* mission to the *Mir* space station. During his 179 days aboard *Mir*, he made two EVAs, the first German to do so. He was trained to command a three-person *Soyuz* crew during its return from space. Selected for the STS-121 mission aboard *Discovery* (July 04-17, 2006), joining 'Expedition 13', he spent another 171 days in space. He returned to Earth aboard *Discovery* on December 22, 2006, ending STS-116. He logged a total of 350 days, five hours and 44 minutes in space. Reiter was named a member of DLR's executive board in August 2007. From April 2011 to December 2015, he served as Director of Human Spaceflight and Operations at ESA. Currently he serves as the ESA Interagency Coordinator and Advisor to the Director General at the ESA. Reiter was awarded the Order of Merit of the Federal Republic of Germany (2007); and the Bavarian Medal Europe (2008). He is married, with two sons.

## Gerst, Alexander. Soyuz TMA-13M, Soyuz MS-09.

Refer above, ESA active astronauts, pages 69-70.

## Maurer, Matthias. SpaceX Crew-3.

Refer above, pages 70, 105.

# ITALY

In chronological order of missions,

## Malerba, Franco E.. STS-46.

Dr. Franco Egidio Malerba was born in October 1946 in Busalla, Metopolitan City of Genoa, Italy. He earned an Electronics Engineer degree from the University of Genoa, Italy, in 1970; and a Doctorate in Physics, specializing in Biophysics in 1974. In 1977, he was chosen by the ESA as one of four European Payload Specialist Candidates for the first 'Spacelab' mission. He then worked in the development, testing and qualification of ES020 – PICPAB, an experiment in Ionospheric Plasma Physics for the first 'Spacelab' payload, flown in 1983. From 1976-89, he held technical and technical management positions with **Digital Equipment** in Italy and in Europe. Selected in 1989 ESA Group, he became a staff member of ASI and was assigned to the NASA JSC for training. Dr. Malerba flew on STS-46 aboard *Atlantis* (July 31-August 08, 1992), and was the first Italian citizen to travel in space, logging seven days, 23 hours and 15 minutes. In 1994, he was elected to the European Parliament on the lists of Forza Italia, serving until 1999.

## Cheli, Maurizio. STS-75.

Lt. Col. Maurizio Cheli was born in May 1959 in Zocca, Modena, Italy. He studied geophysics at the University of Rome; and earned a Master's degree in Aerospace Engineering from the University of Houston. He attended the Italian Air Force Academy, and in 1988 trained as a test pilot at the Empire Test Pilot's School, England. Cheli then trained with the United States Air Force, logged over 4,500 hours flying time in over fifty aircraft types, and was selected in 1992 ESA Group. He flew as a Mission Specialist on STS-75 aboard *Columbia* (February 22-March 09, 1996), and logged 15 days, 17 hours and 40 minutes (over 380 hours) in space. Cheli later joined **Aleni Aeronautica**, then became the Chief Test Pilot for combat aircraft. In 2005, he founded **DigiSky**, an aviation technologies firm. During the 2009 Torino world air games, he piloted the electrically-powered SkySpark experimental aircraft, which logged the world speed record for its class. He is married to fellow former ESA astronaut Marianne Merchez.

## Guidoni, Umberto. STS-75, STS-100.

Dr. Umberto Guidoni was born in August 1954 in Rome, Italy. He is an astrophysicist, science writer, former ESA astronaut and on mission STS-100 became the first European to visit the ISS. He earned a Doctorate in Astrophysics from the University of Rome La Sapienza in 1978 and worked in the Italian Space Agency (ASI) as well as the ESA. One of his research projects was the Tethered Satellite System (TSS), part of the payload of STS-46. He flew on STS-75 aboard *Columbia* in 1996, the second flight of the TSS system. In April 2001, he flew on STS-100 aboard *Endeavour*, carying the *Raffaello* MPLM for its maiden flight, as well as the Space Station Remote Manipulator System (SSRMS), the Canadian robotic arm used for ISS assembly.

Guidoni logged a total of 27 days, 15 hours and 12 minutes in space, and retired from active astronaut status in June 2004. On April 15, 1996, he was appointed Commander of the Order of Merit of the Italian Republic. In May 2001, he received Grand Officer of the Order of Merit of the Italian Republic. In September 2001, Guidoni was assigned to ESA's European Space Research and Technology Center (ESTEC) as crew support for payloads to be developed for the *Columbus* Laboratory. He has received two NASA Space Flight Medals; and the NASA Exceptional Service Medal in 2002. He was a member of the European Parliament (MEP) from June 2004-09, leading a group called PdCI "Unite the left", within the Party of Italian Communists. Guidoni is married, with one child.

## Vittori, Roberto. Soyuz TM-34/33, Soyuz TMA-6/5, STS-134.

Brigadier Roberto Vittori, OMRI was born in October 1964 in Viterbo, Lazio, Italy. He graduated from the Italian Air Force Academy in 1989 and trained as a test pilot in the United States, graduating first in his class from the US Naval Test Pilot School, Maryland, in 1995. He logged over 1,700 flying hours in over forty aircraft types. Selected by ESA in 1998, he trained at JSC in Texas. Vittori flew on the *Soyuz TM-34* taxi flight to the ISS (April 25-May 05, 2002), where he oversaw four European scientific experiments. His second spaceflight was aboard *Soyuz TMA-6* to the ISS on April 15, 2005, returning to Earth aboard *Soyuz TMA-5* on April 24. He became the first European astronaut to visit the ISS twice, and conducted experiments in upper limb fatigue in astronauts, along with germination of herbaceous plant seeds. Vittori was a Mission Specialist on STS-134 aboard *Endeavour* (May 16-June 01, 2011). He logged a total of 35 days, 12 hours and 26 minutes in space. He has three sons.

## Nespoli, Paulo. STS-120, Soyuz TMA-20, Soyuz MS-05.

Major Paulo Nespoli was born in April 1957 in Verano Brianza, northern Italy. He received a B.Sc. in Aerospace Engineering in 1988; and a Master's degree in Aeronautics and Astronautics from Polytech University, New York, in 1989. In July 1998, he was selected by the Italian Space Agency (ASI), and trained at JSC in Houston, Texas. In 2013, Nespoli served as a cavenaut in the ESA CAVES training in Sardinia. On October 23, 2007, he launched as a Mission Specialist on STS-120 aboard *Discovery* to the ISS, delivering the *Harmony* Node 2 module built by **Thales Alenia Space** at its facility in Turin, Italy. On December 15, 2010, he flew aboard *Soyuz TMA-20* to the ISS, returning to Earth on May 24, 2011. His duties included taking part in the docking operations for receiving Europe's second Automated Transfer Vehicle (ATV-2) *Johannes Kepler*, a visiting spacecraft that delivers essential cargo to the ISS. In January 2011, he filmed the majority of the footage for the 2011 documentary film ***First Orbit***, and as a result is credited as its director of photography. Nespoli took part in the arrival of the second uncrewed Japanese *HII Transfer Vehicle* (HTV-2) used to resupply the ISS. He contributed to the scientific exploitation of Europe's *Columbus* laboratory, and participated in some educational activities.

Launched on July 28, 2017, aboard *Soyuz MS-05*, this mission to the ISS was called *VITA*; an acronym for Vitality, Innovation, Technology and Ability. An outreach activity was **Mission-X: Train Like an Astronaut**, where European students run their own code on Raspberry Pi mini computers installed on the ISS. Nespoli completed over sixty expriments, and returned to Earth on December 14, 2017. He logged a total of 313 days, two hours and 36 minutes in space.

## Parmitano, Luca. Soyuz TMA-09M, Soyuz MS-13.

See above, page 71.

## Christoforetti, Samantha. Soyuz TMA-15M, SpaceX Crew-4/Expedition 68.

Capt. Samantha Christoforetti was born in April 1977 in Milan, Italy. At age eighteen years, she participated in a US AFS foreign exchange program, and attended Space Camp. She graduated from the Technical University of Munich with a degree in Mechanical Engineering. Christoforetti became one of the first women to be a lieutenant and fighter pilot in the Italian Air Force. She has logged over 500 hours flying six types of military aircraft. In May 2009, she was selected by the ESA from a pool of 8,000 astronaut applicants, and in 2014 became the first Italian woman in space. In December 2021, ***Politico Europe*** recognized her as one of the "dreamers" in its annual list of the 28 most influential people in Europe.

Launched on November 23, 2014, aboard *Soyuz TMA-15M* to the ISS, she brewed the first cup of espresso in space on May 03, and returned to Earth on June 11, 2015. On July 16, 2015, the Italian President Mattarella awarded her the Order of Merit of the Italian Republic. She commanded NEEMO mission 23 from June 10-22, 2019. Launched with SpaceX 'Crew-4' on April 22, 2022, she made her first EVA of seven hours and five minutes on July 21, and took command of ISS 'Expedition 68' on September 28, 2022; returning to Earth on Crew Dragon *Endurance* on October 14, 2022. She holds the record for the longest uninterrupted spaceflight by a European astronaut (199 days, 16 hours). She has logged a total of 370 days, five hours and 45 minutes in space. Christoforetti can speak Italian, English, German, French, Russian and Chinese. Married to Lionel Ferra, a French engineer, they have a daughter and son.

## Villadei, Walter.

See below, page 102.

# The 2022 European Space Agency Astronaut Group

This is the latest class of the European Astronaut Corps, and the selection includes:

## Career Astronauts:

**Lt. Col. Sophie Adenot** was born in July 1982 in Cosne, Cours-sur-Loire, France. She is a French engineer, helicopter test pilot and astronaut; one of the more than 22,500 candidates who submitted a valid application in 2021 in response to ESA's call for new astronauts. She graduated from Ecole Nationale Superieure de l'Aeronautique et de l'Espace; and from Massachusetts Institute of Technology (MIT). She has served in the French Air and Space Force since 2005. From 2008-20, she served in the Escadron d'Helicopteres 1/67 Pyrenees. The ESA selected her in 2022.

**Pablo Alvarez Fernandez** was born in December 1988 in Leon, Spain. An aeronautical engineer, he holds a Bachelor of Science degree in Aeronautical Engineering from the University of Leon, Spain; and graduated with a Master's degree in Aerospace Engineering from the Warsaw University of Technology (Politechnika Warszawska) in 2011. He is fluent in Spanish, English, Polish and French. He worked as a Project Manager for Airbus operations in Spain, collaborating on industrial projects across different Airbus plants. Fernandez served as the mechanical architect for the ExoMars rover at **Airbus Defense and Space** in the UK from 2017-20. He also worked on the design, development, and testing of various biological seals for the ExoMars rover to prevent biological contamination.

**Dr. Rosemary Theresa Coogan** was born in 1991 in Antrim, Belfast, Northern Ireland, UK. She is an astrophysicist, and her research considers galaxy evolution and space-based telescopes. She graduated from the University of Sussex; and holds two Master's degrees from Durham University. She completed her undergraduate Master's degree in Physics in 2013; and in 2015 received her Master's in Astronomy, where she conducted research on gamma-ray emission from black holes. In 2019, she graduated with a Doctorate in Astronomy from the University of Sussex, UK, then started a postdoctoral research fellowship in astrophysics at the **Max Planck Institute** for Extra-terrestrial Physics in Munich, Germany. Completing this work in 2022, she joined CNES in Paris, France, as a research fellow in space science. Dr. Coogan holds awards for telescope observation time at international observatories ALMA and NOEMA, and for her Ph.D. Thesis from the University of Sussex Astronomy Center. The ESA selected her on November 23, 2022.

**Dr. Raphael Liegeois** was born in January 1988 in Namur, Wallonia, Belgium. He is a Belgian-Luxembourgish biomedical engineer, neuroscientist and astronaut. He graduated from Ecole Centrale Paris in 2009, and earned a Master's degree in Fundamental Physics in 2010. From 2011-15, he completed a Doctorate in Neuroscience at the University of Liege on the functioning of the human brain, with his thesis focusing on dynamic modeling from resting-state brain imaging. Liegeois then held post-doctoral appointments from 2015-17 at the National University of Singapore, and explored neuroimaging markers of neurodegenerative diseases. In 2019, he conducted research as a visiting scholar at Stanford University, California. Since 2021, he has held a permanent position as a Research and Teaching Fellow at the University of Geneva and the Ecole Polytechnique Federale de Lausanne, where he teaches neuroengineering and statistics alongside his research. He was one of five career astronauts selected by the European Astronaut Corps in Cologne in 2022. The basic training of twelve months commenced in Spring of 2023.

**Dr. Marco Alain Sieber** was born in January 1989 in Biel, Kanton Bern, Switzerland. He is a paratrooper and anaesthesiolgist. He graduated from the Gymnasium Burgdorf in Switzerland in 2007. In 2009, he joined paratrooper training with the Swiss Special Forces Commando at the Swiss Army, and achieved the rank of Sergeant. From 2009-15, he worked as an assistant teacher for anatomy and physiology courses at the Faculty of Medicine at the University of Bern. At the same time, he served as a skydiving instructor for pre-military paratrooper courses. Sieber holds a private pilot's license. He received a Doctor of Medicine from the University of Bern, Switzerland, in 2015. His doctoral thesis focused on robotic surgery, and he was awarded the best final exam for his Physician diploma. From 2015-17, Dr. Sieber worked as a general surgery/traumatology and intensive care resident in Interlaken, Switzerland. In 2018, he became the Chief Medical Officer with Swisscoy (Swiss Army) for KFOR (the international peacekeeping mission in Kosovo). Between 2019-21, he pursued an anaesthesiology residency in Interlaken. Since 2020, Dr. Sieber has worked as a helicopter rescue emergency medical doctor, and also joined the hospital in Biel/Bienne, Switzerland, as a urology resident in 2021.

## Project Astronauts:

**John McFall** was born in April 1981 in Frimley, Surrrey, England. He is a British Paralympic sprinter (competing for Wales) and an orthopaedic surgeon. After a serious motocycle accident at the age of nineteen years, McFall's right leg was amputated above the knee. He learned to run with a prosthesis and became a professional track and field athlete. McFall said, ***"I love that sound of air rushing past your ears and the freedom of it. I missed that and wanted to get that back"***. He graduated with an upper second-class Bachelor of Science degree from Swansea University in Summer 2004. In September 2004, he began postgraduate studies in sport and exercise science at the University of Wales Institute, Cardiff; graduating with a Master of Science degree. Representing Great Britain, he completed in the Paralympics, specializing in the 100 meters and 200 meters, and is rated "one of the fastest above-the-knee amputee athletes in the world". At the 2008 Summer Paralympics in Beiging he won a Bronze medal. In 2014, he graduated with a Bachelor of Medicine and Surgery from the Cardiff University School of Medicine, and in 2016 became a member of the Royal College of Surgeons. He is currently a Trauma and Orthopaedic Specialist Registrar. The ESA will do a feasability study on him flying to space and what needs to be adapted for people with disabilities.

## Astronaut with a Disability Program:

**Marcus Wandt** was born in September 1980 in Mora, Kopparburg, Sweden. He is a test pilot, assigned to the Axiom-3 mission in 2024. From 2003-14, he was a fighter pilot in the Swedish Air Force, flying the Saab JAS Gripen. In 2004, he joined the Swedish Air Force Flying Training School, where he received basic fight training. He graduated from Chalmers University of Technology with a Master's degree in Electrical Engineering in 2007. From 2013-14, he trained at the United States Naval Test Pilot School, graduating at the top of his class to become an experimental test pilot. In 2014, he joined **Saab Aeronautics**, and was promoted to Chief Test Pilot and head of Flight Operations in 2020. His wife showed him the ESA application, and in November 2022 Wandt was chosen with 16 others from about 22,500 applicants.

In January 2024, on the *Axiom-3* mission he became the third Swedish astronaut to fly to space. He is the fastest-trained astronaut in history; and logged 21 days, 15 hours and 41 minutes in space. **Saab** was the largest private investor in the mission, which cost 400 million Kroner (about USD 47 million). Experiments included stem cell research led by Uppsala University and improved cognition research led by KTH Royal Institute of Technology.

**Dr. Slawosz Uznanski** was born in April 1984 in Lodz, Poland. He is a radiation effects engineer working at the ESA as a project astronaut since 2023 and formerly at CERN. In 2008, he graduated from the University of Nantes and from the Lodz University of Technology with a Master's degree (with honours). In 2011, he received a Doctorate (with distinction) from the University of the Mediterranean Aix-Marseille for his thesis on radiation-tolerant designs for space applications. He joined the European Organization for Nuclear Research (CERN) in Geneva, Switzerland in 2011. From 2018-20, he was an Engineer in Charge of the Large Hadron Collider. In 2019, Uznanski was a technical expert and reviewer for the Horizon 2020 program, investigating European autonomy and technical competitiveness for the European Research Executive Agency (REA). In August 2023, the Polish Development and Technology Minister announced that a Polish astronaut would fly to the ISS in 2024. Slawosz Uznanski began his ESA training on September 01, 2023. After Miroslaw Hermaszewski in 1978, Uznanski will be the second Pole to go into space.

## Reserve Astronauts:

**Meganne Christian**; a materials scientist.

**Anthea Comellini**; an aerospace engineer.

**Dara Garcia Alonso**; a biomedical scientist.

**Andrea Patassa**; a test pilot.

**Carmen Possnig**; a medical doctor.

**Arnaud Prost**; a flight test engineer.

**Amelie Schoenwald**; an immunologist.

**Ales Svoboda**; a fighter pilot.

**Nicola Winter**; a fighter pilot.

This group of astronauts underwent basic training throughout 2023 at the European Astronaut Center (EAC) facilities in Cologne, Germany, lasting approximately one year. The career astronauts began their training in April 2023, joined by some reserve members later in the year. Notably, Marcus Wandt will be the first from this group to undertake a spaceflight aboard **Axiom-3** in 2024, marking the first commercial mission for an ESA-sponsored astronaut.

# Former ESA astronaut members

There are eighteen former members of the ESA astronaut corps. Some were selected by other European agencies and then enrolled into the ESA Astronaut Corps in 1998.

1. **Hans Schlegel** of Germany. 1987 DLR Group. Flew on missions STS-55, STS-122. Logged a total of 22 days, 18 hours and one minute in space.
2. **Andre Kuipers** of the Netherlands. 1998 ESA Group. *Soyuz TMA-4/4, Soyuz TMA-03M* for Expedition 30/31. Logged 203 days, 15 hours and 50 minutes.
3. **Mr. A. Christer Fugelsang** of Sweden. 1992 ESA Group. STS-116, STS-128. Logged 26 days, 17 hours and 37 minutes.
4. **Leopold Eyarts** of France. 1990 CNES Group. *Soyuz TM-27/26*. STS-122/123 for 'Expedition 16'.
5. **Jean-Francois Clervoy** of France. 1992 ESA Group. STS-66, STS-84, STS-103. Logged 28 days, three hours and four minutes.
6. **Maurizio Cheli** of Italy. 1992 ESA Group. STS-75. Logged 15 days, 17 hours and 41 minutes.
7. **Pedro Duque** of Spain. STS-95, *Soyuz TMA-3/2*. Logged 18 days, 18 hours and 46 minutes.
8. **Reinhold Ewald** of Germany. 1990 DLR Group. *Soyuz TM-25/24*. Logged 19 days, 16 hours and 34 minutes.
9. **Umberto Guidoni** of Italy. 1989 ASI Group. STS-75, STS-100. Logged 27 days, 15 hours and ten minutes.
10. **Claudie Haignere** (nee Andre-Deshays) of France. 1985 CNES Group. *Soyuz TM-24/23, Soyuz TM-33/32*. Logged 25 days, 14 hours and 22 minutes.
11. **Jean-Pierre Haignere** of France. 1985 CNES Group. *Soyuz TM-17/16, Soyuz TM-29* for 'Mir EO-27'. Logged 209 days, 12 hours and 24 minutes.
12. **Ulf Merbold** of Germany. 1978 ESA Group. STS-9, STS-42, *Soyuz TM-20/19* for *EuroMir '94*. Logged 49 days, 21 hours and 36 minutes.
13. **Merchez, Mariane.**
    Dr. Mariane Merchez was born in October 1960 in Uccle, Belgium. She is certified in both aerospace and industrial medicine, holds a Belgian Air Transport Pilot License, and is a former Boeing 737 co-pilot. Selected in May 1992 ESA Group 2, Merchez has not flown in space. She is married to former Italian ESA astronaut Maurizio Cheli, and they co-authored the book *Everything in an Instant: Decisions that Shape a Lifetime*.
14. **Ernst Messerschmid** of Germany. 1982 DLR Group. STS-61-A. Logged seven days, zero hours and 44 minutes.
15. **Paolo Nespoli** of Italy. 1998 ESA Group. STS-120, *Soyuz TMA-20* for 'Expedition 26/27'. *Soyuz MS-05* for 'Expedition 52/53'.
16. **Claude Nicollier** of Switzerland. 1978 ESA Group. STS-46, STS-61, STS-75, STS-103. Logged 42 days, 12 hours and three minutes.
17. **Wubbo Ockels** of the Netherlands. 1978 ESA Group. STS-61-A. Logged seven days, zero hours and 44 minutes.

18. **Phillipe Perrin** of France. 1990 CNES Group. STS-111. Logged 13 days, 20 hours and 35 minutes.
19. **Thomas Reiter** of Germany. 1992 ESA Group. *Soyuz TM-22*/'Mir EO-20', STS-121/116/'Expedition 13/14'. Logged 350 days, five hours and 44 minutes.
20. **Gerhard Thiele** of Germany. 1987 DLR Group. STS-99. Logged 11 days, five hours and 38 minutes.
21. **Michel Tognini** of France. 1985 CNES Group. *Soyuz TM-15/14*, STS-93. Logged 18 days, 17 hours and 45 minutes.
22. **Frank de Winne** of Belgium. 1998 ESA Group. *Soyuz TMA-1/TM-34, Soyuz TMA-15*, for 'Expedition 20/21'. Logged 198 days, 17 hours and 34 minutes.
23. **Roberto Vittori** of Italy. 1998 ESA Group. *Soyuz TM-34/33, Soyuz TMA-6/5*, STS-134.
24. **Peake, Timothy. Soyuz TMA-19M/Expedition 46/47.**

    Maj. Timothy "Tim" Peake was born in April 1972 in Chichester, West Sussex, England. After graduating from the Royal Military Academy at Sandhurst in 1992, he became an officer in the British Army Air Corps. He became a qualified helicopter pilot in 1994 and flew reconnaissance missions in Germany, Northern Ireland, Kenya, Canada, and the Balkans. In 2005, he graduated from the Empire Test Pilots' School, and earned a Bachelor of Science degree (with honors) in Flight Dynamics and Evaluation from the University of Portsmouth in 2006. In a seventeen-year military career, Peake logged over 3,000 hours flying time in helicopters and fixed-wing aircraft. He was selected in 2009 ESA Group, and moved to Cologne, Germany, for basic training; learning Russian language, survival skills, CPR, rescue-diver techniques, and how to move in zero gravity.

    Peake is the first British ESA astronaut, the sixth person born in the United Kingdom to go on the ISS, and the seventh UK-born person in space. He launched on December 15, 2015, aboard *Soyuz TMA-19M* and joined 'Expedition 46/47' on the ISS. After docking, he received messages of support from Queen Elizabeth II and Elton John. On January 15, 2016, he made the first EVA by a British astronaut; to replace a faulty sequential shunt unit on the station's solar arrays. On April 24, he ran the 2016 London Marathon from the ISS treadmill. Tim Peake returned to Earth on June 18, 2016. He logged a total of 185 days, 22 hours and 11 minutes in space. He is an ambassador for the Prince's Trust and the Scout Association in the UK. In October 2016, he received an honorary Doctorate of Science from the University of Leicester. In May 2021, the Institution of Engineering and Technology awarded Peake an Honorary Fellowship for his outstanding contribution to space exploration, engineering and the technology industry. His autobiography is titled *Limitless: The Autobiography, Century 2020.* He is married, with two sons.

# BELGIUM

## De Winne, Frank. Soyuz TMA-1/TM-34, Soyuz TMA-15/Expedition 20/21.

Brig. Gen. Frank, Viscount De Winne was born in April 1961 in Ledeberg, Ghent, Belgium. He graduated from the Royal School of Cadets, Lier, in 1979. He received a Master of Science degree in Engineering (Polytechnics) from the Royal Military Academy in 1984. After graduating from the Belgian Air Component at Goetsenhoven, he flew Dassault Mirage 5 airplanes for the Air Force. In 1991, he completed the Staff Course at the Defence College in Brussels with the highest distinction. He qualified as a Test Pilot from the British Empire Test

Pilot's School in Boscombe Down, UK, also receiving the McKenna Trophy. In August 1998, he became Commander of the 349 Squadron operating from Kleine Brogel. He commanded the Dutch-Belgian **Deployable Air Task Force** during the NATO Operation Allied Force in the Balkans, completing seventeen combat sorties. For this he was awarded the degree of Officer of the Order of Orange-Nassau. He has logged over 2,300 flying hours in Mirage, F-16, Tornado and Jaguar aircraft and serves as the Chairman of the Belgian Armed Forces Flying Personnel Association.

Selected in 1998 ESA Group, in January 2000 he joined the European Astronaut Corps, whose homebase is the European Astronaut Center in Cologne, Germany. Launched on October 30, 2002, as Flight Engineer aboard *Soyuz TMA-1* to the ISS, he carried out a program of twenty-three experiments, returning to Earth aboard *Soyuz TM-34* on November 10. He launched on May 27, 2009, aboard *Soyuz TMA-15* for ISS 'Expedition 20', becoming Commander of 'Expedition 21' on October 11; the first astronaut from the ESA to command a space mission. He returned to Earth in December 2009, having logged a total of 198 days, 17 hours and 34 minutes in space. De Winne has received twelve awards and five Honorary Doctorates. He was made an Officer of the Order of Leopold; Officer of the Order of the Crown; Officer of the Order of Leopold II; received the Military Cross, First Class; the NATO Medal for Kosovo; Member of the Order of Friendship (Russian Federation); and the Joe Bill Dryden Semper Viper Award.

## Frimout, Dirk D.D.. STS-45.

Dr. Dirk Dries David Frimout was born in March 1941 in Poperinge, Belgium. He received a degree in Electrical Engineering from the University of Ghent in 1963; a Ph.D. in Applied Physics from the University of Ghent in 1970; and completed a post-doctorate at the University of Colorado Laboratory for Atmospheric and Space Physics [European Space Research Organization (ESRO) fellow] in 1971-72. As head of instrumentation at the Belgian Institute for Space Aeronomy from 1965-78, he conducted experiments using stratospheric balloons and sounding rockets. He served as Crew Activities Coordinator and Experiment Coordinator for 'Spacelab-1' at the ESA from 1978-84; then worked in the Microgravity Division of ESTEC until 1989.

Frimout flew as a Payload Specialist on STS-45 aboard *Atlantis* (March 24-April 02, 1992), the first 'Spacelab' mission dedicated to NASA's Mission to Planet Earth. The crew operated twelve experiments as part of the 'ATLAS-1' cargo. This mission was the first time an artificial beam of electrons was used to stimulate a man-made auroral discharge. He logged eight days, 22 hours and nine minutes in space. Prince Philippe of Belguim communicated with Frimout during his mission, and a ticker-tape parade welcomed Belgium's first astronaut back home. After his mission, Dirk Frimout was enobled, given the title of Viscount in the Belgian nobility. He currently holds the position of a senior engineer in the Payload Utilization Department of ESA's *Columbus* Directorate. He is married, with two children.

# NETHERLANDS

## Kuipers, Andre. Soyuz TMA-4/3, Soyuz TMA-03M/Expedition 30/31.

Dr. Andre Kuipers was born in October 1958 in Amsterdam, the Netherlands. He received his medical degree from the University of Amsterdam in 1987. Selected to the European Astronaut Corps in 1998, he launched as a Flight Engineer on April 19, 2004, aboard *Soyuz TMA-4* for the *Delta* mission of almost eleven days in space. Padalka, Fincke and Kuipers docked with the ISS two days later, conducted twenty-one exeriments while in orbit, and he returned to Earth with Kaleri and Foale aboard *Soyuz TMA-3* on April 30. He was the second Dutch citizen, third Dutch-born and fifth Dutch-speaking astronaut. On December 21, 2011, he launched with Kononenko and Pettitt aboard *Soyuz TMA-03M* for ISS 'Expedition 30/31', spending another 192 days, 18 hours and 58 minutes in space; returning to Earth on July 01, 2012. He logged a total of 203 days, 15 hours and 50 minutes in space, and retired on November 01, 2018. Dr. Kuipers has received six honors, including Officer of the Order of Orange-Nassau (2004); Knight of the Order of the Netherlands (2012); Recipient of the Order of Friendship (Russian Federation, 2013); and was King of Arms at the inauguration of King Willem-Alexander (2013). He is married, with four children.

## Ockels, Wubbo J.. STS-61-A.

Dr. Wubbo Johannes Ockels (1946-2014) was born in March 1946 in Almelo, Netherlands. He received a Master of Science degree in Physics and Mathematics from the University of Groningen in the Netherlands in 1973, with his work based at the Nuclear Physics Accelerator Institute (KVI) and earning his Ph.D. in 1978. Selected in 1978 by the ESA as one of three candidates for Space Shuttle missions involving 'Spacelab', he was the first Dutch citizen to travel in space. He took basic astronaut training at NASA's JSC from May 1980 to August 1981, and served as a ground communicator and liaison-scientist for the crew on the STS-9/Spacelab-1 mission in 1983. He launched as a Payload Specialist on STS-61-A aboard *Challenger* (October 30-November 06, 1985), the first crew with eight members. Some operations were controlled from the DLR German Space Operations Center, and the crew conducted over seventy-five scientific experiments.

Dr. Ockels was then assigned to the European Space Research and Technology Center in Noordwijk, Netherlands, where he worked on human spaceflight projects until 1996. In 1992, he began teaching at the **Delft University of Technology** in Delft, and in September 2003 he became a full Professor of Aerospace for sustainable engineering and technology. He was a member of the American Physical Society and the European Physical Society. In 2009, he presented a talk arguing that the notion of time is human-constructed as a result of our interpretation of the effects of gravity. Wubbo Ockels died on May 18, 2014, in Amsterdam, Netherlands, aged sixty-eight years. He was married, with two children and two grandchildren. A small planetoid orbiting between Mars and Jupiter was named after him, 9496 Ockels.

# SPAIN

## Duque, Pedro F.. STS-95, Soyuz TMA-3/2.

Pedro Francisco Duque was born in March 1963 in Madrid, Spain. He received a degree in Aeronautical Engineering from the Universidad Politecnica de Madrid (UPM) in 1986, then joined **Grupo Mecanica del Vuelo** (GMV), a Spanish firm providing services for the space industry. Later, he worked at the ESA's European Space Operations Center in Darmstadt, Germany. In May 1992, Duque was selected to join the ESA astronaut corps, and he trained in Cologne, Germany, and at the Yuri Gagarin Cosmonaut Training Center in Star City, Russia. He served as the coordinator between the crew on the *Euromir '94* mission and European scientists. Duque flew as a Mission Specialist on STS-95 aboard *Discovery* (October 29-November 07, 1998), when he focused on studying the Sun and conducted research on weightlessness. He supervised and maintained the ESA experimental modules and scientific equipment. In 2001, he attended advanced training for the International Space Station (ISS).

Launched on October 18, 2003, as Flight Engineer aboard *Soyuz TMA-3* during the *Cervantes* mission to the ISS, he returned to Earth with the 'Expedition 8' crew aboard *Soyuz TMA-2* on October 28. He logged a total of 18 days, 18 hours and 46 minutes in space. From 2008-11, he served as the CEO of **Deimos Imaging**, based in Valladolid, Spain. From 2018-21, he held the position of Minister of Science, Innovation, and Universities; and served as a member of the Congress of Deputies, representing Alicante, Spain, from May 2019 until February 2020. He is married, with three children.

# SWEDEN

## Fuglesang, A. Christer. STS-116, STS-128.

Dr. Arne "Christer" Fuglesang was born in March 1957 in Stockholm, Sweden. He graduated from the Bromma Gymnasium, Stockholm, in 1975; received a Master's degree in Engineering Physics from the Royal Institute of Technology (KTH), Stockholm, in 1981; and a Doctorate in Experimental Particle Physics from Stockholm University in 1987. From 1988, he was a Fellow at CERN and taught mathematics at the Royal Institute of Technology before being selected to join the European Astronaut Corps (EAC) in 1992. He attended an introductory training program at EAC and a four-week training program at the Gagarin Academy (TsPK, Moscow) in Russia, with a view to future ESA-Russian collaboration on the *Mir* Space Station. In May 1993, he and fellow ESA astronaut Thomas Reiter were selected for the *Euromir '95* mission and commenced training in Moscow. In 1996, ESA selected Dr. Fuglesang to train as a Mission Specialist for NASA Space Shuttle missions, and he qualified in April 1998.

Launched on STS-116 aboard *Discovery* (December 10-22, 2006), he became the first Swedish citizen in space. This was an assembly and crew-rotation mission to the ISS. Fuglesang flew on STS-128 (assembly mission 17A) as 'Mission Specialist 4' aboard *Discovery* (August 28-Sept 12, 2009). The primary payload was the Multi-Purpose Module *Leonardo*, containing a collection of experiments for studying the physics and chemistry of microgravity. He logged a total of 26 days, 17 hours and 38 minutes in space; including five EVAs totaling 31 hours and 54 minutes. Fuglesang is a prominent member of the Swedish Skeptics Association and identifies strongly with skeptics and atheists. In 2012, he received the Royal Institute of Technology 2012 Alumni of the Year award. He was awarded H.M. The King's Medal (Stockholm, 2007); NASA Exceptional Service Medal (2010); and two honorary Doctorates. He is married, with three children.

# SWITZERLAND

## Nicollier, Claude. STS-46, STS-61, STS-75, STS-103.

Claude Nicollier was born in September 1944 in Vevey, Switzerland. He was the first astronaut from Switzerland. After graduating from the Gymnase de Lausanne (high school) in 1962 he studied physics at the University of Lausanne and received a Bachelor of Science degree in 1970. From 1970 to 1973 he worked at the Institute of Astronomy at the University and at the Genoa Observatory. Nicollier then obtained a Master of Science degree in astrophysics from the University of Geneva in 1975. In parallel, he became a Swiss Air Force pilot in 1966, in the Fliegerstaffel 5 or Hawker Hunter. He has logged 5,600 hours flying time, including 4,000 hours in jet aircraft. At the end of 1976, he accepted a Fellowship at the European Space Agency's Space Science Department at Noordwijk, the Netherlands, where he worked as a research scientist. Nicollier joined the NASA Astronaut Group 9 selected in

May 1980 for training as a Mission Specialist. He was the first non-American to become a full-time NASA astronaut. In 1988, he graduated as a test pilot from the Empire Test Pilot's School in Boscombe Down, United Kingdom.

Nicollier's first spaceflight was planned to be STS-61-K in October 1986, but it was canceled following the shuttle *Challenger* disaster. He flew on mission STS-46 aboard shuttle *Atlantis* (July 31-August 08, 1992) for almost eight days. His second flight was on STS-61 aboard *Endeavour* (December 02-13, 1993) for ten days and 19 hours. He flew on STS-75 aboard *Columbia*, (February 22-March 09, 1996) which deployed *TSS-1R*, a tethered satellite reflight. This was a follow-up mission to TSS-1 which had been deployed during his mission STS-46. Nicollier's final flight was aboard *Discovery* on mission STS-103 (December 20-28, 1999). He took part in two missions to service the Hubble Space Telescope (missions STS-61 and STS-103). From the spring of 1996 to the end of 1998, he was Head of the Astronaut Office Robotics Branch.

On his fourth mission, Nicollier participated in a spacewalk lasting eight hours and ten minutes, becoming the first European Space Agency astronaut to do so during a Space Shuttle mission. In 2000, he was assigned to the Astronaut Office Extravehicular Activity Branch, while maintaining a position as Lead ESA Astronaut in Houston, Texas. Nicollier retired from ESA in April 2007, after spending forty-two days, twelve hours and five minutes in space. Following his astronaut career, Nicollier delivered a lecture on his experiences, titled "Revisiting Hubble", at the first Starmus Festival in 2011 in the Canary Islands. The talk was published in the book *Starmus: 50 Years of Man in Space.*

He was appointed full professor of Spatial Technology at the **Ecole Polytechnique Federale de Lausanne** on March 28, 2007. Claude Nicollier was an expert board member of **Swiss Space Systems**, until the company's dissolution. He has received twelve Awards and honors, including the Yuri Gagarin Gold Medal from the International Aeronautical Federation (1994); honorary Doctorates from the Swiss Federal Institute of Technology, Lausanne, and the University of Geneva (both in 1994); inducted into the International Space Hall of Fame (1997); the Albert Einstein Medal (1998); and the NASA Distinguished Service Medal (2001).

## Walker, Shannon. Soyuz TMA-19/Expn 24/25, Crew-1/Expn 64/65.

Dr. Shannon Walker was born in June 1965 in Houston, Texas, USA. She began her professional career with the Rockwell Space Operations Company at JSC in 1987 as a robotics flight controller for the Space Shuttle program; working missions STS-27, 32, 51, 56, 60, 61, and STS-66. From 1990-93, Walker took leave of absence from JSC to attend graduate school, where her area of study was the solar wind interaction with the Venusian atmosphere. She received a Bachelor of Arts degree in Physics (1987); a Master of Science (1992) and a Ph.D. (1993) in Space Physics from Rice University. In 1995, she joined the NASA civil service and began working in robotics integration in the ISS program at the Johnson Space Center. From 1999-2000, she moved to Moscow to work with the Russian Space Agency and its contractors in avionics integration and integrated problem solving for the ISS.

Selected in NASA 2004 Group 19, she launched aboard *Soyuz TMA-19* on June 25, 2010, and spent over 163 days on the International Space Station. She launched aboard SpaceX *Crew Dragon* with Crew-1 on November 15, 2020, as a Flight Engineer for ISS 'Expedition 64', and became Commander of 'Expedition 65' for eleven days. Her second mission lasted 167 days. She logged a total of 330 days, 13 hours and 40 minutes in space. Dr. Walker is married to retired NASA astronaut, Australian-born Andy Thomas. She is a member of the Aircraft Owners and Pilots Association and The Ninety-Nines International Organization of Women Pilots. Dr. Walker received the Goethe Institute Scholarship for Study Abroad; the Rice Fellowship for Graduate Study; the Rockwell Sustained Superior Performance Award; and nine Performance Bonus Awards. In 2022, SpaceX's Dragon support vessel *GO Navigator* was renamed *Shannon*. The Dr. Shannon Walker Neighbourhood Library in Westbury, Houston, Texas, is named after her.

# Recent commercial space missions

Space tourism is recreational space travel, either on established government-owned vehicles such as the Russian *Soyuz* and the International Space Station (ISS), or on vehicles funded by private companies. Since the flight of the world's first space tourist, American businessman Dennis Tito, on April 28, 2001, space tourism has gained new prominence as more suborbital and orbital tourism opportunities have become available. The advent of space tourism occurred at the end of the 1990s with a deal between the Russian company **MirCorp** and the American company **Space Adventures Ltd**.

**MirCorp** was a private venture in charge of the Russian space station *Mir*. To generate income for maintenance of the aging space station, **MirCorp** decided to sell a trip to *Mir*, and Tito became its first paying passenger. However, before he could make his trip, space station *Mir* was de-orbited, and the mission was diverted to the ISS. Tito paid USD 20 million for his flight aboard *Soyuz TM-32*, and spent seven days aboard the ISS. In 2002, South African computer millionaire Mark Shuttleworth flew to the ISS, followed by American businessman Gregory Olsen in 2005. Iranian-born American entrepreneur Anousheh Ansari became the first female fee-paying space traveler when she visited the ISS in September 2006. In 2007, American billionaire Charles Simonyi flew aboard *Soyuz TMA-10* to the ISS; and American video game developer Richard Garriott was launched there in October 2008. Since his father, Dr. Owen Garriott was a former NASA astronaut on the *Skylab 3* crew and mission STS-9, Richard became the first second-generation American in space. Cosmonauts Aleksandr Volkov and his son Sergey were the first father-and-son space travelers, and Sergey Volkov was on the ISS when Richard Garriott arrived. A wealthy Canadian entrepreneur, Guy Laliberte flew to the ISS in 2009.

## Olsen, Gregory. Soyuz TMA-7/6.

Dr. Gregory Olsen was born in April 1945 in Brooklyn, New York, USA. He received a Bachelor of Science degree graduating magna cum laude from Fairleigh Dickinson University; a Bachelor of Engineering degree; a Master of Science degree; and later a Ph.D. in Materials Science from the University of Virginia. In 1984, he founded EPITAXX, a fiberoptic detector manufacturer, and sold it in 1990 for USD 12 million. He is a 1992 co-founder and chairman of **Sensors Unlimited, Inc.**, a company developing optoelectronic devices. Olsen trained for nearly two years with the Russian Space Agency, and paid about USD 20 million for his spaceflight.

Launched on October 01, 2005, aboard *Soyuz TMA-7*, he became the third private citizen to make a self-funded trip to the International Space Station with the company **Space Adventures Ltd**. He conducted several experiments in remote sensing and astronomy, and also spoke to students via ham radio from space through the ARISS project. Greg Olsen logged nine days, 21 hours and 15 minutes in space, and returned to Earth aboard *Soyuz TMA-6* on October 10. In 2010, he was elected a member of the National Academy of Engineering, for research and commercilization of optical components for fiber communicationss and national defense. Olsen was elected a member of the Fairleigh Dickinson University Board of Trustees. He has two daughters and six grandchildren.

## Simonyi, Charles. Soyuz TMA-10/9, Soyuz TMA-14/13.

Dr. Charles Simonyi was born in September 1948 in Budapest, Hungary, and left the country at the age of seventeen years to work at the Danish computer company **Regnecentralen**. He graduated from the University of California at Berkeley with a degree in Engineering Mathematics and later earned a Doctorate in Computer Science from Stanford University. After working at the Xerox Palo Alto Research Center from 1972-80, where he created the first WYSIWYG text editor, he joined the **Microsoft Corporation** in 1981. His notable contributions as a software architect include him introducing the graphical user interface to Bill Gates, who later described it as one of the two revolutionary things he experienced in his life. Simonyi played a pivotal role in building the initial versions of Microsoft Office, and co-founded **Intentional Software** in 2002.

Launched aboard *Soyuz TMA-10* to the International Space Station 'Expedition 15' on April 07, 2007, he became the fifth space tourist and the second Hungarian in space. Simonyi paid USD 20 million for his spaceflight. Simonyi arrived at the ISS on April 09, and spent eleven days performing scientific experiments and communicating via amateur radio with high school students. He returned to Earth aboard *Soyuz TMA-9*, landing in the steppes of Kazakhstan on April 21, 2007. Launched on March 26, 2009, aboard *Soyuz TMA-14*, he again visited the ISS, becoming the first repeat space tourist. He returned to Earth on April 08, 2009, aboard *Soyuz TMA-13*.

## Garriott de Cayeux, Richard A. Soyuz TMA-13/12.

Richard Allen Garriott de Cayeux was born in July 1961 in Cambridge, England. He is a British-born American computer-game developer who became the sixth "space tourist" and the first second-generation American to go into space. In December 1993, he bought the *Luna 21* lander and the *Lunokhod 2* rover (both currently on the lunar surface) from the Lavochin Association for USD 68,500 at a Sotheby's auction in New York. He is best known as Lord British, the designer of video games enjoyed for over thirty years. In 2009, he founded the video game development company, **Portalarium**. He was the CEO of Portalarium and creative director of *Shroud of the Aavatar: Forsaken Virtues* until 2018 and relinquished all *Shroud of the Avatar* assets to Catnip Games in 2019. After the sale of Origin Systems, he invested in **Space Adventures Ltd.** and purchased a ticket to become the first private citizen to fly in space, but after the dot-com bubble burst, he was forced to sell his seat to Dennis Tito. Space Adventures later announced that Garriott would fly to the ISS as a self-funded Private Astronaut at a reported cost of USD 30 million.

Launched aboard *Soyuz TMA-13* on October 12, 2008, to the ISS, he worked with the Windows on Earth project, which provides an interactive, virtual view of Earth as seen from the ISS. Garriott returned twelve days later aboard *Soyuz TMA-12*. During his flight he filmed a science fiction movie ***Apogee of Fear***. In 2010, he was featured in a documentary, ***Man on a Mission: Richard Garriott's Road to the Stars***, which covered his spaceflight training and mission in orbit. In January 2021, he was elected President of The Explorers Club. In February 2021, Garriott traveled to the bottom of the Mariana Trench, the deepest oceanic trench on the planet. He married Laeticia Pichot de Cayeux in 2011, and they have two children.

# MISSION:

## Axiom-1 (Ax-1).

Crew Dragon capsule *Endeavour* launched April 08-25, 2022, on a SpaceX *Falcon 9* booster rocket.

Commander: **Capt. Michael Lopez Alegria** (born in May 1958 in Madrid, Spain) is a former NASA astronaut and Axiom vice president. He was the first person to command both a civil and a commercial human spaceflight. He has flown to space four times over a twenty year period. For this first fully private mission, the seat price was a reported USD 50 million. The crew spent eight days aboard the International Space Station. For his full biography, refer Part Three: Astronauts, page 245.

Pilot: **Larry Connor** (January 1950-) is an American real estate and technology entrepreneur and non-profit activist investor. He graduated *summa cum laude* from Ohio University in 1972. From 1982-90, he owned Orlando Computer Corp. He established the Connor Group in 2003, and co-founded Heartland Regional Power in 2004. In 2007, he founded the Connor Group Kids & Community Partners. Connor collaborates with the Mayo Clinic and Cleveland Clinic on research projects and virtually taught students at Dayton Early College Academy, Ohio.

Mission Specialist 1: **Eytan M. Stibbe** (born in January 1958 in Haifa, Israel) is an impact investor and philanthropist. In July 1976, he started the Israeli Air force's fighter pilot training course. His collaboration with the Ramon Foundation and the Israel Space Agency at the Ministry of Science and Technology aimed to inspire children and educators through scientific experiments conducted aboard the International Space Station (ISS). He is married, with three children.

Mission Specialist 2: **Mark Pathy** (July 1969-) is a Canadian businessman, CEO of investment company Mavrik, and a philanthropist. Pathy is the chairman of Stingray Group and the former co-CEO of Fednav. He has an undergraduate degree from the University of Toronto and an M.B.A. from INSEAD. He works on health-related research projects with the Canadian Space Agency and the Montreal Children's Hospital.

Mission duration: Seventeen days, one hour and 49 minutes. Splashdown in the Atlantic Ocean, recovered by *MV Megan*.

## Axiom-2.

Crew Dragon capsule *Freedom* launched May 21-31, 2023 on a SpaceX *Falcon 9* booster.

Commander: **Dr. Peggy Annette Whitson** of NASA. Born in February 1960 in Beaconsfield, Iowa, USA, she received a Bachelor of Science degree from Iowa Wesleyan University; and a Master's degree and Ph.D. from Rice University, Texas. She is an American biochemistry researcher, retired NASA astronaut, former NASA Chief Astronaut, and an active Axiom Space astronaut. Dr. Whitson has logged a total of 675 days, three hours and 49 minutes in space, more than any other American or woman. She was the first female Commander of the ISS. During 'Expedition 50', she made her seventh EVA, setting the record for the most EVAs by a female astronaut.

Pilot: **John Shoffner** of NASA was born in July 1955 in Fairbanks, Alaska, USA. He is an American racing driver, investor and pilot. He had a 21-year career with Kentucky-based telecommunications manufacturer **Dura-Line**, then in 2012 he created his own **J2-Racing** motorsports team. In 2019, he was second in the Nurburgring Endurance Series Carrera Cup Class Championship. He is an air show pilot, and has flown over 8,500 hours in fixed-wing aircraft, decommissioned military aircraft and helicopters. During his stay on the ISS, Shoffner and Dr. Whitson conducted single-cell genome sequencing demonstrations for **10x Genomics**.

Mission Specialist 1: **Ali Al Quarni** was born in March 1992 at Sabt Al-Alaya, Balqarn, Asir Province, Saudi Arabia. He is a Saudi astronaut, selected by the Saudi Space Commission. He holds a Bachelor's degree in Aeronautical Sciences from King Faisal Air Academy. He is a captain in the Royal Saudi Air Force, and has logged over 2,387 hours flying time. He has spent nine days, five hours and 27 minutes in space. He is married, with a daughter.

Mission Specialist 2: **Rayyanah Barnawi** was born in September 1988 in Jeddah, Saudi Arabia. From the Saudi Space Commission, she is the first female Saudi astronaut to go to space. She holds a Bachelor's degree in Biomedical Sciences from the University of Otago; and a Master's degree in Biomedical Sciences from Alfaisal University.

Mission duration: Nine days, five hours and 27 minutes.

## Axiom-3.

Crew Dragon capsule *Freedom*, launched January 18-February 09, 2024.

Commander: **Michael Lopez-Alegria** of Spain/USA. See above.

Pilot: **Walter Villadei** was born in April 1974 in Rome, Italy. He graduated from the Italian Air Force Academy in Pozzouli, Italy, holds a Master's degree in Aerospace Engineering from the University of Naples Federico II; and a specialization in Astronautical Engineering from the University of Rome. Villadei began his spaceflight training at the Yuri Gagarin Cosmonaut Training Center in 2008. He first flew to space in June 2023 on the Virgin Galactic 01 mission aboard *SpaceShipTwo*. He is a regular lecturer at the Sapienza University of Rome and *Tor Vergata* on "Experimentation in Space" and "Space Systems". He graduated from the Senior Courses of the NATO Defense College in 2017. Since September 2017, he has been the head of the Space Policy and Operations Office in the General Directorate of Space of the Air Staff.

Mission Specialist 1: Col. **Alper Gezeravci** (of TSA, Turkey) was born in December 1979 in Silifke, Mersin, Turkey. He is a military pilot and the first Turkish astronaut. He studied Electronic Engineering and graduated from the Turkish Air Force Academy with a Bachelor of Science degree; and from the Air Force Institute of Technology in Dayton, Ohio, USA.

Mission Specialist 2: **Marcus Wandt** of SNSA/ESA was born in September 1980. He is a Swedish test pilot and member of the 2022 ESA Astronaut Group. After A. Christer Fuglesang and Jessica Meir, he became the third Swedish astronaut to visit space. Wandt received a Master's degree in Electrical Engineering from Chalmers Institute of Technology in 2007. From 2013-14, he trained at the US Naval Test Pilot School, graduating at the top of his class. In 2014, he joined **SAAB Aeronautics** as an experimental test pilot, before being promoted to Chief Test Pilot and head of flight operations in 2020. A total of 400 million kroner (USD 47 million) was invested in the 'Axiom-3' mission, with SAAB being the largest private investor. Reseach planned for the mission included stem cell research led by Uppsala University and improved cognition research led by KTH Royal Institute of Technology.

Mission duration: Twenty-one days, 15 hours and 41 minutes.

## Axiom-4.

To be launched no earlier than August/October 2024. The crew will include the winner of the ***Space Hero*** reality television show, a Hungarian (or an Indian) and a Polish astronaut.

Commander:

Pilot:

Mission Specialist 1:

Mission Specialist 2:

Mission duration: days, hours and minutes.

## Crew-1.

Launched aboard Crew Dragon *Resilience*, November 15, 2020-May 02, 2021, for ISS 'Expedition 64/65'.

Commander: Col. **Michael Hopkins** was born in December 1968 in Lebanon, Missouri, USA. He entered the University of Illinois at Urbana-Champaign, where he played defensive back for the Illinois *Fighting Illini* football team. He received a Bachelor of Science degree in Aerospace Engineering in 1991; and a Master of Science degree in Aerospace Engineering from Stanford University in 1992. In 2002, Hopkins was selected as an Olmsted Scholar and was sent to the Defense Language Institute in Monterey, California, studying foreign language. After six months, Hopkins was sent to Parma, Italy, studying Political Science. He was working with the Joint Chiefs when selected in 2009 NASA Group 20. Launched aboard *Soyuz TMA-10M* on September 25, 2013 to the ISS, he joined the 'Expedition 37' crew until March 10, 2014, a flight of 166 days. The 'Crew-1' mission was from November 15, 2020, to May 02, 2021, lasting over 167 days; splashing down in the Gulf of Mexico. Hopkins has logged a total of 333 days, 12 hours and 54 minutes in space. He is married, with two sons.

Pilot: Capt. **Victor J. Glover** was born in April 1976 in Pomona, California, USA. He is a captain and F/A-18 pilot in the US Navy and a graduate of the USAF Test Pilot School. He received a Bachelor of Science degree in General Engineering from the California Polytechnic State University in 1999. From 2007-10, he received three Master's degrees from three different institutions: a Master of Science in Flight Test Engineering from the USAF Air University in Edwards AFB, California; a Master of Science in Systems Engineering from the Naval Postgraduate School in Monterey, California; and a Master of Military Operational Art and Science degree from Air University in Montgomery, Alabama. Glover holds a Certificate in Space Systems from the Naval Postgraduate School, and a Certificate in Legislative Studies from Georgetown University. He has logged 3,000 flying hours in over forty aircraft types, and completed over 400 carrier arrested landings and 24 combat missions. Glover was a crew member of 'Expedition 64' and served as a station systems Flight Engineer. He made four EVAs totaling 26 hours and seven minutes. He is married, with four daughters.

Mission Specialist 1: **Dr. Soichi Noguchi** of JAXA was born in April 1965 in Yokohama, Japan. He received a Bachelor of Science degree in 1989; and a Master's degree in 1991, both in Aeronautical Engineering from the University of Tokyo. In 2020, he obtained a Ph.D. in Advanced Interdisciplinary Studies from the University of Tokyo. Dr. Noguchi worked for **Ishikawajima-Harima Heavy Industries**, on the aerodynamic design of commercial engines. In June 1996, he was selected by the National Space Development Agency of Japan, NASDA (now part of JAXA) and trained at the Gagarin Academy in Russia. He flew on STS-114 aboard *Discovery* (July 26-August 09, 2005) and joined the 'Expedition 11' crew, making three EVAs.

Launched on December 21, 2009, aboard *Soyuz TMA-17* for 'Expedition 22/23', he became the first JAXA astronaut and second Japanese citizen (after Tokyo Broadcasting System reporter Toyohiro Akiyama) to fly on a *Soyuz* spacecraft. With the launch of SpaceX *Dragon 2*, he is the third person, and first non-American, to launch aboard three different spacecraft types. Dr. Noguchi has logged a total of 344 days, nine hours and 33 minutes in space; including four EVAs totaling 27 hours and one minute.

Mission Specialist 2: **Shannon Walker**. Refer page 98.

Mission duration: 167 days, six hours and 29 minutes.

## Crew-2.

Launched aboard Crew Dragon *Endeavour*, April 23, 2021-November 09, 2021, for ISS 'Expedition 65/66'.

Commander: **Shane Kimbrough** of NASA. See pages 234-5 for his biography.

Pilot: **K. Megan McArthur** of NASA. See pages 254-5 for her biography.

Mission Specialist 1: **Akihiko Hoshide** of JAXA was born in December 1968 in Setagaya City, Tokyo, Japan. He received an International Baccalaureate from the United World College of South East Asia, Singapore, in 1987; a Bachelor's degree in Mechanical Engineering from Keio University in 1992; and a Master of Science degree in Aerospace Engineering from the University of Houston Cullen College of Engineering in 1997. He joined NASDA in 1992, and worked in the development of the *H-II* launch vehicle for two years. From 1994-99, he was an astronaut support engineer for the NASDA Astronaut Office. Hoshide has logged 340 days, 11 hours and 41 minutes in space; including four EVAs totaling 28 hours and 17 minutes.

As 'Mission Specialist 4' on STS-124 aboard *Discovery* (May 31-June 14, 2008) he was busy with eight days of operations to install the Japanese Experiment Module *Kibo* (JEM-PM) to the ISS. He launched on July 15, 2012, aboard *Soyuz TMA-05M* for 'Expedition 32/33'. On August 27, Hoshide robotically captured and berthed JAXA's *HTV-3* uncrewed cargo spacecraft. He made his first EVA of eight hours and 17 minutes on August 30; and landed on November 18, 2012, after 127 days in space. From July 21-29, 2014, Hoshide commanded the NEEMO 18 mission to the *Aquarius* Reef Base, undersea off the coast of Key Largo, Florida, USA. He served as Flight Engineer on 'Expedition 65', assuming command of the ISS on April 27, 2021.

Mission Specialist 2: **Thomas G. Pesquet** of ESA was born in February 1978 in Rouen, France. He graduated from the Lycee Pierre Corneille in Rouen in 1996; and received a Master's degree in Space Systems and Space Vehicle Mechanics from Ecole Nationale Superieure de l'Aeronautique et de l'Espace in Toulouse, France, in 2001. He speaks French, English, Spanish, German and Russian. From October 2001, he worked as a spacecraft dynamics engineer on remote sensing missions for **GMV, S.A.** in Madrid, Spain. Pesquet graduated from the Air France Flight School in 2006, and began flying the *Airbus A320*. Selected in May 2009 ESA Group, he completed Astronaut Basic Training in November 2010.

Launched as a Flight Engineer on *Soyuz MS-03* from November 17, 2016, to June 02, 2017, Pesquet was part of 'Expedition 50/51'. He spent 196 days, 17 hours and 49 minutes in space. His second mission was 'Crew-2'. Pesquet has logged a total of 396 days, 11 hours and 34 minutes in space; including six EVAs totaling 39 hours and 54 minutes.

Mission duration: 199 days, 17 hours and 43 minutes.

## Crew-3.

Launched aboard Crew Dragon *Endurance*, November 11, 2021-May 06, 2022, for ISS 'Expedition 66/67'.

Commander: Brig. Gen. **Raja J.V. Chari** of NASA was born in June 1977 in Milwaukee, Wisconsin, USA. He graduated from the USAF Academy, Colorado Springs, with a Bachelor of Science degree in Astronautical Engineering and Engineering Science; earned a Master of Science degree in Aeronautical and Astronautical Engineering from the Massachusetts Institute of Technology in 2001; and from 2007 he attended the US Naval Test Pilot School, at Patuxent River, Maryland. Raja Chari also attended the US Army Command and General Staff College in Fort Leavenworth, Kansas, USA. Chari has logged over 2,000 flying hours.

Selected in 2017 NASA Group 22, Chari was the first NASA rookie to command a spaceflight since Joe Engle on STS-2 in 1981. He has logged a total of 176 days, two hours and 39 minutes in space; including two EVAs totaling 13 hours and 48 minutes. In December 2020, Chari was selected as part of the *Artemis Team*, a group of astronauts *"to help pave the way for the next lunar missions, including sending the first woman and the next man to walk on the lunar surface in 2024"*. Chari has received eight awards, including the Defense Meritorious Service Medal; the Air Force Commendation Medal; the Iraq Campaign Medal; the Korean Defense Service Medal; and the Nuclear Deterrence Operations Service Medal. He is married, with three children.

Pilot: **Thomas Marshburn** of NASA. Refer to Part Three: Astronauts of America, pages 251-2.

Mission Specialist 1: **Dr. Matthias Maurer** of ESA was born in March 1970 in St. Wendl, Saarland, West Germany. He graduated from Gymnasium Wendalinum in Sankt Wendel, Saarland, in 1989. He then began compulsory civilian service as a paramedic with the Malteser Emergency Service. Maurer studied Materials Science and Technology at Saarland University in Saarbrucken; in Nancy, France; the University of Leeds, UK; and at the Polytechnic University of Catalonia, Spain. From 1999-2004, he completed his Engineering Doctorate at the Institute of Materials Sciences of the RWTH Aachen University, Germany. In 2006, Maurer completed a course in Economics at the University of Hagen with a MBA management degree. He applied to the ESA in 2008, and was one of ten to be selected from almost 8,500 candidates.

In 2016, Maurer was part of the NEEMO 21 mission, an underwater training program for future ISS crew, held offshore in Florida; as well as the first ESA PANGEA mission, a field geology and astrobiology training program. He logged a total of 176 days, two hours and 39 minutes in space; including one EVA of six hours and 54 minutes. The SpaceX 'Crew-3' mission launched on November 11, 2021, to join 'Expedition 66'.

Mission Specialist 2: **Kayla J. Barron** of NASA was born in September 1987 in Pocatello, Idaho, USA. She attended the US Naval Academy, earning a Bachelor of Science degree in Systems Engineering in 2010; and a Master of Philosophy degree in Nuclear Engineering in 2011. Her graduate research focused on modeling the fuel cycle for a next-generation thorium-fueled nuclear reactor concept known as accelerator-driven subcritical reactor. After completing her Master's degree, Barron became one of the first women to serve as submarine warfare officers. She was assigned to the Ohio-class submarine *USS Maine*, where she completed three patrols as a division officer. Later, she served as Flag Aide to the Superintendent at the US Naval Academy. She was selected in June 2017 in NASA Group 22, and trained as a Mission Specialist.

Mission duration: 176 days, two hours and 39 minutes.

## Crew-4.

Launched aboard Crew Dragon *Freedom*, April 27, 2022-October 14, 2022, for ISS 'Expedition 67/68'.

Commander: **Kjell Lindgren** of NASA. For these biographies, refer Part Three, pages 241-2.

Pilot: **Bob Hines** of NASA. Refer page 220.

Mission Specialist 1: **Samantha Christoforetti** of ESA. Refer pages 68, 88.

Mission Specialist 2: **Jessica Watkins** of NASA. Refer page 336.

Mission duration: 170 days, 13 hours and three minutes.

## Crew-5.

Launched aboard Crew Dragon *Endurance*, October 05, 2022-March 12, 2023, for ISS 'Expedition 68/69'.

Commander: **Nicole Victoria "Duke" Aunapu Mann** of NASA was born in June 1977 in Petaluma, California, USA. Of Wailaki heritage, she is an enrolled member of the Round Valley Indian Tribes. She attended the US Naval Academy and graduated in 1999 with a Bachelor of Science degree in Mechanical Engineering. After commissioning in the US Marine Corps, she attended graduate school at Stanford University, and received a Master of Science degree in Mechanical Engineering, specializing in Fluid Mechanics. Mann received her Naval Aviator wings in 2002, joined VMFA-251 and deployed twice on the *USS Enterprise* in support of Operation Iraqui Freedom and Operation Enduring Freedom. She was a member of Class 135 at the US Naval Test Pilot School, and served in VX-23 at NAS Patuxent River. Mann logged over 2,500 flying hours in 25 aircraft types, made 200 carrier landings and has flown 47 combat missions in Iraq and Afghanistan.

She was selected as one of the eight members of 2013 NASA Astronaut Group 21. Duke Mann worked on development of the *Orion* spacecraft, Space Launch System, and the Exploration Ground Systems. In August 2018, she was assigned to *Boe-CFT*, the first crewed flight of the Boeing *CST-100 Starliner*, but in October 2021 was moved to fly on SpaceX's 'Crew-5' mission as Commander. Mann is married, with a son. She is a contender to be the first woman on the Moon as part of the Artemis plans for 2025. In a December 2020 interview with *National Geographic*, Mann said she and her son often sit outside looking at the Moon, and *"Hopefully someday, he'll be able to watch Mom fly by and walk on the Moon"*.

Pilot: **Josh A. Cassada** of NASA. Refer page 158.

Mission Specialist 1: **Dr. Koichi Wakata** of JAXA was born in August 1963 in Omiya, Saitama, Japan. He earned a Bachelor of Science degree in Aeronautical Engineering in 1987; a Master of Science degree in Applied Mechanics in 1989; and a Doctorate in Aerospace Engineering from Kyushu University in 2004. He worked as a structural engineer for **Japan Airlines**. He was selected by NASDA (now JAXA) in 1992. In December 2000, he became a NASA robotics instructor astronaut. In July 2006, Dr. Wakata served as commander of NEEMO 10, a seven-day expedition at the NOAA's *Aquarius* undersea laboratory at Key Largo, offshore from Florida, USA.

Dr. Wakata flew on STS-72 aboard *Endeavour* (January 11-20, 1996); on STS-119 aboard *Discovery* (March 15-27, 2009); and on STS-127 aboard *Endeavour* (July 15-31, 2009). During a nearly two-decade career in spaceflight, he has logged more than eleven months in space. During 'Expedition 39', he became the first Japanese commander of the ISS.

Wakata flew on the *Soyuz TMA-11M* for the 'Expedition 38/39' long-duration spaceflight from November 07, 2013 to May 13, 2014. He was accompanied by *Kirobo*, the first humanoid robot astronaut. Dr. Wakata has logged a total of 504 days, 18 hours and 33 minutes in space; including two EVAs totaling 14 hours and two minutes. As of early 2023, he is the world's longest-active astronaut.

Mission Specialist 2: **Anna Kikina** of Roscosmos. Refer Part Four, pages 409-410.

Mission duration: 157 days, ten hours and one minute.

## Crew-6.

Launched aboard Crew Dragon *Endeavour*, March 02, 2023-September 04, 2023, for ISS 'Expedition 68/69'.

Commander: **Stephen Bowen** of NASA. Refer Part Three, pages 145-6.

Pilot: **Warren Hoburg** of NASA. Dr. Warren "Woody" Hoburg was born in September 1985 in Pittsburgh, Pennsylvania, USA. As a teen, he built and launched large-scale rockets, the biggest reaching twenty-one feet tall and weighing over 130 pounds. He participated in the finals of the Team America Rocketry Challenge. Hoburg received a Bachelor of Science degree in Aeronautics and Astronautics from the Massachusetts Institute of Technology in 2008; a Master of Science in 2011; and a Ph.D. in Electrical Engineering and Computer Science at the University of California, Berkeley. He worked in product development at **Boeing** until 2014, when he became an assistant professor at MIT. In December 2020, he was announced as part of the Artemis Program for a lunar mission in 2024. Selected in 2017 NASA Group 22, he underwent two years' training, and logged 185 days, 22 hours and 43 minutes in space on ISS 'Expedition 68/69'. From 2009-13, he was a National Science Foundation research fellow, and is a two-time recipient of the AIAA Aeronautics and Astronautics Teaching Award.

Mission Specialist 1: **Dr. Sultan Al Neyadi** of the Mohammed bin Rashid Space Center (MBRSC) was born in May 1981 in Umm Ghafa, Abu Dhabi. He is an Emirati astronaut, and with Hazza Al-Mansuri is one of the first two astronauts from the United Arab Emirates. Al Neyadi joined the UAE Armed Forces and was sent abroad to study. He received a Bachelor of Science degree in Electronics and Communications Engineering from the University of Brighton, UK; a Master of Science (2008); and a Ph.D. in Information Technology from Griffith University in 2012.

Selected in 2018, Al Neyadi had his space training at the Yuri Gagarin Cosmonaut Training Center, Russia, and further training in Houston, Texas, and Cologne in Germany, as part of partnership agreements with NASA, the ESA, and JAXA. Altogether, he had over ninety courses, with over 1,400 hours of training. He also trained in the Neutral Buoyancy Laboratory, passing evaluations which included using the *Extravehicular Mobility Unit* (EMU), maintenance of the ISS, the Incapacitated Crew Rescue (ICR) and practical training on the T-38 jet aircraft. He is the first Arab to serve on the ISS for a six-month mission, and on April 28, 2023, with NASA's Stephen Bowen, was the first Arab astronaut to perform a spacewalk: of seven hours and one minute. On January 06, 2024, he was appointed as the Minister of State for Youth in the UAE. He is married, with six children.

Mission Specialist 2: **Andrey V. Fedyaev** of Roscosmos.

Maj. Andrey Valerievich Fedyaev was born in February 1981 in Serov, Sverdlovsk Oblast, Russian SFSR, Soviet Union. He is an active Test Cosmonaut, who began training at the Gagarin Academy in 2012. In 2004, he received an Engineering degree in Air Transport and Air Traffic Control from the Balashov Military Aviation School. After graduation, he joined the Russian Air Force in the Three Hundred Seventeenth mixed aviation segment. Before his retirement with the rank of Major in 2013, he logged over 500 hours in Russian aircraft. Selected in TsPK-16 Cosmonaut Group, he was assigned on July 15, 2022, to the SpaceX 'Crew 6' mission to the ISS. He launched as a Mission Specialist on March 02, 2023, from KSC aboard SpaceX's reusable *Crew Dragon* spacecraft atop a *Falcon 9* booster to join 'Expedition 68/69'.

Mission duration: 185 days, 22 hours and 43 minutes.

## Crew-7.

Launched aboard Crew Dragon *Endurance*, October 05, 2023-1Q 2024, for ISS 'Expedition 69/70'.

Commander: **Jasmin Moghbeli** of NASA. Refer pages 265-6.

Pilot: **Dr. Andreas Mogensen** of ESA. Andreas "Andy" Enevold Mogenson was born in November 1976 in Copenhagen, Denmark. He received an International Baccalaureate from Copenhagen International School in 1995. He then went to the Imperial College, London, where he obtained an Master of Engineering degree in Aeronautical Engineering in 1999. From 2001-03, he worked as an engineer in the R & D department of **Vestas Wind Systems** in Denmark, then earned a Ph.D. in Aerospace Engineering at the University of Texas at Austin in 2007. He then worked in Germany for **HE Space Operations** associated with the *SWARM* mission. His position when selected by ESA was Aerospace Engineering at Surrey Space Center, University of Surrey, UK.

In 2013, Mogensen served as an ESA cavenaut, training in Sardinia. He took part in the NEEMO 19 undersea mission from September 07-14, 2014. Launched on September 02, 2015, aboard *Soyuz TMA-16M* to the ISS, he worked up to 9.5 hour days; including remote contol of a robot on Earth, filming Red Sprites and Blue Jets lightnings above thunderclouds, and tried out a new kind of Skinsuit designed to alleviate back-pain astronauts feel due to the lengthening of their spine. He landed on September 12, 2015, just over three hours after departing the ISS. He received the Danish Royal Medal of Recompense. Selected as the pilot of SpaceX 'Crew-7', he launched in August 2023, becoming the first European pilot of a spacecraft, and served as the ISS 'Expedition 70' commander. He returned to Earth on March 12, 2024. As of February 2024, Andy Mogensen had logged a total of 208 days, 22 hours and 34 minutes in space (in progress).

Mission Specialist 1: **Dr. Satoshi Furukawa** of JAXA was born in April 1964 in Yokohama City, Kanagawa, Japan. He received a Doctor of Medicine degree from the University of Tokyo in 1989; and a Doctor of Philosophy degree in Medical Science also from there, in 2000. From 1989-99, Dr. Furukawa worked in the Department of Surgery at the University of Tokyo, and also at JR Tokyo General Hospital, Ibaraki Prefectural Central Hospital and at Sakuragaoka Hospital.

He was selected in February 1999 NASDA Group as one of three Japanese astronaut candidates for the International Space Station (ISS). On October 01, 2003, NASDA merged with the Institute of Space and Astronautical Science (ISAS) and the National Aerospace Laboratory of Japan (NAL) and was renamed the **Japan Aerospace Exploration Agency** (JAXA). In May 2004, Dr. Furukawa completed *Soyuz TMA* Flight Engineer-1 training at the Gagarin Academy in Star City, Russia. In February 2006, he completed NASA Astronaut Candidate Training. In August 2007, he took part in the NEEMO 13 project, held in *Aquarius*, the world's only undersea research laboratory, offshore from Florida. In 2013, he served as a cavenaut in the ESA CAVES training held in Sardinia, Italy. Furukawa's second long-duration flight as a Mission Specialist on the ISS was due to end on March 12, 2024.

Mission Specialist 2: **Konstantin S. Borisov** of Roscosmos was born in August 1984 in Smolensk, Smolensk Oblast, USSR. He received a Bachelor of Economics degree from the Plekhanov Russian University of Economics in June 2007. From September 27, 2006 to December 03, 2007, he studied at the University of Warwick, Coventry, United Kingdom, where he received his Master of Science degree in Operations Research and Systems Analysis. In June 2018, he completed his studies at the Moscow Aviation Institute on the Master Program Life Support Systems for Aircraft, with the qualification "Master" in the field of Aircraft Building. Borisov was selected as a cosmonaut in 2018, and in December 2020, he passed the State exam to qualify for spaceflight assignments. Launched on August 26, 2003 in SpaceX 'Crew-7' for 'Expedition 69/70' as a Flight Engineer, his mission lasted for six months to March 12, 2024, splashing down in the Gulf of Mexico.

Mission duration: 199 days, ten hours and 20 minutes.

## Crew-8.

Launched aboard Crew Dragon *Endeavour* atop a SpaceX *Falcon 9* Block 5 booster on March 03, 2024, for ISS 'Expedition 70/71'. The planned 180-day mission is to conduct research and scientific investigations in the microgravity laboratory.

Commander: **Matthew Dominick** of NASA was born in December 1981 in Wheat Ridge, Colorado, USA. He received a Bachelor of Science degree in Electrical Engineering from the University of San Diego (with minors in Physics and Mathematics) in 2005; and was a member of the Navy ROTC. He completed Primary Flight Training at NAS Pensacola, and was designated a Naval Aviator in 2007. In June 2017, he was selected as a member of NASA Group 22, and began two year's training. His promotion to Navy Commander was approved on June 27, 2019. He is married, with two daughters.

Pilot: **Dr. Michael R. Barratt** of NASA flew as a Mission Specialist on STS-133 aboard *Discovery* (February 24 to March 09, 2011) and has logged a total of 211 days, 11 hours and 46 minutes in space. Refer page 137.

Mission Specialist 1: **Dr. Jeanette Epps** of NASA was born in November 1970 in Syracuse, New York, USA. She received a Bachelor of Science degree from Le Moyne College; a Master of Science and a Ph.D. from the University of Maryland, College Park. After graduating, she worked in research at Ford Motor Company, then as a Technical Intelligence Officer with the Central Intelligence Agency (CIA) for seven years, including deployments to Iraq. From July 21, 2014, Epps served as an aquanaut for nine days during the NEEMO 18 undersea exploration mission. Selected in 2009 NASA Group 20, as of March 2024 she was in space.

Mission Specialist 2: **Alexander Grebenkin** of Roscosmos was born in July 1982. He graduated from the Irkutsk Military Aerospace Engineering Institute with a Technician Diploma in 'Technical Operation of Transport Radio-Electronics' on June 21, 2002. From July 2002 to November 06, 2018, he served at the Kuninka Airbase in the Moscow region. From November 2009 to January 2011, he worked as an engineer and head of the regulations and repair group of a military unit. On March 15, 2011, he graduated in absentia from the Moscow Technical University of Communications and Informatics with an Engineering qualification in Radio Communications, Broadcasting and Television. On March 14, 2017, Grebyonkin applied for recruitment to the cosmonaut corps of the Yuri Gagarin Scientific Research Institute of Cosmonauts. He was accepted on August 10, 2018, and began training on November 06. He passed the State exam on November 24, 2020, and launched as Flight Engineer on March 04, 2024.

**Mission duration:** approximately 180 days.

# Other nations with space travelers

## BELARUS

### Wasilewska, or Vasilevskaya, Marina V.. Soyuz MS-25/24.

Born in September 1990, Marina Vitalyevna Vasilevskaya graduated from Secondary School No. 151 in Minsk, and is a flight attendant for Belavia Airlines in Belarus. In December 2022, she was selected among six candidates from more than 3,000 women to participate in a space flight under the "Belarusian Woman in Space" project on the Russian *Soyuz* spacecraft for a short-duration mission to the ISS. In May 2023, she was selected, with Anastasia Lenkova as her backup, to fly aboard *Soyuz MS-25* in March 2024. Vasilevskaya was designated as a member of the prime crew of ISS mission EP-21. Since July 24, 2023, she has undergone theoretical and practical training at the Gagarin Cosmonaut Training Center in Star City, Russia. From October, she had practical training on the *Soyuz MS* spacecraft simulator, and underwent training in zero-gravity conditions on the Ilyushin Il-76 laboratory aircraft. In December 2023, together with Oleg Novitsky, she conducted training for an emergency landing in a wooded and swampy area. Her planned mission is to spend about twelve days on the ISS, and return to Earth aboard *Soyuz MS-24* with Novitsky and NASA astronaut Loral O'Hara.

## BRAZIL

### Pontes, Marcos. Soyuz TMA-08/07.

Born in March 1963 in Bauru, Sao Paulo, Brazil, Marcos Pontes is a Brazilian Air Force pilot, engineer, astronaut, politician and author. The first Brazilian citizen and first Lusaphone to fly in space, he graduated as a military pilot with a B.Sc. in Aeronautical Technology from the Brazil Air Force Academy in Pirassununga in 1984. He worked as a flight safety officer for fourteen years, investigating aeronautical accidents. Pontes earned a B.Sc. in Aeronautical Engineering from the Aeronautical Institute of Technology in Sao Jose dos Campos in 1993. He completed a test pilot course and gained experience in missile testing, weapons development, and aircraft evaluation. Pontes holds the rank of Lieutenant Colonel, and has logged over 2,000 flying hours in 25 aircraft types. He attended the US Naval Postgraduate School, was selected by NASA in 1998, and trained at the Johnson Space Center (JSC) in Houston, Texas; qualifying as a Mission Specialist in December 2000. Due to the 2003 Space Shuttle *Columbia* disaster, shuttle flights were suspended, but in 2005 Brazil reached a USD 20 million agreement with Russia to send Pontes into space.

He flew to the ISS aboard *Soyuz TMA-08* (March 30-April 06, 2006), and returned aboard *Soyuz TMA-07*. After his mission of nine days, 21 hours and 17 minutes, Pontes returned to technical duties for the ISS Program at JSC, working with FIESP/SENAI-SP on the development and fabrication of the Brazilian parts for the ISS. He entered the reserve of the Brazilian Air Force, and awaits a new space mission. Pontes is an invited researcher at the Institute for Advanced Studies of the University of Sao Paulo at Sao Carlos. He is an ambassador of UNIDO and WorldSkills. In the 2018 Sao Paulo gubernatorial election he was elected as second alternate in the Senate. On October 31, Pontes was announced as the next Minister of Science and Technology, and left the post on March 31, 2022. Pontes is the incumbent Senator for Sao Paulo, Brazil, assuming office on February 01, 2023. He has authored four books, with three on space themes.

# CHINESE TAIKONAUTS

In chronological order of missions,

## Yang Liwei. Shenzhou 5.

Maj. Gen. Dr. Yang Liwei (born in June 1965 in Suizhong, Huludao, Lioaning) is a Chinese Major General, former military pilot, and former taikonaut of the People's Liberation Army (PLA). In 1983, at age eighteen years, Yang enlisted in the Chinese People's Liberation Army where he was chosen to enter the aviation college of the Peoples Liberation Army Air Force (PLAAF). He graduated from the No. 8 Aviation College in 1987 and became a fighter pilot, logging more than 1,350 flight hours. In 1998, he was selected from over 1,500 candidates to enter taikonaut training for China's crewed spaceflight program. With eleven other candidates, he spent five years studying the science and operation of spacecraft and undergoing physical and psychological training. These involved the five feet, six-inch, 160 pounds Yang undergoing months of rigorous physical training at the Star City facility in Russia. In 2003, Yang became the first person sent into space by the Chinese space program, CNSA. The *Shenzhou 5* mission made China the third country to send humans into space.

On October 13, 2003, Lt. Col. Yang was launched aboard *Shenzhou 5* atop a *Chang Zheng 2F* rocket from the Jiuquan Satellite Launch Center in the Gobi Desert in China's Gansu province. He orbited Earth fourteen times, but did not enter the craft's orbital module, which was released to perform a six-month military imaging reconnaissance [spying] mission. Following Yang's return, he was named vice-commander-in-chief of the astronauts' system of China's crewed spaceflight project. He entered Tsinghua University in Beijing for doctoral studies in 2004 and received a Doctor of Philosophy in Management in 2009. In 2008, Yang was promoted to Major General. Yang is currently a vice Chief Designer of **China Manned Space Engineering**. He is married to Zhang Yumei, who also served in China's space program. They have a son.

## Fei Junlong. Shenzhou 6, Shenzhou 15.

Born in May 1965 in Suzhou, Jiangsu province, China, Fei was recruited from high school by the People's Liberation Army Air Force (PLAAF) in 1982 at age seventeen years. He graduated with excellent marks from No. 9 Aviation School, the Changchun No. 1 Flight College, and Flight Training School of the Air Force. Selected for the CNSA astronaut program in 1998, he was in the final five chosen for China's first crewed space flight, *Shenzhou 5*. He was Commander on *Shenzhou 6*, launched on October 12-17, 2005. Fei was a member of the *Shenzhou 15* crew on a six-month mission to the *Tiangong* space station. He launched on November 29, 2022, arriving at the station about seven hours later.

A Major General in the PLASSF, he logged 191 days, two hours and 58 minutes in space; including four EVAs; with the first lasting seven hours and six minutes. He has received eight honors, including Merit Citation Class II, 1996; Aerospace Achievement Medal, 2005; the honorary title of Space Hero, 2005; Top 10 Inspirational Models of China, 2006; and Honorary Professor of Nanjing Audit University, 2012. Asteroid 9512 Feijunlong was named after him. Fei is married, with a son.

## Nie Haisheng. Shenzhou 6, 10, & 12.

Born in October 1964 in Yangdang Town, Zaoyang County, Xiangyang City, Hubei Province, China, Nie joined the PLAAF in June 1983 and became a fighter pilot. He was selected for the spaceflight program in January 1998, and flew as Flight Engineer on *Shenzhou 6*, launched on September 12, 2005. The mission lasted almost five days. A Major General of the Chinese People's Liberation Army Strategic Support Force (PLASSF) and the third commander (Unit Chief) of the PLA Astronaut Corps (PLAAC), he served as Commander on both *Shenzhou 10* (the second crewed mission to the Chinese prototype space station *Tiangong 1*), and the *Shenzhou 12* mission, launched on June 17, 2021. This was the first crewed mission to the new *Tiangong* space station. In 2021, with a combined 111 days, 14 hours and 13 minutes in space (including one EVA of five hours and 55 minutes), he set a new record for longest stay in space by a Chinese astronaut, becoming the first to exceed 100 days and he is one of only two Chinese astronauts to have flown three space missions. Nie is married, with a daughter. The asteroid 9517 Niehaisheng was named after him.

## Zhai Zhigang. Shenzhou 7, Shenzhou 13.

Born in October 1966 in Longjiang County, Qiqihar, Heilongjiang Province, China, Zhai became a fighter pilot and logged over 1,000 hours flying time, rising to the rank of Colonel. He was selected for the Chinese space program in 1998, and spent five years in training. Topics included basic theories of aerospace medicine, geography and meteorology, advanced mathematics, and automatic control. He launched as Commander of *Shenzhou 7* on September 25, 2008, the first mission to have a three-man crew. On Flight Day 2, with a live broadcast to audiences in China, he became the first Chinese astronaut to make a spacewalk of twenty minutes. He wore a Chinese-made *Feitian* space suit. The crew returned to Earth on September 27. He was commander of *Shenzhou 13* on October 15, 2021, the second flight to the *Tiangong* space station, where Zhai and his team spent 182 days in space. He made two EVAs, and returned to Earth on April 16, 2022. Zhai is single.

## Liu Boming. Shenzhou 7, Shenzhou 12.

Born in September 1966 in Nenjiang, Heilongjiang, China, Liu was a fighter pilot in the PLAAF, and was selected for the space program in 1998. Launched aboard *Shenzhou 7* on September 25, 2008, he assisted Zhai in the first Chinese EVA, standing up in the spacecraft to hand him a Chinese flag, while wearing a Russian *Orlan-M* space suit. A Major General of the Chinese People's Liberation Army Strategic Support Force (PLASSF), Liu logged a total of 95 days and 38 minutes in space; including three EVAs totaling 13 hours and three minutes. In 2018, he became a Ph.D. candidate at Beijing Institute of Technology. Liu lauched aboard *Shenzhou 12* on June 17, 2021 to the *Tiangong* space station, one of the three initial crew to enter it. He made EVAs on July 04 and August 20, 2021, returning to Earth in September.

## Jing Haipeng. Shenzhou 7, 9, 11, & 16.

Born in October 1966 in Yungcheng, Shanxi Province, China, Jing became a fighter pilot in the PLAAF. A Major General of the People's Liberation Army Ground Force (PLAGF), and selected in 1998 Chinese Group 1, he was the descent module monitor astronaut on *Shenzhou 7* (September 25-28, 2008), a mission of two days, 20 hours and 27 minutes. This third Chinese human spaceflight mission marked the beginning of the second phase of the Chinese government's Project 921. The *Shenzhou 7* mission won the 2009 Space Achievement Award from the **Space Foundation**. Jing was Commander of *Shenzhou 9* (June 16-29, 2012), which launched atop a *Long March 2F* carrier rocket. The docking with *Tiangong-1* was remotely controlled from a ground station, and marked China's first crewed spacecraft rendezvous and docking. The Capsule landed by parachute in Siziwang Banner, central Inner Mongolia.

He was Commander on *Shenzhou 11* (October 17-November 18, 2016), which landed in Inner Mongolia after 33 days. In July 2017, Chairman Xi Jinping awarded Jing the Order of August First, the highest military award of the People's Republic of China. Jing flew as Commander of *Shenzhou 16* (May 30-October 31, 2023) to the *Tiangong* space station, a mission of over 153 days. The spacecraft came down at the Dongfeng landing site in the Gobi Desert, Inner Mongolia. He has logged a total of 201 days, 17 hours and two minutes in space, holding the Chinese record for most days in space.

## Liu Wang. Shenzhou 9.

Born in March 1969 in Pingyao County, Shanxi Province, People's Republic of China, Liu enrolled at the PLA Air Force Aviation University and became a fighter pilot. A Major General in the PLASSF, he was selected on January 05, 1998, in Chinese Group 2, and flew on *Shenzhou 9* (June 16-29, 2012), logging 12 days, 15 hours and 25 minutes (12.64 days) in space. Liu is married, with one child.

## Liu Yang. Shenzhou 9 & 14.

Born in October 1978 in Zhengzhou, Henan Province, China, Liu was the first Chinese woman in space. She joined the People's Liberation Army in 1997, trained at Changchun No. 1 Flight College, and became a pilot of cargo planes. She has 1,680 hours of flying experience. Liu attained the rank of Major, served as the deputy head of her flight unit, and joined the astronaut corps in 2010. She flew on *Shenzhou 9* (June 16-29, 2012), conducting medical experiments. On mission *Shenzhou 14* TG Expedition 3 (June 05-December 04, 2022), she docked with the *Tianhe* module of the *Tiangong* space station, and spent 182 days, nine hours and 24 minutes in orbit. She assisted with the station's assembly and expansion. On September 01, 2022, she became the second Chinese woman to perform an EVA: six hours and seven minutes. Senior Colonel Liu has logged a total of 195 days, zero hours and 49 minutes in space. She is married, with two children; a son and daughter.

## Wang Yaping. Shenzhou 10 & 13.

Born in January 1980 in Yantai, Shandong Province, China, she graduated from Yantai Yizhong High School in 1997, and was admitted to the Changchun Flight College of the PLAAF. Wang graduated from the Aviation University and flight school in 2001 with the rank of First Lieutenant. As a pilot in Wuhan, she participated in combat readiness exercises, the Wenchuan Earthquake relief effort, and cloud seeding and weather modification for the Beijing 2008 Olympic Games. After 1,600 safe flying hours, she was named an Air Force Class II pilot. A Senior Colonel in the PLASSF, she is a military transport pilot, the second female taikonaut selected to the People's Liberation Army Astronaut Corps (PLAAAC), the second Chinese woman in space, and the first Chinese woman to perform a spacewalk.

Wang flew on *Shenzhou 10* (June 11-26, 2013), docking with the *Tiangong-1* space station. She conducted scientific experiments and taught a physics lesson to Chinese students by live television broadcast. In March 2018, Wang was elected for a five-year term as a deputy to the National People's Congress. Launched on October 15, 2021, the three *Shenzhou 13* crew arrived at the *Tianhe* space station module of the *Tiangong* station, and she made her historic spacewalk of six hours and 25 minutes on November 07. The mission ended on April 16, 2022, by which time she had set a new record for the longest stay in space by a female Chinese astronaut with a cumulative 197 days in space. Wang is studying for a doctoral degree in psychology at Peking University. She is married, with a daughter.

## Zhang Xiaoguan. Shenzhou 10.

Born in May 1966 in Jinzhou, Liaoning, China, to a family of Manchu ethnicity, he was a Squadron Leader in the PLAAF when selected in 1998 in Chinese Group 1 to be an astronaut. As of 2004, he had accumulated 1,000 flight hours. Launched atop a *Long March 2F* carrier rocket from the Jiuquan Satellite Launch Center in Inner Mongolia, *Shenzhou 10* (June 11-26 2013) was the third and final spaceflight to the first Chinese space station, *Tiangong-1*. As pilot operator, Zhang conducted the rendezvous and manual docking. The crew spent twelve days on the station, and their mission lasted 14 days, 14 hours and 29 minutes; the longest Chinese manned mission to date.

## Chen Dong. Shenzhou 11 & 14.

Born in December 1978 in Luoyang, Henan, China, Chen holds a Master's degree in Engineering from Xi'an Jiaotong University. He joined the PLAAF in August 1997, achieving the rank of Senior Colonel. A fighter pilot in the PLAAF, he was selected to be a CNSA taikonaut in 2010 Group 2. Selected to fly on *Shenzhou 11*, he and Jing Haipeng underwent over 3,000 hours of training before launching on October 17, 2016, for a 33-day mission to the *Tiangong-2* space station. They landed on November 18, 2016, in Inner Mongolia. On June 05, 2022, he launched aboard *Shenzhou 14* to the *Tiangong* space station for a six-month mission, and completed three EVAs. On December 26, Chen was awarded the Spaceflight Merit Medal, Third Class; and the honorary title, Hero Astronaut by the Central Military Commission. In December 2022, he set a new record for the longest stay in space by a Chinese astronaut with a cumulative 214 days, 15 hours and 54 minutes. Chen married in 2005, and has twin sons.

## Tang Hongbo. Shenzhou 12 & 17.

Born in October 1975 in Yunhuqiao, Xiangtan County, Hunan, China, Tang joined the PLAAF in September 1995, and graduated from the Changchun Flight College. A Senior Colonel, he accumulated over 1,159 hours of safe flight, and was selected in May 2020 in Chinese Group 2. On the *Shenzhou 12* mission, he visited the *Tiangong* space station, and made the first EVA from there with Liu Boming. He is Commander of *Shenzhou 17* (launched October 26, 2023). As of March 24, 2024, Tang had spent another 141 days in space on the ongoing *Shenzhou 17* mission. He has already logged a total of over 278 days in space; including four EVAs totaling 29 hours and 55 minutes. He is married, with one child.

## Jiang Xinlin. Shenzhou 17.

Born in February 1988 in rural Qiu County, Kaifeng, Henan Province, China, in 2006 he was admitted to the former PLA Army Armored Forces Academy College. He also enlisted in the PLA and became a ground force tank driver. After graduating from the College, he was chosen as an Air Force pilot. Jiang accumulated over 1,000 hours of safe flight, earning the rank of First Class Pilot, later promoted to Lieutenant Colonel. He was selected in Chinese Group 3, and in September 2020 was selected as an aerospace flight engineer. On October 26, 2023, he flew aboard *Shenzhou 17* as a systems operator to the *Tiangong* space station, docking later that same day. As of March 24, 2024, he was in space.

## Tang Shengjie. Shenzhou 17.

Born in December 1989 in Dingxi, Gansu Province, northwest China, Tang enlisted in the People's Liberation army in September 2008, and graduated from the Air Force Aviation University in 2012. He is a First Class PLAAF pilot. As the captain of a flying squadron, in almost thirteen years he gained extensive experience flying six aircraft types, and was promoted to Lieutenant Colonel. Tang was selected in Chinese Group 3 on October 01, 2020. Chosen for the *Shenzhou 17* mission in June 2022, he flew to the *Tiangong* space station on October 26, 2023. He is currently the youngest taikonaut in space, and has logged 137 days, 18 hours and 57 minutes to March 12, 2024. At a pre-launch Press meeting he said, *"It is a great honor and great duty for me, and I feel very lucky and very happy"*.

## Ye Guangfu. Shenzhou 13.

Born in September 1980 in Shuangliu County, Sichuan, China, Ye is a fighter pilot and member of the People's Liberation Army Astronaut Corps (PLAAC). Ye spent four years as an instructor and four years as an Air Force fighter pilot, accumulating 1,100 hours flying time. In 2010, he was selected for Chinese Group 2, and qualified in 2014. Fluent in English and Russian, he participated in the ESA CAVES mission in 2016. He flew on *Shenzhou 13*, literally "Divine Boat Number 13", launched on October 15, 2021, to the *Tiangong* space station. It followed the launch of the *Tianzhou 3* cargo spacecraft. Ye logged 182 days, nine hours and 32 minutes in space. He made his first EVA of six hours and 11 minutes on November 26, 2021. *Shenzhou 13* undocked from *Tiangong* on April 15, 2022, and landed nine hours later on April 16 at the Dongfeng landing site in the Gobi Desert in Inner Mongolia. Ye is married, with a son and daughter.

## Cai Xuzhe. Shenzhou 14.

Born in May 1976 in Shenzhou, Hengshui, Hebei province, China, Cai graduated from the Boading Branch of the Changchun Flight Academy of the Air Force. He served as a fighter pilot in the PLAAF, and was selected to be an astronaut in 2010. A Senior Colonel in the PLASSF, he flew on the *Shenzhou 14* mission to the *Tiangong* space station. Cai logged 182 days, nine hours and 25 minutes in space; including two EVAs totaling nine hours and 46 minutes with Chen Dong. Cai is married.

## Deng Quingming. Shenzhou 15.

Born into a family of farming background in March 1966 in Yihuang County, Jiangxi, China, Deng is a Chinese People's Liberation Army Astronaut Corps (PLAAC) taikonaut selected as part of the Shenzhou program. A fighter pilot in the PLAAF, he was selected in Chinese Group 1 in 1998. After 24 years of training, he lifted off in November 2022 on *Shenzhou 15* to the *Tiangong* station, where he stayed for 183 days, before returning to Earth on June 03, 2023. Deng logged 186 days, seven hours and 24 minutes in space. His daughter was born in 1991 and she now works in the Beijing Aerospace Flight Control Center.

## Zhang Lu. Shenzhou 15.

Born into a fisherman's family in November 1976 in Hanshou County, Hunan Province, China, Zhang joined the Chinese Communist Party in April 1999 and enlisted in the People's Liberation Army in August 1999. He graduated from the PLA Air Force Aviation University in 2000, and became a fighter pilot in the People's Liberation Army Air Force (PLAAF). Zhang was selected as an astronaut on May 05, 2010, in Chinese Group 2 and launched on TG-3 for the *Shenzhou 15* mission to the *Tiangong* space station. In orbit from November 29, 2022 to June 03, 2023, he logged 186 days, seven hours and 24 minutes in space; including four EVAs, the first being of seven hours and six minutes. Zhang married in 2003, and has a daughter.

## Zhu Yangzhu. Shenzhou 16.

Senior Colonel Zhu Yangzhu was born in September 1986 in Pei County, Xuzhou, Jiangsu, China. He registered in the National Defense Science and Technology University in Changsha in September 2005. Also in September 2005, he automatically enlisted in the People's Liberation Army, and joined the Chinese Communist Party (CCP) in December 2006. He studied at the Faculty of Aerospace Enginering Missile Design. He completed his postdoctural work, and graduated from the National University of Defense Technology. His field of expertise had shifted from aviation to space travel, and so in 2017 he joined the Academy of Aerospace Engineering as an Associate Professor at PLA Strategic Support Force Space Engineering University.

Selected in Chinese Group 3, in September 2020 he was inducted into the People's Liberation Army Astronaut Corps. On April 29, 2021, construction of the Chinese *Tiangong* space station began with the launch of the *Tianhe* core module. Zhu's group had relatively little time to prepare for their missions, since the construction time for the station was only eighteen months. The scientists in his group had the advantage over the fighter pilots of the early years in that they already had good prior knowledge of higher mathematics, flight dynamics, etc. In his role as an engineer, Zhu Yangzhu is primarily responsible for repairs, both to the systems of the station and to the laboratory cabinets. While external payloads are typically mounted using a mechanical arm, some spacewalks will still be required after construction is complete. In June 2022, he was selected to the *Shenzhou 16* mission, launched on May 30, 2023. He has logged 153 days, 22 hours and 41 minutes in space.

## Gui Haichao. Shenzhou 16.

Born in November 1986 in Yaoguan Town, Shidian County, Baoshan, Yunnan, China, Dr. Gui is a Chinese pilot, researcher and associate professor. He graduated from Beihang University with a Bachelor of Engineering degree (2009); and Ph.D. (2014) in Aerospace Engineering. From 2014-17, he was a postdoctoral researcher with the Department of Earth and Space Science and Engineering, York University, Toronto, Canada; and the Department of Aerospace Engineering, Ryerson University, Toronto. Since 2017, he has been an associate professor with the School of Astronautics, Beihang University, China.

He was one of 18 astronauts selected in Chinese Group 3 from 2,500 candidates after a three-stage selection process, as part of the Shenzhou program. He is China's first civilian astronaut who is not enlisted in the People's Liberation Army Astronaut Corps. He joined the Chinese Communist Party in June 2022. Gui was appointed to be a Payload Specialist on the *Shenzhou 16* mission, launched on May 30, 2023. He logged a total of 153 days, 22 hours and 41 minutes in space.

# INDIA

## Sharma, Rakesh. Soyuz T-11.

Rakesh Sharma was born in January 1949 in Patiala, present-day East Punjab, India. He graduated from Nizam College, Hyderabad. He joined the National Defence Academy as an Air Force plebe in July 1966, and was commissioned into the Indian Air Force as a pilot in 1970. During the 1971 Bangladesh Liberation War he flew twenty-one combat missions as a MiG-21 pilot. Wing Commander Sharma is the only Indian citizen to travel in space to date, and with his mission India became the fourteenth nation to send a man into outer space. He was selected on September 20, 1982 to become a cosmonaut, and flew on *Soyuz T-11* (April 03-11, 1984) to the *Salyut 7* space stationas part of the Interkosmos program. Sharma logged seven days, 21 hours and 40 minutes in space; including forty-three experimental sessions (mainly in the fields of bio-medicine and remote sensing). He was awarded the title, Hero of the Soviet Union upon his return. India conferred its highest peacetime gallantry award, the Ashoka Chakra on him. In 1987, he joined **Hindustan Aeronautics Limited**, serving as the chief test pilot in the HAL Nashik Division until 1992. He retired from flying in 2001. Sharma is married, with a son and daughter.

# IRAN

## Ansari, Anousheh. Soyuz TMA-9/8.

Born as Anousheh Raissyan in September 1966 in Mashhad, Iran, Ansari is an Iranian American engineer and co-founder and chairwoman of **Prodea Systems,** an Internet of Things company co-founded in 2006. She previously served as co-founder and CEO of **Telecom Technologies, Inc**. Ansari emigrated from Iran to the United States in 1984 as a teenager. She received a Bachelor of Science degree in Electronics and Computer Engineering from George Mason University, Virginia, in 1988; and a Master's degree in Electrical Engineering from George Washington University while working full-time at MCI Communications. In 1993, Ansari, her husband and brother-in-law co-founded Telecom Technologies, Inc., which was acquired by Sonus Networks, Inc. in 2000 for about USD 550 million. Beginning with a multi-million dollar contribution in 2002, the Ansari family is the sponsor of the **Ansari X Prize**. The non-profit XPRIZE Foundation is the world's leader in designing and operating incentive competitions to solve humanity's grand challenges. In 2004, Ansari helped fund the USD ten million Ansari X Prize for private spaceflight.

Launched aboard *Soyuz TMA-9* on September 18, 2006, six days after her fortieth birthday, she became the first Iranian and first Muslim woman in space. Ansari was the fourth overall self-funded "space tourist", and the first self-funded woman to fly to the International Space Station; docking on September 20 for eight days, and returning to Earth aboard *Soyuz TMA-8* on September 29, 2006. In 2009, she received the first Symons Innovator Award given annually by NCWIT to honor successful women entrepreneurs in technology. Her memoir, *My Dream of Stars: From Daughter of Iran to Space Pioneer*, co-written with Horner Hickam, was published by Palgrave Macmillan in 2010.

# JAPAN

In chronological order of missions,

## Akiyama, Toyohiro. Soyuz TM-11/10.

Toyohiro Akiyama was born in July 1942 in Tokyo, Japan. He received a Bachelor of Science degree from the International Christian University, Mitaka, Tokyo. He joined Tokyo Broadcasting System (TBS) as a journalist in 1966, worked for the BBC World Service from 1967-71, and from 1984-88 served as the TBS chief correspondent in Washington, D.C.. In August 1989, Akiyama was selected for a commercial Soviet-Japanese flight. He was not a trained astronaut, scientist nor engineer, but launched on December 02, 1990, aboard *Soyuz TM-11*, the eleventh expedition to Russian space station *Mir*. He docked on December 04, and logged seven days 21 hours and 54 minutes in space. As a reporter his network paid for the flight (possibly USD 14 to 28 million), since they wanted to boost their TV ratings. The spacecraft's launch shroud and its *Soyuz* booster were painted with the Japanese flag and advertisements for Sony, Unicharm, and Otsuka Pharmaceutical. Akiyama was scheduled to make one ten-minute TV broadcast and two 20-minute radio broadcasts each day. He suffered from space sickness, vomited a lot, and craved cigarettes. TBS broadcast his landing aboard *Soyuz TM-10* live from Kazakhstan.

Akiyama returned to TBS and became deputy director of the TBS News Division, but retired in 1995 because he disagreed with the active commercialization of television. From January 1996, he engaged in organic farming in the Abukuma mountains, near Tamura, Fukushima Prefecture. In March 2011, he was forced to abandon his farm because of the Fukushima Daichi nuclear disaster. On November 01, 2011, he became a professor of agriculture at Kyoto University of Art and Design. Akiyama received four awards: the Order of Friendship of Peoples (USSR, 1990); the Tokyo Metropolitan Cultural Honor award (1991); the Japanese Society for Biological Sciences in Space Achievement Award (2000); and the Medal "For Merit in Space Exploration" (Russia, 2011). He was married in 1970, with two children, and divorced in 1995.

## Mukai, Dr. Chiaki. STS-65, STS-95.

Born in May 1952 in Tatebayashi, Gumma Prefecture, Japan, Dr. Mukai became the first Japanese woman and first Asian woman to travel into space. She earned a doctorate in Medicine in 1977; and a doctorate in Physiology in 1988 from Keio University School of Medicine in Tokyo. After working for several years as a Doctor, she became an assistant Professor of the Department of Cardiovascular Surgery at Keoi University. In 1985, she was working as a heart surgeon when the National Space Agency of Japan (NASDA) chose her as a Payload Specialist for the STS-47 Spacelab-J mission, but she did not fly on that mission.

She flew on STS-65 aboard *Columbia* (July 08-23, 1994), taking part in several scientific and medical experiments housed in a 'Spacelab' module in the orbiter's cargo bay. Dr. Mukai's second spaceflight was on STS-95 aboard *Discovery* (October 29-November 07, 1998), when the crew conducted eighty-two experiments in microgravity. She has logged a total of over 22.6 days (566 hours) in space. While serving as a NASDA astronaut at NASA, she also worked at the Baylor College of Medicine in Houston and the Keio University School of Medicine. Dr. Mukai served as the deputy mission scientist for the STS-107 mission, when shuttle *Columbia* disintegrated during re-entry in January 2003. She belongs to the American Aerospace Medical Association; the Japan Society of Microgravity Applications; the Japanese Society for Cardiovascular and Thoracic Surgery, and the United Nations Committee on the Peaceful Uses of Outer Space (COPUOS). She received the highest French award, the Legion of Honor, in 2015.

## Wakata, Koichi. STS-72, STS-92, STS-119/127, Soyuz TMA-11M, SpaceX Crew-5.

See above. Refer pages 106-7.

## Doi, Dr. Takao. STS-87, STS-123.

Born in September 1954 in Minamitama, Tokyo, Japan, Dr. Doi holds a B.Sc., a Master of Engineering degree (1980); a Doctorate in Aerospace Engineering from the University of Tokyo in 1983; and a Ph.D. in Astronomy from Rice University in 2004. As an avid astronomer, he discovered supernovae SN 2002gw and SN 2007aa. He worked for the NASA **Lewis Research Center** as a National Research Council research associate in 1985. Selected in the 1985 NASDA Group, he flew on STS-87 aboard *Columbia* (November 19-December 05, 1997), involved with microgravity experiments and deploying the *SPARTAN-201* satellite.

Dr. Doi was 'Mission Specialist 4' on STS-123 aboard *Endeavour* (March 11-27, 2008), which delivered the Japanese *Kibo* laboratory to the ISS. Takao Doi became the first person to throw a boomerang in space. In September 2009, he retired from astronaut duty and became Chief of the Space Applications Section at the United Nations Office for Outer Space Affairs. Dr. Doi later became a professor at the Unit of Synergetic Studies for Space at Kyoto University, and then at the Graduate School of Advanced Integrated Studies in Human Survivability (GSAIS). He is a member of the Japan Society of Microgravity Applications; the Japan Society for Aeronautical and Space Science; and the American Institute of Aeronautics and Astronautics.

## Mohri, Dr. Mamoru. STS-47, STS-99.

Born in January 1948 in Yoichi, Hokkaido, Japan, Dr. Mamoru "Mark" Mohri is a scientist, a former NASDA astronaut, and the first Japanese who was part of an official Japanese space program. He earned both a Bachelor of Science (1970) and Master's degree in Chemistry (1972) from Hokkaido University; and a Ph.D. in Chemistry from Flinders University in Adelaide, South Australia, in 1976. Most of Dr. Mohri's work has been in the field of materials and vacuum sciences. From 1975-85, he was a member of the nuclear engineering faculty of Hokkaido University, where he worked on nuclear fusion-related projects. Selected in 1985 NASDA Group, he trained as a Payload Specialist for a Japanese materials science payload. Mark Mohri served as 'Payload Specialist 1' on STS-47 aboard *Endeavour* (September 12-20, 1992) for 'Spacelab-J'. This was the fiftieth Space Shuttle mission of the program, and lasted seven days, 22 hours and 30 minutes. Mae Jemison and Mohri were trained to use the autogenic Feedback training Exercise which uses biofeedback and autogenic training to help patients monitor and control their physiology.

He was 'Mission Specialist 4' on STS-99 aboard *Endeavour* (February 11-22, 2000), the Shuttle Radar Topography Mission (SRTM). The SRTM was an international project spearheaded by the National Imagery and Mapping Agency (now the National Geospatial-Intelligence Agency), an agency within the US Department of Defense, and NASA, with participation of the German Aerospace Center, DLR. Dr. Mohri logged a total of 19 days, four hours and nine minutes in space. As of 2001, he was the Chief Executive Director Emeritus for the Miraikan, the **National Museum of Emerging Science and Innovation** in Tokyo. On March 16, 2006, he was appointed an Honorary Member of the Order of Australia (AM), for "service to Australia-Japan education and science relations".

## Noguchi, Soichi. STS-114, Soyuz TMA-17, SpaceX Crew-1.

Refer pages 103-4.

## Hoshide, Akihiko. STS-124, Soyuz TMA-05M, SpaceX Crew-2.

Refer page 104.

## Yamazaki, Naoko. STS-131.

Born in December 1970 in Matsudo City, Chiba Prefecture, Japan, Naoko Sumino received a Bachelor of Science and Master of Science degree in Aerospace Engineering from the University of Tokyo in 1996. She joined NASDA (now JAXA) in 1996 and was part of the development team for the system integration of the Japanese experiment Module (JEM). Selected in NASDA Group 4 on February 10, 1999, she was the second Japanese woman to fly in space. In May 2004, she completed *Soyuz-TMA* Flight Engineer training at the Gagarin Academy in Russia. In June 2004, she arrived at Johnson Space Center to begin Astronaut Candidate Training School, and was selected as a Mission Specialist in 2006.

As 'Mission Specialist 4' on STS-131 (19A) aboard *Discovery* (April 05-20, 2010), she logged 15 days, two hours and 47 minutes in space. Yamazaki supported the primary tasks by operating both the shuttle and space station's Remote Manipulator Systems. As loadmaster, she orchestrated cargo transfer activities between the two spacecraft. From December 2010, she resumed study in the field of aerospace engineering at the University of Tokyo. She retired from the space program on August 31, 2011. From July 2012, Yamazaki was a member of the Space Policy Committee of the Cabinet Office, and from July 2018 is a Representative Director, Space Port Japan Association. Yamazaki is married, with one child.

## Furukawa, Satoshi. Soyuz TMA-02M/Expedition 28/29, SpaceX Crew-7.

Refer page 108.

Born in April 1964 in Yokohama City, Kanagawa, Japan. Flew on *Soyuz TMA-02M* for 'Expedition 28/29'. SpaceX 'Crew-7' for 'Expedition 69/70'. Logged a total of 366 days, eight hours and 34 minutes in space.

## Yui, Kimiya. Soyuz TMA-17M/Expedition 44.

Born in January 1970 in Kawakami, Nagano Prefecture, Japan, after graduating from the National Defense Academy in 1992, Yui enrolled in the Japan Air Self-Defense Force. As a test pilot, he flew the F-15 Eagle fighter. From 2008, he served as a Lieutenant Colonel in the Air Staff Office's Defense Planning Division. In 2009, Yui was selected as an astronaut candidate, the first with a military background. He became a certified ISS astronaut in July 2011, and took part in the NEEMO 16 undersea exploration mission in June 2012. On July 22, 2015, Yui launched on *Soyuz TMA-17M* and joined the 'Expedition 44' crew on the ISS. In August, he robotically supported the arrival of JAXA's *HTV-5* uncrewed resupply spacecraft, and experienced a rare "direct handover" period with nine crew members on board the ISS. During 'Expedition 45' he suported the arrival of Orbital ATK's 'OA-4' mission. Departing the ISS on December 11, 2015, he logged 141 days, 16 hours and nine minutes in space. In November 2016, Yui became chief of the JAXA Astronaut Corps. The 2016 IMAX documentary film ***A Beautiful Planet*** shows scenes of the Earth, featuring Kimiya Yui and other ISS astronauts. He is married, with three children.

## Onishi, Takuya. Soyuz MS-01/Expedition 48/49.

Born in December 1975 in Nerima, Tokyo, Japan, Onishi received a Bachelor of Engineering degree in Aeronautical and Space Engineering from the University of Tokyo in 1998. He then joined **All Nippon Airways** (ANA) and was assigned to the Passenger Service Department, Haneda Airport, Tokyo. He completed two years of basic flight training in Bakersfield, California, and one year of advanced flight training in Tokyo. In October 2003, Onishi was promoted to co-pilot of Boeing 767 airplanes, flying both domestic and international routes. JAXA selected him in February 2009 as an astronaut candidate for the International Space Station. From April 2009, he attended the ISS Astronaut Basic Training domestic program at the Tsukuba Space Center in Japan. From August 2009, he was one of fourteen members of the twentieth NASA astronaut class at JSC, Houston, Texas. Onishi served as an aquanaut aboard the *Aquarius* underwater laboratory during the NEEMO 15 mission of October 20-26, 2011.

Launched on July 06, 2016, aboard *Soyuz MS-01* for 'Expedition 48/49', he logged 115 days, two hours and 22 minutes in space. Asteroid 163153 Takuyaonishi, discovered by a Japanese astronomer in 2002, was named in his honor. The naming citation was published by the Minor Planet Center on February 18, 2011.

## Kanai, Dr. Norishige. Soyuz MS-07/Expedition 54/55.

Born in December 1976 in Tokyo, Japan, Kanai is a Medical Doctor, and a Lieutenant and Diving Medical Officer in the Japan Maritime Self-Defense Force. He graduated from Toho High School, Chiba, in 1995; and received his M.D. from the National Defense Medical College in 2002. From 2002-04, he worked in the Department of Surgery at the National Defense Medical College Hospital. Dr. Kanai qualified as a Diving Medical Officer in 2006. In 2008, he transferred to the Japan Self Defense Force Hospital, Kure, Hiroshima.

Selected in 2009 JAXA Group, he trained at JSC, Texas, as one of the fourteen members of the twentieth NASA astronaut class. Certified as an ISS astronaut in July 2011, Dr. Kanai served as a cavenaut in the ESA CAVES training in Sardinia; and in July 2015 he participated as an aquanaut in the NEEMO 20 crew at the *Aquarius* underwater laboratory, Key Largo, Florida. Launched as 'Flight Engineer 2' aboard *Soyuz MS-07* on December 17, 2017, Dr. Kanai served as a crew member for ISS 'Expedition 54/55' and returned to Earth on June 03, 2018. The crew conducted about 250 science investigations in biology, Earth science, human research, physical sciences, and technology development. He logged 168 days, five hours and 18 minutes in space; including one EVA of five hours and 57 minutes. Dr. Kanai is a member of the Japan Surgical Society, and the Japanese Society of Hyperbaric and Undersea Medicine.

### Hirano, Yozo. Soyuz MS-20.

Born in October 1985 in Imabari, Ehime Prefecture, Japan, Hirano joined **ZOZO, Ltd.**, after graduating from Kyoto Prefectural University with a degree in Casting Director of a photography group. He is now working at the Start Today division of ZOZO as a film producer. Hirano passed the medical examination in May 2021, and from June 15 began preparations for his flight at the Gagarin Cosmonaut Training Center. He was a "space tourist" launched on *Soyuz MS-20* on December 08, 2021, for a mission of 11 days, 19 hours and 34 minutes. Hirano served as a production assistant to document the flight for Yusaku Maezawa, who paid for both seats.

### Maezawa, Yusaku. Soyuz MS-20.

Born in November 1975 in Kamagaya, Chiba, Japan, Maezawa is a billionaire entrepreneur and art collector. While at Waseda Jitsugyo High School he started a hardcore punk band called Switch Style, in which he was the drummer. He founded **Start Today** in 1998 and launched the online fashion retail website **Zozotown** in 2004. In 2012, he founded the Tokyo-based Contemporary Art Foundation, with the goal of "supporting young artists as a pillar of the next generation of contemporary art". In 2018, he introduced **ZOZO**, a custom-fit clothing brand, and resigned from ZOZO in September 2019 after selling a stake of 50.1% in the company for USD 3.7 billion. Maezawa came into the public eye on September 17, 2018, when SpaceX CEO Elon Musk said he would be the first private customer on the SpaceX company's *Starship* rocket to the Moon. On May 13, 2021, Maezawa announced he would be joining **Space Adventures** on a trip to the International Space Station in December. He launched aboard *Soyuz MS-20* on December 08, 2021, for a twelve-day mission, in which he completed the top 100 tasks demanded by the public, including playing a game of badminton. He is divorced, with one child.

## SOUTH AFRICA

### Shuttleworth, Mark R.. Soyuz TM-34/33.

Mark Richard Shuttleworth was born in September 1973 in Welkom, South Africa. He received a Bachelor of Business Science degree in Finance and Information Systems at the University of Cape Town. While studying at the University in 1995, he founded **Thawte**, a consulting firm that became a global leader in Internet security for electronic commerce. He sold the business in 1999 to the US-based company **VeriSign**. With the proceeds, he established a venture capital firm and the **Shuttleworth Foundation**, a non-profit organization dedicated to funding education initiatives in Africa. He lives on the Isle of Man, UK, and holds dual citizenship from South Africa and the United Kingdom. In 2001, at a personal cost of USD 20 million, he embarked on the First African in Space project. Shuttleworth trained in Star City, Russia, and Kazakhstan for a *Soyuz* mission to the ISS. Launched on April 25, 2002 aboard *Soyuz TM-34* with Gidzenko and Vittori, he spent eight days aboard the ISS, conducting scientific experiments related to AIDS and genome research for South Africa. He returned to Earth on May 05 aboard *Soyuz TM-33*. In 2004, he founded **Canonical**, which became the parent company of various ventures, most notably **Ubuntu**, a Linux-based operating system that aims to expand personal computer access, especially in developing countries.

# SOUTH KOREA

## Dr. Yi So-yeon. Soyuz TMA-12/11.

Born in June 1978, in Gwangju, South Korea, Lee graduated from KAIST in Daejeon with a Bachelor of Science and a Master's degree in Mechanical Engineering. In 2006, she was selected as one of two finalists in the Korean Astronaut Program. Her doctorate in Biotech Systems was conferred on February 29, 2008, while she was training in Russia. She launched aboard *Soyuz TMA-12* on April 08, 2008, with two Russian cosmonauts. South Korea is reported to have paid Russia USD 20 million for Yi's space flight. Referred to as a "spaceflight participant", she carried out eighteen science experiments for the Korean Aerospace Research Institute (KARI) and conducted interviews and discussions with media. She observed the Earth, in particular the dust storms from China to Korea. Yi returned to Earth on April 19, 2008, aboard *Soyuz TMA-11*. Due to a malfunction, the craft followed a ballistic trajectory, with up to 10-G's deceleration, and landed 420 kilometres off course in Kazakhstan.

After her flight, Yi worked as a researcher at KARI and as Korea's space ambassador with Ko San. In 2009, she became the first astronaut to attend the **International Space University** (ISU) Space Studies Program at NASA Ames Research Center, held in conjunction with the inaugural class of the Singularity University Graduate Studies Program. In 2010, she enrolled in the MBA program at the Haas School of Business at the University of California. On August 13, 2014, the Korean Aerospace Research Institute announced that Yi had resigned for personal reasons. In an interview, she said she was preparing to marry a Korean-American man, and she wanted to study for an MBA. She then joined the Association of Spaceflight Professionals, and from 2016 taught physics at the Everett Community College in Washington, USA.

# UNITED ARAB EMIRATES

## Al-Mansuri, Hazza. Soyuz MS-15/12.

Born in December 1983 in the Abu Dhabi suburb of Al Wathba as Hazzaa Ali Abdan Khalfan Al Mansoori, of MBRSC is an Emirati astronaut and the first person from the United Arab Emirates in space. He studied aviation at the Khalifa bin Zayed Air College, and joined the UAE Armed Forces in 2004, becoming a military pilot. His skill led him to become a top candidate for the role of a Functional Check Flight (FCF) Pilot F-16B60. Al-Mansuri traveled to Arizona, USA, for three years of training on the F-16 aircraft. He returned to the UAE, and became the youngest military pilot on the F-16, qualified for flight testing and aircraft assessment. Launched on September 25, 2019, aboard *Soyuz MS-15* to the ISS, he returned to Earth aboard *Soyuz MS-12* on October 03, 2019. The mission carried the slogan "Zayed's Ambition", making the UAE the nineteenth country, and the first in the Arab region to visit the ISS. Al-Mansuri logged seven days, 21 hours and one minute in space. He married in July 2007 and has four children.

PART THREE

# Astronauts of America, A-Z

Reading these biographies will give you insights into the lives of 358 brave American astronauts who chose to ride aboard rockets. John W. Young, who flew in space six times said, *"Anyone who sits on top of the largest hydrogen-oxygen fueled system in the world, knowing they're going to light the bottom, and doesn't get a little worried, does not fully understand the situation"*. The tertiary education of American astronauts, their experience before joining NASA, their roles on various space missions, and their lives after traveling to space are outlined. A summary of the awards and honors they received is included. It is known that in about one billion years our Sun will reach the stage in its life where it's so hot that it will cause all the water in Earth's oceans to boil and evaporate, leaving our planet uninhabitable. We can't do anything to stop the evolution of the Sun, but we still might be able to save ourselves by becoming a multiplanetary species that can live beyond this Solar System. Scientists and engineers are already investigating this possibility, and missions to land humans on Mars may occur by 2030 with the **SpaceX** company.

The Apollo missions of the late 1960s and early 1970's to the Moon awakened a sense of awe of the beauty of our home planet, and the need to safeguard its environment. Maj. Bill Anders took the iconic *Earthrise* color photograph from *Apollo 8* in December 1968, an image which became of great significance to the emerging environmental movement. *Earthrise* was later picked by ***Life*** magazine as one of its hundred photos of the century. According to Anders: "***We came all this way to explore the Moon, and the most important thing is that we discovered the Earth***". On conservation of the planet, he said, *"If you can imagine yourself in a darkened room with only one clearly visible object, a small blue-green sphere about the size of a Christmas-tree ornament, then you can begin to grasp what the Earth looks like from space. I think that all of us subconsciously think that the Earth is flat ... Let me assure you that, rather than a massive giant, it should be thought of as the fragile Christmas-tree ball which we should handle with considerable care"*. Unquote.

## Acaba, Joseph M.. STS-119, Soyuz TMA-04M/Expedition 54/55, Soyuz MS-06/Expedition 53/54.

Joseph Michael Acaba was born in May 1967 in Inglewood, California, USA. He is an American educator, hydrogeologist, and NASA astronaut. Seeing 8-mm film of the July 1969 Moon landing intrigued young Joseph about outer space. In 1990, Acaba received his Bachelor's degree in Geology from the University of California, Santa Barbara, and his Master's degree in Geology from the University of Arizona in 1992. Acaba served as a sergeant in the US Marine Corps Reserve for six years. He also worked as a hydrogeologist in Los Angeles, California. After two years in the US Peace Corps, he served as island manager of the Caribbean Marine Research in the Exumas, Bahamas. Returning to the US, he moved to Florida, and taught one year of science in high school and four years at Dunnellon Middle School. In the Fall of 2012, Acaba began coursework in the College of Education at Texas Tech University. He earned his Master of Education, curriculum and instruction from Texas Tech University in 2015.

In May 2004, Acaba became the first person of Puerto Rican heritage to be named as a NASA astronaut candidate, when he was selected as a member of NASA Astronaut Group 19, along with ten others from ninety-nine applicants. He completed his training on February 10, 2006, and was assigned to the Hardware Integration Team in the International Space Station (ISS) branch, working technical issues with European Space Agency (ESA) hardware. Acaba was assigned to Space Shuttle mission STS-119 as an Educator Mission Specialist. Launched on March 15, 2009, aboard shuttle *Discovery*, the crew delivered the final set of solar arrays to the International Space Station, and he performed two Extra-Vehicular Activities (EVA). On May 15, 2012, Acaba was launched from Kazakhstan aboard the *Soyuz TMA-04M* spacecraft to the ISS. He spent almost 125 days in space, returning to Earth on September 17, 2012. Acaba served as the Branch Chief of the ISS Operations branch, responsible for mission preparation and on-orbit support of space station crews. Until being selected as Flight Engineer for 'Expedition 54/55', Acaba served as Director of Operations Russia in Star City, supporting crew training in *Soyuz* and Russian Segment systems.

Acaba returned to Puerto Rico on June 01, 2009, and was presented with a proclamation by then Governor Luis Fortuno. He received the Ana G. Mendez University System Presidential Medal and an honorary Doctorate from the Polytechnic University of Puerto Rico, where he inaugurated a flight simulator on February 07, 2013. Joseph Acaba launched on *Soyuz MS-06* on September 12, 2017, performing a fast six-hour rendezvous with the ISS. On October 20, he and Randy Bresnik performed an EVA of six hours and 49 minutes to continue with lubrication of the new end effector on the robotic arm, and to install new cameras. He logged a total of 306 days and 34 minutes in space; including three EVAs totaling 19 hours and 46 minutes. In February 2023, Acaba became Chief of the Astronaut Office at NASA; replacing Drew Feustel who was acting chief after Reid Wiseman stepped down from the position.

## Acton, Loren W.. STS-51-F.

Dr. Loren Wilber Acton was born in March 1936 in Lewistown, Montana, USA. He is an American physicist, who flew as a Payload Specialist for the Lockheed Palo Alto Research Laboratory. Acton received a Bachelor of Science degree in Engineering Physics from Montana State University in 1959, and a Doctor of Philosophy in Astrophysics from the University of Colorado at Boulder in 1965. Acton was a senior staff scientist with the Space Sciences Laboratory, **Lockheed Palo Alto Research Laboratory**, California. His principal duties included conducting scientific studies of the Sun and other celestial objects using advanced space instruments, and serving as a co-investigator on one of the 'Spacelab 2' solar experiments.

Acton was selected as one of four Payload specialists for 'Spacelab 2' in August 1978, and after seven years of training he flew on mission STS-51-F/Spacelab-2 in 1985. He logged seven days, 22 hours and 45 minutes (over 190 hours) in space. Loren Acton is a retired Research Professor of Physics at Montana State University, and was a principal investigator for the Soft X-ray Telescope (SXT) experiment on the Japan/US/UK Yohkoh mission "YohKoh Legacy Archive". That mission focused on the study of high-energy solar processes, such as solar flares, eruptions, and the heating of the corona. His awards include the 2000 George Ellery Hale Prize, and the 2017 Order of the Rising Sun (by the Japanese government), and he was recognized in February 2020 as Legacy Fellow of the American Astronomical Society.

## Adamson, James C.. STS-28R, STS-43.

Col. James Craig Adamson was born in March 1946 in Warsaw, New York, USA. In 1969, Adamson graduated with a Bachelor of Science degree in Engineering from the US Military Academy at West Point, New York. He later completed a Master of Science degree in

Aerospace Engineering at Princeton University. Adamson served the Army for over two decades as an Army aviator, astronaut and educator. As a military test pilot he flew research aircraft at various locations. During the Vietnam War, Adamson served as a scout pilot, team lead, and air mission commander in the IV Corps area, and in Cambodia with the Air Cavalry. He earned two Distinguished Flying Crosses, 18 Air Medals, and three Vietnamese Crosses of Gallantry.

Adamson's military career included flying with peacetime flight units at Fort Bliss, Texas, at West Point, and at Houston, Texas. While at West Point, he taught courses in fluid mechanics, aerodynamics, aircraft performance, and stability and control. He logged over 3,000 hours in more than thirty types of helicopters and airplanes. James Adamson was employed at the Lyndon B. Johnson Space Center from 1981-92. Following completion of the Space Shuttle operational test flights, he became Guidance Navigation and Control Officer for shuttle missions STS-5 through STS-11. Adamson was selected as an astronaut in 1984.

He flew on Space Shuttle mission STS-28R aboard shuttle *Columbia*, a classified Department of Defense mission; and on STS-43 aboard shuttle *Atlantis*, which deployed the fifth *Tracking and Data Relay Satellite* (TDRS-E). Adamson completed Earth 263 orbits and logged a total of 13 days, 22 hours and 21 minutes (over 334 hours) in space. Following retirement from government service in June 1992, he continued as a management consultant to NASA and the aerospace industry. In September 1994, Adamson joined Lockheed Corporation as Executive Vice President, and was soon promoted to President and CEO. In late 1995, he was selected by Lockheed Martin to start up and become the first COO of the **United Space Alliance** (USA), a joint venture with Rockwell International Corp. In 1999, he was recruited by **Allied Signal Corp** (later merged with Honeywell) until his retirement in 2001. Adamson holds a Chartered Director certification from the Directors College (2010). He currently resides in Fishersville, Virginia.

## Akers, Thomas D.. STS-41, STS-49, STS-61, STS-79.

Col. Thomas "Tom" Dale Akers was born in May 1951 in St. Louis, Missouri, USA. He was the valedictorian of his 29-member 1969 senior class from Eminence, Missouri, USA. He graduated from the University of Missouri at Rolla with Bachelor of Science (1973) and Master of Science degrees in Applied Mathematics (1975). While employed as a school principal, Akers was selected for the astronaut program in 1987. As part of his training with NASA, he taught science classes at Troy State University for a short time. In each of his space flights his role was as a Mission Specialist. Akers has spent more than 800 hours in orbit, including more than 29 hours of extra-vehicular activity (spacewalking).

His first space flight was in 1990 on STS-41, when he was instrumental in deploying the European Space Agency satellite *Ulysses*, as well as tending several secondary payloads and experiments. In 1992, on STS-49 the primary goal was to capture and repair the non-functional *Intelsat VI-F3* satellite, where Akers joined the third attempt, which was successful. This was the first three-person EVA in history, and set an endurance record of eight hours and 29 minutes. In 1993 on STS-61, Akers was one of four astronauts who repaired and upgraded the Hubble Space Telescope on its first servicing mission. He spent almost 13.5 hours in two EVA. On May 18, 1994, Akers appeared on ***Home Improvement*** as himself, along with the rest of the STS-61 crew. His last mission was in 1996 on STS-79, the 17th flight of shuttle *Atlantis*. This was the fourth shuttle flight to rendezvous with the Russian space station *Mir*, and the first to exchange US astronauts with *Mir*, returning Mrs. Shannon Lucid to Earth and leaving John Blaha in orbit.

Tom Akers retired from NASA in 1997 and from the US Air Force in 1999 at the rank of Colonel. He took a position as instructor of Mathematics at the **University of Missouri-Rolla**, which had awarded him an honorary Doctorate of Engineering in 1992, and in 2008 changed its name to the Missouri University of Science and Technology. Akers retired from teaching in 2010. In Series 3, Episode 24, "Reality Bytes", Akers and the Hubble crew appeared as guests on ***Tool Time*** and showed some of the tools they used in space.

## Aldrin, Edwin Eugene "Buzz", Jr.. Gemini XII, Apollo 11.

Dr. Edwin Eugene "Buzz" Aldrin, Jr., was born in January 1930 in Glen Ridge, New Jersey, USA. He received a Bachelor of Science degree in Mechanical Engineering from the United States Military Academy at West Point, New York, in 1951. In 1963 he wrote a dissertation on orbital mechanics to earn a Doctorate in Astronautics from the Massachusetts Institute of Technology. After graduating from the US Military Academy at West Point, Aldrin became a US Air Force pilot. He flew sixty-six combat missions in the Korean War and shot down two MiG-15 jets. Later in 1963, he was chosen as an astronaut. On November 11, 1966, Aldrin made the four-day *Gemini XII* flight as pilot. He made three walks in space (EVA) for a record 5.5 hours, proving that humans can function effectively in the vacuum of space. These EVAs solved the problems of exhaustion and suit overheating encountered on previous flights.

On the second day of the July 1969 *Apollo 11* voyage, the crew aimed a color TV camera back at the Earth and for fifteen minutes made a running commentary on what they were seeing. *"Hey, Houston,"* Buzz said, *"do you suppose you could turn the Earth a bit so we could get a little bit more than just water?"* Aldrin was the Lunar Module pilot on the mission and on July 20 he became the second person to walk on the Moon, landing near the edge of Mare Tranquillitatis (Sea of Tranquillity). About nineteen minutes after Neil Armstrong's first steps on the Moon, Aldrin joined him on the grey lunar surface. Together they planted the flag of the United States. Shortly after, US President Richard Nixon spoke to them from his Oval Office. It was the world's most long-distance telephone call! They spent over two hours gathering rock samples, taking photographs and setting up scientific test equipment. The NASA management had determined that the overall probability of success for the entire mission was roughly 56 percent.

Aldrin retired from NASA in 1971 to become commandant of the Aerospace Research Pilot School at the Edwards Air Force Base in California. In March 1972, he retired from the Air Force to enter private business. In 1988, Edwin Eugene Aldrin, Jr., legally changed his name to Buzz Aldrin. He has received three US patents for his rocket systems and schematics. Since retiring from NASA, Aldrin has worked tirelessly to ensure America's continued leadership in human space exploration, especially toward Mars. In 1998, Buzz founded the **ShareSpace Foundation**, a nonprofit organization to promote the expansion of crewed space exploration.

Buzz Aldrin has written two autobiographies: *Return to Earth* (1973), which told the story of his experience with depression following the *Apollo 11* mission, and *Magnificent Desolation; The Long Journey Home from the Moon* (2009), with Ken Abraham. He also wrote a history of the Apollo program, *Men from Earth* (1989, with Malcolm McConnell), two children's books, and two forward-looking works; *Mission to Mars: My Vision for Space Exploration* (2013), and *No Dream is Too High: Life Lessons from a Man Who Walked on the Moon* (2016). The *Apollo 11* crew received the Collier Trophy (1969), the 1970 Dr. Robert H. Goddard Memorial Trophy, and the Smithsonian Institution's Langley Gold Medal for Aviation (1999). Buzz Aldrin has received nine honors, including the Air Force Distinguished Service Medal (1969) with Oak Leaf Cluster; the Legion of Merit; the NASA Exceptional Service Medal; the NASA Distinguished Service Medal (1970); and was inducted into three Halls of Fame (1982, 1993, 2000 & 2008). The ***Toy Story*** character *Buzz Lightyear* was named in his honor.

## Allen, Andrew M.. STS-46, STS-62, STS-75.

Andrew "Andy" Michael Allen was born in August 1955 in Richboro, Pennsylvania, USA. He graduated from Archbishop Wood Catholic High School in 1973. Allen received his commission in the US Marine Corps at Villanova University in 1977. A 1987 graduate of the US Naval Test Pilot School at Patuxent River, Maryland, he logged over 6,000 hours in more than thirty different aircraft types. Selected in the June 1987 NASA Astronaut Group 12, his technical assignments have included: Astronaut Office representative; Shuttle Avionics Integration Laboratory; Technical Assistant to the Flight Crew Operations Director; lead of the Astronaut Support Personnel team; Special Assistant to the Director of the Johnson Space Center; lead of a Functional Workforce Review at the Kennedy Space Center; and Director of Space Station Requirements at NASA Headquarters. Launched on STS-46 aboard shuttle *Atlantis* on July 31, 1992, Allen was the Pilot. The eight-day mission deployed the *European Retrievable Carrier* (*EURECA*), and demonstrated the *Tethered Satellite System*, a joint project between NASA and the Italian Space Agency.

Launched on STS-62 aboard *Columbia* (March 04-18, 1994, the fourteen-day mission conducted materials processing, biotechnology, and advanced technology experiments, and environmental monitoring. Launched on STS-75 aboard *Columbia* in 1996, he was Commander on a sixteen-day mission with principal payloads being the reflight of the *Tethered Satellite System* and the third flight of the United States Microgravity Payload (USMP-3). The crew worked round the clock performing combustion experiments and research related to USMP-3 investigations used to improve production of medicines, metal alloys, and semiconductors. Allen logged a total of 37 days, 16 hours and 12 minutes in space (over 900 hours). Andy Allen retired from the Marine Corps and left NASA in October 1997. Since 2006, Allen has been CEO/Owner of **Aerodyne Industries**, Cape Canaveral, Florida. In 2022, he received the Entrepreneur of the Year 2022, Florida Award by Ernst & Young. Allen has also received the Defense Superior Service Medal, the Legion of Merit, the DFC, the NASA Outstanding Leadership Medal, and the NASA Exceptional Service Medal.

## Allen, Joseph P.. STS-5, STS-51-A.

Dr. Joseph "Joe" Percival Allen was born in June 1937 in Crawfordsville, Indiana, USA. He won a Fullbright Scholarship (1959-60); received a Bachelor of Arts degree in Mathematics and Physics from DePauw University in 1959; and a Master of Science (1961) and a Doctor of Philosophy degree in Physics from Yale University in 1965. Allen has the Outstanding Flying Award, Class 69-06, Vance Air Force Base (1969). He logged over 3,000 flying time in jet aircraft. Allen was a research associate in the Nuclear Physics Laboratory at the University of Washington prior to his selection as a scientist-astronaut by NASA in August 1967. From August 1975 to 1978, Allen served as NASA Assistant Administrator for Legislative Affairs in Washington, D.C.. He won the NASA Exceptional Scientific Achievement Medal (1973) and the NASA Exceptional Service Medal (1978). Returning to the Johnson Space Center in 1978, he was assigned to the Operations Mission Development Group. He served as a support crew member for the first orbital flight of Space Shuttle *Columbia* in April 1981, and was the CapCom during the re-entry phase for that mission.

Launched on November 11, 1982 aboard shuttle *Columbia* on STS-5, Allen was a Mission Specialist on the five-day flight, and deployed two commercial communications satellites. Launched on November 08, 1984 aboard shuttle *Discovery* on STS-51-A, the crew deployed Canada's *Anik-D2* (aka *Telsat H*) and Hughes' *LEASAT-1* (aka *Syncom IV-1*) satellites, and operated the **3M Company**'s Diffusive Mixing of Organic Solutions experiment. In the first-ever space salvage attempt, the crew performed EVA and retrieved the *Palapa B-2* and *Westar VI* communications satellites. Joe Allen logged a total of 13 days, one hour, and

58 minutes in space (314 hours). Allen served as CEO of Space Industries International, Washington, D.C., and later was chairman of Veridian, until he retired in 2004. He wrote *Entering Space: An Astronaut's Odyssey* (1985). In the 1998 HBO miniseries ***From the Earth to the Moon***, Allen was played by Doug McKeon. In the 1998 movie ***Armageddon***, Allen served as a NASA consultant and cameoed as "Kennedy Launch" during the launching of the X-71 shuttles. Joe Allen was one of three Shuttle astronauts inducted into the US Astronaut Hall of Fame in 2005.

## Altman, Scott D.. STS-90, STS-106, STS-109, STS-125.

Capt. Scott Douglas "Scooter" Altman was born in August 1959 in Lincoln, Illinois, USA. He is a retired US Navy Captain and naval aviator, engineer, test pilot and former NASA astronaut. In 1981, he received a Bachelor of Science degree in Aeronautical and Astronautical Engineering from the University of Illinois. Commissioned as an ensign in the US Navy in August 1981, he received his Naval Aviator wings in February 1983. He was Distinguished Graduate with Test Pilot School Class 97 in June 1990. He was awarded the Air Medal for his role as a strike leader flying over Southern Iraq in support of Operation Southern Watch, then was selected for the NASA astronaut program in December 1994. Altman has logged over 7,000 flight hours in more than forty types of aircraft. He performed many of the stunts in the 1986 film ***Top Gun***.

In 1990, he received a Master of Science degree in Aeronautical Engineering from the US Naval Postgraduate School. In 1996, Altman was initially assigned to work technical aspects of orbiter landing and rollout issues for the Astronaut Office Vehicle Systems Branch. He was the Pilot on STS-90 Neurolab aboard shuttle *Columbia* in 1998, and STS-106 aboard shuttle *Atlantis* in 2000. Altman was the mission Commander on STS-109 aboard *Columbia* in 2002, and on STS-125 aboard *Atlantis* in 2009. This was the fifth and final servicing mission to the Hubble Space Telescope. The Cosmic Origins Spectrograph and the Wide Field Camera 3 were added to Hubble. The mission also replaced a Fine Guidance Sensor, six gyroscopes, and two battery unit modules, enabling Hubble to function to at least 2014.

In total, Scott Altman logged a total of 51 days, 12 hours and 47 minutes in space, and retired from NASA in September 2010 to join ASRC Federal Research and Technology Solutions in Greenbelt, Maryland. Altman was awarded the Defense Superior Service Medal; the Legion of Merit, the DFC; the NASA Distinguished Service Medal, and five other Awards. In 2010, he was honored by the naming of Scott Altman Primary School, in Pekin District, Illinois. In 2018, he was inducted into the US Astronaut Hall of Fame. On November 06, 2021, he was awarded Illinois' highest honor, the Order of Lincoln. As of November 2022, Altman is the president of the Space Operating Group for **ASRC Federal**.

## Anders, William A.. Apollo 8.

Maj. William Alison "Bill" Anders was born in October 1933, in Hong Kong. He earned a Bachelor of Science degree in Electrical Engineering from the US Naval Academy, Annapolis, Maryland, in 1955. He was commissioned a second lieutenant in the United States Air Force, and received his pilots' wings in 1956. Anders became a fighter pilot with the 84th Fighter-Interceptor Squadron, based at Hamilton Air Force Base in California. He then served with the 57th Fighter-Interceptor Squadron in Iceland, where he participated in intercepts of Soviet heavy bombers who at the time were challenging America's air defense borders. Aspiring to become a test pilot, the Air Force Institute of Technology had him instead study nuclear engineering, and he graduated with a Master of Science degree in Nuclear Engineering from the Air Force Institute, Wright-Patterson Air Force Base, Ohio, in 1962. Anders was then assigned to the Air Force Weapons Laboratory at Kirtland AFB in New

Mexico, where he was responsible for the technical management of the nuclear power reactor programs. Anders was not selected by the Aerospace Research Pilots School for training.

On June 05, 1963, NASA announced that it would be recruiting ten to fifteen new astronauts for Project Gemini and Project Apollo. Anders was chosen as one of thirty-four applicants to be interviewed, and, on October 17, he was accepted as a member of the third group of NASA astronauts. At NASA he became involve with dosimetry, radiation effects, and environmental controls. In September 1966, he was the backup pilot for the *Gemini XI* mission. Neil Armstrong and Anders became the first astronauts to fly the Lunar Landing Training Vehicle. The astronauts in his group without test pilot training - Anders, Roger Chaffee, Walter Cunningham, Rusty Schweickart, Gene Cernan and Buzz Aldrin - were assigned to Apollo crews as the Lunar Module (LM) pilot, the lowest-ranking crew member.

On December 22, 1966, Anders was assigned to the third Apollo mission, which was to be commanded by Frank Borman, with Command Module (CSM) pilot Michael Collins (replaced by Jim Lovell in July 1968). Their mission, scheduled for December 1968, was intended to be a second test of the LM in medium Earth orbit, but delivery fell behind schedule. In August 1968, there were reports, including one from the CIA, that the Soviet Union was planning a crewed lunar mission before the end of the year. Since the Apollo CSM would be ready, it was decided to send it to the Moon, and enter lunar orbit before returning to Earth. When Jim McDivitt turned down a mission without the LM, Deke Slatton decided to swap the crews and spacecraft, so Borman, Lovell and Anders' mission became *Apollo 8*. According to Borman, "*Rookie Bill Anders was thirty-five, slightly built, a devout Roman Catholic, and very serious minded. ... Anders was one hell of a worker, a superb technician and all in all a great guy. Some of the guys regarded him as a younger version of Frank Borman in his single-minded concentration on work, his aversion to unnecessary conversation*". Unquote.

In December 1968, he was a member of the crew of *Apollo 8*, the first three people to leave low Earth orbit. Along with fellow astronauts Frank Borman and Jim Lovell, Anders circled the Moon ten times, and broadcast live images and commentary back to Earth. Bill Anders took the iconic *Earthrise* color photograph, which became of great significance to the emerging environmental movement. *Earthrise* was later picked by ***Life*** magazine as one of its hundred Photos of the Century. Anders stated that the *Earthrise* photograph *"really undercut my religious beliefs. The idea that things rotate around the Pope, and up there is a big supercomputer wondering whether Billy was a good boy yesterday? It doesn't make any sense. I became a big buddy of Richard Dawkins"*. The *Apollo 8* Command Module splashed down in the Pacific Ocean on December 27, 1968, after a flight lasting 147 hours and 42 seconds (nearly six days). Their voyage was a record-setting 933,419 kilometres (504,006 miles). The crew landed just 3.7 km (two nautical miles) from their recovery ship, the aircraft carrier *USS Yorktown*. Due to time dilation, Anders had aged about 300 nanoseconds more than people back on Earth. On May 16, 1969, President Richard M. Nixon nominated Bill Anders to become the executive secretary of the **National Aeronautics and Space Council** (NASC). This was the highest government post ever offered to an astronaut up to that time.

In his new role, Anders worked closely with the Office of Science and Technology and the Office of Management and Budget (OMB) and became a personal advisor to the OMB director. He became increasingly pessimistic about the future of the NASC and the space program generally. Anders opposed the development of the Space Shuttle, urging instead that NASA concentrate on developing the *Skylab* space station. Frustrated with the NASC's lack of influence, he recommended in 1972 that it be abolished. This was done on June 30, 1973. On August 06, 1973, Nixon appointed Bill Anders to the five-member **Atomic Energy**

**Commission** (AEC). He also served as the US chairman of the joint US-Soviet Union nuclear fission and fusion power technology exchange program. President Gerald Ford appointed Anders as the first chairman of the **Nuclear Regulatory Commission** (NRC). On April 13, 1976, Anders was appointed Ambassador to Norway, a position he held until June 18, 1977.

In September 1977, he joined **General Electric** as its vice president, and general manager of its Nuclear Products Division. In 1984, Anders left GE to join **Textron** as its executive vice president for aerospace. In 1990, he became vice chairman of **General Dynamics**, and on January 01, 1991, its chairman and CEO. In retirement, Bill Anders bought homes in the San Juan Islands and at Point Loma, California. He established the **William A. Anders Foundation**, a philanthropic organization for educational and environmental issues. In 1996, he founded the **Heritage Flight Museum** in Bellingham, Washington. Anders has received eleven awards and honors. In October, 2018, the International Astronomical Union named a crater seen in the Ander's photo as "Anders' Earthrise".

## Anderson, Clayton C.. STS-117, STS-131.

Clayton Conrad Anderson was born in February 1959 in Omaha, Nebraska, USA. He received a Bachelor of Science degree (cum laude) in Physics at Hastings College, Nebraska in 1981; and a Master of Science degree in Aerospace Engineering at Iowa State University in 1983. Anderson was selected in the June 1998 NASA Astronaut Group. Prior to being assigned to a spaceflight, Anderson served as the lead for the Enhanced Caution and Warning (ECW) system development effort within the Space Shuttle Cockpit Avionics Upgrade (CAU) Project. In 2001, he passed the amateur radio operator Technician Class license exam, and was issued the callsign KD5PLA. In November 2002, he completed training in the EVA skills program. He was backup Flight Engineer for 'Expeditions 12', 13 and 14 to the ISS.

Launched on June 08, 2007 on STS-117 aboard shuttle *Atlantis*, Anderson spent 152 days on the ISS. He remained on board as a member of 'Expedition 16', and returned to Earth aboard *Discovery* on November 07, 2007. Launched on April 05, 2010, on STS-131 aboard *Discovery*, the primary mission payload was a Multi-Purpose Logistics Module (MPLM) loaded with supplies and equipment for the ISS. He logged a total of 166 days, 21 hours and 10 minutes in space. Anderson retired from NASA in January 2013. In May 2022, he became the president and CEO of the **Strategic Air Command & Aerospace Museum**, in his home town of Ashland. He is currently an author, a motivational speaker, and a Professor of Practice at Iowa State University in Ames, Iowa.

## Anderson, Michael P.. STS-89, STS-107.

Michael Phillip Anderson was born in December 1959 in Plattsburgh, New York, USA; an only child. He was a USAF officer and NASA astronaut who served as the Payload Commander and lieutenant colonel in charge of science experiments on the Space Shuttle *Columbia*. He said, "*Like most kids growing up, I had a very wide interest. I tried to take advantage of everything, from the sciences to music to writing to literature*". Unquote. Anderson graduated from Cheney High School in Cheney, Washington, one of four African Americans in a class of 200 students.

In 1981, Michael Anderson earned a Bachelor of Science degree in Physics and Astronomy at the University of Washington, and in 1990 a Master's degree in Physics from Creighton University in Omaha, Nebraska. Anderson joined the USAF in 1982, and served as a pilot and flight instructor from January 1991 to September 1992, before being selected as a NASA astronaut in December 1994. Anderson was one of nineteen candidates selected from a pool of 2,962 applicants. He said, "*Fortunately, I got called down to NASA for an interview.*

*and one thing led to the next, and one day I got that call. I've been here about seven years now and am really enjoying it". "When you launch on a rocket, you're not really flying that rocket. You're just sort of hanging on". "As you look back at your life, there are just a million different things that have happened, just in the right way, to allow you to make your dreams come true, and you know, someone has all that under control".* Unquote. He logged more than 3,000 hours of flight time up to late 1994.

Anderson launched as a Mission Specialist on January 22, 1998, on STS-89 aboard shuttle *Endeavour*, which involved the exchange of crew members with space station *Mir*. Anderson launched on January 16, 2003 on mission STS-107 aboard shuttle *Columbia*, which ended in disaster when the shuttle disintegrated upon re-entry into Earth's atmosphere over Texas on February 01, killing all seven crew. He was aged forty-three years. In acknowledgement of his service as well as sacrifice, Anderson was posthumously awarded the Congressional Space Medal of Honor, the NASA Space Flight Medal, and the Distinguished Service Medal. A school in his hometown of Spokane, Washington was named as a memorial to him, along with a road at the Kennedy Space Center in Florida.

## Antonelli, Dominic A.. STS-119, STS-132.

Dominic Anthony "Tony" Antonelli was born in August 1967 in Detroit, Michigan, USA. He attended the Massachusetts Institute of Technology, where he earned a Bachelor of Science degree in Areonautics and Astronautics. Antonelli earned a Master of Science degree in Aeronautics and Astronautics from the University of Washington. He served as a fleet Naval Aviator and Landing Signal Officer aboard the aircraft carrier *USS Nimitz*, flying F/A-18C Hornets in support of Operation Southern Watch. He accumulated over 3,200 flight hours in forty-one different aircraft types, and completed 273 aircraft carrier arrested landings.

Antonelli was selected as an astronaut candidate by NASA in July 2000. Launched on March 15, 2009, he served as Pilot on mission STS-119 aboard *Discovery*, which delivered the final pair of solar arrays to the ISS. Launched on May 14, 2010, Antonelli was Pilot on STS-132 aboard *Atlantis*, which delivered the Russian Mini-Research Module 1 (MRM-1) to the ISS. Tony Antonelli has received eight awards, including the Navy Meritorious Service Medal; two Unit Battle Efficiency awards; CVW-9 Landing Signal Officer of the Year; and the NASA Superior Accomplishment Award.

## Apt, Jerome "Jay", III.. STS-37; STS-47 (J), STS-59, STS-79.

Dr. Jerome "Jay" Apt, III was born in April 1949 in Springfield, Massachusetts, USA. As a physicist, he worked on the *Pioneer* Venus 1978 space probe project, and used visible light and infrared techniques to study the planets and moons of the Solar System from ground-based observatories. Jay Apt graduated from Shady Side Academy in Pittsburgh, Pennsylvania in 1967. He went on to attend Harvard University, earning a Bachelor of Arts in Physics in 1971. Apt then attended the Massachusetts Institute of Technology and earned a Doctor of Philosophy in Physics in 1976. From 1976-80, he was a staff member of the Center for Earth and Planetary Physics at Harvard, and served as the Assistant Director of Harvard's Division of Applied Sciences from 1978-80. In 1980, he joined the Earth and Space Sciences Division of the NASA Jet Propulsion Laboratory (JPL) as a scientist doing planetary research. From 1982-85 he was a flight controller responsible for Shuttle payload operations at NASA's Johnson Space Center. In 1985, Apt was selected as an astronaut candidate, and qualified after a year of training. He has over 6,000 hours piloting aircraft, has flown on four Space Shuttle missions and logged over 847 hours in space.

Apt flew on mission STS-37 aboard shuttle *Atlantis* (April 05-11, 1991). He made two EVAs with Jerry Ross, manually deploying the Compton Gamma Ray Observatory's radio antenna. On his second EVA, the palm-bar in Apt's right glove punctured the suit, but he was unaware of the problem and sustained only a minor scar. In 1992, Apt flew on STS-47 aboard shuttle *Endeavour* as the Flight Engineer and commander of one of the two shifts in this round-the-clock mission. In April 1994 on STS-59 aboard shuttle *Endeavour*, he was a shift commander of the first Space Radar Laboratory mission. In September 1996, Apt flew on STS-79 aboard shuttle *Atlantis*, and visited the Russian *Mir* space station.

In 2003, Apt joined the faculty of **Carnegie Mellon University** where he is a Full Professor (emeritus). His research and teaching interests are in economics, engineering, and public policy aspects of the electricity industry, and engineering systems design. From 2000-22, he co-directed the Carnegie Mellon Electricity Industry Center. Jerome Apt is the author of the book, *Orbit: NASA Astronauts Photograph the Earth*, published by the National Geographic Society. His book, *Variable Renewable Energy and the Electricity Grid* was published in 2014. Apt received the NASA Distinguished Service Medal in 1997 and the Metcalf Lifetime Achievement Award (for significant contributions to engineering) in 2002. His paper with Ph.D. student Adam Newcomer, "*Near term implications of a ban on new coal-fired power plants in the US*" was cited as one of the top environmental policy papers of 2009 by the American Chemical Society. Jay Apt III is a Fellow of the American Association for the Advancement of Science.

## Archambault, Lee J.. STS-117, STS-119.

Col. Lee Joseph "Bru" Archambault was born in August 1960 in Oak Park, Illinois, USA. He earned a Bachelor of Science degree (1982) and Master of Science degree (with honors) in Aeronautical and Astronautical Engineering from the University of Illinois at Urbana-Champaign in 1984. In January 1985, he received a commission of Second Lieutenant in the USAF, and earned his pilot wings in April 1986. From November 1990 to April 1991, he deployed to Saudi Arabia in support of Operation Desert Shield/Desert Storm and flew twenty-two combat missions during the Gulf War. In August 1992, he was reassigned to Holloman AFB, New Mexico, where he served as an F-117A instructor pilot and operational test pilot for the 57th Wing. From July 1994 to June 1995, he attended the USAF Test Pilot School at Edwards AFB, California.

He has logged over 4,250 flight hours in more than thirty aircraft types. In June 1998, Archambault was selected by NASA as an astronaut candidate. In June 1999, he was assigned to the Astronaut Office Shuttle Operations Branch, where he worked on flight instrument upgrades that were incorporated into the Shuttle in 2003. He was lead Astronaut Support Person for missions STS-111 and STS-114. Lee Archambault was CapCom during daily orbit shifts for STS-121. Launched on June 08, 2007 on STS-117 aboard *Atlantis*, he was the Pilot on the fourteen-day mission, which delivered the S3/S4 truss segment to the ISS. Launched on March 15, 2009, Archambault was Commander for STS-119 aboard *Discovery*, a 13-day mission delivering the final set of solar arrays and batteries to the ISS.

On March 19, 2013, NASA announced that Archambault was leaving after fifteen years' service to join **Sierra Nevada Corporation**. He worked on the *Dream Chaser* lifting body spaceplane, being developed under NASA's Commercial Crew Development program as a potential vehicle for future crewed, and uncrewed cargo missions to the ISS. In early March 2019, Lee Archambault obtained his type rating in the Bombardier Dash 8, and currently flies the Dash 8 part-time. He has received many awards, including the DFC; the Defense Meritorious Service Medal; the Air Medal (2nd Oak Leaf Cluster); the Aerial Achievement Medal (4th Oak Leaf Cluster), the Southwest Asia Service Medal, and the Kuwait Liberation Medal.

## Armstrong, Neil A.. Gemini VIII, Apollo 11.

Lt. Neil Alden Armstrong (1930-2012) was born in August 1930 near Wapakoneta, Ohio, USA. He was a naval aviator, test pilot, aeronautical engineer and the first human being to walk on the Moon. Neil developed a passion for aviation at age six years, when he took his first airplane ride. Over his career he flew more than 200 different models of aircraft. His first flight in a rocket-powered aircraft was on August 15, 1957, in the Bell X-1B airplane, to an altitude of 18.3 km (11.4 miles). Neil became an employee of the **National Aeronautics and Space Administration** (NASA) when it was established on October 01, 1958. Fellow astronaut Michael Collins wrote that Armstrong *"was the very best when it came to understanding the [X-15's] machine's design and how it operated"*. X-15 pilot Bill Dana said Armstrong *"had a mind that absorbed things like a sponge"*. In June 1958, Armstrong was selected for the USAF Man in Space Soonest program. On November 05, 1958, it was superseded by Project Mercury, a civilian project run by NASA. In April 1962, NASA sought applications for the second group of astronauts for Project Gemini. On September 13, Neil Armstrong was asked whether he would be interested in joining the NASA Astronaut Corps as part of what the press dubbed "the New Nine".

Aboard *Gemini VIII* on March 16, 1966, Armstrong became the first American civilian in space. To give astronauts practice in piloting a lunar module on its descent, NASA commissioned Bell Aircraft to build two Lunar Landing Research Vehicles (LLRV). Nicknamed the Flying Bedsteads, they simulated the Moon's one-sixth gravity using a turbofan engine to support 5/6 of the craft's weight. On May 06, 1968, while practicing thirty metres above the ground, Armstrong's controls started to degrade and the LLRV began rolling. He ejected mere seconds before the vehicle struck the ground and disintegrated. If he had ejected a half second later, his parachute would not have opened in time.

A meeting in March 1969 determined that Armstrong, as Commander of the *Apollo 11* mission would be the first person on the Moon, in part because NASA management saw him as a person who did not have a large ego. During the launch, his heartrate peaked at a modest 110 beats per minute. Running short on fuel and with no more than twenty seconds of propellant remaining, Armstrong landed the lunar module at 20:17:40 UTC on July 20, 1969. He confirmed the event with the words, *"Houston, Tranquility Base here. The **Eagle** has landed"*. He set his left boot on the lunar surface at 02:56 UTC on July 21, 1969, then said, ***"That's one small step for [a] man, one giant leap for mankind"***. As he spoke, *Voice of America* was re-broadcast live by the BBC and many other stations worldwide. An estimated 530 million people viewed the event on television, which was 20% of the then-world population of approximately 3.6 billion. On the Moon, Armstrong performed most of the photography with a single Hasselblad camera.

After arriving back on Earth, an official state dinner was held in Los Angeles on August 13, attended by members of Congress, forty-four governors, the Chief Justice of the United States, and ambassadors from eighty-three nations. President Nixon and Vice President Agnew presented the *Apollo 11* astronauts each with a Presidential Medal of Freedom. Armstrong logged a total of eight days, 14 hours and 12 minutes in space; with one EVA on the Moon of two hours and 31 minutes. Shortly after the successful mission, Armstrong stated that he did not plan to fly in space again and he resigned from NASA in 1971. Neil accepted a teaching position in the Department of Aerospace Engineering at the University of Cincinnati. He then completed his master's degree, by reporting on various aspects of Apollo. Armstrong was part of the investigation into the failed *Apollo 13* mission. Neil Armstrong was elected a member to the National Academy of Engineering in 1978, and in 2001 he was elected to the American Philosophical Society.

In 1986, President Reagan asked Armstrong to join the **Rogers Commission** investigating the Space Shuttle *Challenger* disaster. Made Vice Chairman, he helped limit the committee's recommendations to nine, believing that if there were too many, NASA would not act on them. He was appointed to a fourteen-member commission by President Reagan to develop a plan for American civilian spaceflight in the 21st century. After he retired from NASA in 1971, Armstrong acted as a spokesperson for several businesses, and until his retirement in 2002, he served on the boards of directors of several companies.

Armstrong's family described him as a *"reluctant American hero"*. He kept a low profile later in his life, turning down most requests for interviews and public appearances. He was approached by major political parties but declined the offers. Neil Armstrong flew gliders for pleasure, until well into his 70s. Armstrong and his first wife, Janet, separated in 1990 and divorced in 1994 after thirty-eight years of marriage. He met his second wife, Carol, in 1992, and they were married in June 1994. Neil underwent heart bypass surgery in August 2012, but developed complications and died eighteen days later, on August 25, 2012, aged eighty-two years. A financial settlement was made for Armstrong's family. The lunar crater Armstrong, thirty-one miles from the *Apollo 11* landing site, and asteroid 6469 Armstrong are named in his honor. Neil Armstrong's authorized biography, *First Man: The Life of Neil A. Armstrong* was published in 2005.

## Arnold, Richard R., II.. STS-119, Soyuz MS-08/Expedition 55/56.

Richard "Ricky" Robert Arnold, II was born in November 1963 in Cheverly, Maryland, USA. Arnold began working at the US Naval Academy in 1987 as an Oceanographic Technician, then became a science teacher at John Hanson Middle School in Waldorf, Maryland. He completed a Masters program while conducting research in biostratigraphy at the Horn Point Environmental Laboratory in Cambridge, Maryland. In 1993, Arnold joined the faculty at the Casablanca American School in Casablanca, Morocco, teaching Biology and Marine Environmental Science. In 2001, he was hired to teach at the International School of Kuala Kencana in West Papua, Indonesia. NASA selected him as a Mission Specialist-Educator in May 2004 Group 19. After training, he was assigned to the Hardware Integration Team in the Space Station Branch, working on technical issues with JAXA software.

In August 2007, Arnold served as an aquanaut during the NEEMO 13 project, held in *Aquarius*, the world's only undersea research laboratory, located offshore at Key Largo, Florida. Launched on March 15, 2009, mission STS-119 delivered the final set of solar arrays to the ISS. In an EVA on March 19, astronauts Steven Swanson and Arnold plugged in power and data cables to connect new hardware. Their first EVA of six hours and seven minutes installed the ISS's new truss segment and prepared the solar arrays for unfolding. A second EVA with Joseph Acaba on March 23 moved a Crew and Equipment and Translation Aid cart, which also got a new coupler. This EVA also lubricated the end effector capture snares on the ISS's robotic arm. In 2016, Ricky Arnold participated in ESA CAVES training in Sardinia, Italy, spending six nights underground simulating a mission exploring another planet. In March 2017, Arnold was assigned to ISS 'Expeditions 55/56'. On March 21, 2018, he launched aboard *Soyuz MS-08* to the ISS. In a six-hour and ten-minute EVA with Drew Feustel, he installed wireless communications equipment on the station's *Tranquility* module, and swapped out high-definition video cameras on the port truss of the station's backbone. Arnold returned to Earth on October 04, 2018. He logged a total of 209 days, 13 hours and 29 minutes in space; including five EVAs totaling 32 hours and four minutes.

## Ashby, Jeffrey S.. STS-93, STS-100, STS-112.

Capt. Jeffrey Shears Ashby was born in June 1954 in Dallas, Texas, USA. He attended the University of Idaho as a member of the school's joint Navy Reserve Officer Training Corps (NROTC) program, earning a Bachelor of Science degree in Mechanical Engineering in 1976. He earned a Master of Science degree in Aviation Systems from the University of Tennessee in 1993. Ashby is a 1986 graduate of the Navy Fighter Weapons School, also known as TOPGUN, and the US Naval Test Pilot School. He flew the F/A-18 in combat missions as part of Operation Desert Storm and Operation Southern Watch during and after the Gulf War. He was the Navy Attack Aviator of the Year in 1991. During six aircraft carrier deployments in his Navy career, Ashby logged over 7,000 flight hours and 1,000 carrier landings.

He was selected as an astronaut candidate in December 1994 at age forty years. In July 1999, he piloted mission STS-93 aboard shuttle *Columbia* (July 23-28, 1999) and mission STS-100 aboard shuttle *Endeavour* (April 19-May 01, 2001). Ashby commanded STS-112 aboard shuttle *Atlantis* (October 07-17, 2002). He traveled over 11 million miles in space, flew 436 orbits around the Earth, and logged over 27.5 days (660 hours) in space. Jeffrey Ashby received twelve awards, including the Defense Superior Service Medal, the Legion of Merit, the DFC, four Navy Air Medals, two Navy Commendation Medals, and three NASA Space Flight Medals. On July 15, 2019, Ashby sustained severe injuries after falling 500 feet near the summit of La Plata Peak mountain. Hired by Jeff Bezos's private spaceflight company, he currently works for **Blue Origin** as Chief of Mission Assurance.

## Bagian, James P.. STS-29R, STS-40.

Col. Dr. James Bagian, MD, PE was born in February 1952 in Philadelphia, Pennsylvania, USA. Of Armenian descent, he received a Bachelor of Science degree in Mechanical Engineering from Drexel University in 1973, graduating first in his class; and received a Doctor of Medicine degree from Thomas Jefferson University in 1977. Bagian worked as a process engineer for the **3M Company** in Bristol, Pennsylvania, and later, from 1976-78 while studying for his medical degree, as a mechanical engineer at the US Naval Air Test Center at Patuxent River, Maryland. In 1978, he went to work as a flight surgeon and research medical officer at the Johnson Space Center.

James Bagian became a NASA astronaut in July 1980. In 1986, he began serving as an investigator for the STS-51-L accident board, following the Space Shuttle *Challenger* liftoff tragedy. He was responsible for the development program and implementation of the pressure suit used for crew escape and various other crew survival equipment to be used on future shuttle missions. He received his Professional Engineers Certification in 1986, and was board-certified in aerospace medicine by the American College of Preventative Medicine in 1987. In 2003, Bagian served as the Chief Flight Surgeon and Medical Consultant for the shuttle **Columbia Accident Investigation Board**, and currently serves as a member of the NASA Aerospace Safety Advisory Panel. Launched on March 13, 1989, aboard *Discovery*, the five-day mission STS-29 deployed a *Tracking and Data Relay Satellite*. The crew took over 3,000 photographs of the Earth using several camera types, including the IMAX 70-mm movie camera. Launched on June 05, 1991, Bagian was the Lead Mission Specialist aboard shuttle *Columbia*. The nine-day mission STS-40 had crew members performing human health experiments in microgravity, and investigating materials science, plant biology and cosmic radiation. James Bagian logged 337 hours in space.

After leaving NASA in 1995, Dr. Bagian was selected as a member of the **Institute of Medicine**. In 2000, he was elected as member into the **National Academy of Engineering** for the integration of engineering and medical knowledge in applications to aerospace systems, environmental technology, and patient safety. Dr. James Bagian was the founding director of the **VA National Center for Patient Safety**. He has developed numerous patient safety-related tools and programs that have been adopted nationally and internationally. Dr. Bagian is currently the Director of the Center for Healthcare Engineering and Patient Safety at the University of Michigan, and a professor at the University of Michigan with appointments in the Department of Anesthesia and the College of Engineering.

## Baker, Ellen L.. STS-34, STS-50, STS-71.

Dr. Ellen Louise Shulman Baker was born in April 1953 in Fayetteville, North Carolina, USA. She earned a Bachelor of Arts degree in Geology at the University of Buffalo in 1974; a Doctorate in Medicine at Cornell University in 1978; and a Master's degree in Public Health from the University of Texas School of Public Health in 1994. In 1981, Baker joined NASA as a medical officer at the Johnson Space Center. Also in 1981, she graduated from the Air Force Aerospace Medicine Course at Brooks AFB, San Antonio, Texas. Baker was selected in the NASA May 1984 Astronaut Group. She then had a variety of jobs at NASA in support of the Space Shuttle program and International Space Station development.

Baker was a MIssion Specialist aboard *Atlantis* (October 18-23 1989) on STS-34, which deployed the *Galileo* probe to explore Jupiter. Aboard *Columbia* on mission STS-50 (June 25 to July 09 1992), the crew conducted experiments involving crystal growth, fluid physics, fluid dynamics, biological science and combustion science; and flew aboard *Atlantis* on STS-71 (June 27 to July 07 1995), the first shuttle mission to dock with the *Mir* space station. Dr. Ellen Baker logged a total of 28 days, 14 hours and 31 minutes (over 686 hours) in space, then served as Chief of the Astronaut Office Education/Medical Branch.

## Baker, Michael A.. STS-43, STS-52, STS-68, STS-81.

Capt. Michael Allen Baker was born in October 1953 in Memphis, Tennessee, USA. He graduated from Lemoore Union High School, Leemoore, California, in 1971, and received a Bachelor of Science degree in Aerospace Engineering from The University of Texas in 1975. He earned his Wings of Gold as a Naval Aviator in 1977. In 1978, he was assigned to Attack Squadron 56, homeported in Yokosuka, Japan, where he flew the A-7 Corsair II. He attended the US Naval Test Pilot School in 1981 and, after graduation, was assigned to the Carrier Suitability Branch of the Strike Aircraft Test Directorate. In 1983, he returned to the Test Pilot School as an instructor. Baker logged over 5,400 flight hours in about fifty types of aircraft, and completed over 300 aircraft carrier landings.

Selected by NASA in June 1985, he became an astronaut in July 1986 after a one-year training and evaluation program. After working in Shuttle Landing and Deceleration Systems, and the Shuttle Avionics Integration Laboratory, he served as an ascent, entry and orbit spacecraft communicator (CapCom) for eleven shuttle missions from STS-27 to STS-35. From March to October 1995, he served as the Director of Operations for NASA at the Gagarin Cosmonaut Training Center in Star City, Russia, for the joint US/Russian *Shuttle-Mir* program. From October 1997 to August 2001, he was the assistant director of Johnson Space Center for Human Space Flight Programs, Russia. Michael Baker served as pilot on mission STS-43 aboard *Atlantis* (August 02-11, 1991); and STS-52 aboard *Columbia* (October 22 to November 01, 1992), which deployed the Italian *Laser Geodynamic Satellite* (*LAGEOS*), and operated the US Microgravity Payload 1 (USMP-1). He was Commander on mission STS-68 aboard *Endeavour* (September 30 to October 11, 1994), the second flight of the Space Radar Laboratory (SRL) and MAPS (Measurement of Air Pollution from Satellites); and STS-81 aboard *Atlantis* (January 12-22, 1997). In five days of docking with *Mir*, more than three tons of food, water, experiment equipment and samples were moved between the two craft, along with an exchange of crew.

He was the International Space Station Program Manager for International and Crew Operations. In January 2017, Baker left NASA to work in private industry. He has received sixteen awards, including the Defense Superior Service Medal, the Legion of Merit, the DFC, two Defense Meritorious Service Medals, three Navy Expeditionary Medals, the NASA Distinguished Service Medal, four NASA Spaceflight Medals, and was named 1993 Outstanding University of Texas Alumnus.

## Barratt, Michael. Soyuz TMA-14/Expeditions 19/20, STS-133, SpaceXCrew 8.

Dr. Michael Reed Barratt was born in April 1959 in Vancouver, Washington, USA. He graduated with a Bachelor of Science degree in Zoology from the University of Washington in 1981; and an M.D. from Northwestern University in 1988. In 1991, Barratt completed both a residency and a Master of Science degree in Aerospace Medicine jointly run by Wright State University, NASA, and Wright-Patterson AFB. He is Board Certified in Internal and Aerospace Medicine. From July 1995 to July 1998, Barratt served as Medical Operations Lead for the ISS. He worked with counterparts at Star City, Russia, and the Institute of Biomedical Problems; developing medical procedures, training and equipment for use on the ISS. During the eleven-day NEEMO 7 underwater mission of October 2004, the crew tested technologies and procedures for remote surgery, as well as using virtual reality for telemedicine.

Selected by NASA in 2000, and specializing in aerospace medicine, he served as a Flight Surgeon for NASA, and played a role in developing NASA's space medicine programs for both the *Shuttle-Mir* Program (to which he was assigned in January 1994) and the International Space Station. He launched to the ISS in March 2009 aboard *Soyuz TMA-14*, until 'Expedition 20' ended on October 11, 2009. Dr. Barratt spent a total of 211 days, 11 hours and 46 minutes in space aboard the ISS; including two EVAs totaling five hours and six minutes, contributing to station construction. During his extended stay, he participated in experiments, tested technologies, and provided valuable data for future space endeavours. As a Mission Specialist, Dr. Barratt flew aboard shuttle *Discovery* during the STS-133 mission (February 24-March 09, 2011), which delivered the Permanent Multipurpose Module *Leonardo* to the ISS. Barratt was in charge of the robotics activities on the ISS. This mission involved transitioning the ISS crew complement from three to six members. In Spring 2024, Dr. Barratt is slated to pilot the SpaceX 'Crew-8' mission.

## Barry, Daniel T.. STS-72, STS-96/2A.1, STS-105.

Dr. Daniel Thomas Barry was born in December 1953 in Norwalk, western Conneticut, USA. In 1975, he earned a Bachelor of Science degree in Electrical Engineering from Cornell University, in Ithica, New York. He graduated in 1980 with a Ph.D. in Electrical Engineering and Computer Science from Princeton University; and in 1982 he gained a Doctor of Medicine degree from Leonard M. Miller School of Medicine at the University of Miami in Coral Gables, Florida. He has five Patents, over fifty articles in scientific journals, and has served on two scientific journal editorial boards.

Selected by NASA in March 1992, Barry completed one year of training and qualified as a Mission Specialist for shuttle flight crews. The nine-day mission STS-72 aboard *Endeavour* (January 11-20, 1996) was to retrieve the *Space Flyer Unit* (launched from Japan ten months earlier), and deploy and retrieve the *OAST-Flyer.* Barry performed a six-hour and nine minute EVA. Mission STS-96 (May 27-June 06, 1999) aboard *Discovery* was the first mission to dock with the ISS, and delivered four tons of logistics and supplies to prepare for arrival of the first resident crew. Barry perfomed an EVA of seven hours and 55 minutes. Mission STS-105 (August 10-22, 2001) aboard *Discovery* attached the *Leonardo* Multi-Purpose Logistics Module, delivered the 'Expedition 3' crew and transferred over 2.7 tons of supplies and equipment to the ISS. Barry and Patrick Forrester performed two EVAs totaling 11 hours and 45 minutes.

In 2006, Daniel Barry was a contestant on ***Survivor: Panama - Exile Island***. Barry attended Singularity University from 2009 to 2012, where he was the co-chair of the Faculty of Artificial Intelligence and Robotics, and the chair of the graduate summer program. He competed on the televised robot combat tournament ***Battlebots*** revival series on ABC in 2016. Daniel Barry is a co-founder of *Fellow AI*, a telepresence robotics company, and the founder and president of *Denbar Robotics*.

## Bartoe, John-David F.. STS-51-F.

Dr. John-David Bartoe was born in November 1944 in Abington, Pennsylvania, USA. He received his Bachelor of Science degree in Physics in 1966 from Lehigh University, and his Master of Science (1974), and Doctor of Philosophy in Physics from Georgetown University (1976). From 1966-88, Dr. Bartoe worked as an astrophysicist at the Naval Research Laboratory in Washington, D.C., and published over sixty papers in the field of solar physics observation and instrumentation. He served as Chief Scientist for the Space Station from 1987-90. From 1990-94, Dr. Bartoe was Director of Operations and Utilization in the Space Station Office of NASA Headquarters.

As a civilian employee of the US Navy, he launched on July 29, 1985, aboard *Challenger* on mission STS-51-F as a Payload Specialist. Dr. Bartoe logged seven days, 22 hours and 45 minutes (over 190 hours) in space. He became the Research Manager for the International Space Station at NASA's Johnson Space Center. Dr. Bartoe provided oversight for the Program Manager concerning the research capability, research hardware, and research plans of the ISS. John-David Bartoe is a member of the Association of Space Explorers, and is Chairman of the Space Stations Committee of the International Astronautical Federation. His awards include the NASA Exceptional Achievement Medal, the Navy Distinguished Civilian Service Award, the Flight Achievement Award of the American Astronautical Society, and the NASA Skylab Achievement Award.

## Bean, Alan L.. Apollo 12, Skylab 3.

Capt. Alan LaVern Bean (1932-2018) was born in March 1932 in Wheeler County, in the northeastern Texas Panhandle, USA. He graduated from R.L. Paschal High School in Fort Worth, Texas in 1949. Bean earned a degree in Aeronautical Engineering from the University of Texas at Austin in 1955. He received his naval aviator wings in 1956 and served as a fighter pilot. Assigned to the US Naval Test Pilot School at Patuxent River, Maryland, he also took art classes at St. Mary's College. In 1960, Bean graduated from the Test Pilot School, and was ***The New Nine*** astronaut selection finalist in 1962. Bean flew A-4 Skyhawks, logging more than 7,145 hours of flying time, including 4,890 hours in jet aircraft.

Alan Bean was selected by NASA in 1963 as part of Astronaut Group 3. He was the Lunar Module pilot on the *Apollo 12* mission in November 1969, and the fourth person to walk on the Moon. Bean and Pete Conrad landed on the Moon's Ocean of Storms, after a flight of 250,000 miles and a launch that included a harrowing lightning strike thirty-six seconds after liftoff. Bean and Conrad deployed several lunar surface experiments, and installed the first nuclear power generator station on the Moon to provide the power source. Alan Bean commanded the *Skylab 3* mission from July 29 to September 25, 1973. While on *Skylab*, Bean tested a prototype of the *Manned Maneuvering Unit,* performing one EVA. He logged 1,671 hours and 45 minutes in space, of which ten hours and 26 minutes were spent in EVAs on the Moon and in Earth orbit. After eighteen years as an astronaut, Bean resigned in June 1981 for full-time painting the remarkable worlds and space-related sights he had seen. He resigned from the US Navy in October 1975 as a Captain.

Alan Bean's paintings include *Lunar Grand Prix* and *Rock and roll on the Ocean of Storms*, and he used real Moon dust in his works. In July 2009, for the fortieth anniversary of the *Apollo 11* Moon landing, Bean exhibited his lunar paintings at the Smithsonian Institution's National Air and Space Museum in Washington. He was inducted into the International Space Hall of Fame in 1983, the US Astronaut Hall of Fame in 1997, the International Air & Space Hall of Fame in 2010, and the National Aviation Hall of Fame for 2010. Bean died on May 26, 2018 in Houston, Texas, and was interred in Arlington National Cemetery.

## Behnken, Robert "Bob" L.. STS-123, STS-130, SpaceX 'SpX-DM2' Crew Dragon/Expedition 63.

Col. Dr. Robert "Bob" Behnken was born on July 28, 1970 in St. Ann, Missouri, USA. He is a former Chief of the NASA Astronaut Office. He earned Bachelor of Science degrees in Mechanical Engineering and Physics from Washington University in St. Louis in 1992; a Master of Science degree (1993), and a Ph.D. in Mechanical Engineering (1997), both from Caltech. Before entering graduate school, Behnken was an Air Force ROTC student at Washington University, and was later assigned to enter Air Force duty at Eglin AFB, Florida, where he worked as a technical manager and developmental engineer for new munitions systems. After graduating from the USAF Test Pilot School Flight Test Engineer's course at Edwards AFB, California, he was assigned to the F-22 program.

Behnken was selected as an astronaut candidate in the July 2000 NASA Group. Following eighteen months of training and evaluation, he was assigned technical duties in the Astronaut Office Shuttle Operations Branch, supporting launch and landing operations at Kennedy Space Center, Florida. Bob Behnken flew on mission STS-123 aboard *Endeavour* in March 2008 as Mission Specialist, delivering the Japanese Experiment Module and the Special Purpose Dexterous Manipulator to the ISS. He made three EVAs. He launched as a Mission Specialist on STS-130 aboard *Endeavour* on February 08, 2010, when the *Tranquility* module and the *Cupola* observation module were delivered to the ISS. Behnken made three more EVAs.

In August 2018, Behnken was assigned to the first test flight 'SpX-DM2' of the SpaceX *Crew Dragon*. Launched on May 30, 2020, the *Endeavour* spacecraft successfully docked with the ISS on May 31. Behnken and Doug Hurley joined the ISS 'Expedition 63' crew, and returned to Earth in the same *Crew Dragon* capsule on August 02, 2020. Behnken's seat was later occupied by his wife, K. Megan McArthur, who flew in SpaceX 'Crew-2', and returned to Earth on November 08, 2021. Over three space missions, Behnken logged a total of 93 days, 11 hours and 42 minutes (over 708 hours) in space; including fifty-five hours of EVA. Bob Behnken retired from NASA on November 11, 2022. He has received eight awards, including Outstanding Mechanical Engineering Senior, Washington University (1992); National Science Foundation Graduate Research Fellow (1993-96); Distinguished Graduate from the USAF Test Pilot School Program (1999); and the Congressional Space Medal of Honor.

## Blaha, John E.. STS-29R, STS-33, STS-43, STS-58, STS-79.

Col. John Elmer Blaha was born in August 1942 in San Antonio, Texas, USA. He received a Bachelor of Science degree in Engineering Science from the USAF Academy in 1965; and received a Master of Science degree in Aeronautical Engineering from Purdue University in 1966. Blaha received his pilot wings at Williams AFB, Arizona, in 1967. He completed 361 combat missions in Vietnam. In 1973, he was assigned for a three-year tour as a test pilot working with the Royal Air Force at Boscombe Down, United Kingdom. Blaha logged over 7,000 hours flying time in thirty-four aircraft types. In 1976, he attended the USAF Air Command and Staff College. After graduation he was assigned to Headquarters USAF in the Pentagon.

Blaha was selected as an astronaut in May 1980. Launched on March 13, 1989 on mission STS-29R aboard *Discovery*, Blaha served as pilot; supporting the Commander during the launch, orbit, and landing procedures. The main mission objective was to deploy the *Tracking and Data Relay Satellite* (TDRS-D). He launched on November 03, 1989, on STS-33 aboard *Discovery*; and on August 02, 1991, on STS-43 aboard *Atlantis* as Commander. The fifth *Tracking and Data Relay Satellite* (TDRS-E) was deployed. Launched on October 18, 1993, on STS-58 aboard *Columbia*, Blaha was Commander, with the mission emphasis on biomedical studies under the Spacelab Life Sciences (SLS-2) program.

John Blaha began Russian language training in August 1994 at the Defense Language Institute in Monterey, California, and commenced an intensive training program at the Gagarin Cosmonaut Training Center, Star City, Russia, in January 1995. He launched on mission STS-79 on September 16, 1996, and transferred to the *Mir* space station. As 'Board Engineer 2', he spent four months with the 'Mir 22' Cosmonaut crew, conducting material science, fluid science, and life science research. Blaha returned to Earth aboard STS-81 on January 22, 1997. He logged a total of 161 days two hours and 45 minutes in space over five shuttle missions. Blaha has received twenty-seven awards, including the Defense Superior Service Medal, the Legion of Honor, two DFCs, eighteen Air Medals, two NASA Distinguished Service Medals, five NASA Space Flight Medals, the Russian Order of Friendship, the Vietnam Cross of Gallantry, Distinguished Graduate Air Command and Staff College, was inducted into the Astronaut Hall of Fame (May 2008), and his name was given to minor planet 22442 Blaha. Blaha retired from NASA in September 1997, returning to San Antonio, Texas, where he joined the executive management team of the **United Services Automobile Association**.

## Bloomfield, Michael J.. STS-86, STS-97, STS-110.

Col. Michael John "Bloomer" Bloomfield was born in March 1959 in Flint, Michigan, USA. He received his Bachelor's degree in Engineering Mechanics from USAF Academy. He became an F-15 fighter pilot, and then selected as a test pilot. In 1993, he earned his Master's degree in Engineering Management from Old Dominion University. Bloomfield was selected by NASA in December 1994, and from March 1995 he worked as Chief of Safety for the Astronaut Office, Chief Instructor Astronaut, Director of Shuttle Operations, and Chief of the Shuttle Branch. Bloomfield flew as pilot aboard STS-86 aboard *Atlantis* in September 1997, when he docked with space station *Mir*. In December 2000, he piloted STS-97 *Endeavour* to the ISS; and commanded STS-110 aboard *Atlantis* in April 2002. He logged a total of 32 days, 11 hours and two minutes in space. Michael Bloomfield has served as Deputy director of Flight Crew Operations at NASA's Johnson Space Center in Houston, Texas, since 2006, before retiring from NASA in July 2007. He then became Vice President of the Constellation Program for **Alliant Techsystems**, which merged with **Orbital Sciences Corporation** in 2014 to become **Orbital ATK, Inc**. He currently serves on the board of directors at Space Center Houston.

## Bluford, Guion S., Jr.. STS-8, STS-61-A, STS-39, STS-53.

Col. Dr. Guion "Guy" Stewart Bluford, Jr., was born in November 1942 in Philadelphia, Pennsylvania, USA. He became famous in August 1983 when, as a crew member of mission STS-8 aboard *Challenger* he became the first African American to go to space. In 1964, he received an undergraduate degree in Aerospace Engineering from Pennsylvania State University in and was commissioned as an officer in the US Air Force, where he trained as a fighter pilot. Bluford flew 144 combat missions in the Vietnam War, sixty-five of which were over North Vietnam. In 1978 he earned a doctorate in Aerospace Engineering from the Air Force Institute of Technology. Bluford was one of thirty-five individuals selected in 1978 from 10,000 applicants in NASA's first competition to become Space Shuttle astronauts. After Cuban cosmonaut Arnaldo Mendez, he was the second person of African descent in space. In 1987, Bluford received a graduate degree in Business Administration from the University of Houston.

While assigned to NASA, he remained a USAF officer, rising to the rank of Colonel. Guion participated in four Space Shuttle flights between 1983 and 1992. He flew as 'Mission Specialist 1' on STS-8 aboard *Challenger* (August 30-September 05, 1983), with the primary payload of the *INSAT-1B* Indian communications satellite. He was 'Science Mission Specialist' for STS-61-A aboard *Challenger* (October 30-November 06, 1985, carrying the 'Spacelab D-1' module, a scientific laboratory funded by West Germany. He was 'Mission Specialist 3' on STS-39 aboard *Discovery* (April 28-May 06, 1991), which carried unclassified experiments for the Department of Defense (DoD), and he made observations of the atmosphere and gas releases. For Guion Bluford's final mission, he was 'Mission Specialist 1' on STS-53 aboard *Discovery* (December 02-09, 1992), the last space shuttle mission devoted to DoD experiments. The crew deployed a *Satellite Data System-2* classified military communications satellite. Guion Bluford accumulated a total of 28 days, 16 hours and 33 minutes (over 688 hours) in space.

Bluford left NASA in July 1993 for a private sector career in the information technology and engineering services field. He has written several scientific papers in the field of computational fluid dynamics. Guion Bluford is a certified scuba diver, and he has an FAA commercial pilot license. He has received an amazing forty-two honors, including ten Air Force Medals (1967); the Vietnam Campaign Medal (1967); the Defense Superior Service Medal; 1991 Black Engineer of the Year Award; the Legion of Merit (1993); the NASA Distinguished Service Medal (1994); the Ohio Distinguished Service Medal (2020); and thirteen honorary Doctorate degrees. He is a member of seventeen organizations, including ***Who's Who*** Among Black Americans (1975-77); Fellow, American Institute of Aeronautics and Astronautics; Board of Directors, National Inventor's Hall of Fame Foundation (1997-2002); Board of Directors, US Space Foundation (2000-06); and Board of Advisors, Coalition for Space Exploration (2006-10). Married to his wife Linda since 1964, they have two children.

## Bobko, Karol J.. STS-6, STS-51-D, STS-51-J.

Karol Joseph Bobko (1937-2023) was born in Queens, New York, USA, to a family with Polish and Lithuanian roots. He received a Bachelor of Science degree from the USAF Academy in 1959. After receiving his commission and navigator rating, he attended pilot training at Bartow Air Base, Florida, and Vance AFB, Oklahoma; gaining his pilot wings in 1960. He flew F-100 and F-105 aircraft before attending the Aerospace Research Pilot School at Edwards AFB, California, and was assigned as an astronaut to the USAF Manned Orbiting Laboratory (MOL) program in 1966. Bobko logged over 6,600 flight hours in many aircraft types.

After the MOL spy program was canceled, he became part of NASA Astronaut Group 7 in September 1969. Bobko earned a Master of Science degree in Aerospace Engineering from the University of Southern California in 1970. He was a member of the astronaut support crew for the *Apollo-Soyuz Test Project* for 1975, and then a member of the support crew for the Space Shuttle Approach and Landing Tests (ALT) conducted at Edwards AFB, serving alternately as CapCom and prime chase pilot during the ALT flights. Bobko was the first graduate of the USAF Academy to travel in space. Launched on April 04, 1983, Bobko was pilot for STS-6 aboard *Challenger*, a five-day mission which deployed a large communications satellite (TDRS-1). He was mission Commander on STS-51-D aboard *Discovery* (April 12-19, 1985), deploying two communications satellites; and on STS-51-J aboard *Atlantis* (October 03-07, 1985), the second Space Shuttle Department of Defense mission which carried classified payloads. Karol Bobko became the first person to fly on three different Space Shuttles, and logged a total of 386 hours in space.

In 1988, Bobko retired from NASA and the Air Force to join the firm of **Booz Allen Hamilton**, in Houston, Texas. In 2000, he joined **SPACEHAB, Inc**. in Houston, Texas, where he was Vice President for Strategic Programs. In 2005, Bobko joined **Science Applications International Corporation** (SAIC) as Program Manager for the NASA Ames Research Center Simulation Laboratories. Karol Bobko died of an unspecified degenerative disease of the nervous system on August 17, 2023 at age eighty-five years. He received twelve awards, including the DFC; two Meritorious Service Medals; six Johnson Space Center Group Achievement Awards; and was a Fellow of the Society of Experimental Test Pilots (2020).

## Boe, Eric A.. STS-126, STS-133.

Col. Eric Allen Boe was born in October 1964 in Miami, Florida, USA. He is a current, active NASA astronaut. Boe graduated from Henderson High School, Chamblee, Georgia in 1983. He earned a Bachelor of Science degree in Astronautical Engineering from the USAF Academy in 1987, and a Master of Science degree in Electrical Engineering from the Georgia Institute of Technology in 1997. He was a cadet in the Georgia Wing of the Civil Air Patrol. Boe was commissioned from the USAF Academy in 1987, and was later assigned to the Third Tactical Fighter Squadron at Clark AFB in the Philippines as a combat-ready pilot in the F-4E. In 1994, he was assigned to the Thirty-third Fighter Wing's Sixtieth Fighter Squadron at Eglin AFB, Florida. He flew fifty-five combat missions over Iraq in support of Operation Southern Watch after the first Gulf War. Boe later served as a test pilot flying the A/B/C/D and E models of the F-15, and the UH-1N helicopter. He logged over 6,000 flight hours in more than fifty aircraft types.

Boe was selected as an astronaut candidate in July 2000. Following two years of training and evaluation, he was assigned technical duties in the Astronaut Office Advanced Vehicles Branch and Station Operations Branch. His current assignment involves work with displays and controls for future space vehicles. Eric Boe flew as pilot of *Endeavour* on mission STS-126 (November 14-30, 2008); and aboard *Discovery* on STS-133 (February 24-March 08, 2011). In August 2012, Boe was named Deputy Chief of the Astronaut Office. In July 2015, NASA announced Boe as one of the first astronauts for US Commercial spaceflights. He started working with Boeing and SpaceX to train in their Commercial Crew vehicles. In August 2018, he was assigned to *Boe-CFT*, the first test flight of the Boeing *CST-100 Starliner*, but he was unable to fly due to medical reasons. Boe has received eleven honors, including two Air Medals; five Aerial Achievement Medals; three Air Force Outstanding Unit Awards; two National Defense Service Medals; and the War on Terrorism Service Medal.

## Bolden, Charles F., Jr.. STS-61-C, STS-31, STS-45, STS-60.

Maj. Gen. Charles Frank Bolden, Jr. was born in August 1946 in Columbia, South Carolina, USA. He graduated from C.A. Johnson High School in Columbia, South Carolina in 1964; earned a Bachelor of Science degree in Electrical Science from the US Naval Academy in 1968; graduated from the US Naval Academy with the Class of 1968, and was then commissioned a second lieutenant in the US Marine Corps. His flight training was at Pensacola, Florida, at Meridian, Mississippi, and at Kingsville, Texas, before being designated a US Naval Aviator in May 1970. From June 1972 to June 1973, he flew over 100 sorties into North and South Vietnam, Laos and Cambodia. He logged over 6,000 hours flying time. Returning to the US, he began a two-year tour as a Marine Corps officer and recruiting officer in Los Angeles, California, then three years in various assignments at Marine Corps Air Station El Toro, California. Bolden later earned a Master of Science degree in Systems Management from the University of Southern California in 1977.

Bolden became a Marine aviator and test pilot. He was selected as a NASA astronaut candidate in 1980, and was a member of the Astronaut Corps until 1994, when he returned to the Marine Corps. From February to June 1998, he served as Commanding General, I MEF (Forward) in support of Operation Desert Thunder in Kuwait. In July 1998, he was promoted to Major General and became Deputy Commander US Forces Japan. Charles Bolden retired from the military in August 2004. Bolden served as pilot on STS-61-C aboard shuttle *Columbia* (January 12-18, 1986), deploying the *SATCOM* Ku-band satellite; and on STS-31 aboard *Discovery* (April 24-29, 1990), deploying the Hubble Space Telescope. He was the Mission Commander on STS-45 aboard *Atlantis* (March 24-April 02, 1992), the first 'Spacelab' mission dedicated to NASA's "Mission to to Planet Earth"; and on STS-60 aboard *Discovery* (February 03-11, 1994), the first mission involving a Russian cosmonaut, Sergei Krikalev as a Mission Specialist. Bolden logged over 680 hours in space.

After his service as an astronaut, he became Deputy Commandant of Midshipmen at the **Naval Academy**. On May 23, 2009, President Barack Obama announced the nomination of Bolden as **Administrator of NASA**, confirmed by the US Senate on July 15. In a question and answer session in May 2010, he said, *"You know, the Universe is a big place. I'm a practicing Christian, so in my faith, I learn about omnipotent, omnipresent God, which means he's everywhere. ... And I just cannot bring my little pea brain to believe that a God like that would pick one planet of one of millions of suns and say that's the only place in the vast Universe that I'm going to put any kind of life".* Unquote. In a June 2010 interview with ***Al Jazeera***, Bolden said that the three goals he was tasked with by President Obama *were "to help re-inspire children to want to get into science and math, to expand NASA's international relationships, and,* ***"perhaps foremost", "to reach out to the Muslim world*** *... to help them feel good about their historic contribution to science ... and math and engineering".* The White House press secretary disagreed, saying *"Bolden probably misspoke. That was not his task, and that's not the task of NASA".* Unquote.

On October 28, 2015, at the Center for American Progress in Washington, D.C., Bolden presented the next steps for a human journey to Mars. On January 12, 2017, Bolden announced his retirement from NASA. He then served on the United Arab Emirates **Space Advisory Committee**. In 2020, he was elected a member of the **National Academy of Engineering**, for leadership and development of US human spaceflight and space operations programs. He has received twenty-one honors, including eleven honorary Doctorates; the University of Southern California's Alumni Award of Merit (1989); the National Space Trophy (2014); the Wright Brothers Memorial Trophy (2020); and the National Air and Space Museum Trophy for Lifetime Achievement.

## Borman, Frank F., II. Gemini VII, Apollo 8.

Col. Frank Frederick Borman, II (1928-2023) was born in March 1928 in Gary, Indiana, USA. He was an honor student at Tucson High School. Borman learned to fly at age fifteen years, and joined a local flying club. Frank Borman entered the US Military Academy, West Point, New York, on July 01, 1946. In his final year, Borman was a cadet captain, commanding his company, and manager of the varsity football team. He chose to be commissioned as a second lieutenant in the USAF on June 02, 1950. Borman received his pilots wings on December 04, 1951. He became a test pilot, an American USAF Colonel, aeronautical engineer, NASA astronaut, and businessman. In 1950, Borman graduated from West Point with a Bachelor of Science degree, where he was ranked eighth out of 670. He qualified as a fighter pilot and served in the Phillippines. In 1957, he earned a Master of Science degree at California Institute of Technology (Caltech), and then became an assistant professor of thermodynamics and fluid mechanics at West Point. In June 1960, Borman was selected for Class 60-C at the USAF Experimental Flight Test Pilot School at Edwards AFB in California, and became a test pilot. During his Air Force service, Borman logged 3,600 hours of flying time, of which 3,000 was in jet aircraft.

On September 17, 1962, Borman was selected as a NASA astronaut with the second group, known as the *Next Nine*. He moved with his family to Houston, Texas, where the Manned Spacecraft Center (MSC) was still being established. Borman's assignment was to the *Titan II* booster used by Project Gemini, with responsibility for the Emergency Detection System (EDS) developed for an abort situation. Each astronaut was given four months of classroom instruction on subjects such as spacecraft propulsion, orbital mechanics, astronomy, computing, and space medicine. There was also familiarization with the *Gemini* spacecraft, *Titan II* and *Atlas* boosters, and the 26-foot-long *Agena* target vehicle. There was jungle, desert, and water survival training, along with fifty hours of instruction with geology. Borman thought that to be a waste of time. *"I didn't care about picking up rocks"*, he later told an interviewer, *"I wanted to beat the Soviets to the Moon"*. Unquote.

Borman was assigned to *Gemini VII*, planned to be a fourteen-day mission. To keep fit, he and Jim Lovell jogged 3-5 kilometres per day, and played handball after work. They visited the McDonnell Aircraft plant in St. Louis, Missouri, where their spacecraft was being built. At 8,076 pounds (3,663 kg) it was 250 pounds heavier than any previous Gemini craft. A lightwight spacesuit was developed to make the astronauts more comfortable. They launched on December 06, 1965. Wally Schirra and Tom Stafford then launched aboard *Gemini VI-A* on December 15, and at one point the two craft were just thirty centimetres (12 inches) apart in orbit. The re-entry on December 18 was flawless, and *Gemini VII* splashed down 10.3 km (6.4 miles) from the aircraft carrier *USS Wasp*. Borman was awarded the NASA Exceptional Service Medal for this mission, and was promoted to Colonel. At age thirty-seven years, he was the youngest full colonel in the USAF.

Frank Borman served as the only astronaut on the nine-member NASA **AS-204 Accident Review Board** which investigated the AS-204 *Apollo 1* fire of January 27, 1967. A redesigned outward-opening hatch that allowed the astronauts to exit within three seconds (instead of minutes) was one of 1,300 changes costing USD 75 million that were made to the Apollo capsule. The new hatch added 680 kg (1,500 lb) to the weight of the spacecraft, and the wiring was consolidated and coated with Teflon insulation. The parachutes had to be redesigned to ensure they could hold the additional capsule weight, and re-testing them cost USD 250,000. Borman was the commander of *Apollo 8* in December 1968, the first manned mission to fly around the Moon. When *Apollo 8*'s Lunar Module-3 (LM) arrived at the Kennedy Space Center in June 1968, more than 100 significant defects were discovered. This led to Borman's planned *Apollo 9* medium Earth orbit mission being dispensed with, and *Apollo 8* would go to circumnavigate the Moon, but without the Lunar Module. The mission launched at 12:51:00 UTC on December 21, 1968. On the second day, Borman felt ill; he vomited twice and had a bout of diarrhea. Researchers now believe he was suffering from space adaptation syndrome, which affects about a third of astronauts during their first day in space as their vestibular system adapts to weightlessness.

*Apollo 8* made ten lunar orbits in twenty hours before beginning the return journey to Earth. A televised reading from the ***Book of Genesis*** was released for worldwide broadcast on Christmas Eve. Borman later joked, *"One of the things that was truly historic was that we got that good Catholic, Bill Anders, to read from the King James Version of the Bible"*. Their spacecraft splashed down in darkness on December 27, but because the parachutes were not released quickly enough, they dragged the capsule over and left it upside down. Borman inflated the bags in the nose, which righted it, but the crew had to wait forty-five minutes until local sunrise before the frogmen could open the hatch. Borman became seasick, and threw up before he could be taken aboard the aircraft carrier *USS Yorktown*. The crew had survived a mission that they had themselves rated as having a ninety percent chance of surviving, but only a 50:50 chance of succeeding in full. The crew of *Apollo 8* were accorded ticker tape parades in New York, Chicago, and Washington, D.C., where they were awarded the NASA Distinguished Service Medal by President Lyndon B. Johnson. Afterward, Borman was sent on a goodwill tour of Europe, with a secondary objective of finding out more about the space programs of other NATO countries.

After retiring from NASA and the Air Force in 1970, Borman became senior vice president for operations at **Eastern Air Lines**. He became CEO in 1975, and chairman of the board in 1976. After resigning in 1986, he moved to Las Cruces, New Mexico, where he ran a Ford dealership with his son, Fred. In August 1996, Borman published an autobiography, *Countdown*, co-written with Robert J. Serling. In 1999, Borman told an interviewer that "*my reason for joining NASA was to participate in the Apollo Program, the lunar program, and hopefully beat the Russians. I never looked at it for any individual goals. I never wanted to be the first person on the Moon and frankly, as far as I was concerned, when 'Apollo 11' was over the mission was over. The rest was frosting on the cake*". Unquote. In ***From the Earth to the Moon***, a 1998 HBO miniseries, Borman was played by David Andrews. He was interviewed in the 2008 Discovery Channel documentary ***When We Left Earth: The NASA Missions***, and appeared in the 2005 documentary ***Race to the Moon***.

Frank Borman was once described by a NASA psychiatrist as the least complicated man he had ever met. In his business career he was described as autocratic. Borman had a quotation from Tolstoy on a wall in his office: "***The only legitimate happiness is honest hard work and the surmounting of obstacles***". Borman was reunited with Jim Lovell and Bill Anders for celebrations of the fiftieth anniversary of *Apollo 8* in December 2018 at the Museum of Science and Industry in Chicago, where their spacecraft is on display. He declared, "*I have never said it before publicly, but these two talented guys, I'm just proud that I was able to fly with them. It was a tough job done in four months, and we did a good job*". Frank Borman died from a stroke at the Billings Clinic on November 07, 2023, at the age of ninety-five years. He was awarded the Congressional Space Medal of Honor, and many other awards, including ten honorary Doctorates.

## Bowen, Stephen G.. STS-126, STS-132, STS-133, SpaceX Crew-6.

Stephen Gerard "Steve" Bowen (Captain, US Navy, Ret.) was born in February 1964 in Cohasset, Massachusetts, USA. He received a Bachelor of Science degree in Electrical Engineering from the United States Naval Academy in 1986; and received a Master of Science degree in Ocean Engineering from the Massachusetts Institute of Technology in 1993. After submarine training, Bowen spent three years attached to *USS Parche*, and completed qualification in submarines on *USS Pogy*. After the MIT/WHOI Joint Program in Ocean engineering he reported to *USS Augusta* for duty as the Engineering Officer. During this tour, he qualified for command of nuclear powered submarines. In May 2000, Bowen became the first Executive Officer of the Pre-Commissioning Unit *Virginia*, the first of the new *Virginia*-class submarines.

In July 2000, he was the first Submarine Officer to be selected as an astronaut candidate by NASA. Following the completion of two years of training and evaluation, Bowen was initially assigned technical duties in the Astronaut Office Station Operations Branch while awaiting assignment to a Shuttle flight as a Mission Specialist. He served in this role on all his three shuttle missions. Launched on November 14, 2008, mission STS-126 with space shuttle *Endeavour* was the 27th ISS assembly mission. It carried a reusable logistics module that held supplies and equipment. Bowen participated in three EVA, for a total time of 19 hours and 56 minutes. Launched on May 14, 2010, mission STS-132 aboard shuttle *Atlantis* to the ISS saw Bowen conduct two EVAs, for a total time of 14 hours and 34 minutes.

Launched on February 24, 2011, mission STS-133 was aboard shuttle *Discovery*. Again a Mission Specialist, he became the first and only astronaut to fly on consecutive space shuttle missions. Bowen conducted two EVA, for a total time of 12 hours and 48 minutes. On December 16, 2021, NASA assigned Steve Bowen as Commander of the SpaceX 'Crew-6' mission to the ISS, launched on March 02, 2023, using Crew Dragon *Endeavour*. He logged a total of 226 days, eight hours and 43 minutes in space; including ten EVAs totaling 65 hours and 57 minutes. Bowen was awarded the Defense Meritorious Service Medal; US Navy Commendation Medals; and Navy Achievement Medals.

## Bowersox, Kenneth "Ken" D.. STS-50, STS-61, STS-73, STS-82, STS-113/ Expedition 6.

Kenneth Dwane "Sox" Bowersox was born in November 1956 in Portsmouth, Virginia, USA. He is a veteran of five Space Shuttle launches and an extended stay aboard the ISS. Launched on mission STS-50, Bowersox was the Pilot aboard *Columbia* (June 25-July 09, 1992). The primary objective was to conduct experiments in the U.S. Microgravity Laboratory 1 (USML-1), a 'Spacelab' module. For his second spaceflight on STS-61 aboard *Endeavour* (December 02-13, 1993) Bowersox was again the Pilot. The mission installed a new main camera and a Corrective Optics Space Telescope Axial Replacement (COSTAR) to the Hubble Space Telescope.

When he launched on mission STS-73 at the age of 38 years and 11 months, Ken Bowersox became the youngest person to Command a Space Shuttle. On STS-73 aboard *Columbia* (October 20-November 05, 1995), aka the United States Microgravity Laboratory 2 mission, the shuttle carried the 'Spacehab' module, which provided additional laboratory space for experiments. Researchers investigated various phenomena related to fluid behavior, heat transfer, and crystal growth. Life Sciences studies included experiments on the effects of microgravity on living organisms, combustion, and biological processes. Launched as Commander on STS-82 aboard shuttle *Discovery* (February 11-21, 1997) the mission was NASA's second time to service the Hubble Space Telescope. The crew repaired and upgraded Hubble's instruments, significantly enhancing its research capabilities. This mission lasted nine days, 23 hours and 38 minutes.

Launched on November 24, 2002, mission STS-113 aboard shuttle *Endeavour* (November 24-November 07, 2002) was to extend the ISS backbone by attaching the P1 truss, which provided support for the ISS radiators. There was a crew exchange of 'Expedition 5' & '6' members, and three EVAs. Bowersox was a Mission Specialist, and became ISS Commander for 'Expedition 6'. Ken Bowersox retired from NASA on September 30, 2006. In 2009, he was appointed Vice President of Astronaut Safety and Mission Assurance at **SpaceX**, the Space Exploration Technologies Corporation. On June 08, 2010, Bowersox was inducted into the Astronaut Hall of Fame, just four days after SpaceX's successful launch of the *Falcon 9* rocket. He also worked as an independent consultant, advising clients on spacecraft design, proposal development, and providing independent assessments of technical programs.

## Brady, Charles E., Jr.. STS-78.

Capt. Dr. Charles Eldon Brady, Jr. was born in August 1951 in Pinehurst, North Carolina, USA. He studied pre-med at the University of North Carolina at Chapel Hill, graduating in 1971; receiving his medical degree from Duke University in 1975. Dr. Brady specialized in sports medicine and worked as a team physician at several universities (starting in 1978 for Iowa State University in Ames, Iowa) before joining the US Navy and reporting to Carrier Air Wing Two aboard *USS Ranger* in June 1986. He became a Flight Surgeon, and served with the Blue Angels flight demonstration squad from 1988-90. In March 1992, he was selected by NASA and qualified for selection as a Mission Specialist on future Space Shuttle crews.

On STS-78 aboard *Columbia* (June 20-July 07, 1996) he spent 16 days and 21 hours in space, becoming part of the longest Space Shuttle mission to date. The 'Life and Microgravity Spacelab' mission served as a model for future studies on board the ISS. After NASA, Brady returned full-time to the Navy and served as Flight Surgeon at Whidbey Island Naval Air Station in the San Juan Islands before retiring in March 2006 to the Pacific Northwest. Dr. Charles Brady died on July 23, 2006, at his home in Washington State. He had suffered for years from a severe form of arthritis, and died of self-inflicted wounds.

Dr. Brady belonged to the Association of Military Surgeons of the United States; Society of United States Naval Flight Surgeons, the Aerospace Medical Association and Space Medicine Branch, and Ancient Free and Accepted Masons. He had received eleven awards, including the Defense Superior Service Medal; the Navy Commendation Medal with Gold Star; the National Defense Service Medal; the Armed Forces Expeditionary Medal; was Recipient of the Fox Flag at the Naval Aerospace Medical Institute; and the Richard E. Luehrs Memorial Award for Navy Operational Flight Surgeon of the Year (1987).

## Brand, Vance D.. Skylab 3, Apollo-Soyuz Test Project, STS-5, STS-41-B, STS-35.

Maj. Vance DeVoe Brand was born in May 1931 in Longmont, Colorado, USA. He received a Bachelor of Science degree in Business from University of Colorado, Boulder, in 1953; another Bachelor of Science degree in Aeronautical Engineering in 1960; and a Master of Science degree in Business Administration from the University of California, Los Angeles, in 1964. From 1953-57 he served as a commissioned officer with the US Marine Corps, including a 15-month tour in Japan as a jet fighter pilot. In 1963, he graduated from the US Naval Test Pilot School (Class 33), and was assigned as an experimental test pilot on Canadian and German F-104 programs. From 1960-66, Brand worked as a civilian in the **Lockheed Corporation**, until selected for the April 1966 NASA Group 5 as one of nineteen pilot astronauts. Vance Brand was backup Command Module Pilot for *Apollo 15*, and backup Commander for *Skylab 3* and '4' missions. On the *Skylab 3* mission in 1973 he spent fifty-nine days in space. Brand served as Command Module Pilot during the first joint US-Soviet joint *Apollo-Soyuz Test Project* spaceflight from July 17, 1975, which tested a new docking system and demonstrated international cooperation in space; and later as Commander on three Space Shuttle missions. *Apollo* splashed down in the Pacific Ocean near Hawaii, and was recovered by *USS New Orleans*, completing a 217-hour mission.

Brand was Commander on the first fully operational Space Shuttle mission STS-5 aboard *Columbia* (November 11-16, 1982), which first used an upper-stage rocket for payloads, the Payload Assist Module (PAM-D). Satellites were deployed for Satellite Business Systems Corporation of McLean, Virginia, and TELESAT of Ottawa, Canada. Commanding STS-41-B aboard *Challenger* (February 03-11, 1984), the crew deployed two *Hughes HS-376* satellites, which failed to reach the desired orbits due to upper-stage rocket failures. In the first checkout of the *Manned Maneuvering Unit* (MMU), Bruce McCandless II and Robert Stewart performed two untethered EVAs. Commanding STS-35 aboard *Columbia* (December 02-10, 1990), the mission was devoted to round-the-clock astronomical observations. The 13-ton payload consisted of the three ASTRO-1 ultraviolet telescopes and the Broadband X-ray telescope. Brand has logged 9,669 flying hours: including 8,089 hours in jet aircraft; 391 hours in helicopters; 31 days, two hours and two minutes (746 hours) in spacecraft; and checkout in over thirty military aircraft types.

He left the Astronaut Office in 1992 to become Chief of Plans at the National Aerospace Plane (NASP) Joint Program Office at Wright-Patterson AFB. In September 1994, he moved to California to become Assistant Chief of Flight Operations at the Dryden Flight Research Center, then three further roles. Brand retired from NASA in January 2008. Maj. Vance Brand has received thirty-two awards, including two NASA Distinguished Service Medals (1975 & 1992); two NASA Exceptional Service Medals (1974 & 1988); University of Colorado Alumnus of the Century (1976); the AIAA Special Presidential Citation (1977); the Harmon Trophy (Astronaut) (1993); the FAI De la Vaulx Medal (1983); and two FAI V.M. Komarov Diplomas (1983 & 1991). He was inducted into the De Molay Hall of Honor (1989); the International Space Hall of Fame (1996); the US Astronaut Hall of Fame (1997); was awarded an honorary Doctor of Science degree from University of Colorado (2000); the Tsiolkovsky Gold Medal of the International Aeronautical Federation (2005); and Vance Brand Airport in Longmont, Colorado, is named in his honor.

## Brandenstein, Daniel C.. STS-8, STS-51-G, STS-32, STS-49.

Daniel "Dan" Charles Brandenstein was born in January 1943 in Watertown, Wisconsin, USA. He graduated with a Bachelor of Science degree in Mathematics and Physics from the University of Wisconsin-River Falls. He entered active duty with the US Navy in September 1965, and was designated a Naval Aviator in May 1967. From 1968-70, while attached to VA-196 flying A-6 Intruders, he participated in two combat deployments to Southeast Asia, where he flew 192 combat missions in the Vietnam War. Upon graduation from the US Naval Test Pilot School, he conducted tests of electronic warfare systems in various Navy aircraft. During March 1975 to September 1977, he was deployed on *USS Ranger* for nine months. Selected by NASA in January 1978 for Group 8, he became an astronaut in August 1979. He was ascent spacecraft communicator (CapCom) and a member of the astronaut support crew for mission STS-1. Between April 1987 and September 1992, he served as Chief of the Astronaut Office.

Brandenstein was pilot on STS-8 aboard *Challenger* (August 30-September 03, 1983), the first mission with a night launch and night landing. The crew deployed the *Indian National Satellite* (INSAT-1B); operated the Canadian-built Remote Manipulator System (RMS) with the Payload Flight Test Article (PFTA); operated the Continuous Flow Electrophoresis System (CFES) with live cell samples; and activated various Earth resources and space science experiments along with four "Getaway Special" canisters. He was Commander for STS-51-G aboard *Discovery* (June 17-24, 1985), which deployed three communications satellites. He commanded STS-32 aboard *Columbia* (January 09-20, 1990), which deployed the *Syncom IV-F5* satellite, and retrieved the Long Duration Exposure Facility (LDEF) using the RMS. He was Commander for STS-49 aboard *Endeavour* (May 07-16, 1992), when crew made four EVAs to retrieve, repair, and redeploy the *International Telecommunications Satellite* (*INTELSAT*). This was the maiden voyage of *Endeavour*, and the first test of the new drag chute for landing. Daniel Brandenstein logged a total of 32 days, 21 hours and three minutes (over 789 hours) in space.

Daniel Brandenstein has received nineteen awards, including the Defense Superior Service Medal with Oak Leaf Cluster, the Legion of Merit, the DFC, Defense Meritorious Service Medal, 17 Air Medals, the Navy Commendation Medal with Oak Leaf Cluster, two NASA Distinguished Service Medals, the Armed Forces Expeditionary Medal, and the Vietnam Air Gallantry Cross. Daniel Brandenstein was inducted into the Wisconsin Aviation Hall of Fame in 1992. He served as Chief of the Astronaut Office from 1978 to 1992, then became Executive Vice President and COO of **United Space Alliance**.

## Bresnik, Randolf J.. STS-129/Expedition 21, Soyuz MS-05/Expedition 52/53.

Randolf "Randy" James "Komrade" Bresnik was born in September 1967 in Fort Knox, Kentucky, USA. He is an active NASA astronaut. Beresnik earned a Bachelor of Arts degree in Mathematics from The Citadel in 1989; a Master of Science degree in Aviation Systems from the University of Tennessee-Knoxville in 2002; and graduated from the Air War College in 2008. After he graduated from the Naval Reserve Office Training Corps, he attended The Basic School (TBS) and Infantry Officers Course (IOC) at the Marine Corps Base Quantico, Virginia. Designated a Naval Aviator in 1992, he became an F/A-18 Test Pilot. In January 2003, Bresnik was deployed to Kuwait to fly combat missions in support of Operations Southern Watch and Iraqui Freedom. He has logged over 6,000 hours on eighty-one aircraft types.

Selected as a member of NASA Astronaut Group 19 in May 2004, he completed training in February 2006. Mission STS-129 aboard *Atlantis* (November 16-27, 2009) delivered the first two of four ExPRESS Logistics Carriers (ELCs) to the ISS, with two EVAs totaling 11 hours and 50 minutes. Bresnik and five crewmates joined the 'Expedition 21' crew, commanded by Belgian ESA astronaut Frank De Winne. Bresnik launched aboard *Soyuz MS-05* to the ISS for a mission from July 28 to December 14, 2017. He served as Flight Engineer for 'Expedition 52', and commanded 'Expedition 53'. For his third EVA of six hours and 55 minutes, Bresnik and Mark Vande Hei replaced the gripping mechanism on Canadarm2, the latching end effector, or LEE-A. In another EVA of six hours and 49 minutes, he and Mark Vande Hei lubricated the newly installed end effector and replaced cameras. Bresnik logged a total of 149 days, 12 hours and 12 minutes in space; including five EVAs. Bresnik has received the Defense Meritorious Service Medal, three Strike/Flight Air Medals, Navy and Marine Corps Commendation Medal with Combat Valor device, three Navy and Marine Corps Achievement Medals, and the Presidential Unit Citation.

## Bridges, Roy D., Jr.. STS-51-F.

Maj. Gen. Roy Bridges, Jr. was born in July 1943 in Atlanta, Georgia, USA. He is the former director of NASA's John F. Kennedy Space Center (KSC) and Langley Research Center. He earned a Bachelor of Science in Engineering Science from the USAF Academy in 1965; and was commissioned a second lieutenant on June 09. He received a Master of Science degree in Astronautics from Purdue University, Indiana, in 1966. In January 1970, he became a T-37 instructor pilot, 3575th Pilot Training Squadron, Vance AFB, California. From July 1971, he was a test pilot, Air Force Flight Test Center, Edwards AFB, California; and in August 1975 was a distinguished graduate, Air Command and Staff College, Maxwell AFB, Alabama. As a command pilot, he logged over 4,460 flying hours and is a decorated veteran of 262 combat missions during the Vietnam War. Roy Bridges was the Commander, Air Force Flight Test Center, Edwards AFB, California, then served as the Director of Requirements, Headquarters Air Force Material Command, Wright-Patterson AFB, Ohio, from June 1993 until retirement on July 01, 1996.

Selected in the 1980 NASA Group 9, he flew as pilot aboard *Challenger* on mission STS-51-F (July 29 to August 06, 1985). Roy Bridges logged seven days, 22 hours and 45 minutes in space. In May 1986, he was Commander of 6510th Test Wing, Edwards AFB. In March 1989, he became Commander of Eastern Space and Missile Center, Patrick AFB, Florida. In August 1991, he was Commander of Air Force Flight Test Center, Edwards, AFB. Roy Bridges became the Director of NASA's **Langley Research Center** in 2003, and retired from NASA in late 2005. As Director of the KSC for over six years, Bridges was responsible for managing NASA's only site for processing and launch of the Space Shuttle vehicle; processing the payloads flown; and overseeing expendable launch vehicles carrying NASA payloads. He managed a team of about 2,000 NASA civil servants and about 14,000 contractors.

Maj. Gen. Roy Bridges received twenty-four decorations, including the Air Force Distinguished Service ribbon; the DFC ribbon with two Oak Leaf Clusters; the NASA Outstanding Leadership ribbon; the National Defense Service Medal with one Oak Leaf Cluster; and the Presidential Meritorious Executive Award. Currently, Bridges is an executive with **Northrop Grumman** Technical Services.

## Brown, Curtis L., Jr.. STS-47 (J), STS-66, STS-77, STS-85, STS-95, STS-103.

Col. Curtis "Curt" Lee Brown, Jr., was born in March 1956 in Elizabethtown, North Carolina, USA. He received a Bachelor of Science degree in Electrical Engineering from the USAF Academy in 1978, and was commissioned a second lieutenant. Brown completed Undergraduate Pilot Training at Laughlin AFB, Del Rio, Texas, graduated in July 1979 and was assigned to fly A-10 aircraft. In March 1982, he was re-assigned to Davis-Montham AFB as an instructor pilot. In June 1985, he attended USAF Test Pilot School, and, upon graduating in June 1986, was assigned to Eglin AFB, Florida, where he served as a test pilot until his selection for the astronaut program.

Selected in 1987 NASA Group 12, he completed one year of training and evaluation, qualifying for flight assignment as a Pilot. STS-47 aboard *Endeavour* (September 12-20, 1992) was an eight-day co-operative mission between the US and Japan, focused on life science and materials processing experiments in space. STS-66 aboard *Atlantis* (November 03-14, 1994) was the Atmospheric Laboratory for Applications and Science 3 (ATLAS-3) mission, part of a program to determine the Earth's energy balance and atmospheric change over an eleven-year solar cycle. Brown was Pilot for the third time on STS-77 aboard *Endeavour* (May 19-29, 1996) when the crew performed a record number of rendezvous sequences.

He was Commander for STS-85 aboard *Discovery* (August 07-19, 1997), which deployed and retrieved the *CRISTA-SPAS* payload. Curtis Brown was Commander for STS-95 aboard *Discovery* (October 29-November 07, 1998), which deployed the *SPARTAN* solar observing spacecraft; and the Hubble Space Telescope Orbital Systems Test Platform. Commander for the third time on STS-103 aboard *Discovery* (December 20-28, 1999), the crew installed new instruments and upgraded systems on the Hubble Space Telescope. Col. Curtis Brown logged a total of 57 days, 17 hours and seven minutes (over 1,383 hours) in space. He is a member of the United States Air Force Association, the USAF Academy Association of Graduates, and the Experimental Aircraft Association. He has received two Defense Service Medals; three Air Force Medals; and six NASA Space Flight Medals.

## Brown, Mark N.. STS-28R, STS-48.

Col. Mark Neil Brown was born in November 1951 in Valparaiso, Indiana, USA. He graduated with a Bachelor of Science degree in Aeronautical and Astronautical Engineering from Purdue University, West Lafayette, Indiana, in 1973; and a Master of Science degree in Astronautical Engineering from the USAF Institute of Technology, Dayton, Ohio, in 1980. He received his wings at Laughlin AFB, Texas, in 1974, and was assigned to the 87th Fighter Interceptor Squadron, and graduated from the Squadron Officer School in 1975. Brown earned his Master's degree in Astronautical Engineering in 1980. Two years later, he graduated from the Air Command and Staff College at Maxwell AFB, Alabama. In 1980, he was assigned to the Johnson Space Center as an engineer in the Flight Activities Section.

Brown became an astronaut in June 1985, and qualified for assignment as a Mission Specialist. During 1986-87, he was an astronaut member of the solid rocket booster redesign team. He flew on STS-28 aboard *Columbia* (August 08-13, 1989) with a Department of Defense payload, then served as astronaut member on the Space Station *Freedom* program (later canceled). Mission STS-48 aboard *Discovery* (September 12-18, 1991) deployed the *UARS* satellite, designed to provide scientists with their first complete data set on the upper atmosphere's chemistry, winds and energy inputs. Col. Mark Brown was a Distinguished Graduate from Air Force ROTC, and has received eleven honors, including Air Force Command Pilot, the Senior Space Badge, the Defense Superior Service Medal, the Combat Readiness Medal, the Air Force Outstanding Unit Award, and the Aerospace Defense Command "We Point With Pride" Award.

## Buchli, James F.. STS-51-C, STS-61-A, STS-29R, STS-48.

James Frederick Buchli was born in June 1945 in New Rockford, North Dakota, USA. He received a Bachelor of Science degree in Aeronautical Engineering Systems from the US Naval Academy in 1967; and a Master of Science degree in Aeronautical Engineering Systems from the University of West Florida in 1975. He graduated from the Basic Infantry Officer's course, and was sent to the Republic of Vietnam for a one-year tour of duty. Buchli returned to the US in 1969 for naval flight officer training at NAS Pensacola. After earning his wings, he spent two years assigned to Marine Corps stations in Hawaii and Japan. Later, he participated in the Marine Advanced Degree Program at the University of West Florida. In 1977, he was assigned to US Naval Test Pilot School, Patuxent River, Maryland. James Buchli logged over 4,200 flight hours; with 4,000 hours in jet aircraft, including combat in the F-4 Phantom.

Selected as part of Astronaut Group 8, he became an astronaut in August 1979. STS-51-C aboard *Discovery* (January 24-27, 1985) was the first dedicated Department of Defense mission. STS-61-A aboard *Challenger* (October 30-November 06, 1985) performed a West German D-1 'Spacelab' mission, with over 75 scientific experiments. STS-29 aboard *Discovery* (March 13-18, 1989) deployed a *Tracking and Data Relay Satellite*; numerous secondary experiments, including a chromosome and plant cell division; and the crew took over 3,000 photographs of Earth. STS-48 aboard *Discovery* (September 12-18, 1991) deployed the *Upper Atmosphere Research Satellite* (UARS). Buchli logged a total of 20 days, ten hours and 25 minutes in space.

On September 01, 1992, Buchli retired from the Marine Corps and the NASA Astronaut Office to accept a position with **Boeing Defense and Space Group,** at Huntsville, Alabama. He currently serves as Operations and Utilization Manager for Space Station, Boeing Defense and Space Group, at Houston, Texas. Buchli is an associate member of the Naval Academy Alumni, the American Legion, Association of Space Explorers, and American Geophysical Union. He has received the Defense Superior Service Medal; the Legion of Merit; a Purple Heart; the Defense Meritorious Service Medal; the Navy and Marine Corps Commendation Medal; and the Vietnam Gallantry Cross with silver star. He is a 2019 inductee into the US Astronaut Hall of Fame.

## Buckey, Jay C., Jr. STS-90.

Dr. Jay Clark Buckey, Jr. was born in June 1956 in New York City, USA. He holds a Bachelor of Science degree in Electrical Engineering from Cornell University (1977); and an M.D. from Cornell in 1981, interning at New York Hospital-Cornell Medical Center, and completing his residence at Dartmouth-Hitcock Medical Center. Buckey was a Payload Specialist aboard shuttle *Columbia* on STS-90 as part of the 'Neurolab' mission from April 17 to May

03, 1998. The crew performed twenty-six life science experiments focusing on the effects of microgravity on the brain and nervous system. He logged 15 days, 21 hours and 50 minutes (over 381 hours) in space. In 2008, Dr. Buckey briefly ran for the Democratic nomination to challenge New Hampshire Senator John E. Sununu when he was up for re-election. In 2018, Buckey was part of research using virtual reality, at the Australian Antarctic Division's Mawson Station. The research will inform psychological techniques to support long duration spaceflight, such as for astronauts going to Mars. He was a flight surgeon with the USAF Reserve for eight years. Currently, Dr. Jay Buckey is a Professor of Medicine at the **Dartmouth Medical School**.

## Burbank, Daniel C.. STS-106, STS-115, Soyuz TMA-02M/Expedition 29/30.

Capt. Daniel Christopher Burbank was born in July 1961 in Manchester, Conneticut, USA. He graduated from Tolland High School, and attended Fairfield University his freshman year before transferring to the US Coast Guard Academy, where he earned his commission in 1985. He became an instructor pilot, serving at various Coast Guard stations. Burbank has a Master's degree in Aeronautical Science from Embry-Riddle Aeronautical University.

Selected by NASA in April 1996 for two years' training, he then worked technical issues for the Astronaut Office Operations Planning Branch, and the ISS Branch. He served as CapCom for both Space Shuttle and ISS missions. Burbank was a Mission Specialist on STS-106 aboard *Atlantis* (September 08-20, 2000), preparing the ISS for arrival of the first permanent crew. The crew delivered over 6,600 pounds of supplies and installed batteries, power converters, oxygen generators and a treadmill. STS-115 aboard *Atlantis* (September 09-21, 2006) delivered and installed the massive P3/P4 truss and two sets of solar arrays. They performed unprecedented robotics activity using the Shuttle and ISS robotic arms. Burbank made a seven-hour and 11 minute EVA completing the truss installation, activating the Solar Alpha Rotary Joint (SARJ), and enabling the solar arrays to be deployed.

From January 2007 to December 2009, he served as a Professor of Engineering at the US Coast Guard Academy, where he taught astronomy, aerodynamics, and statics and engineering design. Burbank was assigned to ISS 'Expedition 29/30', arriving at the ISS on November 16, 2011, via *Soyuz TMA-02M* for 66 days in space. Burbank then served as Commander of 'Expedition 30', returning to Earth on April 27, 2012, after spending an additional 158 days in space. Daniel Burbank retired from NASA in June 2018. In 2022, he returned to the **US Coast Guard Academy** to teach Mechanical Engineering as a Professor of Practice. He has eleven awards, including the Coast Guard Astronaut Badge; the Defense Superior Service Medal with one Oak Leaf Cluster; the Air Medal; the National Defense Service Medal; the Global War on Terrorism Service Medal; and the NASA Exceptional Service Medal.

## Bursch, Daniel W.. STS-51, STS-68, STS-77, STS-108/Expedition 4.

Capt. Daniel Wheeler Bursch was born in July 1957 in Bristol, Pennsylvania, USA. Bursch received a Bachelor of Science degree in Physics from the US Naval Academy in 1979; and a Master of Science degree in Engineering Science from the US Naval Postgraduate School in 1991. He was designated a Naval Flight Officer in April 1980. In January 1981, he deployed to the Mediterranean aboard *USS John F. Kennedy*, and to the North Atlantic and Indian Oceans aboard *USS America*. He attended the US Naval Test Pilot School in January 1984, and in August 1985 returned to the Test Pilot School as a flight instructor. Redesignated an Aeronautical Engineering Duty Officer (AEDO), he attended the US Naval Postgraduate School, Monterey, California, from July 1989 until his selection to the astronaut program in January 1990. Capt. Bursch has logged over 3,430 flight hours in over thirty-five aircraft types.

He was a Mission Specialist on STS-51 aboard *Discovery* (September 12-22, 1993), which deployed the US *Advanced Communications Technology Satellite* (ACTS), and the *Shuttle Pallet Satellite* (SPAS). The STS-68 Space Radar Lab-2 mission aboard *Endeavour* (September 30-October 11, 1994) was the second flight of three advanced radars called SIR-C/X-SAR; a carbon-monoxide pollution sensor; and MAPS (Measurement of Air Pollution from Satellites). STS-77 aboard *Endeavour* (May 19-29, 1996) included the fourth 'Spacehab' module, with 12 experiments, having an emphasis on commercial space product development. The mission carried the *SPARTAN 207*/Inflatable Antenna Experiment (IAE) and the *Passive Aerodynamically stabilized Magnetically damped Satellite*/Satellite Test Unit (PAMS/STU). On STS-108/STS-111 (December 05, 2001 to June 19, 2002) he served as Flight Engineer for ISS 'Expedition 4' during 196 days in space; including 11 hours and 48 minutes of EVA time. Bursch logged a total of 226 days, 22 hours and 16 minutes in space.

In January 2003, Bursch reported to the US Naval Postgraduate School for a two-year assignment as an instructor in the Space Systems Academic Group. He left NASA in May 2005, and retired from active duty after twenty-six years of service in the US Navy. In July 2005, Bursch joined **The Aerospace Corporation**, and is currently serving as the National Reconnaissance Office (NRO) Chair at the US Naval Postgraduate School. Daniel Bursch has been awarded the Defense Superior Service Medal; NASA Space Flight Medals; the Navy Commendation Medal; and the Navy Achievement Medal.

## Cabana, Robert D.. STS-41, STS-53, STS-65, STS-88/2A.

Robert Donald Cabana was born in January 1949 in Minneapolois, Minnesota, USA. He is a non-flight eligible NASA management astronaut. Cabana graduated from the US Naval Academy, Annapolis, Maryland, with a Bachelor of Science degree in Mathematics in 1971. Initially serving as bombardier and navigator in the Marine Corps, he returned to NAS Pensacola in 1975 for pilot training and was redesignated as a Naval Aviator in September 1976. Cabana graduated from the US Naval Test Pilot School in 1981, and served as the A-6 Intruder program manager, X-29 advanced technology demonstrator project officer, and as a test pilot on A-6 Intruder and A-4 Skyhawk series aircraft. He logged over 7,000 hours in thirty-four aircraft types. Cabana was selected by NASA in June 1985 in Group 11, and completed astronaut training in July 1986. His initial assignment was as the Astronaut Office Space Shuttle flight software co-ordinator until November 1986, when he was assigned as deputy chief of aircraft operations for the Johnson Space Center for 2.5 years. Prior to his assignment to command mission STS-88, Cabana served three years as NASA's Chief of the Astronaut Office from 1994-97.

Mission STS-41 aboard *Discovery* (October 06-10 1990) deployed the *Ulysses* spacecraft, operated the Shuttle Solar Backscatter Ultraviolet Instrument (SSBUV), and activated a controlled "fire in space" experiment. STS-53 aboard *Discovery* (December 02-09, 1992) deployed the classified Department of Defense 'DOD-1' payload, then performed several Military-Man-in-Space and NASA experiments. STS-65 aboard *Columbia* (July 08-23, 1994) conducted the second International Microgravity Laboratory (IML-2) mission, utilizing the long 'Spacelab' module, with eighty-two experiments from fifteen countries. Cabana commanded STS-88 aboard *Endeavour* (December 04-15, 1998), the first International Space Station assembly mission. The US-built *Unity* module was attached to the Russian-built Functional Cargo Block (FCB) *Zarya*. The crew performed IMAX Cargo Bay Camera operations, and deployed two satellites. Cabana logged logged a total of 37 days, 22 hours and 42 minutes (over 1,010 hours) in space.

After mission STS-88, he served as the deputy director of Flight Crew Operations. Joining the ISS program in October 1999, Cabana served as manager for International Operations. Robert Cabana retired from the Marine Corps in August 2000 in the rank of Colonel. From August 2001 to September 2002, he served as director of Human Space Flight Programs, Russia. From November 2002 to March 2004, he served as director of Flight Crew Operations; then was assigned as deputy director of the Johnson Space Center for 3.5 years. In October 2008, he was reassigned as Director of the John F. Kennedy Space Center (KSC), serving for over 12 years. In May 2021, Cabana was appointed as the Associate Administrator of NASA.

He is a Member of the Association of Space Explorers, and an Associate Fellow in the Society of Experimental Test Pilots. Cabana has received thirteen awards, including the Defense Superior Service Medal; the DFC; the National Intelligence Medal of Achievement; the NASA Distinguished Service Medal, two NASA Exceptional Service Medals; the De la Vaulx Medal by the Federation Aeronautique Internationale (FAI, 1994); and was inducted into the Astronaut Hall of Fame (May 2008).

## Caldwell Dyson, Tracy E.. STS-118, Soyuz TMA-18/Expn 23/24, Soyuz MS-25/Expn 70/71.

Dr. Tracy Ellen Caldwell Dyson was born in Arcadia, California, USA. She received a Bachelor of Science degree in Chemistry (1993); and a Ph.D. in Physical Chemistry from the University of California, Fullerton, in 1997. At the University of California, Davis, Caldwell Dyson taught general chemistry laboratory, and began her graduate research. In 1997, she received the Camille and Henry Dreyfus Postdoctoral Fellowship in Environmental Science to study atmospheric chemistry at the University of California, Irvine. In June 1998, she was hired by NASA, and soon began astronaut training. In 1999, Caldwell Dyson was assigned to the Astronaut Office ISS Operations Branch as a Russian Crusader, participating in the testing and integration of Russian hardware and software products developed for the ISS. She served as CapCom during 'Expeditions 4', '5' and '6'. In 2003, she moved to the Astronaut Shuttle Operations Branch and was assigned to flight software verification in the Shuttle Avionics Integration Laboratory. She was lead CapCom for 'Expedition 11' to the ISS.

Caldwell Dyson was selected in the June 1998 NASA Group. On mission STS-118 aboard *Endeavour* (August 08-20, 2007), the crew added a truss segment, a new gyroscope, and external spare parts platform to the ISS. As part of the 'Expedition 23/24' crew, she launched on April 02, 2010, aboard *Soyuz TMA-18* to the ISS for 176 days, one hour and 18 minutes (from April to September 2010). When the STS-131 crew arrived for April 07-17, that was the first and only time four women were together on board the same spacecraft. She completed three EVA, for a total time of 22 hours and 49 minutes. Dr. Caldwell Dyson has logged 188 days, 19 hours and 14 minutes in space. On May 29, 2023, she was officially assigned to 'Expedition 70/71' to fly aboard the *Soyuz MS-25*. As Tracy Dyson, she is the host of a series on ***NASA TV*** called *StationLife*, which focuses on facets of life aboard the ISS. She has received eighteen awards, including the Big West Scholar Athlete (1989-91); the Pro Femina Research Consortium Graduate Research Award (1996); the Outstanding Doctoral Student Award in Chemistry from the University of California, Davis (1997); the NASA Performance Award (2001 & 2002); the NASA Group Achievement Award - Russian Crusader Team (2000); and the NASA Superior Accomplishment Award (2000).

## Camarda, Charles J.. STS-114.

Dr. Charles "Charlie" Joseph Camarda was born in May 1952 in Queens, New York, USA. He graduated from Archbishop Molloy High School in 1970; received a Bachelor's degree in Aerospace Engineering from Polytechnic Institute of Brooklyn in 1974; a Master's degree in Engineering Science from George Washington University in 1980; and a Doctorate in Aerospace Engineering from Virginia Polytechnic Institute and State University in 1990. Dr. Camarda holds seven Patents, including on NASA's Heat-Pipe-cooled Sandwich Panel. NASA named him a Mission Specialist in the 1996 Group, and he served as backup crew for 'Expedition 8'. He launched on STS-114 aboard shuttle *Discovery* on July 26, 2005, and logged 13 days, 21 hours and 32 minutes in space. Following that flight, he was the Director of Engineering at Johnson Space Center, and served as Senior Advisor for Innovation to the Office of Chief Engineer, NASA HQ. He retired from NASA in 2019. Camarda continues to work for NASA, and has pioneered an engineering pedagogical approach called EPIC; that explores how to develop a mission as a collaborative training project.

## Cameron, Kenneth D.. STS-37, STS-56, STS-74.

Col. Kenneth Donald Cameron was born in November 1949 in Cleveland, Ohio, USA. He received a Bachelor of Science degree in Aeronautics and Astronautics from Massachusetts Institite of Technology in 1978; and a Master of Science degree in Aeronautics and Astronautics in 1979. He later earned a Master of Business Administration degree from Michigan State University in 2002. Cameron was commissioned in the US Marine Corps in 1970, and after graduating from The Basic School and the Vietnamese Language School, he was assigned to the Republic of Vietnam for a one-year tour of duty as an infantry platoon commander with the 1st Battalion, 5th Marines. Returning to the US, he served as executive officer with India Company, 3rd Battalion, 2nd Marines at Marine Corps Base Camp Lejeune, North Carolina. He received his Naval Aviator wings in 1973, and flew A-4M Skyhawks. In 1976, Cameron was reassigned to MIT, where he participated in the Marine College Degree and Advanced Degree Programs. He then had a one-year flying duty with Marine Aircraft Group 12 at Marine Corps Air Station Iwakuni, Japan. Graduating from the US Naval Test Pilot School at Patuxent River, Maryland, in 1983, Cameron was assigned as project officer and test pilot with the Systems Engineering Test Directorate at the Naval Air Test Center. He has logged over 4,000 hours flying time in forty-eight aircraft types.

Selected by NASA in May 1984, was CapCom for missions STS-28, 29, 30, 33, and STS-34. In 1994, he served as the first NASA Director of Operations in Star City, Russia; receiving training in Soyuz and Mir spacecraft systems. Cameron returned to the space program in October 2003, taking a founding position as a Principal Engineer in the NASA Engineering & Safety Center, based at Langley Research Center, Virginia. He served as pilot on STS-37 aboard *Atlantis* (April 05-11, 1991), deploying the Gamma Ray Observatory; and was Commander on STS-56 aboard *Discovery* (April 09-17, 1993), carrying ATLAS-2, and conducting solar and atmospheric studies; and STS-74 aboard *Atlantis* (November 12-20, 1995), the second NASA mission to dock with *Mir*, and the first mission to use the shuttle to assemble a module and attach it to a space station. He logged over 561 hours in space.

Following his retirement from NASA on August 05, 1996, Cameron joined **Hughes Training, Inc.**, a subsidiary of **General Motors Corporation** (GM). In September 1997, he transferred to **Saab Automobile, AB**, in Sweden, as Vehicle Line Executive for the Saab 9-3 auto. Returning to the US, he worked at the GM Technical Center in Warren, Michigan. In December 2008, he joined **Northrop Grumman Corporation** as Director of Houston Operations. Kenneth Cameron has received twelve awards, including two DFCs; the Defense Superior Service Medal; the Legion of Merit; the Navy Commendation Medal with Combat Valor device; the NASA Exceptional Achievement Medal; and the Charles Stark Draper Laboratory Fellowship.

## Carey, Duane G.. STS-109.

Lt. Col. Duane Gene "Digger" Carey was born in April 1957 in St. Paul, Minnesota, USA. He graduated from Highland Park High School, St. Paul, Minnesota, in 1975; received a Bachelor of Science degree in Aerospace Engineering and Mechanics (1981); and a Master of Science degree in Aerospace Engineering (1982) from the University of Minnesota. In 1981, he received his commission from the Air Force Reserve Officer Training Corps, and graduated from Undergraduate Pilot Training in 1983. He flew the A-10A Thunderbolt II during tours at the England AFB, Louisiana, and Suwon Air Base, Republic of Korea. In 1991, he was selected to attend the USAF Test Pilot School at Edwards AFB, California. Carey logged over 4,300 flying hours in over thity-five aircraft types. Selected by NASA in April 1996, he qualified as a pilot after two years' training. Duane Carey piloted shuttle *Columbia* on STS-109 (March 01-12, 2002), during the fourth Hubble Space Telescope servicing mission. The service and upgrade was accomplished by four crew during five EVAs on five consecutive days. Carey retired from NASA in October 2004, and has received the DFC with combat Valor Device; and three Air Medals. He is a member of the National Space Society, the American Motorcycle Association, and the Air Force Association.

## Carpenter, M. Scott. Mercury-Atlas "Aurora 7".

Cmdr. Malcolm "Scott" Carpenter (1925-2013) was born in May 1925 in Boulder, Colorado, USA. On February 12, 1943, he enlisted in the US Navy at Lowry Field, near Denver. He returned to Boulder in November 1945 to study Aeronautical Engineering at the University of Colorado. On May 29, 1962, the university granted him his Bachelor of Science degree. Impressed by his flying performance, Carpenter was recommended to the US Naval Test Pilot School. He was part of Class 13, NAS Patuxent River in 1954. He attended the Navy General Line School in California in 1957, then the Naval Air Intelligence School at NAS Anacostia in Washington, D.C.. Selected in the initial group of thirty-two astronaut candidates, he became one of the "Mercury Seven", announced at a Press conference on April 09, 1959. The "original seven" group consisted of Carpenter, Gordon L. Cooper, John H. Glenn, Virgil I. "Gus" Grissom, Walter "Wally" M. Schirra, Alan B. Shepard, Jr., and Donald K. "Deke" Slayton. Carpenter was considered the most physically fit, despite his smoking habit. When an *SM-65D Atlas* rocket exploded a few minutes after liftoff on May 18, 1959, astronaut Alan Shepard turned to John Glenn and said, *"Well, I'm glad they got that out of the way"*.

Carpenter flew aboard *Mercury-Atlas "Aurora 7"* on May 24, 1962, watched by forty million television viewers. He performed five onboard experiments, and became the first American astronaut to eat solid food in space. NASA's official history of Project Mercury says that until the third pass over Hawaii, the mission was "considered the most successful to date". Christopher C. Kraft, Jr. wrote in his 2001 memoir *"He [Carpenter] was completely ignoring our request to check his instruments. ... I swore an oath that Scott Carpenter would never again fly in space"*. At the retrofire event, a Pitch Horizon Scanner malfunctioned, forcing Carpenter to manually control his re-entry, but caused an overshoot of the planned splashdown point by 400 km (250 miles). He inflated his liferaft, climbed into it, and awaited rescue. The sea around him was stained with green dye. Three hour later he was recovered by helicopter, which took him to USS *Intrepid*. Carpenter logged four hours and 56 minutes in space, and retired on August 19, 1967.

When asked in 2012 what his legacy would be, he replied, *"I was an astronaut and an aquanaut"*. He received eighteen awards, including the Legion of Merit; the DFC; the American Campaign Medal; the WWII Victory medal; the Korean Service Medal with two battle stars; the Collier Trophy; the New York City Gold Medal of Honor; and the Numismatica Italiana Award. He was named to the International Space Hall of Fame (1981); the US Astronaut Hall of Fame (1990); and the International Air & Space Hall of Fame

(2008). Carpenter was married four times, divorced three times, and had a total of eight children by three wives. The character of Scott Tracy in the ***Thunderbirds*** TV series was named after him. He wrote four books, including *We Seven: By the Astronauts Themselves* and *For Spacious Skies: The Uncommon Journey of a Mercury Astronaut*, and two science fiction titles. In September 2013, Carpenter suffered a stroke and was hospitalized in Denver, Colorado. He died on October 10, 2013, aged eighty-eight years.

## Carr, Gerald P.. Skylab 4.

Col. Gerard Paul Carr (1932-2020) was born in August 1932 in Denver, Colorado, USA. He was an American mechanical and aeronautical engineer, and NASA astronaut. In 1950, he was appointed a midshipman with the Naval Reserve Officers Training Corps (NROTC. After graduating in 1954, he received his commission in the US Marine Corps, received flight training at Naval Air Stations Pensacola and Kingsville, and was then assigned to VMF (AW)-144. One of nineteen astronauts selected by NASA in April 1966, he served as CapCom for the *Apollo 8* and *Apollo 12* missions, and was involved in the development and testing of the Lunar Roving Vehicle.

Carr was Commander of *Skylab 4* from November 16, 1973 to February 08, 1974, the third and final crewed visit to the *Skylab Orbital Workshop*. The three-man crew completed fifty-six experiments, 26 science demonstrations, 15 subsystem-detailed objectives, and 13 student investigations during their 1,214 Earth orbits. They logged 338 hours of operations of the Apollo Telescope Mount, making extensive observations of the Sun's solar processes. The crew held the record for individual time in space at 84 days and 15 minutes (over 2,017 hours) until March 1978. Carr retired from the Marine Corps as Colonel in September 1975, and from NASA in June 1977. He became manager of corporate development at **Bovay Engineers, Inc.**, a Houston engineering consulting firm, later becoming a senior vice president until leaving in 1981. In 1984, Carr founded **CAMUS, Inc.**, based in Vermont. This was a major contributor as a technical support subcontractor to **The Boeing Company** in the crew systems design of the ISS. Gerald Carr died on August 26, 2020 in Albany, New York, aged eighty-eight years and four days.

## Carter, Manley L., Jr.. STS-33.

Capt. Dr. Manley Lanier "Sonny" Carter, Jr., M.D., (1947-91) was born in August 1947 in Macon, Georgia, USA. He received a Bachelor of Arts degree in Chemistry from Emory University, Atlanta, in 1969; and a Doctor of Medicine from there in 1973. In 1974, Carter entered the US Navy and completed Flight Surgeon school in Pensacola, Florida. He was designated a Naval Aviator on April 28, 1978. As a fighter pilot flying F-4 Phantoms, he completed a nine-month Mediterranean cruise aboard *USS Forrestal* in 1981. In September 1982, he attended US Navy Fighter Weapons School (TOPGUN); and attended the US Naval Test Pilot School, graduating in June 1984. Carter logged 3,000 flying hours and made 160 aircraft carrier landings.

In addition to his intercollegiate athletic career, Carter was an intramural wrestling champion. While attending medical school, he played professional soccer. In 1970, he signed with the Atlanta Chiefs of the North American Soccer League, for which he played three seasons. Selected in 1984 NASA Group 10, he was assigned as Extra-Vehicular Activity representative for the Mission Development Branch of the Astronaut Office when selected to the crew of STS-33 aboard *Discovery* (November 22-27, 1989), the fifth Department of Defense classified mission. Dr. Carter died on April 05, 1991, in the crash of Atlantic Southeast Airlines Flight 2311, in Brunswick, Georgia. At age forty-three years, he was traveling for NASA. He received seven honors, including the Navy Achievement Medal, and the NASA Meritorious Service Medal (1988). He was a member of the Society of US Naval Flight Surgeons, the Marine Corps Aviation Association, the Tailhook Association, and the Society of Experimental Test Pilots.

## Caspar, John. STS-36, STS-54, STS-62, STS-77.

Col. John Caspar was born in July 1943 in Greenville, South Carolina, USA. He graduated with a Bachelor of Engineering Science degree from the USAF Academy (1966); and a Master of Science degree in Aeronautics from Purdue University in 1967. He earned his pilot wings at Reese AFB, Texas, and became a USAF fighter pilot, flying 229 combat missions in the Vietnam War. Selected to attend the USAF Test Pilot School at Edwards AFB, California, he graduated with Class 74A. He later became Commander of the 6513th Test Squadron, where he conducted flight test programs to evaluate and develop tactical aircraft weapons systems. He was then assigned to Headquarters, USAF at the Pentagon and was Deputy Chief of the Special Projects Office. Col. John Caspar logged over 10,000 flying hours in fifty-two aircraft types.

Selected by NASA in May 1984, Caspar logged over 825 hours in space. He was the pilot on mission STS-36 (February 28-March 04, 1990) aboard shuttle *Atlantis*, carrying classified Department of Defense payloads; and Commander on STS-54 (January 13-19, 1993) aboard *Endeavour*, which deployed the USD 200 million NASA *Tracking and Data Relay Satellite* (TDRS-F); STS-62 (March 04-18, 1994) aboard *Columbia*, with primary payloads the US Microgravity Payload (USMP-2) and the Office of Aeronautics and Space Technology (OAST-2), also testing technology for use on future spacecraft; and STS-77 (May 19-29, 1996) aboard *Endeavour,* when the crew performed a record number of rendezvous sequences (one with a *SPARTAN* satellite, and three with a deployed *Satellite Test Unit*).

John Caspar became the Director of Safety, Reliability, and Quality Assurance at the Johnson Space Center, and after the *Columbia* disaster in February 2003, he became the NASA Mishap Investigation Team's deputy for the debris recovery operation; directing the efforts of over 6,000 ground, air, and water search personnel. He was co-chair of the Return-To-Flight Planning Team, then joined the Space Shuttle Program and became Manager of the Management Integration and Planning Office. He is currently Associate Manager of the Space Shuttle Program. Caspar has received twelve different awards, including two Legion of Merit; two DFCs; eleven Air Medals; six Air Force Commendation Medals; two NASA Distinguished Service Medals; and the National Intelligence Medal of Achievement.

## Cassada, Josh A.. SpaceX Crew-5/Expedition 68.

Dr. Josh Aaron Cassada was born in July 1973 in San Diego, California, USA. He graduated summa cum laude from Albion College with a Bachelor of Arts degree in Physics in 1995; and received a Master of Arts (1997) and Ph.D. in Physics (2000), both from the University of Rochester. While at university, he studied high-energy particle physics and conducted research at the Fermi National Accelerator Laboratory. He completed flight training in 2001, deployed in support of Operation Iraqui Freedom and Operation Enduring Freedom, and flew in the humanitarian Operation United Assistance after the 2004 Indian Ocean tsunami. In 2006, Cassada graduated from the US Naval Test Pilot School, and became an instructor test pilot. In 2011, he worked for the Defense Contract Management Agency at Boeing in Seattle.

Prior to his selection to join NASA in 2013, he was serving as a test pilot in the US Navy, and has logged over 3,500 hours in over forty aircraft types; including twenty-three combat missions. In August 2018, Cassada was selected for *CTS-1*, the first operational mission of the Boeing *CST-100 Starliner*. He was subsequently assigned to SpaceX 'Crew-5', and launched on October 05, 2022, aboard Crew Dragon *Endurance*, joining 'Expedition 68'. He logged 157 days, ten hours and one minute in space; including three EVAs totaling 21 hours and 24 minutes. Dr. Cassada has received the Defense Meritorious Service Medal; three Navy and Marine Corps Commendation Medals, including the Combat Valor device; and the Navy and Marine Corps Achievement Medal. He is married, with two children.

## Cassidy, Christopher J.. STS-127, Soyuz TMA-08M/Expedition 35, Soyuz MS-16/Expedition 63.

Capt. Christopher "Chris" John Cassidy was born in January 1970 in Salem, Massachusetts, USA. He is a former US Navy Seal, serving over ten years. Chris Cassidy graduated from BUD/S Class 192 in 1993 as Honor man. He has a Bachelor's degree in Mathematics (1993) from the US Naval Academy, Annapolis, Maryland; a Master's degree in Ocean Engineering from the Massachusetts Institute of Technology (2000); and an honorary Doctorate from Husson University, Bangor, Maine.

Two weeks after the terrorist attacks of September 11, 2001, Cassidy deployed to the Afghanistan region, serving as Ground Assault Force Commander for international and US-only combat missions. Selected by NASA in May 2004, Cassidy became the five hundredth person to fly in space. As a Mission Specialist on STS-127 aboard *Endeavour* (July 15-31, 2009), he helped deliver the Exposed Facility of the Japanese Experiment Module (JEM-EF) to the ISS, and completed three EVAs. Cassidy was a Flight Engineer for 'Expedition 35' , launched on March 28, 2013, flying to the ISS aboard *Soyuz TMA-08M*, and he made two EVA. Chris Cassidy was Commander of the ISS on 'Expedition 63', launched aboard *Soyuz MS-16* on April 09, 2020, with cosmonauts Ivan Vagner and Anatoli Ivanishin. They were joined by NASA astronauts Doug Hurley and Bob Behnken on May 31, with the docking of *Crew Dragon* 'Demo-2' capsule. Cassidy returned to Earth on October 21, 2020. He logged a total of 377 days, 17 hours and 49 minutes in space, and did ten EVAs for a total time of 54 hours and 51 minutes.

Chris Cassidy is a licensed Amateur Radio operator, holding US Technician level license KF5KDR. While on the ISS, he took part in several Amateur Radio on the International Space Station events with students in the USA. He served as CapCom at Johnson Space Center, became the EVA branch chief, leading the ground teams that plan spacewalks and ensure astronaut safety. Cassidy became Chief of the Astronaut Office from July 2015 to June 2017. He has ten Awards and decorations, including the Global War on Terrorism Expeditionary Medal; and retired from the Navy, and from NASA after 17 years' service on May 28, 2021.

## Cernan, Eugene "Gene" A.. Gemini IX, Apollo 10, Apollo 17.

Capt. Eugene "Gene" Cernan (1934-2017) was born in March 1934 in Chicago, Illinois, USA. He graduated from Purdue University in 1956 with a degree in Electrical Engineering and was commissioned in the US Navy that same year. As a fighter pilot, he made some 200 landings on aircraft carriers. In 1963, he earned a Master's degree in Aeronautical Engineering at the US Naval Postgraduate School, Monterey, California. Later in 1963, Cernan was among the fourteen men named to the third NASA group of astronauts. Gene Cernan's first space mission was with Thomas P. Stafford, launched aboard *Gemini IX* on June 03, 1966. Cernan became the second American to walk in space when he undertook more than two hours of EVA in what proved to be a dangerously faulty space suit. During their three-day mission, *Gemini IX* made three rendezvous with an *Agena* target vehicle. On May 18, 1969, Cernan, Tom Stafford and John W. Young began the eight-day mission of *Apollo 10*. As Lunar Module pilot, Cernan brought the craft into a close lunar orbit, within fifteen kilometers (nine miles) of the Moon's surface. They completed a complex series of orbital maneuvers before rejoining John Young in the Command Module. Their mission performed all the necessary functions for a lunar landing, but for the landing itself.

Launched on December 07, 1972, *Apollo 17* reached lunar orbit on December 10. As commander of the mission, Cernan and geologist Dr. Harrison Schmitt alighted from the LM named *Challenger* and explored the Taurus-Littrow Valley region of the Moon's surface. This area was selected for its geologic interest, notably the proximity of relatively young (less than three-billion-year-old) volcanic areas. Cernan and Schmitt traveled 30.5 km (19 miles) in their electrically powered lunar rover, and deployed an experiment package that transmitted data back to Earth for several years. They collected some 115 kg (249 lb) of rocks and surface materials. Gene Cernan was the last person to walk on the Moon in the twentieth century. The CSM pilot was Ronald "Ron" Evans, and splashdown was on December 19, 1972. He logged a total of 23 days, 14 hours and 15 minutes in space; including four EVAs totaling 24 hours and 11 minutes. Cernan later assisted with the *Apollo-Soyuz Test Project* (completed in July 1975), and in 1976 he resigned from the Navy and the space program to enter private business. In 1999, Gene Cernan co-wrote, with journalist Don Davis, *The Last Man on the Moon*. A documentary film of the same name was released in 2014.

## Chaffee, Roger B.. Apollo 1.

Roger Bruce Chaffee (1935-67) was born in February 1935 in Grand Rapids, Michigan, USA, where he became an Eagle Scout. After graduating from Central High School in 1953, he accepted a Naval Reserve Officers Training Corps scholarship. After studies at Illinois Institute of Technology, he transferred to Purdue University in 1954, continuing his involvement in Phi Kappa Sigma and obtaining his private pilot's license on May 24, 1957. After graduating with a Bachelor of Science degree in Aeronautical Engineering (with distinction), Chaffee completed his Navy training, was commissioned as an ensign and began pilot training at Naval Air Station Pensacola, Florida. His time in the Heavy Photographic Squadron included eighty-two missions taking photos of Cuba during the Cuban Missile Crisis of October 1962, earning him the Air Medal.

During his Navy career, Chaffee logged more than 2,300 hours flying time, including more than 2,000 hours in jet aircraft. In 1963, Chaffee was selected, along with thirteen other pilots from an initial pool of 1,800 applicants, to be part of NASA Astronaut Group 3. He served as capsule communicator (CapCom) for the *Gemini 3* and *Gemini 4* missions, and in 1966 received his first spaceflight assignment as the third-ranking pilot for *Apollo 1*. He was assigned to work on flight control, communications, instrumentation, and attitude and translation control systems in the *Apollo* program. The announcement of the three-man crew for *Apollo 1* was made public on March 21, 1966. The two-week flight was intended to test the spacecraft systems and the control and ground tracking facilities.

Chaffee was a member of the crew when a flash fire swept their space capsule during a "plugs out" simulation of a launching scheduled for February 21, 1967. Chaffeee was seated at the right side of the cabin, and his main role was to maintain communications with the blockhouse. A momentary power voltage surge was detected; believed to accompany an electrical short in equipment located on the lower left of the cabin. At 23:31:04 GMT a voice, most likely Ed White's, was heard yelling ***"[We]'ve got a fire in the cockpit***". A moment later, a garbled transmission, possible from Chaffee: *"We've got a bad fire – let's get out ... we're burning up!"* Assigned emergency roles called for Gus Grissom, in the left seat, to open the cabin pressure vent valve, after which Ed White in the center seat was to open the plug door hatch, but Grissom was prevented by the intensity of the fire. The hatch was held closed by the cabin pressure, and increasing pressure burst the inner cabin wall at 23:31:19 GMT. After about twenty-five seconds of being fed by a cabin pressure of pure oxygen at 115 to 200 kPa, the primary fire decreased in intensity and started producing large clouds of smoke. There was a final howl of pain, then silence from the capsule.

Chaffee lost consciousness because of a lack of oxygen, which sent him into cardiac arrest, and he died around 6.31 p.m. local time on January 27 of toxic gas asphyxiation; along with veteran space travelers Virgil I "Gus" Grisson and Edward "Ed" H. White, II. It took about five minutes to get all three hatches open, but it was too late. When Deke Slayton peered in, he saw a blanket of black ash covering the interior. *"It looks like the inside of a furnace"*, he said. The men had suffered serious, but not life-threatening burns. Chaffee was posthumously awarded the Congressional Space Medal of Honor, and a second Air Medal. Chaffee and Grissom were buried in Arlington National Cemetery, while Ed White was buried at West Point Cemetery, New York. Chaffee's widow received USD 100,000 from the life insurance portion of the Contract the astronauts had signed with two publishing firms for exclusive rights to stories and photographs. She also received USD 16,250 p.a. for the life of the insurance Contract.

Roger Chaffee is memorialized in many ways, from the Chaffee Crater on the far side of the Moon, to the Roger B. Chaffee Planetarium in his hometown of Grand Rapids, Michigan. Another memorial is Chaffee Hill on Mars, 14.3 km south-southwest of Columbia Memorial Station. A park in Fullerton, California is named after Roger Chaffee, and other parks are named for his dead comrades. Chaffee's name is included on the plaque left on the Moon in 1971 by the crew of *Apollo 15* with the ***Fallen Astronaut*** statue. The re-designed *Apollo* capsule later had an atmosphere of 60% oxygen and 40% nitrogen for the launch. Astronaut spacesuits, originally made of nylon, were changed to Beta cloth, a non-flammable, highly melt-resistant fabric woven from fiberglass and coated with Teflon. There were many other changes, including replacing 5,000 square inches of flammable Velcro cabin materials with self-extinguishing ones, consolidating the twenty miles of wiring and plumbing and covering it with protective Teflon insulation.

## Chamitoff, Gregory E.. STS-124. STS-134/Expedition 17/18.

Dr. Gregory Errol Chamitoff was born in August 1962 in Montreal, Canada, to a Jewish family of Russian origin. He is a Canadian-born American proponent of "Jews on the Moon", an engineer and former NASA atronaut. He was inspired to become an astronaut by watching the 1969 *Apollo 11* lunar landing. Chamitoff received a Bachelor of Science degree in Electrical Engineering from California Polytechnic State University in 1984; a Master of Science degree in Aeronautical Engineering from California Institute of technology in 1985; a Ph.D. in Aeronautics and Astronautics from Massachusetts Institute of Technology in 1992; and a Master of Science degree in Physical Sciences (Space Science) from the University of Houston at Clear Lake, in 2002. He developed a self-guided robot for his undergraduate thesis project, and his doctoral thesis was titled *"Robust Intelligent Flight Control for Hypersonic Vehicles"* (1992).

From 1993-95, Chamitoff was a visiting professor at the University of Sydney, Australia, where he led a research group in the development of autonomous flight vehicles, and taught courses in flight dynamics and control. In 1995, he joined the Motion Control Systems Group in the Missions Operations Directorate at JSC, where he developed software applications for spaceflight attitude control monitoring, prediction, analysis, and maneuver optimization. Selected by NASA for the Astronaut Class of 1998, he qualified as a Mission Specialist in 2000. He worked in the Space Station Robotics branch, was lead CapCom for ISS 'Expedition 9', and helped develop onboard procedures and displays for Space Station system operations.

As 'Mission Specialist 5' on STS-124 aboard *Discovery* (May 31-June 14, 2008), which delivered *Kibo*, the Japanese Experiment Module (the largest module of the ISS) he spent six months aboard the ISS across 'Expedition 17' and '18' in 2008. *Discovery* carried the replacement parts for a malfunctioning toilet on the ISS. While Richard Garriott was aboard the ISS at the beginning of 'Expedition 18', Chamitoff and Garriott filmed the first magic show in space, and along with Yury Lonchakov, Michael Fincke and Garriott, filmed a science-fiction movie made in space, ***Apogee of Fear***. Gregory Chamitoff brought the first bagels into space (three bags of 18-sesame-seed Montreal-style). He served as 'Mission Specialist 4' on STS-134 aboard *Endeavour* (May 16-June 01, 2011) for 15 days and 17 hours, the last flight of *Endeavour*, which delivered the Alpha Magnetic Spectrometer, and an ExPRESS Logistics Carrier, and completed the US Orbital Segment of the ISS. He made two EVAs, and has lived and worked for 198 days, 18 hours and two minutes in space as a Flight Engineer, Science Officer and Mission Specialist.

## Chang-Diaz, Franklin R.. STS-61-C, STS-34, STS-46, STS-60, STS-75, STS-91, STS-111.

Dr. Franklin Ramon Chang-Diaz was born in April 1950 in San Jose, Costa Rica. He is the first Hispanic former astronaut. He is the sole founder and CEO of **Ad Astra Rocket Company** as well as a member of **Cummins**' board of directors. Franklin Chang-Diaz's parents sent him to the United States in 1967 with only $50 and no English language ability. After living with relatives, he graduated from the University of Connecticut with a Bachelor of Science degree in Mechanical Engineering in 1973; and from the Massachusetts Institute of Technology with a Ph.D. in Applied Plasma Physics and Fusion Technology in 1977. Franklin Chang-Diaz became an American citizen in 1977. He is a veteran of seven Space Shuttle missions, tying the American record, as of 2021 for the most spaceflights.

In 1980, Chang-Diaz was selected for the NASA astronaut program. He flew as 'Mission Specialist 3' on STS-61-C aboard *Columbia* (January 12-18, 1986); as 'Mission Specialist 2' on STS-34 aboard *Atlantis* (October 18-23, 1989) which deployed the *Galileo* spacecraft; on STS-46 aboard *Atlantis* (July 31-August 08, 1992) when he was 'Mission Specialist 4'; on STS-60 aboard *Discovery* (February 03-11, 1994) when he was 'Mission Specialist 3'; and on STS-75 aboard *Columbia* (February 22-March 09, 1996) when he was Payload Commander and 'Mission Specialist 4'. The primary mission objective was to carry the *Tethered Satellite System Reflight* (TSS-1R) into orbit and deploy it spaceward on a conducting tether. On STS-91 aboard *Discovery* (June 02-12, 1998) he was 'Mission Specialist 1' on the final *Shuttle/Mir* Docking Mission, which carried a prototype of the Alpha Magnetic Spectrometer (AMS) (designed to look for dark and missing matter in the Universe). Chang-Diaz was 'Mission Specialist 2' on the STS-111 flight of *Endeavour* (June 05-19, 2002), which carried the MPLM with experiment racks and three stowage and resupply racks, and during which he participated in three EVAs to help repair the robotic arm of the ISS.

Chang-Diaz logged a total of 66 days, 18 hours and 16 minutes in space. He was a visiting scientist at MIT (1983-93), where he led a project that developed plasma propulsion for use in human flights to Mars. He served as director of NASA's Advanced Space Propulsion Laboratory at the Johnson Space Center in Houston from 1993 to 2005, when he retired from the agency. Chang-Diaz then founded and became president of the **Ad Astra Rocket Company**. He also taught at Rice University, Texas, and the University of Houston, Texas. From 2007-23, the English Wikipedia of Franklin Chang-Diaz has received more than 390,532 page views. His biography is available in twenty-nine different languages on Wikipedia.

## Chawla, Kalpana. STS-87, STS-107.

Dr. Kalpana "KC" Chawla was born in March 1962 in Karnal, India. She was an Indian-born American astronaut and aerospace engineer who was the first woman of Indian origin to fly to space. She studied physics, chemistry and math in high school and completed her schooling from Tagore Baal Niketan Senior Secondary School, Karnal. Kalpana Chawla excelled at engineering college in India. Kalpana obtained a Bachelor of Engineering degree from the University of Punjab and moved to the United States at the age of twenty years.

Kalpana Chawla had to overcome gender stereotypes and societal expectations that discouraged women from pursuing careers in science and engineering.

Chawla obtained her Master's degree in Aerospace Engineering from the University of Texas in 1984. Chawla went on to earn a second Master's degree in 1986; and a Ph.D. in Aerospace Engineering in 1988 from the University of Colorado at Boulder. Kalpana held a Certified Flight Instructor rating for airplanes and gliders, and Commercial Pilot licenses for single and multi-engine airplanes and seaplanes. Kalpana Chawla joined the NASA Astronaut Corps in March 1995. She overcame the challenges of rigorous training, long hours, and the risks associated with space travel. Her first mission, STS-87 aboard *Columbia* (November 19-December 05, 1987) was as a Mission Specialist and primary robotic arm operator.

Chawla was a Flight Engineer and Mission Specialist on mission STS-107 aboard Space Shuttle *Challenger*, which broke up on re-entry over Texas on February 01, 2003, killing all seven astronauts aboard. At launch a piece of insulating foam broke off from the Space Shuttle external fuel tank and struck the thermal protection system tiles on the orbiter's left wing. The damage allowed hot atmospheric gases to penetrate the heat shield on re-entry and destroy the internal wing structure, which caused the orbiter to become unstable and break apart. She logged a total of 31 days, 14 hours and 54 minutes in space. Kalpana Chawla was posthumously awarded the Congressional Space Medal of Honor, and several streets, universities and institutions have been named in her honor. Her legacy continues to inspire people around the world to pursue their dreams and overcome adversity.

## Chiao, Leroy. STS-65, STS-72, STS-92, Soyuz TMA-5/Expedition 10.

Dr. Leroy Chiao was born in August 1960 in Milwauke, Wisconsin, USA. He graduated from Monte Vista High School in Danville in 1978; received a Bachelor of Science degree in Chemical Engineering from the University of California, Berkeley; and later earned a Master of Science (1985) and a Ph.D. in Chemical Engineering from the University of California, Santa Barbara, in 1987. Upon graduation, Chiao joined the **Hexcel Corporation** in Dublin, California, from 1987-89. He was involved in process, manufacturing, and engineering research on advanced aerospace materials, and worked on a joint NASA-JPL/Hexcel project to develop a precision segment reflector made entirely of advanced polymer composite materials for future space telescopes. In January 1989, he joined the **Lawrence Livermore National Laboratory**, where he was involved in fabrication of filament-wound and thick-section aerospace composites. An instrument-rated pilot, Chiao has logged over 2,500 flight hours in a variety of aircraft. At age 29, in January 1990 he was selected by NASA (the youngest in Group 13) and became an astronaut in July 1991.

He was a Mission Specialist on STS-65 aboard *Columbia* (July 08-23, 1994), which flew the second International Microgravity Laboratory, and conducted over eighty experiments. STS-72 aboard *Endeavour* (January 11-20, 1996) retrieved the *Space Flyer Unit* (launched from Japan ten months earlier), and deployed and retrieved the *OAST Flyer*. He performed two EVA to demonstrate tools and hardware. Mission STS-92 aboard *Discovery* (October 11-20, 2000) attached the Z1 truss and Pressurized Mating Adapter 3 to the ISS. Chiao is fluent in Mandarin Chinese, and learned Russian to communicate with Russian cosmonauts as part of the ISS program.

Launched aboard *Soyuz TMA-5*, he was Commander of 'Expedition 10', where he lived on the ISS from October 13, 2004 to April 24, 2005. Leroy Chiao logged a total of 229 days, seven hours and 38 minutes in space; including four EVAs totaling over 26 hours. Chiao left NASA in December 2005 to pursue employment and entrepreneurial business ventures in the private sector, both in the US and China. In early 2006, he joined the Atlanta firm of **SpaceWorks Enterprises, Inc.** (SEI) as an affiliate and technical advisor. In July 2006, Chiao accepted a position as the Executive Vice President for Space Operations and a Director of **Excalibur Almaz Limited.** He was responsible for operational aspects of spaceflight, including training for both the *Almaz* capsule and space station. He is now an entrepreneur, motivational speaker and an engineering consultant.

Chiao is currently serving as the chairman of the National Space Biomedical Research Institute User Panel, attached to the Baylor College of Medicine. In July 2007, he joined an expedition to visit Devon Island and conduct five days of webcasts from the Haughton-Mars Project Research Station to illustrate how NASA and other space agencies are learning to live on the Moon and Mars while here on Earth. From 2011-16, he served as the special advisor - human spaceflight for the **Space Foundatio**n. He is a co-founder and CEO of **OneOrbit**, a corporate keynote and training company. Chiao has six awards, including the De la Vaulx Medal (1994); NASA Group Achievement Awards (1995, 1997); NASA Exceptional Service Medals (1996, 2000); the NASA Distinguished Service Medal (2005); and was inducted into the Space Technology Hall of Fame for Mediphan - DistanceDoc and MedRecorder.

## Chilton, Kevin P.. STS-49, STS-59, STS-76.

Gen. Kevin Patrick "Chilli" Chilton was born in November 1954 in Los Angeles, California, USA. He is a retired USAF Four Star General. He graduated from St. Bernard High school, Playa del Ray, California in 1972; received a Bachelor of Science degree in Engineering Sciences from the UASF Academy, Colorado, in 1976; and a Master of Science degree in Mechanical Engineering from Columbia University, New York on a Guggenheim Fellowship in 1977. After receiving his pilot wings at Williams AFB, Arizona, in 1978, he qualified in the RF-4C Phantom II and was assigned to the Fifteenth Tactical Reconnaissance Squadron at Kadena Air Base, Japan. From 1978-80, he served as a combat-ready pilot and instructor pilot in Korea, Japan, and the Philippines. In 1982, Chilton attended the USAF Squadron Officer School, Alabama, and finished as the number one graduate for the year, receiving the Secretary of the Air Force Leadership Award. He served as an F-15 squadron weapons officer, instructor pilot, and flight commander until 1984 when he was selected for the USAF Test Pilot School. He graduated top in his class, to win the Liethen-Tittle Award, as the outstanding test pilot at the School.

In 1985, he completed Air Command and Staff College (by correspondence). In August 1987, Chilton was assigned to NASA and became an astronaut in August 1988. He was lead CapCom for numerous Shuttle flights, and served as Deputy Program Manager for the early ISS program. On STS-49 aboard *Endeavour* (May 07-16, 1992) the crew conducted the maiden test flight, and performed a record four EVA to retrieve, repair and redeploy the *International Telecommunications Satellite* (*INTELSAT*), along with a variety of medical, scientific and operational tests. The landing at Edwards Air Force Base was the first use of the Shuttle drag chute. On STS-59 aboard *Endeavour* (April 09-20, 1994) the SRL mission carried three large radars, which operated in three frequencies and four polarizations. The carbon monoxide (CO) sensor used gas filter radiometry to measure the global distribution of CO in the troposphere. Over 14,000 photographs were taken.

Chilton commanded STS-76 aboard *Atlantis* (March 22-31, 1996) which made rendezvous with *Mir* and transferred a NASA astronaut for a five-month stay on the Russian space station. The 'Spacehab' module carried in the shuttle payload bay was utilized extensively for transfer and return stowage of logistics and science and also carried Biorack, a small multipurpose laboratory used during the mission for research of plant and animal cellular function. STS-76 was also the first flight of Kidsat, an electronic digital camera controlled by classroom students via a Ku-band link fom JSC Mission Control. Chilton logged 29 days, eight hours and 22 minutes (over 704 hours) in space.

Kevin Chilton left NASA in 1998 to become Deputy Director of Political Affairs for Asia, Pacific, and the Middle East on the **Joint Staff**. In 2001, he completed Air War College (by correspondence). In April 2006, he was nominated to become the Commander of the **Air Force Space Command** at Peterson AFB, Colorado. As Commander, he oversees the whole space operations within the USAF jurisdiction, including the Satellites and Ballistic Missile. Gen. Chilton emphasized the important role of satellites to act as eyes for troops, especially those deployed abroad. He was nominated as Commander of **United States Strategic Command** in October 2007. In this capacity, he oversees the whole US nuclear arsenal and the Nuclear Triad. The primary Strategic Deterrence mission is to anticipate an incoming Nuclear-Strike towards the US. During his tenure the Nuclear Missile arsenal was modernised. Since the 2008 cyberattacks on the US, Chilton has emphasized the role of the US Cyber Division; with initiation of Operation Buckshot Yankee. In May 2010, the US Cyber Command was officially formed as the new Unified Combatant Command.

General Kevin Chilton has received eight personal decorations; three unit awards; three NASA awards; two Service awards; three campaign and service medals; and four service, training and marksmanship awards. He also has the National Space Trophy; the NASA "Top Fox" Flight Safety Award; the Commander's Trophy, Undergraduate Pilot Training; and was inducted into the US Astronaut Hall of Fame, May 2012.

## Clark, Laurel B.. STS-107.

Capt. Dr. Laurel Blair Clark (nee Salton) (1961-2003) was born in March 1961 in Ames, Iowa, USA. She was a medical doctor, US Navy Captain, and a NASA astronaut who died with her six fellow astronauts in the Space Shuttle *Columbia* disaster of February 01, 2003. She graduated from William Horlick High School, Racine, Wisconsin, in 1979; received a Bachelor of Science degree in Zoology from the University of Wisconsin-Madison; and a Doctor in Medicine (M.D.) from the University of Wisconsin in 1987. After completing medical school, she underwent postgraduate medical education in pediatrics from 1987-88 at the National Naval Medical Center. After training at the Naval Undersea Medical Institute, she was designated a Radiation Health Officer and Undersea Medical Officer. Clark was then assigned as the Submarine Squadron 14 Medical Department Head in Holy Loch, Scotland. She dove with Navy divers and performed many medical evacuations from United States submarines. After two years of operational experience, she was designated as a Naval Submarine Medical Officer and Diving Medical Officer.

Clark underwent six months of aeromedical training at the Naval Aerospace Medical Institute, Florida, and was designated as a Naval Flight Surgeon. She was assigned to Marine Attack Squadron 211 (VMA-211), which won the Marine Attack Squadron of the Year award for its successful deployment to the Western Pacific, where she flew on multiple aircraft. She was then assigned as the Group Flight Surgeon for the Marine Aircraft Group 13 (MAG-13). Before her selection as an astronaut candidate in April 1996, Clark served as a Flight Surgeon for Training Squadron 86 (VT-86) at NAS Pensacola.

STS-107 aboard *Columbia* was a 16-day dedicated science and research mission, the 113th flight of the Space Shuttle program. Working in two alternating continuous shifts, the crew successfully conducted about 80 experiments, with Clark's including gardening in space. The shuttle disintegrated upon re-entry, just 16 minutes before scheduled landing. Laurel Clark's final message to her friends and family was via an email sent from *Columbia*. She called the planet *"magnificent"*, and and explained that while she spends much of her time working back in 'Spacehab' and away from the sights of Earth, *"whenever I do get to look out, it is glorious"*. She also shared some of the intriguing effects of microgravity on human physiology, such as the constant challenge to stay adequately hydrated due to an *"almost non-existent"* sense of thirst.

Capt. Laurel Clark was a member of the Aerospace Medical Association, and the Society of US Naval Flight Surgeons. She was awarded Naval Astronaut/Flight Surgeon; Submarine Medical Officer; three Navy and Marine Corps Commendation Medals with two gold award stars; the National Defense Service Medal; the Overseas Service Ribbon; and posthumously the Defense Distinguished Service Medal; the Congressional Space Medal of Honor; the NASA Distinguished Service Medal; and NASA Space Flight Medal.

## Cleave, Mary L.. STS-61-B, STS-30R.

Dr. Mary Louise Cleave (1947-2023) was born in February 1947 in Southampton, New York, USA. She received a Bachelor of Science degree in Biological Sciences from Colorado State University in 1969; and a Master of Science degree in Microbial Ecology from Utah State University in 1975. From September 1971 to June 1980, she held graduate research, research psychologist, and research engineer assignments in the Ecology Center and Utah Water Research Laboratory at Utah State University. In 1979, Cleave received a Doctorate in Civil and Environmental Engineering from Utah State University.

Cleave was selected as an astronaut in May 1980. While a Mission Specialist on STS-61-B aboard *Atlantis* (November 26-December 03, 1985) the crew deployed the *MORELOS-B, AUSSAT II,* and *SATCOM K-2* communications satellites. The payload was the heaviest weight carried to orbit by a Shuttle to date. STS-30 aboard *Atlantis* (May 04-08, 1989) deployed the *Magellan* Venus-exploration spacecraft, the first planetary probe to be deployed from a Shuttle. She logged a total of ten days, 22 hours and two minutes in space. In early 1990, Cleave was selected as 'Mission Specialist 3' for the STS-42 mission, but withdrew herself for personal reasons shortly after.

Dr. Cleave left JSC in May 1991 to join NASA's Goddard Space Flight Center in Greenbelt, Maryland. She served as NASA Associate Administrator for the Science Mission Directorate, at NASA HQ from 2005-07. On January 18, 2009, as the inaugural speaker in the Heyden Distinguished Lecture Series, Cleave told students and others at Georgetown University about her education and career, and showed an original film of her Shuttle mission in 1985. Mary Cleave said, *"Looking at the Earth, particularly the Amazon rainforest, the amount of deforestation I could see, just in the five years between my two space flights down there, scared the hell out of me".* Unquote. Cleave was a Member of the Society for Professional engineers; the Association of Space Explorers; Women in Aerospace; and three sororities. She died from a stroke on November 27, 2023 at her home in Annapolis, Maryland, at age seventy-six years.

## Clifford, Michael R. U.. STS-53, STS-59, STS-76.

Lt. Col. Michael Richard "Rich" Uram Clifford (1952-2021) was born in October 1952 in California, USA. He received a Bachelor of science degree from the US Military Academy (USMA) at West Point, New York, in 1974; and a Master of Science degree in Aerospace Engineering from the Georgia Institute of Technology in 1982. Clifford graduated from the US Military Academy, West Point, New York, in June 1974, and was commissioned a second lieutenant in the US Army. In 1976, he was top graduate of his class and was designated an Army Aviator. After completing his graduate training in 1982, he was assigned to the USMA Department of Mechanics as an instructor and assistant professor. In December 1986, he graduated from the US Naval Test Pilot School and was designated an experimental test pilot. As a Master Army Aviator, he logged over 3,400 flying hours in a wide variety of fixed-wing and rotary winged aircraft. In 1995, Rich Clifford retired from the US Army with the rank of Lieutenant Colonel.

As a military officer, he was assigned to JSC in July 1987. Selected in 1990 NASA Group 13, he participated in the design, development, and evaluation of Shuttle payloads and crew equipment having EVA interfaces. From May 1984 to September 1995, he served as lead for Space Station vehicle/assembly issues. On STS-53 aboard *Discovery*, he was the first member of his 1990 astronaut class to reach space. Clifford was responsible for experiments including the Fluid Acquisition and Resupply Experiment (FARE), and conducted the Battlefield Laser Acquisition Sensor Test (BLAST) on a novel laser energy detector for the US military. On STS-59 aboard *Endeavour* (April 09-20, 1994) he and five other astronauts worked around the clock on imaging projects using SRL imaging radars, including Spaceborne Imaging Radar, Synthetic-aperture radar, and a carbon monoxide sensor known as Measurement of Air Pollution from Satellites (MAPS), to collect data on the Earth's surface and atmosphere. A few months after STS-59, at age forty-two years Clifford was diagnosed with Parkinson's disease, with mild symptoms in his right arm.

Mission STS-76 aboard *Atlantis* (March 22-31, 1996) docked with the *Mir* space station, and Clifford performed a six-hour EVA with Linda Godwin to mount experiment packages on the *Mir* docking module. After this mission, he decided he would not fly again, since he did not know how fast his Parkinson's condition would progress. He logged a total of 27 days, 18 hours and 24 minutes in space; including six hours of EVA. Clifford resigned from NASA in January 1997 to join **Boeing**'s Defense and Space Group as Flight Operations Manager for the ISS program. He later oversaw Boeing's shuttle program until it was winding down, and left Boeing in September 2011. Michael Clifford received eight awards, and was a Member of the Association of Space Explorers. In 2012, he received the Award for Public Leadership in Neurology from the American Academy of Neurology. Lt. Col. Michael Clifford died on December 28, 2021, at North Myrtle Beach, South Carolina, aged sixty-nine years.

## Coats, Michael L.. STS-41-D, STS-29R, STS-39.

Michael Lloyd Coats was born in January 1946 in Sacramento, California, USA. He graduated with a Bachelor of Science degree from the US Naval Academy in 1968; received a Master of Science degree in Administration of Science and Technology from George Washington University in 1977; and a Master of Science degree in Aeronautical Engineering from the US Naval Postgraduate School in 1979. Coats served as a naval aviator and test pilot (including as as a flight instructor from April 1976 to May 1977) before joining NASA in 1978 as an astronaut candidate. He logged over 5,000 hours flying time in twenty-eight aircraft types, and made 400 aircraft carrier landings.

Coats was pilot for STS-41-D aboard *Discovery* (August 30-September 05, 1984), when the crew successfully activated the OAST-1 solar cell wing experiment, deployed three satellites, operated the CSEF-III experiment, and used the IMAX motion picture camera. He was Commander of STS-29 aboard *Discovery* (March 13-18, 1989), which deployed a *Tracking and Data Relay Satellite* (TDRS), and took over 3,000 photographs of the Earth. Coats commanded STS-39 aboard *Discovery* (April 28-May 06, 1991), an unclassified Department of Defense mission which included Air Force Program-675 (AFP-675); Infrared Background Signature Survey (IBSS); and Chemical Release Observation. Classified payload was the Multi-Purpose Experiment Canister (MPEC). Crew worked two shift operations to deploy, operate and retrieve the *Shuttle Pallet Satellite-II* (SPAS-II) spacecraft, and conducted various experiments including research of both natural and induced phenomena in the Earth's atmosphere. Michael Coats logged a total of 463 hours in space.

From May 1989 to March 1990, he served as Acting Chief of the Astronaut Office. In 1991, Coats left the Astronaut Corps and retired from the US Navy to work for **Loral Space Information Systems**, serving as Vice President of Avionics and Communications

Operations. From 1996-98, he was Vice President of Civil Space Programs for **Lockheed Martin Missiles and Space** in Sunnyvale, California. From 1998-2005, he was Vice President of Advanced Space Transportation for Lockheed Martin in Denver, Colorado. From December 2005-12, he served as the 10th Director of the Johnson Space Center. He retired from NASA on December 31, 2012. Michael Coats is a Member of the Society of Experimental Test Pilots; and an Associate Fellow, American Institute of Aeronautics and Astronautics. He was awarded the Defense Superior Service Medal; three Navy DFCs; 32 Strike-Flight Air Medals; three Individual Action Air Medals; nine Navy Commendation Medals with Combat Valor device; and three NASA Space Flight Medals.

## Cockrell, Kenneth D.. STS-56, STS-69, STS-80, STS-98, STS-111.

Capt. Kenneth Dale "Taco" Cockrell was born in April 1950 in San Antonio, Austin, Texas, USA. He received a Bachelor of Science degree in Mechanical Engineering from the University of Texas at Austin in 1968, and was commissioned in the US Navy. In 1974, he earned a Master's degree in Aeronautical Systems from the University of West Florida. Cockrell trained as a pilot and was stationed from 1975-78 aboard the aircraft carrier *USS Midway*. He then became a test pilot for several years before serving two tours of duty aboard *USS Constellation* until 1987, when he resigned from the Navy and joined the Aircraft Operations Division of JSC as a research pilot.

Selected by NASA in January 1990 Group, after several technical assignments he served as Assistant to the Chief of the Astronaut Office for Shuttle operations and hardware, and as Chief of the Astronaut Office from 1997-98. He also served one year as the **Director of Operations, Russia**, in Star City, where he was the liaison between the Astronaut Office and the Gagarin Cosmonaut Training Center. Cockrell was a Mission Specialist on STS-56 aboard *Discovery* (April 08-17, 1993), during which the crew conducted atmospheric and solar studies. He was Pilot on STS-69 aboard *Endeavour* (September 07-18, 1995), with the primary objective of deploying and retrieving a *SPARTAN* satellite and the *Wake Shield Facility*. He was Commander on STS-80 aboard *Columbia* (November 19-December 07, 1996), which deployed and retrieved the *Wake Shield Facility*; and *ORFEUS* satellites, which study the origin and makeup of stars.

Kenneth Cockrell was Commander on STS-98 aboard *Atlantis* (February 07-20, 2001), which delivered the US *Destiny* laboratory module to the ISS, with three EVAs to complete its assembly. The crew relocated a docking port, and delivered supplies to the 'Expedition 1' crew. He was Commander on STS-111 aboard *Endeavour* (June 05-19, 2002), which delivered a new ISS resident crew and a Canadian-built mobile base for the robotic arm, and replaced one of the arm's joints. He logged over 1,560 hours in space. Captain Cockrell was reassigned in February 2006 to Flight Crew Operations' Aircraft Operations Division (AOD) at Ellington Field, as the WB-57 High Altitude Research Program Manager.

## Coleman, Catherine G. STS-73, STS-93, Soyuz TMA-20/Expedition 26/27.

Col. Dr. Catherine Grace "Cady" Coleman was born in Charleston, South Carolina, USA. She received a Bachelor of Science degree in Chemistry from Massachusetts Institute of Technology (MIT) in 1983, and was commissioned as a graduate of the Air Force Reserve Officer Training Corps. Coleman entered active duty at Wright-Patterson AFB as a research chemist in 1988. She received a Ph.D. in Polymer Science and Engineering from the University of Massachusetts, Amherst, in 1991. Selected in 1992 NASA Group 14, Cady Coleman flew on mission STS-73 aboard *Columbia* (October 20 to November 05, 1995) on the second United States Microgravity Laboratory scientific mission (USML-2). This focused on materials science, biotechnology, combustion science, the physics of fluids, and numerous other experiments in the pressurized 'Spacelab' module. On STS-93 aboard *Columbia* (July 22-27, 1999) she was Mission Specialist in charge of deploying the Chandra X-Ray Observatory and its Inertial Upper Stage. Col. Coleman served as Chief of Robotics for the Astronaut Office, to include robotic arm operations and training for all Space Shuttle and ISS missions.

She launched aboard *Soyuz TMA-20* on December 15, 2010, to join 'Expedition 26/27' on the ISS. Coleman departed the ISS on May 23, 2011, after logging 159 days on that mission. She logged a total of 180 days, four hours and zero minutes in space. As do many other astronauts, Coleman holds an amateur radio license (callsign: KC5ZTH). In 2019, the Irish postal service **An Post** issued a set of commemorative stamps for the fiftieth anniversary of the Apollo Moon landings. Catherine Coleman is featured, alonside fellow astronauts Neil Armstrong, Michael Collins, and Eileen Collins. She is part of the band ***Bandella,*** which also includes fellow NASA astronaut Stephen Robinson, Canadian astronaut Chris Hadfield, and Micki Pettit (wife of astronaut Donald Pettit). Coleman is a flute player, and took several flutes with her to the ISS, including one from Ian Anderson of ***Jethro Tull***. As of 2015, she was a guest speaker at the Baylor College of Medicine, for 'Saturday Morning Science'.

## Collins, Eileen M.. STS-63, STS-84, STS-93, STS-114.

Col. Eileen Marie Collins was born in November 1956 in Elmira, New York, USA. After graduating from Elmira Free Academy in 1974, she considered enlisting in the USAF, but her father was opposed. Collins earned an Associate degree in Mathematics in 1976 at Corning Community College; a Bachelor of Arts degree in Mathematics and Economics from Syracuse University in 1978, and was commissioned as a second lieutenant in the USAF. She was one of four women chosen for Undergraduate Pilot Training at Vance AFB, Oklahoma. During the American invasion of Grenada in October 1983, and although women were not allowed to fly in combat, her aircraft flew troops from Pope AFB, North Carolina, to Grenada. From 1986-89, she was an assistant professor in Mathematics and a T-41 instructor pilot at the USAF Academy in Colorado. Collins earned a Master of Arts degree in Space Systems Management from Webster University in 1989, and that year became the second woman pilot to attend the USAF Test Pilot School. She logged 6,751 flying hours in thirty aircraft types.

Selected as one of 23 candidates in 1990 NASA Group 13 to be a pilot astronaut, Collins was the first female Space Shuttle pilot on mission STS-63 aboard *Discovery* (February 03-11, 1995), which involved a rendezvous (approach and flyaround) with space station *Mir*. The shuttle carried the 'Spacehab' module and the *SPARTAN-204* astronomy satellite. Collins was pilot for STS-84 aboard *Atlantis* (May 15-24, 1997) and docked with *Mir*. This was the first and only time a launch had occurred without delays. Collins photographed Comet Hale-Bopp, the brightest comet seen for many years.

Eileen Collins became the first woman to command a US spacecraft with Shuttle mission STS-93 (July 23-27, 1999) aboard *Columbia*, (which had to be used because its airlock was inside the crew compartment, instead of the payload bay). On the third launch attempt, one of the main engine controllers failed, and one of the main engines cut out prematurely. Had the controller not been faulty, it might have pumped more liquid oxygen and caused the engine to explode! Crew deployed the Chandra X-Ray Observatory, weighing 30,852 pounds; and the inertial upper stage IUS, weighing 12,930 pounds. This was the heaviest payload ever launched by the Space Shuttle, so the crew was reduced to five members, and the mission was the shortest scheduled since 1990. The crew used the Southwest Ultraviolet Imaging System (SWUIS) to take ultraviolet images of the Earth, Moon, Venus and Jupiter. Collins was awarded the DFC for this mission.

After mission STS-84, Collins was the head of the Vehicle Systems Branch of the Astronaut Office, which she renamed the Spacecraft Systems Branch. After STS-93, she became the chief information officer at the Astronaut Office. Collins commanded STS-114 (July 26-August 09, 2005) aboard *Discovery*, to test safety improvements and resupply the ISS. The mission took 'Expedition 7' to the ISS, and brought 'Expedition 6' crew back. During this return-to-flight mission, Collins became the first astronaut to fly the Space Shuttle orbiter through a complete 360-degree pitch maneuver, so astronauts aboard the ISS could take photographs of its belly to ensure there was no threat from debris-related damage during re-entry. Asked if she felt afraid after the loss of *Columbia*, Collins replied, *"I would have to say no. I want to fly again. I am very mentally ready to go fly again ... when the shuttle is cleared to fly"*. Collins logged a total of 36 days, seven hours and 11 minutes (over 872 hours) in space.

Col. Collins retired from the Air Force in January 2005 and from NASA in May 2006 after a twenty-eight-year distinguished career. She served on the Board of Directors of **USAA** (banking and insurance) from 2006-21. She was chair of the NASA Space Operations Committee from 2007-11. Collins addressed the 2016 Republican National Convention in Cleveland, Ohio on July 20, 2016. She has received 19 awards, including the Defense Meritorious Service Medal; the Air Force Commendation Medal with one Oak Leaf Cluster; the NASA Outstanding Leadership Medal; the Free Spirit Award; the NASA Exceptional Service Award; the National Space Trophy; the Harmon Trophy (2020); and the Wright Brothers Memorial Trophy in 2022.

## Collins, Michael. Gemini X, Apollo 11.

Maj. Gen. Michael Collins (1930-2021) was born in October 1930 in Rome, Italy. He graduated in the Class of 1952 from the United States Military Academy at West Point with a Bachelor of Science degree in Military Science. He joined the USAF, and began basic flight training in the T-6 Texan at Columbus AFB in Columbus, Mississippi, in August 1952, then moved to San Marcos AFB in Texas to learn instrument and formation flying. At James Connally AFB in Waco, Texas, he trained in jet aircraft. In September 1953, he was chosen for advanced day-fighter training flying F-86 Sabre fighters, and moved with the 21st Fighter-Bomber Wing to France in December 1954.

Having served as a fighter pilot and test pilot in the USAF, he was part of the third NASA group of astronauts in October 1963. By this time, Collins had flown over 3,000 hours, of which 2,700 were in jet aircraft. Training began with a 240-hour course on the basics of spaceflight. Collins received his first choice of specialization: pressure suits and extravehicular activities. His first three-day mission was on *Gemini X*, launched on July 18, 1966, when he became the seventeenth American to fly in space. Fifteen scientific experiments were carried. Collins performed a spacewalk on a 15-meter umbilical, the third American to do so. With Command Pilot John Young, he performed orbital rendezvous with two *Agena* target vehicles. This flight set a new altitude record of 764 km (475 miles) above the Earth.

Michael Collins was CapCom for the *Apollo 8* mission, covering the launch phase up to translunar injection. In his preparation for the *Apollo* lunar mission, Collins compiled a 117-page book of 18 different rendezvous schemes for various scenarios; including ones where the Lunar Module did not land, or it launched too early or too late. During the training for *Apollo 11*, Deke Slayton offered to get Collins back into the crew sequence after the flight, but he told Deke Slayton he did not want to travel to space again if *Apollo 11* was successful. Collins wanted to help achieve President Kennedy's goal of landing on the Moon within the decade, and had no interest in further exploration of the Moon once the goal was achieved.

Collins was the Command Module pilot of *Apollo 11*, which had a crowd of about one million spectators for the launch (broadcast live in thirty-three countries), and an estimated TV audience of twenty-five million viewers in the United States alone. The capsule reached orbit in 12 minutes, and, after 1.5 orbits the S-IVB third-stage engine pushed the spacecraft onto its trajectory towards the Moon. About thirty minutes later, Collins performed the transposition, docking, and extraction maneuver. This involved separating the CSM, *Columbia* from the spent S-IVB stage, turning around, and docking with the Lunar Module, named *Eagle*. On July 19 at 17:21:50 UTC, *Apollo 11* passed behind the Moon and fired its service propulsion engine to enter lunar orbit. At 12:52:00 on July 20, Aldrin and Armstrong entered *Eagle* and began final preparations for their lunar descent.

Collins circled the Moon thirty times in *Columbia* while Neil Armstrong and Buzz Aldrin landed in the Lunar Module *Eagle*. In the 48 minutes of each lunar orbit when he was out of radio contact with the Earth, Collins reported feeling not fear or loneliness, but rather *"awareness, anticipation, satisfaction, confidence, almost exultation"*. At 17:54 UTC on July 21, *Eagle* lifted off from the Moon. After rendezvous with CSM *Columbia*, the ascent stage was jettisoned into lunar orbit, and *Columbia* made its way back to Earth, splashing down in the Pacific Ocean 2,660 km east of Wake Island at 05:50 local time on July 24, 1969. The astronauts were recovered by helicopter, and taken to the aircraft carrier *USS Hornet*, then, after a quarantine period, on to Houston. In September, Collins, Armstrong and Aldrin embarked on a 38-day world tour that brought them to twenty-two countries and included visits with world leaders.

After retiring from NASA in 1970, Collins took a job in the Department of State as Assistant Secretary of State for Public Affairs; an unusual choice as he was neither a journalist nor a career diplomat. A year later, he became the director of the National Air and Space Museum, holding the position until 1978, when he stepped down to become Undersecretary of the **Smithsonian Institution**. In 1980, Collins took a job as vice president of **LTV Aerospace**; resigning in 1985 to start his own consulting firm. In a July 2009 interview with ***The Guardian***, Collins said that he was very worried about Armstrong and Aldrin's safety. He was also concerned in the event of their deaths on the Moon, he would be forced to return to Earth alone and, as the mission's sole survivor, be regarded as *"a marked man for life"*.

Maj. Gen. Collins received the Presidential Medal of Freedom in 1969, and the Congressional Gold Medal in 2011. In 1974, he wrote an autobiography titled, *Carrying the Fire: An Astronauts' Journeys*. ***The New York Times*** writer John Wiford wrote that it is *"generally regarded as the best account of what it is like to be an astronaut"*. Collins also wrote *Liftoff: The Story of America's Adventures in Space* (1988); *Mission to Mars* (1990); and *Flying to the Moon and Other Strange Places* (1976). Collins was a long-time trustee of the National Geographic Society, and a fellow of the Society of Experimental Test Pilots and the American Institute of Aeronautics and Astronautics. He was inducted into four halls of fame, and the International Astronomical Union honored him by naming an asteroid after him; 6471 Collins. All three *Apollo 11* astronauts have a lunar crater named after them. Michael Collins died of cancer in Naples, Florida at age ninety years.

## Conrad, Charles, Jr.. Gemini V, Gemini XI, Apollo 12, Skylab 2.

Charles "Pete" Conrad, Jr. (1930-99) was born in June 1930 in Philadelphia, USA, into a well-to-do real estate and banking family. Conrad was a dyslexic boy who loved moving parts and wires, growing up to become one of the most accomplished astronauts. Conrad attended the Haverford School, a private academy in Haverford, Pennsylvania. After failing most of his 11th grade exams and being expelled, his mother placed him in Darrow School in New Lebanon, New York, where he learned to apply a systems approach to learning, and thus found a way to work around his dyslexia. Although he stood only five feet six inches tall, and weighed 135 pounds, Conrad started as the center on his football team and became the Captain.

From age fifteen years, Conrad worked during the summertime at the Paoli Airfield near Paoli, Pennsylvania; doing odd jobs for airplane flights and occasional instruction time. He earned his pilot's certificate before graduating from high school. Admitted on a full Navy ROTC scholarship, he earned his Bachelor of Science degree in Aeronautical Engineering from Princeton University in 1953 - being the first Ivy League astronaut - after completing a 200-page-long senior thesis titled *"The Design of a Turbo-Jet Military Advanced Trainer"*, with R.V. Warden, R.W. Vannata and C.H. Perrine.

Conrad was commissioned an Ensign in the US Navy as a Naval ROTC graduate and joined the US Naval Test Pilot School (Class 20), as a project test pilot. Conrad married Jane DuBose in June 1953, and they had four sons. In September 1954, he was designated a Naval aviator. Conrad logged more than 6,500 hours of flying time, with more than 5,000 hours in jet aircraft. In 1959, he was an astronaut candidate for Project Mercury, but his application was denied with the notation *not suitable for long-duration flight*. Conrad was selected for NASA's second astronaut class in 1962. Regarded as one of the best pilots in the group, with Commander Gordon Cooper he set a seven-day 22-hour and 55-minute space endurance record aboard *Gemini V* in 1965. Conrad commanded *Gemini XI*, flying with Richard Gordon in September 1966, and docked with an *Agena* target vehicle immediately after achieving orbit.

Conrad commanded the *Apollo 12* mission, which was hit with a series of lightning strikes just after liftoff that temporarily knocked out power and guidance in the Command Module. With Alan Bean, he piloted the Lunar Module to touch down near *Surveyor 3*. On November 19, 1969, Conrad was the third person to walk on the Moon. His first words on the lunar surface were, *"Oooh, is that soft and queasy"*. Conrad was the Commander of the *Skylab 2* (first crewed mission) in 1973, repairing significant launch damage to the station during two EVAs. Conrad freed the stuck solar panel using "brute force", and the astronauts erected a parasol-style solar shield to protect the station from intense solar heating. For this, President Jimmy Carter awarded Conrad the Congressional Space Medal of Honor in 1978.

After Conrad retired from NASA and the Navy in 1973, he became a vice president of **American Television and Communications Company**. He went on to work for **McDonnell Douglas**, as a vice president. In 1990, he married his second wife, Nancy Crane. On February 14, 1996, Conrad was part of the crew on on a record-breaking around-the-world flight in a Learjet owned by cable TV pioneer, Bill Daniels. The flight lasted 49 hours, 26 minutes and eight seconds. A month before his death, Conrad appeared on ABC News ***Nightline*** and said, *"I think the Space Shuttle is worth one billion dollars a launch. I think that it is worth two billion dollars for what it does. I think the Shuttle is worth it for the work it does"*. In his last interview, he recommended future missions to Mars and asteroids. He was inducted into several Aviation and Astronaut Halls of Fame, and received twelve other awards, plus three honorary degrees. Pete Conrad died on July 08, 1999, aged sixty-nine years, from injuries sustained in a motorcycle accident in Ojai, California, and was buried with full honors at Arlington National Cemetery. In 2006, NASA posthumously awarded Conrad the Ambassador of Exploration Award for his work for the Agency and for science.

## Cooper, L. Gordon. Mercury-Atlas 9 "Faith 7", Gemini V.

Colonel Leroy Gordon "Gordo" Cooper, Jr. (1927-2004) was born in March 1927 in Shawnee, Oklahoma, USA. He was the youngest of the seven original astronauts in Project Mercury, the first human space program of the United States. Cooper learned to fly as a child, and after brief service in the US Marine Corps during WWII, he was commissioned into the

USAF in 1949. After service as a fighter pilot, he qualified as a test pilot in 1956. Also in 1956, Cooper received a Bachelor of Science degree in Aeronautics from the Air Force Institute of Technology, Wright-Patterson AFB, Ohio, and served as a test pilot at Edwards AFB, California until April 1959, when he was selected as an astronaut.

On May 15-16, 1963, he circled Earth twenty-two times in the Mercury capsule *Faith 7*, completing the sixth and last of the Mercury solo crewed flights, which were launched on *Titan II* boosters. At the end of his 34-hour 20-minute flight, when the automatic control system had broken down, he piloted his craft back to Earth manually and landed just six kilometres (four miles) from the primary recovery ship, *USS Kearsage*. The *Gemini* series helped to develop the techniques for orbital rendezvous and docking with a target vehicle, procedures that were vital to the subsequent *Apollo* Moon-landing program. *Gemini* also provided NASA engineers with an opportunity to improve environmental control and electrical power systems of spacecraft.

As command pilot aboard *Gemini V* in August 1965, he and crewmate Charles "Pete" Conrad were plagued by mechanical problems. They set what was then a world endurance space flight record at 190 hours and 56 minutes (eight days). Their mission traveled 5,331,745 km (3,312,933 miles). Cooper served as backup command pilot for *Gemini XII*, and as backup commander for *Apollo 10*. Gordo Cooper logged a total of nine days, nine hours and 14 minutes (over 222 hours) in space. He retired from the Air Force with the rank of Colonel, and from NASA in 1970. Cooper was vocal about his belief in UFOs, and claimed that the US government was aware of their existence and had kept it secret for decades! Cooper said, "*I believe these extra-terrestrial vehicles and their crews are visiting this planet from other planets*". Most astronauts were reluctant to discuss UFOs. His autobiography, *Leap of Faith: An Astronauts Journey into the Unknown* was published in 2000. Gordo Cooper died aged 77 years on October 04, 2004, in Ventura, California.

## Covey, Richard O.. STS-51-I, STS-26, STS-38, STS-61.

Col. Richard "Dick" Oswalt Covey was born in August 1946 in Fayetteville, Arkansas, USA. He graduated from Choctawhatchee High School, Fort Walton Beach, Florida, in 1964; earned a Bachelor of Science degree in Engineering Science (with a major in Astronautical Engineering) from the USAF Academy in 1968; and a Master of Science degree in Astronautics and Aeronautics from Purdue University in 1969. Covey was an operational fighter pilot from 1970-74, and flew 339 combat missions during two tours in Southeast Asia. Between 1975-78, he was a weapons system test pilot and Joint Test Force Director for electronic warfare testing of the F-15 Eagle. Col. Dick Covey has flown over 5,700 hours in over thirty aircraft types. Selected by NASA in January 1978, he became an astronaut in August 1979.

Mission STS-51-I 1985 aboard *Discovery* (August 27-September 03, 1985) deployed three satellites: the Navy *SYNCOM IV-4*, the Australian *AUSSAT*, and the American Satellite Corporation's *ASC-1*. Covey was pilot on STS-26 1988 aboard *Discovery* (September 29-October 03, 1988), which deployed the *TDRS-C* satellite and operated eleven secondary payloads. He was Commander on STS-38 aboard *Atlantis* (November 15-20, 1990), which conducted Department of Defense operations. Covey was Commander on STS-61 aboard *Endeavour* (December 02-13, 1993), to service and repair the Hubble Space Telescope, for which he and his crew received the National Space Club's Goddard Trophy in 1994. He logged a total of 26 days, 21 hours and nine minutes (over 646 hours) in space.

On August 01, 1994, Col. Covey retired from NASA and the Air Force. After the 2003 shuttle *Columbia* disaster, he headed the Stafford-Covey Commission in returning the Space Shuttle fleet to space. In January 2006, he was selected for the position of Chief Operating

Officer of **United Space Alliance**, and became CEO in September 2007. Covey served on the Board of Directors of **Re/Max Holdings, LLC** from 2005 until 2019. Dick Covey has received twenty-two awards and honors, including two Defense Distinguished Service Medals, five Air Force DFCs, sixteen Air Medals, the Air Force Meritorious Service Medal, the National Intelligence Medal of Achievement, the NASA Distinguished Service Medal, the American Institute of Aeronautics and Astronautics (AIAA) Haley Space Flight Award for 1988; and was inducted into US Astronaut Hall of Fame.

## Creighton, John O.. STS-51-G, STS-36, STS-48.

John Oliver Creighton was born in April 1943 in Orange, Texas, USA. He received a Bachelor of Science degree from the US Naval Academy in 1966; and a Master of Science degree in Administration of Science and Technology from George Washington University in 1978. Creighton received his aviator wings in October 1967, and was with Squadron VF-154 from July 1968 to May 1970, flying F-4J Phantoms, with two combat deployments to Vietnam aboard *USS Ranger*. After graduating from the US Naval Test Pilot school in February 1971, he was assigned as a project test pilot with the Service Test Division at Naval Air Station, Patuxent River, Maryland. He has logged over 6,000 hours flying time, mostly in jet fighters, and has completed over 500 aircraft carrier landings and flew 175 combat missions.

Selected by NASA in January 1978, Creighton held a variety of technical assignments in support of the Space Shuttle Program. He was pilot on STS-51-G aboard *Discovery* (June 17-24, 1985) and deployed satellites for Mexico (Morelos), the Arab League (Arabsat), and the US (A T & T Telstar). He was Commander on STS-36 aboard *Atlantis* (February 28-March 04, 1990), which carried Department of Defense payloads. He commanded STS-48 aboard *Discovery* (September 12-18, 1991) when the crew deployed the *Upper Atmosphere Research Satellite* (UARS). Creighton unsuccessfully ran as a Republican candidate for the Washington State Senate in 1997. He is a member of the Society of Experimental Test Pilots and the Association of Space Explorers. Creighton has received eleven honors, including the Defense Superior Service Medal, the Legion of Merit, the DFC, ten Air Medals, the Vietnam Cross of Gallantry, the NASA Distinguished Service Medal, the NASA Leadership Medal, and the French Legion of Honor.

## Crippen, Robert L.. STS-1, STS-7, STS-41-C, STS-42-G.

Robert "Bob" Laurel Crippen was born in September 1937 in Beaumont, Texas, USA. He became interested in flying and computers at an early age. He received a Bachelor of Science degree in Aerospace Engineering from the University of Texas at Austin in 1960. He earned his wings at the Naval Air Station in Beeville, Texas. From 1962-64, he made two deployments aboard *USS Independence*, flying the A-4 Skyhawk. After graduating from the USAF Aerospace Research Pilot School, he worked as an instructor at Edwards AFB until October 1966, when he was picked for the USAF's Manned Orbiting laboratory (MOL) program, a joint program between the USAF and the National Reconnaissance Office (a follow-on to the X-20 Dyna-Soar program). The secret and primary mission was to perform reconnaissance missions on the Soviet Union and China.

Bob Crippen was selected for astronaut training in October 1966, and decided to stay in the military to work on the MOL program (canceled in June 1969), then transferred to NASA to continue his career; becoming part of Astronaut Group 7 in September 1969. The Skylab Medical Experiment Altitude Test (SMEAT) was a 56-day mission (July 26-September 19, 1972), when astronauts Crippen, Bobko, and Thornton were housed in a vacuum chamber to conduct medical experiments. SMEAT was also to discover and fix the design flaws in the urine handling system for *Skylab*.

Crippen was part of the *Apollo-Soyuz Test Project* (ASTP) support crew member as CapCom; and the Approach and Landing Tests (ALT) for the Space Shuttle for a year from August 12, 1977. He was pilot of mission STS-1 aboard *Columbia* (April 12-14, 1981). He said, *"Living inside the Shuttle was a little like camping out. We ended up sleeping in our seats. You had to pay attention to housekeeping, not get things too dirty"*. Crippen was Commander of STS-7 aboard *Columbia* (June 18-24, 1983), which deployed the *Anik C-2* and *Palapa B-1* satellites. He was the Commander of STS-41-C aboard *Challenger* (April 06-13, 1984), which deployed the Long Duration Exposure Facility (LDEF); retrieved, repaired, and redeployed the *Solar Maximum Mission* (SMM) satellite; and tested the *Manned Maneuvering Units* (MMU) through two EVAs.

Bob Crippen was Commander of STS-42-G aboard *Challenger* (October 05-13, 1984), which deployed the *Earth Radiation Budget Satellite* (*ERBS*); performed scientific observations of the Earth with a Large Format Camera; and demonstrated potential satellite refueling with the Orbital Refueling System (ORS). He has spent more than 6,500 hours in the air, either as a pilot or astronaut. He logged a total of 23 days, 13 hours and 46 minutes in space. In 1986, Crippen participated in the recovery operations for the remains of crew members after the *Challenger* disaster. He was on the Commission responsible for determining the cause of the shuttle accident. After retiring as an astronaut, Bob Crippen worked his way through management at NASA; then went to **Lockheed Martin** and **Thiokol Propulsion** (December 1996 to April 2001) before retiring to private life in Florida. His awards include the Congressional Space Medal of Honor, (2006); and having an elementary school named after him in Porter, Texas. He is a fellow of the American Institute of Aeronautics and Astronautics (AIAA), and the Society of Experimental Test Pilots (SETP).

## Crossfield, Albert Scott. X-15 spaceplane. The "almost astronaut".

Albert "Scott" Crossfield (1921-2006) was born in October 1921 in Berkeley, California, USA. He received a Bachelor of Science degree in Aeronautical Engineering (1949); and a Master of Science from the University of Washington (1950). He graduated from Boistfort High School, attended the University of Washington in Seattle, then worked for **The Boeing Company**. Crossfield served with the US Navy as a flight instructor and fighter pilot during World War II. From 1946-50, he worked in the University of Washington's Kirsten Wind Tunnel while earning his degrees.

In 1958, Crossfield was selected for the X-15 Program, and was the first of twelve pilots to fly the North American X-15, an experimental 15,195 kg rocket-powered spaceplane jointly operated by the USAF and NASA. The first unpowered glide flight was on June 08, 1959, and the landing mass was 6,260 kg. He was the company's Chief Test Pilot, making the first powered flight on September 17, 1959, and the first with the XLR-99 rocket engine on November 15, 1960. Up to December 06, 1960, Crossfield made fourteen of the ninety-nine test flights in the X-15 and D-558-II, gaining more experience with rocket planes than any other pilot in the world by the time he left Edwards AFB to join **North American Aviation** in 1955. On November 20, 1953, Crossfield became the first to fly at twice the speed of sound as he piloted the Skyrocket D-558-II to a speed of 2,078 km/hr (1,291 mph), Mach 2.005. Although it had been his hope to eventually pilot one of the craft into space, the USAF would not allow it, and gave strict orders which basically amounted to ***"stay in the sky, stay out of space"***. The X-15 engine burned about 85 seconds, with the acceleration building to about 4-Gs. The entire flight was about twelve minutes from launch to landing.

With similar flying characteristics, the X-15 proved the Space Shuttle could re-enter the atmosphere and glide to a precision landing; in part relying on a maneuver known as Terminal Area Energy Management, where speed and altitude are carefully controlled so the vehicle can reach the runway instead of falling short or overflying it. The retirement of the X-15 in October 1968 after its record-setting Mach 6.70 (4,520 mph) flight prompted pilot William J. "Pete" Knight to remark that he would have pushed it to even faster speeds if he had known it was the last flight. Crossfield cited the X-15 as one of few aircraft that caused grown men to cry upon its retirement. He remained at **North American Aviation** as systems director of test and quality assurance in the company's Space and Information Systems Division, where he oversaw quality, reliability, engineering and systems test activities for such programs as the Apollo Command and Service Modules and the *Saturn S-II* booster. In 1967, he joined **Eastern Air Lines**, and was subsequently a staff vice president. In 1974-75, he worked for **Hawker Siddeley** as a senior vice president supporting HS146 activities in the United States.

From 1977 to January 1993, Crossfield was a Staff Member, US House of Representatives Committee on Science, Space and Technology. He was the co-author of a book *Always Another Dawn*, about the X-15 program, and authored *"Onward and Upward" Research Airplanes, Act II*. Scott Crossfield received eighteen awards, including the Lawrence Sperry Award (1954); the American Rocket Society Astronautics Award (1960); the Collier Trophy (1962); the NASA Distinguished Public Service Medal (1993); the FAI Gold Air Medal (1994); and the National Air and Space Museum Trophy (2000). Crossfield retired on December 06, 1960, and in 1965 was inducted into the International Aerospace Hall of Fame; the National Aviation Hall of Fame (1983); the International Space Hall of Fame (1988); and the Aerospace Walk of Honor (1990). He died on April 19, 2006, in the crash of his Cessna 210A airplane during a thunderstorm near Ellijay, Ludville, Georgia. A. Scott Crossfield was posthumously awarded the Hoyt S. Vandenberg Award; the Paul Tissandier Diploma; the Victor A. Prather Award; and the Donald D. Engen Award.

## Crouch, Roger K.. STS-83, STS-94.

Dr. Roger Keith Crouch was born in September 1940 in Jamestown, Tennessee, USA. He attended the Alvin C. York Institute; earned a Bachelor of Science in Physics from Tennessee Polytechnic Institute in 1962; a Master of Science (1968) and a Doctor of Philosophy in Physics from Virginia Polytech Institute in 1971. From 1985-96, he was Lead Scientist of the Microgravity Space and Applications Division, serving as Program Scientist on five 'Spacelab' flights. From 1986-97, he was a principal investigator in the MSAD flight program. He has done research in various techniques and types of semiconductor crystal growth, electric and optical properties of materials, electronic devices for remote sensing and flat panel displays, and heat shield protection for atmospheric re-entry of space vehicles.

Crouch flew as a Payload specialist on STS-83 aboard *Columbia* (April 04-08, 1997), the Microgravity Science Laboratory (MSL-1) Spacelab mission (cut short because of problems with one of the shuttle's fuel cell power units); and STS-94 aboard *Columbia* (July 01-17, 1997), logging over 471 hours in space. Crouch was on loan from MIT to NASA HQ as the Senior Scientist for the ISS since 2000. Roger Crouch has received seven special honors, including Distinguished Alumni Achievement, Virginia Tech, 1998; Distinguished Alumnis (1997), Tennessee Technology University; the NASA Exceptional Performance Award, and Certificates for Patents/applications 1975, 85, 86, 87; Certificates for Innovative Technology 1973, 76, 79, 80, 81, 85, 86, 87; and the Floyd Thompson Fellowship 1979-80.

## Culbertson, Frank L., Jr. STS-38, STS-51, STS-105/Expedition 3.

Capt. Frank Lee Culbertson, Jr. was born in May 1949 in Charleston, South Carolina, USA. He is a graduate of the US Naval Academy with a Bachelor of Science in Aerospace Engineering in 1971. After serving aboard *USS Fox* in the Gulf of Tonkin, he reported to flight training in

Pensacola, Florida. Designated a Naval Aviator in May 1973, he flew F-4 Phantom aircraft, with VF-151 aboard the aircraft carrier *USS Midway*; and with the USAF in the 426th Tactical Fighter Training Squadron at Luke AFB, Arizona.

Selected for the Astronaut Corps in May 1984 in NASA Group 10, Culbertson was pilot on STS-38 (November 15-20, 1990) aboard the shuttle *Atlantis* Department of Defense mission. He then became Deputy Chief of the Space Station Crew Support Office and a team member for the early planning for Russian Space Station *Mir* Operations. He was Commander on STS-51 aboard *Discovery* (September 12-22, 1993) which successfully deployed both the *Orbiting and Retrievable Far and Extreme Ultraviolet Spectrograph-Shuttle Pallet Satellite* (ORFEUS-SPAS); and the *Advanced Communications Satellite* with its Transfer Orbital Stage (ACTS-TOS). He then served as Chief of the Astronaut Office Mission Support Branch, followed by assignment as Chief of the Johnson Space Center Russian Projects Office and Deputy Manager of the *Shuttle-Mir* Program. Culbertson became Manager of the *Shuttle-Mir* Program in 1995, then served as Deputy Program Manager of the ISS, until assigned to 'Expedition 3'. He launched on STS-105 aboard *Discovery* (August 10-22, 2001), and spent 129 days aboard the ISS with 'Expedition 3'.

Frank Culbertson was Commander of the ISS for 117 days (almost four months) in 2001. On September 11, he was 400 km above Earth when he saw the huge column of smoke streaming from Lower Manhattan where the Twin Towers fell in the terrorist attack. Later, he could see the gash in the Pentagon building. He logged a total of 143 days, 14 hours and 50 minutes in space, with one EVA. Culbertson resigned from NASA in August 2002, and was employed as a Senior Vice President for **SAIC** for six years, overseeing space and climate-related programs. He was Executive Vice President of **Orbital Sciences Corporation ATK**, a commercial spaceflight company that had in 2017 a contract with NASA to fly eight unmanned resupply missions to the ISS. Culbertson was inducted into the US Astronaut Hall of Fame on June 05, 2010.

## Cunningham, R. Walter. Apollo 7.

Col. Ronnie Walter "Walt" Cunningham (1932-2023) was born in March 1932 in Creston, Iowa, USA. He enlisted in the US Navy in 1951 and transferred to the Marine Corps, where he served as a fighter pilot (1953-56). Cunningham flew fifty-four missions before retiring with the rank of Colonel. He majored in physics at the University of California, Los Angeles, receiving Bachelor's (1960) and Master's (1961) degrees, but abandoning a doctoral program in favor of the crewed space project in 1963. As a civilian participant aboard *Apollo 7*, the first successful crewed mission of October 11-22, 1968, he orbited the Earth 163 times, spending ten days and 20 hours in space. The mission goal was to test the Command and Service Module with a crew, and ensure improvements to Apollo spacecraft safety procedures. As the Lunar Module Pilot, Cunningham was responsible for managing all spacecraft systems except launch and navigation. The crew made the first live television broadcast from an American spacecraft. Despite tensions between the crew and ground controllers, the *Apollo 7* mission was a complete technical success.

After *Apollo 7*, Cunningham moved to management on the first US space station, *Skylab*, which he regarded as his principal contribution to spaceflight. In 1971, he retired from NASA and became a corporation executive. An entrepreneur and venture capitalist, he authored the 1977 book, *The All-American Boys*. He received numerous honors, including the NASA Distinguished Service Medal; the NASA Exceptional Service Medal; the AIAA Haley Astronautics Award (1969); the UCLA Professional Achievement Award (1969); and the Special Trustees Award, National Academy of Television Arts and Sciences (Emmy Award), 1969. Walt Cunningham died on January 03, 2023, aged ninety years, in Houston, Texas.

## Curbeam, Robert L., Jr. STS-85, STS-98, STS-116.

Capt. Robert Lee "Beamer" Curbeam, Jr. was born in March 1962 in Baltimore, Maryland, USA. He had a distinguished career as a Navy pilot. He graduated from Woodlawn High School, Baltimore Co., Maryland in 1980; received a Bachelor of Science degree in Aerospace Engineering from the US Naval Academy in 1984; a Master of Science degree in Aeronautical Engineering from the Naval Postgraduate School in 1990; and a Degree of Aeronautical and Astronautical Engineering from the Naval Postgraduate School in 1991. He commenced Naval Flight Officer training in 1984. In 1986, he was deployed to the Mediterranean and Caribbean Seas. During his tour in XF-11, he also attended Navy Fighter Weapons School (TOPGUN). During the Spring of 2002, he served as Deputy Associate Administrator for Safety and Mission Assurance at NASA Headquarters, Washington.

Selected by NASA in December 1994 Group 15, Curbeam completed a year of training and evaluation and was assigned to the Computer Support Branch in the Astronaut Office. Mission STS-85 aboard *Discovery* (August 07-19, 1997) deployed and retrieved the *CRISTA-SPAS* payload. STS-98 aboard *Atlantis* (February 07-20, 2001) delivered the *Destiny* laboratory module to the ISS. He logged over 19 EVA hours in three spacewalks. STS-116 aboard *Discovery* (December 09-22, 2006) added the P5 spacer truss, and rewired the ISS power system, preparing it to support the addition of European and Japanese science modules. Curbeam logged a total of 37 days, 14 hours and 33 minutes (over 901 hours) in space. He currently holds the record for the most EVAs during a single spaceflight; competing four on mission STS-116.

Capt. Robert Curbeam received the Fighter Wing One Radar Intercept Officer of the Year for 1989, the US Naval Test Pilot School Best Development Thesis (DT-II) award, the Armed Services Expeditionary Medal, the Navy Meritorious Unit Commendation, two Navy Commendation Medals, and the Navy Battle Efficiency Award. He is a member of the US Naval Academy Alumni Association, and the Association of Old Crows.

## Currie-Gregg, Nancy J.. STS-57, STS-70, STS-88/2A, STS-109.

Col. Dr. Nancy Jane Currie-Gregg (nee Decker, then Sherlock) was born in December 1958 in Wilmington, Delaware, USA. She graduated from Troy High School in 1977; received a Bachelor of Arts degree (with honors) in Biological Science from Ohio State University in 1980; a Master of Science degree in Safety Engineering from the University of Southern California in 1985; and a Doctorate in Industrial Engineering from the University of Houston in 1997. Currie-Gregg is an American engineer, US Army officer (serving over twenty-three years), a former helicopter test pilot for the US Army, and a NASA astronaut. As a Master Army Aviator she logged over 3,900 hours in a variety of rotary-wing and fixed-wing aircraft. Currie-Gregg was assigned to Johnson Space Center in September 1987 as a flight simulation engineer on the Shuttle Training Aircraft, modeling flight characteristics of the Shuttle orbiter. Selected in the 1990 NASA Astronaut Group, she has been involved in robotic hardware and procedure development for the Shuttle and space station. She is a consultant to NASA's Space Human Factors Engineering Project.

Currie-Gregg is the shortest person (152 cm, or five feet tall) to have flown in space. She is a very skilled operator of the Canadarm robotic arm, and flew four shuttle missions as a Mission Specialist: STS-57 aboard *Endeavour* (June 21-July 01, 1993) to retrieve the *European Retrievable Carrier* satellite (*EURECA*), and the first flight of 'Spacehab'; STS-70 aboard *Discovery* (July 13-22, 1995), which deployed the final NASA *Tracking and Data Relay Satellite* to complete the constellation of NASA's orbiting communication satellite system; STS-88 aboard *Endeavour* (December 04-15, 1998), the first ISS assembly mission; and STS-109 aboard *Columbia* (March 01-12, 2002), the fourth mission to service the Hubble Space Telescope. She logged a total of 41 days, 15 hours and 32 minutes (1,000 hours) in space.

Currie-Gregg currently holds an appointment as professor of practice in the Department of Industrial & Systems Engineering at **Texas A & M University**. She is a member of the Army Aviation Association of America, Ohio State University and ROTC Alumni Associations, the Institute of Industrial and Systems Engineers, and the Association of Space Explorers. She is an associate fellow of the American Institute of Aeronautics and Astronautics. She has received eighteen awards and honors, including the Arts and Sciences Award for Scholarship, Ohio State University (1980); Honor Graduate of the Army Aviation Office Advanced Course (1986); the NASA Flight Simulation Engineering Award (1988); and the NASA Outstanding Leadership Medal (2005). She was inducted into the Ohio Veteran's Hall of Fame in 1994.

## Dana, William H.. X-15 rocketplane, Flight 174.

William "Bill" Harvey Dana (1930-2014) was born in November 1930 in Pasadena, California, USA. He received a Bachelor of Science degree from the US Military Academy, West Point, in 1952, and served four years as a pilot in the USAF, joining NASA on October 01, 1958, after receiving a Master of Science degree in Aeronautical Engineering from the University of Southern California. From 1960-62, he was a pilot astronaut in the USAF X-20 Dyna-Soar program. Bill Dana flew the hypersonic X-15 rocket-powered research aircraft sixteen times, reaching a top speed of 3,897 mph and a peak altitude of 307,000 feet (nearly 59 miles), thus qualifying as an astronaut. During X-15 Flight 174 on November 01, 1966, he reached 6,035 Km/hr (3,750 mph) at Mach 5.46 and an altitude of 93.5 km (58.1 miles). In Flight 197 on August 21, 1968, he reached 5,541 km/hr (3,443 mph) at Mach 5.01 and an altitude of 81.6 km (50.7 miles). Dana was the pilot on the final (199th) X-15 flight of the ten-year program. In the Northrop HL-10 airplane, he made the highest-ever flight in a lifting body to 90,303 feet.

From 1993-98, Dana was Chief Engineer at NASA's Dryden Flight Research Center, when he retired from NASA after almost forty years' service. In 1997, he received the NASA Distinguished Service Medal for his service as a flight research pilot. Dana was a member of the Society of Experimental Test Pilots, and author of several technical papers. For his contributions to the lifting body program, Dana received the NASA Exceptional Service Medal; and in 1976 the Haley Space Flight Award from the American Institute of Aeronautics and Astronautics for his research work on the M2-F3 lifting body control systems. In 2000, he was awarded the Milton O. Thompson Lifetime Achievement Award by the Dryden Flight Research Center. In 2005, NASA awarded civilian astronaut wings to Bill Dana, and posthumously to McKay and Walker. William "Bill" Dana died on May 06, 2014, of Parkinson's disease in Phoenix, Arizona, aged 83 years.

## Davis, N. Jan. STS-47 (J), STS-60, STS-85.

Dr. Nancy Jan Davis (nee Smotherman) was born in November 1953 in Cocoa Beach, Florida, USA. She received a Bachelor of Science degree in Applied Biology in 1975 from Georgia Institute of Technology, and another in Mechanical Engineering from Auburn University in 1977. Davis received a Master of Science (1983) and a Doctorate (1985), both in Mechanical Engineering at the University of Alabama in Huntsville. In 1977, Davis joined **Texaco** in Bellaire, Texas, working as a petroleum engineer. In 1979, she left to work for NASA's Marshall Space Flight Center as an aerospace engineer. In 1987, she was assigned to be the lead engineer for the re-design of the Shuttle solid rocket booster external tank attach ring. She holds one Patent, has authored several technical papers, and is a Registered Professional Engineer.

Davis became part of NASA Astronaut Group 12 in June 1987. She provided technical support for Shuttle payloads, then was a CapCom for seven missions. On mission STS-47 (September 12-20, 1992) aboard *Endeavour* she was responsible for operating 'Spacehab' and its subsystems. Her then-husband Mark C. Lee was mission commander. She then served as the Astronaut Office representative for the Remote Manipulator system (RMS). Mission STS-60 (February 03-11, 1994) aboard shuttle *Discovery* was the second flight of the 'Spacehab' module, and the first for the *Wake Shield Facility* (WSF). Her prime responsibility was to maneuver the WSF on the Remote Maipulator System (RMS), to conduct thin film crystal growth, and perform scientific experiments in the 'Spacehab'. Dr. Davis was Payload Commander for mission STS-85 aboard *Discovery* (August 07-19, 1997). She deployed and retrieved the *CRISTA-SPAS* payload, and operated the Japanese Manipulator Flight Demonstration (MFD) robotic arm. Davis logged 28 days, two hours and seven minutes (over 673 hours) in space.

She was then assigned to NASA Headquarters as the director of the Human Exploration and Development of Space (HEDS), Independent Assurance Office for the Office of Safety and Mission Assurance. In July 1999, she transferred to the Marshall Space Flight Center as Director of the Flight Projects Directorate, responsible for the ISS Payload Operations Center, ISS hardware and the Chandra X-Ray Observatory Program. After the *Columbia* accident of January 16, 2003, Davis was named head of **Safety and Mission Assurance** at Marshall SFC. Jan Davis secretly married fellow astronaut Mark C. Lee, whom she had met during training. They were the first married couple to fly together in space, but divorced in 1999. Davis is currently married to former Judge Schuyler Richardson. She has received nine awards, including the NASA Outstanding Leadership Medal (1998), the NASA Exceptional Service Medal (1995 & 2002), the NASA Space Flight Medal (1992, 1994 & 1997); and has been inducted into the Alabama Aviation Hall of Fame, the Alabama Engineering Hall of Fame, and received the Presidential Rank Award of Meritorious Executive.

## DeLucas, Lawrence J.. STS-50.

Dr. Lawrence James DeLucas was born in July 1950 in Syracuse, New York, USA. He attended the University of Alabama at Birmingham, Alabama, and received the following degrees: Bachelor of Science in Chemistry (1972), a Master of Science in Chemistry (1974), a Bachelor of Science in Physiological Optics (1979), a Doctorate in Optometry (1981), and a Doctorate in Biochemistry (1981). DeLucas was a Mission Specialist aboard STS-50 on *Columbia* (June 25-July 09, 1992), the United States Microgravity Laboratory-1 (USML-1) Spacelab mission. He logged 13 days, 19 hours and 30 minutes (over 331 hours) in space.

Dr. DeLucas has published 163 research articles in refereed scientific journals, is co-author of two books, and co-inventor on forty-three Patents. From 2001-16, DeLucas was Adjunct Professor, Physiology and Biophysics; from 2016 to present he is Principal Scientist, **The Aerospace Corporation**; and from 2021 to present he is President and Founder, **Soluble Biotech Inc.**. Dr. DeLucas has received twenty-three honors, including co-Chair, Spacehab Science Advisory Board; the NASA Public Service Medal (June 1997); Distinguished Faculty Lecturer, UAB; Distinguished Crystallography Lecturer for Pittsburgh Diffraction Society; four honorary Doctor of Science degrees; and the President's Award for Excellence in Teaching (February 2011). He was inducted into the National Optometry Hall of Fame in October 2002.

## Drew, Alvin. STS-118, STS-133.

Col. Alvin Drew was born in November 1962 in Washington, D.C., USA. Seeing the *Apollo 7* launch on TV at age five years motivated him to become an astronaut. He was named a National Merit Scholar and went on to earn a dual Bachelor of Science degree in Physics and Astronautical Engineering from the USAF Academy in 1984; and a Master of Science degree in Aerospace Science from Embry-Riddle Aeronautical University. He then earned a Master's degree in Political Science from the Air War College. Drew received his commission as a second lieutenant from the USAF Academy in May 1884, and completed Undergraduate Helicopter Pilot Training in 1985. Later assigned to the Air Force Special Operations Command, he flew combat missions in operations Just Cause, Desert Shield, Desert Storm, and Provide Comfort. He completed USAF Fixed Wing Qualification in 1993, and the US Naval Test Pilot School in June 1994. Alvin Drew is a command pilot with over 3,000 hours flying time in over 30 aircraft types. He retired from the Air Force in September 2010 with over 25 years' service.

Drew was selected as a Mission Specialist by NASA in July 2000. He served in technical assignments until he took a sabbatical to the USAF's Air University at Maxwell AFB, Alabama, attending a Master's degree program at their War College. Drew was a Mission specialist on STS-118 aboard *Endeavour* (August 08-21, 2007), then worked as CapCom for the STS-123 mission. On STS-133 aboard *Discovery* (February 24-March 09, 2011), he conducted two EVAs, and logged a total of 25 days and 13 hours in space. Col. Alvin Drew has ten awards, including the Meritorious Service Medal with one Oak Leaf Cluster, the Aerial Achievement Medal with five Oak Leaf Clusters, the Air Force Outstanding Unit Award with three Oak Leaf Clusters, the Armed Forces Expeditionary Medal, and the Southwest Asia Service Medal with three service stars. In October 2020, Drew co-founded the **Patti Grace Smith Fellowship**, a branch of the Brooke Owens Fellowship intended to provide resources for African-American undergraduate students pursuing careers in aerospace.

## Duffy, Brian. STS-45, STS-57, STS-72, STS-92.

Col. Brian Duffy was born in June 1953 in Boston, Massachusetts, USA. He received a Bachelor of Science degree in Mathematics from the USAF Academy in 1975; and completed Undergraduate Pilot Training as part of Class 76-10 at Columbus AFB, Mississippi, in August 1976. He received a Master of Science degree in Systems Management from the University of Southern California in 1981. He flew F-15s at Kadena Air Base, Japan from 1979 to 1982, when he was selected to attend the USAF Test Pilot School at Edwards AFB, California and then became the Director of F-15 Flight Test at Eglin AFB, Florida. Col. Duffy logged over 5,000 hours flight time in over twenty-five aircraft types.

Selected by NASA in June 1985, he participated in the development and testing of displays, flight crew procedures, and computer software to be used on Shuttle flights. He served as CapCom during numerous Space Shuttle missions. Brian Duffy was pilot on mission STS-45 aboard *Atlantis* (March 24-April 02, 1992); pilot on STS-57 aboard *Endeavour* (June 21-July 01, 1993); Commander on mission STS-72 aboard *Endeavour* (January 11-20, 1996); and Commander on STS-92 aboard *Discovery* (October 11-24, 2000). He logged a total of 40 days, 17 hours and 35 minutes in space. Col. Brian Duffy retired from the Air Force and NASA in 2001. From 2004-08, he was Vice President of Mission Operations and Support Services for **Lockheed Martin Corporation**. He is currently the Vice President and JSC Program Manager for Exploration Systems at **Orbital ATK**, which was acquired by and renamed **Northrop Grumman Innovation Systems** in 2018. Duffy has nine awards and Honors, including the DFC; the Air Force Meritorious Service Medal; four NASA Space Flight Medals; and he was inducted into the US Astronaut Hall of Fame, May 14, 2016.

## Dunbar, Bonnie J.. STS-61-A, STS-32, STS-50, STS-71, STS-89.

Bonnie Jeanne Dunbar was born in March 1949 in Sunnyside, Washington, USA. She grew up on a sixteen hectare farm in Outlook, Washington, and could drive a tractor at age nine years. In October 1957, Bonnie and her parents gazed at the night sky, looking for *Sputnik*. She became fascinated with space, reading science fiction novels by H.G. Wells and Jules Verne. Bonnie attended Sunnyside High School, where she took physics, chemistry, math, algebra, trigonometry, analysis and precalculus classes. She played sports and was a cheerleader for three years. After graduating from high school in 1967, she was accepted by the University of Washington. Because she wanted to build spacecraft, her physics teacher advised her to study engineering. The Dean, James I. Mueller of the ceramics engineering department persuaded Bonnie to change her major from aeronautical engineering to ceramic engineering, and she gained her Bachelor of Science degree in Ceramic Engineering in 1971.

After spending two months as an office manager, Bonnie secured a position in the computer service division at **Boeing**, and was trained in COBOL. Eighteen months later, Dean Mueller informed Bonnie he had secured a grant from NASA, and she went on to research the kinetics of ionic diffusion in sodium beta-alumina, leading to her Master of Science degree in Ceramics Engineering from the University of Washington. Dunbar became a senior research engineer in **Rockwell International**'s Space Division, where she designed the equipment and manufacturing processes to fabricate the ceramic tiles used in the Space Shuttle thermal protection system. In 1978, she joined NASA as a flight controller/payload officer, and was a guidance and navigation controller for Skylab during its de-orbiting and re-entry in July 1979. She was selected as one of the nineteen astronaut candidates in NASA Astronaut Group 9 in 1980.

After a year of astronaut candidate training, Bonnie qualified on August 21, 1981. Her technical assignments included assisting with the testing of the Space Shuttle onboard software at the Shuttle Avionics Integration Laboratory (SAIL). She was involved with the remote manipulator system (RMS) development. Dunbar later became the chief of the Mission Development Branch, and was the Astronaut Office interface for "secondary" payloads, and the lead for the Science Support Group. She was a member of the support crews for the STS-2 and STS-9 missions, and worked with Walter Cronkite on the CBS television coverage of the STS-2 mission. Dunbar flew on the STS-61-A mission, where she was primarily responsible for the operation of the 'Spacelab' and its subsystems; STS-32, when she was the principal investigator of the Microgravity Disturbance Experiment; on STS-50, where she was payload commander; STS-71, where she first trained for thirteen months as a cosmonaut in Russia as backup for Norman Thagard, then became 'Mission Specialist 3' to the *Mir* space station; and the STS-89 mission to *Mir*, when she was Payload Commander. Dunbar logged a total of 50 days, eight hours and 24 minutes in space.

Since leaving NASA, she has worked in museums and STEM leadership, and as a professor of aerospace engineering. Dunbar became the president and chief executive officer of the **Museum of Flight** in Seattle, where she was involved in science, technology, engineering, and mathematics education for high school students. From 2013-15 she led the University of Houston's STEM Center, and was a faculty member in the Cullen College of Engineering. In 2016, she became professor of aerospace engineering at **Texas A&M University**. From 2016 to 2020, she was the Director of the **Institute for Engineering Education and Innovation** (IEEI), and is currently Engineering Experiment Station Eminent Research Professor in the aerospace engineering department.

## Duke, Charles M., Jr.. Apollo 16.

Charles "Charlie" Moss Duke, Jr. was born in October 1935 in Charlotte, North Carolina, USA. He graduated from the US Naval Academy in 1957, joined the USAF and completed advanced flight training on the F-86 Sabre at Moody AFB in Georgia, where he excelled as a distinguished graduate. Duke then served as a fighter pilot with the 526th Fighter-Interceptor Squadron at the Ramstein Air Base in West Germany. In September 1965, Duke graduated from the Aerospace Research Pilot School, and stayed on as an instructor.

In April 1966, Duke was selected as one of nineteen astronaut candidates for NASA's April 1966 Group 5. He served as a member of the astronaut support crew for *Apollo 10*. Duke was the CapCom for *Apollo 11*, and the backup LM pilot for *Apollo 13*. He was the Lunar Module pilot of *Apollo 16* launched on April 16, 1972, and on April 21 became the tenth and youngest person to walk on the Moon at the age of 36 years and 201 days. Duke spent over twenty hours on the lunar surface in the Descartes region, conducting three EVAs. Following his retirement, Duke entered the Air Force Reserve and served in various capacities.

## Durrance, Samuel T.. STS-35, STS-67.

Dr. Samuel Thornton Durrance (1943-2023) was born in September 1943 in Tallahassee, Florida, USA. He graduated from Plant High School in 1961; received a Bachelor of Science degree (1972) and a Master of Science degree in Physics (with honors) at California State University in 1974. Durrance received his Doctor of Philosphy degree in Astro-geophysics at the Universty of Colorado at Boulder in 1980. In March 1986, Durrance's first mission was to be STS-61-E, canceled after the *Challenger* mission STS-51-L disaster on January 28. Launched on December 02, 1990, as a Payload Specialist on mission STS-35/Astro-1 aboard *Columbia*; and on March 02, 1995, for STS-67/Astro-2 aboard *Endeavour*, he logged 25 days, 14 hours and 13 minutes (over 615 hours) in space. Durrance was employed by the **Florida Institute of Technology**, a private research university in Melbourne, Florida, serving as a professor in the Department of Aerospace, Physics and Space Sciences.

In 2000, Samuel Durrance was awarded an Honorary Doctor of Science from the University of Colorado, along with eight other astronaut alums. Beginning in 2001, he was the executive director of the **Florida Space Research Institute**, located at the Kennedy Space Center. Durrance was a member of the American Astronomical Society, the American Geophysical Union, the International Astronomical Union, Association of Space Explorers, the Planetary Society, and Phi Kappa Phi. Samuel Durrance was diagnosed with Alzheimer's disease in 2021. He died after battling dementia and Parkinson's disease at a hospice on May 05, 2023, in Viera, Florida, aged seventy-nine years.

## Dutton, James P.. STS-131/Expedition 23.

Col. James Patrick "Mash" Dutton was born in November 1968 in Eugene, Oregon, USA. He graduated from Sheldon High School, joined the Air Force and graduated first in his class from the USAF Academy in 1991 with a Bachelor of Science degree in Astronautical Engineering. He received an unprecedented top military cadet, top academic cadet, top pilot and top engineering student awards. In 1994, Dutton earned a Master of Science degree in Aeronautics and Astronautics from the University of Washington. In 1995, he undertook F-15C training at Tyndall AFB, Florida, and flew as an operational pilot at RAF Lakenheath, UK, from October 1995 to May 1998. Dutton served with the 416th Flight Test Squadron flying the F-16 until June 2002. He joined the F-22 Combined Test Force flying the Raptor at Edwards AFB from August 2002 to June 2004. After retiring from the Air Force in 2014, Col. Dutton had logged over 3,300 flight hours in over thirty aircraft types.

Selected by NASA in May 2004 Group 19, he received his astronaut pin on February 10, 2006. Dutton was initially assigned to the Exploration Branch, working on the development of the Crew Exploration Vehicle (CEV), later named *Orion*. He served as Ascent and Entry CapCom for Space Shuttle missions STS-122 and STS-123. James Dutton flew as Pilot on mission STS-131 aboard shuttle *Discovery* (April 05-20, 2010) to the ISS, joining 'Expedition 23' crew members. He assisted Commander Alan Poindexter during rendezvous and landing, and to maneuver the orbiter during undocking and the fly-around of the ISS. He served as lead shuttle robotic arm operator, and was responsible for airlock operations in preparation for

EVAs. After his retirement from the Air Force, Dutton became an airline pilot with **Southwest Airlines**. In June 2020, Dutton joined **Dynetics, Inc**. as a technical advisor on their design for NASA's Human Landing system (HLS) supporting the *Artemis* program. He is a member of the Society of Experimental Test Pilots and the Officer's Christian Fellowship.

## Edwards, Joe F., Jr. STS-89.

Joe Frank Edwards, Jr. was born in February 1958 in Richmond, Virginia, USA. He earned a Bachelor of Science degree in Aerospace Engineering from the US Naval Academy in 1980; and a Master of Science degree in Aviation Systems from University of Tenessee, Knoxville, in 1994. Designated a Naval Aviator in February 1982, he was assigned to Fighter Squadron 143 in 1983, and flew fighter escort and reconnaisance combat missions over Lebanon in 1983. He graduated from the Navy Fighter Weapons School (TOPGUN) in 1984, and from US Naval Test Pilot School in 1986. From 1992-94, he worked as Operations Officer in the Operations Directorate of the Joint Chiefs of Staff, Washington, D.C.. Edwards has flown 4,000 hours in over twenty-five aircraft types, and logged over 650 aircraft carrier arrested landings.

Selected as an astronaut by NASA in December 1994, Edwards worked on technical issues for the Space Shuttle and International Space Station in the Safety Department of the Astronaut Office. Edwards flew on STS-89 (January 22-31, 1998), the eighth *Shuttle-Mir* docking mission, during which the crew transferred over 9,000 pounds of scientific equipment, logistical hardware and water from shuttle *Endeavour* to *Mir*. Joe Edwards logged eight days, 19 hours and 48 minutes in space. He retired from NASA and the US Navy on April 30, 2000. He is a member of the Society of Experimental Test Pilots, the Association of Naval Aviation, and the US Naval Institute. He has received ten awards, including the DFC; the Defense Superior Service Medal; the Air Medal; the Navy Achievement Medal; and eight Fighter Pilot, Fighter Squadron 142 and 143, and Carrier Air Wing 7 Pilot of the Year Awards.

## Eisele, Donn F.. Apollo 7.

Donn Fulton Eisele (1930-87) was born in June 1930 in Columbus, Ohio, USA. He graduated from West High School in 1948. He received a Bachelor of Science degree from the United States Naval Academy in Annapolis, Maryland, in 1952, and chose a commission in the USAF. After receiving his pilot wings in 1954, Eisele served four years as an interceptor pilot in South Dakota, and in Libya until 1958. Eisele received a Master of Science degree in Astonautics from the USAF Institute of Technology (AFIT), Wright-Patterson AFB, Ohio, in 1960. He graduated from the USAF Experimental Test Pilot School (Class 62A) at Edwards AFB, Califiornia, in 1962. Eisele logged over 4,200 hours flying time; 3,600 of which were in jet aircraft.

Donn Eisele was selected as a NASA astronaut in October 1963. A shoulder injury removed him from the *Apollo 1* crew, then he was named for *Apollo 2*, which was canceled in December 1966. The Apollo capsule was re-designed following the *Apollo 1* fire, and Eisele, Schirra and Cunningham were named to fly the first crewed Apollo mission, *Apollo 7*. Eisele was the Command Module pilot during the *Apollo 7* mission, which launched on October 11, 1968. The crew simulated transposition and docking maneuvers with the upper stage of their *Saturn-IB* launch vehicle. Eisele acted as navigator, taking star sightings and aligning the spacecraft's guidance and navigation platform. The crew completed eight successful test firing maneuvers of the Service Module's propulsion engine, tested the performance of all spacecraft systems, and broadcast the first live TV coverage of their activities. He logged ten days, 20 hours and eight minutes (over 260 hours) in space. Splashdown was in the Atlantic Ocean, only 480 metres from the predicted target, and fifteen kilometres (eight miles) from the aircraft carrier *USS Essex*.

Eisele served as backup Command Module Pilot for the 1969 *Apollo 10* flight. He resigned from the Astronaut Office in 1970 and became technical assistant for crewed spaceflight at the NASA **Langley Research Center**. After retiring from both NASA and the Air Force in 1972, Eisele became the **Peace Corps** country director for Thailand, before moving into private business. Among the honors he received were the NASA Exceptional Service Medal, the Air Force Senior Pilot Astronaut wings, and the Air Force Distinguished Flying Cross. Donn Eisele died of a heart attack while on a business trip to Tokyo, Japan, where he was to attend the opening of a new Space Camp. He was cremated in Japan, and his ashes were buried in Arlington National Cemetery with full military honors. Eisele was inducted into the US Astronaut Hall of Fame in 1997. In the 1998 HBO miniseries ***From the Earth to the Moon***, Eisele was portrayed by John Mese. In the final three episodes of the 2015 ABC television series ***The Astronaut Wives Club***, Eisele was portrayed by Ryan Doom. A family-approved account of Donn Eisele's life appears in the 2007 book *In the Shadow of the Moon*. In 2008, NASA posthumously awarded him the NASA Distinguished Service Medal for his *Apollo 7* mission. Eisele's posthumously discovered memoir, *Apollo Pilot* was published by University of Nebraska Press in 2017.

## England, Anthony W.. STS-51-F.

Professor Dr. Anthony Wayne "Tony" England was born in May 1942 in Indianapolis, Indiana, USA. He received his Bachelor of Science and Master of Science degrees in Earth and Planetary Sciences from the Massachusetts Institute of Technology in 1965; and a Doctor of Philosophy degree in Earth and Planetary Sciences in 1970. Selected by NASA in August 1967 as part of the 6th Astronaut Group, England was among a group of astronauts who served as backups during the Apollo and Skylab programs. He received his wings (with distinguished honors) in April 1969. During the *Apollo 13* emergency, England wrote the procedure to build the vital CO2 scrubber, for which he received the Presidential Medal of Freedom in 1970. He was an EVA CapCom during the *Apollo 16* mission in April 1972, talking to the astronauts exploring the lunar surface. Tony England helped develop and use radars to probe the Moon on *Apollo 17*, and the glaciers in Washington and Alaska. England was formerly dean of the College of Engineering and Computer Science at the University of Michigan's Dearborn campus. He initially retired from NASA in 1972, citing a decline in crewed missions. He was then deputy chief of the Office of Geochemistry and Geophysics for the US Geological Survey, and associate editor for the *Journal of Geophysical Research*.

England returned to the Johnson Space Center in 1979 as a senior science-astronaut (mission specialist), and eventually managed the operation mission development group of the Astronaut Office. He served as a Mission Specialist on STS-51-F 1985, which launched on July 29, and carried 'Spacelab-2'. England logged over 4,000 hours of aircraft flying time, and 188 hours in space. From May 1986 to May 1987, England served as a program scientist for Space Station *Freedom* (which later evolved into the International Space Station). From June to December 1987, he taught Remote Sensing Geophysics at Rice University. In 1988, he joined U-M as a professor of Electrical Engineering and Computer Science, and served as Associate Dean for the Rackham School of Graduate Studies from 1995-98. He retired from NASA in 1988. From 2014-20, Dr. England served as Dean of the College of Engineering and Computer Science at the University of Michigan-Dearborn. England received nine awards and honors, including the NASA Outstanding Scientific Achievement Medal (1973); the NASA Exceptional Service Medal (1988); the Department of Electrical Engineering and Computer Science Exceptional Service Award for 1994; and is a Fellow of the Institue of Electrical and Electronics Engineers (IEEE).

## Engle, Joe H.. X-15 Flights 138, 143 & 153, STS-2, STS-51-I.

Maj. Gen. Joe Henry Engle was born in August 1932 in Chapman, Kansas, USA. He received a Bachelor of Science degree in Aeronautical Engineering from the University of Kansas in 1955. Engle entered flying school in 1957, and received his pilot wings in 1958. After serving as a test pilot in the Fighter Test Group at Edwards AFB, California, Engle was a test pilot in the X-15 research program at Edwards from June 1963 until his assignment to the Manned Spacecraft Center (now the Johnson Space Center). Engle is one of twelve pilots who flew the North American X-15 , an experimental spaceplane jointly operated by the USAF and NASA. As an X-15 pilot, he made three flights (Flight 138 on June 29, 1965, at Mach 4.94; Flight 143 on August 10, 1965, at Mach 5.20; and Flight 153 on October 14, 1965, at Mach 5.08), all above fifty miles (81 km) altitude, thus qualifying for astronaut wings under the American convention for the boundary of space. Engle logged more than 15,000 hours flight time, of which 9,000 were in jet aircraft. He flew over 185 different aircraft types during his career.

In 1966, he was selected for NASA's Astronaut Group 5. He was backup Lunar Module Pilot (LMP) for *Apollo 14* and originally scheduled as LMP for *Apollo 17*, but was displaced by geologist Harrison Schmitt for that mission. According to Engle, Deke Slayton asked him whether he would prefer to fly on *Skylab, Apollo-Soyuz*, or the Space Shuttle. He responded that he "*would prefer the Shuttle, as it was an airplane*". As commander, Joe Engle flew three flights in the Shuttle program's Approach and Landing Tests from June through October 1977. The prototype shuttle *Enterprise* was carried to 25,000 feet atop a Boeing 747 carrier aircraft, then released for its two-minute glide flight to landing. Launched on November 12, 1981, Engle was Commander on the two-day shuttle mission STS-2; and also on STS-51-I, launched on August 27, 1985. From March to December 1982, he was deputy associate **Administrator for Manned Space Flight** at NASA Headquarters. In 1986, he participated in the *Challenger* disaster investigation, and did other consulting work on the Space Shuttle well into the 1990s.

Joe Engle logged nine days, eight hours and 30 minutes (over 225 hours) in space. He retired from NASA on November 28, and from the USAF on November 30, 1986. In November 2001, he was inducted into the US Astronaut Hall of Fame. He was a member of the Society of Experimental Test Pilots, and became a Fellow in 2009. Engle received an impressive twenty-six awards, including the Defense Distinguished Service Medal; two DFC's; Kansan of the Year (1964); USAF Outstanding Young Officer of the Year (1964); the Dr. Robert H. Goddard Memorial Trophy; the Robert J. Collier Trophy; the Harmon International Trophy; Award of Air Force Space and Missile Pioneers (2007); was inducted into the National Aviation Hall of Fame, Dayton, Ohio (2001); and into the International Air & Space Hall of Fame (2014).

## Evans, Ronald E., Jr.. Apollo 17.

Capt. Ronald "Ron" Ellwin Evans, Jr. (1933-90) was born in November 1933 in St. Francis, Kansas, USA. He attended Highland Park High School in Topeka, Kansas, and secured a football scholarship from the Naval Reserve Officers Training Corps (NROTC) to Kansas State University where he played college football, and later played professionally for the Denver Broncos. In his sophomore year, there was an introduction to naval aviation at Naval Air Station Corpus Christi in Texas. This inspired Ron Evans to become a naval aviator. He made his first solo flight on May 21, 1956. Evans graduated with a Bachelor of Science degree in Electrical Engineering from the University of Kansas in June 1956, and joined the US Navy. He served as a fighter pilot and flew combat missions during the Vietnam War, serving two tours of duty with the US Marine Corps. He logged 4,600 hours of flight time, including 2,084 in jet aircraft. In 1964, he received a Master of Science degree in Aeronautical Engineering from the US Naval Postgraduate School. Evans was selected as part of NASA Astronaut Group 5 in 1966. Achieving the rank of Captain, he retired from the Navy on April 30, 1976.

As Command Module Pilot on *Apollo 17* in December 1972, Evans was one of the twenty-four astronauts to fly to the Moon in the 20th century, and one of twelve men to fly to the Moon without landing. Fe, Fi, Fo, Fum, and Phooey were five mice (four male, and one female) who also traveled to the Moon on the *Apollo 17* mission, orbiting it seventy-five times. NASA gave them ID numbers A3305, A3326, A3352, A3356 and A3400. Evans holds the record for the most time spent in lunar orbit: 147 hours and 43 minutes. During the return to Earth, Ron Evans made an EVA of one hour and five minutes, to retrieve film cassettes from the Service Module. It was the third "deep space" EVA, and the one performed at the greatest distance from any planetary body. The splashdown in the Pacific Ocean saw the capsule *America* recovered by *USS Ticonderoga*.

Ron Evans was awarded the NASA Distinguished Service Medal in 1973, the Johnson Space Center Superior Achievement Award in 1970, the Navy Distinguished Service Medal in 1973, Navy Astronaut Wings, eight Air Medals, the Vietnam Service Medal, and the Navy Commendation Medal with Combat Valor device. In 1975, Evans served as backup Command Module Pilot for the *Apollo-Soyuz Test Project* mission. He worked on the development of the Space Shuttle, responsible for launch and ascent phases, before retiring from NASA in March 1977 to become a coal industry executive. He was inducted into the International Space Hall of Fame in 1983; and the US Astronaut Hall of Fame in 1997. Ron Evans died in his sleep of a heart attack at his home in Scottsdale, Arizona, on April 07, 1990, aged fifty-six years.

## Fabian, John M.. STS-7, STS-51-G.

Col. Dr. John McCreary Fabian was born in January 1939 in Goose Creek, Texas, USA. He graduated from Pullman High School, Pullman, Washington, in 1957; received a Bachelor of Science degree in Mechanical Engineering from Washington State University in 1962; a Master of Science degree in Aerospace Engineering from the US Air Force Institute of Technology in 1964; and a Doctorate in Aeronautics and Astronautics from the University of Washington in 1974. After five years' as a KC-135 co-pilot, aircraft commander, and instructor pilot at Wurtsmith AFB, Michigan, he saw action in Southeast Asia, flying 90 combat missions during the Vietnam War. He has logged 4,000 hours flying time, including 3,400 hours in jet aircraft. Selected in January 1978 NASA Group 8, he became an astronaut in August 1979. He worked extensively on satellite deployment and retrieval activities, including development of the Canadian Remote Manipulator System (RMS), the Shuttle's robotic arm.

Dr. Fabian served as Mission Specialist on STS-7 aboard *Challenger* (June 18-24, 1983), and deployed satellites for Canada (*ANIK C-2*) and Indonesia (*PALAPA B-1*). Using the Canadian-built robotic arm, Fabian was the first person to deploy and subsequently retrieve a free-flying satellite. On STS-51-G aboard *Discovery* (June 17-24, 1985), the crew used the RMS to deploy and later retrieve the *SPARTAN* satellite, which performed 17 hours of X-ray astronomy experiments. They conducted a laser tracking experiment as part of the (later abandoned) Strategic Defense Initiative, aka "Star Wars". Fabian logged 13 days, four hours and two minutes (over 316 hours) in space.

John Fabian is a Fellow of the American Institute of Aeronautics and Astronautics; President, Association of Space Explorers - USA; Vice President, International Astronautical Federation; and a Trustee, Washington State University Foundation. He has received twenty-five honors, including the FAI Komarov Diploma; the French Legion of Honor; the NASA Exceptional Public Service Medal; the Air Force Meritorious Service Medal; the Washington State Service to Humanity Award (1983); and the Leonov Medal (Association of Space Explorers). He is a Distinguished Member, Association of Space Explorers (2010); and received the Eleanor Stopps Environmental Leadership Award, Port Townsend Marine Science Center (2017).

## Ferguson, Christopher J.. STS-115, STS-126, STS-135.

Christopher J. "Fergy" Ferguson was born in September 1961 in Philadelphia, Pennsylvania, USA. He received a Bachelor of Science degree in Mechanical Engineering from Drexel University in 1984; and earned a Master of Science degree in Aeronautical Engineering from the US Naval Postgraduate School in 1991. Receiving his Naval Aviator wings in 1986, he trained on the F-14 Tomcat, joined the Red Rippers of VF-11 deploying aboard the aircraft carrier *USS Forrestal*, then attended the Navy Fighter Weapons School (TOPGUN). He was selected for the US Naval Test Pilot School program in 1989 and graduated in 1992. In 1995, he joined the Checkmates of VF-211 and deployed to the Western Pacific and Persian Gulf in defense of the Iraqui no-fly zone on board *USS Nimitz*.

Ferguson was selected for astronaut training in 1998; was deputy chief of the Astronaut Office prior to his selection to the STS-135 crew; and served as CapCom for missions STS-118, STS-128, and STS-129. Christopher Ferguson was pilot on STS-115 aboard *Atlantis* (September 09-21, 2006); delivering a truss segment, a pair of solar arrays, and batteries to the ISS. He was Commander on STS-126 aboard *Endeavour* (November 15-30, 2008), delivering supplies and equipment, servicing the Solar Alpha Rotary Joints (SARJ), and repairing the problem in the starboard SARJ. In 2011, he was assigned as Commander of STS-135 (ISS assembly flight ULF7) aboard *Atlantis* (July 08-21, 2011), the 135th and final Space Shuttle flight, and the only time that a Shuttle flew with a crew of four to the ISS. The primary cargo was the Multi-Purpose Logistics Module (MPLM) *Raffaello*, and a Lightweight Multi-Purpose Carrier (LMC).

Christopher Ferguson retired from NASA in 2011, and currently works for **The Boeing Company** as director of Crew and Mission Operations for Boeing's Commercial Crew Program. In July 2018, Boeing announced his assignment to the first human CST-100 orbital test, known as *Boe-CFT*. He was slated to be capsule commander, but in October 2020 he announced he was stepping down and would not fly, for "personal reasons". Ferguson was elected to the Astronaut Hall of Fame in January 2022.

## Fettman, Martin J.. STS-58.

Dr. Martin Joseph Fettman (B.S., D.V.M., M.S., Ph.D., Diplomate, ACVP) was born in December 1956 in Brooklyn, New York, USA. He has published over one hundred research articles in refereed scientific journals. Fettman graduated from Midwood High School, Brooklyn, New York, in 1973; received a Bachelor of Science degree in Animal Nutrition from Cornell University in 1976; a Doctor of Veterinary Medicine and Master of Science degree from Cornell University in 1980; a Doctor of Philosophy degree in Physiology from Colorado State University in 1982; board certification in Veterinary Clinical Pathology in 1984; and Diplomate of the American College of Veterinary Pathologists. He was named the 1994 Sigma XI honored scientist at Colorado State University, and a Bard College Distinguished Scientist for 1995.

Dr. Fettman's first faculty appointment was from 1982-86 in the Department of Pathology of the College of Veterinary Medicine and Biomedical Science at Colorado State University (CSU). In 1986, he was promoted to Associate Professor, and in 1988 assumed the duties of section chief of Clinical Pathology in the Veterinary Teaching Hospital at CSU. Named by NASA as a candidate in December 1991, he flew on STS-58 aboard *Columbia* as the prime Payload Specialist for 'Spacelab Life Sciences-2' in October 1993. Fettman logged 14 days, zero hours and 12 minutes in space. From 1993 to the present, he has held a joint appointment in the Department of Physiology at **Colorado State University**. Dr. Fettman is a member of the NASA Advisory Council Life and Biomedical Sciences and Applications Advisory Subcommittee.

## Fincke, Michael. Soyuz TMA-4/Expedition 9, Soyuz TMA-13/Expedition 18; STS-134.

Col. Edward Michael "Mike" or "Spanky" Fincke was born in March 1967 in Pittsburgh, Pensylvania, USA. He formerly held the record for the most time in space (381.6 days), until broken by Scott Kelly on October 16, 2015. Immediately after graduating from MIT in 1989, Fincke attended a summer exchange program with the Moscow Aviation Institute in the former Soviet Union, where he studied cosmonautics. After graduating from Stanford University in 1990, Fincke entered the USAF where he was assigned to the Air Force Space and Missiles Systems Center, Los Angeles AFB, California. Upon completion of the USAF Test Pilot School, Edwards AFB, California, Fincke joined the 39th Flight Test Squadron, Eglin AFB, Florida. In January 1996, he reported to the Gifu Test Center, Gifu AFB, Japan, where he was the United States flight test liaison to the Japanese/United States XF-2 fighter program. By 2005, Fincke had accumulated over 800 flight hours in more than thirty varieties of aircraft, and held the rank of Colonel.

Fincke was selected by NASA to be an astronaut in April 1996, and reported to the Johnson Space Center in August. After two years of training and evaluation, he was assigned to technical duties, serving as an International Space Station spacecraft communicator (ISS CapCom), a member of the Crew Test Support Team in Russia, and as the ISS crew procedures team lead. In July 1999, Fincke was assigned as backup crew member for the ISS 'Expedition 4' crew. Additionally, he served as a backup for the ISS 'Expedition 6' crew and is qualified to fly as a left-seat Flight Engineer (co-pilot) on the Russian *Soyuz* spacecraft. Launched aboard *Soyuz TMA-4*, Fincke was the space station Science Officer and Flight Engineer for ISS 'Expedition 9' from April 18 through October 23, 2004. He spent 187 days, 21 hours and 17 minutes (over six months) on the ISS continuing science operations, maintaining station systems and performing four EVAs totaling 15 hours, and 45 minutes.

Fincke was Commander of 'Expedition 18'. He arrived at the ISS aboard the *Soyuz TMA-13* on October 18, 2008, with cosmonaut Yuri Lonchakov and "space participant" Richard Garriott. During his personal time, Fincke (along with Lonchakov, Garriott and Gregory Chamitoff) filmed and starred in a science-fiction movie, ***Apogee of Fear***. On April 08, 2009, Fincke, Lonchakov and space tourist Charles Simonyi returned to Earth. Fincke was a Mission Specialist on STS-134, and made three EVAs during that mission. His total EVA time of 48 hours and 37 minutes places him 11th all time on the list of spacewalkers. On January 22, 2019, NASA announced that Fincke would fly on Boeing's *CST-100 Starliner* Crew Flight Test, scheduled for 2022. On June 16, 2022, NASA confirmed that the CFT would be a two-person flight test consisting of Barry Wilmore and Sunita Williams. Fincke would train as the backup spacecraft test pilot, and remains eligible for assignment to a future mission.

With fellow astronaut Terry Virts, Fincke was a guest star on the final episode of *Star Trek: Enterprise*. He was also featured in the *Star Trek: First Contact* Blu-ray special features. Fincke appeared in ***The Wiggles*** video *Wiggle Around the Clock* (2006), demonstrating a space suit. He voiced himself in the Season 14 ***Arthur*** episode *Buster Spaces Out*. Michael Fincke also appeared in *Man on a Mission: Richard Garriott's Road to the Stars*, a documentary based on Garriott's spaceflight to the ISS, launched on a *Soyuz* spacecraft.

## Feustel, Andrew J.. STS-125, STS-134, Soyuz MS-08/Expedition 55/56.

Dr. Andrew "Drew" Jay Feustel was born in August 1965 in Lancaster, Pennsylvania, USA. He is a former American/Canadian NASA astronaut and geophysicist. He received an AS degree from Oakland Community College in 1985; then received a Bachelor of Science degree in Solid Earth Sciences (1989) and a Master of Science degree in Geophysics from Purdue University in 1991. He then moved to Ontario, Canada, to attend Queen's University, (where he worked as a Graduate Research Assistant and Graduate Teaching Assistant) and received his Ph.D. in Geological Sciences (specializing in Seismology) in 1995. For three years he worked as a Geophysicist for the **Engineering Seismology Group**, Kingston, Ontario. In 1997, Feustel began working for **Exxon Mobil Exploration Company**, in Houston, Texas, as an Exploration Geophysicist.

Selected by NASA in July 2000, he qualified as a Space Shuttle and Space Station Robotic Arm Operator, CapCom, and Instructor Astronaut for EVA training at the Neutral Bouyancy Laboratory. As 'Mission Specialist 5' on STS-125 aboad *Atlantis* (May 11-24, 2009), he spent almost 13 days with three astronauts (Scott Altman, John Grunsfeld, and Michael Massimino) who had previous experience servicing and repairing the Hubble Space Telescope. Feustel performed three EVAs, with a total time of 20 hours and 38 minutes. On mission STS-134 aboard *Endeavour* (May 16-June 01, 2011) he performed another three EVA. Feustel launched on March 21, 2018 on *Soyuz MS-08* for 'Expedition 55/56' to the ISS. With crewmate Ricky Arnold, he installed wireless communications equipment on the *Tranquility* module during a six hour and ten minute EVA. He has logged a total of 225 days, nine hours and 16 minutes in space; including nine EVAs totaling 61 hours and 48 minutes. Dr. Feustel commanded 'Expedition 56' up to October 03, 2018. In November 2022, he became Acting Chief of Astronauts. After working for NASA for 23 years, Drew Feustel announced his retirement. He is a member of the all-astronaut band ***Max Q***.

## Fisher, Anna L.. STS-51-A.

Dr. Anna Lee Fisher (nee Tingle, then Sims) was born in August 1949 in New York City, USA. She received a Bachelor of Science degree in Chemistry (1971); and a Doctor of Medicine from the University of California, Los Angeles, in 1976. A divorcee, in 1977 she met William "Bill" Fisher a fellow intern a year ahead of her, and they were wed in Paseo Del Mar, San Pedro, California, on August 23, 1977. Fisher was selected by NASA in January 1978 as one of the first six American female astronauts. NASA had received 8,079 applicants and chose 208 for further screening. For the first four Space Shuttle missions she was assigned to the Search and Rescue helicopters supporting the flights. For the next four missions, she was involved in the verification of software at the Shuttle Avionics Integration Laboratory (SAIL). During her three-decade career at NASA, Fisher was involved with three major programs: the Space Shuttle, the ISS, and the Orion spacecraft. *"We appreciate all the years that Anna has dedicated to our space program"*, said Chris Cassidy, chief of NASA's Astronaut Office at JSC in Houston. *"Anna, and that first class of astronauts to include women, set the stage for decades of female astronauts after them and provide a tremendous inspiration to young girls"*. Unquote.

Married to Dr. William F. Fisher, in 1984 she became the first mother to fly in space. Anna Fisher was a Mission Specialist on STS-51-A aboard *Discovery* (November 08-16, 1984), deploying two satellites. She was assigned to mission STS-61-H that was planned in 1986, but canceled following the *Challenger* accident, so she resumed technical assignments in the Astronaut Office. Anna Fisher logged 192 hours in space. After a leave of absence to raise her two daughters from 1989 to 1995, Fisher returned to the Astronaut Office, where she worked on procedures and training issues in support of the ISS. From 1996 through 2002, during the early phase of building the ISS, Fisher was Chief of the Space Station branch. From January 2011 through August 2013, Fisher served as an ISS Capsule Communicator (CapCom), and was lead CapCom for 'Expedition 33'. Most recently, she was a management astronaut, working on display development for *Orion*, NASA's new deep-space craft. Anna Fisher was awarded the NASA Space Flight Medal (1984); and two NASA Exceptional Service Medals (1988 & 1998).

## Fisher, William F.. STS-51-I.

Dr. William "Bill" Frederick Fisher was born in April 1946 in Dallas, Texas, USA. He received a Bachelor of Arts degree in Biological Sciences from Stanford University in 1968; a Doctor of Medicine from the University of Florida in 1975; held a Residency in General Surgery at Harbor-UCLA Medical Center (1975-77); entered private practice in Emergency Medicine in 1977; and performed graduate work in engineering at the University of Houston from 1978-80. Fisher logged over 2,000 hours in prop, rotary-wing, jet aircraft and spacecraft. Selected in the 1980 NASA Group, he held fifteen technical assignments from 1980 to 1991, including CapCom for STS-8 and STS-9, Astronaut Office representative for Shuttle Mission Simulator (SMS) development (1983), Member of the USAF Scientific Advisory Board (1986-91), NASA Medicine Policy Board (1987-91), and Astronaut Office representative on space crew selection and retention standards for Space Station (1989-91). Dr. Fisher was a Mission Specialist on STS-51-I aboard *Discovery* (August 27-September 03, 1985), acknowledged as the most successful Space Shuttle mission flown to date. Crew deployed three satellites, and performed a successful on-orbit rendezvous with the ailing 15,400 pound *SYNCOM IV-3* satellite, with two very long EVAs to repair it. He logged seven days, two hours and 17 minutes in space.

In 1990, Dr. Fisher led a study of the design of the planned International Space Station. The ***New York Times*** reported that *study "found the 500-foot structure was so complex and fragile that it **would need up to 3,700 hours of maintenance a year** by space-suited astronauts, as against the designer's original goal of 130 hours".* Fisher was called to Capitol Hill to testify before a number of Committees. His time at NASA coincided with that of his former wife and fellow astronaut Anna Lee Fisher. Dr. William Fisher retired from NASA in 1992 and returned to the full-time practice of medicine.

## Ford, Kevin A.. STS-128, Soyuz TMA-06M/Expedition 33/34.

Col. Dr. Kevin Anthony Ford was born in July 1960 in Portland, Oregon, USA. He received his Bachelor of Science degree in Aerospace Engineering from the University of Notre Dame in 1982; a Master of Science degree in International Relations from Troy State University in 1989; a Master of Science degree in Aerospace Engineering from the University of Florida in 1994; and a Ph.D. in Astronautical Engineering from the US Air Force Institute of Technology in 1997. Ford was commissioned as a second lieutenant in 1982 and completed primary Air Force jet training, and was a Distinguished Graduate of Undergraduate Pilot Training at Columbus AFB in 1984. After spending 1990 as a student, he became a Distinguished Graduate of the USAF Test Pilot School, and flew flight test missions in the in the F-16 Fighting Falcon at Eglin AFB from 1991-94. He has attended the USAF Squadron Office school, the Air Command and Staff College Associate Program, and the Air War College. Ford has 6,100 flying hours and holds FAA Airline Transport Pilot certificates for airplanes, commercial certificates for helicopters and gliders, and a private license for hot air balloons. Ford has logged over 6,100 flying hours, and also holds FAA Airline Transport Pilot certificates for airplanes, helicopters, gliders and balloons.

Selected by NASA in July 2000, he completed two years of training, and later worked on the development and test of the Shuttle Cockpit Avionics Upgrade. Ford served as Director of Operations at the Gagarin Cosmonaut Training Center in Star City, Russia, from January 2004 to January 2005. He was pilot of STS-128 aboard *Discovery* (August 28-September 11, 2005) and then trained for and worked missions as a Space Station CapCom in the Mission Control Center. Col. Ford retired from the US Air Force in June 2008 as a Colonel. He was Flight Engineer of *Soyuz TMA-06M,* launched on October 23, 2012. He served as

ISS Flight Engineer for 'Expedition 33', and from November 18, 2012, to March 16, 2013, was Commander for 'Expedition 34'. He logged a total of 157 days, 13 hours and nine minutes in space. Col. Kevin Ford has received a number awards, including the Legion of Merit; the Air Force Meritorious Service Medal; the Air Force Commendation Medal; the Air Force Longevity Service Award with silver Oak Leaf Cluster; the National Defense Service Medal with service star; and the Armed Forces Expeditionary Medal.

## Foreman, Michael J.. STS-123, STS-129.

Michael James Foreman was born in March 1957 in Columbus, Ohio, USA. He received a Bachelor of Science degree in Aerospace Engineering from the US Naval Academy in 1979; and a Master of Science degree in Aeronautical Engineering from the US Naval Post graduate School in 1986. In 1981, Foreman was designated a Naval Aviator and assigned to Naval Air Station Brunswick, Maine. After deployments to Rota, Spain; Lajes, Azores; Bermuda and Panama, he attended the US Naval Postgraduate School in Monterey, California. Upon selection to the US Naval Test Pilot School in 1989, he moved to NAS Patuxent River, Maryland. He graduated in June 1990, and was assigned to the Force Warfare Aircraft Test Directorate. In 1991, he was reassigned as a flight instructor and the Operations Officer at the US Naval Test Pilot School. He was working as the technical lead for the Advanced Orbiter Cockpit Project team when he was selected for the astronaut program as part of the June 1998 class of twenty-five candidates.

As a Mission Specialist STS-123 aboard *Endeavour* (March 11-26, 2008), he performed three EVA. On STS-129, the first flight of an ExPRESS Logistics Carrier, he performed two EVA, totaling 12 hours and 45 minutes. Michael Foreman logged a total of 26 days, 13 hours and 27 minutes in space; including five EVA totaling 32 hours and 19 minutes. Foreman retired from NASA in July 2015 after a 35-year career in US Government service. He joined **Venturi Outcomes**, a Houston-based construction project management and consulting firm started by his wife. He was elected to the Friendswood Texas City Council in 2016. In 2018, he was elected Mayor of Friendswood, Texas; and re-elected to a second three-year term in 2021. Foreman is a member of the Association of Naval Aviation; US Naval Academy Alumni Association; the Society of Experimental Test Pilots; and the Association of Space Explorers. He has received the Legion of Merit; the Defense Meritorious Service Medal; the Navy Commendation Medal, and various other service awards.

## Forrester, Patrick G.. STS-105, STS-117, STS-128.

Col. Patrick Graham Forrester was born in March 1957 in El Paso, Texas, USA. He received a Bachelor of degree in Applied Sciences and Engineering from the US Military Academy, West Point, New York, in 1979; and a Master of Science degree in Mechanical and Aerospace Engineering from the University of Virginia, Charlottsville, in 1989. Designated an Army Aviator in September 1980, he was assigned as an instructor pilot at the Aviation School, and as the aide-de-camp to the Deputy Commanding General of the US Army Aviation Center. In 1984, Patrick Forrester was assigned to the 25th Infantry Division (Light), Hawaii, where he served as a platoon leader, aviation company operations officer, and an assault helicopter battalion operations officer. In June 1992, he graduated from the US Naval Test Pilot School, and was assigned as an engineering test pilot at the US Army Aviation Technical Test Center, Fort Rucker, Alabama. A Master Army Aviator, Forrester logged over 5,500 hours in over fifty aircraft types.

In July 1993, Forrester was assigned to NASA at the Johnson Space Center as an aerospace engineer. As well as technical assignments, he served as the crew representative for robotics development for the ISS. Forrester was selected in 1996 NASA Group, and after two years' training qualified for flight assignment as a Mission Specialist. He served as the technical assistant to the Director, Flight Crew Operations, then as the Shuttle training and on-board crew procedures representative. Forrester was 'Mission Specialist 1' on STS-105 aboard *Discovery* (August 10-21, 2001), the last mission before *Discovery* was grounded for an refit upgrade of the flight deck to the glass cockpit layout. STS-117 aboard *Atlantis* (June 08-22, 2007) delivered the second starboard truss (S3/S4) and its associated energy systems to the ISS. The mission was extended by two days, adding a fourth EVA and making it the longest flight for *Atlantis*.

He was 'Mission specialist 1' on STS-128 aboard *Discovery* (August 28-September 11, 2009), which carried the Multi-Purpose Logistics Module *Leonardo* as its primary payload. Patrick Forrester logged over 950 hours in space; including four EVA totaling 25 hours and 22 minutes. Forrester retired from the US Army in October 2005. He has received 13 awards, including the Legion of Merit; the Army Achievement Medal; the National Defense Service Medal; The Order of St. Michael (Bronze 2001, Silver 2007); NASA Exceptional Service Medals (2008 & 2010); and was inducted into the US Army Aviation Hall of Fame (2011).

## Fossum, Michael E.. STS-121, STS-124, Soyuz TMA-02M/Expedition 28/29.

Col. Michael Edward Fossum was born in December 1957 in Sioux Falls, South Dakota, USA. He graduated in Mechanical Engineering from Texas A & M University in 1980, and was commissioned as an officer in the USAF. Fossum is a graduate of the USAF Test Pilot School, and received his Master of Science degrees in Systems Engineering and Physical Science (space science) from the University of Houston at Clear Lake. He was selected to attend the Air Force Test Pilot School, from which he flew thirty-four aircraft types. After logging over 2,000 flight hours, Fossum left active duty for the Air Force Reserve in 1992 to work for NASA as a Systems Engineer, where his primary responsibilities were to evaluate the Russian *Soyuz* spacecraft for use as an emergency escape vehicle for the new ISS. In late 1993, he was selected to represent the Flight Crew Operations Directorate in an extensive redesign of the ISS. In 1997, he served as a Flight Test Engineer on the X-38, a prototype crew escape vehicle for the ISS.

Having served as the Astronaut Office Lead for ISS flight software development, Fossum was selected by NASA as an astronaut candidate in June 1998, having applied seven times since 1988. He supported several flights as a CapCom, including Lead Capcom for ISS 'Expedition 6'. Fossum was a Mission Specialist on STS-121 aboard *Discovery* (July 04-17, 2006) ISS Assembly Mission ULF 1.1, with the main purpose to test new safety equipment and procedures for the inspection and repair of the thermal protection system; introduced following the *Columbia* disaster of February 2003. En-route to the ISS, the 15-meter-long Orbiter Boom Sensor System (OBSS) tipped with two types of lasers and a high-resolution TV camera was used to inspect the underside of the Shuttle for damage. Particular attention was paid to the leading edges of the wings. The Multi Purpose Logistics Module *Leonardo* (on its fourth flight) was part of two tons of cargo taken to the ISS.

He was a Mission Specialist on STS-124 aboard *Discovery* (May 31-June 13, 2008), with the primary mission of delivery and installment of the Japanese *Kibo* Laboratory module, which became the largest and most scientifically capable addition to the ISS. He performed three EVAs, and later retired as a Colonel in the USAF Reserve in 2010. Fossum launched on June 07, 2011, aboard *Soyuz TMA-02M* to the ISS as Mission Specialist for 'Expedition 28'. Upon departure of 'Expedition 28' crew, he served as Commander of 'Expedition 29' aboard the ISS from September 09 to November 21, 2011. Fossum performed his seventh EVA, lasting six hours and 31 minutes. He logged a total of 193 days, 19 hours and two minutes in space; with a total EVA time of 48 hours and 32 minutes. He has since served in several capacities, including assistant to the Chief of the Astronaut Office for the ISS. In January 2017, Fossum left NASA and accepted a position with **Texas A & M University** at Galveston as Vice President and COO. In October, 2019, he was appointed as Superintendent of **Texas A & M Maritime Academy**.

In March 2010, Fossum became the youngest person inducted into the Texas A & M University's Corps of Cadets Hall of Honor. A junior high school in McAllen, Texas is named after him. Col. Michael Fossum has received the Meritorious Service Medal with two Oak Leaf Clusters; the NASA Exceptional Service Medal; the NASA Space Flight Medal with two Oak Leaf Clusters; the National Defense Service Medal with service star; the Air Force Longevity Service Award with four Oak Leaf Clusters; and the Air Force Training Ribbon.

## Frick, Stephen N.. STS-110, STS-122.

Capt. Stephen Nathaniel Frick was born in September 1964 in Pittsburgh, Pennsylvania, USA. He graduated from Pine-Richland High School in 1982; received a degree in Aerospace Engineering from the US Naval Academy in 1986; was commissioned as a US Navy officer, and trained as an F/A-18 fighter pilot. Stationed aboard the carrier *USS Saratoga*, he flew combat missions during the Gulf War and then earned a Master's degree in Aeronautical Engineering from the Naval Postgraduate School in 1994.

Frick was selected in the 1996 NASA Group and trained as a Space Shuttle pilot. He was pilot on STS-110 aboard *Atlantis* (April 08-19, 2002), a docking mission with the ISS which installed the S0 truss segment, the backbone of the truss structure on the ISS. Frick was Commander for STS-122 aboard *Atlantis* (February 07-20, 2008), which launched on the third attempt and delivered the European Space Agency's *Columbus* laboratory to the ISS. Stephen Frick logged a total of 23 days, 14 hours and four minutes in space. NASA announced his retirement in July 2015. He has been awarded the DFC; the Defense Meritorious Service Medal; three Navy Commendation Medals; the Air Medal with two Strike-Flight awards; two Southwest Asia Service Medals; the NASA Outstanding Leadership Medal; and NASA Spaceflight Medals.

## Fullerton, C. Gordon. STS-3, STS-51-F.

Col. Gordon Fullerton (1936-2013) was born in October 1936 in Rochester, New York, USA. He graduated from Ulysses S. Grant High School, Portland, Oregon, in 1953; and received a Bachelor of Science degree (1957); and Master of Science degree in Mechanical Engineering from the California Institute of Technology, Pasadena, California, in 1958. After working as a mechanical design engineer for **Hughes Aircraft Company**, he entered the USAF in July 1958. In 1964, he was chosen to attend the Air Force Aerospace Research Pilot School (now the USAF Test Pilot School), Edwards Air Force Base, California, and became a a research pilot at the Dryden Flight Research Facility, Edwards AFB. Upon graduation, Fullerton was assigned as a test pilot with the Bomber Operations Division at Wright-Patterson AFB, Ohio.

In 1966, he was selected as a flight crew member for the MOL program, until its termination in 1969. Fullerton was part of NASA Astronaut Group 7 in September 1969, and served on the support crews for the *Apollo 14*, 15, 16 and 17 lunar missions. In 1977, he was assigned to one of the two-man flight crews which piloted the prototype shuttle *Enterprise* during the Approach and Landing Tests at Dryden. Fullerton was pilot aboard *Columbia* on STS-3 (March 22-30, 1982), which exposed the orbiter to thermal stress, and tested the 50-foot Canadarm used to grapple and maneuver payloads to orbit. He was Commander on the STS-51-F first pallet-only 'Spacelab-2' mission aboard *Challenger* (July 25-August 06, 1985), and the first to operate the Spacelab Instrument Pointing System (IPS). The flight carried 13 major experiments in the fields of astronomy, solar physics, ionospheric science, life science, and a super fluid helium experiment.

Fullerton served as test pilot on the NASA/Convair 990 aircraft, which has been modified as a Landing Systems Research Aircraft to test Space Shuttle landing gear components. He was also project pilot on F-18 Systems Research Aircraft, a test bed to develop new flight control actuators, fiber optic control systems, and other advanced aircraft technology. He flew the B-52 for the first six mid-air launchings of the commercially developed *Pegasus* space vehicle. Fulleron was one of two NASA pilots to fly the Russian-built Tu-144LL supersonic aircraft used in a joint high speed research program. With over 16,000 hours of flying time, he has piloted 135 aircraft types. He received eighteen honors, including the General Thomas D. White Space Trophy (1977); the Iven C. Kinchelow Award from the Society of Experimental Test Pilots (1978); and the Golden Plate Award of the American Academy of Achievement (1982). He is an Honorary Member of the National World War II Glider Pilot Association; and was inducted into the US Astronaut Hall of Fame (2005). Col. C. Gordon Fullerton retired in 2007, suffered a stroke in 2009, and was cared for in a facility in Lancaster, California, where he died on August 21, 2013, aged seventy-six years.

## Gaffney, F. Andrew.. STS-40.

Col. Dr. Francis Andrew "Drew" Gaffney was born in June 1946 in Carlsbad, New Mexico, USA. He received a Bachelor of Arts degree in Psychology from the University of California, Berkeley, in 1968; and received a Doctor of Medicine degree from the University of New Mexico in 1972. Gaffney has more than one hundred publications in the areas of cardiovascular regulation, space physiology and patient safety. A Colonel in the Texas Air National Guard, Dr. Gaffney served as a senior flight surgeon-astronaut for the 147th Fighter Interceptor Group at Ellington Air National Guard Base, Houston, Texas, from 1986-93. From June 1986 to June 1989, he served as a visiting senior scientist with the Life Sciences Division at NASA HQ; and as a reservist at Brooks AFB, San Antonio, Texas until his retirement from the USAF in 2007.

Dr. Gaffney's fifteen years' experience in cardiac research and experience in echocardiography and human physiology led to him being selected as a Payload Specialist on the June 1991 STS-40 'Spacelab Life Sciences' mission (SLS-1), the first Shuttle mission dedicated to biomedical research. He was a co-investigator on an experiment that studied human cardiovascular adaptation to space flight. STS-40 completed over eighteen experiments during a nine-day mission of 218 hours and 14 minutes; bringing back more peer-reviewed, published medical data than any previous NASA flight. Dr. Gaffney then became a member of the Institute of Medicine's Committee on Space Biology and Medicine, serving from 1992 to 2000. He is a Professor of Medicine at **Vanderbilt University**.

## Garan, Ronald J., Jr.. STS-124, Soyuz TMA-21/Expedition 27/28.

Ronald "Ron" John Garan, Jr. was born in October 1961 in Yonkers, New York, USA. He is of Russian Jewish descent, and received a Bachelor of Science degree in Business Economics from the Suny College at Oneonta in 1982; a Master of Science degree in Aeronautics from Embry-Riddle Aeronautical University in 1994; and a Master of Science degree in Aerospace Engineering from the University of Florida in 1996. Garan joined the Air Force, becoming a Second Lieutenant in 1984. From 1986-88, he was a combat-ready F-16 pilot in the 496th Tactical Fighter Squadron. From August 1990 to March 1991, he flew combat missions in Operations Desert Shield and Desert Storm. From January to December 1997, he attended the US Naval Test Pilot School at NAS Patuxent River, and has logged over 5,000 hours flying over thirty aircraft types. Before becoming an astronaut, he was the Operations Officer of Fortieth Flight Test Squadron.

Selected as a pilot by NASA in July 2000, Ron Garan was a Mission Specialist on STS-124 aboard *Discovery* (May 31-June 14, 2008), which delivered the Japanese Experiment Module Pressurized Module (JEM-PM) and the Japanese Remote Manipulator System to the ISS. He completed three EVA for a time of 20 hours and 32 minutes. Launched on April 04, 2011, to the ISS aboard *Soyuz TMA-21* (named *Gagarin* in honor of the 50th anniversary of the first human spaceflight), he logged another six hours and 31 minutes of EVA in the last shuttle-based-spacewalk during the STS-135 mission. He was a member of 'Expedition 27' for six months. Ron Garan returned to Earth on September 16, 2011, describing his experience in a Soyuz capsule *"like going over Niagara Falls in a barrel (that's on fire) followed by a high speed crash"*. He logged a total of over 177 days, 23 hours and 54 minutes in space; including 27 hours and three minutes during four EVA.

Ron Garan is the founder of the **Manna Energy Foundation**, which is assisting the villages of Rwanda to make potable water. He is a member of Engineers Without Borders; International Solar Energy Society; founder of the **Fragile Oasis** project; Vice President of Spaceship Earth Grants; and writer and mentor for Unreasonable, a social entrepreneurship hub. On June 24, 2009, he met Pope Benedict XVI at his general audience. In 2014, Garan retired from NASA to work on communicating what he called the "Orbital Perspective". He has published a book called *The Orbital Perspective - Lessons in Seeing the Big Picture from a Journey of 71 Million Miles*, and is working on a documentary called *Orbital*. On February 23, 2016, **World View Enterprises** announced that Ron Garan will be chief pilot for current robotic flight operations and upcoming human spaceflights via balloon. Garan has received ten Air Force decorations, including the DFC for Combat Valor; the Air Force Outstanding Unit Award with Valor; and two NASA service medals.

## Gardner, Dale A.. STS-8. STS-51-A.

Capt. Dale Allan Gardner (1948-2014) was born in November 1948 in Fairmont, Minnesota, USA. He was Valedictorian of his class from Savanna Community High School, Savanna, Illinois, in 1966; received a Bachelor of Science degree in Engineering Physics from the University of Illinois at Urbana-Champaign in 1970, then joined the US Navy. In October 1970, he began Basic Naval Fight Officer training at Pensacola, was selected a Distinguished Naval Graduate and awarded his wings on May 05, 1971. He flew in the F-14 Tomcat in two cruises while deployed aboard *USS Enterprise*. From December 1976 until July 1978, he was involved in the test and evaluation of Navy fighter aircraft. Dale Gardner logged over 2,300 hours flying time in over twenty aircraft types.

Selected in January 1978 NASA Group 8, he completed one year of training, then flew as a Mission Specialist on STS-8 aboard *Challenger* (August 30-September 05, 1983), which deployed the *Indian National Satellite* (INSAT-1B); and on STS-51-A aboard *Discovery* (November 08-16, 1984), when he captured the malfunctioning *Westar 6* satellite while using the *Manned Maneuvering Unit* (MMU) backpack. Gardner logged 14 days and 52 minutes in space; including two EVAs totaling 12 hours. In October 1986, he returned to his Navy duties and was assigned to the US Space Command, Colorado Springs. He served over two years as Deputy Chief, and, after promotion to Captain in June 1989, became the Command's Deputy Director for Space Control at Peterson AFB. He retired from the US Navy in October 1990 and accepted a position with **TRW Inc.**, Colorado Springs, in their Space and Defense Sector.

Capt. Gardner then became the manager of **Northrop Grumman**'s operations in Colorado Springs. In 2003, he became the Associate Director for Renewable Fuels Science and Technology at the **National Renewable Energy Laboratory**, and retired in January 2013. He received eight honors, including the Humanitarian Service Medal (1979); the Defense Superior Service Medal (1984, 1989, 1990); the DFC (1989); the Lloyds of London Meritorious Service Medal (1984); and the Master Space Badge (1989). Capt. Dale Gardner died after suffering a hemorrhagic stroke at his home on February 19, 2014, in Colorado Springs, Colorado, aged sixty-five years. A bridge connecting Savanna, Illinois and Sabula, Iowa, over the Mississippi River was dedicated to Dale Gardner and US Veterans in November 2017.

## Gardner, Guy S.. STS-27, STS-35.

Col. Guy Spence Gardner was born in January 1948 in Altavista, Virginia, USA. He received a Bachelor of Science degree with majors in Astronautics, Mathematics and Engineering Sciences from the USAF Academy in 1969; and a Master of Science degree in Astronautics from Purdue University in 1970. He completed USAF pilot training at Craig AFB, Alabama, and F-4 Phantom upgrade training at MacDill Air Base, Florida, in 1971, then flew 177 combat missions in Southeast Asia during 1972. In 1977-78, he was an instructor test pilot at the USAF Test Pilot School. In 1979-80, he was Operations Officer of the First Test Squadron at Clark Air Base, Philippines.

Selected as a pilot astronaut in May 1980 NASA Group 9, he was pilot on mission STS-27 aboard *Atlantis* (December 02-06, 1988), which carried a Department of Defense payload. He was pilot on STS-35 aboard *Columbia* (December 02-19, 1990), which carried the ASTRO-1 astronomy laboratory. Gardner logged a total of 13 days, eight hours and ten minutes in space. He left NASA in June 1991 to command the USAF Test Pilot School at Edwards AFB, California. In August 1992, Gardner retired from the Air Force and returned to NASA to direct the joint US and Russian *Shuttle-Mir* Program. In 1995, he joined the **Federal Aviation Administration** as the Director of the William J. Hughes Technical Center, New Jersey. After completing US Government service, Gardner served in various teaching positions, then in 2004 he returned to Purdue University to manage research programs. From 2006-09, he was superintendent of the Riverside Military Academy. From 2009-13, he served as President of the **Williamson College of the Trades**.

Guy Gardner was a Distinguished Graduate of the USAF Academy; the top graduate in pilot training; and the top graduate from the USAF Test Pilot School. He was awarded the Test Pilot School Outstanding Academic Instructor; the Test Pilot School Outstanding Flying Instructor; and the Distinguished Astronaut Engineering Alumnus Award of Purdue University. He has ten decorations, including Defense Distinguished Service Medal; Defense Superior Service Medal; the Legion of Merit, with one Oak Leaf Cluster; the DFC, with two Oak Leaf Clusters; the Air Medal with 13 Oak Leaf Clusters; the National Intelligence Medal of Achievement; the Vietnam Service Medal; and the Republic of Vietnam Campaign Medal.

## Garn, E. Jacob. STS-51-D.

Col. Senator Edwin Jacob "Jake" Garn was born in October 1932 in Richfield, Utah, USA. He is an American politician and member of the Republican Party who served as a US Senator representing Utah from 1974-93. He earned a Bachelor of Science degree in Business and Finance from the University of Utah in 1955. Garn became the first sitting member of the US Congress to fly in space when he flew aboard Space Shuttle *Discovery* as a Payload Specialist. Senator Garn is a former insurance executive. Before election to the Senate, he served on the Salt Lake City commission for four years and was elected Mayor in 1971. He served as chairman of the Senate Banking, Housing, and Urban Affairs Committees, and served on three sub-committees. As chairman of the Senate Banking Committee, Garn was co-author of the **Garn-St. Germain Depository Institutions Act of 1982**, the law that partially deregulated the savings and loan industry and attempted to forestall the looming Savings and Loan crisis.

Garn served in the US Navy as a Martin P5M Marlin pilot, and also as a pilot of the 151st Air Refueling Group of the Utah Air National Guard. He retired as a Colonel in April 1979, when he had flown 17,000 hours in military aircraft, and was promoted to Brigadier General after his Space Shuttle mission. Garn said that flying on the Shuttle would be a fact-finding trip: *"I do really think that it is a necessity that Congressmen check things out that they vote for and make certain that funds are being spent adequately. It might be necessary to have a Senator kick the tire".* Unquote. Mission STS-51-D aboard *Discovery* (April 12-19, 1985) had the primary objective of deploying two communications satellites, and to perform electrophoresis and echocardiograph operations in space, plus other experiments. The space sickness that he experienced was so severe that a scale for space sickness was jokingly based on him, where "one Garn" is the highest possible level of sickness. Astronaut Charles Bolden described Garn as *"the ideal candidate to do it, because he was a veteran Navy combat pilot who had more flight hours than anyone in the Astronaut Office".* Upon his return, Garn co-wrote the 1989 novel *Night Launch*. The Jack Garn Mission Simulator and Training Facility is named after him.

## Garriott, Owen K.. Skylab 3, STS-9.

Dr. Owen Kay Garriott (1939-2019) was born in November 1939 in Enid, Oklahoma, USA. He graduated from Enid High School in 1948, where he served as senior class president and was voted Most Likely to Succeed. He received a Bachelor of Science degree in Electrical Engineering from the University of Oklahoma in 1953; and a Master of Science (1957) and a Ph.D. (1960) from Stanford University; both in Electrical Engineering. From 1953-56, Garriott served as an electronics officer in the US Navy. From 1961-65, he was an assistant professor and associate professor of Electrical Engineering at Stanford University. After obtaining his Doctorate, he performed research and led graduate studies in ionospheric physics; and authored or co-authored more than forty-five scientific papers, mainly in areas of the physical sciences.

As a pre-requisite of the era's scientist-astronaut training, he completed a one-year USAF pilot training program in 1966, qualifying as a pilot in jet aircraft. Garriott was selected in 1965 NASA Group 4. Aboard the *Skylab* space station from July 28, 1973, he set a world space duration record of 60 days, more than double the previous record. Extensive experiments were conducted of the Sun, of Earth sciences, and in various life sciences relating to human adaptation to weightlessness. He then received a NASA fellowship in the Space Station Project Office, advising the project manager concerning the scientific suitability of the space station design. Garriott ten days aboard 'Spacelab-1' during mission STS-9, launched aboard *Columbia* on November 28, 1983. Over 70 separate experiments were conducted in six different disciplines. As an Extra Class amateur radio operator, he operated the world's first amateur radio station in space, and, on December 01, 1983, made the first amateur radio contact from space using a Motorola handheld 2-meter radio while aboard STS-9. He logged a total of 69 days, 18 hours and 56 minutes in space; including three EVAs for a total time of 13 hours and 43 minutes.

After leaving NASA in June 1986, Garriott consulted for various aerospace companies, and served on several committees. From January 1988 until May 1993, he was vice president of space programs at **Teledyne Brown Engineering**. This division provided payload integration for all 'Spacelab' projects at the Marshall Space Center. Later, he accepted a position as adjunct professor in the laboratory for Structural Biology at the University of Alabama, Huntsville. His son, Richard, was launched as a "space tourist" aboard *Soyuz TMA-13* on October 12, 2008, the first American and the second person worldwide to follow a parent into space. Dr. Garriott was one of five Oklahoman astronauts inducted into the Oklahoma Aviation and Space Hall of Fame in 1980, the US Astronaut Hall of Fame (1997), Oklahoma Military Hall of Fame (2000), and the Enid Public Schools Hall of Fame (2001). He recieved seven other awards, including the National Science Foundation Fellowship, 1960-61. Dr. Owen Garriott died on April 15, 2019, at his home in Huntsville, Alabama, aged eighty-eight years.

## Gemar, Charles D.. STS-38, STS-48, STS-62.

Lt. Col. Charles Donald "Sam" Gemar was born in August 1955 in Yankton, South Dakota, USA. He graduated from Scotland Public High School in June 1972; enlisted in the US Army in January 1973, and was assigned to the XVIII Airborne Corps at Fort Bragg, North Carolina, where he received an appointment to the US Military Academy Preparatory School at Fort Belvoir, Virginia, and later an appointment to join the US Military Academy Class of 1979. Gemar received a Bachelor of Science degree in Engineering in 1979, then attended the Infantry Officers Basic Course at Fort Benning, Georgia. In October 1980, he transferred to the 24th Infantry Division, Fort Stewart, Georgia, where he remained until January 01, 1985. Gemar was Distinguished Graduate of his class in Undergraduate Pilot Training, and Distinguished Graduate of his class in fixed-wing and multi-engine pilot training.

Selected by NASA in June 1985, he qualified for assignment as a Mission Specialist, then held a variety of technical assignments in support of the Space Shuttle program. STS-38 aboard *Atlantis* (November 15-20, 1990) conducted Department of Defense operations. STS-48 was aboard *Discovery* (September 12-18, 1991), which carried the classified UARS primary payload. STS-62 aboard *Columbia* (March 04-18, 1994) was a microgravity science and technology demonstrator mission (USMP-2), also carrying the OAST-2 payloads. Sixty experiments or investigations were conducted in many scientific and engineering disciplines. He logged a total of 24 days, five hours and 38 minutes (over 581 hours) in space. Charles Gemar was awarded the Army Achievement Medal; the Defense Superior Service Medal; two National Defense Service Medals; the National Intelligence Medal of Achievement; the NASA Exceptional Service Medal; and an honorary Doctorate of Engineering from the South Dakota School of Mines and Technology.

## Gernhardt, Michael L.. STS-69, STS-83, STS-94 STS-104.

Dr. Michael Landon Gernhardt was born in May 1956 in Mansfield, Ohio, USA. He was manager of the Environmental Physiology Laboratory, and principal investigator of the Prebreathe Reduction Program (PRP) at the Johnson Space Center. Gernhardt graduated from Malabar High School, Mansfield, Ohio, in 1974; received a Bachelor of Science degree in Physics from Vanderbilt University in 1978; a Master of Science degree in Bioengineering from University of Pennsylvania in 1983; and a Doctorate in Bioengineering from University of Pennsylvania in 1991. From 1977-84, Gernhardt worked as a professional diver and project engineer on a variety of subsea oil field construction and repair projects; logging over 700 deep sea dives. From 1984-88, he worked as a Manager, then Vice President of Special Projects, for **Oceaneering International**. In 1988, he founded **Oceaneering Space Systems**, a company formed to transfer subsea technology and operational experience to the ISS program. From 1988-92, he worked on development of new astronaut and robot-compatible tools for performing space station maintenance, and the development of new portable life support systems and decompression procedures for EVA.

Selected in 1992 NASA Group, he reported to JSC in August 1992. Gernhardt flew on STS-69 aboard *Endeavour* (September 07-18, 1995), which deployed and retrieved a *SPARTAN* satellite and the *Wake Shield Facility*, a free-flying experiment. STS-83 aboard *Columbia* (April 04-08, 1997) was the Microgravity Science Laboratory (MSL-1) 'Spacelab' mission, cut short due to problems with one of the Shuttle's three fuel cell power generation units. STS-94 aboard *Columbia* (July 01-17, 1997) was a reflight of the MSL-1 mission. STS-104 aboard *Atlantis* (July 12-24, 2001) was the 100th mission to the ISS, and the crew conducted joint operations with the 'Expedition 2' crew. Gernhardt was one of two astronauts to perform three EVAs to install the joint airlock *Quest*, and to outfit it with four high-pressure gas tanks.

He logged a total of 43 days, seven hours and one minute in space; including four EVAs totaling 23 hours and 16 minutes. Michael Gernhardt received four awards: four NASA Space Flight Medals; the NASA Exceptional Achievement Medal; two NASA Exceptional Service Medals; and the NASA Distinguished Service Medal.

## Gibson, Edward G.. Skylab 4.

Dr. Edward George Gibson was born in November 1936 in Buffalo, New York, USA. From age two to eight years he battled osteomyelitis (soft spots in the bones), until newly-available penicillin cured the disease. He is a retired American engineer and physicist. Gibson earned a Bachelor of Science degree in Engineering from the University of Rochester in 1959; a Master of Science in Engineering (Jet Propulsion) from the California Institute of Technology in 1960; and a Ph.D. in Engineering (with a minor in Physics) from California Institute of Technology (Caltech) in June 1964. While studying at Caltech, Gibson was a research assistant in the field of jet propulsion and classical physics. His technical publications are in the fields of plasma physics and solar physics. From June 1964 he was senior research scientist at Applied Research Laboratories of Philco Corporation, California, where he did research in lasers and the optical breakdown of gases, until moving to NASA in 1965.

Dr. Edward Gibson completed a 53-week course in flight training at Williams AFB, Arizona, and earned his Air Force wings; logging over 4,300 hours flying time, including 2,270 hours in jet aircraft. Selected as a scientist-astronaut in June 1965 NASA Group 4, he served as a member of the astronaut support crew and as a CapCom for the *Apollo 12* lunar mission. Gibson participated in the design and testing of many elements of the *Skylab* space station. He was the science pilot of the *Skylab 4* mission (November 16, 1973 to February 08, 1974), which logged 2,017 hours in space, and made 1,214 Earth orbits. Gibson logged a total of 84 days, one hour and 15 minutes in space, with three EVAs totaling 15 hours and 22 minutes. The three *Skylab* astronaut crews were awarded the 1973 Robert J. Collier Trophy *"For proving beyond question the value of man in future explorations of space and the production of data of benefit to all the people on Earth"*. The FAI awarded the *Skylab 4* crew the De la Vaulx Medal and the V.M. Komarov Diploma for 1974. President Gerald Ford awarded the 1975 Dr. Robert H. Goddard Memorial Trophy to the *Skylab* astronauts. Gibson is the last surviving *Skylab 4* crew member: pilot William Pogue died in 2014, and Commander Gerald Carr died in 2020.

Dr. Gibson resigned from NASA in December 1974 to do research on *Skylab* solar physics data as a senior staff scientist with the **Aerospace Corporation** of Los Angeles, California. In March 1977, he returned to the Astronaut Office as Chief of the Scientist-Astronaut Candidates. Not keen on flying a Space Shuttle mission, he served as CapCom for STS-1. In October 1990, Gibson began his own consulting firm, **Gibson International Corp**. He has performed a significant amount of speaking and writing, and published a text book in solar physics, two novels, and edited *The Greatest Adventure*, a 1994 compilation of stories and pictures on space missions from many astronauts and cosmonauts around the world. He was one of twenty-four Apollo astronauts inducted into the US Astronaut Hall of Fame in 1997.

## Gibson, Robert L.. STS-41-B, STS-61-C, STS-27, STS-47 (J), STS-71.

Capt. Robert Lee "Hoot" Gibson was born in October 1946 in Cooperstown, New York, USA. He is a former test pilot, aeronautical engineer, and graduate of TOPGUN. While assigned to Fighter Squadron 111 (VF-111 & VF-1) from April 1972 to September 1975, he saw duty aboard the aircraft carriers *USS Coral Sea* and *USS Enterprise*, flying combat missions in Southeast Asia in the F-4 and the F-14 Tomcat. On May 30, 1991, he married fellow astronaut Dr. Rhea

Seddon of Tennessee. A retired astronaut, he served as Chief of the Astronaut Office from 1992-94. Robert Gibson has logged 14,000 flying hours in over 140 types of civil and military aircraft. Holding a private pilot rating since age seventeen years, he made over 300 aircraft carrier landings, and holds an airline transport pilot license.

Selected in January 1978 NASA Group 8, he became an astronaut in August 1979. Gibson was pilot on STS-41-B aboard *Challenger* (February 03-11, 1984), which carried the German *Shuttle Pallet Satellite* (SPAS), Remote Manipulator System (RMS), six "Getaway Specials", and materials processing experiments. He was Commander of STS-61-C aboard *Columbia* (January 12-18, 1986), which deployed the *SATCOM* Ku band satellite and conducted experiments in astrophysics and materials processing. He was Commander on STS-27 aboard *Atlantis* (December 02-06, 1988), which carried a Department of Defense payload, and the space shuttle suffered severe damage to its critical heat-resistant tiles during ascent. Gibson was Commander on STS-47 'Spacelab-J' aboard *Endeavour* (September 12-20, 1992), which flew the first Japanese astronaut, Mamoru Mohri; and the first African-American woman, Mae Jemison. The crew focused on over forty life science and materials processing experiments in the 'Spacelab' laboratory.

He commanded STS-71 aboard *Atlantis* (June 27-July 07, 1995), which was modified to carry a docking system compatible with the *Mir* space station. It also carried a 'Spacelab' module in the payload bay, and when the hatch was opened, Gibson and cosmonaut Vladimir Dezhurov shook hands, symbolizing the newly-found cooperation between the USA and the Russian Federation. Robert Gibson logged a total of 36 days, four hours and 15 minutes in space. He left NASA in November 1996 and became a commercial pilot for **Southwest Airlines**, retiring in 2006. In December 2006, he joined the **Benson Space Company** as COO and chief test pilot. From 2009, he flew as a demonstration pilot for **Hawker Beechcraft Corporation** to showcase the Premier 1A light business jet. He has flown 111 aircraft types and today is active as a professional pilot, racing regularly at the annual Reno Air Races. He has eight honors and eleven military Awards, including the FAI Louis Bleriot Medal (1991 & 2004); the FAI Yuri A. Gagarin Gold Medal; three Air Medals; the DFC; the Vietnam Service Medal, the Humanitarian Service Medal; and the Experimental Aircraft Association "Freedom of Flight" Award (1989). He was inducted into the US Astronaut Hall of Fame in 2003; and into the National Aviation Hall of Fame in 2012.

## Glenn, John H., Jr.. Mercury-Atlas 6 "Friendship 7", STS-95.

John Herschel Glenn, Jr., was born in July 1921 in Cambridge, Ohio, USA. He was an American Marine Corps aviator, engineer, astronaut, businessman and politician. In 1941, Glenn obtained a private pilot license after he finished a course in physics and received a credit. In March 1942, he was commissioned as a United States Navy aviation cadet and was trained at the Naval Air Station Olathe, Kansas. During WWII, Glenn was commissioned to the Pacific front, where he flew fifty-nine combat missions. He also flew combat missions in the Korean War. Soon after, he became a test pilot, and in 1957 received a Distinguished Flying Cross (DFC) for his record non-stop supersonic transcontinental flight, Project Bullet, from Los Angeles to New York. He piloted a Vought F8U-1P Crusader in three hours, 23 minutes and 8.3 seconds. In 1958, Glenn applied for the NASA astronaut recruiting program, and was selected.

He was one of the "Mercury Seven" military test pilots selected in 1959 by NASA as the nation's first group of astronauts. His first mission on February 20, 1962, was *Mercury-Atlas 6 Friendship 7*, and it lasted four hours, 55 minutes and 23 seconds. He was the third American in space, and the first American to orbit the Earth, circling it three times. He later said, "*As I hurtled through space, one thought kept crossing my mind – every part of this rocket was supplied by the lowest bidder*". Following his retirement from NASA in 1963, he served from 1974 to January 1999 as a United States Senator (Democrat) from Ohio. He was chair of the Senate Governmental Affairs Committee from 1987-95. In 1976, Glenn made a bid for the Democratic Party vice-presidential nomination, but his keynote address did not impress many delegates. Glenn was a co-sponsor of the **Nuclear Non-Proliferation Act** of 1978, which helped stem the spread of nuclear weapons. In 1980, he won re-election to the Senate, with a landslide winning margin of over 40% of the votes. In addition to his political career, John Glenn served on the board of directors for **Royal Crown Cola** and participated in various other business ventures.

In October 1998, Glenn flew into space at age seventy-seven years, becoming the oldest person in space (at that time). He was a Payload Specialist on STS-95 aboard Space Shuttle *Discovery* (October 29-November 07, 1998). This mission lasted eight days and 21 hours. Glenn took part in observations of health aspects of an older astronaut. He logged a total of nine days, 19 hours and 54 minutes in space. In 2000, he was awarded the US Senator John Heinz Award for Greatest Public Service by an Elected or Appointed Official. Glenn passed away on December 08, 2016, at the age of ninety-five years. He had made the witty saying, "*There is still no cure for the common birthday*". Among the many honors he received were the NASA Distinguished Service Medal (1962); the Congressional Space Medal of Honor (1978); the Presidential Medal of Freedom (2012). He was inducted into the National Aviation Hall of Fame (1976); and into the US Astronaut Hall of Fame in 1990. The NASA **John H. Glenn Research Center** at Lewis Field in Cleveland, Ohio, is named in his honor.

## Godwin, Linda M.. STS-37, STS-59, STS-76, STS-108.

Dr. Linda Maxine Godwin was born in July 1952 in Cape Giradeau, Missouri, USA. She received a Bachelor of Science degree in Mathematics and Physics from Southeast Missouri State University in 1974; a Master of Science degree (1976); and a Doctorate in Physics from the University of Missouri in 1980. An instrument rated private pilot, Godwin joined NASA in 1980, and her assignments included working with flight simulation software verification in the Shuttle Avionics Integration Laboratory (SAIL), and co-ordinating mission development activities for the Inertial Upper Stage, deployable payloads, and 'Spacelab' missions. She was selected by NASA as a candidate in June 1985, becoming an astronaut in July 1986. Godwin logged a total of 38 days, six hours and 13 minutes in space; including over ten EVA hours in two spacewalks. She married fellow astronaut Steven Nagel (who had been her Commander on STS-37), and remained married until his death from melanoma cancer in August 2014.

Dr. Linda Godwin retired in 2010, and accepted the position of professor in the department of Physics and Astronomy at the **University of Missouri**. She is currently the assistant to the director for exploration, Flight Crew Operations Directorate at **Johnson Space Center**. Godwin is a member of the American Physical Society; the Ninety-Nines, Inc.; the Association of Space Explorers; and the Aircraft Owners and Pilots Association. She received the NASA Outstanding Performance Rating; the Sustained Superior Performance Award; the NASA Outstanding Leadership Medal; the NASA Exceptional Service Medal; and the NASA Distinguished Service Medal.

## Good, Michael T. STS-125, STS-132.

Col. Michael Timothy "Bueno" Good was born in October 1962 in Parma, Ohio, USA. He received a Bachelor of Science degree in Aerospace Engineering from the University of Notre Dame in 1984; and a Master of Science degree in Aerospace Engineering from Notre Dame in 1986. Commissioned a second lieutenant in the USAF, he completed his postgraduate degree, and was assigned to the USAF Tactical Air Warfare Center at Eglin

AFB, Florida, as a flight test engineer for the *Gryphon* ground-launched cruise missile (GLCM) program. Good received his wings as a USAF Navigator in 1989. He graduated from the USAF Test Pilot School at Edwards AFB in 1994. He has logged over 3,000 hours flying in more than thirty aircraft types.

Mission STS-125 was aboard *Atlantis* (May 11-24, 2009), a telescope servicing mission when the crew installed new thermal blanket insulating panels, and a soft-capture mechanism that would aid in the safe de-orbiting of the Hubble Space Telescope. The crew used an IMAX 3D camera to document the mission for the March 2010 IMAX film ***IMAX: Hubble 3D***. On mission STS-132 aboard *Atlantis* (May 14-26, 2010), he flew as 'Mission Specialist 2' to the ISS. The shuttle carried the Russian *Rassvet* Mini-Research Module 1, along with the Integrated Cargo Carrier (ICC-VLD2) pallet. Good logged a total of 24 days, 18 hours and six minutes in space, including EVAs totaling 29 hours and 53 minutes. Michael Good's military decorations include the Legion of Merit; four Meritorious Service Medals; two Aerial Achievement Awards; Aircrew of the Year, 77th Fighter Squadron (1991); Air Force Commendation Medal; and the Combat Readiness Medal.

## Gordon, Richard F., Jr.. Gemini XI, Apollo 12.

Richard Francis Gordon, Jr. (1929-2017) was born in October 1929 in Seattle, Washington, USA. He was a former naval officer, aviator, and test pilot. Gordon was selected in October 1963 NASA Group 3. He served as backup pilot for *Gemini VIII*. In September 1966, he piloted *Gemini XI* alongside Pete Conrad, setting an altitude record of 1,369 km (851 miles), which still stands as the highest-apogee Earth orbit. Gordon performed two EVAs, which included attaching a tether to the *Agena* vehicle and retrieving a nuclear emulsion experiment package. Assigned as backup Command Module pilot for *Apollo 9*, he flew as the *Yankee Clipper* Command Module pilot for *Apollo 12* and orbited the Moon 45 times, photographing tentative landing sites for future missions. He logged a total of 315 hours and 53 minutes in space; including two EVAs totaling two hours and 41 minutes. Although he was slated to walk on the Moon as Commander of *Apollo 18,* that mission was canceled due to budget cuts. He then worked in the Astronaut Office, becoming Chief of Advanced Programs in 1971. He worked on the design of the Space Shuttle, and retired from NASA and the US Navy in January 1972.

Richard Gordon then served as executive vice president of the **New Orleans Professional Football Club** (1972-76); was general manager of **Energy Developers, Ltd.** (1977); president of **Resolution Engineering and Development Company** (REDCO) (1978); and following the merger with Amarco Resources, served as vice president for operations, Texas Division; and served as director, **Scott Science and Technology, Inc.**, Los Angeles Division (1981-83). In March 1982, he became president of **Astro Sciences Corporation**. He served as chairman and co-chairman of the Louisiana Heart Fund; was honorary chairman for Muscular Dystrophy; and was on the board of directors for the Boy Scouts of America and Boy's Club of Greater New Orleans. Gordon died in San Marcos, California, on November 06, 2017, aged eighty-eight years, and was buried at Arlington National Cemetery.

Richard Gordon was the author of eight technical papers; including the Gemini XI and Apollo XII Mission Reports. Gordon was a Fellow of the American Astronautical Society, an Associate Fellow of the Society of Experimental Test Pilots, and a member of the Navy League. In 1982, Gordon was inducted into the International Space Hall of Fame; the US Astronaut Hall of Fame in March 1993; and the National Aviation Hall of Fame in 2020. In the 1998 HBO miniseries ***From the Earth to the Moon***, Gordon was played by Tom Verica. He received thirteen awards, including the Bendix Trophy (1961); the NASA Exceptional Service Medal; two Navy DFCs; the Navy Distinguished Service Medal; the Institute of Navigation Award for 1969; and the Manned Spacecraft Center Superior Achievement Award.

## Grabe, Ronald J.. STS-51-J, STS-30R, STS-42, STS-57.

Col. Ronald John Grabe was born in June 1945 in New York, New York, USA. He earned a Bachelor of Science degree in Engineering Science from the USAF Academy in 1966; and studied Aeronautics as a Fulbright Scholar at the Technische Hochschule (Darmstadt, West Germany) in 1967. He returned to the US in 1977 to complete pilot training at Randolph AFB, Texas. In 1969, he was assigned to 3rd Tactical Fighter Wing at Bien Hoa Air Base, in the Republic of Vietnam, where he flew two hundred combat missions. From 1976-79, he served as an exchange test pilot with the RAF at Boscombe Down, UK. He was an instructor at the USAF Test Pilot School, Edwards AFB, when advised of his selection by NASA, becoming an astronaut in August 1981. Grabe logged over 5,500 hours flying time.

Grabe was pilot on STS-51-J, the maiden voyage of shuttle *Atlantis* (October 03-07, 1985), a classified Department of Defense mission. Mission STS-30 aboard *Atlantis* (May 04-08, 1989) deployed the *Magellan* Venus-exploration spacecraft. Ronald Grabe was mission Commander on STS-42 aboard *Discovery* (January 22-30, 1992), which conducted fifty-five major experiments in the International Microgravity Laboratory-1 module. He was Commander on STS-57 aboard *Endeavour* (June 21-July 01, 1993), with the primary mission to retrieve the *European Retrievable Carrier* satellite (EURECA), and the first flight of 'Spacehab'. He logged a total of 26 days, three hours and 38 minutes (over 627 hours) in space. Effective April 11, 1994, Grabe left NASA and the Air Force to join Orbital Sciences Corporation, in Dulles, Virginia. He has received the Liethen-Tittle Award (for outstanding student at the USAF Test Pilot School); the Air Force Distinguished Flying Cross; the Air Medal with seven Oak Leaf Clusters; the Air Force Meritorious Service Medal; and the Royal Air Force Cross.

## Gregory, Frederick D.. STS-51-B, STS-33, STS-44.

Frederick Drew Gregory was born in January 1941 in Washington, D.C., USA, and graduated from Anacostia High School in 1958. He attended the USAF Academy, and received an undergraduate degree in Military Engineering in 1964. He earned his wings after helicopter school, flew 550 combat missions in Vietnam, transitioned to fighter aircraft, attended the Navy Test Pilot School, and then conducted testing for both the Air Force and NASA. During his Air Force service, Gregory logged about 7,000 hours in more than fifty aircraft types as a helicopter, fighter and test pilot. In 1977, Gregory received a Master's degree in Information Systems from George Washington University.

Selected in January 1978 NASA Group 8, Gregory became the first black astronaut to pilot a Space Shuttle on STS-51-B aboard *Challenger* (April 29-May 06, 1985). He was the CapCom during the *Challenger* disaster on January 28, 1986. The first African American to be a shuttle Commander on STS-33 aboard *Discovery* (November 22-27, 1989), he carried the laboratory developed by the European Space Agency (ESA). The crew deployed the *Northern Utah Satellite* (NUSAT). Mission STS-44 aboard *Atlantis* (November 24-December 01, 1991) had the Defense Support Program (DSP) satellite as its main payload. There were extensive studies evaluating medical countermeasures to long duration space flight. The first black man to command two Space Shuttle missions, he logged a total of 18 days, 23 hours and four minutes (over 455 hours) in space.

Frederick Gregory served at NASA HQ as Associate Administrator for the Office of Safety and Mission Assurance (1992-2001), and for the Office of Space Flight (2001-02). On August 12, 2002, he was sworn in as NASA Deputy Administrator. From February 14 to April 20, 2005, he was the NASA Acting Administrator, and resigned on September 09. Frederick Gregory has received twenty-one awards, including the Air Force Legion of Merit; the Defense Meritorious

Service Medal, three DFCs; 16 Air Medals; two NASA Distinguished Service Medals; two NASA Outstanding Leadership Awards; the Presidential Rank Award; the Charles R. Drew University of Medicine and Science President's Medal; and three honorary Doctorates.

## Gregory, William G.. STS-67.

Lt. Col. William "Bill" George "Borneo" Gregory was born in May 1957 in Lockport, New York, USA, to an Albanian American family. He graduated from Lockport High School in 1975, then attended the USAF Academy, graduating with a degree in Engineering Sciences in 1979. He earned a Master's degree in Engineering Mechanics (1980) from Columbia University; and a Master's degree in Management from Troy University in 1984. During 1981-86, Gregory served as an operational fighter pilot, flying the D and F models of the F-111. He attended the USAF Test Pilot School in 1987, then served as a test pilot at Edwards AFB until 1990. He logged over 5,000 hours flying time in over forty aircraft types. Selected by NASA in January 1990, Gregory flew on STS-67 aboard *Endeavour* for the astronomical research mission (March 02-18, 1995). This crew established a new shuttle mission duration record of 16 days, 15 hours, eight minutes and 46 seconds during 262 orbits of Earth; so he logged 400 hours in space. Gregory retired from the Air Force and left NASA in the Summer of 1999. He served as Vice President of Business Development for **Qwaltec, Inc.** in Tempe, Arizona, and retired as a Captain at a US-based air cargo carrier.

## Griggs, S. David. STS-51-D.

Rear Admiral Stanley David Griggs (1939-89) was born in September 1939 in Portland, Oregon, USA. He received a Bachelor of Science degree from the US Naval Academy, Annapolis, in 1962; and a Master of Science degree in Administration from George Washington University in 1970. He entered Naval pilot training in 1962, received his pilot wings in 1964, then was attached to Attack Squadron-72, flying A-4 Skyhawks. In 1967, after two Southeast Asia combat cruises, he entered the US Naval Test Pilot School at Patuxent River, Maryland. In 1970, he affiliated with the Naval Air Reserve, and achieved the rank of Rear Admiral. Griggs logged 9,500 hours flying time, including 7,800 hours in jet aircraft, and flew over forty-five aircraft types, making over 300 aircraft carrier landings.

In July 1970, Griggs was employed at the Johnson Space Center as a research pilot. In 1974, he was assigned duties as the project pilot for the Space Shuttle trainer aircraft; participating in their design, development, and testing until their operational deployment in 1976. He was appointed Chief of the Shuttle Training Aircraft Operations Office in January 1976, and held that position until selected in January 1978 in NASA Group 8. During mission STS-51-D aboard *Discovery* (April 12-19, 1985) he conducted the first unscheduled EVA of the space program. Griggs logged six days, 23 hours and 55 minutes in space. David Griggs died on July 17, 1989, when the vintage WWII-era training aircraft he was piloting crashed near Earle, Arkansas. At that time, he was in training as pilot for STS-33, a dedicated Department of Defense mission, scheduled for launch in September 1989. Griggs was aged forty-nine years, and was buried at Arlington National Cemetery.

## Grissom, Virgil I. "Gus". Mercury-Atlas "Liberty Bell 7", Gemini 3 "Molly Brown", Apollo 1.

Lt. Col. Virgil Ivan "Gus" Grissom (1926-1967) was born in April 1926 in Mitchell, Indiana, USA. In his youth he built model airplanes and developed a passion for aviation. During WWII, he enlisted as an aviation cadet in the US military. After the war ended, he got married and completed his education. He re-enlisted in the USAF and in 1951, officially became a fighter pilot. Grissom was an American mechanical engineer and pilot in the USAF, as well as a member of the original "Mercury Seven" selected by NASA as Project Mercury astronauts. During the Korean War, he flew more than one hundred combat missions and was awarded the Distinguished Flying Cross; and the Air Medal with an Oak Leaf Cluster.

Gus Grissom became the second American in space on July 21, 1961, and the first to fly in space twice, preceded only by Joe Walker with his sub-orbital X-15 rocket-plane flights. The *Liberty Bell 7* capsule, fitted with a trapezoid-shaped window, was the first spacecraft to maneuver in orbit, adjusting the spacecraft's orbital path. There were some issues with the end of his 15-minute 37-second flight, which reached an altitude of 189 km (118 miles). During his re-entry, 486 km downrange, Grissom experienced an uncomfortably heavy 11-G's deceleration, before splashing down into the Atlantic Ocean. The capsule's explosive hatch blew prematurely, and seawater started flooding into the capsule. Grissom's spacesuit began taking on water via the oxygen vent, becoming heavier and waterlogged. Grissom abandoned his craft, struggled for five minutes to remain above the churning waves, and he very nearly drowned. The recovery helicopter was unable to lift the partly submerged capsule, which promptly sank.

As Grissom became the first American to directly view the Earth from space, he marveled at his home planet. "*The view through the window became quite spectacular as the horizon came into view*", he said in his flight report. "*The sight was truly breathtaking. The Earth was very bright, the sky was black, and the curvature of the Earth was quite prominent*". Unquote. A book by George Leopold about Grissom's life, *Calculated Risk: The Supersonic Life and Times of Gus Grissom*, and a recent interview with the head of NASA's recovery options for the Mercury program reveal Grissom as a quick-thinking hero. He reacted decisively in an uncertain situation when otherwise his Mercury mission would have ended in early death.

Gus Grissom commanded the first manned Gemini flight. He flew three Earth orbits with John W. Young in the *Gemini 3* capsule named *Molly Brown* on March 23, 1965. They fired thrusters to change the shape of their orbit and shift their orbital plane slightly. This was the first United States mission with a crew of two astronauts. Grissom and his fellow Gemini astronauts executed the first lifting re-entry of a manned spacecraft, demonstrating the capability to safely return to Earth from orbit. Gus Grisson died in an electrical fire during a pre-launch "plugs out" test on January 27, 1967, at launchpad 34, Cape Canaveral, Florida. There had been a series of problems with the preflight checklist. Constant radio static marred communication between Mission Control and the space craft. Grissom was frustrated, and said, *"Jesus Christ, if we can't communicate across three miles, how the hell are we going to communicate when we're on the Moon?"* Ten minutes before the end of the simulated liftoff, there was a slight surge in voltage. Nine seconds later, one of the crew yelled, *"Hey!"* A moment later came, *"We've got a fire in the cockpit!"*

The three-man crew were unable to escape the capsule pressurised to 16.7 pounds per square inch before being overcome by toxic gases and smoke inhalation causing asphyxia and heart attack. Fourteen seconds after the first shout of alarm, the Command Module's hull ruptured, spewing flames and gases. The twenty-seven men of the pad rescue team rushed across the catwalk and struggled to open the module's hatch, but the heat and smoke drove many of them back. It took five minutes to open the three hatches. The *Apollo 1* tragedy, a fire fueled by a combination of flammable materials in the Command Module's pure oxygen atmosphere, led to a re-designed outward-opening hatch, over 1,300 modifications to the Apollo spacecraft and significant changes in NASA's safety protocols and procedures for future missions. The US government posthumously conferred on V.I. Gus Grissom the Congressional Space Medal of Honor.

## Grunsfeld, John M.. STS-67, STS-81, STS-103, STS-109, STS-125.

Dr. John Mace Grunsfeld was born in October 1958 in Chicago, Illinois, USA. His academic background includes research in high energy astrophysics, cosmic ray physics, and the emerging field of exoplanet studies. He earned a Bachelor of Science degree in Physics in 1980; a Master of Science in Physics from the University of Chicago in 1984; and a Doctor of Philosophy in Physics from the University of Chicago in 1988. Selected by NASA in March 1992, he completed one year of training and qualified for flight selection as a Mission Specialist. He directed an effort to create the Common Display Development Facility (CDDF), responsible for the onboard and MCC displays for the ISS. Following his STS-81 flight, he was assigned as Chief of the Computer Support Branch in the Astronaut Office. After STS-103, he served as Chief of the Extravehicular Activity Branch in the Astronaut Office. Following STS-109, he served as an instructor in the EVA Branch, and worked on the Orbital Space Plane in the Advanced Programs Branch.

STS-67/Astro-2 aboard *Endeavour* (March 02-18, 1995) was a record-setting 16-day mission which studied the far ultraviolet spectra of faint astronomical objects and the polarization of UV light coming from hot stars and distant galaxies. He served as Flight Engineer on STS-81 aboard *Atlantis* (January 12-22, 1997), the fifth mission to dock with *Mir*, and the second to exchange US astronauts. During this flight, Grunsfeld placed a call to ***NPR***'s auto-repair radio show, *Car Talk*. He complained about the performance of his serial-numbered Rockwell-manufactured "government van". To wit, it would run very loud and rough for about two minutes, quieter and smoother for another six and a half, and then the engine would stop with a jolt. He went on to state that the brakes of the vehicle, when applied, would glow red-hot, and that the vehicle's odometer displayed *"about 60 million miles"*. This created some consternation for the hosts, until they noticed the audio of Grunsfeld's voice, being relayed from *Mir* via *TDRS* satellite, sounded similar to that of Tom Hanks in the then-recent film ***Apollo 13***, after which they realized the call was from space and the government van in question was, in fact, the Space Shuttle!

STS-103 aboard Discovery (December 19-27, 1999) upgraded the Hubble Space Telescope, with Grunsfeld performing two EVAs totaling 16 hours and 23 minutes. Grunsfeld served as Payload Commander on STS-109 aboard *Columbia* (March 01-12, 2002), the fourth Hubble servicing mission; installing a new digital camera, a cooling system for the infrared camera, new solar arrays and a new power system. He performed three EVA totaling 21 hours and nine minutes. STS-125 aboard *Atlantis* (May 11-24, 2009) was the final Hubble servicing mission. Grunsfeld performed three of the mission's five back-to-back EVAs, which installed new instruments to the telescope. John Grunsfeld logged a total of 58 days and 30 minutes in space; with eight EVAs. From 2003-04, Grunsfeld served as NASA Chief Scientist assigned to NASA HQ. In January 2012, he became the Associate Administrator for the Science Mission Directorate at NASA HQ in Washington. On April 05, 2016, he announced his retirement from NASA. Grunsfeld has received ten honors, including W.D. Grainger Fellow in Experimental Physics (1988-89); the V.M. Komarov Diploma (1995); the Korolov Diploma (1999, 2002); three NASA Exceptional Service Medals (1997, 1998, & 2000); the NASA Distinguished Service Medal (2002); and he was inducted into the US Astronaut Hall of Fame.

## Gutierrez, Sidney M.. STS-40, STS-59.

Col. Sidney McNeill Gutierrez was born in June 1951 in Albuquerque, New Mexico, USA. He received his Bachelor of Science degree in Aeronautical Engineering from the USAF Academy in 1973, and a Master of Arts degree in Management from Webster University in 1977. After graduating from the Academy, he completed Undergraduate Pilot Training at

Laughlin AFB, and was an instructor pilot there from 1975-77. He attended the USAF Test Pilot School in 1981, then served as primary test pilot for airframe and propulsion testing on the F-16 aircraft. Gutierrez has logged over 4,500 hours flying time in about thirty aircraft types, including sailplanes and balloons.

Selected by NASA in May 1984, he served as commander for the Shuttle Avionics Integration Laboratory. In 1986-87, he participated in the re-certification of the Space Shuttle Main Engines, Main Propulsion System, and External Tank. In 1988, he became the Astronaut Office lead for shuttle software development, verification, and future requirements definition. He was pilot on STS-40 aboard *Columbia* (June 05-14, 1991), the Spacelab Life Sciences (SLS-1) mission, exploring how humans, animals, and cells respond to microgravity, then re-adapt to Earth's gravity. Gutierrez then served as Capcom for STS-42, 45, 46, 49 and STS-52. In 1992, he became the Astronaut Office Branch Chief for Operations Development. He was the Commander on STS-59 aboard *Endeavour* (April 09-20, 1994), the Space Radar Laboratory mission, which imaged over four hundred selected sites with about 14,000 photographs. He logged a total of 20 days, eight hours and three minutes (over 488 hours) in space.

Since retiring from the USAF and NASA in September 1994, Gutierrez has worked in several leadership positions at Sandia National Laboratories, where as a Center Director he was responsible for programs introducing disruptive ideas and technologies into satellites, space payloads and sensors. He retired in February 2015 to become CEO of **Rocket Crafters, Inc.** (now Vaya Space). He serves on the board of directors for Texas-New Mexico Power Company and Goodwill Industries of New Mexico, and is a member of the New Mexico Space Center's Governor's Commission. Sidney Gutierrez has received eleven awards and honors, including Distinguished Graduate of the USAF Academy; the Defense Superior Service Medal; the Air Force Commendation Medal with one Oak Leaf Cluster; Air Training Command Master Instructor; the 1990 Congressional Hispanic Caucus Award; selected in 1994 by ***Hispanic Business*** magazine as one of the 100 Most Influential Hispanics; and was the 2010 Notable New Mexican. He was inducted into the International Space Hall of Fame in 1995, and has a middle school named after him in Roswell, New Mexico.

## Hague, T. Nicklaus. Soyuz MS-10, Soyuz MS-12/Expedition 59.

Col. Tyler Nicklaus "Nick" Hague was born in September 1975 in Belleville, Kansas, USA. He earned a Bachelor of Science degree in Aerospace Engineering from Massachusetts Institute of Technology in 2000. Hague joined the USAF and was commissioned a second lieutenant in May 1998. In 2003, he attended the USAF Test Pilot School in Edwards AFB, and after graduation in 2004 was assigned to the 416th Flight Test Squadron. Hague was deployed for five months in Iraq, supporting Operation Iraqui Freedom, and conducting experimental airborne reconnaissance. In 2006, he began teaching courses in the Department of Astronautics at the USAF Academy, Colorado. In 2009, he received a fellowship for the Air Force Fellows program in Washington, D.C..

From 2012-13, he worked in the Department of Defense as Deputy Chief of the Joint Improvised Explosive Device Defeat Organization. Hague was promoted to Colonel in 2016, and his transfer to the US Space Force was approved in December 2020. Selected by NASA as part of the 2013 astronaut class to be a Flight Engineer on the ISS, his first launch was on *Soyuz MS-10*, which aborted shortly after liftoff at an altitude of 50 kilometers on October 11, 2018. His second launch to the ISS on March 14, 2019, aboard *Soyuz MS-12* was successful, and he joined 'Expedition 59', performing three EVAs totaling 19 hours and 56 minutes. Nick Hague met his wife, Col. Catie Hague in 1996 at the Air Force Academy. He is a ***Doctor Who*** fan. During his service in the Air Force, Hague received seven awards, including two Defense Meritorious Service Medals; Distinguished Graduate, UASF Academy;

Distinguished Graduate in USAF Air Force Test Pilot School Class 03A; six Air Medals; two Aerial Achievement Medals; two Air Force Commendation Medals; the Air Force Combat Action Medal; and the Order of Courage of the Russian Federation (for Expedition 57/58).

## Haise, Fred W., Jr.. Apollo 13.

Capt. Fred Wallace Haise, Jr. was born in November 1933 in Biloxi, Mississippi, USA. He is an American engineer, fighter pilot with the US Marine Corps and Air Force, and former NASA astronaut. He graduated from Perkinston Junior College with a scholarship for journalism in 1952; and joined the Naval Aviation Cadet Program, completing his flight training in 1954. He served as a Marine Corps fighter pilot from March 1954 to September 1956. Haise logged 9,300 hours flying, including 6,200 hours in jet aircraft. After his military service, he graduated with a Bachelor of Science degree (with honors) in Aeronautical Engineering from the University of Oklahoma in 1959, concurrently serving two years in the Oklahoma Air National Guard. He then worked for the newly created NASA, first as a civilian research pilot. He served ten months as a fighter pilot in the USAF during the Berlin Crisis of 1961. Haise completed postgraduate courses at the USAF Aerospace Research Pilot School at Edwards AFB in 1964, then attended the six-week Harvard Business School's Advanced Management Program in 1972. He was one of nineteen candidates selected for NASA Astronaut Group 5. He served as backup Lunar Module Pilot for both *Apollo 8* and *Apollo 11*.

Haise flew as LM pilot on the aborted *Apollo 13* lunar mission in 1970, developing a urinary tract infection and later kidney infections. Haise was in pain for most of the epic journey. Haise later served as backup Commander for *Apollo 16*, then moved to the Space Shuttle program. In 1977, he participated in the Approach and Landing Tests at Edwards AFB. With C. Gordon Fullerton as pilot, Haise was commander for prototype Space Shuttle *Enterprise* in free flight to three successful landings. In June 1979, Haise left NASA to become a test pilot and executive with **Grumman Aerospace Corporation**, until retiring in 1996. On August 22, 1973, Haise was piloting a Convair BT-13, which suffered an undetermined power plant failure that led to a crash landing, when Haise suffered second-degree burns over 50% of his body. He is a fellow of the American Astronautical Society (ASS), and the Society of Experimental Test Pilots. He received eighteen honors, including the City of Houston Medal for Valor (1970); the Mississippi Distinguished Civilian Service Medal (1970); the American Defense Ribbon; an honorary Doctor of Science degree from Western Michigan University (1970); the General Thomas D. White USAF Space Trophy for 1977; the JSC Special Achievement Award (1978); the NASA Exceptional Service Medal; and the Presidential Medal of Freedom.

## Halsell, James D., Jr.. STS-65, STS-74, STS-83, STS-94, STS-101.

James Donald Halsell, Jr. was born in September 1956 in West Monroe, Louisiana, USA. He graduated with a Bachelor of Science degree in Engineering from the USAF Academy in 1978, and from Undergraduate Pilot Training at Columbus AFB in 1979. Halsell received a Master of Science degree in Management from Troy University in 1983; and a Master of Science degree in Space Operations from the Air Force Institute of Technology in 1985. From 1984-85, he was a graduate student at Wright-Patterson AFB, then attended the Air Force Test Pilot School, Edwards AFB, California. He retired from the USAF in July 2004.

Selected by NASA in January 1990, Halsell was pilot on STS-65 aboard *Columbia* (July 08-23, 1994), which flew the second International Microgravity Laboratory. STS-74 aboard *Atlantis* (November 12-20, 1995) was the second Shuttle mission to dock with *Mir*, and crew attached a permanent docking module. He was commander on STS-83 aboard *Columbia* (April 04-08, 1997), a mission cut short due to fuel cell problems. Mission STS-94 aboard

*Columbia* (July 01-17, 1997) was a reflight of STS-83, a 'Spacelab' mission. From February to August 1998, Halsell served as NASA Director of Operations at the Yuri Gagarin Cosmonaut Training Center in Russia. He was Manager, Shuttle Launch Integration, KSC, from July 2000 to January 2003. He was commander on STS-101 aboard *Atlantis* (May 19-29, 2000), the third Shuttle mission devoted to construction of the ISS. Over 5,000 pounds of equipment and supplies were carried to orbit. James Halsell logged over 1,250 hours in space. Halsell retired from NASA in November 2006 for a position with **ATK Launch Systems**, Utah.

Halsell has received five honors: he graduated first in class and was awarded the Liethen/Tittle Trophy (1986); the DFC (1998); four NASA Space Flight Medals (1994, 1995, 1997, 2000); the NASA Distinguished Service Medal (2001); and two NASA Outstanding Leadership Medals (2002 & 2004). In 2021, he pled guilty to two counts of manslaughter and two counts of assault as a result of a motor vehicle accident in 2016. As of May 2021, he was incarcerated in Alabama.

## Ham, Kenneth T.. STS-124, STS-132.

Capt. Kenneth Todd "Hock" Ham was born in December 1964 in Plainfield, New Jersey, USA. He received a Bachelor of Science degree in Aerospace Engineering from the US Naval Academy in 1987; and a Master of Science degree in Aeronautical Engineering from the Naval Postgraduate School in 1996. He was designated a Naval Aviator in October 1989. During two deployments to the Mediterranean sea, he flew combat missions over Bosnia and North Iraq. Ham has over 3,700 flight hours in over forty aircraft types; with over 300 shipboard and 300 land-based arrested landings.

Selected by NASA in June 1998, Ham was pilot on STS-124 aboard *Discovery* (May 31-June 14, 2008) (the 26th flight to the ISS), which delivered the 15,900 kg *Kibo* Japanese Experiment Module-Pressurized Module (JEM-PM) and the six-jointed 33-feet Japanese Remote Manipulator System to the ISS. He was commander on STS-132 aboard *Atlantis* (May 14-26, 2010), and delivered the 6,295 kg Russian-built *Rassvet* Mini Research Module-1 (MRM-1) to the ISS. Hock Ham logged a total of 25 days, 12 hours and 41 minutes in space. Ham left NASA in June 2012. As of July 2013, he was Chair of the Aerospace Engineering Department at the **US Naval Academy** in Maryland. In September 2014, he joined **Bigelow Aerospace** in Las Vegas, to develop a training program that would support Bigelow's proposed space habitats. Kenneth Ham is a member of the Society of Experimental Test Pilots, and the US Naval Academy Association. He was inducted into the New Jersey Aviation Hall of Fame in 2014.

## Hammond, L. Blaine, Jr.. STS-39, STS-64.

Col. Lloyd Blaine Hammond, Jr. was born in January 1952 in Savannah, Georgia, USA. He received a Bachelor of Science degree in Engineering Science and Mechanics from the USAF Academy in 1973; and a Master of Science degree in Engineering Science and Mechanics from Georgia Institute of Technology in 1974. Hammond received his pilot wings in 1975. From 1979-80, he was an instructor pilot at Williams AFB, Arizona. In 1981, he attended the Empire Test Pilot's School, at Boscombe Down, UK. From 1982, he managed projects in the 6512 Test Squadron, until being assigned as an instructor at the USAF Test Pilot School at Edwards AFB. Col. Hammond logged over 4,500 hours in fifteen American and ten RAF aircraft types. He is a Gulfstream airplane test pilot.

Selected by NASA in May 1984, Hammond flew as pilot of STS-39 aboard *Discovery* (April 28-May 06, 1991), with a deploy and rendezvous in support of the Strategic Defense Initiative Office (SDIO). He was pilot on STS-64 aboard *Discovery* (September 09-20, 1994), the first flight of the *Lidar In-Space Technology Experiment* (*LITE*) payload, and the secondary payload was the free-flyer *SPARTAN-201* satellite. Additional life sciences experiments were classified as Detailed Supplementary Objectives (DSOs). Blaine Hammond logged 19 days, six hours and 11 minutes in space. He is a member of the Air Force Academy Association of Graduates; the Air Force Association; and the Order of Daedalians. Hammond has received seven honors, including Distinguished Graduate, USAF Academy; the Commander's Trophy; the Defense Superior Service Medal; two Air Force Commendation Medals; and an upstairs conference room in Gulfstream, Long Beach, California, is named in his honor.

## Harbaugh, Gregory J.. STS-39, STS-54, STS-71, STS-82.

Gregory Jordan Harbaugh was born in April 1956 in Cleveland, Ohio, USA. He received a Bachelor of Science degree in Aeronautical and Astronautical Engineering from Purdue University in 1978; and a Master of Science degree in Physical Science from the University of Houston at Clear Lake in 1986. Selected in 1987 NASA Group, he served as Lead, Data Processing Systems (DPS) Officer for STS-9 and STS-41-D; Orbit DPS for STS-41-B and STS-41-C; and Ascent/Entry DPS for STS-41-G. Harbaugh was 'Mission Specialist 1' on STS-39 aboard *Discovery* (April 28-May 06, 1991), an unclassified DoD mission involving research for the Strategic Defense Initiative. He was 'Mission Specialist 2' on STS-54 aboard *Endeavour* (January 13-19, 1993), which deployed the *TDRS-F* satellite, and Harbaugh made a four-hour 28-minute EVA. STS-71 aboard *Atlantis* (June 27-July 07, 1995) was the first docking with the Russian Space Station *Mir*. Harbaugh was the Flight Engineer (Mission Specialist 2), and was responsible for the modified Shuttle docking system.

STS-82 aboard *Discovery* (February 11-21, 1997) was the second Hubble Space Telescope servicing mission. Five EVAs (two by Harbaugh) upgraded equipment, and installed insulation patches over compartments containing key data processing, electronics and scientific instrument telemetry packages. Hubble was then redeployed and boosted to its highest orbit ever. Harbaugh logged a total of 34 days, one hour and 59 minutes (818 hours) in space; including 18 hours and 29 minutes of EVA, three of which were made in support of repair and refurbishment of the Hubble Space Telescope. After leaving NASA, Harbaugh was Museum Director and COO of **Sun 'n Fun Fly-in**, Lakeland, Florida from 2001-06. He is currently Founder/Chairman/CEO of **Innerspace Consultants**. Harbaugh has received ten awards, including the Purdue University Outstanding Aerospace Engineer and Astronaut Alumnus Awards; the 1999 Rotary National Award for Space Achievement; the NASA Distinguished Service Medal; and the NASA Exceptional Achievement Medal.

## Harris, Bernard A., Jr. STS-55, STS-63.

Dr. Bernard Anthony Harris, Jr. was born in June 1956 in Temple, Texas, USA. He received a Bachelor of Science degree in Biology from the University of Houston in 1978; and a Doctor of Medicine (M.D.) degree from the Texas Tech University Health Science Center of Medicine in 1982. After completing his residency at the Mayo Clinic, he pursued research opportunities at Ames Research Center in California, and later at Johnson Space Center. Selected in the 1990 NASA Group, he was 'Mission Specialist 3' on STS-55 aboard *Columbia* (April 26-May 06, 1993), which carried to orbit the second reusable German 'Spacelab D-2' and demonstrated the shuttle's ability for international cooperation, exploration, and scientific research in space. It augmented the German microgravity research program started by the D-1 mission. This flight marked the one hundredth day of flight time in space for *Columbia*, the fleet's oldest orbiter, on its fourteenth flight.

Harris was 'Mission Specialist 1' on STS-63 aboard *Columbia* (February 03-11, 1993), the second *Shuttle-Mir* Program mission. On February 09, 1995, Harris became the first African American to perform a spacewalk (EVA). There were twenty experiments in 'Spacehab-3', and the crew flew captively, then deployed and retrieved the *Shuttle Pointed Autonomous Research Tool for AstroNomy* (*SPARTAN-204*) payload. This was designed to obtain data in the far ultraviolet region of the spectrum from diffuse sources of light. The closest approach to *Mir* of 11 metres (36 feet) occurred on February 06, 1995. Harris logged a total of 18 days, six hours and eight minutes in space.

He then became an entrepreneur, working first as Vice President of **Spacehab, Inc.** in 1996, where he worked on the commercialization of space exploration. After earning his M.B.A. from the University of Houston at Clear Lake, he became Executive Director of **Versalius Ventures** in 2001. He is the winner of The Challenger Award from the Ronald E. McNair Foundation; Fellow of the American College of Physicians; and recipient of the prestigious Horatio Akger Award. The Dr. Bernard A. Harris Middle School was named in his honor. Harris established the **ExxonMobil Bernard Harris Summer Science Camp**, and the Dream Tour; two STEM programs encouraging minority students to pursue interests in the sciences.

## Hart, Terry J.. STS-41-C.

Lt. Col. Terry Johnathan "TJ" Hart was born in October 1946 in Pittsburgh, Pennsylvania, USA. He received a Bachelor of Science degree in Mechanical Engineering from Lehigh University in 1969; and a Master of Science degree in Electrical Engineering from Rutgers University in 1978. In 1973, he joined the New Jersey Air National Guard, and retired as lieutenant colonel in 1990. Hart logged 3,000 hours flying time, with 2,400 hours in jet aircraft. From 1968-78, Hart was employed on the technical staff of **Bell Labs**, and he received two Patents.

Selected by NASA as an astronaut candidate in January 1978, he was a Mission Specialist on STS-41-C aboard *Challenger* (April 06-13, 1984), where Hart operated the arm of the Remote Manipulator System to deploy the Long Duration Exposure Facility, and retrieve the *Solar Max* satellite, repairing it and then replacing it in orbit. He logged six days, 23 hours and 40 minutes hours in space. After leaving NASA, he was the Director of Engineering and Operations for **AT & T**'s satellite network. Terry Hart retired on June 15, 1984, and was then a member of the engineering faculty at **Lehigh University**. Hart was awarded an honorary Doctorate of Engineering from Lehigh University in 1988.

## Hartsfield, Henry W., Jr. STS-4, STS-41-D, STS-61-A.

Col. Henry Warren Hartsfield, Jr. (1933-2014) was born in November 1933 in Birmingham, Alabama, USA. He received a Bachelor of Science degree in Physics from Auburn University in 1954; and a Master of Science degree in Engineering Science from the University of Tennessee in 1971. Hartsfield entered the USAF in 1955, and was a graduate of the USAF Test Pilot School at Edwards AFB, and was an instructor there prior to his assignment in 1966 to the USAF MOL program as an astronaut. Hartsfield became part of NASA Astronaut Group 7 in September 1969. He was responsible for supporting the development of the Space Shuttle's entry flight control system and its associated interfaces. He logged over 7,400 hours of flying time; including over 6,150 hours in jet aircraft.

Hartsfield was pilot on STS-4 aboard *Columbia* (June 27-July 04, 1982), a mission designed to verify ascent and entry phases of shuttle missions, and perform continued studies of the effects of long-term thermal extremes on the Orbiter subsystems. He was commander on STS-41-D aboard *Discovery* (August 30-September 05, 1984), which deployed three satellites. Henry Hartsfield was commander of STS-61-A aboard *Challenger* (October 30-November 06, 1985), with over seventy-five experiments completed in the areas of physiological sciences, materials processing, biology, and navigation. He logged 20 days, two hours and 50 minutes (over 482 hours) in space. Hartsfield retired from the USAF in

August 1977 with over 22 years of active service, and remained a civilian NASA astronaut. From 1986-87, he served as the Deputy Chief of the Astronaut Office. In 1987, he became the Deputy Director for Flight Crew Operations.

In 1990, Hartsfield accepted a temporary assignment as the Deputy Manager for Operations, Space Station Projects Office, at the Marshall Space Flight Center. In December 1993, he became Manager, International Space Station Independent Assessment. He retired from NASA in 1998, and joined **Raytheon Company** as Vice President of Aerospace Engineering Services. Hartsfield received nine honors, including the Air Force Meritorious Service Medal; the Distinguished Civilian Service Award (DOD, 1982); NASA Distinguished Service Medals (1982, 1988); the NASA Exceptional Service Medal (1988); and was inducted into the US Astronaut Hall of Fame (2006). Hartsfield retired in 2005, and died on July 17, 2014, after back surgery.

## Hauck, Frederick H.. STS-7, STS-51-A, STS-26.

Capt. Frederick "Rick" Hamilton Hauck was born in April 1941 in Long Beach, California, USA. He received a Bachelor of Science degree in Physics from Tufts University in 1962. In 1964, he attended the US Naval Postgraduate school, Monterey, for studies in mathematics and physics. Selected for the Navy's Advanced Science Program, he received a Master of Science degree in Nuclear Engineering from the Massachusetts Institute of Technology in 1966. Hauck received his aviator wings in 1968, became a fighter pilot and graduated from the US Naval Test Pilot School in 1971. In May 1989, he became Director, Navy Space Systems Division, in the Office of the Chief of Naval Operations. He left military duty on June 01, 1990.

Selected in January 1978 NASA Group 8, Hauck was pilot on STS-7 aboard *Challenger* (June 18-24, 1983), which deployed two satellites, and operated the Remote Manipulator System. He commanded STS-51-A aboard *Discovery* (November 08-16, 1984), which deployed two satellites, and retrieved two others for repair. Hauck commanded STS-26 aboard *Discovery* (September 29-October 03, 1988), when the crew deployed the *Tracking and Data Relay Satellite* (TDRS-C) and operated eleven mid-deck experiments. Rick Hauck a total of logged 18 days, three hours and seven minutes in space.

In October 1990, Hauck joined **AXA Space** (formerly INTEC) as President and COO, becoming CEO on January 01, 1993. AXA Space is a world leader in providing property and casualty insurance for the risk of launching and operating satellites. Hauck retired in April 2005. He has received twenty honors, including the Defense Superior Service Medal; the Legion of Merit; the DFC; nine Air Medals; the NASA Distinguished Service Medal; the Navy's Outstanding Test Pilot Award; AIAA Flight Achievement Awards; the Lloyd's of London Silver Medal for Meritorious Service; two FAI Komarov Diplomas; and he is listed in ***Who's Who*** in America.

## Hawley, Steven A.. STS-41-D, STS-61-C, STS-31, STS-82, STS-93.

Dr. Steven Alan Hawley was born in December 1951 in Ottawa, Kansas, USA. He received a Bachelor of Science degree (with highest distinction) from the University of Kansas in 1973; and a Doctor of Philosophy degree in Astronomy and Astrophysics from the University of California in 1977. Prior to his selection by NASA in January 1978, Hawley was a post-doctoral research associate at Cerro Tololo Inter-American Observatory in La Serena, Chile. During 1984-85, he was Technical Assistant to the Director, Flight Crew Operations. From 1987-90, he was the Deputy Chief of the Astronaut Office. In June 1990, he left the Astronaut Office to assume the post of Associate Director of NASA's **Ames Research Center** in California. In August 1992, he returned to the JSC as Deputy Director of Flight Crew Operations, and became Director from October 2001 to November 2002. From 2002-08, he served as Director, Astromaterials Research and Exploration Science Directorate. He became a professor of physics and astronomy and Director of Engineering Physics at the University of Kansas.

Steven Hawley was 'Mission Specialist 2' on STS-41-D aboard *Discovery* (August 30-September 05, 1984), when the crew activated the OAST-1 solar cell wing experiment, and launched three satellites. On STS-61-C aboard *Columbia* (January 12-18, 1986) the crew deployed the *SATCOM* Ku-band satellite, and conducted experiments in astrophysics and materials processing. Mission STS-31 aboard *Discovery* (April 24-29, 1990) deployed the Hubble Space Telescope, and crew operated a variety of cameras. Hawley was 'Mission Specialist 2' on STS-82 aboard *Discovery* (February 11-21, 1997), the second Hubble maintenance mission, when Dr. Hawley's primary role was to operate the Shuttle's 50-foot robotic arm to retrieve and redeploy the telescope, and during five EVAs in which two teams installed new spectrometers and eight replacement instruments. Hubble was then boosted to a higher orbit. He served as Flight Engineer on STS-93 aboard *Columbia* (July 22-27, 1999) to deploy the Chandra X-ray Observatory. Hawley was the primary operator of a second telescope carried in the crew module, used for several days to make broadband ultraviolet observations of a vaiety of Solar System objects. Dr. Hawley logged a total of 32 days, two hours and 42 minutes (over 770 hours) in space.

Hawley married fellow astronaut Sally Ride in 1982, and they divorced in 1987. With Space Shuttle commander Kenneth Bowersox, Hawley appeared on an episode of ***Home Improvement***. He is a member of the American Astronomical Society; the Astronomical Society of the Pacific; and the American Institute of Aeronautics and Astronautics. He has received thirty-four awards, including two Scholarships (1970-73); the Outstanding Physics Major Award (1973); the NASA Outstanding Performance Award (1981); the Group Achievement Award for JSC Strategic Planning (1987); the NASA Exceptional Service Medal (1988, 1991); the NASA Outstanding Leadership Medal (1990); Distinguished Alumni Award, UC Santa Cruz (1991); the Presidential Rank Award (1994 & 1999); and two NASA Distinguished Service Medals (1998, 2000). He was inducted into the Kansas Aviation Hall of Fame (1997); and into the Astronaut Hall of Fame (2007).

## Helms, Susan J.. STS-54, STS-64, STS-78, STS-101/Expedition 2.

Lt. Gen. Susan Jane Helms was born in February 1958 in Charlotte, North Carolina, USA. She was the commander, Fourteenth US Air Force; and commander, Joint Functional Component Command for Space at Vandenberg Air Force Base in California. She graduated with a Bachelor of Science degree in aeronautical engineering from the United States Air Force Academy in 1980. Susan was assigned to Eglin AFB, Florida, as an F-16 weapons separation engineer with the Air Force Armament Laboratory. In 1982 she became lead engineer. In 1985, Helms received a Master of Science in Aeronautics and Astronautics from Stanford University and was assigned as an assistant professor of aeronautics at the US Air Force Academy. In 1987 she attended the Air Force Test Pilot School at Edwards Air Force Base, California.

As a test pilot, she has flown in thirty different types of United States and Canadian military aircraft. While on assignment at Canadian Forces Base Cold Lake in Alberta, she was selected for the astronaut program in January 1990. Helms became an astronaut in July 1991, and in 1993 was the first United States military woman in space. She flew on missions STS-54 (1993); STS-64 (1994); STS-78 (1996); STS-101 (2000); and served aboard the ISS from March 08, 2001, for 163 days (five months) as a member of the ISS 'Expedition 2'. Along with James Voss, her single EVA on 'Expedition 2' was the longest-ever spacewalk of eight hours and 56 minutes duration. She returned to Earth with the STS-105 crew aboard *Discovery* on August 22, 2001. Her total time in space is 210 days, 23 hours and six minutes (5,064 hours); including one EVA. Helms was a crew member on five Space Shuttle missions (a joint US record for five women astronauts).

After a twelve-year NASA career, Helms returned to the USAF in July 2002 to take a position at HQ **USAF Space Command**. She then served as vice commander of the 45th Space Wing at Patrick Air Force Base in Florida. Next, she served as deputy director of operations (Technical Training) at Randolph Air Force Base, Texas. Helms served on the Return to Flight task group after the *Columbia* shuttle accident. In June 2006, she became commander of the 45th Space Wing, and was promoted to Brigadier General that same day. In August 2009 she was promoted to Major General. She was directly responsible to the US Strategic Command commander for the development and implementation of national security policy and guidance; military strategy and guidance; space and weapons employment concepts and policy; and joint doctrine as they apply to the command and the execution of its missions. In January 2011, Helms was promoted to Lieutenant General and assumed duties as Commander, Fourth Air Force (Air Forces Strategic), Air Force Space Command and Commander at US Strategic Command.

Susan Helms retired from the USAF in 2014, with the rank of Lieutenant General. In 2020, she was elected a member of the **National Academy of Engineering** for accomplishments in civil and military space programs. Among her awards are the Command Space Operations Badge; the Air Force Senior Observer Badge with Astronaut Device; the Defense Superior Service Medal with two Oak Leaf Clusters; the Legion of Merit with three Oak Leaf Clusters; the Defense Meritorious Service Medal with two Oak Leaf Clusters; the NASA Distinguished Service Medal; the NASA Outstanding Leadership Medal; the National Defense Service Medal with service star; and the Air Force Longevity Service Award with silver and two bronze Oak Leaf Clusters.

## Henize, Karl G.. STS-51-F.

Lt. Cmdr. Dr. Karl Gordon Henize (1926-93) was born in October 1926 in Cincinnati, Ohio, USA. He became a professor at Northwestern University. Due to World War II, he elected not to finish High School, instead entering the V-12 Navy College Training Program. World War II ended before he received his Naval Commission, so he became a member of the US Naval Reserve. While at the University of Virginia, he received a Bachelor of Arts degree in Mathematics in 1947; and a Master of Arts degree in Astronomy in 1948, while also carrying out research at McCormick Observatory. He was awarded a Doctor of Philosophy in Astronomy in 1954 by the University of Michigan. In 1956, Henize published the Catalogues of Hα-Emission stars and Nebulae in the Magellanic Clouds. The paper references many objects which bear his name, such as the Superbubble Henize 70, and the planetary nebula Henize 3-401.

In 1959, he was appointed associate professor in Northwestern University's Department of Astronomy, and was awarded a professorship in 1964. Henize engaged in studies of ultraviolet optical systems and astronomical programs suited to the crewed space flight program. From 1974-78, he chaired the NASA Facility Definition Team for STARLAB, a proposed 1-meter UV telescope for 'Spacelab'. From 1978-80, he chaired the NASA Working Group for the Spacelab Wide-Angle Telescope. Henize was a member of the astronaut support crew and CapCom for the *Apollo 15* mission. He was also a member of the astronaut support crew for the *Skylab 2*, 3 and 4 missions. He logged 2,300 hours flying jet aircraft. Karl Henize was a Mission Specialist on STS-51-F aboard *Challenger* (July 29-August 06, 1985) the 'Spacelab-2' mission, the first to operate the *Spacelab Instrument Pointing System* (IPS). The mission carried thirteen experiments, of which seven were in the field of astronomy and solar physics, and three were studies of the Earth's ionosphere. Karl Henize logged seven days, 22 hours and 45 minutes in space.

In 1986, he retired as an astronaut and accepted a position as Senior Scientist in the Space Sciences Branch; studying space debris and hazards to the ISS. In the miniseries ***From the Earth to the Moon***, Henize was played by Marc MaCaulay. Henize was presented the Robert Gordon Memorial Award for 1968, and was a recipient of NASA Group Achievement Awards in 1971, 1974, 1975, and 1978. In 1974, he was awarded the NASA Exceptional Scientific Achievement Medal. Henize was a member of the American Astronomical Society; the Royal Astronomical Society; the Astronomical Society of the Pacific; and the International Astronomical Union. He discovered over 2,000 stars, and published seventy-five papers. In October 1993, Henize was on a hiking expedition with three members of High Adventure BVI to Mount Everest, testing the Tissue Equivalent Proportional Counter equipment for NASA at different altitudes, and he reported breathing problems at 22,000 feet altitude. Treatments with oxygen failed, and Henize died in his sleep on October 05, 1993 at base camp, Mount Everest, Nepal, aged sixty-six years. His cause of death was determined to be high altitude pulmonary edema (HAPE).

## Hennen, Thomas J.. STS-44.

Chief Warrant Officer 4 (ret.) Thomas John Hennen was born in August 1952 in Albany, Georgia, USA. Hennen attended Urbana College in Urbana, Ohio, from 1970-72 on both an academic and athletic scholarship. Hennen entered the US Army in 1972 and served over twenty-three years in the Imagery Intelligence (IMINT) field; at both the tactical and national intelligence levels. From 1973-75, he was assigned to the 163rd Military Intelligence Battalion, Fort Hood, Texas. From 1976-78, he was assigned to the 203rd Military Intelligence Detachment; providing imagery collection, exploitation, and intelligence support to the First Cavalry Division, the Second Armoured Division, the Sixth Armoured Cavalry Regiment, and Headquarters III Corps. From 1981 through 1986, Thomas Hennen was assigned to the US Army Intelligence Center, Fort Huachuca, Arizona; responsible for developing all training requirements, concepts and doctrine.

Selected as a Payload Specialist astronaut candidate in September 1988, he reported to NASA in 1990 to begin training. Hennen flew as 'Payload Specialist 1' on STS-44 aboard *Atlantis* (November 24-December 01, 1991), a dedicated Department of Defense mission. The unclassified payload included a Defense Support Program satellite; Military-Man-in-Space (M88-1); Air Force Maui Optical Site; Cosmic Radiation Effects and Activation Monitor (CREAM); and Extended Duration Orbiter Medical Project. He logged six days, 22 hours and 50 minutes in space. Thomas Hennen retired from the US Army in December 1995. In 1996, he co-founded and served as the Executive Director of the **Atlantis Foundation** until 2017. This is both an advocate and service provider for persons with developmental disabilities.

Tom Hennen has received twelve awards, including the Defense Superior Service Medal; the Legion of Merit; the Meritorious Service Medal with two Oak Leaf Clusters; the Army Good Conduct Medal; the Armed Forces Space and Missile Badge (Senior); the National Defense Service Medal; the US Army Military Intelligence Corps Knowlton Award; and he was inducted into the US Army Military Hall of Fame (2022).

## Henricks, Terence T.. STS-44, STS-55, STS-70, STS-78.

Col. Terence Thomas "Tom" Henricks was born in July 1952 in Bryan, Ohio, USA. He received a Bachelor of Science degree in Civil Engineering from the USAF Academy in 1974, and a Master of Public Administration degree from Golden Gate University in 1982. Selected by NASA in June 1985, he had various tecnical assignments, and served as the Assistant for Shuttle to the Chief of the Astronaut Office. He became the first person to log over 1,000 hours as a Space Shuttle pilot/commander.

Henricks was Pilot on STS-44 aboard *Atlantis* (November 24-December 01, 1991), which deployed a Defense Support Program satellite with an Inertial Upper Stage (IUS-14) rocket booster. *Atlantis* made the final Shuttle landing on a dry lake bed runway. Henricks was Pilot on STS-55 aboard *Columbia* (April 26-May 06, 1993), which carried an ambitious eighty-eight experiments in many disciplines, from eleven different nations. Henricks was Commander on STS-70 aboard *Discovery* (July 13-22, 1995) to deploy the sixth and final NASA *Tracking and Data Relay Satellite*. This was the first Shuttle mission controlled from the new Mission Control Center room at JSC in Houston. Henricks was Commander on STS-78 aboard *Columbia* (June 20-July 07, 1996), and the crew performed research into the effects of long-duration spaceflight on human physiology, in preparation for flights on the ISS. He logged a total of 42 days, 18 hours and 38 minutes in space.

Henricks has been named Pilot Training Distinguished Graduate and F-16 Conversion Course Top Gun; awarded the DFC; the NASA Outstanding Leadership Medal; the Defense Superior Service Medal; the Defense Meritorious Service Medal; Air Force Meritorious Service Medals; two Air Force Commendation Medals; four NASA Space Flight Medals; and was inducted into the Ohio Veterans Hall of Fame.

## Herrington, John B.. STS-113.

Cmdr. John Bennett Herrington was born in September 1958 in Wetumka, Oklahoma, USA. He received a Bachelor of Science degree in Applied Mathematics from the University of Colorado in 1983; and a Master's degree in Aeronautical engineering from the US Naval Postgraduate School in 1995. In 1984, he was commissioned as an Ensign at the Naval Air Station in Pensacola, Florida. In 1990, Herrington attended the US Navy Test Pilot school in Maryland. He was selected by NASA in 1996 for two years of astronaut training.

In 2002, he became the first enrolled member of a Native American tribe, the Chickasaw Nation, to fly in space. On STS-113 aboard *Endeavour* (November 23-December 07, 2002) the crew delivered and installed the P1 Truss and transferred cargo. Herrington carried eagle feathers, arrowheads, wooden flutes, and flags of the Chickasaw and Crow nations to the ISS; and made three EVAs totaling 19 hours and 55 minutes. He logged 13 days, 18 hours and 47 minutes in space. In 2005, Herrington left NASA and retired from the Navy, to be hired by **Rocketplane Global** until December 2007. He then left to do public speaking engagements and work with the Chickasaw Nation. In 2013, he was working on a doctoral degree in Education at the University of Idaho. Herrington was inducted into the Chickasaw Nation Hall of Fame in 2002; and the Oklahoma Aviation and Space Hall of Fame in 2007.

## Hernandez, Jose M.. STS-128.

Jose Moreno Hernandez was born in August 1962 in French Camp, California, USA. Of Mexican-American heritage, he received a Bachelor of Science degree in Electrical Engineering from the University of the Pacific in 1984; and a Master of Science degree in Electrical and Computer Engineering from the University of California-Santa Barbara in 1986. From 1990 to 2001, he worked at **Lawrence Livermore National Laboratory** in California and developed (with a commercial colleague) the first full-field digital mammography imaging system, which aids in the early detection of breast cancer.

In 2001, Hernandez joined the JSC in Houston, and later served as chief of the Materials and Processes Branch of JSC. After being turned down eleven times for astronaut training, he was selected in May 2004; working on various technical assignments until his selection in July 2008 as 'Mission Specialist 2' for the STS-128 mission aboard *Discovery* (August

28-September 11, 2009). Flying with John "Danny" Olivas of El Paso, Texas, this was the first time two Hispanic Americans were on the same Shuttle crew. The main payload was the 12,131 kg (26,744 lb) *Leonardo* MPLM FM-1, flown to assist with establishing a six-man crew capacity on the ISS by bringing extra supplies and equipment. The shuttle flew the first test flight of the TriDAR, a 3D dual-sensing laser camera, intended for potential use as an autonomous rendezvous and docking sensor. *Discovery* undertook the testing of a catalytic coating intended for use by the *Orion* spacecraft; in order to understand the entry heating performance. He logged 13 days, 20 hours and 54 minutes in space.

Hernandez supported California's Proposition 30, a ballot measure proposed to increase income taxes on those earning more than USD 250,000.00, and raise the sales tax rate by 0.25 cents for everyone. In October 2011, he ran for Congress as a Democrat in California's 10th Congressional District. He won the Democratic nomination, but lost the 2012 general election. Hernandez is the subject of the 2023 biopic ***A Million Miles Away*** in which he is portrayed by Michael Pena. The film was based on his 2012 autobiography, *"Reaching for the Stars: The Inspiring Story of a Migrant Farmworker Turned Astronaut"*. He has received twelve honors, including Graduate Engineering Minority Fellow (GEM) (1985); the Hispanic Engineer National Achievement Award, Outstanding Technical Contribution (1995); the US Department of Energy "Outstanding Performance Commendation" (2000); the Lawrence Livermore National Laboratory "Outstanding Engineer Award" (2001); and UCLA, Santa Barbara 2015 Distinguished Alumnus. The Jose Hernandez Middle School, in San Jose, California, is named after him. Hernandez currently serves as Regent of the **University of California.**

## Hieb, Richard J.. STS-39, STS-49, STS-65.

Richard James Hieb was born in September 1955 in Jamestown, North Dakota, USA. He received a Bachelor of Arts degree in Mathematics and Physics from Northwest Nazarene College in 1977; and a Master of Science degree in Aerospace Engineering from the University of Colorado at Boulder in 1979. After graduating, Hieb went to NASA/JSC to work in crew procedures development and crew activity planning. He has an extensive background in on-orbit procedures development, particularly in rendezvous and proximity operations.

Selected in 1985 NASA Group 11, he qualified as an astronaut in July 1986. Hieb was 'Mission Specialist 5' on STS-39 aboard *Discovery* (April 28-May 06, 1991) which carried the *Multi-Purpose Release Canister* (MPRC) Department of Defense payload. He was responsible for operating the unclassified *Infrared Background Signature Satellite* (IBSS) from within the payload bay, on the remote manipulator system, and as a free-flying satellite which was retrieved thirty-six hours later. He was 'Mission Specialist 1' on STS-49 aboard *Endeavour* (May 07-16, 1992) which retrieved the 4,215 kg *Intelsat 603* satellite, stranded in an unusable orbit since launch in March 1990. It was equipped with a new perigee kick motor and released to fire into an operational geosynchronous orbit. Crew made the first (and only) EVA involving three astronauts; Hieb, Thuot and Akers.

Hieb was Payload Commander on STS-65 aboard *Columbia* (July 08-23, 1994), the second flight of the International Microgravity Laboratory, and which set a Shuttle flight duration record, with 236 orbits of the Earth. He logged a total of 31 days, 22 hours and 34 minutes (over 750 hours) in space; including three EVAs totaling over 17 hours. After leaving NASA, Hieb worked at **AlliedSignal** and **Orbital Sciences Corporation** before spending 14 years as an executive at **Lockheed Martin**. Richard Hieb is currently a faculty member in the University of Colorado-Boulder Smead Aerospace Engineering Sciences Department.

## Higginbotham, Joan E.. STS-116.

Joan Elizabeth Higginbotham was born in August 1964 in Chicago, Illionois, USA. She received a Bachelor of Science degree from the Southern Illinois University-Carbondale in 1987; and a Master's in Management Science (1992) and in Space Systems (1996), both from the Florida Institute of Technology. She was an accomplished NASA electrical engineer for nearly a decade before becoming an astronaut, working on the development of Space Shuttles and participated in fifty-three launches during her first nine years at KSC.

Selected by NASA in April 1996, Higginbotham logged 12 days, 20 hours and 45 minutes (over 308 hours) in space as 'Mission Specialist 4' on STS-116 aboard *Discovery* (December 09-22, 2006), which delivered the P5 Truss to the ISS, attaching it as the third port truss segment. She was the third African American woman to go into space. Joan Higginbotham has received nine honors, including Commendation of Merit for Service to the Department of Defense Missions; *The Mecklenburg Times* (2014) 50 Most Influential Women; ***World Who's Who*** of Women; the Black Rose Award (2007), awarded by the League of Black Women; Honorary Doctor of Humane Letters (2017), University of New Orleans; and in 2008 she featured in the ***PBS*** special, "Bold Visions: Women in Science & Technology".

## Hilmers, David C.. STS-51-J, STS-26, STS-36, STS-42.

Dr. David Carl Hilmers was born in January 1950 in Clinton, Iowa, USA. He received a Bachelor of Arts degree in Mathematics (summa cum laude) from Cornell College, and, in July 1972, entered active duty with the US Marine Corps. Hilmers received a Master of Science degree in Electrical Engineering (with distinction) in 1977; and a degree in Electrical Engineering from the US Naval Postgraduate School in 1978. He was later assigned to the 1st Marine Aircraft Wing at Marine Corps Air Station Iwakuni, Japan. In 1995, Hilmers received a Doctor of Medicine (M.D.) degree (with honors) from the Baylor College of Medicine; and a Master of Science degree in Public Health (MPH) from the UT Health School of Public Health at the University of Texas Health Science Center, Houston, in 2002.

Selected as a NASA astronaut in July 1980, his early assignments included work on rocket upper stages (such as *PAM, IUS,* and *Centaur*), as well as Shuttle software verification. He was the Astronaut Office training coordinator, worked on various Department of Defense (DoD) payloads, and served as a CapCom on five Shuttle missions. In May 1985, he was named to the crew of STS-61-F, which was to deploy the *Ulysses* spacecraft (eventually launched on STS-41) on an interplanetary trajectory using a *Centaur* upper stage, but the *Shuttle-Centaur* project was ended in July 1986. David Hilmers was 'Mission Specialist 1' on STS-51-J aboard *Atlantis* (October 03-07, 1985), a classified DoD mission, when Hilmers had prime responsibility for a number of on-orbit activities. He was 'Mission Specialist 2' on STS-26 aboard *Discovery* (September 29-October 03, 1988), which deployed the *Tracking and Data Relay Satellite* (TDRS-C), and 11 mid-deck experiments were carried out.

Hilmers was 'Mission Specialist 2' STS-36 aboard *Atlantis* (February 28-March 04, 1990), which carried DoD payloads, launching one satellite (described as *AFP-731*, possibly an *Advanced KH-11* photo-reconnaissance satellite with an all-digital imaging system). While training for his fourth flight, Hilmers attended pre-med courses at night. He was 'Mission Specialist 3' STS-42 aboard *Discovery* (January 22-30, 1992) with fifty-five major experiments in the International Microgravity Laboratory-1 module. He logged a total of 20 days, four hours and 16 minutes (over 493 hours) in space. Hilmers retired from NASA and the US Marine Corps in October 1992, and went on to complete medical school. He currently holds the rank of Professor in the departments of Internal Medicine and Pediatrics, the Center for Space Medicine, and for Baylor Global Initiatives at the **Baylor College of Medicine** in Houston, Texas. Dr. David Hilmers spends much time providing humanitarian medical care in Houston and in developing nations, including disaster relief efforts.

## Hines, Robert "Bob", Jr. SpaceX 'Crew 4'/Expedition 67/68.

Robert "Farmer" Hines, Jr. was born in January 1975 in Fayetteville, North Carolina, USA. He received a Bachelor of Science degree in Aerospace Engineering from Boston University in 1997. Hines has served over twenty-three years in the USAF as an instructor pilot, fighter pilot, and a test pilot. In 1999, he graduated from Air Force Officer Training School and was commissioned as a second lieutenant. In 2008, he attended the USAF Test Pilot School, where he received a Master of Science degree in Flight Test Engineering. In 2010, Hines received a Master of Science degree in Aerospace Engineering from the University of Alabama. He joined the Air Force Reserves in 2011, and worked as a wing plans officer and a F-15E program test pilot. Robert Hines flew seventy-six combat missions, and logged over 3,500 hours in forty-one aircraft types.

In 2017, Bob Hines was selected for NASA Astronaut Group 22, and began two years of training. In February 2021, Hines was assigned as pilot of SpaceX 'Crew-4', the fourth NASA *Commercial Crew* operational flight, which launched on April 27, 2022, to dock with the ISS later that day. Hines, along with Commander K.N. Lindgren, and Mission Specialists Samantha Christoforetti and Jessica Watkins joined the 'Expedition 67/68' team. Their capsule was named *Freedom*, because it "celebrates a fundamental human right, and the industry and innovation that emanate from the unencumbered human spirit". Bob Hines received multiple awards, including the Air Medal; the Aerial Achievement Medal; the Iraq Campaign Medal; the Afghanistan Campaign Medal; and the Nuclear Deterrence Operations Service Medal. He is a member of the Society of Experimental Test Pilots, and the American Institute of Aeronautics and Astronautics.

## Hire, Kathryn P.. STS-90, STS-130.

Capt. Kathryn "Kay" Patricia Hire was born in August 1959 in Mobile, Alabama, USA. She received a Bachelor of Science degree in Engineering and Management from the US Naval Academy in 1981; and a Master of Science degree in Space Technology from the Florida Institute of Technology in 1991. After earning her Naval Flight Officer wings in October 1982, Hire conducted worldwide airborne oceanographic research missions with Oceanographic Development Squadron Eight (VXN-8) based at Patuxent River, Maryland. In January 1989, she resigned her Regular Navy commission, accepted a Reserve commission, and joined the Naval Reserve at NAS Jacksonville, Florida. Kay Hire was the first female in the US military assigned to a combat aircrew. Recalled to full-time active military duty from 2001-03, Hire supported Operation Enduring Freedom and Operation Iraqui Freedom as a member of the staffs at MacDill AFB, Florida. She was promoted to Captain on December 01, 2002.

Selected by NASA in December 1994, Kathryn Hire had a year of training, then flew as 'Mission Specialist 2' on STS-90 aboard *Columbia* (April 17-May 03, 1998), when the crew served as both experimental subjects and operators for 26 life science experiments focusing on the effects of microgravity on the brain and nervous system. She was 'Mission Specialist 1' on STS-130 aboard *Endeavour* (February 09-22, 2010), when the primary payloads were the 13,004 kg *Tranquility* module and the 1,805 kg *Cupola*, a robotic control station providing a 360-degree view around the ISS. As of June 2018, Kay Hire was considered a "NASA Management Astronaut", no longer eligible for spaceflight assignments. On February 28, 2019, Hire retired from NASA. She is a member of Association of Space Explorers; Association of Naval Aviation; Institute of Navigation; Naval Reserve Association; US Naval Academy Alumni Association; and the US Sailing Association. Kay Hire received nine awards, including the Defense Superior Service Medal; three Meritorious Service Medals; the Navy and Marine Corps Commendation Medal; two National Defense Service Medals; the Armed Forces Expeditionary Medal; and the War on Terrorism Service Medal.

## Hobaugh, Charles O.. STS-104, STS-118, STS-129.

Charles Owen Hobaugh was born in November 1961 in Bar Harbor, Maine, USA. He received a Bachelor of Science degree in Aerospace Engineering from the US Naval Academy in 1984. Hobaugh was designated a Naval Aviator in February 1987, and flew combat missions in the Persian Gulf during Operation Desert Storm. In June 1991, he was selected for US Naval Test Pilot School, returning there in July 1994 as an instructor. From January 1993 to August 1994, he attended the University of Tennessee Space Institute. Hobaugh logged over 5,000 flight hours in over forty aircraft types, and made over 200 V/STOL shipboard landings. In September 2010, he retired from the US Marine Corps.

Selected by NASA in April 1996, he was the re-entry and landing CapCom for the STS-107 mission, when the shuttle *Columbia* was destroyed on re-entry over east Texas. Hobaugh was pilot on STS-104 aboard *Atlantis* (July 12-24, 2001), the tenth mission to the ISS, which had the primary objective to deliver and install the 5.9 ton *Quest* Joint Airlock, and perform Station maintenance. He was pilot on STS-118 aboard *Endeavour* (August 08-21, 2007), the first flight of a Mission Specialist Educator (Barbara Morgan), the successor to NASA's Teacher in Space Project. The Shuttle delivered the 1,584 kg starboard S5 truss segment for the ISS, as well as External Stowage Platform 3 (ESP-3) and a replacement Control Moment Gysroscope. The 'Spacehab-SM' in Bays 5-7 had 5,480 kg of cargo.

Charles Hobaugh was commander on STS-129 aboard *Atlantis* (November 16-27, 2009), which focused on staging spare components outside the ISS. This was the second flight to carry two African-American astronauts; Leland Melvin and Robert Satcher. The 6,136 kg EXPRESS Logistics Carrier (ELC-2) was the heaviest item of cargo, which included MISSE-7, an experiment to expose a variety of materials and coatings being considered for future spacecraft to the extreme conditions outside the ISS. He logged a total of 36 days, seven hours and 47 minutes in space. Hobaugh was hired by **FedEx** in 2011. He is a member of the US Naval Academy Alumni Association, and his awards include US Naval Academy Distinguished Graduate; US Naval Test Pilot School Distinguished Graduate; the Strike/Flight Air Medal; the Navy and Marine Corps Achievement Medal; the Combat Action Ribbon; and a Navy Unit Commendation.

## Hoffman, Jeffrey A.. STS-51-D, STS-35, STS-46, STS-61, STS-75.

Dr. Jeffrey Alan Hoffman was born in November 1944 in Brooklyn, New York, USA. He received a Bachelor of Arts degree (summa cum laude) in Astronomy from Amherst College in 1966; a Ph.D. in Astrophysics from Harvard University in 1971; and a Master of Science degree in Materials Science from Rice University in 1988. From 1972-75, during post-doctoral work at the University of Leicester in the UK, he worked on several X-ray astronomy rocket payloads. From 1975-78, he worked in the Center for Space Research at the Massachusetts Institute of Technology as project scientist in charge of the orbiting 'HEAO-1' A4 hard X-ray and gamma ray experiment, launched in August 1977.

Selected by NASA in January 1978, he worked in the Flight Simulation Laboratory at Downey, California; testing guidance, navigation and flight control systems. He also worked on EVA, including the development of an advanced high-pressure spacesuit, and tools for building the ISS. Hoffman was NASA's second Jewish astronaut, and the second Jewish man in space (after Soviet cosmonaut Boris Volynov). He was a co-founder of the Astronaut Office Science Support Group. Hoffman was 'Mission Specialist 3' on STS-51-D aboard *Discovery* (April 12-19, 1985), which deployed two Hughes-built satellites. STS-35 aboard *Columbia* (December 02-10, 1990) was a 'Spacelab' mission that featured the Astro-1 ultraviolet astronomical observatory in the Shuttle's payload bay. Hoffman was 'Mission Specialist 3' on STS-46 aboard *Atlantis* (July 31-August 08, 1992), which deployed two satellites; *EURECA*, and the joint NASA/ASI *Tethered Satellite System* (TSS-1).

He was 'Mission Specialist 3' on STS-61 aboard *Endeavour* (December 02-13, 1993), the first mission to repair the Hubble Space Telescope. Hoffman made three EVAs totaling over 22 hours, and also spun a dreidel for the Jewish holiday of Hanukah, to a live audience watching via satellite. Jeffrey Hoffman was 'Mission Specialist 1' on STS-75 aboard *Columbia* (February 22-March 09, 1996). The principal objective was the reflight of the *Tethered Satellite System* (TSS), and the third flight of the US Microgravity Payload (USMP-3). The TSS demonstrated the ability of tethers to produce electricity, and provided much information on the electrodynamics of tethers and plasma physics (before the tether broke at 19.7 km, just short of the 20.7 km mission goal). During this mission, he became the first person to spend 1,000 hours on Space Shuttle missions. He took with him a small Torah scroll, and read from the ***Book of Genesis*** during the flight. The documentary film, ***Space Torah***, released in 2023, is centered around this event. He logged a total of 50 days, eleven hours and 54 minutes (1,211 hours) in space.

Hoffman left the astronaut program in July 1997 to become NASA's European Representative in Paris, serving until August 2001. He joined the MIT Aeronautics and Astronautics faculty in 2001 as a senior lecturer. Since 2008, he has been a visiting professor at the Department of Physics and Astronomy at the **University of Leicester**. Hoffman is a member of the International Academy of Astronautics; the International Astronomical Union; the American Institute of Aeronautics and Astronautics; and the American Astronomical Society. As of 2005, he was co-director of the Massachusetts Space Grant Consortium and a professor of the practice in the Department of Aeronautics and Astronautics at MIT.

Dr. Hoffman is the author of the book *An Astronaut's Diary* (June 1986), which is accompanied by a cassette tape of original recordings made with a pocket tape recorder. Hoffman has received sixteen awards, including the Amherst College 1963 Porter Prize in Astronomy; the Woodrow Wilson Foundation Pre-Doctoral Fellowship (1966-67); the National Academy of Sciences Post-Doctoral Visiting Fellowship (1971-72); the NATO Post-Doctoral Fellowhip (1973-74); two NASA Exceptional Service Medals; two NASA Distinguished Service Medals; and the FAI V.M Komarov and Sergei P. Korolev Diplomas (1991 & 1994). He was inducted into the Astronaut Hall of Fame in 2007.

## Horowitz, Scott J.. STS-75, STS-82, STS-101, STS-105.

Dr. Scott Jay "Doc" Horowitz was born in March 1957 in Philadelphia, Pennsylvania, USA. He received a Bachelor of Science degree in Engineering from California State University-Northridge in 1978; and a Ph.D. in Aerospace Engineering from the Georgia Institute of Technology in 1982. He then worked for the **Lockheed Company**. He graduated from the USAF Test Pilot School in December 1990, and was selected by NASA in 1992. Scott Horowitz was pilot on STS-75 aboard *Columbia* (February 22-March 09, 1996). Horowitz and Jeffrey Hoffman, both Jewish, had alternating use of the same bunk bed, to which Hoffman attached a mezuzah, using Velcro. A mezuzah is a piece of parchment inscribed with Hebrew verses from the Torah, affixed to the doorpost of a Judaic home. STS-82 aboard *Discovery* (February 11-21, 1997) was the second mission to service the Hubble Space Telescope.

Horowitz was pilot on STS-101 aboard *Atlantis* (May 91-29, 2000), which delivered a 'Spacehab' double module, and an integrated Cargo Carrier Pallet with supplies for the ISS. The crew reboosted the Station from a 370 km to 400 km orbit. On re-entry a damaged tile seam caused a breach which allowed superheated gas to enter the Shuttle's left wing. The gas did not penetrate deeply enough to cause a catastrophic failure, as happened in the *Columbia* disaster. Scott Horowitz was commander on STS-105 aboard *Discovery* (August 10-22, 2001), a mission for ISS supply and crew rotation. He logged a total of

47 days, ten hours and 41 minutes in space. He retired from the USAF and NASA in October 2004, for a position with **ATK Thiokol**, but returned to NASA in September 2005 to become the Associate Administrator for the Exploration Systems Mission Directorate, until leaving on October 01, 2007. Horowitz has been active in advocating NASA's focus to be the human exploration of Mars. He is a member of the steering committee of **The Mars Society**. In 2018, Horowitz became an Advisor to the DAV Foundation.

## Hughes-Fulford, Millie E.. STS-40.

Dr. Millie Elizabeth Hughes-Fulford (nee Hughes, 1945-2021) was born in December 1945 in Mineral Wells, Texas, USA. Millie Hughes received a Bachelor of Science degree in Chemistry and Biology from Tarleton State University in 1968; and a Ph.D. from Texas Women's University in 1972. She then joined the faculty of Southwestern Medical School, University of Texas, as a postdoctoral fellow. She has contributed over 120 papers and abstracts on T-cell activation, bone and cancer growth regulation. From 1971 to 1972, she was both an American Association of University Women Fellow and a MacArthur Foundation Fellow. Hughes-Fulford was a Major in the US Army Reserve Medical Corps, serving from 1981 until 1995.

Selected by NASA in January 1983, she flew as 'Payload Specialist 2' on STS-40 aboard *Columbia* (June 05-14, 1991), the first Spacelab Life Sciences (SLS-1) mission dedicated to biomedical studies, and the first Shuttle crew with three female members. Hughes-Fulford logged nine days, two hours and 14 minutes in space. She was Principal Investigator on a series of SpaceHab/Biorack experiments which flew on STS-76 in March 1996, STS-81 in January 1997, and STS-84 in May 1997. These studies examined the root causes of osteoporosis that occurs in astronauts during spaceflight. Dr. Hughes-Fulford, her team and international colleaguees published a featured article in the *Journal of Leukocyte Biology* showing that microgravity itself is the root cause of T-cell dysfunction. In July 2013, NASA awarded her work as a top discovery on the ISS.

Dr. Hughes-Fulford was a member of the American Association for the Advancement of Science; the American Society for Gravitational Science and Biology; the American Society for Bone and Mineral Research; the American Society for Cell Biology; and the Association of Space Explorers. She was named the International Zontian in 1992; and Marin County Woman of the Year in 1993. She was awarded the 1968-71 National Science Foundation Fellow; the 1984 Presidential Award for Federal Employee, Western Region; the 1987-90 National Research Council, Committee on Space Biology and Space Medicine Award; and 1995-2001 Advisory Board for Marine Biological Laboratory, Sciences Writing Program, at Woods Hole. Dr. Millie Hughes-Fulford died of lymphoma on February 02, 2021, in Mill Valley, California, aged seventy-five years.

## Hurley, Douglas G.. STS-127, STS-135, Crew Dragon 'Demo-2'.

Douglas "Doug" Gerald Hurley was born in October 1966 in Endicott, New York, USA. After graduating from Owego Free Academy in Owego, New York, in 1984, he attended college at Tulane University in New Orleans, Louisiana. Hurley then attended The Basic School at Marine Corps Base Quantico, and later the Infantry Officers Course. He received a Bachelor's degree in Civil Engineering, graduating magna cum laude in 1988. While in college, he was a member of the Naval Reserve Officer Training Corps, where he was commissioned as a second lieutenant in the US Marine Corps. Hurley attended the Naval Air Station, Pensacola, Florida, and became a Naval Aviator in August 1991. He was a distinguished graduate of the US Navy Pilot Training program in December 1997, became a Navy Test Pilot, and was the first Marine to fly a F/A-18 E/F Super Hornet airplane. Hurley logged over 5,500 flying hours in over twenty-five aircraft types.

Selected by NASA in 2000 Group 18, Doug Hurley was pilot on STS-127 aboard *Endeavour* (July 15-31, 2009) on ISS Assembly Mission 2J/A, which delivered the Japanese-built Exposed facility (JEM-EF) and the Experiment Logistics Module Exposed Section (ELM-ES) to the ISS. He was pilot on STS-135 aboard *Atlantis* (July 08-21, 2011), the final flight of the Space Shuttle program, which delivered the MPLM *Raffaello* and a Lightweight Multi-Purpose Carrier (LMC) to the ISS. In August 2014, Doug Hurley became the Assistant Director for the **Commercial Crew Program** following the merger of Flight Operations and Mission Operations. He is married to fellow astronaut Karen Nyberg.

On May 16, 2020, during a meeting of the National Space Council, Douglas Hurley said, *"It's a real honor to just be a part of this program and to launch American rockets from Florida one more time"*. He was Commander of SpaceX Mission *Crew Dragon* 'Demo-2', launched on May 30, 2020, when together with Bob Behnken he became the first astronaut to launch aboard a commercial orbital spacecraft. He logged a total of 92 days, ten hours and 38 minutes in space. On July 16, 2021, NASA announced that Hurley would be retiring after twenty-one years' service. He received the Stephen A. Hazelrigg Memorial Award for best pilot/engineer team, Naval Strike Aircraft Test Squadron. He was awarded the Meritorious Service Medal; two Navy and Marine Corps Commendation Medals; and four NASA Superior Accomplishment Awards (2004, 2005, 2006, & 2007). On January 31, 2023, Hurley was awarded the Congressional Space Medal of Honor for NASA's SpaceX Demonstration Mission-2 *Demo-2* to the ISS in May 2020.

## Husband, Richard D.. STS-96/2A.1, STS-107.

Lt. Col. Richard "Rick" Douglas Husband (1957-2003) was born in July 1957 in Amarillo, Texas, USA. He graduated (with honors) from Amarillo High School in 1975; received a Bachelor of Science degree in Mechanical Engineering from Texas Tech University in 1980; and a Master of Science degree from Fresno State University, then was commissioned as a pilot in the USAF. Transfered to George AFB, California, in December 1985, he became an F-4 instructor. From December 1987, he attended the USAF Test Pilot school, then flew the F-4 and all five models of the F-15 Eagle. He logged over 3,800 hours flight time in over forty aircraft types.

Selected by NASA in December 1994, Husband was pilot on STS-96 aboard *Discovery* (May 29-June 06, 1999), when the crew performed the first docking with the International Space Station (ISS). Rick Husband was commander on STS-107 aboard *Columbia* (January 16-February 01, 2003). He died about 9 a.m. on the morning of February 01, 2003, when Space Shuttle *Columbia* disintegrated during re-entry into Earth's atmosphere over east Texas. This was just sixteen minutes before scheduled touchdown. About 38% of the Shuttle was recovered in small pieces (in a debris field of nineteen kilometers radius), and 35-40% of the scientific experimental data. Some 30% of the total science data was saved and collected through telemetry back to ground stations, with more from samples and hard drives found intact on the ground. For the 'CIBX-2' payload, which studied treatments for cancer, 90% of the data was recovered fully intact. For one of the six FREESTAR experiments, 99% of the data was recovered.

Rick Husband logged a total of 25 days, 17 hours and 33 minutes in space. He had received seven military honors, including the Meritorious Service Medal with two bronze Oak Leaf Clusters; the Air Force Commendation Medal; the National Defense Service Medal; the Air Force Longevity Service Award with two Oak Leaf Clusters; and two NASA Group Achievement Awards. Rick Husband Amarillo International Airport; Rick Husband Drive, El Paso, Texas; Husband Hall, in the Columbia Village apartments, Florida Institute of Technology; Husband Auditorium, Squadron Officer School building, Maxwell AFB, Alabama; the lunar crater Husband; Husband Hill, one of the Columbia Hills on Mars; and Asteroid 51823 Rickhusband were all named in his honor. He was posthumously awarded the Congressional Space Medal of Honor; the NASA Distinguished Service Medal; and the NASA Space Flight Medal.

## Irwin, James B.. Apollo 15.

Col. James "Jim" Benson Irwin (1930-91) was born in March 1930 in Pittsburgh, Pennsylvania, USA. He received a Bachelor of Science degree in Naval Science from the US Naval Academy in 1951; and Master of Science degrees in Aeronautical Engineering and Instrumentation Engineering from the University of Michigan in 1957. Irwin was a developmental test pilot for the Lockheed YF-12, the Mach 3 fighter-interceptor variant which preceded the SR-7 Blackbird. He was one of nineteen astronauts selected by NASA in April 1966. He logged over 7,000 hours flying time; including 5,300 hours in jet aircraft.

As Lunar Module Pilot for *Apollo 15*, he was the eighth person to walk on the Moon, and the first, and youngest, of those astronauts to die. Dave Scott and Irwin spent 66 hours and 54 minutes on the Moon. Their mission was science-based, and they had received intensive geological training; credited with allowing them to make one of the most important discoveries of the Apollo era, the "Genesis Rock". Once the rendezvous procedure was completed with the CSM, Irwin was busy moving rock samples, and preparing the LM for final separation. The crew's vital signs were being monitored back on earth, and the flight surgeons noticed irregularities in Irwin's heart rhythms. He had developed bigeminy, but since the cabin atmosphere was 100% oxygen, and he was in zero gravity, the heart was under no further strain. Irwin logged 295 hours and eleven minutes in space. His EVAs on the Moon's surface amounted to 18 hours and 35 minutes.

James Irwin was probably the only Apollo astronaut who subscribed to creationism. He became a devout born-again Christian after returning from space. He said, *"I felt the power of God as I'd never felt it before"*. After his retirement as a Colonel in 1972, Irwin founded the **High Flight Foundation**. He spent the next twenty years as a "Goodwill Ambassador for the Prince of Peace", stating that *"Jesus walking on the earth is more important than man walking on the Moon"*. He said that his experiences in space had made God more real to him than before, and that his Christian rebirth saved his marriage and made their lives much happier. Beginning in 1973, Irwin led expeditions to Mount Ararat, Turkey, in search of the remains of Noah's Ark. In the book, *More Than Earthlings*, he expressed his view that the Genesis creation narrative was real, literal history.

Jim Irwin survived three major heart attacks; the first occurring less than two years after his Moon mission, when he was aged forty-three years, and another just two months later. He collapsed during a run in June 1986; and on August 08, 1991, he had a fourth heart attack after a bicycle ride, when resuscitation attempts were unsuccessful. He died at Glenwood Springs, Colorado, aged 61 years, and was buried at Arlington National Cemetery. The James Irwin Charter Schools were founded in Colorado in his honor. Irwin received twenty-one awards, including the Air Force Distinguished Service Medal; two Air Force Commendation Medals; the NASA Distinguished Service Medal; the United Nations Peace Medal (1971); the Robert J. Collier Trophy (1971); the Kitty Hawk Memorial Award (1971); the Haley Astronautics Award (1972); and the Arnold Air Society's John F. Kennedy Trophy (1972). He was inducted into the International Space Hall of Fame (1983); and posthumously into the US Astronaut Hall of Fame (1997).

## Ivins, Marsha S.. STS-32, STS-46, STS-62, STS-81, STS-98.

Marsha Ivins was born in April 1951 in Baltimore, Maryland, USA. Of Jewish-American heritage, she received a Bachelor of Science degree in Aerospace Engineering from the University of Colorado at Boulder in 1973. Ivins went to work for NASA's Johnson Space Center and worked mainly on orbiter displays and controls, before being assigned as a Flight Engineer in 1980 and co-pilot on NASA administrative aircraft. Selected by NASA in 1984 Group 10, she was 'Mission Specialist 2' on STS-32R aboard *Columbia* (January 09-20, 1990), which deployed a *Syncom* satellite, and retrieved the 21,400 pound Long Duration Exposure Facility. Ivins was 'Mission Specialist 2' on STS-46 aboard *Atlantis* (July 31-August 08, 1992), which deployed the *EURECA* satellite, and conducted the joint NASA/ASI first *Tethered Satellite System* (TSS-1) test flight.

She was 'Mission Specialist 3' on STS-62 aboard *Columbia* (March 04-18, 1994), which flew the USMP-2 and OAST-2 payloads. These studied the effects of microgravity on materials sciences and other space flight technologies. Ivins was 'Mission Specialist 3' on STS-81 aboard *Atlantis* (January 12-22, 1997), the fifth of nine planned missions to dock with *Mir* and the second to exchange US astronauts. The 'Spacehab' double module provided additional mid-deck locker space for secondary experiments. She was 'Mission Specialist 2' on STS-98 aboard *Atlantis* (February 07-20, 2001), which continued the task of building and enhancing the ISS by bringing up the US *Destiny* laboratory module. Three EVAs were conducted to complete its assembly, and the crew relocated a docking port, also delivering supplies to the resident 'Expedition-1' crew. Marsha Ivins logged a total of 55 days, 21 hours and 48 minutes in space, and retired from NASA on December 31, 2010. She is single, with no children.

## Jemison, Mae C.. STS-47 (J).

Dr. Mae Carol Jemison was born in October 1956 in Decatur, Alabama, USA. She received a Bachelor of Arts degree in African-American Studies and a Bachelor of Science degree in Chemical Engineering from Stanford University in 1977; then an M.D. from Cornell University in 1981. Dr. Jemison was a doctor for the Peace Corps in Liberia and Sierra Leone from 1983-85 and worked as a General Practicioner. Selected by NASA in 1987 Group 12, she became the first African-American woman in space when she served as 'Mission Specialist 4' on STS-47 aboard *Endeavour* (September 12-20, 1992), which mainly involved conducting experiments in life and materials sciences inside 'Spacelab-J', a collaborative laboratory sponsored by NASA and the National Space Agency of Japan (NASDA). She and Mamoru Mohri of JAXA were trained to use the Autogenic Feedback Training Exercise (AFTE), a technique that uses biofeedback and autogenic training to help patients monitor and control their physiology as a possible treatment for motion sickness, anxiety, and stress-related disorders.

Dr. Jemison logged seven days, 22 hours and 30 minutes (over 190 hours) in space. She left NASA in March 1993 and founded a technology research company. Jemison later formed a non-profit educational foundation, and is the Principal of the 100 Year Starship project funded by DARPA. She wrote several books for children, and appeared on television several times; including in a 1993 episode of ***Star Trek: The Next Generation***. Dr. Jemison holds several honorary Doctorates, and has been inducted into the National Women's Hall of Fame and the International Space Hall of Fame.

## Jernigan, Tamara E.. STS-40, STS-52, STS-67, STS-80, STS-96/2A.1.

Dr. Tamara "Tammy" Elizabeth Jernigan was born in May 1959 in Chattanooga, Tennessee, USA. She received a Bachelor of Science degree in Physics from Stanford University in 1981; a Master of Science degree from Stanford in 1983; a Master of Science degree in Astronomy from the University of California, Berkeley, in 1985; and a Ph.D. in Space Physics and Astronomy from Rice University, Texas, in 1988. Jernigan joined NASA in June 1981 while studying at Stanford and Berkeley, initially working as a research scientist in the Theoretical Studies Branch at the NASA **Ames Research Center**. In July 1986, she was selected by NASA in the Astronaut Corps, and flew as 'Mission Specialist 2' on STS-40 aboard *Columbia* (June 05-14, 1991), the first Shuttle mission solely dedicated to biology. Other payloads included twelve "Getaway Special" experiment canisters.

Tammy Jernigan was 'Mission Specialist 3' on STS-52 aboard *Columbia* (October 22-November 01, 1992), which deployed the *LAGEOS-2* satellite and operated the USMP-1 payload. She was 'Mission Specialist 3' on STS-67 aboard *Endeavour* (March 02-18, 1995), which carried the Astro-2 payload: with the Hopkins Ultraviolet Telescope, the Ultraviolet Imaging Telescope, and the Wisconsin Ultraviolet Photo-Polarimeter Experiment (WUPPE). Jernigan was 'Mission Specialist 3' on STS-80 aboard *Columbia* (November 19-December 07, 1996), the longest Space Shuttle mission ever flown at 17 days, 15 hours and 53 minutes. Two satellites were deployed and retrieved; and two EVAs were planned, but were canceled after problems with the airlock hatch prevented Tom Jones and Jernigan from exiting the orbiter. She was 'EVA Specialist 2' on STS-96 aboard *Discovery* (May 27-June 06, 1999), the first Space Shuttle mission to perform a docking maneuver with the ISS. The shuttle carried an Integrated Cargo Carrier with parts for the Russian cargo crane *Strela*, which was mounted to the exterior of the Russian station segment. She performed an EVA of seven hours and 55 minutes.

Dr. Jernigan is one of six women with five Space Shuttle flights to their credit. She logged a total of 63 days, one hour and 24 minutes (over 1,512 hours) in space, and left NASA in 2001. Tammy Jernigan is Deputy Principal Associate Director in the Weapons and Complex Integration (WCI) organization at **Lawrence Livermore National Laboratory**. She has received the NASA Distinguished Service Medal; and the NASA Outstanding Leadership Medal. She is married to former astronaut Peter Wisoff, with two children.

## Jett, Brent W., Jr.. STS-72, STS-81, STS-97, STS-115.

Capt. Brent Ward Jett, Jr. was born in October 1958 in Pontiac, Michigan, USA. He received a Bachelor of Science degree in Aerospace Engineering from the US Naval Academy in 1981; and a Master of Science degree in Aeronautical Engineering from the US Naval Postgraduate School in 1989. Jett was designated a Naval Aviator in March 1983, later assigned to VF-74, and deployed to the Mediterranean Sea and Indian Ocean aboard the aircraft carrier *USS Saratoga*. In July 1986, he was selected for the US Naval Postgraduate School, then attended the US Naval Test Pilot School in June 1989. Jett logged over 5,000 flying hours in over thirty aircraft types, making over 450 aircraft carrier landings.

Selected in March 1992 NASA Group 14, he was pilot on STS-72 aboard *Endeavour* (January 11-20, 1996), when the crew deployed and retrieved the *Office of Aeronautics and Space Technology Flyer* (OAST-Flyer), and performed two EVAs. Jett was pilot on STS-81 aboard *Atlantis* (January 12-22, 1997), the fifth in a series of joint missions between the Space Shuttle and Russian Space Station *Mir*, and the second involving an exchange of US astronauts. From June 1997 to February 1998, Jett was NASA Director of Operations at the Gagarin Center in Russia. He was Commander on STS-97 aboard *Endeavour* (December 01-11, 2000), which delivered a seventeen-ton solar array to the ISS; in addition to batteries, cooling radiators, and a communication system for voice and telemetry. Jett was Commander on STS-115 aboard *Atlantis* (September 09-21, 2006), with the primary objective of delivering and installing the P3/P4 truss and a pair of solar arrays (2A and 4A) to the ISS.

Brent Jett logged a total of 41 days, 18 hours and one minute in space. From November 2007 to February 2011, he was Director of Flight Crew Operations at JSC. From March 2010 to January 2013, he was the Deputy Manager for the NASA Commercial Crew Program. He retired from the US Navy in July 2007, and from NASA in January 2013. Jett belongs to the Association of Space Explorers; the Society of Experimental Test Pilots; and the Association of Naval Aviation. He has received nine honors, including graduating first of 976 students in the Class of 1981 at the US Naval Academy; the DFC; the Defense Superior Service Medal; the Defense Meritorious Service Medal; the Navy Commendation Medal; the NASA Exceptional Service Medal; and a portion of Northeast Fifty Sixth Street in Fort Lauderdale, Florida, has been designated Brent Jett Boulevard in his honor.

## Johnson, Gregory C.. STS-125.

Capt. Gregory Carl "Ray J" Johnson was born in July 1954 in Seattle, Washington, USA. He received a Bachelor of Science degree in Aerospace Engineering from the University of Washington. During college, he obtained his civilian commercial pilot certificate with multi-engine landplane and single-engine seaplane ratings. He was commissioned as an Ensign in the US Navy Reserve through Aviation Officer Candidate School in September 1977, and became a Naval Aviator in December 1978. Assigned to Attack Squadron 52, he completed two deployments in the Western Pacific and Indian Ocean aboard *USS Kitty Hawk*. In 1984, he reported to the USAF Test Pilot School at Edwards AFB, and after graduation reported to Naval Weapons Center, China Lake, California.

G.C. Johnson logged over 10,800 flying hours in over fifty aircraft types, with over 500 aircraft carrier landings. Johnson retired from the US Navy effective October 01, 2007. Selected as a NASA astronaut in June 1998 Group 17, he served as Manager, Launch Integration, for the Space Shuttle program at KSC from June 2004 to November 2005. Johnson piloted STS-125 aboard *Atlantis* (May 11-24, 2009), the final servicing mission for the Hubble Space Telescope. He logged 12 days, 21 hours and 38 minutes in space. Johnson belongs to the Association of Space Explorers; Society of Experimental Test Pilots; American Institute of Aeronautics and Astronautics; and the Tailhook Association. He has received over fourteen awards, including VA-128 Attack Pilot of the Year; three Meritorious Service Medals; two National Defense Service Medals; the War on Terrorism Service Medal; the Armed Forces Reserve Medal; and three Navy and Marine Corps Commendation Medals. "Ray J" Johnson is married, with two children.

## Johnson, Gregory H.. STS-123, STS-134.

Col. Gregory Harold "Box" Johnson was born in May 1962 in South Ruislip, Greater London, UK. He received a Bachelor of Science degree in Aeronautical Engineering from the US Air Force Academy in 1984; a Master of Science degree in Flight Structures Engineering from Columbia University in 1985; and a Master of Business Administration (M.B.A.) from the University of Texas at Austin in 2005. Johnson received his commission from the USAF Academy in May 1984, and attended pilot training at Reese AFB, Texas. He was retained as a T-38A instructor pilot until 1989, when he was selectd for an F-15E Eagle assignment. In December 1990, he deployed to Al Kharj, Saudi Arabia, flying thirty-four combat missions in support of Operation Desert Storm. In December 1992, he flew twenty-seven combat missions in support of Operation Southern Watch. In 1993, he was selected for Air Force Test Pilot School at Edwards AFB.

Selected in June 1998 NASA Group 17, Johnson completed Astronaut Candidate Training in 2000, and was a key player on several "tiger teams" during the investigation into the cause of the Shuttle *Columbia* accident in 2003. He was pilot on STS-123 aboard *Endeavour* (March 11-27, 2008), which delivered the pressurized section of the Japanese Experiment Logistics Module (ELM-PS) aka the *Kibo* module to the ISS. He was the mission's primary robotic arm operator. Johnson was pilot on STS-134 aboard *Endeavour* (May 16-June 01, 2011), which carried the Alpha Magnetic Spectrometer 2 (AMS-02) to the ISS, and it was attached to the S3 truss segment. The ExPRESS Logistics Carrier 3 (ELC3) carried several Orbital Replacement Units (ORU) that were too large or too heavy for other Shuttles to carry to the ISS. The mission delivered the Materials on International Space Station Experiment 8 (MISSE-8) experiments, and returned the completed MISSE-7 experiments to Earth.

Johnson logged a total of 31 days, 11 hours and 50 minutes in space. He retired from the USAF on February 01, 2009, and continued to serve NASA as a civilian. He was awarded the 2005 Stephen D. Thorne Top Fox Safety Award; the 2005 Dean's Award for Academic Excellence, McCombs School of Business; the NASA Superior Service Award; and the 1996 Lieutenant General Bobby Bond Award for the top Air Force Test Pilot. Military decorations include the DFC; two Meritorious Service Medals; four Air Medals; three Aerial Achievement Medals; the Air Force Commendation Medal; and two Air Force Achievement Medals.

## Jones, Thomas D.. STS-59, STS-68, STS-80, STS-98.

Capt. Dr. Thomas David Jones was born in January 1955 in Baltimore, Maryland, USA. He received a Bachelor of Science degree in Basic Sciences from the USAF Academy in 1977; and a Doctorate in Planetary Science from the University of Arizona in 1988. Jones served on active duty as an Air Force officer for six years. As pilot and aircraft commander of a B-52 D Stratofortress, he led a combat crew of six, accumulating over 2,000 hours of jet experience before resigning as a Captain in 1983. From 1983-88, he worked toward a Ph.D. at the University of Arizona. His research interests included the remote sensing of asteroids, meteorite spectroscopy, and applications of space sciences. From 1989-90, he was a program management engineer at the **CIA**'s Office of Development and Engineering. In 1990, he joined **Science Applications International Corporation** (SAIC) in Washington, D.C. as a senior scientist.

Selected by NASA in January 1990 NASA Group 13, he was 'Mission Specialist 4' on STS-59 aboard *Endeavour* (April 09-20, 1994) which flew the Space Radar Laboratory-1 experiments. Ground controllers activated the Spaceborne Imaging Radar-C (SIR-C) and began processing its first images of the Earth. Dr. Jones was 'Mission Specialist 4' on STS-68 aboard *Endeavour* (September 30-October 11, 1994) when the primary payload was the Space Radar Laboratory (SLR-2) making its second flight to study the Earth's environment. On October 01, Tom Jones sent down some video of a volcano erupting in Kamchatka. The latest "burst" from the Kliuchevskoi volcano occurred about eight hours after the Shuttle's launch. On Flight Day 8 (October 07) from an altitude of 206 km, the Spaceborne Imaging Radar and Synthetic Aperture Radar recorded long swaths of interferometric data over central North America, the Amazon forests of central Brazil, and the volcanoes of the Kamchatka Peninsular in Russia.

Produced regularly, these images could eventually provide scientists with insight into movements of the Earth's surface as small as a fraction of an inch. Close monitoring may detect pre-eruptive changes in volcanoes and movements in fault lines that precede earthquakes, providing early warning of imminent natural hazards. Dr. Jones was 'Mission Specialist 2' on STS-80 aboard *Columbia* (November 19-December 07, 1996) which deployed and retrieved two satellites. The *ORFEUS-SPAS II* satellite (with two spectrographs, for far and extreme ultraviolet) was deployed on Flight Day 1, and captured on Flight Day 16, while the *Wake Shield Facility-3* was deployed on Flight Day 4 and recaptured three days later. In 2001, Dr. Tom Jones led three spacewalks to install the Boeing-built 14.5 ton American *Destiny* laboratory module to the ISS. He was 'Mission Specialist 3' on STS-98 aboard *Atlantis* (February 07-20, 2001). *Destiny* was the first NASA lab to be permanently used since the days of *Skylab* nearly thirty years earlier. It is 8.5 metres long and 4.3 metres wide, and three EVAs were made to complete its assembly. Tom Jones logged a total of 53 days, zero hours and 49 minutes (over 1,272 hours) in space. Dr. Jones retired in 2001, and worked as a planetary scientist and consultant in space operations.

He is a senior research scientist at the **Florida Institute for Human and Machine Cognition**, engaged in planning robotic and astronaut expeditions to deep space and near-Earth asteroids. Dr. Jones is an author and speaker, with four adult non-fiction works. His 2006 book, *Sky Walking: An Astronaut's Memoir*, was named one of the top five books on the subject of space by the Wall Street Journal. From 2006-09, he served on the NASA Advisory Council. Thomas Jones is a board member of the Astronauts Memorial Foundation, and was inducted into the US Astronaut Hall of Fame on April 21, 2018. He has received military decorations, including Outstanding Graduate in Basic Sciences at the USAF Academy; and the Air Force Commendation Medal. Dr. Jones was awarded four NASA Space Flight Medals; the NASA Distinguished Service Medal; the NASA Exceptional Service Award; the NASA Exceptional Public Service Medal; and the NASA Outstanding Leadership Medal. The main belt asteroid 1082 TomJones was named in his honor. He appears frequently as a science/space commentator on radio and television.

## Kavandi, Janet L.. STS-91, STS-99, STS-104.

Dr. Janet Lynn Kavandi was born in July 1959 in Carthage, Missouri, USA. She received a Bachelor of Science degree (magna cum laude) in Chemistry from Missouri Southern State College in 1980; a Master's degree from the Missouri University of Science and Technology in 1982; and a Doctorate from the University of Washington in 1990. Selected by NASA in December 1994, after an initial year of training she was assigned to the Payloads and Habitability Branch, where she supported payload integration for the ISS. Dr. Kavandi was a Mission Specialist on STS-91 aboard *Discovery* (June 02-12, 1998), the ninth and final *Shuttle-Mir* docking mission. She was a Mission Specialist on STS-99 aboard *Endeavour* (February 11-22, 2000), the Shuttle Radar Topography Mission. Kavandi then worked in the Robotics Branch, where she trained on both the shuttle and space station robotic manipulator systems.

For her final flight, Kavandi was a Mission Specialist on STS-104/ISS Assembly Flight 7A aboard *Atlantis* (July 12-24, 2001), when the *Quest* joint airlock was installed, and there were joint operations with the 'Expedition 2' crew. Next, Kavandi again served as lead for the Payloads and Habitability Branch, then as the Branch Chief for the ISS. She was also responsible for coordination between international partners for visiting vehicles and associated operations. In 2005, she accepted a position as Deputy Chief of the Astronaut Office, until February 2008. She then became Deputy Director, Flight Crew Operations, then Director of the same. Janet Kavandi logged a total of 33 days, 20 hours and eight minutes in space.

In March 2016, she became the Center Director at the NASA Glenn Research Center in Ohio. In September 2019, Janet Kavandi retired from NASA, and started at **Sierra Nevada Corporation**. She has received nine awards, including National Honor Society (1977); a Presidential Scholarship from Missouri Southern State University (1977); Certificates for Excellence from Boeing Missile Systems (1991); two NASA Exceptional Service Medals (2001 & 2002); and was inducted into the US Astronaut Hall of Fame (2019).

## Kelly, James M.. STS-102, STS-114.

Col. James McNeal "Vegas" Kelly was born in May 1964 in Burlington, Iowa, USA. He received a Bachelor of Science degree in Astronautical Engineering from the USAF Academy in 1986; and a Master of Science degree in Aerospace Engineering from the University of Alabama. Kelly was designated an Air Force pilot in October 1987. After initial F-15A Eagle training, he was assigned to the Sixty-seventh Fighter Squadron, Eighteenth Fighter Wing at Kadena Air Base in Okinawa, Japan. In 1992, he continued flying the F-15 as an instructor and mission commander. Kelly graduated from the Air Force Test Pilot School at Edwards AFB in June 1994. He logged over 2,500 flight hours in more than thirty-five aircraft types.

Selected by NASA as an astronaut candidate in April 1996, James Kelly was pilot on STS-102 aboard *Discovery* (March 08-21, 2001), which carried the MPLM *Leonardo* with five tons of supplies to the ISS. He was pilot on STS-114 aboard *Discovery* (July 26-August 09, 2005) to the ISS. This launch was 907 days (29 months) after the loss of Space Shuttle *Columbia*, and the main focus of the mission was testing and evaluating new flight safety techniques. The *Discovery* landed at Edwards Air Force Base, since poor weather over KSC in Florida prevented use of the primary landing site. James Kelly logged a total of 26 days, 17 hours and 21 minutes (over 641 hours) in space. He has received eight awards and honors, including Distinguished Graduate in the class of 1986 at the USAF Academy, with honors; Top Gun at F-15 initial training at Luke Air Force Base, Arizona; Distinguished Graduate and winner of the Liethen-Tittle Award for Outstanding Graduate of the Air Force Test Pilot School, Class 93B; two Air Force Commendation Medals; two Outstanding Unit Awards; and two Combat Readiness Medals.

## Kelly, Mark E.. STS-108, STS-121, STS-124, STS-134.

Capt. Mark Edward Kelly was born in February 1964 in Orange, New Jersey, USA. An identical twin to Scott J. Kelly, he is an American politician who has served as the junior US Senator from Arizona since 2020. A member of the Democratic Party, he was elected in the special election following the death of Senator John McCain. Mark Kelly was five years old when *Apollo 11* took men to the Moon's surface for the first time. In a NASA 2011 pre-flight interview before mission STS-124, he said that he remembered the later missions more, but it seemed *"completely out of the question"* for *"some kid from New Jersey"*, even though his town was adjacent to another town where *Apollo 11*'s Lunar Module pilot, Dr. Buzz Aldrin, grew up.

Aged eighteen years, Kelly graduated from high school and went to the US Merchant Marine Academy at Kings Point, New York, receiving a Bachelor of Science degree in Marine Engineering and Transportation in 1986. He became a pilot in the US Navy in 1989, and, in 1991, flew 39 combat missions during Operation Desert Storm in the Persian Gulf War. He then attended graduate school and the US Naval Test Pilot School, where he later became an instructor. He said, *"Once I was flying airplanes off the aircraft carrier and later decided to become a test pilot and went to graduate school, I thought,* ***maybe somebody like me could have that opportunity*** *[to become an astronaut], so, I started thinking more seriously about it"*. Unquote. In 1994, Kelly graduated from the US Navy Test Pilot School, Maryland; and received a Master's degree in Aeronautical Engineering from the US Naval Postgraduate School in Monterey, California. Mark Kelly logged over 5,000 hours flying over fifty aircraft types, with over 375 aircraft carrier landings. Both brothers Mark and Scott Kelly were selected by NASA in 1996 on their first application to the Astronaut Corps.

He was pilot on STS-108 aboard *Endeavour* (December 05-17, 2001). The shuttle carried 6,000 small American flags that NASA later distributed to families of victims of the September 11 terrorist attacks in the US. The shuttle returned 1,700 pounds of equipment to Earth. Kelly was pilot on STS-121 aboard *Discovery* (July 04-17, 2006). During and after launch much attention was paid to monitoring the external fuel tank (ET) for the loss of insulation foam. The shuttle was equipped with a number of new cameras, and video was also taken from spotter planes. Close inspection of images revealed several small pieces of debris seen jetting away from the ET, but generally they were seen after the time NASA was most concerned about during launch. En-route to the ISS, the 15-metre-long Orbiter Boom sensor system (OBSS) tipped with two types of lasers and a high-resolution TV camera was used to inspect the underside of the shuttle for damage. Particular attention was paid to the leading edges of the wings.

During an on-orbit press conference, Mark Kelly suggested that the 'wake-up' songs should be made known to the crew before they are woken up by them. *"Sometimes we know ahead of time what the music is going to be and who it's for. Other times it's a complete surprise, and you just get woken up by it and have to scrabble around thinking of something profound to say about it"*. He was Commander on STS-124 aboard *Discovery* (May 31-June 14, 2008), when the main objective was to carry up the Japanese *Kibo* module and berth it to the *Harmony* module. The Japanese Experiment Pressurized Module (PM) [and the Japanese Remote Manipulator system, its' robotic arm] make up the Japanese Experiment Module (JEM), called *Kibo*, on the ISS. The PM is the largest module of the ISS. The lab's logistics module, installed temporarily on the ISS during STS-123, was attached to *Kibo*. It was urgent to bring a new toilet part to the ISS to fix the then-only toilet facility. He logged over fifty-four days in space during this mission.

In a 2011 pre-flight interview for STS-124, Mark Kelly said he still felt like he was a test pilot at NASA, even though the Space Shuttle program was decades old. *"It's still more or less in a test program even though we've been flying it for nearly 30 years now; so it's like the ultimate test job"*. His last six months of training had a major distraction, when his wife was shot in Tucson, Arizona, during an assassination attempt on January 08, 2011. According to his book, *Gabby: A story of Courage and Hope,* after her weeks of rehabilitation, he determined on February 04, 2011, that he could still proceed with the launch. STS-134 launched on May 16; returning on June 01, 2011. It delivered the Alpha Magnetic Spectrometer and an ExPRESS Logistics Carrier to the ISS. After his final mission, Kelly agreed to act as the "control" in a study to see how his identical twin Scott's body reacted to a year in space. The "twin study" was a unique opportunity to see how the genetics of the body could change when flying into space.

On June 21, 2011, Mark Kelly announced that he would leave NASA and the US Navy, effective October 01. He cited his wife Gabby's needs during her recovery as the reason to retire. In 2011, Kelly and Gabby Giffords co-authored *Gabby: A Story of Courage, Love and Resilience,* which describes the assasination attempt and her initial recovery. His second book, *Mousetronaut: Based on a (Partially) True Story,* (2012) is a children's book. In 2014, Giffords and Kelly co-authored *Enough: Our Fight to Keep America Safe From Gun Violence*. In 2015, Kelly and Martha Freeman co-wrote *Astrotwins: Project Blastoff*, a fictional story about twins. Mark Kelly is co-founder and strategic advisor of Tucson-based high-altitude balloon near-space exploration company **World View Enterprises**. On March 28, 2012, SpaceX announced that Kelly would be part of an independent safety advisory panel composed of human spaceflight safety experts.

In January 2013, in response to the Newtown shootings of 2012, Kelly and Giffords started a political action committee, **Americans for Responsible Solutions**; promoting solutions to gun violence with elected officials and the general public. They say, *"It supports the Second Amendment while promoting responsible gun ownership and keeping guns out of the hands of dangerous people like criminals, terrorists, and the mentally ill"*. In March 2013, Kelly said, *"any bill that does not include a universal background check is a mistake. It's the most common-sense thing we can do"*. On February 12, 2019, Mark Kelly announced that he would run as a Democrat in the 2020 US Senate special election in Arizona. He declined to accept campaign contributions from corporate political action committees (PACS), but did accept thousands of dollars from corporate executives and lobbyists. He won the seat on November 04, 2020. Senator Mark Kelly received eleven awards, including two Defense Superior Service Medals; the Legion of Merit; two DFCs; four Air Medals; two Navy Commendation Medals; the Navy Achievement Medal; two Southwest Asia Service Medals; two Sea Service Deployment Ribbons; and the NASA Distinguished Service Medal.

## Kelly, Scott J.. STS-103, STS-118, Soyuz TMA-01M/Expn 25/26, Soyuz 16-M/Expn 43/44.

Capt. Scott Joseph Kelly was born in February 1964 in Orange, New Jersey, USA. He is the identical twin brother of American astronaut and junior Senator Mark Kelly. Scott Kelly is a former American fighter pilot and test pilot, an engineer, retired US Navy Captain and NASA astronaut. Awarded a Naval Reserve Officers Training Corps scholarship, he received a Bachelor of Science degree in Electrical Engineering from the State University of New York Maritime College in 1987. He was then commissioned as an Ensign in the US Navy, graduating as a Naval Aviator in 1989. In September 1990, he was deployed to the North Atlantic and Persian Gulf aboard *USS Dwight D. Eisenhower*. In January 1993, Kelly was selected to attend the US Naval Test Pilot school in Maryland, graduating in June 1994. Scott Kelly received a Master of Science degree in Aviation Systems from the University of Tennessee at Knoxville, in 1996.

Selected by NASA in April 1996, he was pilot on STS-103 aboard *Discovery* (December 19-27, 1999), which upgraded the Hubble Space Telescope. He was then NASA's Director of Operations in Star City, Russia. Scott Kelly was Commander on STS-118 aboard *Endeavour* (August 08-21, 2007) when the shuttle carried about 5,000 pounds of equipment and supplies to the ISS. The crew added a truss segment, a new gyroscope and an external spare parts platform to the ISS. On October 07, 2010, Kelly launched aboard *Soyuz TMA-01M* as a Flight Engineer for 'Expedition 25', and assumed command of 'Expedition 26' once *Soyuz TMA-19* undocked on November 24. After 159 days in space, Scott Kelly landed on the Kazakhstan Steppe aboard *Soyuz TMA-01M* on March 16, 2011. After attaining the rank of Captain in the Navy, Kelly retired from active duty on June 19, 2012, with twenty-five years of naval service. He logged over 8,000 hours in over forty aircraft types, with over 250 aircraft carrier landings.

Scott Kelly and Mikhail Kornienko launched aboard *Soyuz 16-M* on March 27, 2015 for a year-long mission to the ISS, with the goal of understanding how the human body reacts and adapts to the harsh environment of space. He was Flight Engineer for 'Expeditions 43/44'. During the 346-day mission, almost 400 experiments were conducted. Kelly's most recent flight included 5,440 orbits around the Earth, with three EVAs. He returned to Earth in March 2016, and retired from NASA on April 01. He logged a total of 520 days in space, and holds the record for time in orbit by a US Astronaut. Scott Kelly is an Associate Fellow, Society of Experimental Test Pilots; and a Member of the Association of Space Explorers. He has received seventeen honors, including two Defense Superior Service Medals; the Legion of Merit; the DFC; the Southwest Asia Service Medal; two Navy Unit Commendations; the National Defense Service Medal; the Kuwait Liberation Medal; the NASA Outstanding Leadership Medal; and the FAI Korolev Diploma (1999).

## Kilrain, Susan L.. STS-83, STS-94.

Commander Susan Leigh Kilrain (nee Still) was born in October 1961 in Augusta, Georgia, USA. She is the second woman and youngest person to pilot a Space Shuttle. She received a Bachelor of Science degree from Embry-Riddle Aeronautical University in 1982; and a Master of Science degree in Aerospace Engineering from the Georgia Institute of Technology in 1985. In January 1985, she had an opportunity to speak with astronaut Dick Scobee, who advised that she should join the military as a pilot if she wanted to increase her chances of being accepted into the astronaut program. She was commissioned into the US Navy later in 1985, and designated a Naval Aviator in 1987. Still-Kilrain quickly realized that the Navy was not a woman-friendly organization, and were reluctant to accept her.

She said, *"I realized that I was a woman in a man's world, so I was going to be an outsider. My whole philosophy was not to make waves.* ***My goal was to be an astronaut****. I wanted to fit in without accepting unacceptable behaviour".*

She has flown over 3,000 hours in over thirty aircraft types, and was awarded the Defense Superior Service Medal (among ten awards). Selected on December 08, 1994, in NASA Group 15, she flew on STS-83 aboard *Columbia* (April 04-08, 1997), which delivered a collection of 33 microgravity experiments housed inside the Spacelab. She said, *"Looking down at Earth from space the first time ... it's like every dream come true. Even though you've seen pictures of Earth from space, it's not the same as being in space and looking at Earth. It was very rewarding".*

Still-Kilrain flew on STS-94 aboard *Columbia* (July 01-17, 1997) with the same crew as mission STS-83, which was cut short due to a fuel cell problem. This Microgravity Science Laboratory (MSL) flight of 15 days, 16 hours and 45 minutes carried Spacelab Module LM-1. She logged a total of 19 days, 15 hours and 56 minutes (472 hours) in space. Still-Kilrain was an astronaut representative to the Office of Legislative Affairs (Code L), NASA HQ. She retired from NASA in December 2002, went to a Navy base in Puerto Rico, and retired from the Navy with the rank of Commander in June 2005. She now works as a motivational speaker. In 2020, the newspaper ***USA Today*** named her a woman of the century. This was a way to commemorate the one hundredth anniversary of women being granted the right to vote in the USA. In 2022, she participated in Space Camps "Space 2101" in Dubai, and in 2023 at King Abdullah University of Science and Technology. She co-authored a children's book *An Unlikely Astronaut* with space historian Francis French. Susan Still-Kilrain is married to Vice Admiral Colin J. Kilrain, with four children.

## Kimbrough, Robert S.. STS-126, Soyuz MS-02/Expedition 49/50, SpaceX Crew-2.

Col. Robert "Shane" Kimbrough was born in June 1967 in Killeen, Texas, USA. He graduated with a Bachelor of Science degree in Aerospace Engineering from the US Military Academy, West Point, New York, in 1989. He played for the West Point baseball team for four years, and was selected as team Captain in his senior year. Kimbrough served as an Apache helicopter pilot in the first Gulf War, Operation Desert Storm in 1991. He later attended and graduated from Georgia Institute of Technology with a Master of Science degree in Operations Research in 1998. In 2000, he joined NASA and was assigned as a Flight Simulation Engineer on the Shuttle Training Aircraft. NASA selected Kimbrough for astronaut training in 2004. He was 'Mission Specialist 4' on STS-126 aboard *Endeavour* (November 15-30, 2008), and conducted two EVAs with a total time of 12 hours and 52 minutes. Four EVAs were devoted to servicing and repair of the Solar Alpha Rotary Joints (SARJ). In addition to lubricating both port and starboard bearings, the remaining eleven trundle bearings in the starboard SARJ were replaced. *Endeavour* carried the *Leonardo* MPLM on its fifth spaceflight; holding over 6,400 kg of supplies and equipment.

Kimbrough launched aboard *Soyuz MS-02* on October 19, 2016, for 'Expedition 49/50'. With the departure of *Soyuz MS-01* on October 28, he became commander of 'Expedition 50', which ended in April 2017; and performed four more EVAs. On December 25, 2016, the crew celebrated Christmas by floating in microgravity, and opening presents recently delivered on a Japanese cargo spacecraft. In July 2020, NASA announced that Kimbrough would fly to space for the third time as Commander of SpaceX 'Crew-2'; along with NASA's Megan McArthur, JAXA's Akihiko Hoshide, and ESA's Thomas Pesquet. Crew-2 launched and docked with the ISS on April 24, 2021, beginning their six-month mission. Kimbrough as 'Flight Engineer 3' performed three EVAs to install the iROSA solar arrays in the P6 Truss.

He conducted experiments including Earth observations, protein crystal growth and cultivating cotton and peppers. Crew-2 splashed down off the coast of Florida on November 09, 2021, after 199 days in space. He logged a total of 388 days, 17 hours and 28 minutes in space. Robert Kimbrough retired from NASA on July 31, 2022.

## Kerwin, Joseph P.. Skylab 2.

Capt. Dr. Joseph Peter Kerwin was born in February 1932 in Oak Park, Illinois, USA. He was the first physician to be selected for astronaut training, and the first doctor from the United States to enter space. He received a Bachelor of Arts degree in Philosophy, magna cum laude, from the College of the Holy Cross in 1953; and a Doctor of Medicine (M.D.) from the Feinberg School of Medicine of Northwestern University in 1957. Kerwin completed his internship at the District of Columbia General Hospital in Washington, D.C., and attended the US Navy School of Aviation Medicine at NAS Pensacola; before being designated a Naval Flight Surgeon in December 1958. Kerwin was a Captain in the Navy Medical Corps, commissioned in July 1958. He earned his Flight Surgeon's wings at Beeville, Texas, in 1962, and logged 4,500 hours flying time. While serving as a pilot and flight surgeon for the Navy, he was selected in NASA astronaut Group 4 in June 1965. Dr. Kerwin was one of the CapComs for the *Apollo 13* mission in 1970.

Kerwin served as the science pilot for the *Skylab 2* mission (May 25-June 22, 1973), the first crewed mission to an American orbital space station. *Skylab 2* was launched on an *Apollo* Command and Service Module by a *Saturn-IB* rocket. He logged 28 days, zero hours and 50 minutes in space; including EVA of three hours and 23 minutes. Joseph Kerwin was then in charge of the on-orbit branch of the Astronaut Office, where he coordinated astronaut activity involving rendezvous, satellite deployment and retrieval, and other Space Shuttle payload operations. He was part of the NBC broadcasting team for coverage of the launch of *Columbia* on mission STS-1 on April 12, 1981. From 1982-83, he served as NASA's senior science representative in Australia; liaising with Australia's Commonwealth Scientific and Industrial Research Organization. From 1984-87, Kerwin served as Director of Space and Life Sciences at JSC. He was responsible for the direction and coordination of medical support to operational crewed spacecraft programs. In 1986, he issued a report on the January 28 deaths of the crew killed in the Space Shuttle *Challenger* disaster.

Joseph Kerwin retired on March 31, 1987, and joined **Lockheed**, where from 1988-90 he managed the Extravehicular Systems Project, providing hardware for Space Station *Freedom* (later known as the ISS). He was co-inventor of the *Simplified Aid For EVA Rescue* (SAFER) jetpack. In 1994-95, he led the Houston liaison group for Lockheed Martin's FGB contract, the procurement of the Russian "space tug". Kerwin served on the NASA Advisory Council from 1990-93. In June 1996, he joined **Systems Research Laboratories** (SRL) to serve as Program Manager of the team which bid to win the Medical Support and Integration Contract at JSC. He was recruited in April 1997 by competitor **KRUG Life Sciences** to replace its retiring President. On March 16, 1998, KRUG became the Life Sciences Special Business Unit of **Wyle Laboratories** of El Segundo, California, from which he retired in Summer 2004.

Joseph Kerwin serves on the Board of directors of the National Space Biomedical Research Institute (NSBRI) as an industry representative. The Secretary of the Navy awarded the *Skylab 2* crew the Navy Distinguished Service Medal in 1973; and the three *Skylab* crews were awarded the 1973 Robert J. Collier Trophy *"For proving beyond question the value of man in future explorations of space and the production of data of benefit to all the people on Earth"*. Kerwin is a Fellow of the Aerospace Medical Association; a member of the Aircraft Owners and Pilots Association; and was one of 24 *Apollo* astronauts inducted into the US Astronaut Hall of Fame in 1997. Kerwin was portrayed by Jack Hogan in the 1974 TV movie ***Houston, We've Got a Problem***. He is co-author with astronaut Owen K. Garriott of *Homesteading Space*, published in 2008; a history of the Skylab program. He appears as himself in the 2018 documentary film ***Searching for Skylab***.

## Knight, William J.. X-15 rocketplane.

Col. William J. "Pete" Knight (1929-2004) was born in November 1929 in Noblesville, Indiana, USA. He attended Butler University and Purdue University. Knight enlisted in the USAF in 1951, and completed pilot training in 1953. Flying an F-89D for the 438th Fighter-Interceptor Squadron, he won the prestigious Allison Jet Trophy Race in September 1954. After receiving a Bachelor of Science degree in Aeronautical Engineering from the Air Force Institute of Technology in 1958, he attended the USAF Test Pilot School at Edwards AFB, then remained at Edwards as project test pilot on five test programs.

In 1960, Knight was one of six test pilots selected to fly the X-20 Dyna-Soar, designed for lifting re-entries and conventional landings. After that program was canceled in 1963, Knight completed the astronaut training curriculum at the new USAF Aerospace Research Pilot School at Edwards AFB in 1964 and was selected to fly the North American X-15 rocketplane. On June 29, 1967, he was climbing through 107,000 feet at Mach 4.17 when the plane suffered a total electrical failure and all onboard systems shutdown. Flying by "the seat of his pants" he glided down to a safe landing and earned the DFC for his remarkable feat of airmanship. On Flight 188 on October 03, 1967, he piloted the modified X-15A-2 to 31,120 metres altitude (102,100 feet, or 19.35 miles) and reached 7,274 km/hr or 4,520 mph (Mach 6.7), still the highest speed attained in an airplane. On October 17, 1967, in the X-15-3 he climbed to 280,500 feet (53.12 miles), qualifying him as an astronaut. In 1968, Knight went to Southeast Asia where he completed 253 combat sorties in the F-100 during the Vietnam War. He then served as test director for the F-15 System Program Office at Wright-Patterson AFB, Ohio. Following an assignment as director of the Fighter Attack System Program Office, he returned to Edwards AFB as vice commander of the AFFTC in 1979.

Col. Pete Knight remained an active test pilot in the F-16 combined Test Force. Knight retired from the USAF in 1982; after logging over 6,000 hours flying in over one hundred aircraft types. In 1988, he was elected Mayor of Plamdale, California. In 1992, Knight was elected to the California State Assembly, and since 1996 served as a State Senator representing California's 17th Senate District. Knight gained statewide attention in 2000 as the author of **Proposition 22**, aka the Knight Initiative, the purpose of which was to ban same-sex marriage. It passed with 61.4% approval, but was later found to be unconstitutional in ***In re Marriage Cases*** (2008). Knight had a younger brother who died of AIDS-related complications in 1995 at age 60; and a son, David Knight who is gay. Among Pete Knight's many honors are the Legion of Merit with one Oak Leaf Cluster; the DFC with two Oak Leaf Clusters; the Air Force Medal with ten Oak Leaf Clusters; the Harmon International Trophy; the Octave Chanute Award; and the Air Force Association Citation of Honor. He was inducted into the National Aviation Hall of Fame (1988); the Aerospace Walk of Honor (1990); and the International Space Hall of Fame (1998). In 2003 in Palmdale, California, the Pete Knight High School was opened. He died on May 07, 2004.

## Kopra, Timothy L.. STS-127/Expedition 20/STS-128, Soyuz TMA-19M/Expedition 46/47.

Col. Timothy "Tim" Lennart Kopra was born in April 1963 in Austin, Texas, USA. He received a Bachelor of Science degree from the US Military Academy, West Point in 1985; a Master of Science degree in Aerospace Engineering from Georgia Institute of Technology in 1995; a Master of Strategic Studies degree from the US Army War College in 2006; and a Master of Business Administration (M.B.A.) degree from the London Business School and Columbia Business School in 2013. Kopra was designated as an Army Aviator in August 1986, then completed a three-year assignment at Fort Campbell, Kentucky. In 1990, Kopra was assigned to the 3rd Armoured Division in Hanau, Germany. He was deployed to the Middle East in support of Operations Desert Shield and Desert Storm. Timothy Kopra retired from the US Army in November 2010.

Selected by NASA in July 2000, Kopra completed the initial two years of intensive space shuttle and ISS training, scientific and technical briefings, and T-38 flight training. He was 'Mission Specialist 5' on STS-127 aboard *Endeavour* (July 15-31, 2009), which launched to the ISS for 'Expedition 20' on the sixth attempt. Pieces of foam were observed falling off the External Tank during the ascent, however the shuttle received only minor scuffs to its heat shield. The primary objective was to deliver and install on the ISS the final two components of the Japanese Experiment Module. When *Endeavour* docked with the ISS, it set a record for the most humans in space at the same time in the same vehicle: thirteen. Tim Kopra served almost sixty days aboard the ISS, returning to Earth aboard *Discovery* on STS-128 on September 11, 2009.

Kopra returned to space aboard *Soyuz TMA-19M* as 'Flight Engineer 5' to be part of 'Expedition 46' (December 15, 2015, to March 2016), and Commander of 'Expedition 47' from March 02 to June 18, 2016. During an EVA, Kopra's spacesuit began to leak water into his helmet, causing the EVA to be cut short. After 186 days in space, he returned to Earth aboard *Soyuz TMA-19*. He logged a total of 244 days, one hour and one minute in space; including three EVAs totaling 13 hours and 31 minutes. In 2020, he was announced as Vice President of Robotics and Space Operations at **MDA Corporation**. Col. Kopra is a member of the Society of Experimental Test Pilots; the Army Aviation Association of America; American Helicopter Society; the US Military Academy Association of Graduates; and the West Point Society of Greater Houston. He has received honors including the Empire Test Pilot School Award; the Silver and Bronze Order of St. Michael, Army Aviation Awards (1999, 2009); the Bronze Star Medal; the Southwest Asia Service Medal; the National Defense Service Medal with one service star; two Meritorious Service Medals; the Air Medal; the Army Commendation Medal; the Army Achievement Medal; and the NASA Distinguished Service Medal.

## Kregel, Kevin R.. STS-70, STS-78, STS-87, STS-99.

Col. Kevin Richard Kregel was born in September 1956 in Amityville, New York, USA. He is a former member of the Space Launch Initiative Project at the Johnson Space Center. He received a Bachelor of Science degree in Astronautical Engineering from the USAF Academy in 1978; and a Master of Science degree in Public Administration in 1988. Kregel earned his pilot wings at Williams AFB, Arizona, in August 1979. From 1980-83, he was assigned to F-111 aircraft at RAF Lakenheath. Aboard the *USS Kitty Hawk*, he made sixty-six aircraft carrier landings during a cruise of the Western Pacific. Upon graduation from the US Naval Test Pilot School in Maryland, he conducted weapons and electronic systems testing on the F-111, F-15, and F-15E aircraft. Kregel resigned from active duty in 1990 to work for NASA. He logged over 5,000 flight hours in thirty aircraft types.

Selected in March 1992 NASA Group 14, he completed one year of training. Kregel was pilot on STS-70 aboard *Discovery* (July 13-22, 1995), when the crew deployed the sixth *TDRS* satellite. He was pilot on STS-78 aboard *Columbia* (June 20-July 07, 1996), a Life and Microgravity 'Spacelab' mission, with studies sponsored by ten nations and five space agencies. Kregel was Commander on STS-87 aboard *Columbia* (November 19-December 05, 1997), the fourth US Microgravity payload flight (USMP-4). The *SPARTAN 201-04* was a Solar Physics spacecraft, deployed on orbit 18 and retrieved on orbit 52. Thirteen experiments included the Sodium Sulfur Battery Experiment, which characterized the performance of four 40 ampere-hour sodium-sulfur battery cells, the first test of this technology in space.

Kregel was Commander on STS-99 aboard *Endeavour* (February 11-22, 2000) when the crew worked dual shifts to support payload operations. The Shuttle Radar Topography Mission mapped more than 47 million miles of the Earth's land surface. It was the most complete high-resolution digital topographic database of the Earth; gathering around eight terabytes of data. Col. Kregel logged a total of 52 days, 18 hours and 19 minutes in space, and retired on June 27, 2002. He received six honors: the Defense Meritorious Service Medal; the Air Force Meritorious Service Medal; the Air Force Commendation Medal; the Navy Commendation Medal; the NASA Exceptional Service Medal; and four NASA Space Flight Medals.

## Lawrence, Wendy B.. STS-67, STS-86, STS-91, STS-114.

Capt. Wendy Barrien Lawrence was born in July 1959 in Jacksonville, Florida, USA. She received a Bachelor of Science degree in Ocean Engineering from the US Naval Academy in 1981 (as a distinguished Flight School graduate); and a Master of Science degree in Ocean Engineering from MIT and the Woods Hole Oceanographic Institution in 1988, as part of a joint program between the two schools. Wendy Lawrence was designated a Naval Aviator in July 1982. She logged over 1,500 hours of flight time in six different helicopter types, and made over 800 shipboard landings. Lawrence was the first female graduate of the US Naval Academy to fly into space, and in September 1996 she began training as a crew member for a four-month mission to the Russian Space Station *Mir*, but was replaced by David Wolf due to concerns over minimum size requirements for the Russian *Orlan* EVA suit.

Selected by NASA in March 1992, she completed one year of training. Lawrence was 'Mission Specialist 2' on STS-67 aboard *Endeavour* (March 02-18, 1995), the second flight of the ASTRO observatory. STS-86 aboard *Atlantis* (September 25-October 06, 1997) was the seventh mission to rendezvous and dock with *Mir*, and transferred 10,400 pounds of science and logistics. STS-91 aboard *Discovery* (June 02-12, 1998) was the ninth and final *Shuttle-Mir* docking mission, and marked the conclusion of the joint US/Russian Phase 1 Program. She was 'Mission Specialist 4' on STS-114 aboard *Discovery* (July 26-August 09, 2005), which saw the first in-flight repair to the orbiter during a spacewalk. She logged a total of 51 days, three hours and 56 minutes (over 1,200 hours) in space.

Capt. Wendy Lawrence is married to Cathy Watson, a former NASA scientist. She belongs to the Association of Naval Aviation; Women Military Aviators; Naval Helicopter Association; and The Mars Generation. Wendy Lawrence was awarded the Legion of Merit; the Defense Meritorious Service Medal with two clusters; the Meritorious Service Medal with star; the National Defense Service Medal with star; the Global War on Terrorism Service Medal; and the Sea Service Deployment Ribbon with star.

## Lee, Mark C.. STS-30R, STS-47 (J), STS-64, STS-82.

Col. Mark Charles Lee was born in August 1952 in Viroqua, Wisconsin, USA. He received a Bachelor of Science degree in Civil Engineering from the USAF Academy in 1974, then spent 2.5 years at Kadena Air Base in Okinawa, Japan, in the 25th Tactical Fighter Squadron. In 1979, he studied for a Master of Science degree in Mechanical Engineering at MIT, where he specialized in graphite/epoxy advanced composite materials. In 1982, he served as Executive Officer for the 388th Tactical Fighter Wing Deputy Commander for Operations, and as flight commander in the Fourth Tactical Fighter Squadron at Hill AFB, Utah. Lee was selected by NASA in May 1984, and completed a one-year training and evaluation program.

Mark Lee was 'Mission Specialist 1' on STS-30 aboard *Atlantis* (May 04-08, 1989), when crew launched the *Magellan* probe, and made experiments involving life sciences and crystals. He was 'Payload Commander' and 'Mission Specialist 1' on STS-47 aboard *Endeavour* (September 12-20, 1992), the fiftieth space shuttle mission, which carried forty-four Japanese and US life science and materials processing experiments. His wife at that time, N. Jan Davis was 'Mission Specialist 3' on the same flight. They had met in January 1991 during training for the flight, married in secret, but disclosed their marriage to NASA shortly before their flight, when it was too late to train a substitute. They divorced in 1999. NASA since changed the rules, and will not now allow married astronauts on the same flight.

Mark Lee was 'Mission Specialist 4' on STS-64 aboard *Discovery* (September 09-20, 1994). Lee logged six hours and 51 minutes of EVA to test a self-rescue jetpack *SAFER*, undertaking the first untethered spacewalk in ten years. He was 'Mission Specialist 4' on STS-82 aboard *Discovery* (February 11-21, 1997), the second Hubble Space Telescope maintenance mission. Mark Lee logged a total of 32 days, 21 hours and 46 minutes in space. He retired from the Air Force and NASA on July 01, 2001. Among his awards are the DFC; the Legion of Merit; the Defense Superior Service Medal; the Air Force Meritorious Service Medal; the Defense Meritorious Service Medal; two Air Force Commendation Medals; the NASA Distinguished Service Medal; the NASA Outstanding Leadership Medal; and the NASA Public Service Group Achievement Award.

## Leestma, David C.. STS-41-G, STS-28R, STS-45.

Capt. David Cornell Leestma was born in May 1949 in Muskegon, Michigan, USA. He received a Bachelor of Science degree in Aeronautical Engineering from the US Naval Academy in 1971 (graduating first in his class); and a Master of Science degree in Aeronautical Engineering from the Naval Postgraduate School in 1972. He received his Naval Flight Officer wings in October 1973. Leestma made three overseas deployments to the Mediterranean and North Atlantic areas aboard the aircraft carrier *USS John F. Kennedy*. He was an operational test director with the F-14A in 1977, using new tactical software, and evaluating new avionics. He logged over 3,500 hours of flight time, including nearly 1,500 hours in the F-14A airplane.

Selected by NASA in 1980 Group 9, Leestma was 'Mission Specialist 3' on STS-41-G aboard *Challenger* (October 05-13, 1984), when crew deployed the *Earth Radiation Budget Satellite (ERBS)* satellite, operated the OSTA-3 payload, and conducted a satellite refueling demonstration. Leestma and Kathryn Sullivan made a 3.5 hour EVA to demonstrate the satellite Orbital Refueling System (ORS). He was 'Mission Specialist 2' on STS-28 aboard *Columbia* (August 08-13, 1989), which carried classified Department of Defense payloads. Leestma was 'Mission Specialist 2' on STS-45 aboard *Atlantis* (March 24-April 02, 1992), when crew operated twelve experiments that made up the ATLAS-1 cargo; contributing significantly to our understanding of Earth's climate and atmosphere. He logged a total of 22 days, four hours and 32 minutes (532.7 hours) in space; including three hours and 29 minutes of EVA.

David Leestma is an Associate Fellow, American Institute of Aeronautics and Astronautics (AIAA); and a Life Member, Association of Naval Aviation. Leestma has been awarded the DFC; the Legion of Merit; the Defense Superior Service Medal; the Defense Meritorious Service Medal; the National Defense Service Medal; the Navy Achievement Medal; four NASA Exceptional Service Medals; two NASA Outstanding Leadership Medals; and the Presidential Rank of Meritorious Executive Award (1998 & 2004).

## Leslie, Frederick W.. STS-73.

Dr. Frederick "Fred" Weldon Leslie was born in December 1951 in Ancon, Panama. He received a Bachelor of Science degree in engineering Science from the University of Texas in 1974; and a Master's degree (1977) and Doctor of Philosophy degree in Meteorology (with a minor in fluid mechanics) from the University of Oklahoma (1979). Leslie then served as a post-doctoral research associate at Purdue University studying fluid vortex dynamics. In 1980, Dr. Leslie worked for the Universities Space Research Association as a visiting scientist at the Marshall Space Flight Center. He is an instrument rated commercial pilot with over 900 hours in various aircraft.

Since 1983, he served as a co-investigator for the Geophysical Fluid Flow Cell experiment, which examines spherical rotating and convection relevant to the atmosphere of stars and planets. The experiment flew on 'Spacelab-3' and is also a part of the US Microgravity Laboratory-2 (USML-2) payload. In 1987, Dr. Leslie became Chief of the Fluid Dynamics Branch, where he directed and conducted both laboratory and theoretical research with over twelve scientists. He was a Payload Specialist on STS-73 aboard *Columbia* (October 20-November 05, 1995), a mission which focused on materials science, biotechnology, combustion science, and fluid physics. He logged 15 days, 21 hours and 34 minutes in space. Fred Leslie served on the American Institute of Aeronautics and Astronautics Fluid Dynamics Technical Committee. He is a member of the American Meteorological Society; and the US Parachute Association.

## Lichtenberg, Byron K.. STS-9, STS-45.

Lt. Col. Dr. Byron Kurt Lichtenberg was born in February 1948 in Stroudsburg, Pennsylvania, USA. He received a Bachelor of Science degree in Aerospace Engineering from Brown University in 1969; a Master of Science degree in Mechanical Engineering from MIT in 1975; a Doctor of Science degree in Biomedical Engineering from MIT in 1979; and an honorary Doctor of Science degree from Westminster College. He was a USAF fighter pilot for 23 years. Lichtenberg flew 138 combat missions during the Vietnam War, and received two DFCs, ten Air Medals, and many other decorations. He flew as a Captain for **Southwest Airlines**; and was a founder of **Payload Systems, Inc.**, a company that has provided hardware and flight support for MODE and MACE experiments for the Space Shuttle and ISS.

Selected by NASA in 1978, Lichtenberg was a researcher at MIT from 1978-84 on 'Spacelab' experiments; and a co-principal investigator for the Mental Workload and Performance experiment flown on IML-1. Byron Lichtenberg was 'Payload Specialist 2' on STS-9 aboard *Columbia* (November 28-December 08, 1983), a mission devoted entirely to 'Spacelab-1', a joint NASA/ESA program to demonstrate the ability to conduct advanced scientific research in space. STS-45 aboard *Atlantis* (March 24-April 02, 1992) carried the first Atmospheric Laboratory for Applications and Science (ATLAS-1), placed on Spacelab pallets mounted in the orbiter's payload bay. It was equipped with twelve instruments from the US, France, Germany, Belgium, Switzerland, the Netherlands and Japan. He logged a total of 19 days, five hours and 56 minutes in space. Lichtenberg is a founding member of the Association of Space Explorers; the X-Prize Foundation; and a member of the User Panel for National Space Biomedical Research Institute. He received two NASA Space Flight Medals; the AIAA Haley Space Flight Award; and the FAI's V.M. Komarov Award. He is now a professor at **LeTourneau University** in Longview, Texas.

## Lind, Don L.. STS-51-B.

Cmdr. Dr. Don Leslie Lind (1930-2022) was born in May 1930 in Midvale, Utah, USA. He received a Bachelor of Science degree in Physics (with high honors) from the University of Utah in 1953; and a Ph.D. in High-energy Nuclear Physics from the University of California-Berkeley, in 1964. Lind was a Naval Aviator and attained the rank of Commander in the US

Naval Reserve. He worked at NASA's Goddard Research Center from 1964-66. Selected in 1966 Astronaut Group 5, he helped develop the *Apollo 11* EVA activities [including the Apollo Lunar Surface Experiments Package (ALSEP)], and he served as CapCom for the *Apollo 11* and *Apollo 12* missions. During a leave of absence from NASA, Dr. Lind conducted post-doctoral research at the University of Alaska's Fairbanks Geophysical Institute from 1975-76. He logged over 4,500 hours flight time, including 4,000 in jet aircraft. He served as backup science pilot for *Skylab 3* and '4' missions.

Lind flew as the lead Mission Specialist and Payload Commander on STS-51-B aboard *Challenger* (April 29-May 06, 1985). With payload experiments consisting primarily of microgravity research and atmospheric measurement, with the help of his Alaska postdoctoral group he designed and conducted an experiment to capture Earth's aurora. NASA needed only to purchase three rolls of film for USD 36.00; Lind described it as *"the cheapest experiment that has ever gone into space"*. He logged seven days, zero hours and eight minutes (over 168 hours) in space. A space program afficionado has speculated that Lind's science-dominant assignment was a *"reward ... for sticking around so long"*, in contrast to the majority of early STS missions that were centered around routinized satellite deployments.

In 1986, Dr. Don Lind retired from NASA on the 20th anniversary of his selection in 1966. For the next nine years he served as Professor of Physics and Astronomy at Utah State University, until his retirement in 1995. He was a member of the American Geophysical Union; the American Association for the Advancement of Science; received the NASA Exceptional Service Medal (1974); and the NASA Space Flight Medal. He served as a member of the lay ecclesiastical hierarchy of the Church of Jesus Christ of Latter Day Saints, and with his wife served as public affairs missionaries in the Europe West Area of the Church. Cmdr. Don Lind died on August 30, 2022, in Logan, Utah, aged ninety-two years, and surrounded by family.

## Lindgren, Kjell N.. Soyuz TMA-17M/Expedition 44/45, SpaceX Crew 4/ Expedition 67/68.

Dr. Kjell (pronounced Chell) Norwood Lindgren was born in January 1973 in Taipei, Taiwan. He is an active American astronaut who was selected in June 2009 as one of fourteen members of NASA Astronaut Group 20. He received a Bachelor of Science degree in Biology from the USAF Academy in 1995; and a Doctorate of Medicine from the University of Colorado in 2002. He is Board Certified in Emergency Medicine. In 2006, he completed a National Library of Medicine Post-Doctoral Fellowship and Master of Health Informatics at the University of Minnesota. In 2007, he completed a Master of Public Health at the University of Texas Medical Branch (UTMB) in Galveston, Texas. As part of his Masters studies at CSU, he conducted cardiovascular countermeasure research in the Space Physiology Lab at NASA Ames Research Center, California.

When selected to the astronaut corps, he was serving as the Deputy Crew Surgeon for STS-130 and 'Expedition 24'. Lindgren launched aboard *Soyuz TMA-17M* on July 22, 2015, to the ISS as part of 'Expedition 44/45' to December 11, 2015. He logged 141 days in space, participating in two EVAs of 15 hours and four minutes in total. Lindgren was Commander of NASA's SpaceX 'Crew-4' mission to the ISS, launched on April 22, 2022, as part of 'Expedition 67/68'. He logged another 170 days in space, and returned to Earth on October 14, 2022. Dr. Kjell Lindgren has logged a total of 312 days, five hours and 11 minutes in space.

He is a Fellow of the Aerospace Medical Association; Fellow of the American Academy of Emergency Medicine; Member of the Association of Space Explorers; American Medical Informatics Association; Space Medicine Association; and the Christian Medical and Dental

Associations. Among Lindgren's eleven awards are: the US Air Force Achievement Medal (1998); Edgar and Marion Adler Scholar, University of Colorado School of Medicine (2001, 2002); the Hippocrates Award, University of Colorado School of Medicine (2002); the USAF Primary Flight Surgeon Course (2007); the NASA Exceptional Achievement Medal (2013); and the NASA Distinguished Service Medal (2016).

## Lindsey, Steven W.. STS-87, STS-95, STS-104, STS-121, STS-133.

Steven Wayne Lindsey was born in August 1960 in Arcadia, California, USA. He received a Bachelor of Science degreee in Engineering Sciences from the USAF Academy in 1982; and a Master of Science degree in Aeronautical Engineering from the Air Force Institute of Technology in 1990. After receiving his pilot wings in 1983, he was assigned to the 12th Tactical Reconnaissance Squadron at Bergstrom AFB, Texas. From 1984-87, he served as a combat-ready pilot, instructor pilot, and academic instructor at Bergstrom. In 1989, Lindsey attended the USAF Test Pilot School at Edwards AFB. In August 1993, he was selected to attend the Air Command and Staff College at Maxwell AFB, Alabama. Lindsey has logged over 7,000 hours flying time in over fifty aircraft types.

In March 1995, he was assigned to NASA as an astronaut candidate. Steven Lindsey was pilot on STS-87 aboard *Columbia* (November 19-December 05, 1997), which carried experiments designed to study how the weightless environment affects various physical processes, and also made observations of the Sun's outer atmospheric layers. Lindsey piloted the first flight of the *AERCam Sprint*, a free-flying robotic camera. He was pilot on STS-95 aboard *Discovery* (October 29-November 07, 1998), which deployed several research payloads; including deployment and retrieval of the *SPARTAN* solar-observing spacecraft, and the Hubble Space Telescope Orbital Systems Test Platform. Aboard was 77-year-old US Senator John H. Glenn, on his second spaceflight. He remains the oldest person to reach Earth orbit. The mission inaugurated ASTC HDTV broadcasting in the US, with live coast-to-coast coverage of the launch.

Lindsey then served as Deputy for Shuttle Operations and Co-Chairman of the Space Shuttle Cockpit Council; responsible for designing, testing, and implementing crew interfaces and displays for the USD 400 million Shuttle Cockpit Avionics Upgrade. He was Commander on STS-104 aboard *Atlantis* (July 12-24, 2001), a mission which installed the *Quest* airlock to the ISS, and outfitted it with four high-pressure gas tanks. STS-121 aboard *Discovery* (July 04-17, 2006), ISS Assembly Mission ULF 1.1 had the main purposes to test new safety and repair techniques following the February 2003 *Columbia* disaster. Lindsey was Commander on STS-133 aboard *Discovery* (February 24-March 09, 2011), which transported the Permanent Multi-Purpose Module *Leonardo*, and left it docked to the ISS to add much-needed storage space. Other cargo was the third of four ExPRESS Logistics Carriers, as well as a humanoid robot called *Robonaut*. He logged a total of 62 days, 22 hours and 33 minutes (over 1,500 hours) in space.

Steven Lindsey served as Chief of the Astronaut Office from September 2006 until October 2009. He has received thirteen awards, including Distinguished Graduate of the Air Force Undergraduate Pilot Training (1983); the Liethen-Tittle Award for outstanding test pilot (1989); the DFC; the Defense Superior Service Medal; the Defense Meritorious Service Medal; the NASA Exceptional Service Medal; the NASA Outstanding Leadership Medal; the Air Force Achievement Medal; the Air Force Commendation Medal; and the Air Force Distinguished Service Medal. He is a member of the Society of Experimental Test Pilots; the USAF Academy Association of Graduates; and the Association of Space Explorers.

## Linenger, Jerry M.. STS-64, STS-81.

Capt. Dr. Jerry Michael Linenger was born in January 1955 in East Detroit, Michigan, USA. He received a Bachelor of Science degree in Bioscience from the US Naval Academy in 1977; a Doctorate in Medicine from Wayne State University School of Medicine in 1981; a Master of Science degree in Systems Management from the University of Southern California in 1988; a Master of Public Health degree in Health Policy from the University of North Carolina-Chapel Hill in 1989; and a Doctor of Philosophy degree in Epidemiology from the University of North Carolina in 1989. Selected in NASA Group 14 in August 1992, Linenger flew as 'Mission Specialist 1' on STS-64 aboard *Discovery* (September 09-20, 1994), when mission highlights included the first flight of the *Lidar In-space Technology Experiment* (*LITE*), using lasers for environmental research; deployment and retrieval of a solar science satellite; robotic processing of semiconductors; and the first untethered spacewalk in ten years (since STS-51-A in 1984), to test a self-rescue jetpack called *Simplified Aid For EVA Rescue* (*SAFER*). STS-64 was the first mission to use the new full-pressure Advanced Crew Escape Suit, which eventually replaced the partial-pressure Launch Entry Suit.

In January 1995, Linenger began training in the Russian language at the Gagarin Training Center for a long-duration stay aboard Space Station *Mir*. He also trained as chief scientist to conduct the entire US science program, consisting of over 100 planned experiments in various disciplines. Dr. Linenger launched aboard *Atlantis* on STS-81 on January 12, 1997; remained on *Mir* with two cosmonauts, and returned on STS-84 aboard *Atlantis* on May 24, 1997. He spent an American male mission record-setting 132 days, four hours and one minute in space. He and his crewmates endured the worst on-orbit fire in the history of human spaceflight, and multiple systems failures aboard *Mir*, but all mission goals were accomplished; including EVA, flyaround, and all planned US science experiments. He logged a total of 143 days, two hours and 50 minutes in space. Dr. Linenger retired from NASA and the US Navy in January 1998.

Jerry Linenger received ten awards, including the Navy Commendation Medal with gold star; the Navy Unit Commendation; the Navy Battle Effectiveness Award; two National Defense Service Medals; and the Distinguished Alumni Award from Wayne State University School of Medicine. He is a member of the Association of Naval Aviation; the US Navy Flight Surgeons Association; the Aerospace Medicine Association; the American Medical Association; the American College of Preventative Medicine; the Society of US Navy Preventative Medicine Officers; and the American College of Sports Medicine.

## Linnehan, Richard M.. STS-78, STS-90, STS-109, STS-123.

Capt. Dr. Richard Michael Linnehan was born in September 1957 in Lowell, Massachusetts, USA, and was raised by his paternal grandparents. He received a Bachelor of Science degree in Animal Sciences from the University of New Hampshire in 1980; and a Doctor of Veterinary Medicine from the Ohio State University College of Veterinary Medicine in 1985. In 2009, Dr. Linnehan received the Master of Public Administration degree from Harvard Kennedy School at Harvard University. Selected in March 1992 NASA Group 14, he completed one year of astronaut training, qualifying for Space Shuttle flights as a Mission Specialist.

Richard Linnehan flew as 'Mission Specialist 1' on STS-78 aboard *Columbia* (June 20-July 07, 1996), the Life Sciences and Microgravity Spacelab (LMS) mission. Dr. Linnehan flew as 'Mission Specialist 1' on STS-90 Neurolab aboard *Columbia* (April 17-May 03, 1998), his second 'Spacelab' mission. It was a joint venture of six space agencies and seven US research agencies, and thirty-one microgravity studies were completed. He was 'Mission Specialist 3' on STS-109 aboard *Columbia* (March 01-12, 2002), Hubble Space Telescope

Servicing Mission 3B. Dr. Linnehan was 'Mission Specialist 3' on STS-123 aboard *Endeavour* (March 11-27, 2008), which delivered the Japanese Logistics Module and the Canadarm robotic *Dextre* SPDM to the ISS. Linnehan performed three of the five scheduled EVAs. He logged a total of 59 days, 20 hours and 49 minutes in space; including six EVAs totaling 42 hours and 11 minutes.

As of 2020, Linnehan was a management astronaut; splitting his time between the Astronaut Office Exploration and Integration branches and the NASA Institutional Review Board, JSC Institutional, and NASA Flight, Animal Care and Use Committees (IACUCs). Dr. Linnehan belongs to the American Association of Zoo Veterinarians; the International Association of Aquatic Animal Medicine; the Association of Space Explorers; and is Adjunct Professor at the North Carolina State University College of Veterinary Medicine in Raleigh, North Carolina. He received ten awards, including the Navy Group Achievement Award (1991); the American Veterinary Medical Association's President's Award; The University of New Hampshire Distinguished Alumni Award (1999); the NASA Outstanding Leadership Award (1999); the NASA Exceptional Service Award (2003); the NASA Distinguished Service Medal (2009); and two honorary Ph.D's. He is single.

## Linteris, Gregory T.. STS-83, STS-94.

Dr. Gregory "Greg" Thomas Linteris was born in October 1957 in Englewood, New Jersey, USA. He received a Bachelor of Science degree in Chemical Engineering from Princeton University in 1979; a Master of Science degree in Mechanical Engineering from Stanford University in 1984; and a Ph.D. in Mechanical and Aerospace Engineering from Princeton University in 1990. He conducts research to understand the detailed mechanisms of chemically acting fire suppressants. Linteris is an instructor in the part-time engineering program at the John Hopkins University, where he teaches a graduate course in combustion.

Selected in February 1996, he was 'Payload Specialist 2' on STS-83 aboard *Columbia* (April 04-08, 1997), the planned 15-day, 16-hour Microgravity Science Laboratory (MSL-1) mission that was aborted to three days and 23 hours, due to a technical problem with the electrical power generating Fuel Cell #2 (of three). Despite troubleshooting , the anomaly appeared to worsen, and Linteris later described the mission as *"an exercise in crisis management"*. The same crew flew the reflight (STS-94) three months later. He was 'Payload Specialist 2' on STS-94 aboard *Columbia* (July 01-17, 1997), a mission which focused on materials and combustion science research in microgravity. Dr. Greg Linteris logged a total of 19 days, 15 hours and 56 minutes (over 471 hours) in space. He retired from NASA on July 17, 1997. Dr. Greg Linteris is a member of the American Institute of Aeronautics and Astronautics; the American Physical Society; and the Combustion Institute. He has authored or co-authored fifty refereed and eighty-three un-refereed scientific publications in the areas of combustion, chemical kinetics, spectroscopy and heat transfer. Dr. Linteris is currently a staff scientist in the Fire Research Division of the Engineering Laboratory at the **National Institute of Standards and Technology** in Gaithersburg, Maryland. He is single.

## Lockhart, Paul S.. STS-111, STS-113.

Col. Paul Scott "Paco" Lockhart was born in April 1956 in Amarillo, Texas, USA. He received a Bachelor of Arts degree in Mathematics from Texas Tech University in 1978; a Master of Science degree in Aerospace Engineering from the University of Texas at Austin in 1981, then was commissioned into the US Air Force. He also studied at the University of Innsbruck and the University of Vienna Summer School from 1978-79 on a Rotarian Fellowship. Lockhart is a distinguished graduate of both ROTC and the Air Force Squadron Officer School. He has logged over 5,000 hours flying time in over thirty aircraft types.

A test pilot for the F-16 aircraft, he was selected in 1996 NASA Group 16, and was pilot on STS-111 aboard *Endeavour* (June 05-19, 2002) to the ISS, a resupply flight which brought up the 'Expedition 5' crew. Col. Lockhart was Pilot on STS-113 aboard *Endeavour* (November 22-December 07, 2002), an ISS assembly mission to deliver the P1 truss segment, which provides structural support for the Space Station radiators. He logged a total of 27 days, 15 hours and 23 minutes in space. After his service with NASA, Lockhart was assigned to and graduated in 2004 from the Royal College of Defense Studies in London, UK. He retired from the USAF in January 2007, and returned to NASA in an administrative position. Col. Lockhart belongs to the Society of Experimental Test Pilots; and the Order of Daedalians (Fraternal Order of Military Pilots). He has received seven awards, including the Defense Superior Service Medal; the Defense Meritorious Service Medal; the Outstanding Unit Award with Valor device; and the National Defense Service Medal.

## Lopez-Alegria, Michael E.. STS-73, STS-92, STS-113, Soyuz TMA-9/Expn 14, Axiom-1, Axiom-3.

Capt. Michael Lopez-Alegria was born Miguel Eladio Lopez Alegria in May 1958 in Madrid, Spain. He is a commercial astronaut with dual nationality (American and Spanish). He earned a Bachelor of Science degree in Systems Engineering from the US Naval Academy in 1980; and a Master of Science degree in Aeronautical Engineering from the US Naval Postgraduate School in 1988. Designated a Naval Aviator in 1981, he was the first EP-3 pilot to be selected for US Naval Test Pilot school. He is a graduate of Harvard Kennedy School's Program for Senior Executives in National and International Security.

Selected in 1992 NASA Group 14, he was 'Mission Specialist 2' on STS-73 aboard *Columbia* (October 20-November 05, 1995), the second US Microgravity Laboratory (USML-2) 'Spacelab' mission. Lopez-Alegria was 'Mission Specialist 4' on STS-92 aboard *Discovery* (October 11-24, 2000), the one hundredth mission of the Space Shuttle, which brought the Z1 truss (the first exterior framework installed, that allowed the first US solar arrays to be temporarily installed on *Unity* for early power during Flight 4A), Control Moment Gyros, Pressurized Mating Adapter-3, and two DDCU heat pipes to the ISS. In October 2001, he served as an aquanaut aboard the first NASA Extreme Environment Mission Operations (NEEMO) crew aboard the *Aquarius* underwater laboratory in Key Largo, offshore from Florida. He was 'Mission Specialist 1' on STS-113 aboard *Endeavour* (November 22-December 07, 2002), an ISS assembly mission which delivered the P1 truss segment. Capt. Lopez-Alegria and John Herrington performed three EVAs to activate and outfit the P1 truss. This mission delivered the 'Expedition 6' crew to the ISS; and returned the 'Expedition 5' crew after 185 days in space.

Launched from Baikonur, Kazakhstan on September 18, 2006, aboard *Soyuz TMA-9*, he docked with the ISS as Commander of 'Expedition 14', and performed five spacewalks before returning to Earth on April 21, 2007. On a privately funded mission, Lopez-Alegria commanded *Axiom-1* which launched on April 08, 2022, in a SpaceX *Crew Dragon* spacecraft from **Axion Space**'s Mission Control Center in Houston, Texas; and he spent 17 days, one hour and 48 minutes in Earth's orbit. The other three seats were reserved for space tourists paying USD 55 million each. On his sixth spaceflight, Lopez-Alegria was Commander on *Axiom 3* aka *Ax-3*, which launched on January 18, 2024, on a *Falcon 9* booster. The flight lasted for 21 days (including 18 days, three hours and 38 minutes docked with the *Harmony* forward port of the ISS), then splashed down in the Atlantic Ocean on February 09. It was privately operated by **Axiom Space** and used a SpaceX *Crew Dragon* space craft. He holds the all-time American record for ten EVAs, presently holding the record for the second longest EVA duration record (67 hours and 18 minutes: first among NASA astronauts) and having the fifth-longest spaceflight of any American at 215 days; on board the ISS from September 18, 2006, to April 21, 2007. Lopez-Alegria has logged a total of 296 days, 16 hours and 16 minutes in space.

## Lounge, John M.. STS-51-I, STS-26R, STS-35.

John Michael "Mike" Lounge (1946-2011) was born in June 1946 in Denver, Colorado, USA. He received a Bachelor of Science degree from the US Naval Academy in 1969; and a Master of Science degree in Astrophysics from the University of Colorado at Boulder in 1970. Lounge served in various assignments in the US Navy, including a nine-month Southeast Asia cruise with VF-142 aboard *USS Enterprise*, flying ninety-nine combat missions in Vietnam, and a seven-month Mediterranean cruise aboard *USS America*. In 1974, Mike Lounge returned to the US Naval Academy as an instructor in the Physics department. In 1978, he resigned his regular navy commission and joined the Naval Air Reserve. Later, he transferred to the Navy Space Project Office in Washington, D.C., where he contributed to space-related projects.

Selected by NASA in May 1980 Group 9, he served as a member of the launch support team at KSC for several missions, including STS-1, STS-2 and STS-3. Mike Lounge flew as 'Mission Specialist 2' on STS-51-I aboard *Discovery* (August 27-September 03, 1985), which deployed the *AUSSAT-1* comms satellite, the Navy's *SYNCOM IV-4* and American Satellite Company's *ASC-1* satellite. The crew also performed a successful on-orbit rendezvous and repair of the 6,990 kg *SYNCOM IV-3* satellite. He was 'Mission Specialist 1' on STS-26R aboard *Discovery* (September 29-October 03, 1988), which deployed the *Tracking and Data Relay Satellite* (TDRS-C), which was then carried to orbit by the Inertial Upper Stage (IUS) rocket. Lounge was Flight Engineer on STS-35 aboard *Columbia* (December 02-10, 1990), a mission dedicated to astronomy. Observations of the Universe were collected by the ASTRO-1 ultraviolet telescope and by the Broad Band X-ray Telescope. He logged a total of 20 days, two hours and 23 minutes (over 482 hours) in space.

Mike Lounge served as Chief of the Space Station Support Office from 1989, and retired on June 20, 1991. He was an asociate fellow of the American Institute of Aeronautics and Astronautics. Mike Lounge received six different awards, including six Navy Air Medals; three Navy Commendation Medals with Combat Valor device; the Johnson Space Center Superior Achievement Award (for service as a member of the *Skylab* Re-entry Team); and three NASA Exceptional Service Medals. Part of Eighteenth Street in his hometown of Burlington, Colorado was renamed Mike Lounge Drive in his honor. Michael Lounge died of liver cancer on March 01, 2011, in Houston, Texas, aged sixty-four years. He was married, with three children.

## Lousma, Jack R.. Skylab 3, STS-3.

Jack Robert Lousma was born in February 1936 in Grand Rapids, Michigan, USA. He received a Bachelor of Science degree in Aeronautical Engineering from the University of Michigan in 1959; and a Master of Science degree in Aeronautical Engineering from the US Naval Postgraduate School in 1965. He received his aviator wings in 1960 and was assigned to VMA-224, Second Marine Aircraft Wing as an attack pilot, and later served at Marine Corps Air Station Iwakuni, Japan. He was a reconnaissance pilot with VMCJ-2 in North Carolina, before being assigned to the Manned Spacecraft Center in Houston, Texas. Lousma logged over 7,000 hours flying time; including 700 hours in general aviation aircraft, 1,619 hours in space, 4,500 hours in jet aircraft and 240 hours in helicopters.

Lousma was one of 19 candidates selected in April 1966 NASA Astronaut Group 5, was pilot of the *Skylab 3* second crew (July 28-September 25, 1973), and made two EVAs. The crew repaired nine major experiment or operational equipment items, and installed a two-pole sunshade for thermal control. They devoted 305 man-hours to extensive solar observations from above the atmosphere, and acquired 16,000 photographs and 29 km of magnetic tape documenting Earth resources observations. Lousma served as backup docking module pilot for the US flight crew for the *Apollo-Soyuz Test Project* (ASTP) mission of July 1975. He may have been considered as a Lunar Module Pilot for the *Apollo 20* mission, which was canceled.

Jack Lousma was Commander on STS-3 aboard *Columbia* (March 22-30, 1982), when objectives included exposing the orbiter to extremes in thermal stress, and the first use of the 15 meter (49 foot) Remote Manipulator System (RMS) to grapple and maneuver a payload in space. After a one-day delay due to bad weather, they landed at White Sands, New Mexico, the only Shuttle flight to land there. Lousma logged a total of 67 days, 11 hours and 13 minutes in space; including two EVAs totaling 11 hours and one minute. He retired from NASA on October 01, 1983. As a Republican, Lousma lost the 1984 US Senate election in Michigan with 47% of the vote. He had survived a bitter primary fight against former Republican congressman Jim Dunn to capture the nomination with 63% of the vote, but was hurt late in the campaign when video surfaced of him telling that his family owned a Japanese-made Toyota car.

The three *Skylab* crews were awarded the 1973 Robert J. Collier Trophy, and Gerald P. Carr accepted the 1975 Dr. Robert H. Goddard Memorial Trophy from President Ford (awarded to all the *Skylab* astronauts). Lousma was awarded thirteen honors, including the Johnson Space Center Certificate of Commendation (1970); the NASA Distinguished Service Medal (1973); the Navy Distinguished Service Medal; the City of Chicago Gold Medal (1974); the FAI V.M. Komarov Diploma for 1973; the AIAA Octave Chanute Award for 1975; and the Department of Defense Distinguished Service Medal (1982). He received four honorary Doctorates, and was inducted into four Halls of Fame. Lousma Drive in Wyoming, Michigan, is named after him (off Roger B. Chaffee Memorial Boulevard). Jack Lousma is married, with four children.

## Love, Stanley G.. STS-122.

Dr. Stanley Glen Love was born in June 1965 in San Diego, California, USA. Love worked at the University of Oregon as a computer programming instructor during the summer of 1984, and as an assistant in physics and chemistry laboratories from 1985-87. He received a Bachelor of Science degree in Physics from Harvey Mudd College, California; a Master's degree (1989) and a Ph.D. in Astronomy from the University of Washington in 1993. His thesis was titled *"The Source of Interplanetary Dust"*. He moved to the University of Hawaii in 1994 for postdoctoral research. In 1995, he was awarded a postdoctoral prize from the California Institute of Technology, and in 1997 he went to work as a staff engineer at JPL.

Selected in 1998 NASA Group 17, he served as a CapCom for ISS Expeditions 1 through 7, and for STS-104, STS-108 and STS-132. Love was 'Mission Specialist 4' on STS-122 aboard *Atlantis* (February 07-20, 2008), which carried the ESA's *Columbus* Laboratory and installed it to the ISS. Love made two EVAs to help prepare the Laboratory for installation, to add two science payloads to the outside of *Columbus*, and to carry a failed ISS gyroscope to the Shuttle for return to Earth. He logged 12 days, 18 hours and 22 minutes (over 306 hours) in space. Dr. Love took part in the October 2011 NEEMO 15 undersea exploration mission from the DeepWorker 2000 submersible; an underwater stand-in for the Space Exploration Vehicle, which might someday be used to explore the surface of an asteroid.

As of May 2019, Stanley Love was a management astronaut, working as the Deputy Chief of the Astronauts Office's Rapid Prototyping Laboratory; developing cockpit displays and controls for *Orion* and other forthcoming spacecraft. He belongs to the American Astronomical Society; the Division for Planetary Science; the American Geophysical Union; is an Associate Fellow of the American Institute for Aeronautics and Astronautics; a member of the Meteorological Society; and of the Harvey Mudd College Alumni Association. He is married, with two children.

## Lovell, James A., Jr.. Gemini VII, Gemini XII, Apollo 8, Apollo 13.

Capt. James "Jim" Arthur Lovell, Jr. was born in March 1928 in Cleveland, Ohio, USA. He is a mechanical engineer who graduated from the US Naval Academy, Annapolis, Maryland, in 1952. From October 1952 to February 1954, he had flight training at NAS Pensacola, then was assigned to VC-3 at Moffett Field, near San Francisco. Lovell deployed in the Western Pacific aboard *USS Shangri-La*, completing 107 aircraft carrier deck landings. In January 1958, he entered a six-month test pilot training course at NAS Patuxent River, Maryland, and graduated at the top of Class 20. In 1960, he became the Navy's McDonnell Douglas F-4 Phantom II program manager. Lovell completed Aviation Safety School at the University of Southern California. He was serving as a flight instructor and safety officer when selected in 1963 by NASA for the crewed space program.

The astronaut group became known as the New Nine, and the task of supervising their training fell to Gus Grissom. Each astronaut was given four months of classroom instruction on spacecraft propulsion, orbital mechanics, astronomy, computing and space medicine. There was familiarization with the *Gemini* spacecraft, the *Titan II* and *Atlas* boosters and the *Agena* target vehicle. Classes were for six hours per day, two days per week, with each astronaut assigned a special area to develop expertise that could be shared with the other candidates. Lovell became responsible for spacecraft recovery systems. Jim Lovell flew on *Gemini VII*, launched on December 04, 1965, with Commander Frank Borman for fourteen days. The crew made rendezvous with *Gemini VI-A*, which was launched eleven days later. The crews were awarded the Harmon International Trophy for 1966. Lovell flew with Buzz Aldrin on *Gemini XII*, launched on November 11, 1966, in orbit for four days. Aldrin and Lovell also received the Harmon International Trophy in 1967.

Capt. James Lovell was Command Module Pilot of *Apollo 8* (launched December 21, 1968), when he, Frank Borman and William Anders became the first three astronauts to fly to and orbit the Moon. They made ten orbits in 20 hours and ten minutes. On Christmas Eve, the crew broadcast black-and-white television pictures of the lunar surface back to Earth, and took turns reading the Biblical creation story in the ***Book of Genesis***. The crew splashed down after 147 hours (nearly six days) of flight; just 4.8 km from the aircraft carrier recovery ship *USS Yorktown*. Mount Marilyn on the lunar surface was named for Lovell's wife. Lovell was the first person to fly in space four times, and the first to fly to the Moon twice. Capt. Jim Lovell was Commander of *Apollo 13*, with Fred W. Haise and John "Jack" L. Swigert, Jr., launched on April 11, 1970. At the time of the mission Lovell had logged more time in space than anyone in the world.

Some 330,000 km (205,000 miles) from Earth, an explosion ruptured an oxygen tank in the *Apollo 13* Service Module. No one was aware that the tank contained a wrong voltage-rated thermostat switch, and turning on the heaters to purge the oxygen on the launch pad had removed teflon insulation from the copper wiring. The resulting shortage of power and oxygen forced the abandonment of the Moon mission, and in the emergency the crew changed course to loop once around the Moon and return to Earth; returning safely on April 17. Capt. Jim Lovell logged a total of 29 days, 19 hours and five minutes (over 700 hours) in space, and in 1971 became a deputy director of the Johnson Space Center (JSC) in Houston.

Lovell retired from NASA and the Navy on March 01, 1973, and remained in Houston as a corporation executive until 1991. He worked at the **Bay-Houston Towing** company, and became president of **Fisk Telephone Systems** in 1977, and later worked for **Centel Corporation** in Chicago. He later moved to Illinois and in 1999 opened a successful restaurant in Lake Forest. Capt. Jim Lovell received seven military, federal service and foreign awards, including the Navy Distinguished Service Medal with Gold Star; the DFC with Gold Star; the NASA Exceptional Service Medal with star; the Congressional Space Medal of

Honor; and the Presidential Medal of Freedom. He received twelve other awards, including the Henry H. Arnold Trophy (1969); the Institute of Navigation Award (1969); FAI De la Vaulx Medal and Gold Space Medals (1971); and the Space Foundation's General James E. Hill Lifetime Space Achievement Award (2003). Lovell co-authored the 1994 book *Lost Moon*, on which the 1995 film ***Apollo 13*** was based. He is married, with four children.

## Low, G. David. STS-32, STS-43, STS-57.

George "David" Low (1956-2008) was born in February 1956 in Cleveland, Ohio, USA. He received a Bachelor of Science degree in Physics from Washington and Lee University in 1978; a Bachelor of Science degree in Mechanical Engineering from Cornell University in 1980; and a Master of Science degree in Aeronautics and Astronautics from Stanford University in 1983. In the early 1980s, Low worked at the Jet Propulsion Laboratory (JPL), which is part of the California Institute of Technology. He was involved with preliminary planning for planetary missions and systems engineering design for the *Galileo* spacecraft. Selected by NASA in 1984, he was 'Mission Specialist 3' on STS-32 aboard *Columbia* (January 09-20, 1990), which deployed the *Syncom IV-F5* communications satellite, and retrieved the 21,400 pound Long Duration Exposure Facility (LDEF) using the Canadarm robotic arm. David Low was 'Mission Specialist 2' and Flight Engineer on STS-43 aboard *Atlantis* (August 02-11, 1991), which deployed the *TDRS-E* satellite, and conducted thirty-two physical, material, and life science experiments.

Tracking and Data Relay Satellites are used by NASA and other US government agencies for communications to and from independent User Platforms; such as satellites, balloons, aircraft, the ISS and remote scientific bases. They are designed to function in geosynchronous orbit, 35,786 km (22,236 miles) above the surface of the Earth, and are managed by NASA's Goddard Space Flight Center. David Low was 'Mission Specialist 1' and Payload Commander on STS-57 aboard *Endeavour* (June 21-July 01, 1993) with the primary objective to retrieve the *European Retrievable Carrier* satellite (*EURECA*) using the Canadarm. This mission featured the first flight of 'SpaceHab', a commercially provided module for the conduct of microgravity experiments. Astronauts Low and Peter Wisoff made an EVA of five hours and 20 minutes to evaluate various EVA tools and techniques for use on future missions. He logged a total of 29 days, 18 hours and five minutes (over 714 hours) in space.

David Low left NASA in 1996 to pursue a career in the private sector with **Orbital Sciences Corporation**'s Launch Systems Group in Dulles, Virginia. He received three NASA Space Flight Medals; the NASA Outstanding Leadership Medal; the NASA Exceptional Service Medal; and an honorary Doctorate of Engineering from Renssaeler Polytechnic Institute. David Low was an Associate Fellow of the American Institute of Aeronautics and Astronautics. He was married, with three children, and died on March 15, 2008. In his memory, the Cygnus Orb-D1 spacecraft was named the *S.S. G. David Low*.

## Lu, Edward T.. STS-84, STS-106, Soyuz TMA-2/Expedition 7.

Dr. Edward "Ed" Tsang Lu was born in July 1963 in Springfield, Massachusetts, USA. He received a Bachelor of Science degree in Electrical Engineering from Cornell University where he lettered in wrestling; a Master of Science degree and a Ph.D. in Applied Physics from Stanford University in 1989. Dr. Lu became a specialist in solar physics and did postdoctoral work at the Institute for Astronomy in Honolulu, Hawaii, before being selected on December 08, 1994, in NASA Group 15. He was 'Mission Specialist 3' on STS-84 aboard *Atlantis* (May 15-24, 1997) for the sixth *Shuttle-Mir* docking mission when 3,318 kg of water and logistics were transferred, including an Elektron oxygen-generating unit. Dr. Lu was

'Mission Specialist 1' on STS-106 aboard *Atlantis* (September 08-20, 2000), which utilized the 'SpaceHab' Double Module and the Integrated Cargo Carrier to bring supplies to the ISS. Lu made an EVA of six hours and 14 minutes to route and connect nine cables between the *Zvezda* and *Zarya* modules, as well as install the six-foot-long magnetometer.

Ed Lu flew aboard *Soyuz TMA-2* as Flight Engineer for ISS 'Expedition 7' (April 26-October 28, 2003) for nearly 185 days. While on orbit, Lu communicated with Yang Liwei, then flying aboard *Shenzhou 5*. He congratulated the Chinese for the achievement and said in Putonghua, *"Welcome to space. Have a safe journey"*. Dr. Lu logged a total of 205 days, 23 hours and 18 minutes in space. In 2002, he co-founded the **B612 Foundation**, dedicated to protecting the Earth from asteroid strikes. As of 2020, he is the Executive Director. Dr. Ed Lu retired from NASA in August 2007 to become the program manager of **Google**'s Advanced Projects Team. In September 2011, he joined **Liquid Robotics** as Chief of Innovative Applications where his work includes outreach to promote new applications for ocean science. In 2012, he joined **Hover, Inc.** as its Chief Technology Officer. He is married, with two children.

## Lucid, Shannon M. Wells. STS-51-G, STS-34, STS-43, STS-58, STS-76.

Dr. Shannon Matilda Wells Lucid was born in January 1943 in Shanghai, People's Republic of China. She is a biochemist who made a prolonged 188-day mission aboard the space station *Mir* in 1996; the only American woman to have stayed on *Mir*. As a youngster, Shannon was fascinated with American frontier history and its pioneering spirit. A biography of rocket pioneer Dr. Robert Goddard inspired her to be a space explorer. *"People thought I was crazy, because that was long before America had a space program"*, she said. Shannon Wells earned her pilot's license and purchased an old Piper Clipper airplane, in which she flew her father to church revival meetings. *"The Baptists wouldn't let women preach"*, she has said, *"so I had to become an astronaut to get closer to God than my father"*. Shannon Wells earned a Bachelor of Science degree in Chemistry from the University of Oklahoma in 1963, followed by a Ph.D. in Biochemistry in 1973. Shannon Wells married Michael Lucid, a chemist, in 1968, and they have two daughters and one son. In 1978, she was a member of NASA's first class of female astronauts.

Her first space mission was on STS-51-G aboard the Space Shuttle *Discovery* in 1985, with three more missions before her record-breaking mission aboard the *Mir* station. In addition to maintenance and organizational duties aboard *Mir*, Lucid carried out scientific experiments which involved documenting how a candle burns in space, how protein crystals grow, and how quail embryos developed in zero gravity. When the first space shuttle scheduled to bring her back to Earth was plagued with safety problems and a second flight was stalled by Hurrican Fran, Shannon was unflappable, although she did issue an urgent request for a fresh stock of M&Ms and more books. An avid reader, Shannon read close to fifty books while in orbit. Upon landing at the Kennedy Space Center in Florida, Dr. Lucid surprised everyone by walking off the aircraft. Admitting to feeling a bit *"wobbly and woozy"*, she nonetheless made it twenty-five feet to the transporter.

From 1996 to 2007, Shannon Wells Lucid held the record for the longest duration spent in space by an American and by a woman. She orbited the Earth more than 3,000 times, covering 75 million miles while aboard the *Mir* station. She holds the worldwide space record for the most flight hours for a woman astronaut, and the most flight hours in orbit by any non-Russian. Colonel Richard M. Mullane, a veteran of three space shuttle missions said, *"The 'Right Stuff' nowadays is being able to thrive in situations a lot of us couldn't have imagined – like six months on a space station with two Russians"*. Shannon Wells Lucid flew on STS-51-G aboard *Discovery* in June 1985; on STS-34 aboard *Atlantis* in October

1989; on STS-43 aboard *Atlantis* in August 1991; on STS-58 aboard *Columbia* in October 1993; and finally on mission STS-76 aboard *Atlantis*, which launched on March 22, 1996, taking her to *Mir*, while the return flight of shuttle *Atlantis* mission STS-79 on September 26, 1996, ended her final of five spaceflights, a joint female US record. Her accumulated time in space is 223 days, two hours and 50 minutes. Dr. Shannon Wells Lucid was awarded the Congressional Space Medal of Honor in December 1996, making her the tenth person and the first woman accorded the honor.

## Magnus, Sandra H.. STS-112, STS-126, STS-135.

Dr. Sandra Hall Magnus was born in October 1964 in Belleville, Illinois, USA. She earned a Bachelor of Science degree in Physics and Electrical engineering from University of Missouri-Rolla (now known as the Missouri University of Science and Technology); and a Master of Science degree and Ph.D. in Materials Science and Engineering from the Georgia Institute of Technology in 1996. During the 1980s, Dr. Magnus worked on stealth aircraft design as an engineer for **McDonnell Douglas**, and contributed to the propulsion system for the A-12 Avenger II (canceled by the US Navy in 1991).

Selected in May 1996 NASA Group 16, in 2000 she was CapCom for the International Space Station (ISS). Magnus served as 'Mission Specialist 2' on STS-112 aboard *Atlantis* (October 07-18, 2002), with the objective to install the 28,000 pound S1 truss section, and deliver consumables to the ISS. She operated the station's robotic arm during the three EVAs required to install and activate the S1 truss. From September 16-22, 2006, she served as Commander of NASA's NEEMO 11 mission, located aboard *Aquarius* undersea off the coast of Florida. Dr. Magnus was 'Mission Specialist 5' on STS-126 aboard *Endeavour* (November 14-30, 2008) for ISS 'Expedition 18', returning to Earth with the crew of STS-119 aboard *Discovery* on March 28, 2009, after 134 days in orbit. She was 'Mission Specialist 1' on STS-135 aboard *Atlantis* (July 08-21, 2011), the final flight of the Space Shuttle program. The primary cargo was the MPLM *Raffaello* and a Lightweight Multi-Purpose Carrier (LMC).

Dr. Sandra Magnus logged a total of 157 days, eight hours and 42 minutes in space. In September 2012, she became deputy Chief of the Astronaut Office. She soon retired from NASA on October 21, 2012, and from 2012 until January 2018 was the executive director of the **American Institute of Aeronautics and Astronautics**. In 2019, she became the deputy director for Engineering within the **Office of the Undersecretary of Defense** for Research and Engineering. On February 09, 2021, **Virgin Galactic** announced that Dr. Sandra Magnus (along with former CSA astronaut Chris Hadfield and Cubic Corporation's David A. Whelan) would be joining their Space Advisory Board. Dr. Magnus is divorced, with no children.

## Marshburn, Thomas H.. STS-127, Soyuz TMA-07M/Expn 34/35, SpaceX Crew 3/Expn 66.

Dr. Thomas Henry Marshburn was born in August 1960 in Statesville, North Carolina, USA. He received a Bachelor of Science degree in Physics from Davidson College in 1982; a Master of Science degree in Engineering Physics from the University of Virginia in 1984; an M.D. from Wake Forest University in 1989; and an M.M.S. Degree from the University of Texas Medical Branch in 1997. After completing medical school, he trained in emergency medicine and also worked as a flight physician. He then worked as an emergency physician in Seattle, before being accepted into the first class of the NASA/UTMB Space Medicine Fellowship in Galveston, Texas. Dr. Marshburn joined NASA in November 1994, serving as a flight surgeon at JSC in Houston, Texas. From February 1996 to May 1997, he served as a flight surgeon for NASA personnel deployed to the **Yuri Gagarin Cosmonaut Training Center**, Star City, Russia. From July 1997 to August 1998, he was co-chair of Medical Operations for the *Shuttle-Mir* Program.

From 1998 to 2000, Dr. Marshburn was deputy Flight Surgeon for Neuronal (STS-98) and lead Flight Surgeon for the STS-101 mission to the ISS. He was the lead Flight Surgeon for 'Expedition 7' to the ISS in 2003. Selected in 2004 NASA Group 19, he was 'Mission Specialist 3' on STS-127 aboard *Endeavour* (July 15-31, 2009), which delivered the Japanese-built Exposed Facility (JEM-EF), and he made three EVAs. In May 2010, he served as an aquanaut during the NEEMO 14 mission, living underwater for fourteen days. He served as Flight Engineer on 'Expedition 34/35' to the ISS, launched aboard *Soyuz TMA-07M* on December 19, 2012, and returning to Earth on May 13, 2013. On November 11, 2021, Dr. Marshburn launched on the SpaceX 'Crew-3' flight as part of the 176-day long-duration 'Expedition 66' mission on the ISS. He performed his fifth EVA with Kayla Barron, and took over command of the ISS on March 29, 2022; landing on May 06, 2022. Dr. Thomas Marshburn logged a total of 337 days, nine hours and 43 minutes in space; including five EVAs totaling 31 hours and one minute. In 2019, he served as backup 'Flight Engineer 1' for the *Soyuz MS-13* and *Soyuz MS-15* long-duration flights to the ISS. After leaving NASA, Dr. Marshburn joined **Sierra Space** in late 2022 as the Chief Medical Officer in the company's Human Spaceflight Center and Astronaut Training Academy.

## Massimino, Michael J.. STS-109, STS-125.

Dr. Michael James Massimino was born in August 1962 in Oceanside, New York, USA. He received a Bachelor of Science degree in Industrial Engineering from Columbia University in 1984; a Master of Science degree in Mechanical Engineering (1988); and a Master of Science degree in Technology and Public Policy from Massachusetts Institute of Technology (MIT, 1988); then earned a Degree of Mechanical Engineer (MIT, 1990) and a Ph.D. in Mechanical Engineering from MIT in 1992. After completing his B.Sc., Massimino worked for **IBM** as a systems engineer in New York City from 1984-86. He then entered graduate school at MIT and conducted research on human operator control of space robotic systems in the MIT Mechanical Engineering Department's Human-Machine Systems Laboratory. During the summers of 1988 and 1989, he worked as a research fellow in the Man-Systems Integration Branch at the NASA **Marshall Space Flight Center**. He worked at **McDonnell Douglas Aerospace** in Houston, where he developed laptop computer displays to assist operators of the Space Shuttle remote manipulator system.

From 1992-94, he was an adjunct assistant professor in the Mechanical Engineering and Materials Sciences Department at Rice University. In September 1995, Massimino joined the faculty of the **Georgia Institute of Technology** as an assistant professor in the School of Industrial and Systems Engineering. Selected in May 1996 NASA Group 16, he flew as 'Mision Specialist 5' on STS-109 aboard *Columbia* (March 01-12, 2002), the fourth Hubble Space Telescope servicing mission. The crew accumulated a total of 35 hours and 55 minutes of EVA time to install the Advanced Camera for surveys, new rigid solar arrays, a new Power Control unit, a Reaction Wheel Assembly (which stabilizes a spacecraft axis to point in a nearly-fixed direction), and an experimental cryocooler for the Near Infrared Camera and Multi-Object Spectrometer (NICMOS).

Massimino was 'Mission Specialist 4' on STS-125 aboard *Atlantis* (May 11-24, 2009), the final Hubble Space Telescope servicing mission. The Shuttle carried two new instruments for Hubble: the Cosmic Origins Spectrograph, and the Wide Field Camera 3. Other parts were replaced to allow the telescope to function at least through 2014. Massimino became the first person to use ***Twitter*** in space, writing, *"From orbit: Launch was awesome! I am feeling great, working hard, & enjoying the magnificent views, the adventure of a lifetime has begun!"* He performed two EVAs, and logged a total of 23 days, 19 hours and 47 minutes in space. Beginning in 2012, Massimino appeared as himself in a recurring role on the CBS TV sitcom ***The Big Bang Theory***, launching aboard a Russian *Soyuz* to the ISS. On July 29, 2014, Dr. Massimino announced that he would be leaving NASA to pursue a teaching career at **Columbia University**, in New York City.

His book *Spaceman: An Astronaut's Unlikely Journey to Unlock the Secrets of the Universe* was published by Crown Archetype on October 04, 2016. He is a frequent guest on television news and talk show programs, including **NBC**'s *Today Show,* **ABC**'s *Good Morning America*, **CNN**, and **Fox News**. He often appears on Neil DeGrasse Tyson's podcast ***Star Talk***. Massimino is the senior advisor of space programs at the **Intrepid Sea, Air & Space Museum.** In early 2024, he was a guest speaker on the podcast ***Finding Mastery***, called an astronaut's guide to achieving the impossible. Dr. Mike Massimino has received six awards, including the NASA Graduate Student Researchers Program Fellowship; the Christopher Award; and the Communications Award of the National Space Club. He is married, with two children from a previous marriage.

## Mastracchio, Richard A. STS-106, STS-118, STS-131, Soyuz TMA-11M/ Expedition 38/39.

Richard Alan Mastracchio was born in February 1960 in Waterbury, Conneticut, USA. He received a Bachelor of Science degree in Electrical Engineering/Computer Science from the University of Conneticut in 1982; a Master of Science degree in Electrical Engineering from Rensselaer Polytechnic Institute in 1987; and a Master of Science degree in Physical Science from the University of Houston at Clear Lake in 1991. From 1982-87, he worked for **Hamilton Standard** in Conneticut as an engineer in the system design group. He moved to Houston, Texas, to work for the **Rockwell** Shuttle Operations Company at the JSC. In 1990, he joined NASA as an engineer in the Flight Crew Operations Directorate. From 1993-96, Mastracchio worked as an ascent/entry Guidance and Procedures Officer in Mission Control; supporting seventeen missions as a flight controller.

Selected in 1996 NASA Group 16, he flew on three shuttle missions as a Mission Specialist. STS-106 aboard *Atlantis* (September 08-20, 2000) delivered over 6,600 pounds of supplies to the ISS to prepare for the first resident crew. Masstrachio was the ascent/entry Flight Engineer, the primary robotic arm operator, and responsible for transfer of logistics. STS-118 aboard *Endeavour* (August 08-21, 2007) was when crew added another truss segment, a new gyroscope and an external spare parts platform to the ISS. STS-131 aboard *Discovery* (April 05-20, 2010) was an ISS resupply mission with over 27,000 pounds of hardware, supplies and equipment transferred. As lead EVA, Mastracchio performed three EVAs totaling 20 hours and 17 minutes.

He was Flight Engineer aboard *Soyuz TMA-11M* for the ISS 'Expedition 38/39' long-duration mission, and helped repair the malfunctioning main cooling system; returning to Earth on May 13, 2014, after nearly 188 days in space. Mastracchio had performed another three EVAs. He logged a total of 227 days, 13 hours and 38 minutes in space; including nine EVAs totaling 53 hours and four minutes. On June 19, 2017, he was appointed Senior Director of Operations for Commercial Resupply Services at **Orbital Sciences ATK**. He is currently Director of Business Development at **Northrop Grumman Space Systems**. Mastracchio is a member of the Institute of Electrical and Electronics Engineers, and he is married.

## Mattingly, Thomas K.. Apollo 16, STS-4, STS-51-C.

Rear Admiral Thomas Kenneth "Ken" Mattingly (1936-2023) was born in March 1936 in Chicago, Illinois, USA. He recalled that his *"earliest memories ... all had to do with airplanes".* He received a Bachelor of Science degree in Aeronautical Engineering from Auburn University in 1958. On September 10, 1965, NASA began the selection process for Group 5. They picked 159 candidates from a pool of 351 applicants. The candidates had to be less than six feet tall, have at least 1,000 hours of flight time in jet aircraft, and a Bachelor's degree in engineering or the physical or biological sciences. Lt. Mattingly was a student at the USAF Aerospace Research Pilot School at Edwards AFB when NASA selected him in April 1966.

He was initially in the support crew for *Apollo 8*, then trained for the *Apollo 13* mission, but was exposed to German measles (which he did not contract) and was replaced three days before launch by backup Command Module pilot Jack Swigert. Mattingly played a large role in helping the *Apollo 13* crew solve the problem of power conservation during their re-entry. Ken Mattingly flew on *Apollo 16* (April 16-27, 1972), the fifth crewed lunar landing mission. While in lunar orbit, the instruments aboard the CSM, named *Casper* extended the photographic and geochemical mapping of a belt around the lunar equator. During the return to Earth, he performed an EVA to retrieve film cassettes from the exterior of the spacecraft Command and Service Module. As of 2023, it remains one of only three "deep space" EVAs.

He commanded STS-4 aboard *Columbia* (June 27-July 04, 1982), the fourth and final Space Shuttle orbiter test flight. An experiment called the Continuous Flow Electrophoresis System was designed to investigate the separation of biological materials in a fluid according to their surface electrical charge. Mattingly and pilot Henry Hartsfield, Jr. were greeted by President Ronald Reagan after the landing. Ken Mattingly commanded STS-51-C aboard *Discovery* (January 24-27, 1985), the first Space Shuttle classified **Department of Defense** mission, which included deployment of a modified Inertial Upper Stage (IUS) vehicle from the orbiter. He logged a total of 21 days, four hours and 34 minutes in space; including one EVA of one hour and 23 minutes. Mattingly retired from NASA in 1985, and from the Navy in 1986 with the two-star rank of Rear Admiral. He worked as a director in **Grumman**'s Space Station Support Division, then headed the Atlas booster program for **General Dynamics** in San Diego, California.

Rear Adm. Ken Mattingly was a member of many organizations: an associate fellow, American Institute of Aeronautics and Astronautics; a fellow of the American Astronautical Society, a member of the Society of Experimental Test Pilots, and the US Naval Institute. He received at least eighteen awards, including the JSC Certificate of Commendation (1970); the Navy Distinguished Service Medal; the National Defense Service Medal; the AIAA Haley Astronautics Award for 1973; the FAI V.M. Komarov Diploma (1973); two NASA Distinguished Service Medals; and the Department of Defense Distinguished Service Medal (1982). He was portrayed by Gary Sinese in the 1995 movie ***Apollo 13***, and by Zelko Ivanek in the 1998 miniseries ***From the Earth to the Moon***. Mattingley died on October 31, 2023, in Arlington, Virginia, aged eighty-seven years. He was married, with one child.

## McArthur, K. Megan. STS-125, SpaceX Crew 2/Expedition 65/66.

Dr. Katherine "Megan" McArthur was born in August 1971 in Honolulu, Hawaii, USA. She received a Bachelor of Science degree in Aerospace Engineering from the University of California, Los Angeles, in 1993; and a Master of Science degree and a Ph.D. in Oceanography from Scripps Institution of Oceanography at the University of California, San Diego. She served as chief scientist during at-sea data collection operations, and has planned and led diving operations during sea-floor instrument deployments and sediment-sample collections.

Selected in 2000 NASA Group 18, she was CapCom for STS-116 in 2006, and the EVA CapCom for STS-117 in 2007. McArthur was 'Mission Specialist 2' on STS-125 aboard *Atlantis* (May 11-24, 2009), the fifth mission to service the Hubble Space Telescope, upgrading it to its most technologically advanced state since its launch in April 1990. She was pilot on the SpaceX 'Crew-2' mission aboard Crew Dragon *Endeavour,* launched on April 23, 2021 for 'Expedition 65/66' on the same booster as SpaceX 'Crew-1'. Returning to Earth and splashing down in the Gulf of Mexico in the evening of November 09, 2021, the mission set a record for the longest spaceflight by a crewed US spacecraft: 199 days. Dr. Megan McArthur logged a total of 212 days, 15 hours and 21 minutes in space. In 2019, she was appointed Deputy Chief of the Astronaut Office. In 2022, SpaceX's Dragon support vessel *GO Searcher* was renamed *Megan*. She is married to fellow astronaut Robert "Bob" Behnken, and on her 'Crew-2' mission used the same seat her husband had used in the SpaceX *Demo-2* mission. They have one son.

## McArthur, William S., Jr.. STS-58, STS-74, STS-92, Soyuz TMA-7/ Expedition 12.

Col. William Surles McArthur, Jr. was born in July 1951 in Laurinburg, North Carolina, USA. He received a Bachelor of Science degree in Applied Science and Engineering from the US Military Academy in 1973; and a Master of Science degree in Aerospace Engineering from the Georgia Institute of Technology in 1983. In 1987, he attended the US Naval Test Pilot School, then was assigned to a post as a test flight engineer at NASA. He logged over 9,000 flying hours in forty-one aircraft types. Selected by NASA in January 1990 Group 13, he has held various assignments within the Astronaut Office, including working on issues relating to the Solid Rocket Booster, redesigned solid rocket motor, and the advanced solid rocket motor. He was 'Mission Specialist 2' on STS-58 aboard *Columbia* (October 18-November 01, 1993), a mission devoted to life sciences research. Crew conducted experiments aimed at understanding bone tissue loss and the effects of microgravity on sensory perception. McArthur was 'Mission Specialist 3' on STS-74 aboard *Atlantis* (November 12-20, 1995), and docked to the *Kristall* module of space station *Mir* on November 15. Along with water, supplies and equipment, two new solar arrays were transferred to *Mir*.

He was 'Mission Specialist 2' on STS-92 aboard *Discovery* (October 11-24, 2000), the one hundredth mission of the Space Shuttle, which brought the Z1 truss, Control Moment Gyros, Pressurized Mating Adapter-3, and two DDCU (heat pipes) to the ISS. William McArthur launched via *Soyuz TMA-7* for ISS 'Expedition 12' (October 03, 2005 to April 08, 2006), and tested a new method of preparing for spacewalks by spending the night in the *Quest* airlock at a reduced pressure; instead of breathing pure oxygen for several hours to purge the body of nitrogen. He logged a total of 224 days, 22 hours and 19 minutes in space, and served as Director of Safety and Mission Assurance for the JSC until his retirement in June 2017. Col. McArthur's awards include the Army Distinguished Service Medal; the Defense Superior Service Medal; the Defense Meritorious Service Medal with Oak Leaf Cluster; the National Defense Service Medal with one star; and the NASA Exceptional Service Medal. In May 2009, William McArthur was awarded an honorary Doctorate from the University of Strathclyde, UK, in recognition of his contribution to space science and engineering, as well as his outreach work with the Scottish Space School. He is married, with two daughters.

## McBride, Jon A.. STS-42-B.

Capt. Jon Andrew McBride was born in August 1943 in Charleston, West Virginia, USA. He attended West Virginia University from 1960-64; received a Bachelor of Science degree in Aeronautical Engineering from the US Naval Postgraduate School in 1971; and pursued graduate studies in Human Resource Management at Pepperdine University. He began flight training in 1965 at NAS Pensacola, Florida, and earned his pilot wings in August 1966, then joined Fighter Squadron 1010 for F-4 Phantom II aircraft training. During the Vietnam War, he deployed to Southeast Asia, flying sixty-four combat missions. Jon McBride attended USAF Test Pilot School, Class 75A, at Edwards AFB, California. He piloted over forty aircraft types of military and civilian airplanes, and logged over 8,800 hours flying time, including 4,700 hours in jet aircraft and he made over 600 aircraft carrier landings. During airshows in 1976, 1977 and 1978, he showcased the Navy "Spirit of '76" bi-centennial-painted F-4J Phantom.

Selected by NASA in January 1978, he was CapCom for missions STS-5, STS-6 and STS-7. McBride was pilot on STS-42-B aboard *Challenger* (February 03-11, 1984), which deployed the *Earth Radiation Budget Satellite (ERBS)*, conducted scientific investigations of the Earth with the OSTA-3 pallet and Large Format Camera, and demonstrated potential satellite refueling with an EVA and associated hydrazine transfer. He logged eight days, five hours and 23 minutes in space. Jon McBride was slated to pilot STS-61-E, which was one of several missions canceled following the *Challenger* disaster in January 1986. From September 1987 to March 1989, he served as Assistant Administrator for Congressional Relations at NASA HQ, and retired on May 12, 1989.

He was President and CEO of the **Flying Eagle Corporation** in Lewisburg, West Virginia, and President of the Constructor's Labor Council of West Virginia. In 1996, he unsuccessfully vied for the Republican nomination for Governor of West Virginia. By 2000, he was retired and living near Cocoa, Florida. In September 23, 2011, the NASA Independent Verification and Validation Facility in Fairmont, West Virginia dedicated a NASA software laboratory to Jon McBride. He belonged to eleven organizations and received seventeen awards, including the Legion of Merit; three Air Medals; the Defense Superior Service Medal; the National Defense Service Medal; the Vietnam Service Medal; the Navy Commendation Medal with Combat Valor device; Kanawha County, West Virginia's "Famous Person Award" (1988); and Daughters of the American Revolution "Medal of Honor" (1993). McBride is married, with three children and eight grandchildren.

## McCandless, Bruce II.. STS-41-B, STS-31.

Capt. Bruce McCandless, II (1937-2017) was born Byron Willis McCandless in June 1937 in Boston, Massachusetts, USA. McCandless was a third-generation US Navy officer, with his father and a grandfather being Medal of Honor recipients. He received a Bachelor of Science degree from the US Naval Academy in 1958; a Master of Science degree in Electrical Engineering from Stanford University in 1965; and a M.B.A. from the University of Houston at Clear Lake in 1987. In March 1960, McCandless was designated a US Naval Aviator, and from December 1960 to February 1964, he was assigned to Fighter Squadron 102. He saw duty aboard *USS Forestal* and *USS Enterprise*, including the latter's participation in the 13-day October 1962 Cuban Missile Crisis. During Naval service, McCandless gained flying proficiency in eight airplanes and the Bell 47G helicopter. He logged over 5,200 hours flying time, including 5,000 hours in jet aircraft. At age twenty-eight years, he was the youngest of nineteen astronauts selected in 1966. He served as the voice of Mission Control during the *Apollo 11* mission in 1969 and was a CapCom for *Apollo 14* in 1971.

McCandless was 'Mission Specialist 1' on STS-41-B aboard *Challenger* (February 03-11, 1984), which deployed two Hughes-built communications satellites, and he completed the first untethered spacewalk by using the *Manned Maneuvering Unit* (MMU). He reached a maximum distance of ninety-eight metres from the orbiter, while Robert Stewart tested the "work station" foot restraint at the end of the Remote Manipulator System. McCandless described his experience, *"I was grossly over-trained. I was just anxious to get out there and fly. I felt very comfortable ... It got so cold my teeth were chattering and I was shivering, but that was a very minor thing.... I'd been told of the quiet vacuum you experience in space, but with three radio links saying, "How's your oxygen holding out?", "Stay away from the engines!" and "When's my turn?", it wasn't that peaceful ...* ***It was a wonderful feeling, a mix of personal elation and professional pride****: it had taken many years to get to that point"*. Unquote.

Bruce McCandless was 'Mission Specialist 1' on STS-31 aboard *Discovery* (April 24-29, 1989), when the crew deployed the Hubble Space Telescope from their record-setting altitude of 610 km (380 miles). He logged a total of 13 days, zero hours and 31 minutes (over 312 hours) in space; including EVA of six hours and 17 minutes, plus five hours and 55 minutes of MMU time. **Lockheed Martin** later developed the McCandless Lunar Lander and named it after him. This reflected the fact that the MMU spacewalk was facilitated by the jetpack developed by Lockheed Martin. John McCain, who graduated from the US Naval Academy with McCandless in the Class of 1958, stated after McCandless' death: *"The iconic photo of Bruce soaring effortlessly in space has inspired generations of Americans to believe that there is no limit to the human potential"*.

After retiring from NASA in 1990, Bruce McCandless worked for **Lockheed Martin Space Systems**. He belonged to the US Naval Academy Alumni Association; the US Naval Institute; Institute of Electrical and Electronics Engineers (IEEE); the American Institute of Aeronautics and Astronautics (AIAA); the Association for Computing Machinery; and the National Audubon Society. McCandless was awarded a Patent for the design of a tool tethering system that was used during Space Shuttle spacewalks, and he received twelve awards, including the NASA Exceptional Service Medal (1974); the National Defense Service Medal (1974); the Armed Forces Expeditionary Medal (1974); the Defense Distinguished Service Medal (1985); the NASA Exceptional Engineering Achievement Medal (1985); the National Aeronautic Association Collier Trophy (1984); and the Legion of Merit (1988). He was inducted into the International Space Hall of Fame (1995); and the US Astronaut Hall of Fame (2005). Bruce McCandless died on December 21, 2017, in Los Angeles, California, aged eighty years. He was married, with two children.

## McClain, Ann C.. Soyuz MS-11/Expedition 58/59.

Col. Ann Charlotte McClain was born in June 1979 in Spokane, Washington, USA. She received a Bachelor of Science degree in Mechanical Engineering from the US Military Academy, West Point, New York, and was commissioned as an Army officer in 2002. In receipt of a Marshall Scholarship, she received a Master's degree in Aerospace Engineering at the University of Bath, UK, in 2004; and a Master's degree in International Relations from the University of Bristol, UK, in 2005. McClain is an avid rugby player who has competed in the women's premiership in England, and represented the US women's national Rugby Union team. She qualified as a Bell OH-58 Kiowa Warrior attack helicopter pilot, and was deployed to the Second Battalion, Sixth Cavalry Regiment at Wheeler Army Airfield, Hawaii. During her deployment as part of Operation Iraqui Freedom, she flew 800 hours and completed 216 combat missions.

In 2013, she graduated from the US Naval Test Pilot School in Maryland, and was selected as a NASA astronaut. Launched on December 03, 2018, aboard *Soyuz MS-11*, she served as a Flight Engineer on ISS 'Expedition 58/59', performing two EVAs, and spending 204 days in space to June 25, 2019. In 2020, McClain was chosen for the team of eighteen astronauts who will be eligible to fly on the first missions of the Artemis program. For the *Artemis III* mission, scheduled for 2026, two astronauts would land on the lunar surface and spend over six days in the Moon's south polar region.

## McCool, William C.. STS-107.

Capt. William "Willie" Cameron McCool (born William Cameron Graham, 1961-2003) was born in September 1961 in San Diego, California, USA. His parents divorced when he was young, and in 1969 his mother married Barent McCool, a Navy aviator. He received a Bachelor of Science degree in Applied Science from the US Naval Academy in 1983; a Master of Science degree in Computer Science from the University of Maryland, College Park, in 1985, and was designated a Naval Aviator in August 1986. In 1991, he graduated from the US Naval Test Pilot School at NAS Patuxent River, Class 101, and received a Master of Science degree in Aeronautical

Engineering from the US Naval Postgraduate School in 1992. He was responsible for the management and conduct of a wide variety of projects, ranging from airframe fatigue life studies to numerous avionics upgrades. His efforts were primarily dedicated to flight test of the Advanced Capability (ADVCAP) EA-6B. McCool logged over 2,800 flying hours in twenty-four aircraft types and made over 400 aircraft carrier arrested landings.

Selected in April 1996 NASA Group 16, he was initially assigned to the Computer Support Branch, and also served as technical assistant to the director of flight crew operations, and worked Shuttle cockpit upgrade issues for the Astronaut Office. He was pilot on STS-107 aboard *Columbia* (January 16-Feb 01, 2003), when the crew worked two alternating shifts, successfully conducting about 80 experiments. According to NASA, McCool said of the unique view he and his crewmates had from *Columbia*, *"From our orbital vantage point, we observe an Earth without borders, full of peace, beauty and magnificence, and we pray that humanity as a whole can imagine a borderless world as we see it, and strive to live as one in peace"*. A quote from January 29, 2003, aboard *Columbia*. He logged 15 days, 22 hours and 20 minutes in space.

Capt. William McCool died on February 01, 2003, over Texas when the Space Shuttle *Columbia* disintegrated during re-entry, just 15 minutes before scheduled landing. Following the disaster, all seven crewmembers were posthumously awarded the NASA Space Flight Medal; and the Congressional Space Medal of Honor. The lunar crater McCool, asteroid 51829 WilliamMcCool, and McCool Hill in the Columbia Hills on Mars are named for him; as are four school facilities, and three memorials. He had received the NASA Distinguished Service Medal; the Defense Distinguished Service Medal; and four special honors: McCool was presented "Outstanding Student" and "Best DT-II Thesis" awards as a graduate of the US Naval Test Pilot School, Class 101; two Navy Commendation Medals; two Navy Achievement Medals; and was posthumously promoted to Captain by the Secretary of the Navy, Carlos Del Toro on October 07, 2023.

## McCulley, Michael J.. STS-34.

Capt. Michael "Mike" James McCulley was born in August 1943 in San Diego, California, USA. He was the first submariner in space. He received a Bachelor of Science degree and Master of Science degree in Metallurgical Engineering from Purdue University, both in 1970. He then received his Officer's commission in the US Navy, and following flight training he served tours of duty in A-4 Skyhawk and A-6 Intruder aircraft, then was selected to attend the Empire Test Pilot's School in Great Britain. McCulley served in several test pilot positions before returning to sea duty on *USS Saratoga* and *USS Nimitz*. He has flown over fifty aircraft types, logged over 5,000 flying hours and made nearly 400 aircraft carrier landings.

Selected in 1984 NASA Group 10, he was pilot on STS-34 aboard *Atlantis* (October 18-23, 1989), which deployed the *Galileo* spacecraft onto its journey toward Jupiter, and operated the Shuttle Solar Backscatter Ultraviolet Instrument (SSBUV) to map atmospheric ozone. Mike McCulley logged four days, 23 hours and 39 minutes in space. He retired in October 1990, and was employed by **Lockheed Martin Space Operations**. He served as Vice President and Deputy Launch Site Director for the Kennedy Space Center (KSC), being promoted to Director in November 1995. From November 1999, he was named Chief Operating Officer (COO) of **United Space Alliance** (USA), the prime contractor for the Space Shuttle program. From May 15, 2003 to September 28, 2007, Michael McCulley was President and CEO of **United Space Alliance**. He received nine awards, including the Legion of Merit; the Defense Superior Service Medal; the National Defense Service Medal; the Armed Forces Expeditionary Medal; the Meritorious Unit Commendation; the Small Arms Expert Ribbon; and Child Advocate of the Year Award for the State of Florida. Mike McCulley is married, with six children and twelve grandchildren.

## McDivitt, James A., Jr.. Gemini 4, Apollo 9.

Brig. Gen. James "Jim" Alton McDivitt, Jr. (1929-2022) was born in June 1929 in Chicago, Illinois, USA. In June 1957, he entered the University of Michigan under the USAF Institute of Technology (AFIT) program. He graduated first in his class with a Bachelor of Science degree in Aeronautical Engineering from the University of Michigan in 1959, qualified as a test pilot at the Air Force Experimental Flight Test Pilot School (Class 59C) and Aerospace Research Pilot School (Class I), and joined the Manned Spacecraft Operations Branch. McDivitt joined the USAF in 1951 and flew 145 combat missions in the Korean War. By September 1962, he had logged over 2,500 flying hours, including over 2,000 hours in jet aircraft. Selected by NASA in September 1962 as part of the second astronaut class, dubbed The Next Nine, he was known for his sense of teamwork and gutsy leadership, as well as his sense of humor.

McDivitt was Command Pilot of the *Gemini 4* mission (June 03-07, 1965), when Ed White performed the first US spacewalk. The hatch was difficult to open and relatch after the EVA, which could have compromised their re-entry. Jim McDivitt became the first US astronaut to command a crew on his first spaceflight. Their first objective was to attempt the first space rendezvous with the spacecraft's spent *Titan II* launch vehicle's upper stage, and *Gemini 4* came within 61 metres of it. A computer failed in the final stages of the mission, and McDivitt was ordered to perform a rolling atmospheric re-entry, rather than the lifting bank angle entry that the computer had intended for them. Their spacecraft overshot the landing point by eighty kilometres, but they were soon reached by the *USS Wasp* recovery ship.

Jim McDivitt was Commander of *Apollo 9*, the first crewed flight of the Apollo lunar program and the complete set of Apollo flight hardware. Launched on March 03, 1969, this was a ten-day Earth orbital Lunar Module test session. The CSM was named *Gumdrop*, and the LM was *Spider*. The probe-and-drogue docking assembly worked well, and the two docked craft were successfully maneuvered with the Service Propulsion System engine. The LM's propulsion systems were tested, then the craft separated to test the descent engine, and later used the LM's ascent engine to dock with the CSM. After 151 Earth orbits, the crew splashed down in the Atlantic Ocean, and were recovered by *USS Guadalcanal*. Jim McDivitt logged a total of 14 days, two hours and 56 minutes in space. He later became the manager of NASA lunar landing operations, and was the Apollo spacecraft program manager from 1969 to June 1972, when he left NASA and retired from the USAF. McDivitt became Executive Vice President, Corporate Affairs for **Consumers Power Company**. In March 1975, he joined **Pullman, Inc.**; and in October became President of the Pullman Standard Division. In January 1981, Jim McDivitt joined **Rockwell International**, until retiring in 1995. He belonged to the Society of Experimental Test Pilots; the American Institute of Aeronautics and Astronautics; the Atlantic Council on Foreign Diplomacy; and the Advisory Council-University of Michigan.

Brig. Gen. James McDivitt received sixteen awards, including two Air Force Distinguished Service Medals; two NASA Distinguished Service Medals; four DFCs; five Air Medals; the Sword of Loyola; four honorary degrees; and was inducted into four Halls of Fame. In the 1998 miniseries ***From the Earth to the Moon***, McDivitt was played by Conor O'Farrell. He appeared as himself on ***The Brady Bunch*** in a 1974 fifth-season episode about UFOs, *Out of This World*, when he talked about his UFO experience. He married in 1965 and had four children; was later divorced and remarried with two stepchildren. Jim McDivitt died in his sleep on October 13, 2022, in Tucson, Arizona, aged ninety-three years.

## McNair, Ronald E.. STS-41-B, STS-51-L.

Dr. Ronald Erwin McNair was born in October 1950 at Lake City, South Carolina, USA. He received a Bachelor of Science degree in Engineering Physics (magna cum laude) from North Carolina Agricultural & Technical State University; and a Master's degree and a Ph.D. in Physics from the Massachusetts Institute of Technology in 1976. He received four honorary Doctorates, and many fellowships and commendations. Selected in 1978, McNair was one of thirty-five candidates from a pool of 10,000 applicants as NASA tried to increase the number of minority and female astronauts.

On February 1984, he became the second African American to reach space, serving as 'Mission Specialist 3' on STS-41-B aboard the Space Shuttle *Challenger* (February 03-11, 1984). About eight hours after launch, two communications satellites were deployed: *Westar 6* for America's Western Union, and *Palapa B2* for Indonesia. However, both the satellite Payload Assist Modules malfunctioned, placing them into a lower-than-planned orbit. [Both satellites were successfully retrieved during STS-51-A in November 1984]. The mission carried five "Getaway Special" canisters, a Cinema-360 camera and a continuation of the Continuous Flow Electrophoresis System and Monodisperse Latex Reactor experiments. Ronald McNair logged seven days, 23 hours and 15 minutes in space.

At the launch of STS-51-L aboard *Challenger*, the seven-person crew were killed on January 28, 1986, following the explosion of a Solid Rocket Booster which caused the Shuttle to disintegrate when nine miles high, just 73 seconds after liftoff. McNair died over the Atlantic Ocean, aged thirty-five years. Ronald McNair Boulevard is named in his honor, and lies near other streets named for astronauts who perished in the *Challenger* disaster. His brother, Carl, wrote Ronald McNair's official biography, *In the spirit of Ronald E. McNair – Astronaut: An American Hero*. Ron McNair and all the crew members lost in the *Challenger* and *Columbia* disasters were posthumously awarded the Congressional Space Medal of Honor in 2004. Thirty-five buildings are named in his honor. McNair was portrayed by Joe Morton in the 1990 TV movie, ***Challenger***. He is buried at Ronald E. McNair Memorial Park, Lake City, South Carolina. He was married, with two children.

## McMonagle, Donald R.. STS-39, STS-54, STS-66.

Col. Donald Ray McMonagle was born in May 1952 in Flint, Michigan, USA. He received a Bachelor of Science degree in Astronautical Engineering from the USAF Academy in 1974; a Master of Science degree in Mechanical Engineering from California State University, Fresno, in 1985; and a Master of Business Administration from the University of Michigan Ross School of Business in 2003. He completed pilot training at Columbus AFB, Mississippi, in 1975, and as an F-4 pilot he served at Kunsan Air base, South Korea, and Holloman AFB, New Mexico. McMonagle was an F-15 instructor pilot at Luke AFB, Arizona. In 1981, he excelled at the USAF Test Pilot School at Edwards AFB, California. From 1982-85, he was the operations officer and project test pilot for the Advanced Fighter Technology Integration F-16 aircraft. He logged over 5,000 hours in aircraft including the T-38, F-4, F-15 and F-16 jets.

Selected by NASA in June 1987, Donald McMonagle was 'Mission Specialist 2' on STS-39 aboard *Discovery* (April 28-May 06, 1991), a Department of Defense mission when the seven-man crew deployed, operated and retrieved a remotely controlled spacecraft, and conducted research of both natural and induced phenomena in the Earth's atmosphere. McMonagle was Pilot on STS-54 aboard *Endeavour* (January 13-19, 1993) which featured the deployment of a *Tracking and Data Relay Satellite* (TDRS-F), and the collection of data about celestial X-rays. He was Commander on STS-66 aboard *Atlantis* (November 03-14,

1994) on the ATLAS-3 mission. He logged a total of 25 days, five hours and 34 minutes (over 605 hours) in space. In January 1996, McMonagle was assigned the task to establish a new Extra-Vehicular Activity (EVA) Project Office for managing all NASA resources associated with space suits and tools used to conduct spacewalks.

On August 15, 1997, Col. Donald McMonagle became the Manager, Launch Integration, at KSC. In this capacity, he oversaw final Shuttle preparation, launch execution, and orbiter return following landings at any location other than KSC. As of June 2006, he was the Vice-President of Quality and Mission Effectiveness of the Missile Systems business unit of **Raytheon** in Tucson, Arizona. McMonagle received the DFC; the Air Medal; the Meritorious Service Medal; three Air Force Commendation Medals; the Defense Meritorious Service Medal; the Liethen-Tittle Award; three NASA Space Flight Medals; the NASA Exceptional Service Medal; and the NASA Outstanding Leadership Medal. He is married, with two children.

## Meade, Carl J.. STS-38, STS-50, STS-64.

Col. Carl Joseph Meade was born in November 1950 at Chanute Air Force Base, Illinois, USA. He received a Bachelor of Science degree in Electronics Engineering (with honors) from the University of Texas in 1973; and a Master of Science degree in Electronics Engineering from the California Institute of Technology in 1975. Meade worked as a Hughes Fellow at the **California Institute of Technology** and as an electronics design engineer at **Hughes Aircraft Company** in Culver City, California, before joining the US Air Force (USAF) at Laughlin AFB, Texas. He excelled as a distinguished graduate of Undergraduate Pilot Training, and was assigned to the Three Sixty-Third Tactical Reconnaissance Wing at Shaw AFB, South Carolina. Meade was then selected for the USAF Test Pilot School, Class 80B, where he received the Liethen-Tittle Award as the Outstanding Test Pilot. His next assignment was with the Test Wing 6510 at Edwards AFB, where he contributed to flight testing. He logged over 4,800 hours of jet time in twenty-seven aircraft types, and was selected by NASA in June 1985.

Prior to STS-38 and after STS-50, Carl Meade was the Astronaut Office representative at the Solid Rocket Booster Program and the Space Shuttle Main Engine Program at the **Marshall Space Flight Center**. He served as a Mission Specialist on three Space Shuttle flights: STS-38 aboard *Atlantis* (November 15-20, 1990) was a classified Department of Defense mission; STS-50 aboard *Columbia* (June 25-July 09, 1992) carried the first US Microgravity Laboratory (USML). This mission was the first to utilize the Extended Duration Orbiter capabilities of the newly-modified Orbiter *Columbia*. At 13 days, 19 hours and 30 minutes, this was the longest Shuttle flight to date. In 1994, Meade performed the first untethered spacewalk in ten years: on STS-64 aboard *Discovery* (September 09-20, 1994). The objective was to flight test a self-rescue jetpack. Meade logged six hours and 54 minutes in EVA, including 3.6 hours piloting the *SAFER* jetpack.

Col. Carl Meade logged a total of 29 days, 16 hours and 14 minutes (over 712 hours) in space. He has served as the Deputy Division Chief of the Crew and Thermal Systems Division and as the Chief of the Flight Support Branch of the Astronaut Office. Carl Meade has received twelve awards, including the DFC; the Intelligence Medal; the National Defense Service Medal; the Air Force Commendation Medal; the Defense Superior Service Medal; two Air Force Outstanding Unit Awards; the NASA Group Achievement Award; and the NASA Exceptional Service Medal. Meade currently works for **Northrop Grumman Integrated Systems** as an engineering manager on the Crew Exploration Vehicle project. He is married, with two sons.

## Meir, Jessica U.. Soyuz MS-15/Expedition 61/62.

Dr. Jessica Ulrika Meir was born in July 1977 in Caribou, Maine, USA. Of Israeli-Swedish descent, she received a Bachelor of Arts degree in Biology from Brown University in 1999; a Master of Science degree in Space Studies from the International Space University, Strasbourg, France, in 2000; and a Doctorate in Marine Biology from Scripps Institution of Oceanography (UCSD) in 2009. She is an American Jewish university professor, scientist, explorer and astronaut. Meir was previously an assistant professor of anesthesia at Harvard Medical School, Massachusetts General Hospital, Boston, following postdoctoral research in comparative physiology at the University of British Columbia. In September 2002, she served as an aquanaut for five days of saturation diving on the NEEMO 4 crew. In 2013, she was selected in NASA Group 21; and in 2016 participated in ESA CAVES, a training course in which international astonauts train in a space-analogue cave environment in Sardinia.

On September 25, 2019, she launched aboard *Soyuz MS-15* to the ISS, where she served as a Flight engineer during 'Expedition 61/62'. Meir made history with Christina Koch in the first all-woman spacewalk of seven hours and 17 minutes on October 18, 2019. She returned to Earth on April 17, 2020. Meir has logged 204 days, 15 hours and 19 minutes in space; including three EVAs totaling 21 hours and 44 minutes. In 2020, she was one of the eighteen astronauts chosen to be in the *Artemis* program. She is trained to help direct the first human mission to an asteroid and then to Mars. She is a member of the science advisory board of **Adventurers and Scientists for Conservation**. Meir has received eight awards, including the Philanthropic Educational Organization Scholar Award (2008); the Lockheed Martin Space Operations Special Recognition Award (2002); the Sir Charles Sherrington Prize lecture and Medal (2021); and the President's Lecture and Medal (2021), of The Physiological Society of the United Kingdom & Republic of Ireland.

## Melnik, Bruce E.. STS-41, STS-49.

Cmdr. Bruce Edward Melnik was born in December 1949 in New York City, USA. He received a Bachelor of Science degree in Engineering (with honors) from the US Coast Guard Academy in 1972; a Master of Science degree in Aeronautical Systems from the University of West Florida in 1975; and an honorary Doctorate from the University of West Florida in 2001. He spent twenty years in the **US Coast Guard**, rising to the rank of Commander. His assignments included serving as operations officer and chief test pilot at **Coast Guard** Aircraft Program Office in Grand Prairie, Texas. Melnik logged over 5,000 hours flying time. Selected by NASA in 1987, he holds the distinction of being the first Coast Guard aviator to go into space.

Melnik was 'Mission Specialist 1' on STS-41 aboard *Discovery* (October 06-10, 1990), with the primary goal to deploy the ESA-built *Ulysses* spacecraft, which observed the Sun's activity across different latitudes on a five-year mission. This was part of the International Solar Polar Mission. Additional crew activities included experimenting with a voice command system (VCS) to control onboard television cameras, and monitoring ionizing radiation exposure to the crew within the orbiter cabin. He was 'Mission Specialist 2' on STS-49 aboard *Endeavour* (May 07-16, 1992), with the primary mission to capture and repair the stranded 4,215 kg (9,292 pound) *Intelsat VI* (aka *Intelsat 603*) satellite, which failed to leave low Earth orbit two years before. A new upper stage perigee kick motor was attached, to relaunch it to its intended geosynchronous orbit. This mission featured the first three-person EVA, an unplanned solution to challenges encountered during the satellite capture, and set a record for the longest-duration EVAs at that time (until STS-102 in 2001). He logged a total of 12 days, 23 hours and 27 minutes (over 300 hours) in space.

Following retirement from NASA and the US Coastguard in July 1992, Bruce Melnik entered the aerospace industry. From 1992-94, he was director of process improvement technology at **Lockheed Space Operations Company**. He then served as a vice president with **The Boeing Company**'s Integrated Defense Systems group, overseeing Boeing's Florida operations at the J.F. Kennedy Space Center (KSC). Bruce Melnik belongs to fourteen organizations, including the American Institute of Aeronautics and Astronautics; the Association of Space Explorers; the Coastal Conservation Association; Missile, Space and Range Pioneers; the Naval Aviation Museum Foundation; and the United States Space Foundation. He has served on seven boards of directors; five boards of advisors; and is on the board of governors of Brevard Community College Foundation. Bruce Melnik has received two Defense Distinguished Service Medals; two DFCs; and the Secretary of Transportation Heroism Award. He retired in 2008. Melnik is married, with two children.

## Melroy, Pamela A.. STS-92, STS-112, STS-120.

Pamela Ann Melroy was born in September 1961 in Palo Alto, California, USA. She received a Bachelor of Arts degree in Physics and Astronomy from Wellesley College in 1983; and after completing a Master of Science degree in Earth and Planetary Science from the Massachusetts Institute of Technology in 1984, she attended Undergraduate Pilot Training at Reese AFB, Lubbock, Texas. She is a veteran of the US invasion of Panama, and Operation Desert Shield/Desert Storm, with over 200 combat and combat support hours. In June 1991, she attended the USAF Test Pilot School at Edwards AFB, where she served as a test pilot until her selection for the astronaut program. On May 18, 2008, she received an honorary degree from Iona College in New Rochelle, New York.

Selected in December 1994 NASA Group 15, Melroy was pilot on STS-92 aboard *Discovery* (October 11-24, 2000), which delivered and attached the Z1 truss and Pressurized Mating Adapter 3 to the ISS. The Shuttle's robotic arm and four EVAs configured these elements. She was pilot on STS-112 aboard *Atlantis* (October 07-18, 2002), which delivered and installed the S1 truss (the third piece of the space station's 11-piece Integrated Truss Structure). Pamela Melroy was Commander on STS-120 aboard *Discovery* (October 23-November 07, 2007) to the ISS during 'Expedition 16' (commanded by Dr. Peggy Whitson), which delivered the Node 2 *Harmony* element, providing for future international laboratories to be added. STS-120 marked the first time that two female mission commanders were in orbit at the same time. Melroy logged a total of 38 days, 20 hours and six minutes (over 924 hours) in space. She retired from the US Air Force in February 2007, and left NASA's astronaut program in August 2009. On June 16, 2021, the Senate Commerce Committee confirmed her nomination, and she is now serving as the deputy **Administrator of NASA**. Pamela Melroy is married to geologist Douglas Hollett.

## Melvin, Leland D.. STS-122, STS-129.

Leland Devon Melvin was born in February 1964 in Lynchburg, Virginia, USA. He received a Bachelor of Science degree in Chemistry from the University of Richmond in 1989; and a Master of Science degree in Materials Science Engineering from the University of Virginia about 1993. From 1982-85, he was a wide receiver on the University of Richmond football team, and a team captain during his senior season. Melvin caught at least one pass in every game he played as a Richmond Spider. He began working in the Nondestructive Evaluation Sciences Branch at NASA's **Langley Research Center** in 1989. In 1994, he was selected to lead the Vehicle Health Monitoring team for the NASA/Lockheed Martin *X-33* Reusable Launch Vehicle program.

Selected in 1998 NASA Group 17, he was 'Mission Specialist 1' on STS-122 aboard *Atlantis* (February 07-20, 2008), the twenty-fourth shuttle mission to the ISS, which delivered and installed the ESA's *Columbus* Laboratory Module. The Shuttle also carried the Solar Monitoring Observatory (SOLAR), the European Technology Exposure Facility (EuTEF), and a new nitogen tank assembly. Melvin was 'Mission Specialist 1' on STS-129 aboard *Atlantis* (November 16-29, 2009), which delivered two ExPRESS Logistics Carriers (ELC racks) to the ISS; about 30,000 pounds of replacement parts for systems that provide power to the station to keep it from overheating, and maintain proper orientation in space. Leland Melvin logged a total of 23 days, 13 hours and 28 minutes (over 565 hours) in space.

Leland Melvin was elected to the University of Richmond Athletic Hall of Fame (1996-97); and selected for the All-UR Stadium Team in 2009. Melvin is a TV personality: a guest judge in the twelvth episode of ***Top Chef*** (season 7); on ***The Dog Whisperer***; and was host of ***Child Genius*** (season 1 and 2). Since retiring from NASA, Melvin has given several STEM lectures of his experiences in space, as well as his football career in the NFL. He has published two books, *Chasing Space: An Astronaut's Story of Grit, Grace, and Second Chances* and *Chasing Space: Young Reader's Edition.*

## Metcalf-Lindenburger, Dorothy M.. STS-131.

Dorothy "Dottie" Marie Metcalf-Lindenburger was born in May 1975 in Colorado Springs, Colorado, USA. She received a Bachelor of Arts degree in Geology (with honors, and cum laude "with praise") from Whitman College, Washington, in 1997; teaching certification from Central Washington University in 1999; and a Master of Science degree in Geology from the University of Washington, Seattle, in 2016. Selected in 2004 NASA Group 19, she flew on STS-131 aboard *Discovery* (April 05-20, 2010), a resupply mission to the ISS which carried the *Leonardo* MPLM to deliver over 27,000 pounds of hardware, supplies and equipment; including a tank full of ammonia coolant that required three EVAs to hook up, new crew sleeping quarters and three experiment racks. This was the first mission with four women in space at the same time: Dottie Metcalf-Lindenburger, Tracy Caldwell Dyson, Stephanie Wilson and Naoko Yamazaki of JAXA. Dottie logged 15 days, two hours and 47 minutes in space. Metcalf-Lindenburger commanded the NEEMO 16 mission aboard the *Aquarius* underwater laboratory offshore from Florida for twelve days from June 11, 2012.

She belongs to the International order of the Rainbow for Girls; National Education Association; and The Mars Generation. She has received seven awards, including 1995-96 NAIA Academic All-American in Cross Country and Track; the 1997 Whitman College Leed's Geology Award; was 1999 Outstanding Teacher Preparation Candidate at Central Washington University; and being a 2007 Space Camp Hall of Fame Inaugural Inductee. Dottie Metcalf-Lindenburger retired from NASA on June 13, 2014. She is a lead singer with the all-astronaut rock band, ***Max Q***. Dorothy Metcalf-Lindenburger is married, with a daughter.

## Mitchell, Edgar D.. Apollo 14.

Capt. Dr. Edgar "Ed" Dean Mitchell (1930-2016) was born in September 1930 in Hereford, Texas, USA. He first learned to fly at age 13, and received his private pilot license at age 16 years. Edgar Mitchell received a Bachelor of Science degree in Industrial Management from Carnegie Institute of Technology (now Carnegie Mellon University) and entered the US Navy in 1952; a Bachelor of Science degree in Aeronautical Engineering from the US Naval Postgraduate School in 1961; and a Ph.D. in Aeronautics and Astronautics from the Massachusetts Institute of Technology in 1964.

Mitchell completed flight training in July 1954, was designated a Naval Aviator, having received the highest overall marks during flight training. He was assigned to Patrol Squadron 29, flying land-based patrol planes, deployed to Okinawa, Japan. From 1957-58, he transitioned to carrier-based jet aircraft, and deployed aboard *USS Bon Homme Richard* and *USS Ticonderoga*. Mitchell qualified as a research pilot and flew with Air Development Squadron Five until 1959. From 1964-65, he served as Chief, Project Management Division of the Navy Field Office for the Manned Orbiting Laboratory (MOL). From 1965-66, he attended the USAF Aerospace Research Pilot School for certification as a test pilot, graduating first in his class. Ed Mitchell logged 5,000 hours flying time, including 2,000 hours in jet aircraft.

Selected in 1966 NASA Group 5, he was a part of the *Apollo 13* Mission Operations Team, for which he was awarded the Presidential Medal of Freedom by President Richard Nixon in 1970. Mitchell flew as Lunar Module Pilot of *Apollo 14* in 1971, in which he and Alan Shepard landed on February 05, 1971, and explored the hilly upland Fra Mauro Highlands region on the Moon. They stayed on the Moon for 33 hours, deployed and activated lunar surface scientific equipment and experiments, and collected almost 100 pounds of lunar samples. The sixth person to walk on the Moon, he logged 216 hours and 42 minutes in space, and retired from NASA on October 01, 1972. Ed Mitchell wrote, *"You develop an instant global consciousness, a people orientation, an intense dissatisfaction with the state of the world, and a compulsion to do something about it. From out there on the Moon, international politics look so petty. You want to grab a politician by the scruff of the neck and drag him a quarter of a million miles out and say, "**Look at that, you son of a bitch**"*. Unquote.

On his way back to Earth, he had a powerful savikalpa samadhi experience, and he claimed to have conducted private ESP experiments with his friends on Earth. The results were published in the *Journal of Parapsychology* in 1971. He retired from the US Navy with the rank of Captain, and founded **Edgar D. Mitchell & Associates** of Monterey, California, a "commercial organization promoting ecologically-pure products and services designed to alleviate planetary problems". After moving to Atherton, California, he became founding chairman of the **Institute of Noetic Sciences** (IONS) in Palo Alto, California, in 1973, for the purpose of consciousness research and other "related phenomena". He co-founded the Association of Space Explorers in 1983, and later served as chairman of the **Mitchell Communications Company**. He publicly said he was *"ninety percent sure that many of the thousands of unidentified flying objects, or UFOs, recorded since the 1940s, belong to visitors from other planets"*.

Dr. Edgar Mitchell's 1996 book, *The Way of the Explorer*, discusses his journey into mysticism and space. He said, *"We all know that UFOs are real: now the question is where they come from"*. A NASA spokesman stated: *"NASA does not track UFOs. **NASA is not involved in any sort of cover-up** about alien life on this planet or anywhere in the Universe"*. Dr. Edgar Mitchell died on February 04, 2016, in West Palm Beach, Florida, aged eighty-five years. He was married from 1951 to 1972, with two children. He married his second wife from 1973-84, adopted her three children, and has a child with his third wife of 1989-99.

## Moghbeli, Jasmin. SpaceX Crew 7/Expedition 70.

Jasmin Moghbeli was born in June 1983 in Bad Nauheim, West Germany. She graduated with a Bachelor of Science degree in Aerospace Engineering with Information Technology from the Massachusetts Institute of Technology (MIT); a Master's degree in Aerospace Engineering from the US Naval Postgraduate School in California; and graduated from the US Naval Test Pilot School at Patuxent River, Maryland. She was commissioned as an officer in the US Marine Corps in 2005, and trained to become an AH-1 Super Cobra pilot; deploying overseas three times and completing one hundred and fifty combat missions, including sorties in Afghanistan. As of 2019, Moghbeli had over 2,000 hours flying time.

Selected in June 2017 NASA Astronaut Group 22, she graduated along with thirteen others in January 2020. In March 2022, she was assigned as Commander for SpaceX 'Crew 7', launched aboard *Crew Dragon* on August 26, 2023 and became Flight Engineer to the ISS for 'Expedition 69/70'. 'Expedition 70' began with the departure of *Soyuz MS-23* on September 27, 2023, and will end on April 02, 2024. Moghbeli is part of the NASA *Artemis Team*, focusing on missions to the Moon, and eventually to Mars. She speaks Persian, English, and learned Russian during her astronaut training. Jasmin Moghbeli has received four Air Medals; a Navy and Marine Corps Commendation Medal; three Navy and Marine Corps Achievement Medals; the US Navy Test Pilot School Class 144 Outstanding Developmental Phase II Award; and the Commander Willie McCool Outstanding Student Award as the Class 144 Honor graduate. She is married to Sam Wald, and they have twin girls.

## Morgan, Andrew R. Soyuz MS-13/Expedition 60/61/62/Soyuz MS-15.

Dr. Andrew "Drew" Richard Morgan was born in February 1976 in Morgantown, West Virginia, USA. He received a Bachelor of Science degree in Environmental Engineering from the US Military Academy, West Point, New York, in 1998, and was commissioned into the US Army as a medical officer. He volunteered for the US Army Special Operations Command and served as a physician at Fort Bragg. He deployed to Iraq, Afghanistan, and Africa as part of special operations assignments. He later earned an MD from the Uniformed Services University of Health Sciences. In 2005, he completed his residency in emergency medicine at the Madigan Army Medical Center near Tacoma, Washington. In 2013, Dr. Morgan completed a fellowship in primary care sports medicine at Virginia Commonwealth University.

Morgan was preparing for a tour of duty in Germany when selected by NASA as part of Group 21 (aka The Eight Balls) in June 2013. He underwent two years of basic training, including Russian language, robotics, ISS operations, T-38 flight operations, survival and spacewalk training. He launched aboard *Soyuz MS-13* on July 20, 2019, and served as a Flight Engineer for ISS 'Expedition 60/61/62'. He logged 271 days, 12 hours and 47 minutes in space, returning to Earth aboard *Soyuz MS-15* on April 17, 2020. Dr. Morgan has made seven EVAs totaling 45 hours and 48 minutes. Since his spaceflight, he served as the Mission Support Branch Chief of the NASA Astronaut Office and Increment Lead for Expedition 67. He is married, with four children.

## Morgan, Barbara R.. STS-118.

Barbara Radding Morgan was born in November 1951 in Fresno, California, USA. She received a Bachelor of Science degree in Human Biology from Stanford University in 1973; and obtained her teaching credential from Notre Dame de Namur University, Belmont, in 1974. She began teaching on the Flathead Indian Reservation in Arlee, Montana. From 1975-78, she taught remedial reading and math and second grade at McCall-Donnelly Elementary School in Idaho. From 1978-79 she taught in Quito, Ecuador; and from 1979-98, she taught second, third and fourth grades at McCall-Donnelly Elementary School. Selected in the 1985 Teacher in Space Project, she was backup to Christa McAuliffe for the 1986 *Challenger* mission STS-51-L, which ended in disaster on January 28, 1986. Following McAuliffe's death, Morgan assumed the duties of Teacher in Space Designee; which included public speaking, educational consulting, curriculum design, and serving on the National Science Foundation's Federal Task Force for Women and Minorities in Science and Engineering.

Selected in 1998 NASA Group 17, which consisted of thirty-two candidates including eight pilots, seventeen Mission Specialists, and seven international Mission Specialists, she completed two years of training. Barbara Morgan flew on STS-118 aboard *Endeavour* (August 08-21, 2007), serving as robotic arm operator and transfer coordinator. Over 2,300 kg

(5,000 pounds) of cargo went to the ISS, and over 1,400 kg was returned to Earth aboard the Shuttle. Morgan passed the technician class amateur radio operator license in 2003, and participated in 20-minute question-and-answer sessions with young people at the Discovery Center of Idaho, and other centers. She logged 12 days, 17 hours and 53 minutes in space.

On June 28, 2008, Morgan announced that she would leave NASA for a teaching job at Boise State University (BSU). She became a distinguished educator in residence; a dual appointment to BSU's colleges of engineering and education. On July 04, 2008, she received the "Friend of Education" award from the National Education Association. In August 2008, the Barbara R. Morgan Elementary School opened in McCall, Idaho. In 2008, she received the Adler Planetarium Women in Space Science Award. Actress Mary Chris Wall portrayed Morgan in the 1990 television film ***Challenger***. Barbara Morgan appeared on the 2020 Netflix documentary miniseries ***Challenger: The Final Flight***. She is married to Clay Morgan, with two sons.

## Morin, Lee M.E.. STS-110.

Capt. Dr. Lee Miller Emile Morin was born in September 1952 in Manchester, New Hampshire, USA. He received a Bachelor of Science degree in Mathematical/Electrical Sciences from the University of New Hampshire in 1974; a Master of Science in Biochemistry in 1978, and a Doctor of Medicine (1981) and Doctor of Microbiology from New York University School of Medicine in 1982; a Master of Public Health from the University of Alabama, Birmingham in 1988; and a Master of Science degree in Physics from the University of Houston, Clear Lake, in 2008.

Selected in 1996 NASA Group 16, he was 'Mission Specialist 3' on STS-110 aboard *Atlantis* (April 08-19, 2002) with the main purpose to install the S0 Truss segment on the ISS to the *Destiny* Laboratory Module, and to deliver the 885 kg (1,950 pound) Mobile Transporter assembly. He logged ten days, 19 hours and 42 minutes in space; including two EVAs totaling 14 hours and seven minutes. Capt. Lee Morin belongs to the Aerospace Medical Association; the Force Recon Association; the Undersea and Hyperbaric Medical Society; and the Society of United States Naval Flight Surgeons. He has received eleven awards, including the Meritorious Service Medal; two Meritorious Unit Commendation; the Overseas Service ribbon; the National Defense Service Medal; the Kuwait Liberation Medal; the Southwest Asia Service Medal; and three US Navy badges. He was a recipient of the 1994 Chairman of the Joint Chiefs of Staff Award for Excellence in Military Medicine; and the 1996 Sustaining Membership Lecture Award for the Association of Military Surgeons of the United States. Dr. Lee Morin is married, with two children and three grandchildren.

## Mullane, Richard M.. STS-41-D, STS-27, STS-36.

Col. Richard Michael "Mike" Mullane was born in September 1945 in Wichita Falls, Texas, USA. He received a Bachelor of Science degree in Military Engineering from the US Military Academy, West Point, in 1967; and a Master of Science degree in Aeronautical Engineering from the USAF Institute of Technology in 1975. From January to November 1969, he completed 134 combat missions as an RF-4C weapons systems officer in Vietnam, then served a four-year tour of duty in England. Upon completing the USAF Flight Test Engineer Course at Edwards AFB, he was assigned as a flight test weapons system officer to Test Wing 3246 at Eglin AFB, Florida.

Selected by NASA in January 1978, he flew as a Mission Specialist on three Space Shuttle missions: STS-41-D aboard *Discovery* (August 30-September 05, 1984), when the crew activated the OAST-1 solar cell wing experiment, deployed three satellites, and used an IMAX motion picture camera; STS-27 aboard *Atlantis* (December 02-06, 1988) with a classified Department of Defense payload; and STS-36 aboard *Atlantis* (February 28-March 04, 1990), which carried Department of Defense and several secondary payloads. In 2006, Mike Mullane published an autobiography, *Riding Rockets: The Outrageous Tales of a Space Shuttle Astronaut*, which expresses critical views of the NASA Astronaut Corps. He received ten awards, including six Air Medals; the DFC; the Meritorious Service Medal; NASA Space Flight Medals; the Vietnam Campaign Medal; the National Defense Service Medal; the National Intelligence Medal of Achievement; and was a Distinguished Graduate of the USAF Navigator Training School. Col. Mike Mullane is a member of the Air Force Association, and of the Leadership Board of **For all Moonkind, Inc.**

## Musgrave, F. Story. STS-6, STS-51-F, STS-33, STS-44, STS-61, STS-80.

Dr. Franklin "Story" Musgrave was born in August 1935 in Boston, Massachusetts, USA. He is an athlete, photographer and poet. Musgrave is the most formally educated astronaut, having six academic degrees. He is a public speaker and consultant to both Disney's Imagineering group and Applied Minds in California. He received a Bachelor of Science degree in Mathematics and Statistics from Syracuse University in 1958; a Master of Business Administration (M.B.A.) in Operations Analysis and Computer Programming from the University of California, Los Angeles, in 1959; a Bachelor of Arts degree in Chemistry from Marietta College in 1960; an M.D. from Columbia University College of Physicians and Surgeons in 1964; a Master of Science degree in Physiology and Biophysics from the University of Kentucky; and an Master of Arts degree in Literature from the University of Houston at Clear Lake in 1987. He has flown 17,700 hours in 160 different types of civilian and military aircraft, including 7,500 hours in jet aircraft. Musgrave has made over 800 parachute jumps, including 100 experimental free-falls to study human aerodynamics. He has written or co-authored twenty-five scientific papers in the areas of aerospace medicine and physiology, temperature regulation, exercise physiology and clinical surgery.

Selected as a scientist-astronaut by NASA in August 1967 as a member of Group 6, Story Musgrave completed flight and academic training, then worked on the design and development of the *Skylab* program. He participated in the design and development of all Space Shuttle EVA equipment. He served as a CapCom on two *Skylab* and five Space Shuttle missions. In 1996, he became the second astronaut to fly on six spaceflights. He was 'Mission Specialist 1' on STS-6 aboard *Challenger* (April 04-09, 1983), when a *Tracking and Data Relay Satellite* (IUS/TDRS) was deployed. Musgrave was 'Mission Specialist 2' on STS-51-F aboard *Challenger* (July 29-August 06, 1985), the first pallet-only 'Spacelab' mission, and the first to operate the *Spacelab Instrument Pointing System* (IPS). There were thirteen major experiments in astronomy, astrophysics and life sciences. The Shuttle had to perform an "Abort to Orbit" emergency procedure when its Engine 1 Center SSMME engine shut down. He was 'Mission Specialist 2' on STS-33 aboard *Discovery* (November 22-27, 1989) with payloads for the US Department of Defense.

Story Musgrave was 'Mission Specialist 2' on STS-44 aboard *Atlantis* (November 24-December 01, 1991), which deployed a Department of Defense Support Program (DSP) satellite with an Inertial Upper Stage (IUS) rocket booster. He was 'Mission Specialist 4' on STS-61 aboard *Endeavour* (December 02-13, 1993), the first Hubble Space Telescope servicing mission, when Musgrave performed three of five EVAs. He was 'Mission Specialist 1' on STS-80 aboard *Columbia* (November 19-December 07, 1996), when the crew deployed and retrieved the *Wake Shield Facility* and the *Orbiting Retrievable Far and Extreme Ultarviolet Spectrometer* (ORFEUS) satellites. At 17 days, 15 hours and 53 minutes, this was the longest Shuttle mission ever flown.

Story Musgrave logged a total of 53 days, nine hours and 55 minutes (over 1,281 hours) in space; including nearly 27 hours of EVA. Since he first flew with the US Marines, he has spent 18,000 hours in various air crafts. In 1997, he retired from NASA after thirty years as an astronaut. He has made cameo appearances on several documentary TV programs, as well as the movie ***Mission to Mars*** (2000), and ***Home Improvement*** (Series 3, Episode 24). In 2012, he appeared at Chicon 7, the seventieth World Science Fiction Convention, as the Astronaut Guest of Honor. Story Musgrave has received seven awards, including a USAF Post-doctoral Fellowship (1965-66); the Reese Air Force Base Commander's Trophy (1969); the Flying Physicians Association Airman of the Year Award (1974, 1983); and the NASA Exceptional Service Medal. He was inducted into the International Space Hall of Fame (1995); and the National Aviation Hall of Fame (2022). He had seven children, one of whom is deceased.

## Nagel, Steven R.. STS-51-G, STS-37, STS-55.

Col. Steven Ray Nagel (1946-2014) was born in October 1946 in Canton, Illinois, USA. He received a Bachelor of Science degree in Aeronautical and Astronautical Engineering (with high honors) from the University of Illinois in 1969; and a Master of Science degree in Mechanical Engineering from California State University, Fresno, in 1978. Steven Nagel received his commission in 1969 through the Air Force Reserve Officer Training Corps (AFROTC) program at the University of Illinois. He completed Undergraduate Pilot Training at Laredo AFB, Texas, in February 1970. From October 1970 to July 1971, he was an F-100 pilot at England AFB, Louisiana, then served a tour of duty as a T-28 instructor for the Laotian Air Force at Udon RTAFB, Thailand. He logged 12,600 hours flying time, including 9,640 hours in jet aircraft.

Selected in 1978 NASA Group 8, Nagel was 'Mission Specialist 2' on STS-51-G aboard *Discovery* (June 17-24, 1985), when crew deployed satellites for Morelos of Mexico, the Arab League, and AT & T 'Telstar' of the USA. Col. Nagel was Pilot on STS-61-A aboard *Challenger* (October 30 to November 06, 1985), the Deutschland-1 mission (funded and directed by West Germany) which carried the NASA/ESA 'Spacelab' module with seventy-six experiments on board. Payload operations were controlled from the German Space Operations Center in Oberpfaffenhofen. Col. Nagel was Commander on STS-37 aboard *Atlantis* (April 05-11, 1991), which deployed the Compton Gamma Ray Observatory to explore gamma ray sources throughout the Universe. He commanded STS-55 aboard *Columbia* (April 26-May 06, 1993), when eighty-nine experiments were performed in materials processing, life sciences, robotics, technology, astronomy and Earth mapping.

Steven Nagel logged a total of 30 days, one hour and 34 minutes (over 721 hours) in space. After NASA, he worked at the **University of Missouri** College of Engineering as an instructor in its Mechanical and Aerospace Engineering Department. He retired from the USAF on February 28, 1995. From March 01, 1995, he became Deputy Director for Operations Development, Safety, Reliability, and Quality Assurance Office, JSC, Houston, Texas. In September 1966, he transferred to Aircraft Operations Division, where he performed duties as a Research Pilot. Col. Nagel received fourteen awards, including the DFC; the Air Medal with seven Oak Leaf Clusters; the Orville Wright Achievement Award; the Air Force Meritorious Service Medal (1978); two NASA Exceptional Service Medals (1988, 1989); the NASA Outstanding Leadership Medal (1992); and the NASA Distinguished Service Medal (1994). Col. Steven Nagel died in Columbia, Missouri, on August 21, 2014, aged sixty-seven years, from advanced melanoma. He was married to fellow astronaut Linda M. Godwin, and they had two daughters.

## Nelson, George D.. STS-41-C, STS-61-C, STS-26.

Dr. George Driver "Pinky" Nelson was born in July 1950 in Charles City, Iowa, USA. He received a Bachelor of Science degreee in Physics from Harvey Mudd College in 1972; and a Master of Science (1974) and a Ph.D. in Astronomy from the University of Washington in 1978. Dr. Nelson performed astronomical research at the Sacramento Peak Solar Observatory; the Astronomical Institute at Utrecht, Netherlands; the University of Gottingen Observatory, West Germany; and at the Joint Institute for Laboratory Astrophysics, Boulder, Colorado. Selected in January 1978 NASA Group 8, he was 'Mission Specialist 3' on STS-41-C (formerly STS-13) aboard *Challenger* (April 06-13, 1984), which marked the first direct ascent trajectory for a Space Shuttle mission. During the ascent phase, the Mission Control Center main computer failed, as did the backup, and for about an hour the controllers had no data on the orbiter. The Shuttle crew captured and repaired the Solar Maximum Mission *Solar Max* satellite, and deployed the Long Duration Exposure Facility (LDEF) experimental apparatus, a passive, retrievable, twelve-sided experimental cylinder. The crew tested the *Manned Maneuvering Unit* (MMU) in two EVAs.

Pinky Nelson was 'Mission Specialist 1' on STS-61-C aboard *Columbia* (January 12-18, 1986), with the primary objective to deploy the *Satcom K-1* communications satellite. The Shuttle also carried a large number of small experiments, including thirteen "Getaway Special" (GAS) canisters devoted to investigations involving the effect of microgravity on materials processing, seed germination, chemical reactions, egg hatching, astronomy and atmospheric physics. He was 'Mission Specialist 3' on STS-26 aboard *Discovery* (September 29-October 03, 1988), with the primary payload of the *Tracking and Data Relay Satellite* (TDRS-C). Eleven scheduled mid-deck scientific and technological experiments were carried out. During this mission, *Discovery* became the first spacecraft to fly with a Voice Control Unit (VCU), a computer capable of recognizing and responding to human speech. Voice templates created prior to liftoff were found to have less than 60% recognition for one crew member, and it was found that weightless conditions caused a fundamental change in human speech, so new templates were retrained. Nelson logged a total of 17 days, two hours and 43 minutes (411 hours) in space; including two EVAs totaling ten hours and six minutes.

Dr. George Nelson left NASA in June 1989, became an assistant provost at the University of Washington, and now directs the Science, Mathematics and Technology Education program at **Western Washington University** in Bellingham. He served as faculty on the Fall 2013 Semester at Sea Voyage, where he taught introductory astronomy and celestial navigation. He has received six awards, including the NASA Exceptional Engineering Achievement Medal; the NASA Exceptional Service Medal; the AIAA Haley Space Flight Award; and the FAI V.M. Komarov Diploma. In 2009, he was inducted into the US Astronaut Hall of Fame. Nelson is an elected member of Washington State Academy of Science and and an Elected Fellow of the American Association for the Advancement of Science. He is married, with two daughters.

## Nelson, [Senator] Clarence William, II.. STS-61-C.

Capt. Dr. Clarence William "Bill" Nelson, II was born in September 1942 in Miami, Florida, USA. He is an American politician and attorney, serving since 2021 as the Administrator of the National Aeronautics and Space Administration (NASA). A member of the Democratic Party, he served in the Florida House of Representatives from 1972-78, and was the US Senator from Florida from 2001-19. He received a Bachelor of Arts in Political Science from Yale University in 1965; a Doctorate of Law degree (J.D.) from the University of Virginia School of Law in 1968; and two honorary Doctorates. Nelson served in the US Army Reserve from 1965-71, saw active duty for two years (1968-70), and earned the rank of Captain.

While in the US Congress, he established a reputation as a liberal. He took an interest in climate change, especially its effects on his home state of Florida. In 1986, he was the second sitting member of Congress to travel into space. Senator Bill Nelson began his astronaut training in September 1985 at JSC. He flew as a Payload Specialist on STS-61-C aboard *Columbia* (January 12-18, 1986), when the crew launched a communications satellite and performed several experiments in materials processing and astrophysics. He logged six days, two hours and three minutes in space, and retired from the Astronaut Corps on January 18, 1986. He is married, with two children.

## Newman, James H.. STS-51, STS-69, STS-88, STS-109.

Dr. James Hansen Newman was born in October 1956 in Truk, Palau, the Trust Territory of the Pacific Islands (now the Federated States of Micronesia). He received a Bachelor of Science degree from Dartmouth College; and a Master's degree and a Ph.D. from Rice University. After graduating from Rice University in 1984, he did an additional year of postdoctoral work at Rice. His research interests are in atomic and molecular physics. Selected by NASA in January 1990, Dr. Newman was detailed to the Space Shuttle Program Office from March 1999 to March 2001, serving as the Remote Manipulator System (RMS) Integration Manager responsible for the Orbiter Canadian robotic arm and the Space Vision System.

He was 'Mission Specialist 1' on STS-51 aboard *Discovery* (September 12-22, 1993), when the crew deployed the *Advanced Communications Technology Satellite* (ACTS) and the *ORFEUS/SPAS satellite* and its IMAX camera. Newman was responsible for the operation of the *Shuttle Pallet Satellite* (SPAS), was the backup operator for the RMS, and on flight day five made a seven-hour five-minute EVA with Carl Walz. The crew tested a Global Positioning System (GPS) receiver to determine real-time Shuttle positions and velocities. Newman was 'Mission Specialist 2' on STS-69 aboard *Endeavour* (September 07-18, 1995), when the crew deployed and retrieved a *SPARTAN-201* satellite (on its third flight) and the *Wake Shield Facility* (WSF). The purpose of the WSF was to grow thin films in a near perfect vacuum created by the wake of the Shuttle as it moved through space. Also on board was the International Extreme Ultraviolet Hitchhiker payload, numerous secondary payloads, and medical experiments.

He was 'Mission Specialist 3' on STS-88/2A aboard *Endeavour* (December 04-16, 1998), the first ISS assembly mission, when the *Unity* module was mated with the *Zarya* module. Dr. Newman performed three EVAs totaling 21 hours and 22 minutes. They crew connected power and data cables between the Node, PMAs and the FGB. An S-band communications system was established to enable US flight controllers to monitor the systems on the ISS. Two satellites, the *Mighty Sat 1*, sponsored by the US Air Force, and *SAC-A* from Agentina were deployed.

James Newman was 'Mission Specialist 4' on STS-109 aboard *Columbia* (March 01-12, 2002), the fourth Hubble Space Telescope servicing mission. A cooler was installed to reactivate an old infrared camera. New equipment was installed with five EVAs in five consecutive days. Newman performed two EVAs totaling 14 hours and 46 minutes. He logged a total of 43 days, ten hours and seven minutes in space; including six EVAs totaling 43 hours and 13 minutes. From December 01, 2002, to January 2006, he served as NASA's Director, Human Space Flight Programs, Russia. He oversaw NASA's mission operations in Korolev, and crew training at the Gagarin Center in Star City, Russia. In March 2006, he was detailed to the Naval Postgraduate School, Monterey, California, as a NASA Visiting Professor in the NPS Space Systems Academic Group. Dr. James Newman is a member of the American Physical Society; and the American Institute of Aeronautics and Astronautics. He is married, with three children.

## Noriega, Carlos I.. STS-84, STS-97.

Lt. Col. Carlos Ismael Noriega was born in October 1959 in Lima, Peru. He received a Bachelor of Science degree in Computer Science from the University of Southern California in 1981; and a Master of Science degree in Computer Science plus a Master of Science degree in Space Systems from the US Naval Postgraduate School in 1990. He received his commission in the US Marine Corps in 1981. From 1983-85, he flew CH-46 Sea Knight helicopters with HMM-165 in Hawaii. In 1986, Noriega was transferred to MCAS Tustin, California, where he served as the aviation safety officer and instructor pilot with HMT-301. In September 1990, he was assigned to **United States Space Command** in Colorado Springs, Colorado. He was ultimately the command representative for the development and integration of the major space and missile warning computer system upgrades for Cheyenne Mountain Air Force Station. He has logged about 2,200 flight hours in various fixed-wing and rotary-wing aircraft.

Selected in December 1994 NASA Group 15, Lt. Col. Noriega completed a year of training and evaluation. He was 'Mission Specialist 2' on STS-84 aboard *Atlantis* (May 15-24, 1997), the sixth Space Shuttle mission to dock with Russian Space Station *Mir*, when 3,318 kg (7.315 pounds) of supplies, including 465 kg of water, 1,168.6 kg (2,576 pounds) of Russian logistics and 383.2 kg of US science experiment equipment were transferred between the craft. Noriega was 'Mission Specialist 3' on STS-97 aboard *Endeavour* (December 01-11, 2000), the fifth Shuttle mission dedicated to assembly of the ISS. The crew delivered the P6 truss, installed the first set of US solar arrays, performed three spacewalks, and delivered supplies and equipment. Carlos Noriega logged a total of 20 days, one hour and 18 minutes in space; including three EVAs totaling 19 hours and 20 minutes. Lt. Col. Noriega has received six awards, including two Defense Meritorious Service Medals; the Air Medal with Combat Valor device; the Air Medal (Strike Flight Award); and the Navy Achievement Medal.

## Nowak, Lisa M.. STS-121.

Commander Lisa Marie Nowak (nee Caputo) was born in May 1963 in Washington, D.C., USA. While growing up, she followed the Space Shuttle program, particularly the introduction of female astronauts in 1978, and paid frequent visits to the National Air and Space Museum. She graduated from the US Naval Academy on May 22, 1985, with a Bachelor of Science degree in Aeronautical Engineering, and was commissioned as an Ensign in the US Navy. In June 1987, she qualified as a Naval Flight Officer (NFO). On April 06, 1988, she married an Annapolis classmate, Richard T. Nowak. In 1990, Lisa Nowak entered the US Naval Postgraduate School in Monterey, California, where she earned both a Master of Science degree in Aeronautical Engineering and a degree in Aeronautical and Astronautical Engineering in September 1992.

She graduated from the US Naval Test Pilot school at NAS Patuxent River in June 1994, and served as a Naval flight officer and test pilot, participating in the development of the F/A-18 Hornet and EA-6B Prowler. During her Navy career, Lisa Nowak logged over 1,500 hours flying time in over thirty aircraft types, and was awarded the Defense Meritorious Service Medal; the Navy Commendation Medal; and the Navy Achievement Medal. On June 15, 1995, NASA announced it was selecting a new group of astronauts. Over 2,400 applications were received, and Nowak was one of 150 finalists deemed highly qualified. Thirty-five were selected in 1996 NASA Group 16, and she qualified as a Mission Specialist in robotics. In October 2001, she gave birth to twin daughters.

In January 2004, Lisa Nowak took part in an eleven-day cold weather survival training course in Canada with five other space trainees, including William Oefelein. After returning to Houston, they began an extramarital affair, which they attempted to conceal. As serving Navy officers, they could have been charged under the Uniform Code of Military Justice. Nowak was 'Mission Specialist 2' on STS-121 aboard *Discovery* (July 04-17, 2006), when the crew continued to test new equipment and procedures for the inspection and repair of the shuttle's thermal protection system. Two tons of cargo were delivered to the ISS, including the MPLM *Leonardo* on its fourth flight. The Minus Eighty Degree Laboratory Freezer (MELFI) and the European Modular Cultivation System (EMCS) for biological experiments were included. She logged 12 days, 18 hours and 36 minutes in space.

Lisa Nowak was involved in an ongoing adulterous relationship with William Oefelein. She was arrested on February 05, 2007, when she used pepper spray and attempted to kidnap fellow astronaut Bill Oefelein's current girlfriend from her car at Orlando International Airport. There was widespread public reaction to her arrest, concerns being expressed about NASA's astronaut selection and screening processes. A psychiatic examination diagnosed her with obsessive-compulsive personality disorder, Asperger syndrome, and a single episode of major depressive disorder. She later pleaded guilty to felony burglary and misdemeanor battery, and was sentenced to time served plus one year's probation. Nowak and Oefelein became the first astronauts ever dismissed from NASA, which then created its first astronaut Code of Conduct. A new Astronaut Code of Professional Responsibility was issued. Lisa Nowak remained a Navy Captain until 2010, when a Naval Board of Inquiry voted unanimously to reduce her rank to Commander, and to unfavorably discharge her after twenty-five years of service.

## Nyberg, Karen L.. STS-124, Soyuz TMA-09M/Expedition 36/37.

Dr. Karen LuJean Nyberg was born in October 1969 in Pakers Prairie, Minnesota, USA. Of Norwegian ancestry, she graduated *summa cum laude* "with high praise" with a Bachelor of Science degree in Mechanical Engineering from the University of North Dakota (UND) in 1994; a Master of Science degree in Mechanical Engineering from the University of Texas at Austin in 1996; followed by a Ph.D. in Mechanical Engineering from the University of Texas in 1998. Selected in July 2000 NASA Group 18, she took part in NEEMO 10, a deep-sea training and simulation exercise at the *Aquarius* underwater laboratory offshore from Florida. Karen Nyberg became the fiftieth woman in space on her first mission in 2008. She was 'Mission Specialist 1' on STS-124 aboard *Discovery* (May 31-June 14, 2008), which delivered the Pressurized Module of the Japanese Experiment Module (JEM), called *Kibo,* which was berthed to the *Harmony* module of the ISS. The Japanese Remote Manipulator system was also delivered and attached to *Kibo*. Nyberg was lead for all the robotic arm operations.

She was 'Flight Engineer 2' on *Soyuz TMA-09M*, [spacecraft 11F732A47 No. 709, launched with a *Soyuz-FG* carrier rocket] for 'Expedition 36/37' (May 28-November 11, 2013). The *Soyuz* docked with the nadir port of the *Rassvet* module, less than six hours after launch, as part of a fast rendezvous profile designed to minimise the time the crew must spend in the cramped *Soyuz* spacecraft. On June 16, 2013, the fiftieth anniversary of *Vostok 6*, the first spaceshot by a woman (Valentina Tereshkova of the USSR), there were two women in space: Karen Nyberg of NASA, and Wang Yaping of China aboard *Tiangong-1* on the *Shenzhou 10* mission. Nyberg returned to Earth on *Soyuz TMA-09M* after a flight of 166 days, six hours and 18 minutes. She logged a total of 180 days in space. From 2013 to her retirement in March 2020, Dr. Nyberg worked for NASA in the Space Shuttle branch, the Exploration branch, and as Chief of the Robotics branch. She has received numerous awards, including the D.J. Robertson Award of Academic Achievement (1992); the Space Act Award (1993); the Joyce Medalen Society of Women Engineers Award (1993-94); the UND Young Alumni Achievement Award (2004); and the NASA Distinguished Service Medal (2014). Karen Nyerg is married to fellow astronaut Doug Hurley, and they have a son.

## Ochoa, Ellen L.. STS-56, STS-66, STS-96, STS-110.

Dr. Ellen Lauri Ochoa was born in May 1958 in Los Angeles, California, USA. She received a Bachelor of Science degree in Physics from San Diego State University in 1980; and a Master of Science (1981) and a Ph.D. from Stanford University Department of Electrical Engineering in 1985. As a doctoral student at Stanford, and later as a researcher at Sandia National Laboratories and the NASA Ames Research Center, Ochoa investigated optical systems for performing information processing. She is a co-inventor on three Patents for an optical inspection system, an optical object recognition method, and a method for noise removal in images. She wanted to help computers "see". NASA recognized her methods in optics could be used in fabricating: such as inspecting for flaws, or it might be used on a space vehicle such as the Rover.

Selected in January 1990 NASA Group 13, she flew as 'Mission Specialist 3' on STS-56 aboard *Discovery* (April 08-17, 1993), becoming the first Hispanic woman to go to space. The primary payload was ATLAS-2; one element of NASA's Mission to Planet Earth program. The purpose was to collect data on the relationship between the Sun's energy output and Earth's middle atmosphere, and study how these factors affect the Earth's ozone layer. Ochoa is a classical flutist, and took a flute to space with her. She was 'Mission Specialist 1' and Payload Commander on STS-66 aboard *Atlantis* (November 03-14, 1994) with the ATLAS-03 primary payload. This mission made the first detailed measurements from the Shuttle of the Northern Hemisphere's middle atmosphere in late Fall. The timing of the flight allowed scientists to study possible effects of the ozone hole on mid-latitudes.

Ellen Ochoa was Mission Specialist and Flight Engineer on STS-96/2A.1 aboard *Discovery* (May 27-June 06, 1999), a logistics and resupply mission to the ISS. The Shuttle carried an Integrated Cargo Carrier (ICC) with parts for the Russian cargo crane *Strela*, which was mounted to the exterior of the Russian station segment. Dr. Ellen Ochoa was 'Mission Specialist 2' and Flight Engineer on STS-110 aboard *Atlantis* (April 08-19, 2002), whose main purpose was to attach the stainless steel S0 Truss segment to the ISS and the Destiny Laboratory Module. The S0 truss forms the backbone to the station, to which the S1 and P1 truss segments were attached on the following STS-112 and STS-113 missions. This mission also delivered the Mobile Transporter (MT), an 885 kg assembly that glides down rails on the Space Station's integrated trusses. The MT thus allows the Canadarm2 to travel down the length of the installed truss structure. Dr. Ochoa logged a total of 41 days, 19 hours and 35 minutes (over 950 hours) in space.

From 2007, Dr. Ochoa served as Deputy Director of NASA's Johnson Space Center (JSC), helping to manage and direct the Astronaut Office and Aircraft Operations. She was Director from 2013. Ellen Ochoa has received many awards, including the NASA Outstanding Leadership Medal (1995); the NASA Exceptional Service Medal (1997); and the NASA Distinguished Service Medal (2015). She is a Fellow of the American Association for the Advancement of Science; the American Institute of Aeronautics and Astronautics, the National Academy of Inventors, and Optica. She was named Vice Chair of the **National Science Board** for the 2018-20 term. She currently chairs the committee evaluating nominations for the National Medal of Technology and Innovation. Dr. Ellen Ochoa is married to Coe Miles, an intellectual property attorney, and they have two sons.

## O'Connor, Bryan D.. STS-61-B, STS-40.

Col. Bryan Daniel "O.C." O'Connor was born in September 1946 in Orange, California, USA. He received a Bachelor of Science degree in Engineering from the US Naval Academy in 1968; and a Master of Science degree in Aeronautical Systems from the University of West Florida in 1970. He graduated from the Naval Safety School at the US Naval Postgraduate School, Monterey, California, in 1972 and from the US Naval Test Pilot School, NAS Patuxent River, Maryland, in 1976. Selected on May 19, 1980 in NASA Group 9, he was pilot on STS-61-B aboard *Atlantis* (November 26-December 03, 1985), which deployed the *Aussat-2, Morelos-2* and *Satcom-K2* satellites; all equipped with Payload Assist Modules to boost them to geostationary orbits. A key mission objective was EASE/ACCESS, an experiment in assembling large structures in space. The astronauts judged that performing six-hour spacewalks every other day over a five or six-day period was feasible, and recommended glove changes to reduce hand fatigue.

O'Connor was Commander on STS-40 aboard *Columbia* (June 05-14, 1991) for 'Spacelab Life Sciences-1', the fifth 'Spacelab' mission and the first dedicated solely to biology. These were the most detailed and interrelated physiological measurements in space since the 1973-74 *Skylab* missions. The subjects involved were humans (studying six body systems), thirty rodents (for seven experiments) and thousands of tiny jellyfish. Bryan O'Connor logged a total of 15 days, 23 hours and 18 minutes in space. He retired from NASA on July 29, 1991. Col. O'Connor has received thirteen awards, including two Defense Superior Service medals; the DFC; two NASA Outstanding Leadership Medals; two NASA Exceptional Service Medals; the Navy Meritorious Service Medal; the AIAA System Effectiveness and Safety Award; the AIAA Barry M. Goldwater Education Award; *Aviation Week & Space Technology* Laureate (Space and Missiles); and was inducted into the US Astronaut Hall of Fame in 2008. He is married, with two children.

## Oefelein, William A.. STS-116.

William "Bill" Anthony Oefelein was born in March 1965 in Fort Belvoir, Virginia, USA. He is an American freelance adventure writer and photographer and former NASA astronaut. He received a Bachelor of Science degree in Electrical Engineering from Oregon State University in 1988; and a Master of Science degree in Aviation Systems from the University of Tennessee Space Institute in 1998. He then became an Ensign in the US Navy and was designated a Naval Aviator in September 1990. An early assignment was to the Persian Gulf as a strike fighter pilot in the F/A-18 Hornet. After attending the US Navy Strike Fighter Tactics Instructor program (aka TOPGUN), he graduated in 1995 from the US Naval Test Pilot School at NAS Patuxent River, Maryland, where he later served as an instructor.

In June 1998, Oefelein was selected in NASA Group 17. He was pilot on STS-116 aboard *Discovery* (December 10-22, 2006), when crew added the P5 spacer truss segment during the first of four EVAs. The fourth EVA was added to allow the crew to coax and retract a solar panel to fold up accordion-style into its box. There was a major re-wire of the ISS's power system, and exchange of 'Expedition 14' personnel. Over two tons of equipment and supplies were delivered to the ISS, and almost two tons of items no longer needed were returned with STS-116. Bill Oefelein logged 12 days, 20 hours and 44 minutes (over 308 hours) in space. Following the arrest of fellow astronaut Lisa Nowak in February 2007 (his married former girlfriend from 1998), he admitted to a two-year affair with her, and they became the first astronauts dismissed from NASA. On May 23, 2007, NASA terminated Oefelein's service, and he was reassigned to the US Navy. He retired from the Navy in late 2008, returning to Alaska to start a business called **Adventure Write** as a freelance writer and photographer with USAF Captain Colleen Shipman, then his fiancee and current wife. Oefelein and his first wife had two children, and divorced in 2005. He and Shipman married in Summer 2010, and they have a son. In September 2011, he survived a small plane crash in rural Alaska, near where they live.

## O'Hara, Loral A.. Soyuz MS-24/Expedition 69/70.

Loral Ashley O'Hara was born in May 1983 in Houston, Texas, USA. She received a Bachelor of Science degree in Aerospace Engineering from the University of Kansas in 2005, then worked for **Rocketplane Ltd.** in Oklahoma City. In 2008, she was awarded a National Science Foundation Graduate research fellowship, and earned a Master of Science degree in Aeronautics and Astronautics from Purdue University in 2009. She then began working at the Woods Hole Oceanographic Institution. Selected by NASA in June 2017, she graduated from the astronaut corps in January 2020. O'Hara launched on September 15, 2023, aboard *Soyuz MS-24* to join ISS 'Expedition 69/70'. She has made one EVA of six hours and 42 minutes. She is a private pilot, certified Emergency Medical Technician, and a wilderness First Responder. In 2015, she was an invited speaker at TEDx New Bedford. As of March 28, 2024, she was in space.

## Olivas, John D.. STS-117, STS-128.

Dr. John Daniel "Danny" Olivas was born in May 1966 in North Hollywood, California, USA. Of Mexican descent, he received a Bachelor of Science degree in Mechanical Engineering from the University of Texas at El Paso in 1989; a Master of Science degree in Mechanical Engineering from the University of Houston in 1993; and a Doctorate in Mechanical Engineering and Materials Science from Rice University in 1996. Selected on June 04, 1998, in NASA Group 17, from 1999 to 2002 he was assigned technical responsibilities within the Robotics Branch as lead for the Special Purpose Dexterous Manipulator Robot and the Mobile Transporter. In July 2002, he served as an aquanaut during the NEEMO 3 mission aboard the *Aquarius* underwater laboratory, offshore at Key Largo, Florida. In April 2005, he was a crew member on the NEEMO 8 underwater training mission.

Dr. Danny Olivas was 'Mission Specialist 3' on STS-117 aboard *Atlantis* (June 08-22, 2007), the twenty-first mission to visit the ISS, delivering the second starboard truss segment, the third set of US solar arrays, batteries, and associated equipment. This mission entailed the first on-orbit EVA repair to the Space Shuttle. Olivas spent 14 hours and 13 minutes in two EVAs. He was 'Mission Specialist 3' on STS-128 aboard *Discovery* (August 29-September 12, 2009). Three EVAs removed and replaced a materials processing experiment outside the ESA's *Columbus* module, and returned an empty ammonia tank assembly. Dr. John Olivas logged a total of 27 days, 17 hours and five minutes in space; including EVAs totaling 34 hours and 28 minutes. He retired on May 25, 2010. In 2013, he joined the University of Texas at El Paso as Director of the Center for the Advancement of Space Safety and Mission Assurance Research (CASSMAR). In 2019, he appeared as a contestant on ***Nickelodeon***'s revival of *Are You Smarter than a Fifth Grader?* He is married, with five children.

## Oswald, Stephen S.. STS-42, STS-56, STS-67.

Rear Adm. Stephen Scot Oswald was born in June 1951 in Seattle, Washington, USA. He received a Bachelor of Science degree from the US Naval Academy in 1973. He was designated a Naval Aviator in September 1974, and flew aboard the aircraft carrier *USS Midway* from 1975-77. In 1978, he attended the US Naval Test Pilot School at Patuxent River; and remained there to conduct flying qualities, performance, and propulsion flight tests until 1981. In November 1984, he joined NASA Group 11 as an aerospace engineer and instructor pilot, and was selected as an astronaut candidate in June 1985. He later resigned from active Navy duty and joined **Westinghouse Electric Corporation** as a civilian test pilot. In 1988, he transferred to the Naval Reserve space community. Oswald logged over 7,000 flight hours in over forty aircraft types.

He was pilot on STS-42 aboard *Discovery* (January 22-30, 1992), the International Microgravity Laboratory-1 mission where the crew conducted experiments on the human nervous sytem's adaptation to low gravity, and the effects of microgravity on other life forms including shrimp eggs, lentil seedlings, fruit fly eggs, and bacteria. Oswald was pilot on STS-56 aboard *Discovery* (April 08-17, 1993), the second Atmospheric Laboratory for Applications and Science (ATLAS-2) mission. He commanded STS-67 aboard *Endeavour* (March 02-18, 1995), the second flight of the Astro Observatory (Astro II). Oswald logged a total of 33 days, 22 hours and 30 minutes in space.

He was then assigned to NASA HQ in Washington, D.C., as deputy Associate Administrator for Space Operations. He was responsible for Space Shuttle, expendable launch vehicles, and space communications for the agency. Oswald returned to the Astronaut office in July 1998, and retired from NASA in January 2000. Stephen Oswald is a member of the Society of Experimental Test pilots; the Association of Space Explorers; the Naval Reserve Association; and the American Institute of Aeronautics and Astronautics. He received seven awards, including the Legion of Merit; the DFC; the Meritorious Service Medal; two Navy Commendation Medals; the NASA Outstanding Leadership Medal; two NASA Exceptional Service Medals; and various service awards. Stephen Oswald is married, with three children.

## Overmyer, Robert F.. STS-5, STS-51-B.

Col. Robert "Bob" Franklyn Overmyer (1936-96) was born in July 1936 in Lorain, Ohio, USA. He received a Bachelor of Science degree in Physics from Baldwin-Wallace College in 1958; and a Master of Science degree in Aeronautics (with a major in Aeronautical Engineering) from the US Naval Postgraduate School in 1964. Overmyer entered active duty with the US Marine Corps in January 1958. In 1962, he was assigned to study aeronautical engineering, served one year with Marine Maintenance Squadron 17 at MCAS Iwakuni, Japan, then was assigned to Edwards AFB, California. He logged over 7,500 flying hours, with over 6,000 in jet aircraft. Selected by the USAF as an astronaut for its Manned Orbiting Laboratory (MOL) in 1966, he transferred to NASA in 1969 and served support crew duties for the Apollo program, Skylab program (1969-71), and the *Apollo-Soyuz Test Project* (1973-75).

In 1976, Overmyer was assigned to the Space Shuttle program for the Approach and Landing Tests (ALT) program, flying as pilot on STS-5 in 1982 and as commander on STS-51-B in 1985. He was selected as a lead investigator into the shuttle *Challenger* disaster in 1986, and was involved in the recovery of the crew remains. STS-5 aboard *Columbia* (November 11-16, 1982) was the first four-man crew, and marked the first use of the Payload Assist Module (PAM-D) and its new ejection system. Numerous flight tests were performed to document shuttle performance during launch, boost, orbit, atmospheric re-entry and landing phases. STS-51-B aboard *Challenger* (April 29-May 06, 1985), the 'Spacelab-3' (SL-3) mission, was the first to launch a small payload from a "Getaway Special" canister. A broad range of experiments were conducted: from space physics to the suitability of animal holding facilities. He logged a total of 12 days, two hours and 22 minutes in space.

Col. Robert Overmyer retired from NASA and the Marine Corps in May 1986. He commenced his own consulting business, **Mach Twenty Five International, Inc**. In March 1988, he joined the Space Station Team at **McDonnell Douglas Aerospace**, where he led crew and operations activities for seven years, until April 1995. He then continued his aerospace consultation work, as well as speaking engagements and writing. In November 1995, he joined **Cirrus Design**, a Minnesota-based aircraft manufacturer as a test pilot, helping pioneer the use of glass cockpits, composite materials and ballistic parachutes in the light production aircraft industry. Col. Overmyer was awarded the Air Force Meritorious Service Medal (1969); the Marine Corps Meritorious Service Medal (1978); US Naval Postgraduate

School Distinguished Engineers Award (January 1983); the DFC (1983); and an Honorary Doctor of Philosophy degree from Baldwin-Wallace College (December 1982). He was a member of the Society of Experimental Test Pilots; the Experimental Aircraft Association; and the Aircraft Owners and Pilots Association. Robert Overmyer died in a plane crash on March 22, 1996, while testing the Cirrus VK-30 homebuilt aircraft, and was buried at Arlington National Cemetery. He was married, with three children.

## Pailes, William A.. STS-51-J.

Col. William Arthur Pailes was born in June 1952 in Hackensack, New Jersey, USA. He received a Bachelor of Science degree in Computer Science from the USAF Academy in Colorado Springs, Colorado, in 1974; and a Master of Science degree in Computer Science from Texas A & M University in 1981. From 1974-75, he attended pilot training at the Williams AFB, Arizona, where he trained as a HC-130 rescue pilot in the Aerospace Rescue and Recovery Service. In 1978 he attended the Squadron Officer School. Selected in September 1982, he was a Payload Specialist on STS-51-J aboard *Atlantis* (October 03-07, 1985), the second Space Shuttle Ministry of Defense mission. He logged four days, one hour and 44 minutes (over 97 hours) in space. From January 1983, he was a manned spaceflight engineer at USAF Manned Spaceflight Engineering Program, and retired from the astronaut corps in July 1987. From August 2002 to 2011 he was the instructor of the AFJROTC Program, Corsicana, Texas; and became a senior aerospace science instructor from August 2011-19. He is married, with one child, and has been a deacon and treasurer of his church in El Segundo, California.

## Parazynski, Scott E.. STS-66, STS-86, STS-95, STS-100, STS-120.

Dr. Scott Edward Parazynski was born in July 1961 in Little Rock, Arkansas, USA. In March 1969, he traveled to Florida to witness the launch of *Apollo 9*, an event he claims inspired him to pursue a career in science. Of Polish descent, he received a Bachelor of Science degree in Biology (1983); and an MD (with honors) from Stanford University Medical School in 1989. He has logged over 2,500 flying hours. Selected in 1992 NASA Group 14, he was 'Mission Specialist 4' on STS-66 aboard *Atlantis* (November 03-14, 1994) with the primary payload being the Atmospheric Laboratory for Applications and Sciences-3 (ATLAS-3), continuing the series of 'Spacelab' flights. This mission deployed and retrieved the *Cryogenic Infrared Spectrometer Telescope for Atmosphere* (CRISTA), designed to explore the variability of Earth's atmosphere. Parazynski was 'Mission Specialist 2' on STS-86 aboard *Atlantis* (September 26-October 06, 1997), the seventh space station *Mir* docking mission. The Shuttle carried a double module, from which four tons of materials (with 771 kg of water) were transferred. The first joint US-Russian EVA during a Shuttle mission was made by cosmonaut Vladimir Titov and Parazynski on October 01, lasting five hours and one minute. Mission STS-95 aboard *Discovery* (October 29-November 07, 1998) used the 'SpaceHab' module to perform life-sciences experiments. The *SPARTAN-201* spacecraft was released to study the acceleration of the solar wind that originates in the Sun's solar corona.

Parazynski was 'Mission Specialist 3' on STS-100 aboard *Endeavour* (April 19-May 01, 2001), which had the main objectives of installing, activating and checking out the **Canadarm2** robotic arm on the ISS. The *Rafaello* Multi-Purpose Logistics Module (MPLM) was berthed, unloaded, and returned to the orbiter's payload bay. The shuttle also boosted the space stations' altitude, and performed a flyaround of the complex; recording views of the ISS with an IMAX cargo bay camera. STS-120 aboard *Discovery* (October 23-November 07, 2007) was highlighted by a dramatic, unplanned EVA to repair a live solar array. The Shuttle first delivered the US *Harmony* (Node 2) module to the ISS, and installed it to the port node of the *Unity* module. On flight day ten, the crew took time to gather in the *Harmony* module and speak with former President George H.W. Bush and his wife, Barbara, who were visiting Johnson Space Center. Astronauts used five "cufflinks" assembled by crew to take the structural load off the damaged hinges of a solar array. On flight day twelve, Parazynski and Wheelock managed to fully deploy the faulty P6 4B solar array, then made additional inspections of the 2B and 4B arrays. Parazynski's accumulated EVA time of 47 hours and five minutes now placed him fifth overall for total EVA duration.

Scott Parazynski logged a total of 57 days, 15 hours and 34 minutes in space; including seven EVAs. He retired from NASA in March 2009 to pursue entrepreneurial private opportunities, and is currently the CEO of a technology start-up. Dr. Scott Parazynski is a Fellow of the Aerospace Medical Association and The Explorers Club. He is also a member of the American Society for Gravitational and Space Biology; the Wilderness Medical Society; the American Alpine Club; the Association of Space Explorers; the Experimental Aircraft Association; and the Aircraft Owners and Pilots Association. On May 20, 2009, he summitted the 8,848 meter (29,029 feet) Mount Everest. Parazynski describes his life's experiences in his memoir, *The Sky Below*. In 2014, he was appointed as a professor of practice at Arizona State University, and subsequently founded **Fluidity Technologies**. He has received twenty-two awards, including the NASA Graduate Student Researcher's Award (1988), and asteroid (534299) 2014 SP311 was renamed Parazynski (2021). He is married, and has two children with his first wife.

## Parise, Ronald A.. STS-35, STS-67.

Dr. Ronald "Ron" Anthony Parise (1951-2008) was born in May 1951 in Warren, Ohio, USA. By the age of eleven years, he was a licensed amateur radio operator. He became a pilot during his teenage years, attended Western Reserve High School and graduated in 1969. Parise received a Bachelor of Science degree in Physics (with minors in mathematics, astronomy and geology) at Youngstown State University in 1973; and a Master's degree (1977) and a Ph.D., both in Astronomy, from the University of Florida in 1979. He joined **Operations Research, Inc.**, where he worked on avionics requirements and failure mode analyses for several NASA missions. In 1980, he began work at **Computer Sciences Corporation**, in the International Ultraviolet Explorer (IUE) operations center. In 1981, he began work on the development of a new Spacelab experiment called the Ultraviolet Imaging Telescope (UIT). Ron Parise was selected by NASA on June 20, 1984, for the newly formed Astro mission series, beyond which he played a crucial role in advanced planning and communications engineering support for projects including Mir, the ISS, and the X-38 vehicle.

He was 'Mission Specialist 3' on STS-35/Astro-1 aboard *Columbia* (December 02-10, 1990), which used three telescopes to simultaneously record spectral data, polarimetric data and imagery of faint astronomical objects in the far ultraviolet. STS-67/Astro-2 aboard *Endeavour* (March 02-18, 1995) was the second flight of the Astro observatory. The crew worked round-the-clock to study the far ultraviolet spectra of faint astronomical objects and the polarization of ultraviolet light coming from hot stars and distant galaxies. Parise brought amateur radio equipment to the Space Shuttle, and operated on air during his missions. He logged a total of 25 days, 14 hours and 13 minutes (over 614 hours) in space, and retired from NASA on March 18, 1995. At the end of his career, Ronald Parise supported the Goddard Space Flight Center, Networks and Mission Services Project, in the area of advanced communications planning for human spaceflight missions. He died from a brain tumor on May 09, 2008, in Silver Spring, Maryland, aged fifty-six years. He was married, with two children.

## Parker, Robert A.R.. STS-9, STS-35.

Dr. Robert Allan Ridley Parker was born in December 1936 in New York City, USA. He received a Bachelor of Science degree in Astronomy and Physics from Amherst College in 1958; and a Master of Science and a Ph.D. from the California Institute of Technology in 1962. He logged over 3,500 hours in jet aircraft. Selected as a scientist-astronaut in August 1967 NASA Group 6, Parker served on the astronaut support crews for *Apollo 15* and *Apollo 17* and later was program scientist during the three manned *Skylab* flights. From March 1988 to March 1989, he served as director of the NASA HQ Space Flight/Space Station Integration Office.

Dr. Robert Parker flew as 'Mission Specialist 2' on STS-9 Spacelab-1 aboard *Columbia* (November 28 to December 08, 1983), which carried the first 'Spacelab' laboratory into orbit. This was a joint NASA/ European Space Agency (ESA) program designed to demonstrate the ability to conduct advanced scientific research in space. The crew worked in two twelve-hour shifts in the 'Spacelab' module, and coordinated their efforts on seventy-two experiments with scientists at the Marshall Space Flight Center (MSFC) and Payload Operations Control Center (POCC). Four hours before re-entry, one of the flight control computers crashed, followed by a second a few minutes later, but was successfully rebooted. Commander John Young delayed the landing, and later testified: *"Had we then activated the Backup Flight Software,* ***loss of vehicle and crew would have resulted****".* Later analysis revealed the General Purpose Computers failed when the reaction Control System thrusters knocked a piece of solder loose, which shorted out the CPU circuit board.

Parker flew as 'Mission Specialist 3' on STS-35 aboard *Columbia* (December 02-10, 1990), which featured the ASTRO-1 ultraviolet astronomy laboratory consisting of four telescopes. The primary objectives were round-the-clock observations of the celestial sphere in ultraviolet and X-ray spectral wavelengths. The Shuttle Amateur Radio Experiment (SAREX-II) communicated in line-of-sight of the orbiter in one of four transmission modes. He logged a total of 19 days, six hours and 52 minutes (over 463 hours) in space.

Dr. Robert Parker was director of the Division of Policy and Plans for the Office of Space Flight at NASA HQ from January to December 1991. After two interim posts, he was named Director of the NASA Management office at the Jet Propulsion Laboratory, Pasadena, California. Parker retired from NASA on August 31, 2005. He is a member of the American Astronomical Society and of the International Astronomical Union. He was awarded the NASA Exceptional Scientific Achievement Medal (1973); and the NASA Outstanding Leadership Medal (1974). In the 1998 miniseries ***From the Earth to the Moon***, Parker was portrayed by Chris Ellis. He is married, with five children and nine grandchildren.

## Patrick, Nicholas J.M.. STS-116, STS-130.

Dr. Nicholas James MacDonald Patrick was born in November 1964 in Saltburn-by-the-Sea in the North Riding of Yorkshire, England, UK. He grew up in both London and Rye, New York, USA, and became a US citizen in 1994. Patrick learned to fly during his university years as a member of the RAF's Cambridge University Air Squadron. After Cambridge, he worked for four years as an engineer for the Aircraft Engines Division of **General Electric**, in Lynn, Massachusetts, USA. He received a Bachelor of Science degree in Engineering from Trinity College, Cambridge, UK in 1986; and a Master of Science degree (1990) and a Ph.D. from Massachusetts Institute of Technology (MIT) in 1996, both in Mechanical Engineering. He then joined **The Boeing Company**'s Commercial Airplane Group in Seattle, USA. Selected in June 1998 NASA Group 17, in July 2004, Dr. Nicholas Patrick served as an aquanaut during the NEEMO 6 mission, living and working underwater in the *Aquarius* laboratory offshore from Florida for ten days. He occasionally works as a CapCom, e.g., for the Orbit 2 team for STS-120.

Patrick flew as 'Mission Specialist 1' on STS-116 aboard *Discovery* (December 10-22, 2006), which made him the fourth person born in the United Kingdom to go into space. On EVA one, the crew added the P5 spacer truss to the ISS. EVAs two and three rewired the station's power system, preparing it to support the addition of European and Japanese science modules. Dr. Patrick was 'Mission Specialist 3' on STS-130 aboard *Endeavour* (February 08-22, 2010), which launched at night, carrying the ISS's final permanent modules: *Tranquility* (Node 3), the life support hub of the ISS, and the *Cupola* with seven hemispheric windows. He logged a total of 26 days, 14 hours and 52 minutes in space. Nicholas Patrick retired from NASA in May 2012, and joined the **Blue Origin** space company; currently holding the positions of New Shepard Flight Director and Senior Director of Human Integration. He is married, with three children.

## Pawelczyk, James A.. STS-90.

Dr. James "Jim" Anthony Pawelczyk was born in September 1960 in Buffalo, New York, USA; the first American astronaut of full-blooded Polish descent. In 1982, Pawelczyk earned a Bachelor of Arts degree in Biology; and another BA in Psychology from the University of Rochester, New York. In 1985, he received a Master of Science degree in Physiology; and a Ph.D. in Biology (with a major in physiology) from the University of North Texas in 1989. He continued studies with a post-doctoral fellowship in cardiovascular neurophysiology at the University of Texas Southwestern Medical Center from 1989-92; and became an assistant Professor of Medicine (cardiology) there from 1992-95. Since 1995, he has been an associate professor of physiology and kinesiology at Penn State University in University Park, Pennsylvania.

In 1998, Dr. Jim Pawelczyk took leave from Pennsylvania State University to fly as a Payload Specialist on STS-90 (Neurolab) aboard *Columbia* (April 17-May 03, 1998), the second longest Space Shuttle mission. The crew served as both experiment subjects and operators for twenty-six individual life science experiments focusing on the effects of microgravity on the human brain and nervous system. He logged 15 days, 21 hours and 50 minutes in space. In 1999, Pawelczyk and three other STS-90 crew were guests of the Republic of Poland. He presented the Polish president with the national flag that flew with him to space. Dr. James Pawelczyk is co-editor of *Blood Loss and Shock*, published in 1994. He has more than twenty refereed journal articles and three invited book chapters in the areas of cardiovascular regulation and cardiovascular physiology. In 2003, he testified before the US Senate Committee on Commerce, Science and Transportation, advocating for strengthening research on board the International Space Station. He is married, with a daughter and son.

## Payton, Gary E.. STS-51-C.

Col. Gary Eugene Payton was born in June 1948 in Rock Island, Illinois, USA. He attended Bradley University in Peoria, Illinois, but after one year he entered the USAF Academy, Colorado Springs; graduating with a Bachelor of Science degree in Astronautical Engineering in 1971. In 1972, he graduated from Purdue University of Aeronautics and Astronautics in West Lafayette, Indiana, with a Master of Science degree in Astronautical and Aeronautical Engineering. Payton completed pilot training at Craig AFB, Alabama, in 1973. Selected in 1979 USAF Group, he flew as 'Payload Specialist 1' on STS-51-C aboard *Discovery* (January 24-27, 1985), the first Department of Defense classified mission, which deployed a *Magnum* satellite with an Inertial Upper Stage booster. Gary Payton logged three days, one hour and 33 minutes in space.

He was Director of NASA's Space Transportation Division from 1995-97. From 1997 to 2000, Payton served as NASA's Deputy Associate Administrator for Space Transportation Technology where he initiated, planned and led the Reusable Launch Vehicle technology demonstration program. After leaving NASA, he joined the ORBIMAGE as its Vice President for Engineering and Operations. From 2005-10, Col. Gary Payton served as Deputy Undersecretary of the Air Force for Space Programs under the Bush and Obama Administrations. He is currently distinguished visiting professor in the Schriever Chair in Astronautics at the **United States Air Force Academy**, teaching astronautical engineering. Col. Gary Payton is married, with a daughter.

## Peterson, Donald H.. STS-6.

Col. Donald Herod Peterson (1933-2018) was born in October 1933 in Winona, Mississippi, USA. He enrolled in the US Military Academy at West Point, New York, and received a Bachelor of Science degree in 1955, electing to join the Air Force. After working with Air Training Command until 1960, he was asked if he would be interested in getting a degree in nuclear engineering so he could join a program involving aircraft powered by a nuclear reactor! He enrolled in nuclear engineering at the USAF Institute of Technology of Air University, but six months before graduation, the program was canceled. Peterson earned his Master of Science degree in 1962; and also worked towards his Ph.D. at the University of Texas. He was a graduate of the Aerospace Research Pilot School at Edwards AFB, California, and was originally selected as one of the 1967 Group 3 astronauts for the USAF Manned Orbiting Laboratory (MOL) program, but when the MOL was canceled, he became a NASA astronaut in September 1969 Group 7. He logged over 5,300 hours flying time, including over 5,000 hours in jet aircraft.

Peterson served on the astronaut support crew for *Apollo 16*, and was a Mission Specialist on STS-6 aboard *Challenger* (April 04-09, 1983). The crew of Paul Weitz, Karol Bobko, Peterson and Dr. Story Musgrave had a combined 111 years of flying experience and an average age of 48 years and five months. They were dubbed The Geritoll Bunch for their high exerience. They deployed the first *Tracking and Data Relay Satellite*, and activated three "Getaway Special" (GAS) canisters. Peterson and Musgrave conducted the Shuttle program's first EVA (four hours and 15 minutes) to test the new suit, the shuttle airlock, and new tools and techniques for construction and repair outside a spacecraft. Peterson logged five days and 23 minutes in space; including one EVA of four hours and 15 minutes. He resigned from the Astronaut Corps in November 1984 and worked as a consultant in the area of crewed aerospace operations. He was awarded the Air Force Commendation Medal; the Meritorious Service Medal; and the JSC Group Achievement Award (1972). Col. Donald Peterson died on May 27, 2018, in El Lago, Harris County, Texas, of Alzheimer's disease, aged eighty-four years. He was married, with three children.

## Pettitt, Donald R.. STS-113/Expedition 6/Soyuz TMA-1, STS-126, Soyuz TMA-22/Expedition 31.

Dr. Donald Roy Pettit was born in April 1955 in Silverton, Oregon, USA. He received a Bachelor of Science degree in Chemical Engineering from Oregon State University in 1978; and a Ph.D. from the University of Arizona in 1983. Pettitt worked as a scientist at the **Los Alamos National Laboratory** until 1996 when he was selected as an astronaut candidate. He was a junior advisor to the Synthesis Committee of the **Space Exploration Initiative** on its May 1991 report *America at the Threshold*, recommending plans for a human mission to Mars. He is renowned for his orbital astrophotograhy and in-space inventions, including the microgravity coffee cup, which holds the distinction of being the first patented object invented in space. The cup uses the wetting angle to carry fluid along a crease to permit drinking and avoid the necessity of a straw. This zero-g cup was featured in the May 2009 issue of ***National Geographic*** magazine.

Donald Pettitt was 'Mission Specialist 5' on STS-113 aboard *Endeavour* (November 24-December 07, 2002) and the crew extended the ISS backbone with the Port 1 truss. Carried aboard the Shuttle was the *Micro-Electromechanical System* (MEMS) -based Pico Satellite Inspector (MEPSI). This payload deployed two small satellites which were connected via a 15-meter tether. He became 'Flight Engineer 2' for ISS 'Expedition 6' from November 25, 2002, to May 03, 2003. During his six-month stay, he performed two EVAs to help install external scientific equipment. In his free time, Pettitt conducted demonstrations showing how fluids react in microgravity in a series he called Saturday Morning Science. The 'Expedition 6' mission was extended by about two months, following the shuttle *Columbia* disaster in February 2003. Instead of returning on a Space Shuttle, the crew returned on a Russian capsule, *Soyuz TMA-1*. The capsule touched down about 444 km (276 miles) from its planned landing zone.

From November 2006 through January 2007, Pettitt joined the Antarctic Search for Meteorites (ANSMET), spending six weeks in the Antarctic summer collecting meteorite samples, including a lunar meteorite. He estimated Antarctic glacial ice to contain roughly one micrometeorite per litre. During the expedition, he was called on to perform emergency electrical repairs to a snowmobile and emergency dental surgery. He was 'Mission Specialist 1' on STS-126 aboard *Endeavour* (November 14-30, 2008), a mission to deliver equipment and supplies to the ISS. He performed experiments related to the clumping of solid particles in microgravity. When confined to a volume of four litres, various material (with sizes of one micrometer to six millimetres) naturally clumped together when the container had a few grams of the materials. The cause was theorized to be electrostatic: which presents a plausible mechanism for the initial stages of planetary formation, since particles of this size do not have sufficient gravity to cause this phenomenom.

Donald Pettitt launched to the ISS aboard *Soyuz TMA-22* on December 21, 2011, as part of the 'Expedition 31' crew, beginning on April 27, 2012, with the departure of *Soyuz TMA-22* from the ISS, which returned the 'Expedition 30' crew to Earth. On May 25, 2012, he and Andre Kuipers operated the Canadarm-2 to grapple the first **SpaceX** *Dragon-1* capsule and berth it to the *Harmony* module. During the capture, he was quoted saying, *"Houston, Station [calling], we've got a dragon by the tail"*. This was the first time a private spacecraft had ever rendezvoused and brought supplies to the ISS. Dr. Donald Pettitt has logged a total of 369 days, 16 hours and 41 minutes in space to date; including two EVAs totaling 13 hours and 17 minutes. As of 2023, at age sixty-eight years, he is NASA's oldest active astronaut. In September 2024, he will fly on *Soyuz MS-26* with Aleksey Ovchinin and Ivan Vagner. He is married, with twin sons.

## Phillips, John L.. STS-100, Soyuz TMA-6/Expedition 11, STS-119.

Capt. Dr. John Lynch Phillips was born in April 1951 in Fort Belvoir, Virginia, USA. He received a Bachelor of Science degree in Mathematics from the US Naval Academy in 1972 (graduating second in a class of 906 people); a Master of Science degree in Aeronautical Systems from the University of West Florida in 1974; a Master of Science degree from UCLA (1984); and a Ph.D. in Geophysics and Space Physics from UCLA in 1987. Phillips was designated a Naval Aviator in November 1974, and made overseas deployments with Attack Squadron 155 aboard the *USS Oriskany* and *USS Roosevelt*. He logged over 4,500 flight hours and made 250 aircraft carrier landings. He served as a Navy Reservist from 1982 to 2002, and retired with the rank of Captain, USNR.

In 1982, Phillips enrolled as a graduate student at UCLA and carried out research involving observations by the *Pioneer* Venus spacecraft. After completing his Doctorate in 1987, he was awarded an Oppenheimer Fellowship at **Los Alamos National Laboratory** (LANL) in New Mexico. From 1993-96, he was the Principal Investigator for the Solar Wind Plasma Experiment aboard the *Ulysses* spacecraft as it executed a unique trajectory over the poles of the Sun. Dr. Phillips has authored 156 scientific papers dealing with the plasma environments of the Sun, Earth, other planets, comets and spacecraft.

Selected in 1996 NASA Group 16, Dr. Phillips has flown on three different spacecraft types: the Space Shuttle, Soyuz and ISS. He was Ascent/Entry Flight Engineer on STS-100 aboard *Endeavour* (April 19-May 01, 2001), when the crew delivered and installed the **Canadarm-2** Robotic Arm. It was the maiden flight of the *Raffaello* Multi-Purpose Logistics Module (MPLM). Launched with Sergei Krikalev and Roberto Vittori on *Soyuz TMA-6* for 'Expedition 11' (April 15-October 10, 2005) he served as NASA Science Officer and Flight Engineer aboard the ISS. On June 14, Dr. John Phillips became the first person to testify before Congress from the ISS, or outer space in general. Via a live video feed, he gave testimony to the US House of Representatives Science Space Subcommittee. On August 18, he made an EVA of four hours and 58 minutes. He returned to Earth aboard *Soyuz TMA-6* on October 11, 2005.

On STS-119 aboard *Discovery* (March 15-28, 2009) he operated the Canadarm-2 robotic arm to unberth and install the S6 Integrated Truss Segment and the fourth set of solar arrays and batteries to the ISS. He also served as Loadmaster, Rendezvous Mission Specialist and Crew Medical Officer. Dr. Phillips logged a total of 203 days, 17 hours and 22 minutes in space. He was then detailed to the US Naval Postgraduate School in Monterey, California, from 2009-11 as the Smith/McCool NASA Chair Professor. Capt. Phillips has received numerous awards, including the NASA Distinguished Service Medal; the National Defense Service Medal with two service stars; the Gagarin Medal; the Russian Medal of Merit for Space Exploration; the Los Alamos National Laboratory Distinguished Performance Award; and various military awards such as the Navy Sea Service Deployment Ribbon, and Recruiting Service Ribbon. He is married, with two children.

## Poindexter, Alan G.. STS-122, STS-131.

Capt. Alan Goodwin "Dex" Poindexter (1961-2012) was born in November 1961 in Pasadena, California, USA. He received an Associate degree in Engineering from Pensacola Junior State College; a Bachelor of Science degree in Aerospace Engineering (with highest honors) from Georgia Institute of Technology in 1986; and a Master of Science degree in Aeronautical Engineering from the US Naval Postgraduate School in 1995. He was designated a Naval Aviator in 1988, was assigned to Fighter Squadron 211 and made two deployments to the Persian Gulf, including Operation Southern Watch. In 1993, he was selected to attend the Naval Postgraduate School/US Naval Test Pilot School Cooperative Program. After graduation in December 1995, he was assigned as a test pilot and Project Officer at NAS Patuxent River, Maryland. Later, he reported to Fighter Squadron 32 at NAS Oceana, Virginia, where he was serving as a department head when he was selected for astronaut training. He logged over 4,000 hours in over thirty aircraft types, and made over 450 aircraft carrier landings.

Selected in 1998 NASA Group 17, Capt. Poindexter was pilot on STS-122 aboard *Atlantis* (February 07-20, 2008), which delivered the *Columbus* Laboratory (built by the ESA) to the ISS, along with the Biolab, Fluid Science Laboratory (FSL), and transferred crew for 'Expedition 16'. He served as lead CapCom for STS-125 in May 2009, as well as launch CapCom for STS-127 in July 2009. Poindexter was Commander on STS-131 aboard *Discovery* (April 05-20, 2010) with the primary payload being the *Leonardo* Multi-Purpose Logistics Module (MPLM). Upon arrival at the ISS, crew performed three EVAs to replace an empty ammonia tank for the ISS Thermal Control System. They transferred over 13,000 pounds of hardware, supplies and equipment. The MPLM was then packed with over 6,000 pounds of hardware, and scientific and technical research return samples. He logged a total of 27 days, 21 hours and nine minutes in space.

Following his retirement from NASA in 2010, Alan Poindexter served in the US Navy as Dean of Students and executive director of programs at the **Naval Postgraduate School**. He belonged to the Society of Experimental Test Pilots; and the Association of Space Explorers. He received nineteen awards, including the Legion of Merit; the DFC; the Navy and Maritime Corps Commendation Medal with Combat Valor device; a Navy Meritorious Unit Commendation; the National Defense Service Medal; the Southwest Asia Service Medal; the Kuwait Liberation Medal; and Top Ten Carrier Aviator, Carrier Air Wing Nine. Following a water jet-scooter accident in Pensacola Beach, Capt. Alan Poindester died on July 01, 2012, aged fifty years, in Pensacola Beach, Florida. On **Orbital ATK**'s July 2016 *CRS 1-5* mission, Cygnus *CRS OA-5*'s space capsule was named *S.S. Alan Poindexter* in his honor. He was married, with two sons.

## Polansky, Mark L.. STS-98, STS-116, STS-127.

Mark Lewis "Roman" Polansky was born in June 1956 in Paterson, New Jersey, USA. He received the nickname *Roman* as a joke, because he shares a last name with film director Roman Polanski. Of Jewish/Korean descent, he received a Bachelor of Science in Aeronautical and Astronautical Engineering (1978); and Master of Science degree in Aeronautics and Astronautics (1978), both from Purdue University. Polansky earned his pilot wings in January 1980 at Vance AFB, Oklahoma. From 1980-83, he was assigned to Langley AFB, Virginia, then transitioned from the F-15 to the F-5E aircraft and served as an Aggressor Pilot, where he trained tactical aircrews to defeat enemy air attacks. In 1986, he was selected to attend USAF Test Pilot School, then was assigned to Eglin Air Force Base, Florida, where he conducted weapons and systems testing in aircraft. He left active duty in August 1992 to join NASA, having logged over 5,000 flying hours in over thirty aircraft types.

Selected in April 1996 NASA Group 16, Roman Polansky was Pilot on STS-98 aboard *Atlantis* (February 07-20, 2001), which delivered and attached the *Destiny* Laboratory Module to the ISS. He was then assigned as CapCom, and was Chief of the CapCom Branch from April to December 2002. From April 2003 to Janauary 2004, he served as Chief Instructor Astronaut. He was Commander on STS-116 aboard *Discovery* (December 10-22, 2006), which added the P5 Truss Segment during the first of four EVAs. More than two tons of equipment and supplies were taken aboard the ISS, and nearly two tons of items were returned to Earth. Mark Polansky commanded STS-127 aboard *Endeavour* (July 15-31, 2009), an assembly flight to the ISS when crew installed the Japanese Exposed Facility onto the ISS. Polansky began posting public updates and replies to *Twitter* on May 07, 2009, and continued throughout the mission.

Mark Polansky logged a total of 41 days, ten hours and 50 minutes in space, and retired from the astronaut Corps on June 30, 2012, when Peggy Whitson, Chief of the Astronaut Office said, *"Mark is a remarkably talented individual. His skills as an aviator coupled with his engineering expertise were a valuable contribution to our team. We wish him well in his future endeavours"*. Polansky is a member of the Society of Experimental Test Pilots; the American Institute of Aeronautics and Astronautics; and the Aircraft Owners and Pilots Association. He is a Distinguished Graduate of the USAF Test Pilot School (1987); a Distinguished Graduate of USAF Undergraduate Pilot Training (1980); Recipient of the USAF Flying Training Award (1980); and was awarded the Air Force Meritorious Service Medal; and the Air Force Commendation Medal with two Oak Leaf Clusters. In June 2006, his hometown of Edison, New Jersey, unveiled a sign for Mark Polansky Park, on Grove Ave. He is married, with two children; a daughter and a son.

## Pogue, William R.. Skylab 4.

Col. William "Bill" Reid Pogue (1930-2014) was born in January 1930 in Okemah, Oklahoma, USA. He was a teacher, public speaker and author. Pogue received a Bachelor of Science degree in Education from Oklahoma Baptist University in 1951; and a Master of Science degree in Mathematics from Oklahoma State University, Stillwater, in 1960. He first flew an airplane while in high school, and served with the Fifth Air Force from 1953-54 in the Korean War, with forty-three combat missions in fighter bombers. From 1955-57, he was a member of the USAF Thunderbirds as an aerobatics pilot. He logged 7,200 hours flying time in over fifty aircraft types, including 4,200 hours in jet aircraft and 2,000 hours in space flight.

Col. Bill Pogue was an instructor at the USAF Aerospace Reseach Pilot School from October 1965, until selected in April 1966 NASA Group 5. He was pilot for the *Skylab 4* mission, the third and final crewed visit to the Skylab Orbital Workshop; from November 16, 1973 to February 08, 1974. The crew of Commander Gerald Carr, Pogue, and science pilot Edward Gibson conducted 56 in-orbit research experiments, 26 science demonstrations, and 13 student investigations. They set a space flight duration record of 84 days, which was unbroken in NASA for over twenty years. After about six weeks on *Skylab*, there were disagreements between the crew and Ground Control, with the crew turning off radio transmission on December 28, 1973. Pogue later commented that the team was *"studying the Sun, the Earth below, and ourselves"*. When an agreement was reached for the flight to continue, the tensions diminished significantly. He commented in 1985 that the flight had made him more empathetic, saying *"I try to put myself into the human situation, instead of trying to operate like a machine"*.

Bill Pogue logged 84 days, one hour and 15 minutes in space; including two EVAs totaling 13 hours and 34 minutes. After he retired from NASA on September 01, 1975, Pogue was self-employed as an aerospace consultant and a producer of general-interest videos about space flight. He wrote four books: *Astronaut Primer* (1985); *How Do You Go To The Bathroom In Space?* (1991); *Space Trivia* (2003); and *But for the Grace of God: An Autobiography of an Aviator and Astronaut* (2011). With Ben Bova, he co-authored *The Trikon Deception*, a science-fiction novel. Over a forty-year career, he continuously presented lectures, working at more than 400 schools and 100 civic clubs.

William Pogue and his crew members received many awards, including the NASA Distinguished Service Medal (presented by President Richard Nixon in 1974); the FAI De la Vaulx Medal and the V.M. Komarov Diploma (1974); the City of Chicago Gold Medal; the American Astronautical Society's 1975 Flight Achievement Award; the AIAA Haley Astronautics Award (1975); the Dr. Robert H. Goddard Memorial Trophy (1975) and were inducted into three Halls of Fame. As a member of the USAF Thunderbirds he won four military awards. Bill Pogue died at home in Cocoa Beach, Florida, on March 03, 2014, aged eighty-four years. His ashes were sent into orbit using *Celestis*, a memorial rocket service launch on a *Falcon Heavy* booster on June 25, 2019. He was married three times, with three children from his first marriage.

## Precourt, Charles J.. STS-55, STS-71, STS-84, STS-91.

Col. Charles Joseph Precourt was born in June 1955 in Waltham, Massachusetts, USA. He was heavily involved with Russian/US Space relations as well as the International Space Station collaboration. He received a Bachelor of Science degree in Aeronautical Engineering from the USAF Academy in 1977; a Master of Science degree in Engineering Management from Golden Gate University in 1988; and a Master of Arts degree in National Security Affairs from the US Naval War College in 1990. Precourt attended the French Air Force Academy in 1976 as part of an exchange program. He is fluent in French and Russian. Selected in 1990 NASA Group 13, he was 'Mission Specialist 2' on STS-55 (aka Deutschland 2) aboard *Columbia* (April 26-May 06, 1993), a multinational 'Spacelab' flight involving eighty-eight experiments from eleven different nations. It carried to orbit the second reusable 'SpaceLab D-2', and crew worked around the clock in two shifts for investigations into fluid physics, materials sciences, life sciences, biological sciences, technology, Earth observations, atmospheric physics and astronomy; building on the work of the D-1 mission.

Charles Precourt was pilot on STS-71 aboard *Atlantis* (June 27-July 07, 1995) the first Space Shuttle docking to Russian space station *Mir*. It was the first time a Shuttle crew switched members with the crew of a space station, and the one hundredth crewed launch by the United States. The crews conducted various life science investigations with 'Spacelab', along with the Shuttle Amateur Radio Experiment-II (SAREX-II) experiment. He was Commander on STS-84 aboard *Atlantis* (May 15-24, 1997), the sixth *Shuttle/Mir* Docking Mission, which saw the transfer of 3,400 kg of water and logistics to *Mir*, and about 1,123 kg of material returned to Earth. For this mission the orbiter was equipped with a 4,187 kg SpaceHab Double Module and a 1,922 Orbiter Docking System. Col. Precourt commanded STS-91 aboard *Discovery* (June 02-12, 1998). The transfer of astronaut Andy Thomas from *Mir* marked a total of 907 days spent by seven US astronauts aboard the Russian space station as long-duration crew members. Over the next four days, 500 kg of water and almost 2,130 kg of cargo, experiments and supplies were exchanged.

Charles Precourt logged a total of 38 days, 20 hours and 16 minutes in space. He retired as an Air Force colonel in 2000. From 1998 to 2002, he served as Chief of the Astronaut Office. He retired from NASA in 2004, and became a vice president at **Orbital ATK** in Utah, USA. He is Vice President of the Association of Space Explorers; an Associate Fellow of the Society of Experimental Test Pilots; and a Member and on the board of directors of the Experimental Aircraft Association. Col. Charles Precourt has received eight military awards, including the Legion of Merit; the DFC; the National Defense Service Medal; and four NASA awards: the NASA Distinguished Service Medal; the NASA Exceptional Service Medal; the NASA Outstanding Leadership Medal; the NASA Space Flight Medal (with three Oak Leaf Clusters); and was inducted into the NASA Astronaut Hall of Fame (2012). He is married, with three daughters and two grandsons.

## Pudwill Gorie, Dominic L.. STS-91, STS-99, STS-108, STS-123.

Capt. Dominic Lee Pudwill Gorie was born in May 1957 in Lake Charles, Louisiana, USA. His father, Capt. Paul Pudwill, was killed in an airplane crash when Dominic was six years of age. He received a Bachelor of Science degree in Ocean Engineering from the US Naval Academy in 1979; and a Master of Science degree in Aviation Systems from the University of Tennessee Space Institute, Tullahoma, in 1990. Designated a Naval Aviator in 1981, he flew with Attack Squadron 46 aboard *USS America* from 1981-83. In 1987, he trained as a test pilot and worked at the Naval Air Test Center until the Gulf War, where, stationed aboard the carrier *USS Theodore Roosevelt*, he flew thirty-eight combat missions in Operation Desert Storm. He was en-route to his command tour of VFA-37 when he was selected in 1994 NASA Group 15. Capt. Dom Gorie logged over 6,000 flying hours in over 35 aircraft types, and made over 600 aircraft carrier landings.

He was Pilot on STS-91 aboard *Discovery* (June 02-12, 1998), the ninth and final *Shuttle/Mir* Docking Mission. The shuttle carried a prototype of the Alpha Magnetic Spectrometer, designed to look for dark and missing matter in the Universe. Gorie was Pilot on STS-99 aboard *Endeavour* (February 11-22, 2000), the Shuttle Radar Topography Mission, with the objective of obtaining the most complete high-resolution digital topographic database of the Earth. Around eight terabytes of data was gathered, to produce high-quality 3-D images of the Earth's surface. He was Commander on STS-108 aboard *Endeavour* (December 05-17, 2001), the twelvth Shuttle flight to the ISS, which carried the Italian cargo module *Raffaello* that was temporarily attached to the *Unity* module of the ISS for unloading of three tons of equipment and supplies. *Endeavour* delivered the 'Expedition 4' crew, and later returned the 'Expedition 3' crew to Earth.

Capt. Pudwill Gorie was Commander on STS-123 aboard *Endeavour* (March 11-27, 2008), which delivered the pressurized section of the Japanese Experiment Logistics Module (ELM-PS) as well as the Special Purpose Dexterous Manipulator (SPDM) *Dextre* to the ISS. Dom Gorie logged a total of 49 days, zero hours and six minutes in space. He retired from the US Navy in September 2005. Capt. Dom Pudwill Gorie received six awards: the Defense Superior Service Medal; the Legion of Merit; two DFCs; two Air Medals; the Defense Meritorious Service Medal; and two Navy Commendation Medals with Combat Valor device.

## Readdy, William F.. STS-51, STS-79.

Capt. William Francis Readdy was born in January 1952 in Quonset Point, Rhode Island, USA. He is a former Associate Administrator of the Office of Space Flight at NASA HQ. He received a Bachelor of Science degree in Aerospace Engineering (with honors) from the US Naval Academy in 1974; and was a distinguished graduate of the US Naval Test Pilot School in 1980. He joined Attack Squadron 85, and from 1976-80 he deployed to the North Atlantic and Mediterranean aboard *USS Forestal*. In 1986, he transferred into the Naval Reserve to join NASA, and served as an instructor pilot and unit commander until his naval retirement in August 2000. Selected on June 05, 1987, in NASA Group 12, he was 'Mission Specialist 2' on STS-42 aboard *Discovery* (January 22-30, 1992), whose main goal was to work in two teams to study the effects of microgravity on a variety of organisms and on materials processing.

Readdy was Pilot on STS-51 aboard *Discovery* (September 12-22, 1993), which deployed the *Advanced Communications Technology Satellite* (ACTS) on flight day one. The mission also featured the deployment and retrieval of the *SPAS-ORFEUS* satellite and its IMAX camera, which captured spectacular footage of *Discovery* in space. STS-51 was the first shuttle mission to fly a Global Positioning System receiver. He was Commander on STS-79 aboard *Atlantis* (September 16-26, 1996), which docked with the fully completed *Mir* space station, to deliver equipment, 4.7 tons of supplies, and relief personnel. The Shuttle carried the 1,821 kg Orbiter docking system. This mission marked the second flight of the 'SpaceHab' module, and the first flight of its Double Module configuration. Capt. William Readdy logged a total of 28 days, zero hours and 43 minutes in space, and retired from NASA on October 14, 2005.

William Readdy currently serves on the board of directors of the **Challenger Center for Space Science Education** as the Treasurer; and on the board of directors of **Astrobotic Technology**, a Carnegie-Mellon University spinoff company that is competing for the Google Lunar X Prize. Capt. Readdy was US Naval Test Pilot Instructor of the Year (1984) and has received 19 awards, including the Legion of Merit; the DFC; two NASA Outstanding Leadership Medals; two NASA Exceptional Service Medals; the Armed Forces Expeditionary Medal; and received the NASA Space Flight Safety Award. He is a member of the Association of Space Explorers; a Fellow of the Society of Experimental Test pilots; the Explorers Club; the Royal Astronautical Society; and an Associate Fellow of the AIAA. He is married, with two sons and a daughter.

## Reightler, Kenneth S., Jr.. STS-48, STS-60.

Capt. Kenneth Stanley Reightler, Jr. was born in March 1951 in Patuxent River, Maryland, USA. He received a Bachelor of Science degree in Aerospace Engineering from the US Naval Academy in 1973; a Master of Science degree in Aeronautical engineering from the US Naval Postgraduate School in 1984; and a Master of Science degree in Systems Management from the University of Southern California. He was designated a Naval Aviator in August 1974. After deployments to Keflavik, Iceland, and Sigonella, Sicily, he attended the US Naval Test Pilot school at NAS Patuxent River, Maryland. After serving as a flight test instructor and safety officer, in June 1981 Reightler was assigned to the aircraft carrier *USS Dwight D. Eisenhower* as communications officer and carrier on-board delivery pilot.

He then attended the US Naval Postgraduate School in Monterey, California, and was redesignated an Aerospace Engineering Duty Officer (AEDO). From March 1985, he served as the chief flight instructor, until selected on June 05, 1987, in NASA Group 12. Capt. Reightler logged over 5,000 hours flying in over sixty aircraft types. He was pilot on STS-48 aboard *Discovery* (September 12-18, 1991), the first Space Shuttle flight in support of Mission to Planet Earth. The crew deployed the *Upper Atmosphere Research Satellite* (UARS) and conducted numerous secondary experiments. Capt. Reightler was pilot on STS-60 aboard *Discovery* (February 03-11, 1994), the first joint US/Russian Space Shuttle Mission, the first flight of the *Wake Shield Facility*, and the second flight of the 'SpaceHab'. He logged a total of 13 days, 15 hours and 36 minutes (over 327 hours) in space.

Capt. Reightler retired from the Navy and left NASA in July 1995 to join **Lockheed Martin** as the Program Manager for the Engineering, Test, and Analysis contract at the Johnson Space Center. He was then promoted to Vice President of the Science, Engineering, Analysis and Test (SEAT) Operation, providing engineering and science services to NASA. In 2001, Reightler was promoted to Senior Vice President (COO) of **Lockheed Martin Space Operations**, and in 2004 he was promoted to President of Lockheed Martin Space Operations. In February 2011, he left Lockheed Martin to become the Vice President, Engineering Services for **Orbital Sciences**, Space Systems Division. In August 2012, he joined the faculty of the **US Naval Academy**. Kenneth Reightler is married, with two daughters.

## Reilly, James F., II. STS-89, STS-104, STS-117.

Dr. James Francis Reilly II was born in March 1954 at Mountain Home Air Base, Idaho, USA. He has three degrees from the University of Texas at Dallas: a Bachelor of Science degree in Geosciences (1977); a Master of Science degree in Geosciences (1987); and a Ph.D. in Geosciences (1995). In graduate school, he was selected to participate in the 1977-78 scientific expedition to Marie Byrd Land, West Antarctica, as a research scientist specializing in stable isotope geochronology. In 1979, he began work as an exploration geologist with **Santa Fe Minerals, Inc.** in Dallas, Texas. From 1980 to December 08, 1984 (when he was selected for the astronaut program), he was an oil and gas exploration geologist for **Enserch Exploration, Inc.** in Dallas, Texas, becoming Chief Geologist of the Offshore Region.

Reilly flew as 'Mission Specialist 1' on STS-89 aboard *Endeavour* (January 23-31, 1988), the eighth of nine planned missions to space station *Mir*. More than 3,175 kg of experiments, supplies and hardware were transferred between the craft, which were docked for four days, 20 hours and 41 minutes. STS-104 aboard *Atlantis* (July 12-21, 2001) had the primary purpose to deliver and install the 5.9 metric ton *Quest* airlock; two cylindrical chambers joined end-to-end by a connecting bulkhead and hatch. With Michael Gernhardt, Reilly made three EVAs totaling 16 hours and 30 minutes. The mission was the first to fly with a Block II Space Shuttle Main Engine. Post-launch analysis indicated an anomaly occurred when the Main Engine was shut down; the cause was determined and corrected for STS-108 in November 2001.

Reilly worked on both the *Mir* and ISS, then was assigned as the Astronaut Office lead on Shuttle training. He was 'Mission Specialist 4' on STS-117 aboard *Atlantis* (June 08-22, 2007), which delivered the second stainless steel starboard truss segment (S3/S4) and associated energy systems to the ISS. At 16,183 kg (35,678 pounds), this was the heaviest payload ever carried by a Space Shuttle. James Reilly logged a total of 35 days, ten hours and 34 minutes (over 517 hours) in space; including three EVAs totaling 16 hours and 30 minutes. Reilly retired from the space program in May 2008. From January 2010 to May 2014, he was the American Public University System's Dean of the School of Science and Technology. In January 2018, President D.J. Trump nominated Reilly to be the seventeenth Director of the **United States Geological Survey** (2018-21). After leaving government service in 2020, Reilly later joined defense contractor **Booz Allen Hamilton** as an executive advisor. Reilly received a US Navy ROTC Scholarship; the Antarctic Service Medal (1978); and became the Seventh Honorary US Marshall (2001). He is married, with three children.

## Reisman, Garrett E.. STS-123/Expedition 16/17/STS-124, STS-132.

Dr. Garrett Erin Reisman was born in February 1968 in Morristown, New Jersey, USA, to a Jewish family. He went to the Wharton School and the School of Engineering and Applied Science. He received a Bachelor of Science degree in Economics, as well as a B.Sc. in Mechanical Engineering and Applied Mechanics from the University of Pennsylvania, both in 1991; a Master of Science degree (1992); and a Ph.D. (both in Mechanical Engineering) from the California Institute of Technology (Caltech) in 1997. In 2003, he served as an aquanaut during the 14-day NEEMO 5 mission aboard the *Aquarius* underwater laboratory in Florida.

Selected in 1998 NASA Group 17, Reisman was a backup crew member for 'Expedition 15'. He launched as 'Mission Specialist 5' on STS-123 aboard *Endeavour* on March 11, 2008, and joined 'Expedition 16' aboard the ISS for a short time before becoming a member of 'Expedition 17'. He was the first Jewish crew member on board the ISS, and sent a greeting from space to the people of Israel during the celebration of Israel's sixtieth Independence Day in May 2008. He also did an entertaining high-definition video of "A day in the life of a space station crew member", as well as demonstrating in the large, and at the time empty, newly-arrived *Kibo* section, that humans cannot swim in the microgravity of orbital space. A self-proclaimed member of the Colbert Universe, Reisman was interviewed live from space on the May 08, 2008 episode of ***The Colbert Report***. He returned to Earth on June 14, 2008, on STS-124 aboard *Discovery*. On July 24, he appeared in person on ***The Colbert Report*** show as that night's feature guest. Reisman flew as 'Mission Specialist 1' on STS-132 aboard *Atlantis* (May 14-26, 2010), with the primary payload being the Russian *Rassvet* Mini-Research Module. Reisman logged a total of logged 107 days, three hours and 15 minutes in space; including three EVAs totaling 21 hours and 21 minutes.

On March 04, 2011, **SpaceX** announced that Garrett Reisman would be joining the company as a senior engineer working on astronaut safety and Mission Assurance. He was later promoted to Director of Crew Operations, but on May 31, 2018 he announced he would leave that position. He now operates as a **SpaceX** part-time consultant and Professor of Astronautics Practice at the **University of Southern California**'s Viterbi School of Engineering. In July 2019, it was reported that producer Ron Moore had created his "alternative reality" series ***For All Mankind*** after he and Reisman had discussed the possibility of a timeline in which the Soviet Union landed people on the Moon before the United States did. Reisman also served as technical advisor for the show, and guest appeared as Commander of a Moon Space Shuttle in Season 2, Episode 2.

## Resnik, Judith A.. STS-41-D, STS-51-L.

Dr. Judith Arlene Resnik (1949-86) was born in April 1949 in Akron, Ohio, USA. She was the second American woman in space (and the fourth worldwide), and the first Jewish woman of any nationality to fly in space, spending 145 hours in orbit. Resnik had a paternal grandfather born in Preluke, Ukraine, who was a Rabbi, and she grew up in an observant Jewish home, celebrating her Bat Mitzvah in 1962. From age 17 years, she studied Electrical Engineering at Carnegie Institute of Technology (now called Carnegie-Mellon University), earning her Bachelor of Science degree in 1970, before working in the missile and surface radar division of RCA as a senior systems engineer for **Xerox Coroporation**. In July 1970 she married Michael Oldak, but they divorced in 1975 since she did not want to have children.

Judith Resnik obtained her Ph.D. in Electrical Engineering from the University of Maryland in 1977. She also qualified as a pilot in 1977 and made research contributions to biomedical engineering as a research fellow at the National Institutes of Health. Astronaut Jerome Apt described her as *"an excellent pilot"*. Fellow astronaut Rhea Seddon said, *"I thought she was really really bright, obviously a very beautiful person, flirtatious, funny. She was just a live wire. We would do the happy hours, or we'd go on these NASA trips, and **Judy was just a star attraction**"*. In January 1978, Resnik became a NASA astronaut at the age of twenty-nine years; one of six women accepted into the US space program. She developed the software and operating procedures for the Space Shuttle's Remote Manipulator System (RMS). Dr. Judith Resnik also developed the deployment systems for the tethered satellite systems and worked on orbiter development. She avoided television interviews when possible, and resented intrusive questions about her private life, such as questions about her divorce. Instead, Dr. Sally Ride was chosen as the first American woman to fly in space, since that event would attract an unusual amount of public interest.

Dr. Resnik flew STS-41-D aboard *Discovery* (August 30-September 05, 1984), which launched on the fourth attempt. On mission day one, Resnik and Mike Mullane deployed the first of three commercial communications satellites: *SBS-1;* then *Syncom IV-2* the next day; and *Telstar 302* the following day. On day three, Resnik deployed the OAST-1 solar array wing, considered a potential future way of generating more electrical power. After the six-day and 56 minute mission, astronaut Mullane wrote, *"I was happy to be crewed with Judy ... She was smart, hard working, and dependable, all the things you would want in a fellow crewmember"*.

Dr. Resnik was 'Mission Specialist 2' on STS-51-L aboard *Challenger*, launched on January 28, 1986. A faulty O-ring seal on a Solid Rocket Booster led to destruction of the Orbiter by aerodynamic forces just seventy-three seconds after liftoff. None of the seven crew survived when the debris crashed into the ocean at 207 mph. Following the disaster, examination of the recovered vehicle cockpit revealed that the Personal Egress Air Packs (PEAP) were activated for pilot Mike Smith and two other crew members. The location of Smith's activation switch on the back of his seat means either Judith Resnik or Ellison Onizuka likely activated it for him. Mike Mullane wrote, *"The fact that Judy or El had done so for Mike Smith made them heroic in my mind. They had been able to block out the terrifying sights and sounds and motions of Challenger's destruction ... It was the type of thing a true astronaut would do - maintain their cool in the direst of circumstances"*. Unquote. This is the only evidence that shows Onizuka and Resnik were alive after the cockpit separated from the vehicle. Her remains were recovered by navy divers from the *USS Preserver*, and they were cremated and scattered over the water.

Dr. Judith Resnik was posthumously awarded the Congressional Space Medal of Honor, and the NASA Space Flight Medal for STS-41-D. The Fairlawn Elementary School which she had attended was renamed Judith A. Resnik Community Learning Center in her honor, as was Judith A. Resnik Middle School, established in 2016, in San Antonio, Texas. A dormitory at her alma mater,

Carnegie-Mellon University, a crater on the Moon, one on Venus (where all features are named after women), and an asteroid, 3356 Resnik, were also named after her. A memorial to all the crew of the Space Shuttle *Challenger* disaster was dedicated in Seabrook, Texas, and she was buried at Arlington National Cemetery. Dr. Judith Resnik is also commemorated on the Space Mirror Memorial at KSC. The IEEE Judith A. Resnik Award was established in 1986 by the Institute of Electrical and Electronics Engineers and is presented annually in recognition of outstanding contributions to space engineering in areas of relevance to the IEEE. Julie Fulton portrayed Judith Resnik in the 1990 made-for-TV movie ***Challenger***.

## Richards, Paul W.. STS-102.

Paul William Richards was born in May 1964 in Dunmore, Scranton, Pennsylvania, USA. He received a Bachelor of Science degree in Mechanical Engineering from Drexel University in 1987; and a Master of Science degree in Mechanical Engineering from the University of Maryland, College Park, in 1991. He is the inventor of the Pistol Grip Tool (PGT), a computer controlled, battery powered 3/8-inch drive power tool that can apply 38 foot-pounds of torque; and has been used on every NASA EVA since 1996. From 1983-87, he was employed in the Department of the Navy, Naval Ship Systems Engineering Station. He was senior EVA tool development engineer for the Hubble Space Telescope Servicing Project.

Selected in April 1996 NASA Group 16, he flew as 'Mission Specialist 2' on STS-102 aboard *Discovery* (March 08-19, 2001), the eighth shuttle mission to visit the ISS; delivering the 'Expedition-2 ' crew, and returning the 'Expedition-1' crew to Earth. The *Leonardo* Multi-Purpose Logistics Module (MPLM), built by the Italian Space Agency delivered its first supplies to the ISS, and was brought back for reuse. Paul Richards logged 12 days, 19 hours and 49 minutes in space; including an EVA of six hours and 21 minutes. He received sixteen awards, including the Department of Defense Superior Service Medal; the Silver Snoopy Award (1994); US Patent for Computerized Pistol Grip Power Tool; Hubble Space Telescope First Servicing Mission Crew Aids and Tools Development Team; Certificate of Outstanding Performance (1990, 1993, & 1994); the NASA Exceptional Service Medal; and NASA Outstanding Performance Awards (1992, 1994, & 1995).

Paul Richards retired from NASA in February 2002 to pursue private interests, but returned in 2004 as the Observatory Manager for the Geostationary Operational Environmental Satellite (GOES-R Series), an advanced weather satellite. In 2014, he was promoted to Deputy Project Manager for the Laser Communications Relay Demonstration (LCRD) Project. In 2018, he was detailed to NASA HQ as a Program Executive for NASA's Satellite and Communications and Navigation (ScaN) Program, finally retiring from NASA in 2019. He is a member of the American Society of Mechanical Engineers; the National Society of Professional Engineers; the American Institute for Aeronautics and Astronautics; the American Society of Naval Engineers; and the United States Naval Reserves.

## Richards, Richard N.. STS-28R, STS-41, STS-50, STS-64.

Capt. Richard "Dick" Noel Richards was born in August 1946 in Key West, Florida, USA. He studied chemical engineering, receiving a Bachelor of Science degree from the University of Missouri in 1969; and a Master of Science degree from the University of West Florida in 1970. He was designated a Naval Aviator in August 1970 and until 1973 flew support missions while assigned to Tactical Electronic Warfare Squadron 33 at Norfolk Naval Air Station, Virginia. With Fighter Squadron 103, he deployed to the North Atlantic and Mediterranean aboard the aircraft carriers *USS America* and *USS Saratoga*. Selected for test pilot training, Richards reported to NAS Patuxent River, Maryland, in 1976. He served for three and a half years as a project test pilot for automatic carrier landing systems development work in F-4 and A-7 aircraft.

Selected in 1980 NASA Group 9, he served as an ascent CapCom for STS-51-F, passing up the "Abort ATO" call when *Challenger*'s centre engine went down. Richards was pilot on STS-28R aboard *Columbia* (August 08-13, 1989), which carried classified Department of Defense payloads. He commanded STS-41 aboard *Discovery* (October 06-10, 1990), and crew deployed the *Ulysses* spacecraft to investigate the polar areas of the Sun. Dick Richards was Commander on STS-50 aboard *Columbia* (June 25-July 09, 1992), the first United States Microgravity Laboratory mission, and the first extended duration Orbiter flight (ten days and 22 hours). He commanded STS-64 aboard *Discovery* (September 09-20, 1994), where mission highlights included: the first use of a space based laser for environmental research; deployment and retrieval of a spacecraft in support of solar wind and corona studies; robotic processing of semiconductors; maneuvering the robotic arm in proximity to over 100 Shuttle reaction control system jet firings to measure forces imparted to a plume detection instrument; and the first untethered spacewalk in ten years (to test a self-rescue jetpack).

Richard Richards logged a total of 32 days, 21 hours and 29 minutes in space on four Space Shuttle missions. He left NASA in August 1998 to join **The Boeing Company**. From 1998 to 2007, he supported NASA via Boeing's Space Shuttle Sustaining Engineering Support Contract. In 2007, he retired as the Deputy Program Manager for Space Shuttle. Dick Richards received nine awards, including Naval Air Test Center Test Pilot of the Year for 1980; the DFC; the Defense Superior Service Medal; two National Defense Service Medals; the Vietnam Service Medal; and the NASA Exceptional Service Medal. He is married.

## Ride, Sally K.. STS-7, STS-41-G.

Dr. Sally Kristen Ride (1951-2012) was born in May 1951 in Los Angeles, California, USA. Sally grew up in the Van Nuys and Encino neighborhoods of Los Angeles, and became an American astronaut and physicist. In 1960 her family spent a year traveling in Europe. In Spain, Sally played tennis for the first time, and by 1963 she was ranked number 20 in Southern California for girls aged twelve years and under. Sally was admitted on a full scholarship to Swarthmore College in Pennsylvania. Later, she entered the University of California, Los Angeles, where she enrolled in courses in Shakespeare and quantum mechanics, earning A's in both subjects. She was the only woman majoring in physics. Sally was romantically involved with the teaching assistant, until he went to Moscow to conduct research. She also concluded that she did not have what it took to be a professional tennis player. At Stanford, Sally renewed her acquaintance with Molly Tyson, and the two played doubles together. To earn money, ride and her then-girlfriend Tyson gave tennis lessons.

Ride was a graduate of Stanford University, where she earned a Bachelor of Science degree in Physics and a Bachelor of Arts degree in English Literature in 1973, a Master of Science degree in Physics in 1975, and a Doctor of Philosophy in Physics in 1978 for research on the interaction of X-rays with the interstellar medium. She joined NASA in 1978 and was selected as a Mission Specialist with NASA Astronaut Group 8, the first class to include women. She served as the first female ground-based Capsule Communicator (CapCom) for the second and third Space Shuttle flights and helped develop the Space Shuttle's robotic arm. On June 18, 1983, aboard mission STS-7 she became the first American woman [and the third woman, after Valentina Tereshkova (in June 1963) and Svetlana Savitskaya (in August 1982)] to fly in space. At the age of thirty-two years, she was also the youngest American astronaut to have flown in space. Many of the people attending the launch wore T-shirts bearing the words *Ride, Sally Ride*. NASA wanted a woman to fly on the mission, and factors in her favor included her agreeable personality and ability to work with others. There were over five hundred requests for private interviews, all of which were declined.

Sally's second mission was STS-41-G in 1984, also on board the shuttle *Challenger*. During 1985, Ride began an affair with Tam O'Shaughnessy. Sally was appointed to the Rogers Commission, to investigate the shuttle *Challenger* disaster of January 28, 1986; the only Space Shuttle astronaut and the only current NASA employee on the Commission. Next, she was assigned to NASA headquarters in Washington, D.C., where she led NASA's first strategic planning effort. After retiring from NASA in 1987, she founded **Sally Ride Science** and the **Sally Ride Science Festivals**. She also wrote several science books for children, including *To Space and Back* (with co-author Sue Okie). In May 1987, Ride took up a two-year fellowship at the Stanford University Center for International Security and Arms Control. In June 1987 she divorced her husband, fellow astronaut Steven A. Hawley (they married in July 1982; he was a veteran of five Space Shuttle flights up to STS-93), and Ride continued her relationship with Tam O'Shaughnessy until her own death in 2012. Dr. Sally Ride logged a total of 14 days, seven hours and 46 minutes in space. She retired from NASA on August 15, 1987.

On July 01, 1989, Dr. Ride became a professor of physics at the University of California, San Diego (UCSD), director of the California Space Institute. From the mid-1990s until her death, Ride led two public-outreach programs for NASA – the ISS EarthKAM and GRAIL MoonKAM projects. Not wanting to leave California, where she was now living with O'Shaughnessy, Ride turned down offers from President Bill Clinton to become the NASA Administrator. She did agree to serve on the President's Committee of Advisors on Science and Technology (PCAST). In 2003, Ride served on the Space Shuttle **Columbia Accident Investigation Board**. In 2008, she endorsed Barack Obama for US President. Ride served on the board of the National Math and Science Initiative in 2007 and the Educate to Innovate initiative in 2009.

Ride was inducted into the National Women's Hall of Fame, the Astronaut Hall of Fame and was twice awarded the NASA Space Flight Medal. Two US elementary schools were named after her. Having been married to astronaut Steven Hawley during her spaceflight years and in a private, long-term relationship with former Women's Tennis Association player Tam O'Shaughnessy, she is the first astronaut known to have been LGBT. When Ride drew up her Will in 1992, she ensured that Tam O'Shaughnessy would inherit her estate. They registered their domestic partnership on August 15, 2011. Dr. Sally Ride died of pancreatic cancer on July 23, 2012. A National Tribute to Sally Ride was held at the John F. Kennedy Centre for the Performing Arts in Washington, D.C., on May 20, 2013. That day, President Obama announced that Ride would posthumously receive the Presidential Medal of Freedom, the highest civilian award in the US. It was presented to her partner of twenty-seven years, Tam O'Shaughnessy. In 2014, Sally Ride was inducted into the Legacy Walk, an outdoor public display in Chicago that celebrates LGBT history and people.

## Robinson, Stephen K.. STS-85, STS-95, STS-114, STS-130.

Dr. Stephen "Steve" Kern Robinson was born in October 1955 in Sacramento, California, USA. He received a Bachelor of Science degree in Mechanical and Aeronautical Engineering from the University of California, Davis, in 1978; a Master of Science degree in Mechanical Engineering from Stanford University in 1985; and a Doctorate in Mechanical Engineering (with a minor in Aeronautics and Astronautics) from Stanford University in 1990. After graduation in 1979, he joined **NASA Ames** as a research scientist in the fields of fluid dynamics, aerodynamics, experimental instrumentation, and computational scientific visualization. In 1990, he was selected as Chief of the Experimental Flow Physics Branch at NASA's **Langley Research Center** in Hampton, Virginia, where he led a group of thirty-five engineers and scientists engaged in aerodynamics and fluid physics research. He has logged over 1,400 hours in aircraft. He began applying to become an astronaut in 1983, and was selected to join the NASA Astronaut Corps Group 15 on October 26, 1995. He served as backup Flight Engineer for the ISS 'Expedition 4'.

Dr. Robinson was 'Mission Specialist 3' on STS-85 aboard *Discovery* (August 07-19, 1997), the fourth mission in a cooperative venture between the German Space Agency (DARA) and NASA, when the *CRISTA-SPAS-2* satellite made its second flight. STS-95 aboard *Discovery* (October 29-November 07, 1998) was noted for inaugurating ATSC HDTV broadcasting in the US, with live coast-to-coast coverage of the launch, which was witnessed by President Bill Clinton and his wife, Hillary Rodham-Clinton. The mission's objectives involved investigating life-sciences experiments, using the 'SpaceHab' module to perform those experiments on Senator John H. Glenn, who at age 77 years, returned to low Earth orbit as the oldest astronaut to date. This age record was unbroken for twenty-three years. The *SPARTAN-201* spacecraft was released, flying free from the Shuttle to study the acceleration of the solar wind.

STS-114 aboard *Discovery* (July 26-August 09, 2005) was launched 907 days after the loss of shuttle *Columbia*. Analysis of the launch footage showed debris separating from the External Tank during ascent; a serious concern since this was the issue that had set off the *Columbia* disaster. As a result, NASA decided to postpone future Space Shuttle flights (until July 04, 2006) pending additional modifications to the flight hardware. Steve Robinson made the first podcast from space on August 07. This mission flew three members of 'Expedition 7' to the ISS, and returned with three from 'Expedition 6'. The major focus of STS-114 was testing and evaluating new Space Shuttle flight safety techniques, which included new inspection and repair techniques. The first (of three) EVAs demonstrated repair techniques on the Shuttle's Thermal Protection System. Stephen Robinson served as 'Mission Specialist 2' and Flight Engineer on STS-130 aboard *Endeavour* (February 08-22, 2010). As Intravehicular Officer, he choreographed the three EVAs involved with the installation and activation of the Node 3 *Tranquility* module and the ISS *Cupola*.

Dr. Steve Robinson logged a total of 48 days, nine hours and 52 minutes in space; including three EVAs totaling 20 hours and five minutes. He retired from the space program on June 30, 2012, to take a teaching position as a full-time professor in the Mechanical and Aerospace Engineering department at the **University of California** at Davis. Dr. Stephen Robinson is a member of the American Institute of Aeronautics and Astronautics; the Aerospace Medical Association; and the Experimental Aircraft Association. He was awarded the NASA Ames Honor Award for Scientists in 1989; the American Institute of Aeronautics and Astronautics Outstanding Technical Paper Award (1992); and the NASA/Space Club G.M. Low Memorial Engineering Fellowship in 1993. He is single, and plays lead guitar in ***Max Q***, an all-astronaut rock band.

## Rominger, Kent V.. STS-73, STS-80, STS-85, STS-96, STS-100.

Kent Vernon "Rommel" Rominger was born in August 1956 in Del Norte, Colorado, USA. He received a Bachelor of Science degree in Civil Engineering from Colorado State University; and a Master of Science degree in Aeronautical Engineering from the US Naval Postgraduate School in 1987. He was commissioned throught the Aviation Reserve Officer Candidate (AVROC) Program in 1979, and was designated a Naval Aviator in September 1980. From October 1981 to January 1985 he flew the F-14 Tomcat aboard the aircraft carriers *USS Ranger* and *USS Kitty Hawk*. While assigned to VF-2, he attended the Navy Fighter Weapons School (TOPGUN). In 1987, he completed the Naval Postgraduate School/Test Pilot School Cooperative Program, and was assigned to NAS Patuxent River, Maryland. In September 1980, he reported to Fighter Squadron 211 (VF-211), where he served as Operation officer and completed a Desert Storm deployment to the Persian Gulf aboard *USS Nimitz*. He logged over 5,000 flying hours in over thirty-five aircraft types, and made 685 aircraft carrier landings.

Selected in March 1992 NASA Group 14, he completed one year of training to qualify for assignment as a pilot on future Space Shuttle flight crews. Rominger was Pilot on STS-73 aboard *Columbia* (October 20-November 05, 1995), the second US Microgravity mission. He was pilot on STS-80 aboard *Columbia* (November 19-December 07, 1996), when the crew deployed and retrieved the *Wake Shield Facility* and the *Orbiting Retrievable Far and Extreme Ultraviolet Spectrometer* (*ORFEUS*) satellites. The free-flying WSF created a super vacuum in its wake and grew thin film wafers for use in semiconductors and other high-tech electrical components. The *ORFEUS* instruments studied the origin and makeup of stars. Rominger was pilot on STS-85 aboard *Discovery* (August 07-19, 1997) when the crew deployed and retrieved the *CRISTA-SPAS* satellite, operated the Japanese Manipulator Flight Demonstration (MFD) robotic arm, studied changes in the Earth's atmosphere and tested technology destined for use on the future International Space Station (ISS).

Kent Rominger was Commander on STS-96 aboard *Discovery* (May 27-June 06, 1999), when four tons of logistics and supplies were delivered to the ISS in preparation for the first resident crew. STS-100 aboard *Endeavour* (April 19-May 01, 2001) was when the crew (from the US, Russia, Canada and Italy) installed the Canadian-built Robotic Arm and the *Rafaello* Logistics Module to the ISS. Rominger logged a total of 67 days, two hours and 58 minutes (over 1,600 hours) in space, a record for STS Orbiter crew members. From 2002-06, Kent Rominger served as Chief of the Astronaut Office and professional head of the NASA Astronaut Corps. He is a member of the Association of Space Explorers; the Society of Experimental Test Pilots; the American Institute of Aeronautics and Astronautics; the Association of Naval Aviation; and The Mars Generation. He has received ten awards, including the Defense Superior Service Medal; the DFC; the NASA Distinguished Service Medal; and was the Naval Air Test Center Test Pilot of the Year in 1988.

## Roosa, Stuart A.. Apollo 14.

Col. Stuart Allen Roosa (1933-94) was born in August 1933 in Durango, Colorado, USA. He received a Bachelor of Science degree in Aeronautical Engineering (with honors) from the University of Colorado, Boulder, in 1960. He began his career as a smokejumper with the **United States Forest Service**, then was a graduate of the Aviation Cadet Program at Williams AFB, Arizona. He also attended the USAF Aerospace Research Pilot School (Class 64C) and was an experimental test pilot at Edwards AFB in California, before being one of 19 astronauts selected by NASA in April 1966 Group 5. He logged 5,500 hours flying time; including 5,000 hours in jet aircraft. Roosa was the CapCom at the Launch Complex 34 blockhouse during the *Apollo 1* fire on January 27, 1967. He was a member of the astronaut support crew for the *Apollo 9* mission in 1969.

Stuart Roosa was the Command Module Pilot for the *Apollo 14* mission (January 31-February 09, 1971), which saw the third lunar landing by Commander Alan Shepard and Lunar Module pilot Edgar Mitchell. Roosa orbited the Moon thirty-four times in 33 hours while conducting experiments in the CSM *Kitty Hawk*. He was backup Command Module Pilot for *Apollo 16* and *Apollo 17*, and was assigned to the Space Shuttle program until his retirement as a colonel from the Air Force in 1976. Roosa logged nine days, zero hours and one minute (217 hours) in space, and retired from the space program on February 01, 1976. He attended Harvard Business School's six-week Advanced Management Program in 1973, and later held a number of positions in international and US businesses.

Stuart Roosa received nine awards, including the Arnold Air Society's John F. Kennedy Award (1971); the City of New York Gold Medal (1971); an honorary LL.D. from the University of St. Thomas, Houston (1971); and an elementary school in Claremore, Oklahoma, is named in his honor. He was one of five Oklahoman astronauts inducted into the Oklahoma Aviation and Space Hall of Fame in 1980; was inducted into the International Space Hall of Fame in 1983;

and posthumously inducted into the US Astronaut Hall of Fame in 1997. In the HBO miniseries ***From the Earth to the Moon*** (1998), Roosa was played by George Newbern. The book *Smoke Jumper, Moon Pilot* (pub. 2011), by Willie E. Moseley, is a biography on Roosa's life. He was the founder and president of **Gulf Coast Coors** in 1981, until his death on December 12, 1994, in Falls Church, Virginia, aged sixty-one years, from complications of pancreatitis. Roosa was survived by his wife, three sons and a daughter, and seven grandchildren. He was buried in Arlington National Cemetery.

## Ross, Jerry L.. STS-61-B, STS-27, STS-37, STS-55, STS-74, STS-88, STS-100.

Jerry Lynn Ross was born in January 1948 in Crown Point, Indiana, USA. He graduated from Crown Point High School in 1966; received a Bachelor of Science degree in 1970 and a Master of Science degree in 1972, both in Mechanical Engineering from Purdue University. He then entered active duty with the US Air Force and was assigned to the Ramjet Engine Division of Air Force Aero-Propulsion Laboratory at Wright-Patterson Air Force Base, in Ohio. From June 1974 to July 1975, he was the Laboratory Executive Officer and Chief of the Management Operations Office. Ross graduated from the USAF Test Pilot School's Flight Test Engineer Course in 1976 and was then assigned to the 6510th Test Wing at Edwards AFB, California.

In February 1979, Ross was assigned to the Payload Operations Division at the **Lyndon B. Johnson Space Center**, responsible for the flight operations integration of payloads into the Space Shuttle. He was selected as an astronaut in May 1980, and mainly served as a Mission Specialist, but was the Payload Commander on STS-55 in 1993. He spent over 1,393 hours (almost two months) in space and held the American record for nine EVAs. Jerry is a veteran of seven US Space Shuttle missions, and first set the individual world record (now held jointly with Franklin Chang-Diaz) for the most spaceflights flown. These were missions STS-61-B, in November 1985; STS-27, December 1988; STS-37, April 1991; STS-55, April 1993; STS-74, November 1995; STS-88, December 1998; and STS-100, launched on April 08, 2002. Ross spent a total of fifty-eight days, zero hours and 52 minutes in space; including nine EVAs, which totaled 57 hours and 55 minutes.

Ross has flown in twenty-one different types of aircraft, holds a private pilot's license, and has logged over 3,900 flying hours, the majority in military aircraft. He retired from the Air Force in March 2000. Ross served as Chief of the Vehicle Integration Test Office at the **Johnson Space Center**, Texas until he retired in January 2012. His papers, photographs, and many personal items are in the Barron Hilton Flight and Space Exploration Archives at Purdue University. Ross was inducted into the Astronaut Hall of Fame during ceremonies in May 2014. Jerry Ross is the author of *Spacewalker. My Journey in Space and Faith as NASA's Record-Setting Frequent Flyer* (Purdue Press, 2013), written with John Norberg. Fellow astronaut, Chris Hadfield, describes Ross as *"the embodiment of the trustworthy, loyal, courteous and brave astronaut archetype"*. Jerry Ross is a lifetime member of the **Association of Space Explorers**, the Purdue Alumni Association, and a corresponding member of the International Academy of Astronauts.

## Rubins, Kathleen H.. Soyuz MS-01/Expedition 48/49, Soyuz MS-17/ Expedition 63/64.

Dr. Kathleen "Kate" Hallisey Rubins was born in October 1978 in Fermington, Conneticut, USA. She received a Bachelor of Science degree in Molecular Biology from the University of California at San Diego; and a Masters' degree and a Ph.D. in Cancer Biology from Stanford University. Her Thesis was titled *A Genome-Wide Analysis of the Host and Viral Responses during Poxvirus Infection*, 2005. Selected in July 2009 NASA Group 20, she participated in the fourth edition of the ESA PANGEA training program in September to November 2021, which prepared astronauts and space engineers to identify planetary geological features.

Dr. Rubins became the sixtieth woman to fly in space when she launched on July 17, 2016, aboard *Soyuz MS-01* and joined 'Expedition 48/49', where in August she became the first person to sequence mouse, virus and bacteria DNA in space. She returned to Earth on October 30, 2016, after 115 days in space. Launched on October 14, 2020, aboard *Soyuz MS-17* to join 'Expedition 63/64', she continued research for the Cardinal Heart experiment, and returned to Earth on April 17, 2021. She logged a total of 300 days, one hour and 31 minutes in space; including four EVAs totaling 26 hours and 46 minutes. Rubins is among the group of astronauts for the *Artemis* program that will put the first American woman on the Moon. She has received five awards, including UCSD Emerging Leader of the Year (1998); a National Science Foundation Predoctural Fellowship (2000); and ***Popular Science***'s Brilliant Ten (2009).

## Runco, Mario, Jr.. STS-44, STS-54, STS-77.

Lt. Cmdr. Mario Runco, Jr. was born in January 1952 in The Bronx, New York, USA. He received a Bachelor of Science degree in Meteorology and Physical Oceanography from the City College of New York in 1974; a Master of Science in Meteorology from Rutgers University in 1976; and an honorary Doctorate in Science (1999). Working at various positions even through his school years, Runco worked continuously from age nine until his retirement on December 31, 2017; eventually accumulating fifty-seven years of continuous employment and almost forty-eight years of federal service with four different agencies. In 1977, he joined the New Jersey State Police, working as a State Trooper until he entered the US Navy in June 1978. In September 1978, he was commissioned and assigned to the Naval Oceanographic and Atmospheric Research Laboratory in Monterey, California. He held two roles before serving as Commanding Officer of Oceanographic Unit 4 from December 1985 to December 1986.

Selected by NASA in 1987 Group 12, Mario Runco qualified as a Mission Specialist in August 1988 and is believed to be the first person of Italian origin to fly in space. Runco served as 'Mission Specialist 3' on STS-44 aboard *Atlantis* (November 24-December 01, 1991), which deployed a Defense Support Program (DSP) satellite and conducted two Military Man-in-Space Earth Observation experiments, on which he was the lead. The mission was cut short (to six days and 22 hours) when an Inertial Measurement Unit was powered up for redundancy and failed immediately. STS-54 aboard *Endeavour* (January 13-19, 1993) had the primary objective of deploying a NASA *Tracking and Data Relay Satellite* (TDRS-F), and Runco performed an EVA. Greg Harbaugh (EVA1) said, *"Mario was the most naturally skilled EVA guy I ever saw. He moved with ease and was able to accomplish everything he was challenged to do. ... He even caught a small tool case when it came loose after my tether hook failed to close"*.

Runco was 'Mission Specialist 3' on STS-77 aboard *Endeavour* (May 19-29, 1996), which featured the fourth flight of a (double) SpaceHab module, and Runco was the Robotic Arm operator who deployed two satellites. He captured some video Physics of Toys scenes for a sequel to the original STS-54 educational video, demonstrating how everyday toys behave in space to an interactive audience of elementary school students across the USA. Runco subsequently made several appearances on the children's television show ***Sesame Street*** in 1998 for the *Slimey to the Moon* series during Season 29. Mario Runco logged a total of 22 days, 23 hours and eight minutes (over 551 hours) in space; with one EVA of four hours and 28 minutes. He was a lead for Spacecraft Window Optics and Utilization for the ISS's windows, and assisted with window design. Mario Runco received twelve awards, including the Defense Superior Service Medal; the NASA Exceptional Service Medal; two Navy Sea Service Ribbons; the Navy Pistol Expert Medal; and the City College of New York's Townsend Harris Medal (1993). He is married, with two children.

## Rushworth, Robert A.. X-15 rocketplane Flight 87.

Maj. Gen. Robert "Bob" Aitken Rushworth (1924-93) was born in October 1924 in Madison, Maine, USA. He was a World War II, Korean War and Vietnam War pilot. He was one of the twelve pilots who flew the North American X-15, an experimental spaceplane jointly operated by the US Air Force and NASA. In six years, Rushworth flew thirty-four of the program's 199 flights, more than any other pilot. He earned his pilot wings in September 1944, after graduating from the aviation cadet training program. From February 1945, he flew C-47 Skytrain transport combat missions from India throughout the Burma Campaign. After five years with the Reserve and Air National Guard, during which he received a Bachelor's degree from the University of Maine, Rushworth was recalled to active duty in 1951. He served as a F-80C Shooting Star pilot at Dow Air Force Base during the Korean War.

Rushworth studied mechanical and aeronautical engineering, and after graduating from the USAF Institute of Technology in 1954, he stayed at Wright-Patterson AFB in Dayton, Ohio, where he specialized in the development and flight testing of experimental automatic flight control systems. In 1957, he graduated from Air Force Experimental Flight Test Pilot School (Class 56C). Bob Rushworth was selected in the 1958 Man in Space Soonest (MISS) Group. He qualified as an astronaut on X-15 Flight 87, June 27, 1963, when he reached an altitude of 285,000 feet (above fifty miles). However, this altitude did not surpass the Karman line, the internationally accepted boundary of space at sixty-two miles (100 km) altitude. On a later flight, he was awarded a DFC for successfully landing an X-15 after its nose wheel extended while flying at nearly Mach 5 speed. He made his final X-15 flight on July 01, 1966, then returned to regular Air Force duties. In March 1968, he was sent to Cam Ranh Bay Air Base, Republic of Vietnam, and flew 189 combat missions.

From April 1969 to January 1971, Rushworth was program director of the AGM-65 Maverick program, and in February 1971 became Commander of the newly organized Test Wing 4950. At Edwards AFB, he served as Commander of the major test programs, including the F-5, A-10, F-15, YF-16, YF-17, and B-1, and he was commander of the Air Force Test and Evaluation Center at Kirtland AFB, New Mexico. From October 1976 until his retirement on June 01, 1981, he was vice commander, Aeronautical Systems Division, at Wright-Patterson AFB; having served in the USAF since 1944. He logged over 6,500 flying hours in over fifty aircraft types. Maj. Gen. Rushworth received twelve military and NASA awards, including the World War II Victory Medal; the Legion of Merit; the DFC with two Oak Leaf Clusters; the Vietnam Service Medal; the NASA Exceptional Service Medal; and the James H. Doolittle Award (1975). Bob Rushworth died of a heart attack on March 18, 1993, aged sixty-eight years in Camarillo, California. He was married, with a daughter.

## Sacco, Albert, Jr.. STS-73.

Dr. Albert "Al" Sacco, Jr. was born in May 1949 in Boston, Massachusetts, USA. He received a Bachelor of Science degree in Chemical Engineering (with honors) from Northeastern University, Boston, in 1973; and a Ph.D. in Chemical Engineering from the Massachusetts Institute of Technology (MIT), Cambridge, Massachusetts, in 1977. He joined the faculty of the **Worcester Polytechnic Institute** (WPI), where he became a full professor and rose to the position of department head in July 1989. He flew on STS-73 aboard *Columbia* (October 20-November 05, 1995), a mission focused on materials science, biotechnology, combustion science, and fluid physics within the 'Spacelab' module. He was the Principal Investigator of the Zeolite Crystal Growth experiment.

Dr. Sacco was initially selected as the Payload Specialist for STS-107 and trained alongside the other six astronauts for two months until the US government decided to replace him with Israeli Payload Specialist Ilan Ramon. That mission ended in disaster when *Columbia* broke up on re-entry on February 01, 2003, and all seven astronauts were killed. As of December 1995, Dr. Sacco was a professor and head of the Chemical Engineering Department at Worcester Polytechnic Institute. Sacco served as the Dean of the **Edward E. Whitacre, Jr. College of Engineering** at Texas Tech University from 2011-22. Dr. Al Sacco is a member of seven professional organizations, and has over seventy publications. He is married, with four children.

## Satcher, Robert L., Jr.. STS-129.

Dr. Robert "Bobby" Lee Satcher, Jr. was born in September 1965 in Hampton, Virginia, USA. He received a Bachelor of Science degree in 1986; a Ph.D. in Chemical Engineering from the Massachusetts Institute of Technology in 1993; then went on to study medicine at Harvard Medical School, and received his M.D. in 1994 through the Harvard-MIT Program in Health Sciences and Technology. Prior to his entry into the astronaut program, Dr. Satcher was the Assistant professor at Feinberg School of Medicine, Northwestern University, in the Department of Orthopaedic Surgery from 2001. He was a member of the Robert H. Lurie Comprehensive Cancer Center and the Institute for Bioengineering and Nanotechnology in Advanced Medicine at Northwestern University.

Selected by NASA in May 2004, he completed Astronaut Candidate Training in February 2006. As 'Mission Specialist 4' on STS-129 aboard *Atlantis* (November 16-27, 2009), he was the first Black male astronaut and physician to act as the crew's medical doctor to the ISS. During his EVAs, he repaired ISS exterior robotic arms, and installed an S-band antenna sub-assembly to improve satellite communication and file updates on Twitter. Dr. Bobby Satcher logged ten days, 19 hours and 16 minutes (over 259 hours) in space; including two EVAs totaling 12 hours and 19 minutes. In 2011, he joined The University of Texas **M.D. Anderson Cancer Center** as a surgical oncologist and assistant professor. Dr. Bobby Satcher was a National Merit scholar, and received the Monsanto Award and the Albert G. Hill Award from MIT, and Fellowhips from both the Robert Wood Johnson Foundation and UNCF/Merck Research department. He received two honorary Doctorates in 2012. He is married, with a son and daughter.

## Schirra, Walter M., Jr.. Mercury-Atlas 8 "Sigma 7", Gemini VI-A, Apollo 7.

Capt. Walter "Wally" Marty Schirra, Jr. (1923-2007) was born in March 1923 in Hackensack, New Jersey, USA. He became one of "the Original Seven" astronauts chosen for Project Mercury. He graduated with a Bachelor of Science degree from the US Naval Academy, was commissioned as an Ensign on June 06, 1945, and served at sea during the final months of World War II aboard the large cruiser *USS Alaska*. In 1948, he became a naval aviator, and flew ninety combat missions in the Korean War, shooting down two MiG-15 jets. In 1958, Schirra graduated from the US Naval Test Pilot School, Class 20.

Selected in February 1959, Schirra was the first astronaut to go into space three times, and the only astronaut to fly in the Mercury, Gemini, and Apollo programs. He launched on October 03, 1962, on the *Mercury-Atlas 8* mission in a spacecraft he nicknamed *Sigma 7*, becoming the fifth American and nineth human to travel into space. Schirra demonstrated manually positioning and maneuvering his spacecraft using a reaction control system. After six orbits, he manually aligned his craft and performed retro-fire; landing eight kilometres (five miles) from the *USS Kearsarge*. He then flew on *Gemini VI-A* and made the first orbital rendezvous (just five hours after launch) with *Gemini VII* (December 04-18, 1965) on December 15-16. The two craft flew within thirty centimetres of each other. Schirra and Tom Stafford splashed down southeast of Cape Canaveral and were recovered by *USS Wasp*.

Launched on October 11, 1968, he commanded *Apollo 7* for an eleven-day shakedown test of the *Apollo* Command/Service Module; the first crewed launch of the Apollo program. After reaching orbit, the crew performed space rendezvous and docking exercises with the *Saturn-IVB* booster stage to simulate retrieving the Apollo Lunar Module. During the mission Schirra became sick with a head cold, which he passed to Donn Eisele. Because of congestion, he proposed to Mission Control that they would not wear their helmets during re-entry, defying Chris Kraft and Deke Slayton in Mission Control. They landed southeast of Bermuda on October 22, 1968. The cold medicine carried aboard *Apollo* missions was a combination of pseudoephidrene decongestant with triprolidine antihistamine: when this later became available as Actifed, the makers hired Schirra as a TV commercial spokesperson, based on the notoriety of his *Apollo 7* in-space head cold.

Schirra logged a total of 12 days, seven hours and 12 minutes in space. In 1969, he retired from the US Navy with the rank of Captain, and became a consultant to **CBS News** in the network's coverage of the following *Apollo* flights. Wally Schirra joined Walter Cronkite as co-anchor commentator for all seven of NASA's Moon landing missions. In January 1979, he founded **Schirra Enterprises**, and worked as a consultant until 1980. From 1971-84, he worked for the Belgian Consulate for Colorado and New Mexico, and was a board member of several corporations including **Electromedics, Finalco, Kimberly-Clark, Net Air International, Rocky Mountain Airlines**, and **Johns-Manville Corporation**. From 1980-81, he was president of the energy development company **Prometheus**.

Along with the rest of the Mercury Seven astronauts, he co-authored the 1962 book, *We Seven: By the Astronauts Themselves*. With Richard N. Billings, he released his autobiography, *Schirra's Space* in 1988. In 1995, he co-authored the book, *Wildcats to Tomcats: The Tailhook Navy*. In 2005, he co-authored the book, *The Real Space Cowboys* with Ed Buckbee. He was a contributor to the 2007 book, *In the Shadow of the Moon*. Walter Schirra died of a heart attack on May 03, 2007, aged eighty-four years, in San Diego, California. He was having treatment for abdominal cancer. Wally Schirra was married, with two children.

## Schmitt, Harrison H.. Apollo 17.

Dr. Harrison "Jack" Hagan Schmitt was born in July 1935 in Santa Rita, New Mexico, USA. He is the most recent living person to have walked on the Moon. He received a Bachelor of Science degree in Geology from the California Institute of Technology in 1957; then spent a year studying geology at the University of Oslo in Norway as a Fulbright Scholar; and received a Ph.D. in Geology from Harvard University in 1964. He then worked at the **United States Geological Survey**'s Astrogeology Center at Flagstaff, Arizona, developing field techniques that would later be used by the Apollo crews.

Selected by NASA in June 1965 along with a group of scientist-astronauts - the first group not to be test pilots - he was part of the Apollo space program's last flight in 1972. His first year of astronaut training was spent at Air Force UPT learning to become a jet pilot. Schmitt then trained Apollo crews to be geologic observers for when they were in lunar orbit, and to be competent geologic workers when they were on the lunar surface. After each of the landing missions, he took part in the examination and evaluation of the returned lunar samples and helped the crews with the scientific aspects of their mission reports. He was backup Lunar Module pilot for *Apollo 15* and was to be on the *Apollo 18* crew, but when that flight was canceled, he was moved up to *Apollo 17* so that a professionally trained geologist could fly the mission.

On the *Apollo 17* mission, launched December 07, 1972, he was the first person originally trained as a geologist to walk on the Moon. He collected the rock sample designated Troctolite 76535, which has been called *"without doubt the most interesting sample returned from the Moon"*. It is the central piece of evidence suggesting that the Moon once possessed an active magnetic field. Jack Schmitt logged 12 days, 13 hours and 52 minutes in space. He made three lunar EVAs totaling 20 hours and 35 minutes.

Dr. Jack Schmitt resigned from NASA on August 30, 1975, to run as a Republican candidate to the US Senate. In the 1976 election, he defeated Democratic incumbent Joseph Montoya. Schmitt was in office from January 03, 1977 to January 03, 1983, then being defeated by Democrat Jeff Bingaman; criticized for not paying enough attention to local matters, but impacted by the deep recession. Notably, Schmitt was the chairman of the Science, Technology and Space subcommittee of the US Senate Committee on Commerce. He then worked as a consultant in Business, geology, space, and public policy. He is an adjunct professor of engineeering physics at the University of Wisconsin-Madison, and has long been a proponent of lunar resource utilization; particularly for helium-3 as a fuel for notional nuclear fusion reactors.

In 2006, Dr. Schmitt wrote a book, *Return to the Moon: Exploration, Enterprise, and Energy in the Human Settlement of Space*. Jack Schmitt was chair of the NASA Advisory Council from November 2005 until his abrupt resignation on October 16, 2008. In November 2008, he quit The Planetary Society over policy advocacy differences: he said that going back to the Moon would speed progress toward a crewed Mars mission. Additionally, he criticized the Society's stance on global warming, writing in his resignation letter that the " ***'global warming scare'*** *is being used as a political tool to increase government control over American lives, incomes and decision making"*. He spoke at the March 2009 International Conference on Climate Change, an anthropogenic climate change denier event, where he said that climate change was a ***"stalking horse for National Socialism!"*** In January 2011, he was appointed as secretary of the New Mexico Energy, Minerals and Natural Resources department, but was forced to give up the position the following month after refusing to submit to a required background investigation.

On April 29, 2018, the Schmitt Space Communicator SC-1x (named in his honor) was carried aboard the **Blue Origin** *New Shepard* crew capsule in a project partly funded by NASA. It launched the first commercial two-way data and wi-fi hotspot service in space, and sent the first commercial *Twitter* message from space. Jack Schmitt has received eight awards, including the NASA Distinguished Service Medal (1973); was made an honorary fellow of the Geological Society of America for his efforts in geoscience in 1984; and was the 2015 recipient of the Leif Erikson Exploration Award for his work on the surface of the Moon in 1972, and for his part in the geology training of all the astronauts that walked on the Moon before him. He was a contributor to the 2006 book, *NASA's Scientist-Astronauts* by David Shayler and Colin Burgess; and he is one of the astronauts featured in the 2007 documentary ***In the Shadow of the Moon***.

## Schweickart, Russell L.. Apollo 9.

Russell "Rusty" Louis Schweickart was born in October 1935 in Neptune Township, New Jersey, USA. He is a research scientist, and former business and government executive. He received a Bachelor of Science degree in Aeronautical Engineering in 1956; and a Master of Science degree in Aeronautics and Astronautics from the Massachusetts Institute of Technology in 1963. From 1956-63, he served in the USAF and Massachusetts Air National Guard, logging over 4,000 hours flying time as a fighter pilot, including 3,500 hours in jet aircraft. His thesis for a Master's degree at MIT concerned stratospheric radiance.

Selected in October 1963 for NASA's third astronaut group, Schweickart was the youngest in the group, and became Lunar Module pilot on *Apollo 9* in 1969. He performed the first in-space test of the portable life support system used by the Apollo astronauts who walked on the Moon. He was backup Commander of the first crewed *Skylab* mission in 1973, then served as Director of User Affairs in NASA's Office of Applications. Schweickart logged ten days, one hour and zero minutes (241 hours) in space; including one EVA of one hour and 17 minutes. During a five-minute pause tethered outside his spacecraft, he felt he underwent a metaphysical experience as he stared at the Earth, contemplating its place in the Universe. He subsequently practised Trancendental Meditation based on his experience. Following the loss of *Skylab*'s thermal shield during launch, he assumed responsibility for the development of hardware and procedures for erecting an emergency solar shade and deploying a jammed solar array that saved the space station.

Schweickart left NASA in 1977 to serve for two years as California Governor Jerry Brown's assistant for science and technology. He was then appointed to California's **Energy Commission** for five and a half years, serving as Chairman for three years. In 1984-85, he co-founded **The Association of Space Explorers** and later in 2002 co-founded the **B612 Foundation**, a non-profit organization dedicated to defending Earth from asteroid impacts. His satellite and telecommunications work involved him in the development of international communications regulations and policies, including participation in the 1992 and 1995 World Radiocommunication Conferences of the International Telecommunication Union (ITU). From 1996-98 he was President and Chief Executive of **ALOHA Networks, Inc.**, a data communications company. Schweickart has spoken and taught at the Esalen Institute, and currently serves as the B612 Foundation's Chair Emeritus. Rusty Schweickart received seven awards, including a 1969 Emmy for transmitting the first live TV pictures from space; and the NASA Exceptional Service Medal. Main-belt asteroid 17033 Rusty was named in his honor. He was married twice, and has seven children.

## Scobee, Francis R.. STS-41-C, STS-51-L.

Lt. Col. Francis Richard "Dick" Scobee (1939-86) was born in May 1939 in Cle Elum, Washington, USA. He graduated from the University of Arizona with a Bachelor of Science degree in Aerospace Engineering in 1965. He was a reciprocating engine mechanic for the USAF, earned his pilot wings in 1966, and served as a combat aviator in the Vietnam War. He then attended the USAF Aerospace Research Pilot School (Class 71B) at Edwards AFB. Upon graduation in 1972, he became an Air Force test pilot; logging thousands of flying hours in dozens of aircraft. Selected for the NASA astronaut corps in January 1978, he served as an instructor pilot for the Shuttle's 747 carrier aircraft. On April 06, 1984, he piloted *Challenger* on STS-41-C, a mission of six days and 23 hours, which successfully deployed one satellite and repaired the malfunctioning *Solar Max* satellite.

Dick Scobee was killed on January 28, 1986, while commanding the Space Shuttle *Challenger*, which suffered a catastrophic booster O-ring failure over the North Atlantic Ocean during the launch of the STS-51-L mission. Scobee's last recorded words were, *"Roger, go at throttle up"*. The shuttle broke up at an altitude of 14.6 km (48,000 feet). Some experts, including one of NASA's lead investigators, Robert Overmyer, believed most if not all of the crew were alive and possibly conscious during the entire descent until the high-speed impact with the ocean. After the disaster investigation, Overmyer stated, *"I know Scobee did everything he could to save his crew. Scob fought for any and every edge to survive"*. Dick Scobee logged six days, 23 hours and 40 minutes in space.

Richard Scobee is buried at Arlington National Cemetery. He was posthumously awarded the Congressional Space Medal of Honor and was inducted into the US Astronaut Hall of Fame. On July 09, 1994, the San Antonio College Planetarium was rededicated The Scobee Planetarium. After the *Challenger* disaster, a number of schools, streets, and municipal facilities in the US were renamed in his honor. Scobee made an appearance in the 1985 IMAX documentary ***The Dream is Alive***, shot during the STS-41-C mission. He was portrayed in the 1990 TV film ***Challenger*** by Barry Bostwick. Dick Scobee was married, with two children.

## Scott, David R.. Gemini VIII, Apollo 9, Apollo 15.

Col. David "Dave" Randolph Scott was born in June 1932 at Randolph Field, in San Antonio, Texas, USA. He was the seventh person to walk on the Moon. Selected as part of NASA's third group of astronauts, he's now the only living human being who commanded an Apollo mission and walked on the Moon. In 1950, he received an invitation to attend West Point, New York. Dave Scott graduated with a Bachelor of Science degree in Military Science from the United States Military Academy at West Point. He was fifth in his class of 633, and was commissioned to join the US Air Force. Scott began his primary pilot training in July 1954. From April 1956 to July 1960, he flew with the Thirty Second Tactical Fighter Squadron at Soesterberg Air Base, Netherlands. After serving as a fighter pilot in Europe during the Cold War, he graduated from the Air Force Experimental Test Pilot School (Class 62C) and the Aerospace Research Pilot School (Class IV). Dave Scott logged over 5,600 hours flying time. Scott achieved both a Master of Science degree in Aeronautics and Astronautics, and a degree in Aeronautics and Astronautics Engineering from the Massachusetts Institute of Technology in 1962.

Col. Scott was selected as one of fourteen NASA astronauts late in 1963. On March 16-17, 1966, he was pilot of *Gemini VIII* with Commander Neil Armstrong, spending just under eleven hours in low Earth orbit. They made the first US orbital space docking with an *Agena* target rocket. The Gemini's Orbit Attitude and Maneuvering System (OAMS) thrusters were firing unexpectedly, and needed to be shut down. To maintain control they had to fire the Reaction Control System thrusters. Since these were to be used for re-entry, the mission rules said that if they were activated early, the mission had to be aborted and the crew return to Earth. According to Franciolin French and Colin Burgess in their book on NASA and the Space Race, *"Scott, in particular, had shown incredible presence of mind during the unexpected events of the Gemini VIII mission. Even in the middle of an emergency, out of contact with Mission Control, he had thought to re-enable ground control command of the Agena before the two vehicles separated"*. Along with Armstrong, he received the NASA Exceptional Service Medal; and a DFC from the Air Force, who promoted him to lieutenant colonel.

Launched on March 03, 1969, he was Command Module Pilot of *Apollo 9*, making extensive tests of the Apollo spacecraft during a ten day flight. The Lunar Module (LM) named *Spider* and Command Service Module (CSM) *Gumdrop* navigated to a space rendezvous, simulating a return from the lunar surface. If the LM had failed to return, Col. Scott would have had to run the entire spacecraft for re-entry (normally a three-person job). He would also have to rescue the LM if it could not perform the rendezvous, and if it could not dock, would have to assist Jim McDivitt and Russell "Rusty" Schweickart in performing an EVA back to the safety of the CSM. Scott was named as backup Commander of *Apollo 12* and flew as Commander of *Apollo 15*, launched on July 26, 1971, the fourth crewed lunar landing mission. With James Irwin, he was on the lunar surface for three days; splashing down in the Pacific Ocean north of Honolulu on August 07, 1971. Scott left a lunar memorial to the astronauts and cosmonauts who had died to advance space exploration. He logged a total of 22 days, 18 hours and 54 minutes in space. On August 25, 1971, he and the *Apollo 15* crew were jointly awarded the first United Nations Peace Medal.

In his role with the *Apollo-Soyuz Test Project*, David Scott traveled to Moscow, Russia, leading a team of technical experts. He met Alexei Leonov, the Commander of the Soviet part of the mission, with whom he would later write a joint autobiography, *Two Sides of the Moon: Our Story of the Cold War Space Race*, published in 2006. After serving as director of NASA's Dryden Flight Research Center in California, Scott retired from NASA on September 30, 1977, then worked on a number of space-related projects, and served as a consultant for several films about the space program, including ***Apollo 13*** and the 1998 miniseries ***From the Earth to the Moon*** and the 3D IMAX film ***Magnificent Desolation*** (2005).

Dave Scott founded **Scott Science and Technology, Inc.** He worked on several government projects in the late 1970s and 1980s; including four years on the Commercial Space Transportation Advisory Committee. He was a commentator for British television for the first Space Shuttle flight (STS-1) in April 1981. From 2003-04, Scott was a consultant on the BBC TV series ***Space Odyssey: Voyage to the Planets***. He received nine awards and two honorary doctorates. He is a fellow of the American Astronautical Society; an associate Fellow of the American Institute of Aeronautics and Astronautics; and a member of the Society of Experimental Test Pilots.

## Scott, Winston E.. STS-72, STS-87.

Winston "Win" Elliott Scott was born in August 1950 in Miami, Florida, USA. He received a Bachelor of Arts degree in Music from Florida State University in 1972; and a Master of Science Degree in Aeronautical Engineering (with an emphasis in Avionics) from the US Naval Postgraduate School in 1980. He entered Aviation Officer Candidate School in December 1972, and as a Naval Aviator he flew both rotary-wing then fixed-wing aircraft. He flew a four-year tour of duty with Helicopter Anti-Submarine Squadron Light 33 at the Naval Air Station North Island, California, flying the SH-2F LAMPS helicopter. In 1978, Win Scott was selected to attend the Naval Postgraduate School at Monterey, California. In June 1986, he became an Aerospace Engineering Duty Officer, then served as a production test pilot at Naval Aviation Depot Jacksonville, flying the F/A-18 Hornet and the A-7 Corsair II aircraft.

Winston Scott logged over 5,000 hours flying time in twenty different military and civilian aircraft types; and made over 200 shipboard landings. He was an associate instructor of electrical engineering at **Florida A & M University** and **Florida Community College at Jacksonville**. Selected in NASA Group 14 on March 31, 1992, he flew on STS-72 aboard *Endeavour* (January 11-20, 1996), an eight-day 22-hour flight which retrieved the Space Flyer Unit launched from Japan ten months earlier, deployed and retrieved the *OAST-Flyer satellite*, and crew conducted two EVAs. Scott was a Mission Specialist on STS-87 aboard *Columbia* (November 19-December 05, 1997), the fourth US microgravity payload flight, and Scott performed two EVAs. He logged a total of 24 days, 14 hours and 34 minutes in space; including three EVAs totaling 19 hours and 26 minutes.

Winston Scott retired from NASA on July 31, 1999, and from the US Navy in September 1999 to accept a position at his alma mater as Vice President for Student Affairs. Scott was then Executive Director of the **Florida Space Authority**. He currently serves as the Vice President of **Florida Institute of Technology**. Winston Scott wrote a book about his experiences in space, *Reflections From Earth Orbit*, published by Apogee Books. He belongs to the American Institute of Aeronautics and Astronautics; the National Naval Officers Association; the Naval Helicopter Association; and the Bronze Eagles Association of Texas. He plays trumpet with a Houston-based big band. Scott is married, with two children; one of whom has followed in his footsteps as a Naval Aviator.

## Scully-Power, Paul D.. STS-41-G.

Paul Desmond Scully-Power (AM GOSE FRAeS) was born in May 1944 in Sydney, New South Wales, Australia. He is an Australian-American oceanographer, technology expert and business executive. He attended schools in London, and Saint Ignatius College, Riverview and St. Pius X College, Chatswood, in Sydney. Paul studied applied mathematics at the University of Sydney and graduated with a Bachelor of Science (with honours) in 1966; and a Diploma of Education in 1967. In January 1967, he was approached by the Royal Australian Navy to set up the first oceanographic group within the Navy. From July 1972 to March 1974, he was an Australian Navy Exchange Scientist, United States Navy. He worked at the US Naval Underwater Systems Center, New London, Connecticut, and at the Office of Naval Research, Washington, D.C.. During this period, he was invited to assist the Earth Observations team on the Skylab Project and has worked in space oceanography for each manned spacecraft mission since that time.

In October 1977, Scully-Power emigrated to the United States and was offered a position at the Naval Underwater Systems Center. He became a US citizen in 1982. In June 1984, Paul was chosen by NASA to be a Payload Specialist on the thirteenth Space Shuttle mission, which would study Earth Sciences. The *Challenger* STS-41-G mission of October 05-13, 1984, was the first with a seven-person crew, and the first to demonstrate American orbital fuel transfer. The flight logged over 197 hours in space and traveled over 3.4 million miles (about 5,471,770 km) in 133 Earth orbits. In October 1984, while a civilian employee of the United States Naval Undersea Warfare Center, he flew on NASA mission STS-41-G as a Payload Specialist. He was the first Australian-born person to journey into space, and the first astronaut with a beard. His time in space was eight days, five hours and 23 minutes. He was able to confirm the existence of spiral eddies, observe them with the naked eye and photograph them.

Paul Scully-Power went on to work in private industry. In 2004 he became the CTO of **Tenix Group** (Australia's largest Defence and Technology contractor), and in 2007 he was appointed Chairman and CEO of **SensorConnect Inc**. Scully-Power has extensive commercial, government and academic experience in Australia, New Zealand, the United Kingdom and the United States. He is widely recognized in the fields of defence and national security, aviation and aerospace, marine science, communications and systems analysis, and education. Paul Scully-Power is considered a world expert in remote sensing: visible, infrared, radar and acoustic. He was the first president of the UN International Commission on Space Oceanography. He is a Fellow of the Royal Aeronautical Society, a Liveryman of the Guild of Air Pilots and Air Navigators, and a Freeman of the City of London. In the 2004 Australia Day Honours, he was appointed a Member of the Order of Australia. He has received twelve honors, including a US Presidential Letter of Commendation.

## Searfoss, Richard A.. STS-58, STS-76, STS-90.

Col. Richard "Rick" Alan Searfoss (1956-2018) was born in June 1956 in Mount Clemens, Michigan, USA. He received a Bachelor of Science degree in Aeronautical Engineering from the USAF Academy in 1978; and a Master of Science degree in Aeronautics from the California Institute of Technology on a National Science Foundation Fellowhip in 1979. In the Air Force, he attended Squadron Officer School, Air Command and Staff College, and Air War College. Searfoss graduated from Undergraduate Pilot Training at Williams AFB, Arizona, in 1980. In 1988, he attended the US Naval Test Pilot School, Patuxent River, Maryland. He was a flight instructor at the USAF Test Pilot School at Edwards AFB when selected by NASA. He logged over 6,000 hours flying time in 77 aircraft types, and over 939 hours in space.

Selected by NASA in January 1990, he was Pilot on STS-58 aboard *Columbia* (October 18-November 01, 1993), and the crew performed sixteen engineering tests and twenty Extended Duration Orbiter Medical Project experiments. Rick Searfoss was Pilot on STS-76 aboard *Atlantis* (March 22-31, 1996), the third docking with *Mir*, and crew transferred nearly two tons of supplies and scientific equipment. He was Commander on STS-90 (April 17-May 03, 1998), an almost 16 day 'Spacelab' flight with twenty-six life science experiments. This was the last and most complex of the twenty-five 'Spacelab' missions, and the results have broad applicability both in preparing for future long-duration human space missions and in clinical applications on Earth. Searfoss logged a total of 39 days, three hours and 18 minutes in space.

After leaving NASA, Rick Searfoss served as a judge for the **Ansari X-Prize**, where he officially declared *SpaceShipOne* to have won the competition on October 04, 2004, after completing two flights within a two-week period. He was a test pilot instructor at the National Test Pilot School at the Mojave Spaceport. In 2008, he was a featured guest in a Volkswagen commercial. From the mid-2000s until the company closed, Searfoss worked with **XCOR Aerospace** in the development of their rocket-powered aircraft. In September 2013, he was Chief Test Pilot at XCOR, and made a total of fifty-two rocket-powered flights. Rick Searfoss received sixteen awards, including USAF Squadron Officer's School Commandant's Trophy as top graduate; selected for Outstanding Young Men of America (1987); the Defense Meritorious Service Medal; the NASA Exceptional Service Medal; the DFC; the Orbit Award; and Asteroid 13157 Searfoss (discovered in 1995) was named in his honor. In January 2017, he was a featured guest on the first ***Star Trek*** Cruise, where he gave a presentation about the history and future of space flight. Richard Searfoss died on September 29, 2018.

## Seddon, Margaret "Rhea". STS-51-D, STS-40, STS-58.

Dr. Margaret "Rhea" (pronounced "Ray") Seddon was born in November 1947 in Murfreesboro, Tennessee, USA. She received a Bachelor of Arts degree in Physiology from the University of California in 1970; and a Doctor of Medicine from the University of Tennessee, Memphis, in 1973. Selected in 1978 NASA Group 8, part of the first group of 35 astronauts to include eight US women. At 157 cm tall (62 inches) she was just tall enough to meet the minimum height requirement of 60 inches. Pilot training was not required of Mission Specialist candidates, but they were given training in how to handle emergencies while flying in the back seat of NASA's Northrop T-38 Talon jets. As an astronaut candidate, she drew a civil service salary of about USD 22,000 (equivalent to $99,000 in 2022), which was more than she made as a surgical resident.

In February 1981, Seddon became engaged to fellow astronaut Robert L. "Hoot" Gibson. They were married on May 30, and honeymooned in Hawaii. She then resumed her role with search and rescue for the upcoming ST-2 mission. Their first child, Paul Seddon Gibson, was born in July 1982; the first child born to a US astronaut couple. Dr. Seddon was a rescue helicopter physician for the early Space Shuttle flights and a support crew member for STS-6, launched April 04, 1983. She served as a member of NASA's Aerospace Medical Advisory Committee, and as a CapCom in the Mission Control Center. Dr. Rhea Seddon flew as a Mission Specialist on STS-51-D, which deployed the *ANIK-C* and *Syncom IV-3* satellites. Rhea Seddon used her surgical skills to operate a bone saw to help build home-made repair tools for the *Syncom* satellite, but the launch sequence did not commence, and the satellite was left in low Earth orbit.

She was a Mission Specialist on STS-40 aboard *Columbia* (launched June 05, 1991). During the nine-day two-hour mission, the crew performed experiments that explored how humans, animals and cells respond to microgravity and re-adapt to Earth's gravity on return. There were tests of hardware proposed for the Space Station *Freedom* Health Maintenance Facility. From September 1991 to July 1992, Dr. Seddon was a CapCom in the Mission Control Center, handling the STS-42 and STS-45 missions. Rhea Seddon was Payload Commander on STS-58 aboard *Columbia* (launched October 18, 1993). During the 14-day flight, the crew performed neurovestibular, cardiovascular, cardiopulmonary, metabolic and musculoskeletal medical experiment on themselves and forty-eight rats. She logged a total of 30 days, two hours and 21 minutes (over 722 hours) in space.

After her third child was born in June 1995, Seddon became the Assistant to the Director of Flight Crew Operations for *Shuttle-Mir* Payloads, a new position, which involved travel to Russia. In 1996, she was detailed by NASA to Vanderbilt University Medical School in

Tennessee, where she assisted in the preparation of cardiovascular experiments that flew on the STS-90 Neurolab Spacelab flight in April 1998. Dr. Seddon retired from NASA in November 1997 and became Assistant Chief Medical Officer of the **Vanderbilt Medical Group**. Dr. Rhea Seddon had begun writing her memoirs in December 1993, and the book, titled *Go for Orbit* was published in 2015. It won the Independent Book Publishers Ben Franklin Gold Award for Best Autobiography/Memoir. Rhea Seddon received two NASA Exceptional Service Medals (1988, 1992); and the NASA Outstanding Leadership Medal (1994). She was inducted into the Tennessee Aviation Hall of Fame (2005); and the US Astronaut Hall of Fame and Tennessee Women's Hall of Fame (2015). In 2017, she was named as one of the University of Tennessee's Centennial Top 100 Alumni.

## Sega, Ronald M.. STS-60, STS-76.

Maj. Gen. Dr. Ronald "Ron" Michael Sega was born in December 1952 in Cleveland, Ohio, USA. He received a Bachelor of Science degree in Mathematics and Physics from the US Air Force Academy in 1974; a Master of Science degree in Physics from Ohio State University in 1975; and a Ph.D. in Electrical Engineering at the University of Colorado at Boulder in 1982. Dr. Ron Sega completed pilot training in 1976 and served as an instructor pilot at Williams AFB, Arizona. From 1987-88, he served as Technical Director, Lasers and Aerospace Mechanics Directorate of the **Frank J. Seiler Research Laboratory** at the USAF Academy. He has authored or co-authored over 100 technical publications. Selected by NASA on January 17, 1990, in Group 13, he became an astronaut qualified as a Mission Specialist in July 1991. Dr. Sega was a co-principal investigator and mission director for the *Wake Shield Facility* which flew on missions STS-60 and STS-76.

He served as Flight Engineer on STS-60 aboard *Discovery* (February 03-11, 1994), the first joint space mission for Russia and the US. He operated the robotic arm, berthing the *Wake Shield Facility* onto its payload bay carrier on four separate occasions. From November 1994 to March 1995, Sega was the NASA Director of Operations, Star City, Russia. This involved building an organization and infrastructure to support astronaut and cosmonaut mission and science training for flight on the Russian Space Station *Mir*. He was a Mission Specialist on STS-76 aboard *Atlantis* (March 22-31, 1996), the third docking mission to *Mir*. As Payload Commander, Dr. Sega was responsible for planning and on-orbit operations, including extensive transfer of logistics and science. Some 4,800 pounds of science and mission hardware, food, water and air went aboard *Mir*, and over 1,100 pounds of US and ESA science and Russian hardware were returned to Earth. Ron Sega logged a total of 17 days, 12 hours and 25 minutes (over 420 hours) in space and retired as an astronaut on July 01, 1996 to become Dean of the College of Engineering and Applied Science, **University of Colorado** at Colorado Springs.

In 2001, he became acting Assistant to the **Secretary of Defense** for Nuclear, Chemical and Biological Programs; becoming Assistant Secretary from August 2001 to August 2005. From August 2005 to August 2007, he served as Under Secretary for the **US Air Force**. He belongs to six professional organizations. Maj. Gen. Ronald Sega has received eighteen awards, including Distinguished Graduate of the USAF Academy (1974); Officer of the Year in the Department of Physics USAF Academy (1980); Outstanding Faculty Award, University of Colorado (1985); the NASA Superior Achievement Award - NASA Director of Operations, Russia (1995); an IEEE Fellow, for contributions to material research in space and regional engineering education (2002); and the Air Force Commander's Award for Public Service (November 14, 2016). Currently he is professor of systems engineering and Vice President for Energy and the Environment at the **Colorado State University Research Foundation**. Dr. Sega also holds a similar position at **Ohio State University**. He was married to fellow astronaut Bonnie J. Dunbar and is now married to Ann Sega. They have two sons.

## Shaw, Brewster H., Jr.. STS-9, STS-61-B, STS-28R.

Col. Brewster Hopkinson Shaw, Jr. was born in May 1945 in Michigan, USA. He received a Bachelor of Science degree in Engineering Mechanics from the University of Wisconsin-Madison in 1968; and a Master of Science degree in Engineering Mechanics from University of Wisconsin-Madison in 1969. He entered the US Air Force in 1969 after completing Officer Training School and undergraduate pilot training at Craig AFB, Alabama. In April 1973, he reported to George AFB, California, for F-4 instructor duties. From July 1975 he attended the USAF Test Pilot School at Edwards AFB, then remained in California as an operational test pilot; serving as an instructor from August 1977 to July 1978. Col. Shaw served as a US Air Force pilot in the Vietnam War. He logged over 5,000 hours flying time in thirty aircraft types.

Selected by NASA in January 1978, he served on loan from the US Air Force. He was pilot on STS-9 aboard *Columbia* (November 28-December 08, 1983), the first flight to carry Payload Specialists. The crew conducted over 70 multi-disciplinary experiments; returning the heaviest-ever payload to Earth in the Shuttle's cargo bay. Shaw was Commander on STS-61-B aboard *Atlantis* (November 26-December 03, 1985), which deployed communications satellites, and the crew conducted two six-hour EVAs. He was Commander on STS-28R aboard *Columbia* (August 08-13, 1989), which carried classified **Department of Defense** payloads and a number of secondary payloads. Brewster Shaw logged a total of 22 days, five hours and 51 minutes (over 533 hours) in space. Following the Columbia accident in 1986, he supported the **Rogers Presidential Commission** investigations.

He left JSC in October 1989 to become deputy director, Space Shuttle Operations at KSC. He was the final authority for the launch decision, and chaired the Mission Management Team. Shaw worked as a manager in Space Shuttle Operations at NASA until 1996 when he left the agency, retired from the Air Force and went to work at **Rockwell**, which was acquired by **The Boeing Company** in December 1996. He was then an aerospace executive at Boeing. In mid-2003, Brewster Shaw left Boeing and became the COO of **United Space Alliance (USA)**, with overall management of its 10,000 employees in Florida, Texas, Alabama and Russia. In January 2006, he returned to **The Boeing Company**'s Houston campus; and retired from Boeing on August 26, 2011. Shaw was inducted into the US Astronaut Hall of Fame on May 06, 2006. He received twenty-eight medals in the Vietnam War; the Defense Superior Service Medal; the Air Force Distinguished Flying Cross; and the Defense Meritorious Service Medal. Brewster Shaw is married, with three children.

## Shepard, Alan B., Jr.. Mercury-Atlas "Freedom 7", Apollo 14.

Rear Adm. Alan Bartlett Shepard, Jr. (1923-98) was born in November 1923 in Derry, New Hampshire, USA. He was the first American and the second man in space (after cosmonaut Yuri Gagarin), piloting the Mercury spacecraft *Freedom 7* on a suborbital flight on May 05, 1961. Shepard also became the fifth person (and the oldest, at age 47 years) to walk on the Moon in February 1971. Alan was a brilliant student in school, skipping both the sixth and eighth grades. He completed his schooling privately at the Pinkerton Academy in Derry. As a student he was fascinated by planes, often cycling to Manchester Airfield to do odd jobs in exchange for occasional rides and informal flying lessons. While Shepard passed the entrance exam of the United States Naval Academy in 1940, he was underage; joining in 1941. There, he won prizes for swimming and sailing. He graduated from the United States Naval Academy at Annapolis as an ensign in June 1944 and received a Bachelor of Science degree from the Naval Academy.

Alan Shepard served in the Pacific during World War II. His first post from August 1944 was on the destroyer *USS Cogswell* as a gunnery officer. His flight training started at the Naval Air Station, Corpus Christi in Texas in January 1946, with later training in Florida. Following six perfect landings on the *USS Saipan* he earned his naval aviator wings in 1947. After his flying training, he joined the Fighter Squadron 42 (VF-42) to fly the Vought F4U Corsair. He attended the United States Naval Test Pilot School, becoming a qualified test pilot. In 1950, Shepard experimented with high-altitude aircraft, in-flight fueling systems, and landings on angled carrier decks. His next assignment was with the VF-193 night-fighter squadron that was deployed for the early 1950s Korean War. Alan Shepard was one of the seven last pilots to be selected to be part of NASA's space initiative and he became a national hero after his maiden flight aboard *Mercury-Atlas Freedom 7* on May 05, 1961. He was awarded the NASA Distinguished Service Medal by President John F. Kennedy.

He was designated as the commander of the first manned *Gemini* mission, which involved launching a crew of two into space, but because of health issues (Meniere's disease, causing extreme dizziness and nausea) he had to be taken off flying duties. He was appointed as the Chief of the Astronaut Office in November 1963, being responsible for training United States astronauts and planning their missions. After an operation in 1969 he was again declared fit for flying. Alan Shepard was the Commander of the *Apollo 14* mission, which took place from January 31 to February 09, 1971. This was the third manned lunar landing and the first to land in the lunar highlands. Shepard piloted the Lunar Module *Antares*. During his time on the Moon, he famously used the head of a six iron to hit two golf balls on its surface. Shepard was the only one of the "Mercury Seven" astronauts to walk on the Moon. He logged a total of nine days, zero hours and 57 minutes in space; including two lunar EVAs totaling nine hours and 23 minutes. After retiring from NASA in 1974, Shepard worked in private business, investing in banks, real estate and oil exploration. He also did volunteer work to support education and to help people learn about spaceflight.

In 1977, Alan Shepard was inducted into the National Aviation Hall of Fame, the International Space Hall of Fame in 1981, and the United States Astronaut Hall of Fame in 1990. He also received several awards, including the Congressional Space Medal of Honor. In recognition of his contribution to the United States space mission, Shepard was appointed as a delegate to the 26th **United Nations General Assembly**. Shepard was promoted to the rank of Rear Admiral in the United States Navy before he retired in 1971. He made a fortune in banking and was part of the board of directors of several institutions. Alan spent the rest of his life close to his family. In 1994, he published a book titled *Moon Shot: The Inside Story of America's Race to the Moon*, which was made into a TV mini-series. Alan Shepard died of leukemia on July 21, 1998, at Pebble Beach, California. He was aged seventy-four years.

At Shepard's funeral, fellow astronaut John Glenn said this about his friend: *"Alan Shepard was many things, a patriot, leader, a competitor – a fierce competitor - a hero, and to us here today, a close friend. When we seven were chosen as America's first astronauts, Al stood out. I first knew Al as test pilots when we were at the Naval Test Center, Patuxent River, Maryland. But it wasn't until the Mercury program that I saw, firsthand, the determination and toughness of Alan Shepard. Al named his spacecraft "Freedom 7", based on his view of where this program stood, the openness of it around the world and what we stood for. In fact, as he sat in "Freedom 7" through another delay in countdown before his historic Mercury mission, he was very anxious to get going. We heard an anxious Al say, **"Why don't you fix your little problem and light this candle!"** America has lost one of its true adventurers, someone whose entire life was dedicated to America's questing spirit. Scott, Wally, Gordon and I have lost more than a friend, we've lost another brother".* Unquote.

## Shepherd, William M.. STS-27R, STS-41, STS-52, Soyuz TM-31/Expedition 1.

William "Bill" McMichael Shepherd was born in July 1949 in Oak Ridge, Tennessee, USA. He served as Commander of 'Expedition 1', the first resident crew on the ISS, and is a recipient of the Congressional Space Medal of Honor. He received a Bachelor of Science degree in Aerospace Engineering from the US Naval Academy in Annapolis, Maryland, in 1971. He completed Basic Underwater Demolition/SEAL (BUD/S) training in 1972, and later qualified as a Navy SEAL. He obtained a Bachelors degree in Ocean Engineering and a Master of Science degree in Mechanical Engineering, both in 1978, from the Massachusetts Institute of Technology. Selected in 1984 NASA Group 10, Shepherd was the first military non-aviator in astronaut training. In 1986, his Navy SEAL training proved unexpectedly useful to NASA as he helped to direct the underwater salvage operations of the destroyed Space Shuttle *Challenger*.

He then served as a Mission Specialist on three Space Shuttle flights: STS-27R aboard *Atlantis* (December 02-06, 1988), when the vehicle's heatshield tiles were substantially damaged during lift-off, and intense heat damage needed to be repaired after landing. The mission launched a Department of Defense satellite. He flew on STS-41 aboard *Discovery* (October 06-10, 1990), which deployed the ESA's *Ulysses* probe as part of the International Solar Polar Mission. STS-52 aboard *Columbia* (October 22-November 01, 1992) deployed the *Laser Geodynamics Satellite 2* (*LAGEOS-2*) and operated the US Microgravity Payload-1 (USMP-1). From March 1993 to January 1996, Shepherd was assigned to the ISS Program, serving as Deputy, then Program Manager. In November 1995, he was selected to command the first crew of the ISS, but a long series of political, financial and technical problems caused significant delays. Shepherd flew as Flight Engineer on *Soyuz TM-31*, launched October 31, 2000, the first *Soyuz* spaceflight to dock with the ISS, and it carried the members of 'Expedition 1', the first long-duration ISS crew. After being ISS Commander, he returned to Earth aboard *Discovery*, STS-102, on March 19, 2001, which saw the first use of the Multi Purpose Logistics Module *Leonardo* (MPLM) to bring supplies to the ISS. William Shepherd logged a total of 159 days, seven hours and 49 minutes in space.

He was then assigned to the staff of Commander, **Naval Special Warfare Command**, to assist with the development of new capabilities and programs for the Navy's SEAL and Special Boat units. He retired from the Navy in January 2002. Capt. Shepherd also served as US Special Operations Commands's first Science Advisor from 2008-11. He has received eleven awards, including NASA's Steve Thorne Aviation Award; the Federation Aeronautique Internationale (FAI) V.M. Komarov Diploma; the FAI Yuri Gagarin Gold Medal; the Spirit of St. Louis Medal; the Dr. Robert H. Goddard Memorial Trophy; and in 2009, the ISS Program team (which he led both on the ground and in space) was awarded the Collier Trophy.

## Shriver, Loren J.. STS-51-C, STS-31, STS-46.

Col. Loren James Shriver was born in September 1944 in Jefferson, Iowa, USA. He graduated from the USAF Academy with a Bachelor of Science degree in Aeronautical Engineering in 1967; and from Purdue University with a Master of Science degree in Astronautical Engineering in 1968. From 1975, he attended the USAF Test Pilot School at Edwards AFB, California, and was later assigned to the Test Squadron 6512. Selected in January 1978 NASA Group 8, Shriver was pilot on STS-51-C aboard *Discovery*, launched on January 24, 1985, a classified DoD mission which may have launched the *Magnum* signals intelligence system. He was Commander on STS-31 aboard *Discovery*, launched on April 24, 1990, which deployed the Hubble Space Telescope into a 380-statute mile low-Earth orbit.

Shriver was Commander on STS-46 aboard *Atlantis*, launched July 31, 1992, which had problems with the *Tethered Satellite System* (TSS-1). Col. Loren Shriver logged a total of 16 days, two hours and four minutes (over 386 hours) in space. He served as the Space Shuttle program Manager for Launch Integration to 1997. Before his retirement, Col. Shriver was Deputy Director for Launch and Payload Processing at KSC. Shriver was the vice-president of engineering and integration and Chief Technology Officer with **United Space Alliance** until he retired from NASA in 2011. He is married, with three daughters and a son.

## Slayton, Donald K.. Apollo-Soyuz Test Project.

Donald Kent aka D.K. "Deke" Slayton (1924-93) was born in March 1924 in Sparta, Wisconsin, USA. He went on to become NASA's first **Chief of the Astronaut Office** and Director of Flight Crew Operations. Slayton joined the United States Army Air Forces during World War Two and flew in Europe and the Pacific, including seven combat missions over Japan. He received a Bachelor of Science degree in Aeronautical Engineering from the University of Minnesota in 1949, and later joined the Minnesota Air National Guard after working for **Boeing** as an aeronautical engineer. Slayton joined the USAF and attended the United States Air Force Test Pilot School in 1955. While a test pilot, one of his pilot colleagues was also named Don. To avoid confusion in radio communications, Slayton was referred to by his initials, D.K., which eventually shortened to "Deke". In 1959, he applied to, and was selected as one of the "Mercury Seven" astronaut candidates on April 02. He was scheduled to pilot the second US crewed orbital spaceflight, but was grounded in 1962 by idiopathic atrial fibrillation, an irregular heart rhythm.

In early 1962, Slayton was selected to serve as the senior manager of the astronaut office. One of his first roles was to select the Group 2 astronauts. He was tasked with making future crew assignments. In October 1963, Slayton became assistant director of Flight Crew Operations. In November 1963 he resigned from his commission in the Air Force and became a civilian executive for NASA, becoming director for Flight Crew Operations in 1966. He oversaw the process to determine which astronauts would fly on the *Gemini* and *Apollo* missions. In 1971, Slayton was examined at the Mayo Clinic after a long period without heart fibrillation and was determined not to have a coronary condition. On March 13, 1972, he was medically cleared to fly and was named the docking module pilot of the 1975 *Apollo-Soyuz Test Project* (ASTP).

Slayton remained in a managerial role throughout the *Skylab* program and resigned as Director of Flight Crew Operations in February 1974 in preparation for his upcoming spaceflight of July 1975. At age fifty-one years, he was the oldest astronaut to fly at that time. Prior to the ASTP flight, Slayton was assigned to manage the Approach and Landing Tests of the Space Shuttle program. While working on the Space Shuttle, he also assisted in developing the Shuttle Carrier Aircraft, a modified Boeing 747. The ALT program ended in late 1977 and Slayton agreed to manage the Space Shuttle's Orbital Test Flight Tests. He officially retired in 1980 but continued to serve in an advisory role for mission STS-1 and flew a T-38 chase plane during the landing of Space Shuttle mission STS-2.

Deke Slayton continued to work at NASA until 1982. After his retirement, he served as president of **Space Services Inc.**, a Houston-based company founded to develop rockets for small commercial payloads. He also served on the Department of Transportation's Commercial Space Advisory Committee. In 1991, Slayton began working with space historian Michael Cassutt to write his autobiography, titled *Deke!: US Manned Space from Mercury to the Shuttle*, published in 1994, a year after his death. With fellow astronaut Alan Shepard he co-wrote the 1994 book, *Moon Shot: The Inside Story of America's Race to the Moon.* Slayton was awarded the Distinguished Flying Cross, the NASDA Distinguished Service Medal, the Collier Trophy, and the James H. Doolittle Award. In 1992, Deke Slayton was diagnosed with a malignant brain tumor, and he died from a brain cancer on June 13, 1993, aged sixty-nine years.

## Smith, Steven L.. STS-68, STS-82, STS-103, STS-110.

Steven Lee Smith was born in December 1958 in Phoenix, Arizona, USA. He graduated from Stanford University with a Bachelor of Science degree in Electrical Engineering (1981), a Master of Science in Electrical Engineering (1982), and a Master's degree in Business Administration (M.B.A.) in 1987. From 1982-85, he worked for **IBM** in the Large Scale Integration (Semiconductor) Technology Group in San Jose as a technical group lead. Following a leave to pursue graduate studies, Smith returned to IBM's Hardware and Systems Management Group as a product manager until he joined NASA in 1989 as a payload officer responsible for pre-flight payload integration.

Selected in 1992 NASA Group 14, he underwent a year of astronaut candidate training, and in September 1993 became the first member of the 1992 class to receive a flight assignment. Smith served as a Mission Specialist on STS-68 aboard *Endeavour* (launched September 30, 1994), when he was responsible for Space Radar Lab-2 (SRL-2). He was a Mission Specialist on STS-82 aboard *Discovery* (launched February 11, 1997) to service the Hubble Space Telescope. Steven Smith was Payload Commander on STS-103 aboard *Discovery* (launched December 19, 1999) when crew performed three EVAs to upgrade several sub-systems for Hubble. After STS-103, he served as the Deputy Chief Astronaut for a year. Smith was lead spacewalker on STS-110 aboard *Atlantis* (April 08-19, 2002), when crew installed the S0 Truss on the ISS. A week was spent in joint operations with the 'Expedition 4' crew. He is a veteran of four spaceflights covering 16 million miles, and has performed seven EVAs totaling 49 hours and 25 minutes.

From 2000-03, Steven Smith was the director of **Special Olympics Texas**. He served as the NASA ISS Program Liaison to the European Space Agency until mid-2015, then he went to the NASA **Ames Research Center** in Silicon Valley. Since retiring from NASA, he is a Keynote Speaker, giving talks for companies like IBM and LinkedIn. He has also joined several boards, including that of The Charles A. and Anne Morrow Lindbergh Foundation, and the Blue Sky Network.

## Spring, Sherwood C.. STS-61-B.

Col. Sherwood Clark Spring was born in September 1944 in Hartford, Connecticut, USA. He received a Bachelor of Science degree in General Engineering from the US Military Academy at West Point in 1967; and a Master of Science degree in Aerospace Engineering from the University of Arizona in 1974. He graduated from the US Naval Test Pilot School in 1976; and from the Defense Systems Management College in 1989. Spring logged 3,500 hours in twenty-five military and civilian aircraft types, including over 1,500 hours in jet aircraft. Selected in May 1980 NASA Group 9, Spring served as a Mission Specialist on STS-61-B aboard *Atlantis* (November 26-December 03, 1985). He was responsible for launching three communications satellites. His EVA included a Time and Motion study for comparison between Earth training and space performance, with the EASE/ACCESS experiment. He logged six days, 21 hours and four minutes (165 hours) in space; including two EVAs totaling 12 hours and 14 minutes.

After the Space Shuttle *Challenger* accident, Col. Spring participated in the Kennedy Space Center (KSC) **Tiger Team** for accident investigation, then led the Astronaut Office EVA program until his retirement from NASA in August 1988. Spring then spent five years directing the Army Space Program Office in Washington, D.C.. He retired from the Army in July 1994 and worked as a contractor on Defense and Intelligence programs in Washington, D.C.. Spring is currently a professor at the **Defense Acquisition University** (DAU), living in San Diego, California. He has received ten awards, including the DFC; three Army Commendation Medals; nine Air Medals; the National Defense Service Medal; Vietnam Service Medals; and three honorary Doctorate degrees. Sherwood Spring is married, with two children.

## Springer, Robert C.. STS-29R, STS-38.

Col. Robert "Bob" Clyde Springer was born in May 1942 in St. Louis, Missouri, USA. He received a Bachelor of Science degree from the US Naval Academy, Annapolis, Maryland, in 1964; and a Master of Science degree from the Naval Postgraduate School in 1971. A decorated aviator in the US Marine Corps, he flew over 300 F-4 combat sorties from Chu Lai, South Vietnam. He flew another 250 combat missions in O-1 Bird Dogs and UH-1 Iroquois "Huey" helicopters in the Vietnam War; logging 4,500 hours flying time, including 3,500 hours in jet aircraft. Springer flew UH-1Es in 1972 at Camp Pendleton, California, then went to Okinawa, Japan, to fly with HML-367, First Marine Aircraft Wing. In 1975, he graduated from the US Naval Test Pilot School at NAS Patuxent River, Maryland, and served as a test pilot for over 20 types of fixed and rotary-winged aircraft. He graduated from the Armed Forces Staff College in Norfolk, Virginia, in 1978.

Springer was serving as aide-de-camp for the Commanding General, Fleet Marine Force, Atlantic, when advised of his selection in May 1980 NASA Group 9. Col. Springer flew as a Mission Specialist on STS-29R aboard *Discovery* (March 13 to 18, 1989), when the crew deployed a *Tracking and Data Relay Satellite* (TDRS-D), and took over 4,000 photographs of Earth. Springer also flew as a Mission Specialist on STS-38 aboard *Atlantis* (November 15-20, 1990) for Department of Defense operations. After eighty Earth orbits, the Shuttle landed back at KSC, the first Shuttle recovery in Florida since 1985. Robert Springer logged a total of nine days, 21 hours and 32 minutes (over 237 hours) in space. He is married, with three children.

## Stafford, Thomas P.. Gemini VI-A, Gemini IX, Apollo 10, Apollo-Soyuz Test Project.

Lt. Gen. Thomas "Tom" Patten Stafford was born in September 1930 in Weatherford, Oklahoma, USA. He is one of the twenty-four US astronauts who flew to the Moon. He received a Bachelor of Science degree (with honors) from the US Naval Academy in 1952. He was the first member of his Naval Academy Class to wear the first, second and third stars of a General Officer. In September 1953, he received his pilot wings at Connally AFB, Waco, Texas. He later attended the USAF Experimental Test Pilot School and received the A.B. Honts Award as the Outstanding Graduate. Tom Stafford is the co-author of the *Pilot's Handbook for Performance Flight Testing* and the *Aerodynamics Handbook for Performance Flight Testing*. Selected in NASA's second group of astronauts in 1962, he flew on the December 1965 *Gemini VI-A* and June 1966 *Gemini IX* space rendezvous missions, then was Commander of *Apollo 10* in May 1969. In the first flight of the Lunar Module, he descended to nine miles above the Moon, performing the entire lunar landing mission, except the actual landing. General Stafford was cited in the *Guinness Book of World Records* for the highest speed ever attained by a man that occurred during the *Apollo 10* re-entry, when the spacecraft attained 24,791 statute miles per hour velocity.

In June 1969, he was assigned as head of the Astronaut Group; responsible for the selection of flight crews for projects *Apollo* and *Skylab*. In June 1971, he was assigned as Deputy Director of Flight Crew Operations at the NASA Manned Space Flight Center. Tom Stafford also commanded the *Apollo-Soyuz Test Project* mission from July 15-24, 1975. Gen. Stafford was promoted to Lt. General on March 15, 1978, and on May 01 assumed duties as Deputy Chief of Staff, Research, Development, and Acquisition, HQ USAF, Washington, D.C.. During this time he personally initiated the F-117A Stealth Fighter. Stafford also initiated the AGM-129 Stealth Cruise Missile. He retired from the Air Force in November 1979. He has flown over

127 types of aircraft and helicopters, and four different types of spacecraft. He has flown six rendezvous in space and logged a total of 21 days, three hours and 42 minutes (over 507 hours) in space. Lt. Gen. Stafford is a Fellow of the American Astronautical Society; the American Institute of Aeronautics and Astronautics; and the Society of Experimental Test Pilots. He has two daughters from his first marriage, and two adopted sons.

## Stefanyshyn-Piper, Heidemarie M.. STS-115, STS-126.

Capt. Heidemarie Martha Stefanyshyn-Piper was born in February 1963 in St. Paul, Minnesota, USA. She has Bachelor of Science (1984) and Master of Science (1985) degrees in Mechanical Engineering from the Massachusetts Institute of Technology. Heidemarie Stefanyshyn received her commission from the Naval ROTC Program at MIT in June 1985, then completed training as a Navy Basic Diving Officer and Salvage Officer. She is an experienced salvage officer, whose major projects included de-stranding the tanker *Exxon Houston* on Oahu, Hawaii, and developing the plan for the Peruvian Navy salvage of the Peruvian submarine *Pacocha*. She has received numerous medals, including the Defense Superior Service Medal, and two Legion of Merit medals.

Selected in April 1996 NASA Group 16, she flew on STS-115 aboard *Atlantis* (September 09-21, 2006) and became the eighth woman to perform an EVA (out of then 180 total spacewalkers). During her pre-flight interview, she described her philosophy about human exploration of space: *"To me, **exploring space is just a natural progression of where humans are going.** As we become more advanced, we have more technology to go farther. Thousands of years ago, people would go just beyond the next hill ... then it was "OK, let's go into the skies". ... So looking into the skies and looking at the stars and at the planets and thinking, what's out there? ... We're curious. We, as humans always want to know what's out there. ... We're still developing the means to go out there farther. ... **what else can we learn, and how can we go there?** Maybe we can learn something that we can bring back here and help solve some of the problems we have on Earth"*. Unquote.

Stefanyshyn-Piper was commander of NEEMO 12, a NASA program for studying human survival in the *Aquarius* underwater laboratory, Key Largo, Florida, in preparation for future space exploration. She flew as a Mission Specialist on STS-126 aboard *Endeavour* (November 14-30, 2008). While making her first EVA on November 18, a grease gun exploded in her large tool bag, and while cleaning up she accidentally pushed aside the apparently untethered bag. She said, *"It was very disheartening to watch it float away"*. The bag entered low Earth orbit as space debris, and burned up on August 03, 2009. The estimated value of the equipment lost was USD 100,000. Stefanyshyn-Piper logged a total of 27 days, 15 hours and 36 minutes in space; including five EVAs totaling 33 hours and 42 minutes. This placed her in twenty-fifth place for total time in EVA.

In July 2009, she retired from the Astronaut Corps to return to her Navy duties. Steven Lindsay, Chief of the Astronaut Office at JSC said, *"Heide has been an outstanding astronaut, contributing significantly to the Space Shuttle and Space Station programs. In particular, her superb leadership as lead spacewalker during the STS-126 mission resulted in restoring full power generation capability to the International Space Station. We wish her best of luck back in the Navy – she will be missed"*. Unquote. In August 2009, she became Chief Technology Officer at **Naval Sea Systems Command**. On May 20, 2011, she became Commander of the Carderock Division of the **Naval Surface Warfare Center** in Maryland, USA. On July 01, 2015, she retired from the US Navy after thirty years of active duty. She married Glenn A. Piper, and they have a son.

## Stewart, Robert L.. STS-51-J.

Brigadier General Robert Lee Stewart was born in August 1942 in Washington, D.C., USA. He received a Bachelor of Science degree in Mathematics from the University of Southern Mississippi in 1964; and a Master of Science degree in Aerospace Engineering from the University of Texas at Arlington in 1972. He entered active service with the US Army in May 1964, and was designated an Army Aviator in July 1966. He flew 1,035 hours of combat time in the Vietnam War from August 1966-67. He served in Seoul, South Korea, from 1972-73 as a Battalion Operations Officer and Battalion Executive Officer, then attended the US Naval Test Pilot school at NAS Patuxent River, Maryland, completing the Rotary Wing Test Pilot Course in 1974.

Later, Stewart participated with Sikorsky Aircraft Corporation test pilots in developing an electronic automatic flight control system for the new Army transport helicopter, the UH-60A Black Hawk. Stewart has military and civilian experience in thirty-eight types of airplanes and helicopters, logging about 6,000 hours total flying time. Selected in January 1978 NASA Group 8, he served as Mission Specialist on STS-41-B aboard *Challenger* (February 03-11, 1984) for seven days and 23 hours, and made two EVAs to conduct first flight evaluations of the *Manned Maneuvering Unit* (MMU). For achieving man's first untethered operations from a spacecraft in flight, he was the first to be awarded the Army Astronaut Badge.

Robert Stewart flew on STS-51-J aboard *Atlantis* (October 03-07, 1985), the second classified Department of Defense mission, for four days and one hour and 44 minutes (almost 98 hours). Stewart logged a total of 12 days, zero hours and 49 minutes in space; including two EVAs totaling 12 hours and 12 minutes. He retired from the Army in 1992, and was employed as Director of Advanced Programs at the **Nichols Research Corporation** in Colorado Springs, Colorado. Robert Stewart retired to Huntsville, Alabama, and makes appearances at the **US Space & Rocket Center**, greeting visitors and signing autographs. He is a member of the Society of Experimental Test Pilots; the Association of Space Explorers; and the Scabbard and Blade (a military honor society). Stewart is married, with two children.

## Sturckow, Frederick W.. STS-88, STS-105, STS-117, STS-128.

Frederick "Rick" Wilfred Sturckow was born in August 1961 in La Mesa, California, USA. He received a Bachelor of Science degree in Mechanical Engineering from California Polytechnic State University in 1984; and a Master of Science degree in Mechanical Engineering from Florida Institute of Technology in 2000. He was commissioned into the US Marine Corps in December 1984, and earned his aviator wings in April 1987. While assigned to VMFA-333, he made an overseas deployment to Japan, South Korea and the Philippines, then was selected to attend the Navy Fighter Weapons School (TOPGUN) in March 1990. Deployed to Sheik Isa Air Base, Bahrain for eight months from August 1990, he flew forty-one combat missions during Operation Desert Storm. In January 1992, he attended the USAF Test Pilot School at Edwards AFB. He flew a wide variety of projects and classified programs as an F/A-18 tst pilot.

Selected in December 1994 in NASA-15 Group, Sturckow flew four missions to the ISS: as Pilot for STS-88/2A (December 04-16, 1998), the first ISS assembly mission, and STS-105 (August 10-22, 2001), and he was Commander of STS-117 (June 08-22, 2007) and STS-128 (August 28-September 12, 2009). Sturkow is one of three American astronauts to visit the ISS four times. He logged a total of 51 days, nine hours and 34 minutes in space; and was later assigned to JSC as a CapCom. Sturckow retired from the Marine Corps as a Colonel in September 2009, after 25 years' active service. He logged over 6,500 flying hours in over sixty aircraft types. Rick Sturckow left NASA in March 2013 after eighteen years, to become a commercial pilot for **Virgin Galactic**. He flight tested their *SpaceShipTwo* spacecraft and the *White Knight Two* aircraft. On December 13, 2018, he reached outer space with *VSS Unity*. Sturckow is married, with two children.

## Stott, Nicole M.. STS-128/Expedition 21/STS-129, STS-133.

Nicole Marie Passonno Stott was born in November 1962 in Albany, New York, USA. She graduated from St. Petersburg College, earned her Bachelor of Science degree in Aeronautical Engineering from Embry-Riddle Aeronautical University in 1987; and received her Master of Science degree in Engineering Management from the University of Central Florida in 1992. Stott began her career in 1987 as a structural design engineer with **Pratt and Whitney Government Engines** in West Palm Beach, Florida. In 1988, she joined NASA at KSC, Florida as an Operations Engineer in the Orbiter Processing Facility (OPF). In 1998, she joined the **Johnson Space Center** (JSC) team in Houston, Texas, as a member of the NASA Aircraft Operations Division, where she served as a Flight Simulation Engineer (FSE) on the Shuttle Training Aircraft.

Stott began her two-year astronaut candidate training in August 2000, then performed crew evaluations of station payloads while working in the Astronaut Office Station Operations Branch. In April 2006, she was a crew member on the NEEMO 9 mission, spending eighteen days aboard the *Aquarius* undersea research habitat. Launched on August 28, 2009, as a Mission Specialist aboard *Discovery* on STS-128 to the ISS for 'Expedition 20/21', she participated in the first EVA on that mission, and returned to Earth on STS-129 aboard *Atlantis* on November 26, 2009. Stott was the first person to paint with watercolor in space. Her current artistic works often relate to astronomy, including her Earth Observation collection and Spacecraft collection. She flew on STS-133 aboard *Discovery* (February 24-March 09, 2011), a mission of 12 days, 19 hours and four minutes.

After twenty-seven years of working at NASA, the space agency announced her retirement effective June 01, 2015. Nicole Stott's book, *Back to Earth*, is described as *What Life in Space Taught Me About Our Home Planet and Our Mission to Protect It*. In 2022, she provided the narration to a piece performed by the Schenectady Symphony Orchestra: Glen Cortese's ***Voyager: A Journey to the Stars***. She is married to Christopher Stott, a Manx (Isle of Mann)-born American space entrepreneur.

## Sullivan, Kathryn D.. STS-41-G, STS-31, STS-45.

Lt. Cmdr. Dr. Kathryn Dwyer Sullivan was born in October 1951 in Paterson, New Jersey, USA. She received a Bachelor of Science degree in Earth Sciences from University of California-Santa Cruz in 1973; and a Doctor of Philosophy degree in Geology from Dalhousie University and the Bedford Institute of Oceanography in Nova Scotia, Canada, in 1978. In early 1977, she saw a NASA advertisement in a science journal, decided to apply, and was invited to a week of interviews and physical examinations in November 1977. Sullivan was successful, being selected as one of six women among thirty-five members of NASA Astronaut Group 8, the first group to include women. She qualified as an astronaut on August 31, 1979, and John W. Young presented the 35 newbies with a silver NASA astronaut pin. Sullivan then helped develop systems management checklists for the first Space Shuttle flights. After nine months, she became a mission manager at NASA's High Altitude Research Project; setting an unofficial sustained American aviation altitude record for women of 19,000 metres (63,000 feet) aboard a NASA WB-57F aircraft.

In July 1983, she joined the Mission Development group, which organised and supervised the development of payloads for future missions. She was assigned the Office of Space and Terrestrial Applications *OSTA-3* satellite and the Orbital Refueling System (ORS), to demonstate that a Shuttle could be used to refuel a satellite in orbit. Sullivan was launched aboard *Challenger* on mission STS-41-G on October 05, 1984. On October 11, she performed the first EVA by an American woman. Of 3.5 hours duration, it was to operate the ORS to refuel the ageing *Landsat 4* satellite, transferring 59 kg (130 lb) of hydrazine. The crew deployed the *Earth Radiation Budget Satellite*, several in-cabin experiments, and activated eight "Getaway Special" canisters. The mission ended on October 13, 1984. Sullivan was then CapCom for STS-26, 27, and 29. Sullivan was 'Mission Specialist 3' on STS-31 aboard *Discovery* (April 24-29, 1990), which deployed the Hubble Space Telescope, designed to be maintained and serviced in orbit for up to fifteen years. The flight apogee of 617 km (333 nautical miles) above Earth was the highest yet achieved by a Shuttle orbiter.

On STS-45 aboard *Discovery* (March 24-April 02, 1992) the crew operated the 12 experiments that constituted the Atmospheric Laboratory for Applications and Science cargo; studying the compostion of the mid-atmosphere over an eleven-year solar cycle. Sullivan had special responsibility for a dose-response experiment that involved firing an electron pulse into the upper atmosphere and recording the luminosity induced with a special camera. She left NASA in 1993, having logged 532 hours in space. Kathryn Sullivan became an adjunct Professor of Geology at Rice University, Houston, in 1985. Later that year she became a direct commission officer in the US Naval Reserve, with the rank of Lieutenant Commander. In October 1990, she assumed command of a small specialized unit of oceanographers and meteorologists based at Naval Air Station Dallas. Sullivan retired from the USNR in 2006 with the rank of Captain.

Kathryn Sullivan became the chief scientist of the **National Oceanic and Atmospheric Administration** (NOAA), confirmed by the Senate on May 28, 1993. She oversaw a diverse portfolio which included ranging research into climate change, the use of satellites for oceanography, and marine biodiversity. From 1996 to 2006, she was President and CEO of the **COSI Columbus**, an interactive science center. From 2006-11, she was Director for Ohio State University's Battelle Center for Mathematics and Science Education Policy. In 2009, Sullivan was elected to a three-year term as the chair of the Section on General Interest in Science and Engineering for the American Association for the Advancement of Science. She was named the 2018 Charles A. Lindbergh Chair of Aerospace History, a competitive twelve-month fellowship at the National Air and Space Museum. Her book *Handprints on Hubble: an Astronaut's Story of Invention* was released from MIT Press in November 2019. She was inducted into the Astronaut Hall of Fame (2004); received honorary Doctor of Science degrees in 2013 and 2015; was elected to the National Academy of Engineering (2016); the American Academy of Arts and Sciences (2017); and the National Aviation Hall of Fame (2023).

## Swanson, Steven R. STS-117, STS-119, Soyuz TMA-12M/Expedition 39/40.

Dr. Steven Roy Swanson was born in December 1960 in Syracuse, New York, USA. He received a Bachelor of Science degree in Engineering Physics from the University of Colorado in 1983; a Master of Applied Science degree in Computer Systems from Florida Atlantic University in 1986; and a Doctorate in Computer Science from Texas A & M University in 1998. After graduation, he worked for **GTE** in Phoenix, Arizona, as a software engineer. Swanson joined NASA in 1987 as a systems engineer and a Flight Engineer working on the Shuttle Training Aircraft (STA). In 1989, he became a flight simulation engineer on the STA; working on improvement of the STA's navigation and control systems and the incorporation of a real-time wind determination algorithm. Swanson was selected as an astronaut candidate in May 1998, and after training was assigned to the Astronaut Office Space Station Operations Branch. He completed advanced training for EVA, the Shuttle and ISS robotic arms, and Shuttle rendezvous. Swanson has served as a CapCom for both ISS and Space Shuttle missions.

He was 'Mission Specialist 2' and Flight Engineer on STS-117 aboard *Atlantis* (June 08-22, 2007), the 250th orbital human spaceflight, which delivered to the ISS the second starboard truss segment (S3/S4) and its associated energy systems. A two-day extension of the mission allowed a fourth EVA to repair the gap in the Orbital Maneuvering System thermal blanket. As 'Mission Specialist 2', he launched on March 25, 2014, on STS-119 aboard

*Discovery* (March 15-28, 2009) to the ISS, which delivered and installed the 14,088 kg fourth starboard integrated Truss Segment (S6) and the fourth set of solar arrays and batteries. The mission carried experiments including the Shuttle Ionospheric Modification with Pulsed Local EXhaust (SIMPLEX), Shuttle Exhaust Ion Turbulence Experiments (SEITE), and Maui Analysis of Upper Atmospheric Injections.

Dr. Swanson was 'Flight Engineer 5' as a member of 'Expedition 39/40', which began on March 11, 2014. Launched from Kazakhstan aboard *Soyuz TMA-12M* on March 25, he installed the "Veggie" project on the ISS, and returned to Earth on September 11, 2014. Steven Swanson logged 195 days, 20 hours and 47 minutes in space; including five EVAs totaling 28 hours and five minutes. NASA announced his retirement in August 2015, and since then he worked at **Boise State University** as a distinguished educator in residence, assisting with programs such as the NASA SUITS Program. Dr. Steven Swanson has received the NASA Exceptional Achievement Medal; and the Johnson Space Center *Certificate of Accommodation*.

## Swigert, John "Jack" L., Jr.. Apollo 13.

John "Jack" Leonard Swigert, Jr. (1931-82) was born in August 1931 in Denver, Colorado, USA. He was most famous for his role as the Command Module Pilot of the fraught *Apollo 13* mission in April 1970. Swigert received a Bachelor of Science degree in Mechanical Engineering from the University of Colorado-Boulder in 1953. He later earned a Master of Science degree in Aerospace Engineering from Rensselaer Polytechnic Institute (Hartford campus) in 1965; and a Master of Business Administration degree from the University of Hartford in 1967.

Upon graduating from the USAF Pilot Training program and Gunnery School at Nellis AFB, he was assigned as a fighter pilot in Japan and South Korea. In 1953, he survived his plane crashing into a radar unit on a Korean airstrip. Before joining NASA, Swigert was a civilian test pilot for **North American Aviation** and fighter pilot in the Air National Guard. He logged over 7,200 flying hours; including more than 5,725 hours in jet aircraft. Accepted into NASA Astronaut Group 5 in April 1966, he joined the *Apollo 13* crew just three days before the launch, replacing Ken Mattingly, who was exposed to German Measles (the rubella virus). He flew on the *Apollo 13* mission (April 11-17, 1970) in which an intended Moon landing was canceled because of a ruptured fuel-cell oxygen tank in the Service Module. It was Swigert who first announced, ***"Houston, we've had a problem here"***. The crew of Jim Lovell, Fred Haise and Swigert went into emergency mode, conserved power and improvised a CO2 scrubber, traveled around the Moon in a "slingshot" maneuver. They safely returned home, each receiving the Presidential Medal of Freedom the next day.

Jack Swigert took a leave of absence from NASA in April 1973 and went to Washington D.C. to become Executive Director of the Committee on Science and Astronautics, US House of Representatives. He was received the AIAA Haley Astronautics Award (1971); the 1972 Antonian Gold Medal; and was inducted into the International Space Hall of Fame (1983); and the Colorado Aviation Hall of Fame (1988). After leaving NASA, he ran for Senate in 1978, but lost heavily in a primary election. Later, he ran for Congress in 1982, but was diagnosed with a nasal tumor. He won the election for Colorado's new Sixth District in 1982 with 64% of the popular vote, but died of bone marrow cancer and respiratory failure before being sworn in.

Jack Swigert died on December 27, 1982, in Washington, D.C., aged fifty-one years. He was a Fellow of the American Astronautical Society; an Associate Fellow of the Society of Experimental Test Pilots; and the American Institute of Aeronautics and Astronautics. In 1997, Swigert was posthumously inducted into the US Astronaut Hall of Fame. His statue was placed on display in the US Capitol Building, and as of December 2008, it was on display in the Emancipation Hall in the US Capitol Visitor Center. In 2004, the Space Foundation launched the John L. "Jack" Swigert Jr. Award for Space Exploration, presented annually.

## Tani, Daniel M.. STS-108, STS-120/Expedition 16.

Daniel "Dan" Tani was born in February 1961 in Ridley Park, Pennsylvania, USA. He received a Bachelor of Science degree (1984) and a Master of Science degree in Mechanical Engineering from Massachusetts Institute of Technology (MIT) in 1988. After gaining his Bachelor's degree, Tani worked at **Hughes Aircraft Corporation** in El Segundo, California, as a design engineer in the Space and Communications group. In 1986, he returned to MIT, and after graduation worked for **Bolt, Beranek and Newman** in Cambridge, Massachusetts, in their experimental psychology department. In 1988, Tani joined **Orbital Sciences Corporation** in Dulles, Virginia, and later became the Mission Operations Manager for the Transfer Orbit Stage. He then moved to the air-launched *Pegasus* rocket program at OSC as launch operations manager.

In April 1996, Dan Tani was selected by NASA as an astronaut candidate. He performed technical duties in the Astronaut Office Computer Support Branch and Extra-Vehicular Activity Branch, and served as a crew support astronaut for 'Expedition 2'. He was 'Mission Specialist 2' on mission STS-108 aboard *Endeavour* (December 05-17, 2001). The crew transferred over three tons of supplies, logistics and science experiments from the *Raffaello* MPLM to the ISS. Tani performed an EVA of over four hours to wrap thermal blankets around the ISS Solar Array Gimbals. Tani was assigned to 'Expedition 16' as 'Mission Specialist 5' and ISS Flight Engineer, and was launched to the ISS on STS-120 aboard *Discovery* on October 23, 2007, which delivered the 14,288 kg *Harmony* module. Tani completed one EVA with the STS-120 crew, and four more during his time aboard the ISS. He returned to Earth on STS-122 on February 20, 2008.

Daniel Tani logged a total of 131 days, 18 hours and 29 minutes in space; including six EVAs. He left NASA in August 2012 to become the Vice President of Mission and Cargo Operations in the Advanced Programs Group of **Orbital Sciences Corporation** (OSC). In August 2016, he left OSC to join the High School Faculty at **The American School** in Japan, where he taught science, engineering and design. Tani has received the Orbital Sciences Corporation Outstanding Technical Achievement Award (1993); and the Medal "For Merit in Space Exploration" (Russia, April 12, 2011) for *"outstanding contribution to the development of international cooperation in crewed space flight"*.

## Tanner, Joseph R. STS-66, STS-82, STS-97, STS-115.

Joseph "Joe" Richard Tanner was born in January 1950 in Danville, Illinois, USA. After graduating as a Mechanical Engineer from the University of Illinois in 1973, Tanner joined the US Navy; receiving his Naval Aviator wings in 1975, then serving aboard *USS Coral Sea*. He finished his active service as an advanced jet instructor pilot. Tanner has logged almost 9,000 hours in military and NASA aircraft, and over 1,000 hours in space; including over 46 hours in EVA. Tanner is unusual among astronauts as he did not have a background in flight test, nor did he earn any advanced academic degrees. His path followed operational military flying, and then into NASA for operational jet training, before being selected into the Astronaut Corps in 1992. He was 'Mission Specialist 2' on STS-66 aboard *Atlantis* (November 03-14, 1994), the Atmospheric Laboratory for Applications and Science (ATLAS-3) mission, which deployed the *CRISTA-SPAS* satellite. Tanner was 'Mission Specialist 1' on STS-82 aboard *Discovery* (February 11-21, 1997), which serviced the Hubble Space Telescope, when Tanner made two EVAs totaling 14 hours and one minute.

STS-97 aboard *Endeavour* (November 30-December 11, 2000) was the fifth Shuttle mission dedicated to assembly of the ISS. The crew installed the first set of solar arrays, and delivered supplies and equipment to the ISS's first resident crew. Tanner performed three EVAs totaling 19 hours and 20 minutes. His final mission was STS-115 aboard *Atlantis* (September 09-21, 2006), when Tanner participated in the five-hour, 26 minute EVA to connect the P3/4 truss to the ISS. A photo he took during his spacewalk was later listed on ***Popular Science's*** photo gallery of the best astronaut selfies. He logged a total of 43 days, 13 hours and 15 minutes in space; including seven EVAs totaling 46 hours and 29 minutes. Currently he is a Senior Instructor in the **University of Colorado** Aerospace Engineering Sciences Department in Boulder, Colorado; teaching and mentoring undergraduate and graduate students in space systems design. He is also a self-employed aerospace systems consultant. Joe Tanner has eight awards and honors, including Captain of the Swimming team and "Top 100 Seniors" Award at University of Illinois; Distinguished Graduate from Navy Flight training; the NASA Exceptional Service Medal; and the JSC Superior Achievement Award.

## Thagard, Norman E.. STS-7, STS-51-B, STS-30R, STS-42, Soyuz TM-21/Mir EO-18/STS-71.

Capt. Dr. Norman Earl Thagard was born in July 1943 in Marianna, Florida, USA. His Alma mater is Florida State University. In September 1966, Thagard entered active duty with the US Marine Corps Reserve. He achieved the rank of Captain in 1967, was designated a Naval Aviator in 1968, and flew 163 combat missions in Vietnam from January 1969-70. Thagard resumed his academic studies in 1971, pursuing additional studies in Electrical Engineering, and a Doctor of Medicine (M.D.) degree from the University of Texas Southwestern Medical School in 1977. Before joining NASA, Thagard was interning in the Department of Internal Medicine at the Medical University of South Carolina. Norman Thagard is a licensed Physician. He is also a pilot, with over 2,200 hours flying time, mostly in jet aircraft.

Selected in 1978 NASA Group 8, he was the first American to ride to space aboard a Russian vehicle. Thagard's first mission was STS-7 aboard *Challenger* (June 18-24, 1983), the first mission with five crew, which deployed satellites for Canada and Indonesia, operated the Remote Manipulator System (RMS) to perform the first deployment and retrieval exercise with the *Shuttle Pallet Satellite* (SPAS-01), and activated seven "Getaway Specials". STS-51-B aboard *Challenger* (April 29-May 06, 1985) deployed the *NUSAT* satellite, as well as animal care for the twenty-four rats and two squirrel monkeys contained in the Research Animal Holding Facility (RAHF). STS-30 aboard *Atlantis* (May 04-08, 1989) deployed the *Magellan* spacecraft, which arrived at Venus in mid-1990 and radar-mapped the entire surface. Thagard was Payload Commander on STS-42 aboard *Discovery* (January 22-30, 1992) with fifty-five major experiments conducted in the International Microgravity Laboratory-1 module. Biological experiments on the effects of zero gravity on plants, tissues, bacteria, insects and human vestibular response were conducted.

He launched aboard *Soyuz TM-21* on March 14, 1995, for the 'Mir EO-18' mission, returning on *Atlantis* at the end of mission STS-71 on July 07, 1995. Twenty eight experiments were conducted during the 115-day flight. He logged a total of 140 days, 13 hours and 24 minutes in space. Thagard is a Tenured Professor at the FAMU-FSU College of Engineering; a Member of the Board of Directors of EMS Technologies, Inc.; co-founder and Executive Director, Challenger Learning Center of Tallahassee; Member, Editorial Review Board of the *Journal of the Society for Human Performance in Extreme environments*; Distinguished Lecturer for the American Institute of Aeronautics and Astronautics; and an Advisor for Space Adventures, Ltd., a company offering aerospace experiences, including orbital flight. He was technical advisor for ***Virus***, a 1999 film starring Jamie Lee Curtis and Donald Sutherland; and for ***Armageddon*** (1998) starring Bruce Willis, Ben Affleck, and Liv Tyler.

Capt. Norman Thagard belongs to the American Institute of Aeronautics and Astronautics; the Aerospace Medical Association; and the Tallahassee Symphony Orchestra. Among his thirty-two awards are: The Marine Corps "E" Award; eleven Air Medals; the Vietnam Service

Medal; ***Aviation Week & Space Technology***'s 1995 Laurels Award; and Recipient of Russia's Order of Friendship (1996). He was inducted into New Mexico's International Space Hall of Fame (1996); listed in ***Who's Who in America,*** Who's Who in Science and Engineering; received the NASA Distinguished Service Medal; The Society of NASA Flight Surgeons' 1997 W. Randolph Lovelace Award; was Erskine Visiting Fellow, University of Canterbury, New Zealand (2001); and he was inducted into the United States Astronaut Hall of Fame (2004).

## Thomas, Andrew S.W.. STS-77, STS-89, STS-102, STS-114.

Dr. Andrew "Andy" Sydney Withiel Thomas was born in December 1951 in Adelaide, South Australia. He has dual nationality; he became a US citizen in December 1986, hoping to gain entry to the US astronaut program. Andy Thomas is married to fellow NASA astronaut Shannon Walker. He received a Bachelor of Engineering degree (with First Class Honors) from the University of Adelaide, Australia, in 1973; and a Ph.D. in Mechanical Engineering in 1978. After completing his studies, Thomas accepted an offer from **Lockheed** in Atlanta, USA. By 1990, he was the organization's principal aerodynamic scientist.

Andy Thomas was selected in March 1992 NASA Group 14, and while awaiting space flight assignment he supported Shuttle launch and landing operations at KSC. He also provided technical support to the Space Shuttle Main Engine project, the Solid Rocket Motor project, and the External Tank project at the Marshall Space Flight Center. On STS-77 aboard *Endeavour* (May 1996) he was the first Australia-born professional astronaut to enter space. Thomas next trained at the Gagarin Cosmonaut Training Center in Star City, Russia, in preparation for a long-duration flight. In 1998, he served as 'Board Engineer 2' aboard the Russian Space Station *Mir* for 130 days. From August 2001 to November 2003, he served as Deputy Chief of the Astronaut Office.

STS-77 aboard *Endeavour* (May 19-29, 1996) deployed two satellites, tested BEAM (a large inflatable space habitation structure) on orbit, and conducted experiments in the 'Spacehab' laboratory module. STS-89 launched aboard shuttle *Endeavour* on January 22, 1998, and docked with *Mir*. Andy Thomas returned to Earth with the crew of STS-91 aboard *Discovery* on June 12, 1998. STS-102 aboard *Discovery* (March 08-21, 2001) was the eighth shuttle mission to visit the ISS, resupplying logistics with the *Leonardo* MPLM; also delivering the 'Expedition 2' crew, and returning 'Expedition 1' crew to Earth. Thomas performed an EVA of 6.5 hours to install components to the exterior of the ISS. STS-114 aboard *Discovery* (July 26-August 09, 2005) was a mission to resume assembly of the ISS. Andy Thomas tested and evaluated new procedures for flight safety, and inspection and repair techniques for the Shuttle's thermal protection system. He logged a total of 177 days, nine hours and 14 minutes in space.

In 2000, Andrew Thomas was appointed an Officer of the Order of Australia (AO) *"for service to science and technology ... as an astronaut, and for contributions to the human exploration of space"*. In 2001, he received the Centenary Medal. Thomas worked with issues for the Exploration Branch of the Astronaut Office until his retirement from NASA in February 2014, after twenty-two years' service. He was awarded Life Membership of the Space Industry Association of Australia (2018). On November 20, 2020, the **Andy Thomas Foundation** was launched, with the purpose of advancing space education, raising space awareness, and contributing to the national space community.

## Thomas, Donald A.. STS-65, STS-70, STS-83, STS-94.

Dr. Donald "Don" Thomas was born in May 1955 in Cleveland, Ohio, USA. He received a Bachelor of Science degree (with honors) in Physics from Case Western Reserve University in 1977; a Master of Science degree (1980); and a Doctorate in Materials Science from

Cornell University in 1982. His dissertation involved evaluating the effect of crystalline defects and sample purity on the superconducting properties of niobium. Thomas joined **AT & T Bell Laboratories** in New Jersey, where until 1987 he was a Senior Member of the Technical Staff. He worked on development of advanced materials and processes for high-density interconnections of semiconductor devices. Thomas holds two Patents and has authored several technical papers. In 1987, he began work for **Lockheed Engineering and Sciences Company** in Houston, Texas, where his responsibilities involved reviewing materials used in Shuttle payloads. In 1988, he joined the JSC as a materials engineer, where his work involved lifetime projections of advanced composite materials for use on the Space Station.

Dr. Donald Thomas joined NASA in 1988 as a materials engineer working on the Space Shuttle program, and was selected for NASA Group 13 in 1990; being accepted on his fourth application. He was a Mission Specialist on STS-65 aboard *Columbia* (July 08-23, 1994), which set a new Shuttle flight duration record of 353 hours and 55 minutes. STS-70 aboard *Discovery* (July 13-22, 1995) deployed the sixth and final *Tracking and Data Relay Satellite*. STS-83 aboard *Columbia* (April 04-08, 1997) was the Microgravity Science Laboratory (MSL-1) 'Spacelab' mission; cut short by a problem with a Shuttle fuel cell power generation unit. STS-94 aboard *Columbia* (July 01-17, 1997) was a reflight of the Microgravity Science Laboratory (MSL-1) 'Spacelab' mission, and focused on materials and combustion science research in microgravity.

Dr. Don Thomas logged a total of 43 days, eight hours and 13 minutes (1,040 hours) in space. He is a private pilot with over 250 hours in single-engine land aircraft and gliders, and over 800 hours flying in NASA T-38 jet aircraft. He is head of the Willard Hackerman Academy of Mathematics and Science at Towson University in Towson, Maryland. In 2013, Dr. Thomas co-wrote *Orbit of Discovery: The All-Ohio Space Shuttle Mission*, referencing the STS-70 flight. He is a recipient of the NASA Sustained Superior Service Award (1989); four NASA Space Flight Medals; four NASA Group Achievement Medals; two NASA Exceptional Service Medals; and the NASA Distinguished Service Medal.

## Thornton, Kathryn R. C.. STS-33, STS-49, STS-61, STS-73.

Dr. Kathryn Ryan Cordell Thornton was born in August 1952 in Montgomery, Alabama, USA. She received a Bachelor of Science degree in Physics in 1974 from Auburn University; a Master of Science degree in Physics (1977); and a Ph.D. in Physics, also from the University of Virginia, in 1979. After earning her doctorate, Thornton was awarded a NATO Postdoctoral Fellowship to continue her research at the Max Planck Institute for Nuclear Physics in Heidelberg, Germany. In 1980, she became a physicist at the US Army Foreign Science and Technology Center in Charlottesville, Virginia. Selected in 1984 NASA Group 10, she was 'Mission Specialist 3' on STS-33 aboard *Discovery* (November 22-27, 1989), which carried Department of Defense payloads. Thornton was 'Mission Specialist 4' on STS-49 aboard *Endeavour* (May 07-16, 1992) when the crew performed a record four EVAs to retrieve, repair and redeploy the *International Communications Satellite* (INTELSAT), which was stranded in an unusable orbit, and then equipped with a new perigee kick motor which fired it into a usable geosynchronous orbit.

She was 'Mission Specialist 1' on STS-61 aboard *Endeavour* (December 02-13, 1993) for a Hubble Space Telescope servicing and repair mission, over 3.5 years since its launch. There were a record five spacewalks by four astronauts, including Thornton. Dr. Kathryn Thornton was Payload Commander on STS-73 aboard *Columbia* (October 20-November 05, 1995), the second US Microgravity Laboratory (USML-2) mission. She logged a total of 40 days, 15 hours and 14 minutes (over 975 hours) in space; including 21 hours of EVA. Kathryn Thornton left NASA on August 01, 1996, and became the chairwoman of the Board of Directors at the **Space Foundation**. She was inducted into the 2010 Class of the US Astronaut Hall of Fame. On October 16, 2019, she completed a through-hike of the Appalachian Trail. Dr. Kathryn Thornton is a member of the American Physical Society; and the American Association for the Advancement of Science. She was the Associate Dean for graduate programs at the **University of Virginia School of Engineering and Applied Science**, and is currently a Professor of Mechanical and Aerospace Engineering.

## Thornton, William E.. STS-8, STS-51-B.

Dr. William Edgar Thornton (1929-2021) was born in April 1929 in Faison, North Carolina, USA. He received a Bachelor of Science degree in Physics from the University of North Carolina in 1952; and a Doctorate in Medicine also from UNC in 1963. Selected in 1967 NASA Group 6, he was physician crew member on the highly successful Skylab Medical Experiments Altitude Test (SMEAT) - a 56-day simulation of a *Skylab* mission enabling crewmen to collect medical experiments baseline data and evaluate equipment, operations, and procedures. He was a member of the astronaut support crew for the *Skylab 2*, 3 and 4 missions, and principal investigator for *Skylab* experiments. He first documented the shift and loss of fluid changes in body posture size and shape, including increase in height and the rapid loss of muscle strength and mass in space flight. During Shuttle operations he continued physiological investigations in the cardiovascular, musculoskeletal and neurological areas. He developed several on-board exercise devices.

William Thornton flew on STS-8 aboard *Challenger* (August 30-September 05, 1983) with the primary payload of the *INSAT-1B* Indian communications and weather observation satellite. A four-ton dummy payload was used to test the Canadarm remote manipulator system. There were studies of biological materials in electric fields under microgravity, and research into space sickness. Dr. Thornton was 'Mission Specialist 3' on STS-51-B aboard *Challenger* (April 29-May 06, 1985). He logged a total of 13 days, one hour and 16 minutes (over 313 hours) in space, and retired from NASA on May 31, 1994. Dr. Thornton had a significant influence in the evolution of space life sciences, holding patents on a Shuttle flight exercise treadmill and an improved waste collection facility. He held over sixty Patents; from military weapons systems through the first real-time EKG computer analysis.

Dr. William Thornton was later a Clinical Assistant Professor in the Department of Medicine, University of Texas Medical Branch, Galveston, Texas, and an adjunct professor at the University of Houston at Clear Lake. He received ten awards, including the Air Force Legion of Merit (1956); the NASA Exceptional Scientific Achievement Award (1974); the Association of Military Surgeons of the United States Kern Award (1986); and the NASA Exceptional Engineering Achievement Medal (1988). William Thornton died on January 11, 2021 at his home, Fair Oaks Ranch, Texas, aged ninety-one years.

## Thuot, Pierre J.. STS-36, STS-49, STS-62.

Capt. Pierre Joseph Thuot was born in May 1955 in Groton, Conneticut, USA. He received a Bachelor of Science degree in Physics from the US Naval Academy in 1977 (graduating 30th in his class); and a Master of Science degree in Systems Management from the University of Southern California in 1985. After training as a Radar Intercept Officer, he was deployed to the Mediterranean and Caribbean Seas aboard the aircraft carriers *USS John F. Kennedy* and *USS Independence*. Thuot attended the Navy Fighter Weapons School (TOPGUN), then was selected to attend the US Naval Test Pilot School in May 1982. He worked as a project test flight officer until June 1984, when he returned to the Test Pilot School as an instructor. He logged over 3,500 flight hours in over fifty aircraft types, with over 270 aircraft carrrier landings.

Selected by NASA as an astronaut in June 1985, Thuot served in a varity of technical assignments. He participated in the design, development and evaluation of Space Shuttle payloads, crew equipment and crew procedures. He served as Chief of the Astronaut Office Mission Support Branch, as well as supervising candidate training for the Class of 1995. Thuot was a Mission Specialist on STS-36 aboard *Atlantis* (February 18-March 04, 1990), which carried Department of Defense payloads. On STS-49 aboard *Endeavour* (May 07-16, 1992) he performed three EVAs with Richard Hieb, to capture and repair the stranded *Intelsat VI F3* comms satellite. STS-62 aboard *Columbia* (March 04-18, 1994) was a microgravity science and technology demonstration mission carrying the USMP-2 payload. Over sixty experiments or investigations were comnducted in scientific and engineering disciplines. He logged a total of 27 days, six hours and 51 minutes in space; including over 17.7 hours of EVA. From 1992 until 2001, he held the US and absolute world records for total time spent on EVA: eight hours and 29 minutes; breaking the *Apollo 17* record.

Thuot was the first member of NASA Astronaut Group 11 to fly a Shuttle mission. He retired from the US Navy in 1998 and took a job with **Orbital Sciences Corporation** as a vice president in the Space Systems Group. Thuot attended Harvard Business School's six-week Advanced Management Program in 2004. He is currently a Senior Systems Engineer at the **John Hopkins University** Applied Physics Laboratory. Pierre Thuot is a member of the US Naval Academy Alumni; the Association of Naval Aviation; the Association of Space Explorers; the American Astronautical Society; and an Associate Fellow of the American Institute of Aeronautics and Astronautics (AIAA). He has received eleven honors, including the Legion of Merit; the National Intelligence Medal of Achievement; the National Defense Service Medal; the AAS Flight Achievement Award and Victor A. Prather Award for 1993; and two NASA Exceptional Service Medals.

## Tito, Dennis A.. Soyuz TM-32/31.

Dennis Anthony Tito was born in August 1940 in Queens, New York City, USA. He received a Bachelor of Science degree in Astronautics and Aeronautics from New York University in 1962; and a Master of Science degree in Engineering Science from the Renssaeler Polytechnic Institute satellite campus in Hartford, Conneticut. He was a former scientist at the NASA Jet Propulsion laboratory. In 1972, Tito founded **Wilshire Associates**, a leading provider of investment management, consulting, and technology services based in Santa Monica, California. In a project first managed by **MirCorp**, Tito was accepted by the Russian Federal Space Agency as a candidate for a commercial spaceflight. MirCorp, Daniel Goldin (then the Administrator of NASA) and Tito are profiled in the documentary film ***Orphans of Apollo***. When he arrived at JSC for training on the American portion of the ISS, he and his two fellow cosmonauts were sent home by Robert Caban, NASA manager, stating *"... we are not willing to train with Dennis Tito"*. Later, through an arrangement with **Space Adventures, Ltd.**, he joined the *Soyuz TM-32* mission to the ISS.

He made history in mid-2001 by becoming the first "space tourist" to fund his own journey at a reported cost of USD 20 million, spending seven days, 22 hours and four minutes in orbit as a crew member of ISS mission EP-1. Launched aboard *Soyuz TM-32* on April 28, 2001, he returned to Earth on *Soyuz TM-31* on May 06, 2001. In May 2002, he received an honorary Doctorate of Engineering from Renssaeler Polytechnic Institute. He testified at the Senate Committee on Commerce, Science and Transportation Subcommittee Joint Hearing on 'Commercial Human Spaceflight' on July 24, 2003. In February 2013, Tito announced his intention to send a privately financed spaceflight to Mars by 2018, stating that the technology is already in place. However, in November 2013, Tito and other **Mars Inspiration** team members admitted that their plan was impossible without significant levels of assistance and funding from NASA. On October 12, 2022, SpaceX announced that Dennis and his wife, Akiko Tito will be on the crew of the second commercial spaceflight of **SpaceX's** *Starship* around the Moon. Tito has married four times, and has three children with his first wife.

## Trinh, Eugene H.. STS-50.

Dr. Eugene Huu-Chau "Gene" Trinh was born In September 1950 in Saigon, South Vietnam. Trinh came to the United States at age eighteen years, and later became an American citizen. He is a biochemist who flew aboard NASA mission STS-50, becoming the second Vietnamese in space. Trinh graduated from Lycee Michelet in Paris, France, in 1968 with a baccalaureate degree. He received a Bachelor of Science degree in Mechanical Engineering (Applied Physics) from Columbia University in 1972; a Master of Science degree from Yale University in 1974; a Master of Philosophy degree in 1975; and a Doctor of Philosophy in Applied Physics in 1977. Dr. Trinh started with NASA in 1979 as a senior research scientist at JPL. He served as the director of the physical sciences research division in the Biological and Physical Research Enterprise at NASA HQ. For twenty years, he conducted experimental and theoretical research in fluid dynamics, fundamental materials science, and levitation technology.

Dr. Gene Trinh was 'Payload Specialist 2' aboard *Columbia* on STS-50/United States Microgravity Lab-1 (USML-1) flight (June 25-July 09, 1992), with thirty-one microgravity experiments in material science (the study of materials solidification and crystal growth), fluid physics (the study of how liquids and gases respond to the application or absence of differing forces), and biotechnology (the study of phenomena related to products derived from living organisms). Secondary experiments were: Investigations into Polymer Membrane Processing, Shuttle Amateur Radio Experiment, and Ultraviolet Plume Instrument. He logged 13 days, 19 hours and 30 minutes in space. In May 2004, Dr. Trinh received the Golden Torch Award from the Vietnamese American National Gala in Washington, D.C.

## Truly, Richard H.. STS-2, STS-8.

Vice Admiral Richard Harrison Truly was born in November 1937 in Fayette, Mississippi, USA. He received a Bachelor of Science degree in Aeronautical Engineering from Georgia Institute of Technology in 1959. Being a member of the Navy ROTC at Georgia Tech, Truly entered the US Navy, where he was ordered to flight school and was designated a Naval Aviator on October 07, 1960. He flew F-8 Crusaders aboard *USS Intrepid* and *USS Enterprise*, making over 300 carrier landings. From 1963-65, Truly was first a student and later an instructor at the USAF Aerospace Research Pilot School at Edwards, AFB. In 1965, he was among the first military astronauts selected to the USAF's Manned Orbiting Laboratory (MOL) program in Los Angeles. He joined NASA Astronaut Group 7 in August 1969. Truly flew three times on one of the two-man crews for the 1977 Approach and Landing Tests of the prototype Space Shuttle *Enterprise*.

Richard Truly was Pilot on STS-2 aboard *Columbia* (November 12-14, 1981). The mission marked the first time that a crewed, reusable orbital vehicle returned to space. The crew tested the Shuttle Imaging Radar (SIR), the Shuttle robotic arm, and six experiments. There were various tests on the Orbital Maneuvering system (OMS) including starting and restarting the engines while in orbit, and various adjustments to orbit. Truly was Commander on STS-8 aboard *Challenger* (August 30-September 05, 1983), whose flight was marred by the subsequent discovery that a solid-fuel rocket booster had almost malfunctioned catastrophically during the launch. The primary payload was *INSAT-1B*, an Indian communications and weather observation satellite. Scientific experiments included the environmental testing of new hardware and materials designed for future spacecraft.

Richard Truly left NASA to become the first commander of the Naval Space Command. Three weeks after the *Challenger* disaster of January 28, 1986, Truly returned to NASA to become Associate Administrator for Space Flight. In March 1986, Truly noted in a memo that there were several actions NASA needed to accomplish before launching another shuttle flight.

They included *"Solid Rocket Motor joint redesign, Critical Items review, and Operations and Maintenance Instructions review"*. It took Truly and NASA's Return to Flight program thirty-one months before Space Shuttle *Discovery* successfully flew on September 29, 1988. Truly retired from the Navy as a Vice Admiral before becoming the eighth NASA Administrator in May 1989, holding that position until May 1992. He became Vice President and Director of the **Georgia Tech Research Institute** from 1992-97. Truly then served as Director of the **Department of Energy**'s National Renewable Energy Laboratory and as Executive Vice President of the **MRIGlobal** (formerly Midwest Research Institute) from 1997-2005.

Richard Truly was awarded the Legion of Merit with one star; the DFC; the NASA Outstanding Leadership Medal; the Johnson Space Center Superior Achievement Award (1972); the American Astronautical Society's Flight Achievement Award (1977); the General Thomas D. White USAF Space Trophy for 1981; the Society of Experimental Test Pilots James H. Doolittle Award (1988); and he was inducted into the Georgia Aviation Hall of Fame (1995).

## Vande Hei, Mark T.. Soyuz MS-06/Expedition 53/54, Soyuz MS-18/Expn 64/65/Soyuz MS-19.

Col. Mark Thomas Vande Hei was born in November 1966 in Falls Church, Virginia, USA, from Dutch parents. He earned a Bachelor of Science degree (with a major in Physics) from St. John's University in 1989; and a Master of Science degree in Applied Physics from Stanford University in 1999. He was commissioned in the US Army through the ROTC program in 1989 and became a combat engineer, serving in Iraq in Operation Provide Comfort. After earning his Master's degree in 1999, he became an assistant professor of Physics at the US Military Academy in West Point, New York. In 2003, he became part of the Army's 1st Space Battalion at Peterson AFB. Vande Hei served again in Iraq in Operation Iraqi Freedom. He retired in October 2016 at the rank of Colonel.

Vande Hei began working at JSC in 2006 as part of the US Army contingent there. In June 2009, he was selected as a member of NASA Astronaut Group 20. On June 10, 2014, NASA announced that he would serve as an aquanaut aboard the *Aquarius* underwater laboratory during the NEEMO 18 undersea exploration mission, which ran for nine days from July 21, 2014. Later, he was assigned to ISS 'Expedition 53/54' scheduled to launch in 2017 aboard *Soyuz MS-06*. Launched on September 12, 2017, the crew performed the fast rendezvous with the ISS, and docked automatically after about six hours. As a Flight Engineer, he performed an EVA of six hours and 55 minutes on October 05. A second EVA on October 10 lubricated the end effector, and replaced cameras. The mission ended on February 27, 2018.

In March 2021, it was confirmed that Vande Hei would be a Flight Engineer aboard *Soyuz MS-18* and be part of 'Expedition 64/65'. On September 14, his six-month stay was extended by another six months. This meant he would break the record for the longest spaceflight by an American astronaut with 355 days. On January 06, 2022, Vande Hei and Dubrov completed 273 days on the ISS, surpassing Andrew R. Morgan's record. Vande Hei returned to Earth with *Soyuz MS-19* on March 20, 2022, after 355 days in space.

## van den Berg, Lodewijk. STS-51-B.

Dr. Lodewijk van den Berg (1932-2022) was born in March 1932 in Sluiskil, Netherlands. He studied crystal growth and flew on STS-51-B aboard *Challenger* (April 29-May 06, 1985). He attended the Technical University Delft from 1949, graduating with a Master of Science degree in Chemical Engineering. He then moved to the US and attended the University of Delaware, obtaining a Master of Science degree in Applied Science in 1972, then a Ph.D. in Applied Science in 1975. Van den Berg was then offered a job at **EG & G Corporation** Energy

Measurements in Goleta, California, working in crystal growth. Since EG & G was a defense contractor of the US government, he was required to become a naturalized US citizen. He worked many years in research, acquiring management experience in the preparation of crystalline materials, and the investigation of associated defect chemistry and electronic properties. Van den Berg became an international authority on vapor growth techniques with an emphasis on mercuric iodide crystals and its application in the nuclear industry as gamma ray detectors.

While employed at EG & G, he asked NASA for permission to conduct crystal growth experiments in space, and NASA agreed. Van den Berg and his colleages designed the 'EG & G Vapor Crystal Growth System' experiment apparatus for a Space Shuttle flight. In 1983, he started to train as an astronaut, and six months before the launch was told he would be the prime astronaut. He was fifty-three years old when he went to space, becoming one of the oldest rookie astronauts. Lodewijk van den Berg was launched on STS-51-B aboard *Challenger* on April 29, 1985, the 'Spacelab-3' mission. While a co-investigator of the Vapor Crystal Growth System (VCGS) experiment, he supervised crystal growth aspects of the experiment.

Returning to Earth on May 06, 1985, van den Berg continued working at **EG & G Corporation**, becoming head of the materials science section. He then moved to Florida to become a chief scientist at the Constellation Technology Corporation. At age seventy-two years in 2004, he continued to work full-time on crystal growth experiments, a process he compared to gardening. He was the subject of a short 2004 documentary named *The Forgotten Astronaut,* by the TV-show ***Netwerk*** on the Dutch channel **Nederland 2**. A secondary school has been named after him, the Lodewijk College in Terneuzen. Van den Berg logged seven days, zero hours and eight minutes in space. He died on October 16, 2022, aged ninety years.

## van Hoften, James D.A.. STS-41-C, STS-51-I.

Dr. James Dougal Adrianus "Ox" van Hoften was born in June 1944 in Fresno, California, USA. He earned a Bachelor of Science degree in Civil Engineering from the University of California, Berkeley, in 1966; a Master of Science (1968); and Doctor of Philosophy degree (1976), both in Hydraulic Engineering from Colorado State University. From 1969-74, van Hoften was a pilot in the US Navy. While assigned to *USS Ranger* in 1972, he participated in two cruises to Southeast Asia, where he flew about sixty combat missions during the Vietnam War. In September 1976, he accepted an assistant professorship of Civil Engineering at the University of Houston, and until his selection as an astronaut in January 1978 taught fluid mechanics and conducted research on biomedical fluid flows concerning flows in artificial internal organs and valves. From 1977-80, van Hoften flew as a member of the Texas Air National Guard with the 147th Fighter-Interceptor Group at Ellington Field. He logged a total of 3,300 hours flying time, mainly in jet aircraft.

Selected by NASA in January 1978, he was a Mission Specialist on STS-41-C aboard *Challenger* (April 06-13, 1984), which deployed the Long Duration Exposure Facility; and retrieved, repaired and replaced in orbit the ailing *Solar Maximum Mission* satellite. STS-51-I aboard *Discovery* (August 27-September 03, 1985) successfully deployed three communications satellites; and the salvage of the ailing Navy *Syncom IV-3* satellite. This required two EVAs, with van Hoften attached to the Remote Manipulator System (RMS) performing the first manual grapple and manual deployment of a satellite in orbit. Dr. van Hoften logged a total of 14 days, one hour and 57 minutes in space; including four EVA totaling 21 hours and 57 minutes.

Dr. van Hoften has published a number of papers on turbulence, waves, and cardiovascular flows. He is a former senior vice president and partner of the **Bechtel Corporation**. In the early 1990s, he was the program manager of the USD 23 billion Hong Kong Airport Core

Program. He later acted as Director of Projects for the UK National Air Traffic Services. Dr. van Hoften has received the Meritorious Service Medal; two Navy Air Medals; the National Defense Service Medal; the Vietnam Service Medal; and two NASA Space Flight Medals. He is a member of the American Institute of Aeronautics and Astronautics.

## Veach, Charles L.. STS-39, STS-52.

Charles Lacy Veach (1944-95) was born in September 1944 in Chicago, Illinois, USA. He received a Bachelor of Science degree in Engineering Management from the USAF Academy in 1966; received his pilot wings at Moody AFB, Georgia, in 1967, then attended fighter gunnery school at Luke AFB, Arizona. For the next 14 years he served as a USAF fighter pilot on many assignments, including a 275-mission combat tour in South Vietnam. In 1976-77, he was a member of the USAF Air Demonstration Squadron, the *Thunderbirds*, flying the T-38 Talon. He left active duty in 1981, but continued flying with the 147th Fighter Group; logging over 5,000 flying hours. Veach came to work for NASA in January 1982 as an engineer and research pilot at JSC in Houston. His primary duty was as an instructor pilot in the Shuttle Training Aircraft. He was selected as an astronaut candidate in May 1984.

On STS-39 aboard *Discovery* (April 28-May 06, 1991) Veach was responsible for operating a group of instruments including an ultraviolet astronomical camera, an X-ray telescope, and a liquid helium-cooled infrared telescope which preformed landmark observations of the Earth's atmosphere and Aurora Australis. Mission STS-52 aboard *Columbia* (October 22-November 01, 1992) deployed the *Laser Geodynamic Satellite* (*LAGEOS*), a joint Italian-American project. He logged a total of 18 days, four hours and 18 minutes (436.3 hours) in space. Charles Veach received the DFC with two Oak Leaf Clusters; the Air Medal with 13 Oak Leaf Clusters; the Air Force Commendation Medal with one Oak leaf Cluster, and the Purple Heart. Lacy Veach died of cancer on October 03, 1995, in Houston, Texas.

## Virts, Terry W., Jr.. STS-130, Soyuz TMA-15M/Expedition 42/43.

Col. Terry W. Virts was born in December 1967 in Baltimore, Maryland, USA. He graduated from Oakland Mills High School in 1985; earned a Bachelor of Science degree in Mathematics (with a French minor), graduating with distinction from the USAF Academy in 1989; and a Master of Aeronautical Science degree from Embry-Riddle Aeronautical University in 1997. While at the Air Force Academy, he attended the Ecole de l'Air in 1988 on an exchange program. Commissioned a second lieutenant, he earned his pilot wings at Williams AFB, Arizona, as a distinguished student. After training with the F-16 Fighting Falcon, he was assigned to the 31st Tactical Fighter Wing at Homestead AFB, Florida. Virts was later assigned to the 36th Fighter Squadron at Osan Air Base, Republic of Korea; and the 22nd Fighter Squadron at Spangdahlem Air Base, Germany. He was selected for Test Pilot School in 1997 at Edwards AFB, then was an Experimental Test Pilot at the F-16 combined Test Force. Col. Virts logged over 5,300 flight hours in 40 aircraft types.

Selected in 2000 NASA Group, his career was noted for a series of important technical assignments and mission roles; particularly in serving as the lead astronaut for the NASA T-38 program, where he contributed to the training and preparation of fellow astronauts. Virts also played a crucial role as a member of the Shuttle Avionics Integration Laboratory (SAIL) test crew, participating in critical systems testing and integration efforts. In 2004, he supported 'Expedition 9', and took on the role of Capcom. He contributed to the Space Launch System (SLS) program, where he served as lead astronaut. Terry Virts was pilot on STS-130 aboard *Endeavour* (February 08-22, 2010), the final assembly flight for the ISS. On November 23, 2014, he launched aboard *Soyuz TMA-15M* from Baikonur Cosmodrome in Kazakhstan, and docked at the ISS about six hours later. On February 28, 2015, Virts paid tribute to the late ***Star Trek*** actor Leonard Nimoy, who portrayed Spock. Virts tweeted an image of himself striking the Vulcan salute as the ISS flew over Boston, MA, Nimoy's birthplace. In March 2015 (before *Soyuz TMA-14M* departed), Virts assumed the role of Commander of the ISS, leading 'Expedition 43' until June 11, 2015, when he returned to Earth. Virts retired from NASA on August 23, 2016. He logged a total of 213 days, ten hours and 48 minutes in space; including three EVAs totaling 19 hours and two minutes.

In recent years Virts has become a noticeable figure in the fields of public speaking, podcast appearances, business consultancy, literature, film-making and screenwriting. He wrote *View From Above: An Astronaut Photographs the World* (2017), *Apollo: To the Moon and Back* (2019), and *How to Astronaut* (2020). Terry Virts and other ISS astronauts feature in the 2016 IMAX documentary film ***A Beautiful Planet***. In 2019, he was part of a pioneering crew of eight aviation explorers who achieved the fastest circumnavigation of Earth via both geographic poles by airplane; in a world-record time of 46 hours, 40 minutes and 22 seconds. On December 11, 2020, Virts made an appearance on *"The Joe Rogan Experience"* podcast. He has received the Meritorious Service Medal; the Air Medal; the Aerial Achievement Medal; and a United States Air Force Commendation Medal.

## Voss, James S.. STS-44, STS-53, STS-69, STS-101, STS-102.

James Shelton Voss was born in March 1949 in Cordova, Alabama, USA. He graduated with a Bachelor of Science degree in Aerospace Engineering from Auburn University in 1972, and was commissioned a second lieutenant in the US Army. He graduated with a Master of Science degree in Aerospace Engineering Sciences from the University of Colorado in 1974, then attended the Infantry Basic Course, where he was listed as a Distinguished Graduate. Voss taught in the Department of Mechanics at the **US Military Academy** at West Point, New York. He participated in a NASA Summer Faculty Research Fellowship and received the William P. Clements Jr. Award for Excellence in Education as the outstanding professor at the US Military Academy. Voss attended the US Naval Test Pilot School, graduating with the Outstanding Student Award in 1983. He then attended the Armed Forces Staff College, and was subsequently assigned as a Flight Test Engineer/Research and Development Coordinator with the US Army Aviation Engineering Flight Activity. He worked at NASA's JSC from November 1984 to June 1987, when he was selected as an astronaut candidate.

James Voss was a Mission Specialist on STS-44 aboard *Atlantis* (November 24-December 01, 1991), whose primary objectice was the deployment of a Defense Support Program satellite with an Inertial Upper Stage booster rocket. The crew conducted two Military-Man-in-Space experiments, and numerous medical tests to support longer duration spaceflight. He was a Mission Specialist on STS-53 aboard *Discovery* (December 02-09, 1992), which deployed a classified Department of Defense payload; a *Satellite Data System-2* military communications satellite. Voss was a Mission Specialist and Payload Commander on STS-69 aboard *Endeavour* (September 07-18, 1995), which deployed a *SPARTAN* satellite and the *Wake Shield Facility*. Voss and Michael Gernhardt performed a six-hour EVA to test space suit modifications and to evaluate tools and procedures for future use on the ISS. He then spent two years training at the Yuri Gagarin Cosmonaut Training Center, and served as backup on the STS-84, STS-86 and STS-89 missions.

Jim Voss was 'Mission Specialist 3' and 'Flight Engineer 1' on STS-101 aboard *Atlantis* (May 19-29, 2000), a resupply mission to the ISS, whose launch was delayed three times in April due to high winds. Voss and Jeff Williams performed a six-hour EVA to finish installation of a Russian crane, replace a faulty antenna, and install handrails on the outside of the ISS. Voss and his 'Expedition 2' (ISS EO-2) crewmates launched as Mission Specialists on STS-102 aboard *Discovery* on March 08, 2001. He and Susan Helms conducted an eight-hour and 56 minute EVA, the longest to date. He returned on STS-105, after having spent 163 days aboard the ISS, conducting eighteen scientific experiments, maintenance, and outfitting. He was the first person to operate the Space Station Robotic Manipulator System, *Canadarm-2*. Voss was one of the cinematographers and featured astronauts of the first 3D film shot in space, ***Space Station 3D***; released in theatres in 2002, and on DVD in 2005. He returned to Earth aboard *Discovery* on August 20, 2001.

After retiring from NASA in 2003, Voss became Associate Dean of Engineering for External Affairs at **Auburn University**. In 2005, Voss became VP of Space Exploration Systems for **t/Space**, leaving in November 2007 to become the Vice President of Engineering at **SpaceDev** (later acquired by Sierra Nevada Corporation), where he became Director of their Advanced Programs and the VP of Space Exploration Systems and program executive of the *Dream Chaser* spaceplane. James Voss has received nineteen awards, including the Army Distinguished Service Medal; the NASA Exceptional Service Medal; the Russian Spaceflight Medal of Achievement; Most Outstanding Faculty Member, Auburn University; and the National Aeronautic Association Gagarin Gold Medal. He was inducted into the Alabama Engineering Hall of Fame (2001); and the Alabama Aviation Hall of Fame.

## Voss, Janice E.. STS-57, STS-63, STS-83, STS-94, STS-99.

Dr. Janice Elaine Voss (1956-2012) was born in October 1956 in South Bend, Indiana, USA. She first became involved with NASA in 1973, and worked as co-op at **Johnson Space Center** until 1975. Voss has cited Madeline L'Engle's novel *A Wrinkle in Time* (published in 1962) as one of her primary inspirations for becoming an astronaut. She claimed that powerful female roles were the norms she accepted in life. Voss received her Bachelor of Science in Engineering Science from Purdue University in 1975; her Master of Science degree in Electrical Engineering from MIT in 1977; and her Ph.D. in Aeronautics and Astronautics from MIT in 1987. For her Ph.D. work, Janice Voss focused on developing algorithms to identify frequencies, damping, and mode shapes for the International Space Station.

From 1977-78, she completed work in space physics at Rice University, Texas. In 1983, Voss became a Draper Fellow while continuing her graduate studies at MIT. She worked on developing software for the Space Shuttle program. She then began working at **Orbital Sciences Corporation**, including support for a rocket upper stage called the *Transfer Orbit Stage* (TOS). This launched the *Advanced Communications Technology Satellite* in September 1993. In 1990, Voss was selected by NASA as an astronaut candidate, qualifying in 1991. Voss served as a Mission Specialist on STS-57, (1993); STS-63, (1995); STS-83, (1997); STS-94, (1997) and STS-99 in February 2000. She participated in the first shuttle rendezvous with the *Mir* space station on mission STS-63; and on mission STS-99 she worked in shifts to produce what was at the time the most accurate digital topographic map of the Earth.

Voss logged forty-nine days, three hours and 49 minutes in space, traveled 18.8 million miles (about 30,255,667 km) in 779 Earth orbits, and all her missions included at least one other woman. She holds the joint record for five spaceflights by an American woman; along with Dr. Shannon Wells Lucid (achieved in March 1996); Bonnie Dunbar (January 1998); Tamara Jernigan (May 1999); Marsha Ivins (February 2001) and Susan Helms (achieved in August 2011). From October 2004 to November 2007, Dr. Janice Voss was the Science Director for NASA's **Kepler Space Observatory**. At the Astronaut Office Branch Station, she served as the Payloads Lead. Voss also worked for **Orbital Sciences Corporation** in flight operations support. She was awarded the NASA spaceflight medal (1993 and 1995). Janice Voss died of breast cancer in Arizona on February 06, 2012. She was aged fifty-six years. The Cygnus *CRS Orb-2* spacecraft was named *SS Janice Voss* in her honor. The VOSS Model is a scaled model of the Solar System, dedicated to Janice Voss, and located at Purdue University's Discovery Park in West Lafayette, Indiana.

## Walheim, Rex J.. STS-110, STS-122, STS-135.

Rex Joseph Walheim was born in October 1962 in Redwood City, California, USA. He received a Bachelor of Science degree in Mechanical Engineering from the University of California, Berkeley, in 1984; and a Master of Science degree in Industrial Engineering from the University of Houston in 1989. In May 1984, he was commissioned a second lieutenant in the Air Force, and worked as a missile warning operations crew commander. In October 1986, he was reassigned to the JSC, where he worked as a mechanical systems flight controller, and was the lead operations engineer for the Space Shuttle landing gear, brakes, and emergency runway barrier. In August 1989, he was transferred to **Air Force Space Command** HQ in Colorado Springs, where he was manager of a program upgrading missile warning radars. He was selected for USAF Test Pilot School in 1991, and after graduation was assigned to the F-16 Combined Test Force at Edwards AFB. In January 1996, Walheim became an instructor at USAF Test Pilot School, where he served until he began astronaut training.

Selected in 1996 NASA Group 16, he was 'Mission Specialist 1' on STS-110 aboard *Atlantis* (April 08-19, 2002), whose main purpose was to install the S0 Truss segment to the ISS. It formed the backbone to which the S1 and P1 truss segments were attached on following missions. Walheim made two EVAs. The Shuttle also delivered the Mobile Transporter (MT), an 885 kg assembly that glides down rails on the Space Station integrated trusses. During mission STS-111, the Mobile Base System was mounted to the MT, allowing the Canadarm2 to travel down the length of the installed truss structure. On STS-122 aboard *Atlantis* (February 07-20, 2008) the main payload was the 12,077 kg (26,625 lb) *Columbus* Orbital Facility laboratory module; along with the Biolab, Fluid Science Laboratory (FSL); European Drawer Rack (EDR); and the European Physiology Module (EPM).

During an address at the Marshall Space Flight Center on November 16, 2010, NASA Administrator Charles Bolden said that the agency needed to fly mission STS-135 in 2011, due to possible delays in the development of commercial rockets and spacecraft designed to transport cargo to the ISS. Walheim was assigned as Mission Specialist and Flight Engineer on STS-135 aboard *Atlantis* (July 09-21, 2011), the final Space Shuttle mission, when the primary cargo was the Multi-Purpose Logistics Module *Raffaello* and a Lightweight Multi-Purpose Carrier (LMC). He logged a total of 36 days, eight hours and 31 minutes (over 566 hours) in space; including 36 hours and 23 minutes of EVA, and retired from NASA on July 27, 2020. Rex Walheim's awards include Distinguished Graduate, Reserve Officers Training Corps; Distinguished Graduate in USAF Test Pilot School Class 92A; the Meritorious Service Medal; two Air Force Commendation Medals; and the Aerial Achievement Medal.

## Walker, Charles D.. STS-41-D, STS-51-D, STS-61-B.

Charles "Charlie" David Walker was born in August 1948 in Bedford, Indiana, USA. He graduated from Purdue University with a Bachelor of Science degree in Aeronautical and Astronautical Engineering in 1971, then worked as a civil engineering technician, land acquisition specialist and forest firefighter for the US Forest Service. He was a design engineer with the **Bendix Aerospace Company**; working on aerodynamic analysis, missile subsystem design, and flight testing. He joined **McDonnell Douglas Corporation** in 1977 as a test engineer on the Aft Propulsion Subsystem for Space Shuttle orbiters. From 1979-86, Walker was chief test engineer and payload specialist for the McDonnell Douglas EOS commercialization project, leading the EOS laboratory test and operations team developing biomedical products. McDonnell Douglas proposed that Walker fly as a Payload Specialist to operate the complex Continuous Flow Electrophoresis System (CFES) payload himself. NASA calculated that would cost McDonnell Douglas USD 40,000 per flight, and in May 1983 he was assigned to mission STS-41-D. His flight was part of a NASA effort in the 1980s to fly civilians on the Shuttle. Walker later stated that his experience showed that a "working passenger" could fly after an abbreviated training program of a few months.

Charles Walker flew as a Payload Specialist STS-41-D aboard *Discovery* (August 30-September 05, 1984) for McDonnell Douglas Corporation. During the June 26 launch attempt, a stuck valve prevented proper flow of LOX into the combustion chamber of the Shuttle's center engine, which led to a launch abort, followed by an invisible hydrogen fire around the engine nozzles while on the pad. Mission Specialist Steve Hawley was reported as saying, *"Gee, I thought we'd be a lot higher at MECO (Main Engine Cut-Off)!"* If the astronauts had used the normal emergency escape procedure across the service arm to the (then untested by a human) slidewire escape baskets, they would have run into the fire. At launch on the fourth attempt, the combined cargo weighed over 18,681 kg (41,185 pounds), a record for a Shuttle payload up to that time. The crew deployed three commercial satellites: *SBS-4, Telstar 302* and *Syncom IV-2* aka *LEASAT-2*.

On STS-51-D aboard *Discovery* (April 12-19, 1985) Charlie Walker performed early protein crystal growth experiments and participated as a test subject in numerous medical studies. The *Telesat-1* aka *Anik C1* and *Syncom IV-3* (aka *LEASAT-3*) satellites were deployed. The Continuous Flow Electrophoresis System III (CFES-III) payload was flying for the sixth time. During the shuttle landing at KSC, extensive brake damage was suffered, and a landing gear tire ruptured. More effective nose wheel steering was later implemented to reduce risks during landing. On STS-61-B aboard *Atlantis* (November 26-December 03, 1985) the crew deployed three communications satellites, and tested techniques of constructing structures in orbit. Walker logged a total of 19 days, 21 hour and 56 minutes in space. Since 1986, Charlie Walker has served in various NASA study and review team capacities; including as a member of the NASA Microgravity Material Science Assessment Task Force, the NASA Space Station office Quick-is-Beautiful/Rapid Response Research Study Group, the NASA Space Station Operations Task Force, and the ISS Strategic Roadmap committee.

Walker retired in 2005 as the director for NASA Systems Government Relations with the Washington D.C. Operations Office of **The Boeing Company**. In 2018, he joined **Solstar** as an advisor; he is currently a consultant, author and lecturer. Walker makes an occasional public appearance to sign memorabilia at KSC, and appears for the complex's *Lunch with an Astronaut* program. Charles Walker has received nine awards: US Patent No. 4,394,246 Electrophoresis Apparatus with Flow Control, issued July 19, 1983; three NASA Space Flight Medals (1984, and twice in 1985); Doctor of Science, honoris causa, St. Louis College of Pharmacy (1985); the Aerospace Laurels Award, *Aviation Week & Space Technology* Magazine (1986); the Lindbergh Award, American Institute of Aeronautics and Astronautics - St. Louis Section (1986); the NASA Group Achievement Award (1987-88); the Engineering Astronaut Alumnis Award, Purdue University (September 1989); and was made Kentucky Colonel, Commonwealth of Kentucky (May 1990).

## Walker, David M.. STS-51-A, STS-30R, STS-53, STS-69.

David Mathieson Walker (1944-2001) was born in May 1944 in Columbus, Georgia, USA. He received a Bachelor of Science degree and graduated from the US Naval Academy at Annapolis, Maryland, in 1966, then received flight training from the Naval Air Training Command at bases in Florida, Mississippi, and Texas. Designated a Naval Aviator in December 1967, he was assigned to the aircraft carriers *USS Enterprise* and *USS America*. From December 1970 to 1971, he attended the USAF Test Pilot school at Edwards AFB, and was assigned in January 1972 as an experimental and engineering test pilot at NAS Patuxent River, Maryland. In 1975, Walker was assigned to Fighter Squadron 142 and was deployed to the Mediterranean Sea twice aboard *USS America*. Walker logged over 7,500 hours flying time, with over 6,500 hours in jet aircraft. He was one of thirty-five candidates selected by NASA in January 1978 for the new Space Shuttle program, and he became an astronaut in August 1979.

David Walker was pilot on STS-51-A aboard *Discovery* (November 08-16, 1984), which deployed two satellites and retrieved for return to Earth the *Palapa B-2* and *Westar VI* satellites. He was Commander of STS-30 aboard *Atlantis* (May 04-08, 1989), which deployed the *Magellan* Venus-exploration spacecraft. STS-53 aboard *Discovery* (December 02-09, 1992) deployed a classified Department of Defense payload DOD-1, and then performed several Military-Man-in-Space and NASA experiments. STS-69 aboard *Endeavour* (September 07-18, 1995) deployed and retrieved a *SPARTAN* satellite and the *Wake Shield Experiment*. David Walker logged a total of 30 days, four hours and 27 minutes (almost 725 hours) in space.

In April 1996, Walker left NASA and retired from the Navy to become Vice President for sales and marketing for **NDC Voice Communications** in San Diego, California. In 1998, Walker served as space technology consultant on the film ***Deep Impact***. He joined **Ultrafast, Inc.**, of Malvern, Pennsylvania, in April 1999 as Vice President of Aerospace Sales, and later retired to McCall, Idaho. David Walker was a Senior Member of the American Institute of Aeronautics and Astronautics; the Association of Space Explorers; was awarded the Defense Superior Service Medal; the DFC; the National Intelligence Medal of Achievement; the Legion of Merit; two Defense Meritorious Service Medals; two NASA Distinguished Service Medals; the NASA Outstanding Leadership Medal; and the Vietnamese Cross of Gallantry. He died from cancer on April 23, 2001, in Houston, Texas, aged fifty-six years, and was interred at Arlington National Cemetery.

## Walker, Joseph "Joe" A.. X-15 rocketplane Flight 77, Flight 90, Flight 91.

Capt. Joseph "Joe" Albert Walker (1921-66) was born in February 1921 in Washington, Pennsylvania, USA. He was an American World War II pilot, experimental physicist, NASA test pilot and astronaut who was the first person to fly an airplane to space. On October 27, 1961, Walker became the first person in the mesosphere when piloting USAF X-15 Flight 35, and in 1963, Walker made three flights above fifty miles altitude, thereby qualifying as an astronaut according to the US definition of the boundary of space. Flights 90 and 91 also surpassed the Karman line, the internationally accepted space boundary of 100 kilometers altitude (62.14 miles). He was the only X-15 pilot to fly above one hundred kilometres altitude. Making the latter flights immediately after the completion of the *Mercury* and *Vostok* programs, Walker became the first person to fly to space twice.

Joe Walker was the first American civilian to make any spaceflight, and he flew the X-15 rocket plane 25 times. On X-15 Flight 77 (January 17, 1963) he exceeded eighty km (50 miles) altitude. Walker flew at his highest speed in the X-15A at 6,605 km/hr (Mach 5.92) during Flight 59 on June 27, 1962. On X-15 Flight 90 (July 19, 1963) he reached 106 km (66 miles) altitude, and on X-15 Flight 91 (August 12, 1963) he reached 108 km (67 miles). Walker became the first test pilot of the Bell Lunar Landing Research Vehicle (LLRV), used to develop piloting and operational techniques for lunar landings. On October 30, 1964, he reached an altitude of about ten feet and a total flight time of just under one minute for the maiden flight. Walker made thirty-five LLRV flights. It was later flown by Neil Armstrong in preparation for the *Apollo 11* mission.

Joe Walker was a charter member and one of the first Fellows of the Society of Experimental Test Pilots. He received the Robert J. Collier Trophy; the Harmon International Trophy for Aviators; National Pilots Association named him Pilot of the Year (1963); the Iven C. Kincheloe Award; the John J. Montgomery Award; and the Octave Chanute Award. Walker was inducted into the Aerospace Walk of Honor (1991); and the International Space Hall of Fame (1995). Joe Walker Middle School in Quartz Hill, California, is named in his honor, as well as the Joe Walker Elementary School in Washington, Pennsylvania. Capt. J.A. Walker died on June 08, 1966, in a group formation flying accident. ***Star Trek*** starship designer John Eaves created the *Walker*-class starships named for him that first appeared in the 2017 TV series ***Star Trek: Discovery***. On August 23, 2005, NASA posthumously conferred Astronaut Wings on Joseph Walker.

## Walz, Carl E.. STS-51, STS-65, STS-79, STS-108/Expedition 4.

Col. Carl Erwin Walz was born in September 1955 in Cleveland, Ohio. He received a Bachelor of Science degree in Physics from Kent State University, Ohio, in 1977; and a Master of Science degree in Solid State Physics from John Carroll University, Ohio, in 1979. From 1979-82, Walz was responsible for analysis of radioactive samples from the US Atomic Energy Detection System at the 1155th Technical Operations Squadron, McClellan AFB, California. After a year at the USAF Test Pilot School, from January 1984 to June 1987 he served as a Flight Test Engineer to the F-16 Combined Test Force at Edwards AFB, California. From July 1987 to June 1990, he served as a Flight Test Manager at Detachment 3, Air Force Flight Test Center.

Selected by NASA in January 1990, Walz was a Mission Specialist on STS-51 aboard *Discovery* (September 12-22, 1993), which deployed the US *Advanced Communications Technology Satellite* (ACTS). Walz made a seven hour EVA. He was Orbiter Flight Engineer on STS-65 aboard *Columbia* (July 08-23, 1994) and conducted over eighty experiments on materials and life sciences in microgravity. Walz was a Mission Specialist on STS-79 aboard *Atlantis* (September 16-26, 1996), which docked with space station *Mir*, delivered supplies and experiments, and exchanged crew members. When Walz launched on STS-108 aboard *Endeavour* and docked with the ISS on December 07, 2001, he became 'Flight Engineer 1' for 'Expedition 4' (December 05, 2001 to June 19, 2002) spending 196 days on the ISS. Wearing the Russian *Orlan* spacesuit, Walz logged six hours of EVA time. He also logged another five hours and 52 minutes of EVA in the US *EMU* spacesuit, preparing the ISS for the S0 truss. Carl Walz logged a total of 230 days, 13 hours and four minutes in space.

Walz retired from NASA on December 05, 2008, and is currently working for **Orbital Sciences Corporation ATK**'s Advanced Programs Group as vice president for Human Space Flight Operations. He enjoys piano and vocal music, sports, and was notably the lead singer for ***Max Q***, an all-astronaut rock and roll band. Carl Walz graduated Summa Cum Laude from Kent State University; was awarded Distinguished Graduate from the USAF Test Pilot School, Class 83A; the Defense Superior Service Medal with one Oak Leaf Cluster; the USAF Legion of Merit; the Meritorious Service Medal with one Oak Leaf Cluster; the NASA Exceptional Service Award; the Gagarin Award (2003); and he was inducted into the Ohio Veterans Hall of Fame.

## Wang, Taylor G.. STS-51-B.

Dr. Taylor Gun-Jin Wang was born in June 1940 in Shanghai, Republic of China. His family moved to Taiwan in 1952, and Wang graduated from The Affiliated Senior High School of National Taiwan Normal University in Taipei, Taiwan, and later moved to Hong Kong. He received a Bachelor of Science degree in Physics from UCLA in 1967; and his Master of Science degree in 1968; with a Doctorate in Low Temperature Physics -Superfluid and Solid State Physics in 1972. After completing his doctorate, Wang joined the **California Institute of Technology**'s Jet Propulsion Laboratory (JPL) in 1972 as a senior scientist.

He gained US citizenship in 1975, and published a paper on the dynamic behaviour of rotating spheroids in zero gravity the following year. This received attention in NASA, and in June 1983 Dr. Wang was selected for the 'Spacelab-3' mission. While an employee of the JPL, Wang was a Payload Specialist on the Space Shuttle *Challenger* mission STS-51-B (April 29-May 06, 1985); the first person of Chinese origin to go into space. The seven-man crew conducted investigations in crystal growth, drop dynamics leading to containerless material processing, atmospheric trace gas spectroscopy, solar and planetary atmospheric simulation, cosmic rays, laboratory animals and human medical monitoring. Dr. Taylor Wang logged seven days, zero hours and eight minutes in space.

Using insights gained from his mission, Dr. Wang has developed an immunoisolation encapsulation system that protects cellular transplants, and sustains cell function - without immuno-suppression drugs and their side effects. Dr. Wang later became a Centennial Professor at **Vanderbilt University** in Nashville, Tennessee. He has written about two hundred journal articles and holds twenty-eight US Patents. Dr. Wang has received various honors and awards, including the NASA Exceptional Scientific Achievement Medal (1987); the Asian Pacific American Achievement Award (1989); the Llewellyn J. Evans Distinguished Scientific, Engineering and Management Award (1994); the Educational Award, Vanderbilt University Alumni League (1996); and he was awarded the Asian American Engineer of the Year Distinguished Science and Technology Award, CIE-USA, National Engineer's Foundation (2007).

## Watkins, Jessica A.. SpaceX 'Crew 4'/Expedition 67/68.

Dr. Jessica Andrea "Watty" Watkins was born in May 1988 in Gaithersburg, Maryland, USA. She received a Bachelor of Science degree in Geological and Environmental Sciences from Stanford University; and earned a Master of Science degree and a Ph.D. in Geology from the University of California, Los Angeles, researching landslides on both Earth and Mars. Prior to her selection as an astronaut, Watkins was a postdoctoral fellow at the California Institute of Technology, where she was also an assistant coach to the women's basketball team. Watkins began playing rugby during her freshman year at Stanford. In both 2008 and 2010, she became a member of First Team Collegiate Rugby All-American. She played for the USA Eagles as a semi-finalist in the 2009 Rugby World Cup Sevens.

As an undergraduate, Watkins worked at the Ames Research Center to support the Mars Phoenix lander and prototype Mars drill testing. As a graduate student, she worked at the JPL on the NEOWISE project to survey near-Earth asteroids. Watkins also worked on planning for the Mars rover *Curiosity*. In 2009, she was chief geologist for the NASA Spaceward Bound Crew 86 at the Mars Desert Research Station. She served as a planner for the Mars 2020 rover and a Mars sample-return mission. As a collaborator on the Mars Science Laboratory Science Team, she participated in daily planning of the Mars rover activities, and uses its image data combined with orbital data to investigate the stratigraphy, geology, and geomorphology of Mars.

In December 2020, Watkins was selected to be a part of the *Artemis* Team to return humans to the Moon. In November 2021, she became the fourth astronaut of Group 22. When SpaceX 'Crew 4' launched on April 27, 2022, she was a Mission Specialist on the maiden flight of the *Crew Dragon* spacecraft *Freedom*. Joining 'Expedition 67/68' she became the first Black woman to complete an ISS long-term mission (170 days, 13 hours and three minutes). On June 09, she became the African American woman with the most time in space, surpassing Stephanie Wilsons's 42 day, 23 hour and 46 minute record. Splashdown was on October 14, 2022, in the Atlantic Ocean. Dr. Watkins has received thirteen awards, including USA Rugby Collegiate All-American (2008-10); the Chancellor's Prize, UCLA (2010); a Graduate Research Fellowship in Geosciences, National Science Foundation (2012); the Caltech Division of Geological and Planetary Sciences Chairs' Postdoctoral Fellowship (2015); the NASA Group Achievement Award, Mars Science Laboratory (2015); and Jessica Watkins Day, April 19, 2022, City of Lafayette, Colorado, was in her honor.

## Weber, Mary E.. STS-70, STS-101.

Dr. Mary Ellen Weber was born in August 1962 in Cleveland, Ohio, USA. She received a Bachelor of Science degree in Chemical Engineering (with honors) from Purdue University (1984); a Ph.D. in Physical Chemistry from the University of California, Berkeley in 1988; and a M.B.A. from Southern Methodist University in 2002. As an undergraduate, Weber was a chemical engineering intern at **Ohio Edison, Delco Electronics**, and at **3M**. In her doctoral research, she explored the physics of chemical reactions involving silicon. At **Texas Instruments** she researched new processes and revolutionary equipment for making computer chips. Dr. Weber holds one Patent, and has published nine papers in scientific journals.

Selected by NASA in 1992 Group 14, she assisted in the processing and launch of Shuttles at KSC, and in payload and robotics development. She was 'Mission Specialist 3' on STS-70 aboard *Discovery* (July 13-22, 1995) which delivered to orbit a critical USD 200 million NASA communications satellite *TDRS-G* to its 22,000-mile orbit above the Equator. Dr. Weber performed pioneering biotechnology experiments, growing colon cancer tissues, later becoming a leading NASA spokesperson in this field. Weber was 'Mission Specialist 1' on STS-101 aboard *Atlantis* (May 19-29, 2000), the third Shuttle mission devoted to ISS construction. This was the first mission to the ISS in over a year and batteries were failing. The crew repaired and installed electrical and life-support components, and boosted the Station to a safe orbit. Weber drove the 60-foot robotic arm to maneuver spacewalk crewmembers along the Station exterior, and directed the logistics and transfer of over 3,000 pounds of equipment. She also developed new crew checklists for engine failures during ascent and new procedures for robotic arm operations. Weber was among the youngest to fly in space, and logged a total of 18 days, 18 hours and 30 minutes.

Weber resigned from NASA in December 2002. For nine years, Dr. Weber was vice president for Government Affairs and Policy for the **University of Texas Southwestern Medical Center** in Dallas, Texas. She is currently with **Stellar Strategies, LLC.**, providing consulting services for operations in high-stakes business ventures, technology communications, and legislative strategy. Dr. Mary Weber is an active skydiver, with nearly 6,000 jumps, and thirteen silver and bronze medals to date at the US National Skydiving Championships. She has received the NASA Exceptional Service Medal.

## Weitz, Paul J.. Skylab 2, STS-6.

Paul Joseph Weitz (1932-2017) was born in July 1932 in Erie, Pennsylvania, USA. He graduated with a degree in Aeronautical Engineering from Pennsylvania State University in 1954; and later earned a Master of Science degree in Aeronautical Engineering from the US Naval Postgraduate School in Monterey, California, in 1964. Weitz served in various naval aircraft squadrons, and became a test pilot before being selected as an astronaut. He logged over 7,700 flying hours; including 6,400 hours in jet aircraft.

Selected by NASA in April 1966 as one of nineteen candidates, Weitz flew as Pilot of the three-man crew on *Skylab 2* (May 25-June 22, 1973), the first crewed *Skylab* mission. With Pete Conrad and Joseph Kirwin, he performed extensive and unprecedented repairs to serious damage sustained during the uncrewed launch of *Skylab*. The three *Skylab* astronaut crews were jointly awarded the 1973 Robert J. Collier Trophy. In 1976, Weitz retired from NASA and went back to the Navy, but returned to NASA at age fifty years to be Commander of STS-6, the maiden flight of Space Shuttle *Challenger* (April 04-09, 1983). The crew tested a variety of support systems and equipment in preparation for future spacewalks, and carried three "Getaway Special" canisters. Weitz logged a total of 33 days, one hour and 13 minutes (793 hours) in space; including one EVA of two hours and 11 minutes.

Weitz was Deputy Director of JSC when he retired from NASA in May 1994. He received thirteen awards, including the Navy Distinguished Service Medal; five Air Medals; the Navy Commendation Medal (for combat flights in Vietnam); the NASA Distinguished Service Medal; the Los Angeles Chamber of Commerce Kitty Hawk Award (1973); the AIAA Haley Astronautics Award for 1974; the 1974 Harmon International Aviation Trophy for Astronaut (1975); and the 1984 Harmon International Award (1989). Weitz was a Fellow, American Astronautical Society; and Master Mason, Lawrence Lodge 708, Erie, Pennsylvania. He died on October 22, 2017, from myelodysplastic syndrome in Flagstaff, Arizona, aged eighty-five years.

## Wetherbee, James D.. STS-32, STS-52, STS-63, STS-86, STS-102, STS-113.

James "Jim" Donald "Wxb" Wetherbee was born in November 1952 in Flushing, New York, USA. He graduated from the University of Notre Dame with a Bachelor of Science degree in Aerospace Engineering in 1974. He was designated a Naval Aviator in 1976, and after training in the A-7E was assigned to Attack Squadron 72 from August 1977 to November 1980 aboard *USS John F. Kennedy*, logging 125 night aircraft carrier landings. After attending the US Naval Test Pilot School in Patuxent River, in 1981 he was assigned to the Systems Engineering Test Directorate. From January 1984 while assigned to Strike Fighter Squadron 132, he flew operationally in the F/A-18 until his selection by NASA. Wetherbee logged over 7,000 flight hours in twenty aircraft types; including 345 aircraft carrier landings.

Selected by NASA in 1984, he is the only American to have commanded five spaceflight missions. Jim Wetherbee was pilot on STS-32 aboard *Columbia* (January 09-20, 1990), which deployed the *Syncom IV-F5* satellite and retrieved the 21,400 pound Long Duration Exposure Facility (LDEF) using the Remote Manipulator System. He was Commander on mission STS-52 aboard *Columbia* (October 22-November 01, 1992), which deployed the *Laser Geodymnamic Satellite* (LAGEOS), and operated the first US Microgravity Payload (USMP). STS-63 aboard *Discovery* (February 02-11, 1995) was the first joint flight of the Russian-American Space Program; and included a rendezvous with *Mir* space station, operation of 'Spacehab', and the deployment and retrieval of the *SPARTAN-204* satellite. STS-86 aboard *Atlantis* (September 25-October 06, 1997) was the seventh mission to dock with *Mir*, transferring 10,400 pounds of science and logistics, and the return of experimental hardware and results to Earth.

Wetherbee served as Deputy Director of Johnson Space Center (JSC) from August 1995 to April 2000; Director of the Flight Crew Operations Directorate (April 2000-2002); and Technical Assistant to the Director of JSC's Safety and Mission Assurance Directorate (April 2003 to June 2004). Wetherbee commanded STS-102 aboard *Discovery* (March 08-21, 2001), the eighth shuttle mission to visit the ISS and delivered the contents of the *Leonardo* Multi-Purpose Logistics Module (MPLM). STS-113 aboard *Endeavour* (November 23-December 07, 2002) delivered the P1 Truss segment, which provides support for the ISS radiators. He logged a total of 66 days, ten hours and 23 minutes in space. James Wetherbee retired from the US Navy in 2003 and left NASA in 2005 to become a consultant. He joined **BP** in 2006, as a safety auditor, and retired in 2014.

Wetherbee has received ten honors, including the DFC; the Navy Achievement Medal; two Meritorious Unit Commendations; two NASA Outstanding Leadership Medals; and four NASA Distinguished Service Medals. He was inducted into the Long Island Air and Space Hall of Fame; and into the US Astronaut Hall of Fame. He is a Lifetime Member of the Society of Experimental Test Pilots; and as a drummer is an Honorary Member, Musician's Union, Local 47, American Federation of Musicians.

## Wheelock, Douglas H.. STS-120, Soyuz TMA-19/Expedition 24/25.

Col. Douglas "Doug" Harry "Wheels" Wheelock was born in May 1960 in Binghamton, New York, USA. He attended the US Military Academy at West Point, New York, graduating with a Bachelor of Science degree in Applied Science and Engineering. Wheelock was commissioned a second lieutenant in the US Army's infantry branch and entered flight school the next year. In September 1984, Lt. Wheelock graduated at the top of his flight class and was designated an Army Aviator. He served in the Pacific, laterly as commander of an air cavalry troop in the 9th Cavalry. He was selected as a member of Class 104 at the US Naval Test Pilot School and, upon completion was assigned as an experimental test pilot with the Army

Aviation Technical Test Center (AATTC). While on active duty, Wheelock earned a Master of Science degree in Aerospace Engineering at Georgia Institute of Technology in 1992. Selected in 1998 NASA Group, Wheelock had two years of intensive training, then was assigned to the Astronaut Office ISS Operations Branch as a Russian Liaison, participating in the testing and integration of Russian hardware and software products developed for the ISS. Wheelock worked extensively with the **Energia Aerospace Company** in Moscow, Russia.

In 2001, he assumed duties as the crew support astronaut for ISS 'Expedition 2' crew; and for 'Expedition 4' crew, in orbit from December 2001 to June 2002. In August 2002, he was assigned as a CapCom at Houston, Texas, being lead CapCom for the 194-day 'Expedition 8'. Wheelock was 'Mission Specialist 1' on STS-120 aboard *Discovery* (October 23-November 07, 2007), which delivered and outfitted the 14,288 kg *Harmony Node 2* module (made by **Thales Alenia Space** in Torino, Italy) to the ISS. He made an EVA for an emergency repair to a solar array on the P6 truss. Launched on June 15, 2010, aboard *Soyuz TMA-19* to the ISS, he joined 'Expedition 24'. This was the 106th crewed flight of a *Soyuz* spacecraft. He made three EVAs to repair a failed pump module. On September 22, 2010, Wheelock took command of the ISS at the beginning of 'Expedition 25', the first US Army officer to do so. He shared pictures of the earth, station and views of space via ***Twitter***. In September 2010, he tweeted photos of Hurricane Earl, and returned to Earth on November 26. He logged a total of 178 days, nine hours and 34 minutes in space.

In 1989, Wheelock was selected by the US Jaycees as one of "Ten Outstanding Young Men of America". In 1990, the Veterans of Foreign Wars selected him as an Outstanding Spokesman for Freedom. Wheelock is a member of the Society of Experimental Test Pilots; the Society of American Military Engineers; the Association of the United States Army; and the Army Aviation Association of America. Doug Wheelock has received twenty-three awards, including the Air Assault Badge; the Defense Superior Service Medal; the Meritorious Service Medal with Oak Leaf Cluster; the Army Achievement Medal with Oak Leaf Cluster; the NASA Space Flight Medal with Oak Leaf Cluster; the NASA Superior Accomplishment Award (2002 & 2004); the Order of St. Michael, Bronze (2007); the National Defense Service Medal with two award stars; the Global War on Terrorism Service Medal; and the Army Overseas Service Ribbon. On July 21, 2011, Doug Wheelock announced that he would be returning to active duty with the US Army in support of Operation Enduring Freedom. He is currently working with NASA to test the *Orion* spacecraft at the Glenn Research Center in Plum Brook, Ohio. Today, he stands as the only Active Duty Army member to command the ISS.

## White, Edward "Ed" H., II.. Gemini 4, Apollo 1.

Lt. Col. Edward "Ed" Higgins White (1927-67) was born in January 1927 in San Antonio, Texas, USA. He graduated from Western High School, Washington; and from West Point in 1952 with a Bachelor of Science degree (ranked 128th out of 523 in the class of 1952). He was then was sent to flight training, received his pilots wings in 1953, and was assigned to the Twenty-Second Fighter Day Squadron at Bitburg Air Base, West Germany. In 1958, he enrolled in the University of Michigan (under Air Force sponsorship) to study aeronautical engineering, receiving his Master of Science degree in 1959. White received test pilot training at Edwards AFB, graduating in July 1959, then was assigned as a test pilot for the Aeronautical Systems Division at Wright-Patterson AFB, Ohio. During his Air Force career, Ed White logged over 3,000 flight hours; including about 2,200 hours in jet aircraft.

Selected in 1962 NASA Group 2 as the so-called Next Nine, White was pilot during the four-day *Gemini-Titan 4* mission of June 03, 1965. The two-man crew attempted to rendezvous with the *Titan II* upper stage booster, but this was unsuccessful and the objective had to be abandoned since the allocated propellant had been used up. On the third revolution, Ed White became the first American to do a spacewalk. During his EVA, White used an oxygenpropelled gun called the Hand Held Maneuvering Unit (HHMU), and found the experience so exhilarating he was reluctant to terminate after the allotted twenty minutes. Ordered back inside the craft, he said, *"I'm coming back in ... and it's the saddest moment of my life"*. A devout Methodist, White carried three pieces of religious jewelry to space: a gold cross, a St. Christopher's Medal and a Star of David. He commented, *"I felt while I couldn't take one for every religion in the country, I could take the items most familiar to me"*.

Ed White was assigned as Senior Pilot for *Apollo 1*, but died in a Block I capsule flash fire on launch pad 34 during a "plugs out" simulated launch routine on January 27, 1967 at Cape Kennedy, Florida. He was buried at West Point Cemetery, New York. White logged four days, one hour and 56 minutes in space, with one EVA. Lt. Col. Edward White was awarded the NASA Distinguished Service Medal for his flight in *Gemini 4*, and was posthumously awarded the Congressional Space Medal of Honor. The *Apollo 1* fire resulted in the capsule manufacturer, **North American Aviation** undergoing a management shakeup, and they incurred heavy financial penalties. NASA was determined to build a safer, more reliable spaceship, and implement improved safety practices and higher levels of quality control. NASA investigators compiled a list of 8,000 potential problems that needed to be resolved, and the A-204 Command Module design underwent 1,300 changes. The hatch was redesigned so that it opened outwards, easily and in three seconds, flammable materials were replaced with fireproof Beta cloth, and the twenty miles of wiring were consolidated and insulated with Teflon. Manager Deke Slayton said, *"Gentlemen, we won't make the same mistake twice"*.

## White, Robert M.. X-15 rocketplane Flight 62.

Maj. Gen. Robert "Bob" Michael White (1924-2010) was born in July 1924 in New York City, USA. He entered active military service as a USAF aviation cadet in November 1942, receiving his pilot wings in February 1944. During World War II, he served with the 355th Fighter Group in Europe. White flew P-51 Mustangs from July 1944 to February 1945, when he was shot down on his fifty-second combat mission and became a prisoner of war. In December 1945 he became a member of the USAF Reserve, while studying Electrical Engineering at New York University; earning his Bachelor of Science degree in 1951; and a Master of Business Administration from George Washington University in 1966. Robert White was recalled to active duty on May 1951 for the Korean War, serving with the Troop Carrier Wing 514 at Mitchel AFB. In February 1952, he was assigned to the 40th Fighter-Interceptor Squadron, based at Johnson Air Base, Japan. He attended the USAF's Experimental Flight Test Pilot school at Edwards AFB, California, and became a test pilot, flying five advanced aircraft models.

White attended the Air Command and Staff College at Maxwell AFB, Alabama, graduating in 1959. He was one of twelve pilots who flew the North American X-15, an experimental rocket-powered spaceplane jointly operated by the USAF and NASA. His first X-15 flight was on April 15, 1960, when the craft was fitted with two interim 71 kN XLR-11 thrust rocket engines. Four months later, he flew to an altitude of 41.5 km (136,000 feet) above Rogers Dry Lake. In February 1961, he set an unofficial air speed record of 3,661 km/hr (2,275 mph) using the XLR-99 engine. White was the first human to fly at Mach 4 speed, and later at Mach 5 over the next eight months. On November 09, 1961, he became the first pilot to fly a winged aircraft at six times the speed of sound, at 6,587 km/h (4,093 mph), Mach 6. President John F. Kennedy conferred the most prestigious award in American aviation, the Robert J. Collier Trophy, jointly to White and three of his fellow X-15 pilots: Joseph A. Walker, Forrest S. Peterson, and Scott Crossfield.

On July 17, 1962, Robert White flew the X-15 on Flight 62 to an altitude of 96 km (314,750 feet), which is above fifty miles, thereby qualifying him as an astronaut, according to the US definition of the boundary of space. He was the first "winged" astronaut; one of a few who have flown into space without a conventional spacecraft. Major Bob White was featured with a cover story in the August 03, 1962, issue of ***Life*** magazine. Commenting on his high altitude X-15 flights, White said, *"My flights to 217,000 feet (66 km) and 314,750 feet (96 km) were very dramatic in revealing the Earth's curvature ... at my highest altitude I could turn my head through a 180-degree arc and wow! - the Earth is really round. At my peak altitude I was roughly over the Arizona/California border in the area of Las Vegas, and this was how I described it: Looking to my left I felt I could spit into the Gulf of California; looking to my right I felt I could toss a dime into San Francisco Bay".* Unquote.

In October 1963, White returned to Germany, where he served as Operations Officer for the 22nd Tactical Fighter Squadron, 36th Tactical Fighter Wing, and from July 1964 to August 1965 as Commander of the wing's Tactical Fighter Squadron 53. Returning to the US in August 1965, he attended the Industrial College of the Armed Forces, Washington, D.C.. After graduation, he was Chief of the Tactical Systems Office at Wright-Patterson AFB, Ohio, from September 1966 to May 1967. White was then assigned as Deputy Commander for Operations of Tactical Fighter Wing 355, based at Takhli Royal Thai AFB, Thailand. He flew seventy combat missions over North Vietnam, and was awarded the Air Force Cross.

Effective February 12, 1975, White was promoted to Major General. In March he became Chief of Staff of the fourth Allied Tactical Air Force. Robert White served with the Air Force from 1942-46 and 1951-81, retiring on February 01, 1981. In 1992, he was inducted into the Aerospace Walk of Honor; and in July 2006 into the National Aviation Hall of Fame in Dayton, Ohio. Robert White received thirty decorations, including the Legion Of Merit; a Presidential Unit Citation; the World War II Victory Medal; the DFC with four clusters; the Outstanding Unit Award with cluster and Combat Valor device; the National Defense Service Medal with one star; the United Nations Korea Medal; the Vietnam Campaign Medal; the Air Medal with sixteen clusters; the Air Force Longevity Service Award, with eight clusters; the Armed Forces Honor Medal; and the NASA Distinguished Service Medal. Robert White died on March 17, 2010, aged eighty-five years.

## Wilcutt, Terrence W.. STS-68, STS-79, STS-89, STS-106.

Terrence Wade Wilcutt was born in October 1949 in Russellville, Kentucky, USA. He graduated from Southern High School in 1967; received a Bachelor of Arts degree in Mathematics from Western Kentucky University in 1974. Wilcutt then taught high school math for two years before entering the US Marine Corps. He was commissioned in 1976, earned his aviator wings in 1978, and became a test pilot. He logged over 6,600 flight hours in over thirty aircraft types. Selected in 1990 NASA Group 13, Wilcutt was pilot for STS-68 aboard *Endeavour* (September 30-October 11, 1994). STS-79 aboard *Atlantis* (September 16-26, 1996) made rendezvous with space station *Mir* 240 miles above the Earth, and the crew transferred over 3.5 tons of supplies. Wilcutt was Commander for STS-89 aboard *Endeavour* (January 22-31, 1998), the eighth *Shuttle- Mir* docking mission, transferring over 9,000 pounds of scientific equipment, logistical hardware and water. Mission STS-106 aboard *Atlantis* (September 08-20, 2000) to the ISS prepared the station for the arrival of the first permanent crew, delivering over 6,600 pounds of supplies and installing batteries and power converters. An EVA was made to connect power, data and communications cables to the *Zvezda* Service Module. Wilcutt logged a total of 42 days, zero hours and five minutes in space.

Terrence Wilcutt served as NASA's Chief of Safety and Mission Assurance until his retirement from NASA in December 2020. He is currently serving as Director of Safety and Mission Assurance at JSC in Houston, Texas. He has received thirteen honors, including the Navy Commendation Medal (1990); the Defense Superior Service Medal (1995); the Defense Meritorious Service Medal (1997); the American Astronautical Society Flight Achievement Award (1997); the DFC (1998); the NASA Exceptional Service Medal (1998); the NASA Outstanding Leadership Medal (2000); and the V.M. Komarov Diploma.

## Williams, Donald E.. STS-51-D, STS-34.

Capt. Donald Edward Williams (1942-2016) was born in February 1942 in Otterbein, Indiana, USA. He graduated from Otterbein High School in 1960, received a Bachelor of Science degree in Mechanical Engineering from Purdue University in 1964, then received his commission through the Naval ROTC program. After flight training at three locations, he received his Naval Aviator wings in May 1966. After A-4 Skyhawk training, he made two Vietnam War deployments aboard *USS Enterprise*. Williams served as a flight instructor for two years and transitioned to the A-7 Corsair II aircraft, then made two more Vietnam deployments aboard *USS Enterprise*, completing 330 combat missions. He became a test pilot.

Selected by NASA in January 1978, Williams became an astronaut in August 1979 and qualified for assignment as a Shuttle pilot. From July 1985 through August 1986, he was the Deputy Chief of the Aircraft Operations Division at JSC, and from September 1986 through December 1988, he served as Chief of the Mission Support Branch in the Astronaut Office. Donald Williams was pilot on STS-51-D aboard *Discovery* (April 12-19, 1985). He was Commander on STS-34 aboard *Atlantis* (October 18-23, 1989), which deployed the *Galileo* spacecraft, and operated the Shuttle Solar Backscatter Ultraviolet Instrument (SSBUV) to map atmospheric ozone. Donald E. Williams logged a total of 287 hours and 35 minutes in space.

On March 01, 1990, Williams retired from the US Navy as Captain, and left NASA. He joined **Space Applications International Corporation** (SAIC), until his retirement in April 2006. Donald Williams was a member of the Society of Experimental Test Pilots, the Association of Space Explorers, and the National Aeronautic Association. He received fourteen special honors, including Legion of Merit, the DFC, the Defense Superior Service Medal, two Navy Commendation Medals with Combat Valor device, the Meritorious Unit Commendation, the National Defense Service Medal, the Armed Forces Expeditionary Medal; the NASA Outstanding Leadership Medal; the Vietnam Service Medal, with four stars; and the Vietnam Campaign Medal. Capt. Donald Williams died on February 23, 2016, aged seventy-four years.

## Williams, Jeffrey N. STS-101, Soyuz TMA-8/Expedition 13, Soyuz TMA-6/ Expedition 21/22/Soyuz TMA-16, Soyuz TMA-19M/Expedition 47/48.

Jeffrey Williams was born in January 1958 in Superior, Wisconsin, USA. He received a degree in Applied Science and Engineering from the US Military Academy in 1980, receiving his commission in the US Army. He served with the Army at JSC from 1987-92 before training as a test pilot. He was selected by NASA as an astronaut candidate in 1996. Williams was 'Mission Specialist 2' and Lead Spacewalker for STS-101 aboard *Atlantis* (May 19-29, 2000), a resupply mission to the ISS, and the first to fly with a glass cockpit (featuring an array of digital flight instrument displays, typically large LCD screens driven by flight management systems, rather than traditional analog dials and gauges). The Shuttle carried a 'Spacehab' double module and an Integrated Cargo Carrier pallet. Williams and James Voss performed an EVA, then boosted the ISS from 370 km to a higher 400 km orbit.

Launched aboard *Soyuz TMA-8* on March 30, 2006, to dock with the ISS *Zarya* nadir port, Williams was Flight Engineer and Commander for 'Expedition 13', returning to Earth on September 28, 2006, after six months and over 2,800 Earth orbits. Launched aboard *Soyuz TMA-6* on September 30, 2009, he was Flight Engineer for 'Expedition 21' and assumed command of 'Expedition 22' in November 2009. He landed with *Soyuz TMA-16* on March 18,

2010, after a 167-day stay aboard the ISS. Williams wrote the book *The Work of His Hands: A View of God's Creation From Space*, which, in his words, reflects the *"vivid lessons about the meticulous goodness of divine providence, God's care for His creation, and His wisdom in ordering the Universe"*.

Jeffrey Williams returned to space on March 02, 2016, aboard *Soyuz TMA-19M* as 'Flight Engineer 5', to join 'Expedition 47/48'. Upon the departure of *Soyuz TMA-19M* (which landed on June 18, 2016), he became Commander of 'Expedition 48', returning to Earth on September 06, 2016. Jeffrey Williams formerly held the American record for most days spent in space (534 days), which was surpassed by Dr. Peggy Whitson in April 2017. He has received eight awards, including the Defense Superior Service Medal with Oak Leaf Cluster; the Legion of Merit with Oak Leaf Cluster; the Army Commendation Medal; the National Defense Service Medal with award star; the Army Service Ribbon; and the NASA Exceptional Service Medal.

## Williams, Sunita L.. STS-116/Expedition 14/15, Soyuz TMA-05M/ Expedition 32/33.

Sunita Lyn Williams (nee Pandya) was born in September 1965 in Euclid, Ohio, USA, the second American astronaut of Indian heritage to go into space. She is a US Navy officer, NASA astronaut, and former record holder for most spacewalks by a woman and most time for a woman outside the spacecraft. Williams is nicknamed "Suni" in the US and "Soncka" in Slovenia. She graduated from Needham High School in 1983; entered the US Naval Academy at Annapolis, Maryland, in 1983; received a Bachelor of Science degree in Physical Science from the US Naval Academy in 1987; was made an Ensign, and reported for aviator training at the Naval Aviation Training Command. In July 1989, she began combat helicopter training, then flew in helicopter support squadrons during the preparations for the Persian Gulf War and the establishment of no-fly zones over Kurdish areas of Iraq, as well as in relief missions during Hurricane Andrew in 1992 in Miami, Florida. In 1993, Suni Williams became a naval test pilot and later became a test pilot instructor, flying more than thirty aircraft types and logging over 2,770 flight hours.

Sunita Williams completed a Master of Science degree in Engineering Management from the Florida Institute of Technology, Melbourne, in 1995, and entered astronaut training in 1998. She traveled to Moscow where she received training in robotics and other ISS operational technologies while working with the Russian Federal Space Agency (Roscosmos) and with crews preparing for expeditions to the ISS. As Flight Engineer for 'Expeditions 14/15', she flew on STS-116 aboard *Discovery*, launched on December 09, 2006. Williams made four EVAs totaling over 29 hours, and set a female astronaut record for over 195 days in space (broken by Samantha Cristoforetti in 2015 with over 199 days). Suni Williams returned to Earth with the crew of STS-117 on June 22, 2007.

Williams launched aboard *Soyuz TMA-05M* on July 15, 2012, as a Flight Engineer on 'Expedition 32', and on September 16 she became Commander of 'Expedition 33'. Williams made three more EVA totaling over 21 hours. She completed a triathlon in orbit using a treadmill, a stationary bicycle, and a weightlifting machine to simulate the swimming portion of the race. Williams returned to Earth on November 11 after nearly 127 days in space. Her two spaceflights logged a total of over 321 days in space, ranking her second, after American astronaut Dr. Peggy Whitson, for the most time spent in space by a woman. Suni Williams trained for the Boeing *CST-100 Starliner* spacecraft, the Crew Flight Test (CFT), initially scheduled to launch to the ISS in 2021. On June 16, 2022, NASA confirmed that CFT would be a two-person flight test, consisting of Barry Wilmore and Williams.

## Wilmore, Barry E.. STS-129, Soyuz TMA-14M/Expedition 41/ Soyuz TMA-13M.

Capt. Barry Eugene "Butch" Wilmore was born in December 1962 in Murfreesboro, Tennessee, USA. He graduated from Mount Juliet High School in Tennessee; received a Bachelor of Science and a Master of Science degree in Electrical Engineering from Tennessee Technological University. He was a member, letterman and team captain of the University football team. Wilmore is a graduate of the US Naval Test Pilot School. He completed four operational deployments; including missions in support of Operation Desert Shield, Desert Storm (flying twenty-one combat missions) and Southern Watch over the skies of Iraq, as well as missions over Bosnia in support of US and NATO interests. As a Navy test pilot, he participated all aspects of the initial development of the T-45 Goshawk jet trainer. Wilmore has logged over 6,200 hours flying time and made 663 aircraft carrier landings; all in tactical jet aircraft.

Selected as a pilot by NASA in July 2000, Wilmore completed two years' training, and then represented the Astronaut Office on all propulsion systems issues including the Space Shuttle Main Engines, solid rocket motor, external tank, and served on the astronaut support team. Wilmore was pilot on STS-129 aboard *Atlantis* (November 16-27, 2009), which focused on staging spare components outside the ISS. This was the first flight of a 6,136 kg ExPRESS Logistics Carrier, with two aboard, having volumes of thirty cubic metres each; along with an 1,800 kg Orbiter docking system (EMU 3010/3008). Wilmore launched on September 25, 2014 aboard *Soyuz TMA-14M* to join 'Expedition 41', and the crew used the ISS's 3-D printer to print a ratchet wrench tool with a design file transmitted from the ground. He returned to Earth aboard *Soyuz TMA-13M* on November 10, 2014. Wilmore made an EVA of six hours and 34 minutes. He logged a total of 178 days, one hour and zero minutes in space.

Barry Wilmore has received numerous honors, including the Navy Meritorious Service Medal; five Air Medals; six Navy Commendation Medals; two Navy Achievement Medals; the Aviation Office Candidate School "Distinguished Naval Graduate" award; the US Atlantic Fleet "Light Attack Wing One - Pilot of the Year" (1991); and US Atlantic Fleet "Strike Fighter Aviator of the Year" (1999). In 2003, Wilmore was inducted into the Tennessee Technological University Sports Hall of Fame. He trained as Commander for the first flight of the Boeing *CST-100 Starliner* spacecraft, the Crew Flight Test (CFT), initially scheduled to launch to the ISS in 2021. On June 16, 2022, NASA confirmed that CFT would be a two-person flight test, consisting of Wilmore and Sunita Williams.

## Wilson, Stephanie D.. STS-121, STS-120, STS-131.

Stephanie Diana Wilson was born in September 1966 in Boston, Massachusetts, USA. She is the second African-American woman to go into space. She graduated from Yaconic High School, Pittsfield, Massachusetts, in 1984; received a Bachelor of Science degree in Engineering from Harvard University in 1988; and a Master of Science degree in Aerospace Engineering from the University of Texas in 1992. Wilson has returned to Harvard as a member of the Harvard Board of Overseers. From 1988-90, Wilson worked for the former **Martin Marietta Astronautics Group** in Denver as a Load and Dynamics engineer for the *Titan IV* rocket. After her graduate work, she began working for the JPL in Pasadena, California, in 1992.

Selected by NASA in April 1996, Wilson completed two years' training and qualified as a Mission Specialist. Wilson flew as 'Mission Specialist 3' on STS-121 aboard *Discovery* (July 04-17, 2006), a return-to-flight and ISS assembly flight. Wilson supported robotic arm operations for vehicle inspection, Multi-Purpose Logistics Module installation and EVAs, and was responsible for the transfer of over 28,000 pounds of supplies and equipment to the ISS. Wilson was 'Mission Specialist 2' on STS-120 aboard *Discovery* (October 23 to November 07, 2007), which delivered the 14,288 kg *Harmony* (Node 2) connecting module to the ISS. The crew conducted four EVAs and performed a previously untested repair method in the station's solar array. STS-131 aboard *Discovery* (April 05-20, 2010) was a resupply mission to the ISS; delivering over 27,000 pounds of hardware, supplies, experiments and equipment, including a tank of ammonia coolant that required three EVA and robotics to install. Wilson was responsible for robotics for spacewalking support, and for robotic removal of the *Leonardo* MPLM from the payload bay. Stephanie Wilson has logged a total of 42 days, 23 hours and 46 minutes in space.

On October 18, 2019, Wilson was ground controller at Houston for the first all-woman EVA by Christina Koch and Jessica Meir. On December 09, 2020, NASA announced that Stephanie Wilson was among the eighteen candidates for the *Artemis* program that is working to send the first woman and next man to the Moon by 2025, and if selected, she could be both the first woman and the first African-American on the Moon. Wilson has been awarded the Harvard College Women's Professional Achievement Award (2008); the NASA Distinguished Service Medal (2009 & 2011); an honorary Doctorate of Science from Williams College (2011); the Johnson Space Center Innovation Group Achievement Award (2013); the Johnson Space Center Director Commendation Award (2013); and the Salem Trailblazer Award (2019).

## Wisoff, Peter J.K.. STS-57, STS-68, STS-81, STS-92.

Dr. Peter Jeffrey "Jeff" Kelsay Wisoff was born in August 1958 in Norfolk, Virginia, USA. He graduated from Norfolk Academy in 1976; received a Bachelor of Science degree in Physics (with Highest Distinction) from the University of Virginia in 1980; a Master of Science degree (1982); and a Doctorate in Applied Physics from Stanford University in 1986. Wisoff's graduate work at Stanford (as a National Science Foundation Graduate Fellow) focused on development of short wavelength lasers. He later joined the faculty of the Electrical and Computer Engineering Department at Rice University, Houston, Texas. His research focused on the development of new vacuum ultraviolet and high intensity laser sources. Wisoff has collaborated with researchers at Rice University on new techniques for growing and evaluating semiconductor materials using lasers.

Selected by NASA in January 1990, he qualified as a Mission Specialist, and was lead for the Payloads and Habitability Branch of the Astronaut Office. Mission STS-57 was aboard *Endeavour* (June 21-July 01, 1993), when the first flight of 'Spacehab' carried twenty-two individual flight experiments in materials and life sciences research. STS-68 aboard *Endeavour* (September 30-October 11, 1994) was the Space Radar Lab-2 mission. STS-81 aboard *Atlantis* (January 12-22, 1997) was the fifth mission to dock with *Mir*, and carried the 'Spacehab' double module. STS-92 aboard *Discovery* (October 11-24, 2000) attached the Z1 truss and Pressurized Mating Adaptor 3 to the ISS, and required four EVAs to achieve that. Dr. Wisoff made two EVAs totaling 13 hours and 16 minutes. He logged a total of 42 days, 56 hours and two minutes in space; including three EVAs totaling 19 hours and 53 minutes.

Dr. Jeff Wisoff retired from NASA in September 2001 to join the Lawrence Livermore National Laboratory, where he serves as Deputy Program Manager for Systems Engineering at the National Ignition Facility. On October 01, 2013, he was named the interim Principal Associate Director for the NIF and Photon Sciences Directorate at **Lawrence Livermore National Laboratory**. He was married to fellow astronaut Tamara E. Jernigan. Dr. Wisoff was awarded the Physics Prize and Shannon Award from the University of Virginia (1980); the NCR Faculty Award of Excellence (1989); and NASA Space Flight Medals (1993, '94 & '97).

## Wolf, David A.. STS-58, STS-86/Mir EO-24, STS-112, STS-127.

Dr. David Alexander Wolf was born in August 1956 in Indianapolis, Indiana, USA. He graduated from North Central High School; received a degree in Electrical Engineering from Purdue University, where he graduated with distinction; earned a medical degree from Indiana University School of Medicine in 1982, and subsequently trained as a Flight Surgeon with the USAF. Wolf joined the staff of JSC in 1983 and investigated the physiological effects of microgravity. Selected in 1990 NASA Group, he completed eighteen months training and began his career in the Medical Sciences Division at JSC in Houston. He was responsible for engineering development and spacecraft avionics integration of the American Flight Echocardiograph for investigating cardiovascular physiology in microgravity. At Kennedy Space Center (KSC) in Florida, he was involved in Orbiter vehicle processing and testing.

Dr. Wolf is an expert in extravehicular activity, spacesuit design, and rendezvous navigation. He was assigned as Chief Engineer for design of the Space Station medical facility. During his training for a *Mir* mission, he lived in Star City, Russia, and became fluent in the Russian language. Wolf was 'Mission Specialist 3' on STS-58 aboard *Columbia*, (October 18-November 01, 1993), a Spacelab Life Sciences-2 mission to study DNA, regulatory physiology, cardiovascular, neurovestibular and musculoskeletal systems. There were experiments aimed at understanding bone tissue loss, space motion sickness and the effects of microgravity on sensory perception. On October 23, 1993, the payload crew devoted time to metabolic studies of the forty-eight rodents on board the 'Spacelab' science workshop. Blood plasma volume was measured, to see how weightlessness may be affecting the red blood cell count of the caged rats. Wolf was a CapCom during the historic Shuttle *Atlantis* docking with *Mir* in 1995.

He flew on STS-86 aboard *Atlantis* (September 1997), docking with *Mir* for mission 'EO-24' on September 27, when he transfered to the Russian space station for 128 days. There were numerous systems failures, including multiple failed spacecraft systems (atmospheric life support, three total power systems outages, loss of attitude control, primary computer system failure, airlock hatch failure, and humidity separation system loss). While aboard *Mir*, David Wolf became the first American to vote from space, casting a ballot in a 1997 local election. *Endeavour* launched on January 22, 1998, for STS-89 docked and with *Mir* on January 24. He touched down aboard *Endeavour* on January 31. Mission STS-112 aboard *Atlantis* (October 07-18, 2002) delivered the S1 truss segment to the ISS. Wolf made three EVAs for a total of 19 hours and one minute. Mission STS-127 aboard *Endeavour* (July 15-31, 2009) delivered the Japanese Experiment Logistics Module (ELM-ES). Wolf made another three EVAs totaling 18 hours and 24 minutes. He logged a total of 168 days, eight hours and 57 minutes in space; including seven EVAs totaling 41 hours and 17 minutes.

Dr. David Wolf has received many awards, including the NASA Exceptional Engineering Achievement Medal (1990); and NASA Inventor of the Year (1992). He has received fifteen US Patents and over twenty Space Act Awards, primarily for 3-dimensional tissue engineering technologies, earning the Texas State Bar Patent of the Year in 1994. He has published more than forty technical papers. Wolf is a member of the Institute of Electrical and Electronics Engineers; the Aerospace Medical Association; the Experimental Aircraft Association; the International Aerobatic Club; and Indiana National Guard (retired).

## Worden, Alfred M.. Apollo 15.

Col. Alfred "Al" Merrill Worden (1932-2020) was born in February 1932 in Jackson, Michigan, USA. He graduated from Jackson High School, where he became the Student Council President. Worden attended the University of Michigan for one year, before securing an appointment to the US Military Academy at West Point, New York, in July 1951. He later stated, *"There was no way I was going to live the rest of my life on a farm. That kind of got me started down the path that led to NASA".*

Graduating in 1955 from West Point with a Bachelor of Science degree in Military Science (finishing 47th out of 470 in his class), he elected to be commissioned in the USAF, although he had no piloting experience. Worden received primary flight training at Moore AFB, Texas, trained on Lockheed jet fighters at Laredo AFB, Texas; and went on to Tyndall AFB for Air Defense Command training. While serving as a pilot at Andrews AFB from March 1957 to May 1961, he was his squadron's armament officer. He became a test pilot before his selection in the NASA April 1966 Group 5, then served as a member of *Apollo 9's* support crew, and was backup Command Module Pilot for *Apollo 12*. On March 26, 1970, Worden, Scott and Irwin were publicly named as the crew for *Apollo 15*; originally scheduled to be an H-type mission with a limited stay of 33 hours on the Moon. The cancellation of two Apollo flights in mid-1970 meant the flight would be a J-type mission, with three Moonwalks during a three-day stay, and with the first Lunar Roving Vehicle. For the first time, observations from lunar orbit were made a formal mission objective.

Launched on July 26, 1971, the mission arrived in lunar orbit and Dave Scott and James Irwin entered the LM *Falcon*, but the two craft failed to separate. Worden went into the docking tunnel and reconnected a loose umbilical cable, fixing the problem. As CM Pilot, he orbited the Moon seventy-four times in the Command Module *Endeavour*, taking photographs of the Moon and operating a suite of scientific instruments in the SIM bay that probed the Moon. Seeing many more stars, and more intensely, than Earthbound observers, he concluded *"it was naive to believe Earth had the only life in the Universe"*, and he wondered *"if space exploration was part of humanity's survival instinct to avoid being trapped in a single solar system"*. The mission left the Moon on August 02 with seventy-seven kilograms of rock and soil samples, splashed down in the Pacific Ocean, and the crew were recovered by the amphibious assault ship *USS Okinawa*. Al Worden logged 12 days, seven hours and 12 minutes (over 295 hours) in space.

During the return to Earth, Worden was the first human to do a deep-space EVA to retrieve film cassettes from the exterior of the spacecraft. He has been listed in the ***Guiness Book of World Records*** as the "most isolated human being" during his time alone in *Endeavour*. While orbiting the Moon, he was at his greatest distance from Scott and Irwin in *Falcon*; 3,597 km (2,235 miles) away from any other human beings. After their return, the crew were involved in a controversy over 100 postal covers they had taken to the Moon (in exchange for about $7,000 to each astronaut). In the end, they returned the money. Astronauts were forbidden by standards of conduct issued in 1967 from using their position for financial gain for themselves or other people. They were reprimanded by NASA for poor judgement, and did not fly in space again.

Worden remained at NASA until 1975 at the **Ames Research Center** as a Senior Aerospace Scientist, then entered the private sector. He founded **Alfred M. Worden, Inc.**, then served as the Director of Energy Management Programs at the **Northwood Institute** in Midland, Michigan. In 1982, he ran in the Republican primary in Florida's 12th Congressional District. He held executive positions with **Jet Electronics and Technology, Inc.**, and with **B.F. Goodrich** prior to his retirement in 1996. In 2011, his autobiography, *Falling to Earth: An Apollo 15 Astronaut's Journey to the Moon* made the top twelve books of the ***Los Angeles Times*** Bestseller list. He wrote four other books, including *Hello Earth; Greetings from Endeavour* (1974), and *The Light of Earth: Reflections on a Life in Space* with Francis French (2021).

Al Worden had a long involvement with the Astronaut Scholarship Foundation. He made many public appearances, promoting a renewed space program and science education. He was inducted into the US Astronaut Hall of Fame (1997); and the International Air & Space Hall of Fame (2016). He received the NASA Distinguished Service Medal (1971); and the NASA Ambassador of Exploration Award (2009). Alfred Worden died on March 18, 2020, at Sugar Land Assisted Living Center, Texas, aged eighty-eight years.

## Young, John W.. Gemini 3, Gemini X, Apollo 10, Apollo 16, STS-1, STS-9.

John Watts Young (1930-2018) was born in September 1930 in San Francisco, California, USA. He is the only astronaut to fly on four different classes of spacecraft; and the only astronaut to walk on the Moon and fly on the Space Shuttle. Young attended the Georgia Institute of Technology on a Naval ROTC scholarship. He completed a midshipman cruise aboard *USS Missouri*, and another aboard *USS Newport News*. In 1952, Young graduated second in his class with a Bachelor of Science degree in Aeronautical Engineering and was commissioned as an Ensign in the US Navy on June 06, 1952. He applied to become a naval aviator, but was selected to become a gunnery officer aboard *USS Laws*, and completed a Pacific deployment as a fire control and division offer in the Sea of Japan during the Korean War.

In May 1953, Young received orders to flight school at NAS Pensacola, Florida. He graduated and received his aviator wings in December 1954. He flew during the Suez Crisis, but not in combat. In September 1958, he deployed with the sixth Fleet on *USS Forrestal* to the Mediterranean Sea. In January 1959, Young was selected to be in Class 23 at the US Naval Test Pilot School. He graduated second in his class in 1959. In 1962, he was assigned to Fighter Squadron 143 until his selection as an astronaut in September 1962. John Young was pilot for *Gemini 3* (March 23, 1965), a three-orbit mission with Commander Gus Grissom to test the ability of the spacecraft to perform orbital maneuvers. Biological experiments were assigned to test the effects of radiation on human blood and microgravity on cell division, and an experiment to test re-entry communications was created. They splashed down eighty-four kilometres (52 miles) short of the target area, and after 30 minutes were recovered aboard *USS Intrepid*. After the flight it was discovered that Young had smuggled a corned beef sandwich aboard, which he and Grissom shared while testing food!

Young was Commander for *Gemini X* (July 18-21, 1966) with the primary mission to dock with an *Agena* target vehicle (ATV) and use its engines to maneuver (a failed objective of the *Gemini VIII* and *Gemini IX* missions). Once in orbit, the crew attempted to navigate to their first rendezvous using celestial navigation, but were unable and required inputs from Mission Control. The higher-than-expected fuel consumption during the midcourse corrections resulted in additional planned practice dockings to be canceled. Using the *Agena*'s engines, *Gemini X* set another new altitude record of 764 km (475 miles). The capsule then undocked from its *Agena* and performed two maneuvers to rendezvous with the *Gemini VIII* Agena, which was successful. Young held the capsule three metres from the *Agena* while Collins conducted an EVA to retrieve a micrometeorite experiment package. Young conducted the retrofire burn and manually flew the re-entry to splash down in the western Atlantic Ocean 5.4 km from their recovery ship, *USS Guadalcanal*.

John Young was Command Module Pilot on *Apollo 10* (May 18-26, 1969), the only F-type mission, which entailed crewed entry into lunar orbit and testing of the Lunar Module, but without a landing. The crew selected the call sign *Charlie Brown* for the Command Module and *Snoopy* for the LM. Young took celestial navigation measurements en-route to the Moon as a contingency for loss of communication. Stafford and Cernan flew the LM down to 14.447 km (8.977 miles) above the Moon, while Young in the Command Module prepared to maneuver to the LM in the event that its ascent engine did not work. Once the LM rendezvoused with the CM, Young successfully docked the spacecraft. Young tracked landmarks in preparation for a lunar landing, then flew the trans-Earth injection (TEI) maneuver. The capsule splashed down 690 km from Samoa, and six km (3.7 miles) from the recovery ship, *USS Princeton*.

On March 13, 1971, Young was assigned as Commander of *Apollo 16*, along with Duke and Mattingly. The mission's science objective was to study material from the lunar Descartes Highlands, as they were believed to contain older volcanic material. To prepare for their

EVAs, Young and Duke participated in field exercises in geological research. Their mission was from April 16-27, 1972. On April 21, Young was the first to exit the LM, and his first words on the surface were, *"I'm glad they got ol' Brer Rabbit here, back in the briar patch where he belongs"*. He became the ninth person to walk on the Moon, with an EVA of seven hours. The astronauts deployed the lunar rover and the Apollo Lunar Surface Experiments Package (ALSEP). They conducted a seismic experiment using pneumatic hammers, and set up a geology station at Flag crater, 1.4 km west of the landing site. Their second seven-hour EVA on April 22 experienced a failure of the rover's navigation system. For their third EVA of five hours, they drove to North Ray crater and collected rock samples from the rim, and inside the crater. With ninety-four kilograms (207 pounds) of samples they docked with the CM on April 24, returned to Earth and splashed down about 350 km (220 miles) southeast of Christmas Island on April 27, 1972. The crew were recovered aboard *USS Ticonderoga*.

In January 1973, John Young was made Chief of the Space Shuttle Branch of the Astronaut Office. From 1974-87, he was **Chief of the Astronaut Office.** In March 1978, he was selected to be Commander of STS-1, with Robert Crippen as Pilot. The development of *Columbia* was delayed because of the longer than predicted installation time of the Space Shuttle thermal protection system. Young commanded STS-1 aboard *Columbia* on its maiden flight (April 12-14, 1981). Vice President George H.W. Bush called the crew to congratulate them. The crew inspected their thermal tiles and determined that some had been lost during launch, so a *KH-11 KENNEN* satellite was used to image the orbiter, and it was determined the craft could safely re-enter the atmosphere. Young and Crippen tested the orbital maneuvering capabilities of the Shuttle, as well as its mechanical and computer systems.

Young commanded STS-9 aboard *Columbia* (November 28-December 08, 1983), which carried the first 'Spacelab' module into orbit. Young tested a new portable onboard computer, and attempted to photograph Russian airfields. Prior to re-entry, two of the four primary General Purpose Computers failed, causing a delay in landing at Edwards AFB. After STS-9, Young was critical of NASA management following the *Challenger* disaster, and blamed it on the lack of safety culture within the Space Shuttle program. He testified before the **Rogers Commission**, and suggested improvements for the safety program. In May 1987, Young was reassigned as Special Assistant to Johnson Space Center's Director Aaron Cohen for Engineering, Operations and Safety. He oversaw the redesign of the solid rocket boosters to prevent a repeat of the *Challenger* disaster. In February 1996, he was assigned as the Associate Director (Technical) of Johnson Space Center, where he was involved in the development of the *Shuttle-Mir* program and the design process for the ISS.

John Young retired from NASA on December 31, 2004, after forty-two years' service. During his career, Young flew more than 15,275 hours; including more than 9,200 hours in T-38 jets and 835 hours in spacecraft during six spaceflights. Additionally, he spent over 15,000 hours in training to prepare for eleven primary and backup crew positions. Young then worked as a public speaker; advocating for the importance of asteroid impact avoidance, colonization of the Moon, and climate engineering. In November 2011, Young and Crippen met with the crew of STS-135, the final Space Shuttle mission. In 2012, John Young and James R. Hansen co-authored his autobiography, *Forever Young*.

In 1981, NASA and the developers of the Space Shuttle won the Collier Trophy. Young was inducted into the International Space Hall of Fame (1982); the National Aviation Hall of Fame (1988); the International Air & Space Hall of Fame (1995); and the Georgia Aviation Hall of Fame (2001). He was awarded the American Astronautical Society Space Flight Award (1993); the Exceptional Engineering Achievement Award (1985); the Golden Plate Award of the American Academy of Achievement (1993); and the Philip J. Klass Award for Lifetime Achievement (1998). John Young was a fellow of the American Institute of Aeronautics and Astronautics; the American Astronautical Society; and the Society of Experimental

Test Pilots. Named in his honor are: John Young Parkway, Florida State Road 423; John Young Elementary School; and the Orlando Science Center planetarium. John Young died of complications from pneumonia, on January 05, 2018, at his home in Houston, Texas, aged eighty-seven years. He was buried at Arlington National Cemetery. In 2018, Northrop Grumman announced that Cygnus NG-10, their 10th cargo resupply mission to the ISS, would be named *S.S. John Young*.

## Zamka, George D.. STS-120, STS-130/Expedition 22.

Col. George David "Zambo" Zamka was born in 1962 in Jersey City, New Jersey, USA, of Colombian and Polish ancestry. He received a Bachelor of Science degree in Mathematics from the US Naval Academy in 1984, and was commissioned a second lieutenant in the US Marine Corps. In 1990, he trained as an F/A-18D Hornet pilot and was then assigned to VMFA (AW)-121. Overseas deployments were to Japan, Korea, Singapore and Southwest Asia. Zamka flew 66 combat missions over occupied Kuwait and Iraq during Operation Desert Storm. In December 1994, he graduated from the USAF Test Pilot School, then served as an F/A-18 Hornet test pilot and operations officer. In 1997, Zamka earned a Master of Science degree in Engineering Management from the Florida Institute of Technology. In 1998, he returned to VMFA (AW)-121 and deployed to MCAS Iwakuni, Japan. As a test pilot and Colonel in US Marine Corps, he logged over 3,500 flight hours in over thirty aircraft types.

Selected in June 1998 NASA Group, he served as lead for the Space Shuttle training and procedures division and as supervisor for the Astronaut Class of 2004. Zamka was pilot on STS-120 aboard *Discovery* (October 23-November 07, 2007), which delivered the *Harmony* module to the ISS, and reconfigured a portion of the space station in preparation for future assembly missions. He was Commander on STS-130 aboard *Endeavour* (February 08-21, 2010) when the primary payloads were the 13,004 kg *Tranquility* module, the 410 kg Canadarm 201 and the 1,805 kg *Cupola*, a robotic control station providing a 360-degree view around the station. On Flight Day 11, the joint 'Expedition 22/STS-130' crew received a conference phone call from US President Barack Obama and several schoolchildren.

Col. George Zamka logged a total of 28 days, 20 hours and 32 minutes in space. In March 2008, he visited Colombia's Planetarium of Bogota with the crew of STS-120, to share their experience as NASA astronauts. In August 2010, Zamka retired from the Marine Corps after almost thirty years of distinguished service. As of May 2018, he is chairman of the board of the **National Association of Spaceports**. Zamka has received 14 awards and honors, including the Legion of Merit; the DFC; the Defense Meritorious Service Medal; six Navy Strike Air Medals; the Navy Commendation Medal with Combat Valor device; Distinguished Graduate, US Naval Academy; the NASA Outstanding Leadership Medal; four NASA Superior Accomplishment Awards; the GEM Award; and the Officer's Cross of the Order of Merit of the Republic of Poland.

PART FOUR

# Cosmonauts of the Soviet Union and Russia

## Preamble

The space program of the Soviet Union was extremely secretive. Successful missions were not announced until after the event, and failures were not mentioned at all. However, Western observers were able to determine when the USSR had made a rocket launch, and ground stations were able to independently track its spacecraft orbits. Soviet media releases were calculated to inspire the population with sanitised accounts of the brave exploits of cosmonauts, while any misdemeanours such as drunkeness or speaking publicly about forthcoming missions were punished by banishment from the space program. Nikolai Kamanin, director of cosmonaut training, had read in the American media that female pilots were training to be astronauts. In early 1962, he wrote in his diary, *"We cannot allow that the first woman in space will be American. This would be an insult to the patriotic feelings of Soviet women"*. Approval was then given to place five female cosmonauts into the next group, selected on February 16, 1962, to begin training for space flights.

The flight of Valentina Tereshkova for almost three days from June 16, 1963, was a deliberate effort to steal a march on the Americans by being first to send a woman into space. While in Earth orbit, she spoke directly with Premier Nikita Krushchev, reporting that *"All systems are working perfectly"* and that she felt *"excellent"*. In reality, she was experiencing space sickness i.e. nausea and physical discomfort for much of her seventy-one-hour mission. The Soviets worked hard with programs that were often underfunded to achieve many significant milestones in space exploration.

On April 12, 1967, Vladimir Komarov rode *Soyuz 1* into orbit. Over 203 design flaws were known to be unresolved, but the Soviet authorities wanted to stage a successful mission to boost national celebrations marking the fiftieth anniversary of 1917 Russian Revolution. The spacecraft was rushed to launch before it was safe to operate. On orbital insertion, the solar panels failed to fully deploy, preventing the craft from being fully powered. Komarov reported, *"The high frequency communications are not working. I cannot orient the spacecraft to the Sun. I tried orienting the spacecraft manually ... "*. He tried for over five hours to regain control, but plagued by technical issues, his craft began tumbling about its axis, an enlarged main parachute (jammed in a too-small container) failed to open, and Komarov's emergency re-entry ended in disaster with a high-speed fiery crash at 140 km (about 87 miles/hr). Komarov was the first space fatality.

The June 30, 1971, tragedy which cost the lives of three cosmonauts was caused by the Soviet regime's fixation to put the second three-man crew into orbit. Georgy Dobrovolsky, Vladislav Volkov and Viktor Patsayev had spent twenty-two days aboard the *Salyut 1* orbital space station as the inaugural crew. On their return flight, and at an altitude of about 104 miles, an air pressure equalization valve opened prematurely, causing their capsule to lose its' atmosphere. Within two minutes, the crew had died of asphyxiation, being exposed to the vacuum of space. Their spacecraft had been modified to fit a third cosmonaut aboard, at the expense of having room for pressure suits and a life support system. Following this event, the Soyuz vehicle underwent a major redesign process.

**From an article by Richard Hollingham, April 08, 2021:** While the faces of NASA's "Mercury Seven" astronauts were splashed across the world's media, Russia's cosmonauts trained in secret, hidden from public view. On April 13, 1961, Soviet newspaper ***Izvestia***'s special correspondent Georgi Ostroumov met the first man to fly in space. A day after returning to Earth, *"space pilot"* Yuri Gagarin is, reported Ostroumov, *"in high spirits, hale and hearty ... a wonderful smile illumines his face". "Every now and then dimples appear on his cheeks"*, Oustromov wrote. *"He appreciates the curiosity with which he is pressed for details of what he saw and experienced during the one and a half hours he spent outside the Earth"*. In a booklet published to commemorate the flight, ***Soviet Man in Space,*** the interview with Gagarin continued for several pages. The cosmonaut described the experience: *"The horizon presents a very unique and unusually beautiful sight"*. He praised the Soviet Union: *"I dedicate my flight to ... all our people who are marching in the forefront of humanity and building a new society"*.

In a political system where journalism tended towards propaganda, rather than a realistic portrayal of events, it is easy to argue that Gagarin's quotes were made-up. But, although they might have been refined by censors, there is a good chance they were the cosmonaut's actual words. A jet fighter pilot who had grown up in a small Russian village, Gagarin was a well-liked family man. He was indeed good looking, personable, and, crucially, a loyal card-carrying member of the Communist Party. Although the drama of NASA's early human space program played out in public, only recently has the full story of how the Soviet Union selected and trained its cosmonauts emerged. The Communist empire was keen to encourage the view that selection was open to all, and that these first men in space – and the first woman, Valentina Tereshkova – were volunteers. But that is not strictly true.

After qualifying as a fighter pilot, Gagarin was stationed at a remote airfield on the Russian border with Norway, flying MiG-15 jet fighters on the western frontier of the Cold War. In late Summer of 1959, two medical doctors arrived at the base to interview a pre-selected group of aviators. After starting with about 3,500 potential candidates, they had already narrowed their search to some 300 pilots across the west of Russia. *"The guys being interviewed really had no clue as to why they were being interviewed"*, says Stephen Walker (the author of ***Beyond***), who has spent years scouring Russian archives to piece together the full story of Gagarin's mission. The doctor's interviews consisted of a seemingly casual chat about career, aspirations and family. Some interviewees were invited back for a second conversation. Although the doctors hinted that they were looking for candidates to fly a new type of machine, at no point did they reveal their true purpose. Walker said, *"What they were looking for was basically a human version of a dog – somebody who could sit there and endure the mission.* ***They were looking for military pilots, people who had already signed up for the possibilty of killing themselves for their country****, which is really what we're dealing with here, because the chances of coming back alive were not necessarily that great"*. Unquote.

While NASA recruited military test pilots as its first astronauts to fly its more complex *Mercury* spacecraft, the Soviet capsule, named *Vostok*, was designed to be controlled remotely from Earth. Except in an emergency, the Soviet pilots would not get to do much flying. *"They were not looking for people that had much experience"*, said Walker. *"They wanted someone who could ... deal with the acceleration forces and get back alive"*. Much like the "space dogs" that Soviet rocket scientists had been launching for more than a decade, their cosmonauts would have to be fit, obedient, and small enough to fit into the cramped capsule. In the end, one hundred and thirty-four selected individuals – all young pilots, all under 168 centimetres tall – were given the opportunity to "volunteer" for the new, top-secret assignment. Some were told this would involve training to fly a spacecraft; others believed it was a new model of helicopter. None of them were allowed to discuss the intriguing offer with their colleagues, or to consult with their families.

Meanwhile, in April 1959, the United States publicly announced the names of its first seven *Mercury*-series astronauts. Their candidates had been put through a series of gruelling physical, medical and psychological tests; as detailed in Tom Wolfe's book ***The Right Stuff***.

When asked at a Press conference which of the tests they liked the least, US candidate John Glenn replied: *"It's difficult to pick one, because if you can figure out how many openings there are on the human body, and how far you can go into any one of them ... you answer which one would be the toughest for you"*. Unquote. With many questions unanswered about how humans would cope with the rigours of spaceflight – the accelerations, weightlessness and isolation – there was every reason to pick the most physically and psychologically capable. Vladimir Yazdovsky, a professor at Moscow's Institute of Aviation and Space Medicine, was put in charge of testing the Soviet space candidates. He had overseen the "space dog" program, and was described by colleagues (in private!) as a harsh and arrogant man. Walker said, *"He's a kind of terrifying James Bond horror figure, and he was ... brutal with those guys"*.

In almost every case, the Soviet tests were longer, tougher and more rigorous than those endured by the US astronaut candidates. Over a month, the cosmonaut volunteers were injected, probed and prodded. They were put in rooms with the temperature raised to 70 degrees Celsius, into chambers where they were progressively starved of oxygen, and on vibrating seats to simulate a rocket launch. Some of the men collapsed, while others just walked away from their ordeal. All through the Soviet process, the men were forbidden to tell their families or friends what they were doing. Even during that punishing month, there were still candidates who did not know what they were being tested for. Twenty of the young men made it through to training at a new Cosmonaut Center. Initially, it was just a few military huts in a forest near Moscow, but later it became Star City, Russia. There was no Press conference or public announcement – officially, the Soviet's human spaceflight program did not exist! *"If they left the base, they were told not to tell anyone what they were doing, or why they're there. If anyone asked, they were to say they're part of a sports team"*, said Walker. *"**Everything was controlled**, everything was secret. Everything was behind closed doors"*. Unquote.

The Soviet space training program was similar to the American system, but with less emphasis on controlling the spacecraft. Just like the "space dogs" that went before them, the men were spun at dizzying accelerations on large centrifuges, sealed in sound-proofed isolation chambers for multiple days (to see how they coped with no reference to the passing of time), and were subjected to almost constant physical and psychological evaluation. One significant difference from the American program was the amount of parachute training the Russians received. That was because they would need to eject at an altitude of around seven kilometres from their spacecraft as it plummeted towards the ground, to avoid being seriously injured by the impact on landing in the remote steppes of Kazakhstan.

The fact that the Soviet capsule and its pilot landed separately was another secret carefully concealed until years later. With several Russians failing their evaluations, an initial six candidates (the "Soichi Six") were selected for the first Soviet flights. When NASA publicly declared that it hoped to launch its first astronaut in the Spring of 1961, Sergei Korolev (1907-66, the head of the Soviet space program) knew he had but a narrow window of opportunity to stage a propaganda coup by getting in first.

On April 05, 1961, the cosmonauts arrived at what is now called the Baikonur Cosmodrome; in the remote Kazakh desert steppe, where Korolov's huge *R-7* rocket was being prepared. To conceal its' true location, Baikonur was named for a town many miles away. No-one knew who would be chosen to be first in space. Only a few days before the scheduled launch, Yuri Gagarin was given the task. Not until Gagarin was in Earth orbit did any but those closest to the space program know his name. Assured of success, an official broadcast via ***Izvestia***'s special correspondent, Ostroumov, on the morning of April 12, 1961, reported Gagarin gave *"a last wave to the friends and comrades down below [the rocket], then he stepped inside the spaceship. A few seconds [sic] later the command was given ... the gigantic ship rose up out of a fiery cloud towards the stars"*. Yuri Gagarin would return to Earth within one hour and forty-eight minutes, the poster child for the Soviet Union – the space pilot with the Russian ***Right Stuff***. He was awarded the title, Hero of the Soviet Union!

# Soviet cosmonaut selections

**1960 March 07** - Air Force Group 1 (USSR)

Ivan Anikeyev, Pavel Belyayev, Valentin Bondarenko, Velery Bykovsky, Valentin Filatyev, Yuri Gagarin, Viktor Gorbatko, Anatoli Kartashov, Yevgeny Krunov, Vladimir Komarov, Alexei Leonov, Grigori Nelyubov, Pavel Popovich, Mars Rafikov, Georgi Shonin, Gherman Titov, Valentin Varlamov, Boris Volynov, and Dmitri Zaikin.

**1962 March 12** - Female Group (USSR)

Tatyana Kuznetsova, Valentina Ponomaryova, Irina Solovyova, Valentina Tereshkova, and Zhanna Yorkina.

**1963 January 10** - Air Force Group 2

Yuri Artyukhin, Eduard Buinovski, Lev Dyomin, Georgy Dobrovolsky, Anatoly Filipchenko, Aleksei Gubarev, Vladislav Gulyayev, Pyotr Kolodin, Eduard Kugno, Anatoli Kuklin, Aleksandr Matinchenko, Vladimir Shatalov, Lev Vorobyov, Anatoly Voronv, and Vitaly Zholobov.

**1964 January 25** - Air Force Group 2 Supplemental

Georgi Beregovoi

**1964 May 26** - Voskhod Group - Medical Group 1

Vladimir Benderov, Georgy Katys, Vasili Lazarev, Boris Polyakov, Aleksei Sorokin, and Boris Yegorov.

**1964 June 11** - Civilian Specialist Group 1

Konstantin Feoktistov

**1965 June 01** - Journalist Group 1

Yaroslav Golovanov, Yuri Letunov, and Mikhail Rebrov.

**1965 June 01** - Medical Group 2

Yevgeni Illyin, Aleksandr Kiselyov, and Yuri Senkevich (all subsequently canceled).

**1965 October 28** - Air Force Group 3

Boris Belousov, Vladimir Degtyarov, Anatoli Fyodorov, Yuri Glazkov, Vitali Grishchenko, Veygeni Khludeyev, Leonid Kizim, Pyotr Klimuk, Gennadi Kolesnikov, Aleksandr Kramarenko, Mikhail Lisun, Aleksandr Petrushenko, Vladimir Perobrazhhensky, Valery Rozhdestvensky, Gennadi Sarafanov, Ansar Sharafutdinov, Vasili Shcheglov, Aleksandr Skvortsov, Eduard Stepanov, Valeri Voloshin, Oleg Yakolev, and Vyacheslav Zudov.

**1966 May 23** - Civilian Specialist Group 2

Sergei Anokhin, Vladimir Bugrov, Gennadi Dolgopolov, Georgi Grechko, Valeri Kubasov, Oleg Makarov, Vladislav Volkov, and Aleksei Yeliseyev.

**1966 September** - Military Cosmonaut Group

Pavel Popovich, Alexei Gubarev, Yuri Artyukhin, Vladimir Gulyaev, Boris Belousov, and Gennady Kolesnikov.

**1967 January 31** - Civilian Specialist Group 2 Supplemental

Nikolai Rukavishnikov and Vitali Sevastyanov.

**1967 February** - Soviet crewed lunar programs cosmonauts in two training groups

Group 1, commanded by Vladimir Komarov (Gagarin, Nikolayev, Bykovsky, Khrunov; Engineer-Cosmonauts: Gorbatko, Grechko, Sevastyanov, Kubasov, Volkov).

Group 2, commanded by Alexei Leonov (Popovich, Belyayev, Volynov, Klimuk; Engineer-Cosmonauts: Makarov, Voronov, Rukavishnikov, Artyukhin).

**1967 May 07** - Air Force Group 4

Vladimir Alekseyev, Vladimir Beloborodov, Mikhail Burdayev, Sergei Gaidukov, Vladimir Isakov, Vladimir Kovalyonok, Vladimir Kozelsky, Vladimir Lyakhov, Yuri Malyshev, Viktor Pisarev, Nikolai Porvatkin, and Mikhail Sologub.

**1967 May 22** - Academy of Sciences Group

Mars Fathulin, Rudolf Gulyayev, Ordinard Kolomitsev, Vsevolod Yegorov, and Valentin Yershov.

**1968 May 27** - Civilian Specialist Group 3

Vladimir Fartushny, Viktor Patsayev, and Valeri Yazdovsky.

**1969 September 10** - Civilian Engineer Group

Anatoli Demyanenko, Valeri Makrushin, and Dmitri Yuyukov.

**1970 April 27** - Air Force Group 5

Anatoli Berezovoi, Aleksandr Dedkov, Vladimir Dzhanibekov, Nikolai Fefelov, Valeri Illarianov, Yuri Isaulov, Vladimir Kozlov, Leonid Popov, and Yuri Romanenko.

**1971 February 25** - 1971 Scientific Group

Gurgen Ivanyan

**1972 March 22** - Civilian Specialist Group 4

Boris Andreyev, Valentin Lebedev, and Yuri Ponomaryov.

**1972 March 22** - Medical Group 3

Georgi Machinski, Valeri Polyakov, and Lev Smirenny.

**1973 March 27** - Civilian Specialist Group 5

Vladimir Aksyonov, Vladimir Gevorkyan, Aleksandr Ivanchenkov, Valeri Romanov, Valey Ryumin, and Gennady Strekalov.

**1974 January 01** - Physician Group

Zyyadin Abuzyarov.

**1976 August 23** - Air Force Group 6 - Space Shuttle Buran crew

Leonid Ivanov, Leonid Kadenyuk, Nikolai Moskalenko, Sergei Protchenko, Yevgeni Saley, Anatoly Solovyev, Vladimir Titov, Vladimir Vasyutin, and Alexander Volkov.

**1976 November 25** - 1976 Intercosmos Group

Miroslaw Hermaszewski (Poland), Zenon Jankowski (Poland), Sigmund Jahn (East Germany), Eberhard Kollner (East Germany), Oldrich Pelcak (Czechoslovakia), and Vladimir Remek (Czechoslovakia).

**1977 July 12** - The first group of test pilots for Buran - Gromov Flight Research Institute group

Igor Volk, Oleg G. Kononenko, Anatoly Levchenko, Nikolai Sadovnikov, Rimantas Stankevicius, and Alexander Schukin.

**1978 March 01** - 1978 Intercosmos Group

Aleksandr P. Aleksandrov (Bulgaria), Dumitru Dediu (Romania), Jose Lopez Falcon (Cuba), Bertelan Farkas (Hungary), Maidarjavyn Ganzorig (Mongolia), Jugderdemidiin Gurrachaa (Mongolia), Georgi Ivanov (Bulgaria), Bela Magyari (Hungary), Arnaldo Tamayo Mendez (Cuba), Dumitru Prunariu (Romania).

**1979 April 01** - 1979 Intercosmos Group

Tuan Pham (Vietnam), and Thanh Liem Bui (Vietnam).

**1980 July 30** - LII-1/IMBP-3/MAP/NPOE-5/AN-2 Cosmonaut Group

LII-1: Anatoly Levchenko, Alexandr Shchukin, Rimantas Stankevicius, Igor Volk.

IBMP: Galina Amelkina, Yelena Dobrokvashina, Larisa Pozharskaya, and Tamara Zakharova.

MAP: Svetlana Savitskaya.

NPOE: Yekaterina Ivanova, Natalya Kuleshova, and Irina Pronina.

AN-2: Irina Latysheva.

**1982 September 11** - 1982 Intercosmos Group

Reinhard Furrer, and Ernst Messerschmid.

**1983 April 25** - The second group of test pilots for project Buran

Ural Sultanov and Magomed Tolboev.

**1984 February 15** - NPOE-6 Cosmonaut Group

Aleksandr Kaleri and Sergei Yemelyanov.

**1984 June 12** - The third group of test pilots for project Buran

Victor Zabolotski.

**1985 September 02** - GKNII-2/NPOE-7 Cosmonaut Group

GKNII: Viktor Afansyev, Anatoly Artsebarsky, and Gennadi Manakov.

NPOE: Sergei Krikalyov, and Andrei Zaytsev.

**1985 September 30** - Intercosmos Group (Syria)

Muhammed Ahmed Faris, and Munir Habib Habib.

# Russian space organizations:

## The Yuri Gagarin Air Force Academy

The Gagarin Air Force Academy is a Russian military aviation academy located in Morino, Moscow Oblast, Russia. The Academy prepares high-ranking military personnel for the Russian Aerospace Forces. Among the alumni are around 700 Heroes of the Soviet Union (the highest award in the USSR), more than ten cosmonauts, and over 2,000 military specialists from twenty-one foreign countries. The Academy was founded in 1940; named the Air Force Academy in 1946; and became the **Yuri Gagarin Air Force Academy** in 1968. In Soviet times, only the officers with a primary military education and holding the rank of Major could study there. According to tradition, after the end of active military service as teachers of the Academy, the officers became the guides in the museum. The school provides Regiment and Division-level commanding officers to fill commanding, staff, navigation, logistics, communications and radar-support positions.

Alternative names for the Academy in the English-language literature include **Yuri Gagarin Military Air Academy** and **Yuri Gagarin Air Force Academy**. In conversational speech it is often simply referred to as **Gagarin Academy** or **Morino Academy**. By late 2008, this Academy and the Professor N.E. Zhukovsky Aviation Engineering Academy merged to become the **Zhukovsky-Gagarin Military Combined Air Force Academy**, but it still retained its Monino campus. It was now a federal government military educational institution of higher education run by the Russian Ministry of Defence. Since 2010, the full name is **Russian Air Force Military Educational and Scientific Center**. Later in 2011, the Monino campus was closed.

## Yuri Gagarin Cosmonaut Training Center

Star City, romanized as Zvyozdny gorodok, lit. *"starry townlet"* is a common name of an area in Zvyozdny gorodok, Moscow Oblast, Russia, which has since the 1960s been home to the **Yuri A. Gagarin State Scientific Research-and-testing Cosmonaut Training Center** (GCTC); a Russian training facility responsible for training cosmonauts for their space missions. The facility was originally known only as **Military Unit 26266**, and was a secret training base for Soviet Cosmonaut candidates. The site at latitude 55 degrees, 52 minutes, 38 North and longitude 38 degrees, six minutes, and 48 East was chosen for its proximity to Moscow and other infrastructure that would be essential to its function, namely Chkalovsky Airport and the Yaroslavl railroad. The densely forested area was originally a radar range with some existing infrastruture. Initially, cosmonaut candidates were housed at the nearby Frunze Central Airfield (Moscow), followed by an apartment block in Chkalovsky, before being moved to the newly-built apartments on site where they could remain with their families while training. In the 2010 Census, the Star City population was 6,332 persons.

This small town didn't appear on any map in past decades, since it was a secret, guarded and severely restricted place. Officially, the area was known as "closed military townlet No. 1", and at various times had also been designated as **Shchyolkovo-14** and **Zvozdny**. The Star City facility has its own Post Office, high school, shops, child day care, movie theatre, sports and recreation facilities, railway station, and a museum of space travel and human exploration. In Star City there is a statue of the "space dog" Laika, the first animal to orbit the Earth. The "Prophylactory" (aka the "Prophy"), was a three-story dormitory built by the Russians to house American astronauts taking part in the 1975 *Apollo-Soyuz Test Project*. NASA leased the second floor for team housing and for the Office of the Director of Operations Russia. For the 'Shuttle-Mir' program, NASA built several duplexes for astronaut housing. During an oral history interview in 1998, NASA Flight Surgeon Michael Barratt said, *"Star City has always been kind of felt as the "forbidden city" or the hidden city. It wasn't on any maps, certainly, by then.* ***It was a secret cosmonaut training base****. Of course, everyone knew where it was, but it was considered a closed and secure city"*. Unquote.

Until April 2009, the Yuri Gagarin Training Center was owned and operated by the Ministry of Defence (Russia), in cooperation with the Russian Federal Space Agency (Roskosmos). Russian President Dmitry Medvedev then signed a presidential decree; transferring the centre on October 01, 2008, from the Defence Ministry to the Russian Federal Space Agency. The facility contains infrastructure essential for the training of cosmonauts across a wide range of experiences, including g-loads; mission specific/suit training; medical observation/testing and astronavigation. Key GCTC facilities include full-size mock-ups of all major spacecraft developed since the Soviet era; zero-gravity training aircraft for simulating weightlessness; a medical observation clinic and testing facility; and the original office of Yuri Gagarin and a number of monuments and busts to him and other cosmonauts. The first head of the Cosmonaut Training center was Yevgeny Karpov (1960-63), and Maksim Kharlamov has held that office since June 2021.

# Bauman Moscow State Technical University

The Bauman Moscow State Technical University (BMSTU), sometimes colloquially referred to as the **Bauman Schoo**l, or Baumanka, is a public technical university (polytechnic) located in Moscow, Russia. Established in 1830, it is the second-oldest educational institition in Russia (after Lomonosov Moscow State University, 1755). Bauman University is regularly ranked first in official government and business press rating. In 2011/12 QS World University rankings, BMSTU ranked 379th overall, and 229th in Engineering and IT. The admissions process includes exams on math, physics and Russian. The academic year begins on September 01, and the Engineer Specialist degree requires six years of study, with students specializing from their third year.

Thirty-eight notable faculty and alumni include: Sergei Pavlovich **Korolev**, lead Soviet rocket engineer and spacecraft designer, the founder of Soviet cosmonautics; Vladimir Grigorievich **Shukhov**, Russian engineer, scientist and architect, inventor of the Shukhov cracking process; Vladimir **Solovyov**, Soviet cosmonaut, and first man on the *Mir* orbital space station; Andrey Nikolaevich **Tupolev**, pioneering Soviet aircraft engineer, chief designer of the "Tupolev" aircraft; Pavel Osipovich **Sukhoi**, Soviet aerospace engineer, chief designer of the "Sukhoi" military aircraft; Pafnuty Lvovich **Chebyshev**, Russian mathematician and mechanician; Vladimir Mikhailovich **Myasichchev**, Soviet aircraft designer; Georgy **Malenkov**, Prime Minister of the Union of Soviet Socialist Republics; Nikolai Yegorovich **Zhukovsky**, Russian scientist and mechanician, founder of modern aerodynamics; Vladimir Nikolayevich **Chelomei**, Soviet mechanics scientist, aviation and missile engineer; Nikolay Alekseevich **Pilugin**, Soviet chief designer of rocket guidance systems; Semyon Alekseyevich **Lavochkin**, Soviet aircraft designer, founder of the Lavochkin Aircraft Design Bureau; Vladimir Mikhailovich **Petlyakov**, Soviet aeronautical engineer; Sergey Alekseyevich **Lebedev**, Soviet scientist in the fields of electrical engineering and computer science, designer of the first Soviet computers; Ivan Georgievich **Petrovsky**, Soviet mathematician; Vladimir Evgenievich **Zotikov**, Russian and Soviet scientist and textile engineer; Lev Iakolevich **Karpov**, Russian chemist and Bolshevik revolutionary; Sergei Ivanovich **Vavilov**, Soviet physicist, the President of the USSR Academy of Sciences;

Arkady Grigoryevich **Mordvinov**, Soviet architect and construction manager; Aleksandr **Aleksandrovich Mikulin**, Soviet aircraft engine designer and chief designer in the Mikulin Design Bureau; Sergey Alexandrovich **Afanasyev**, Soviet engineer and statesman, space and defence industry executive, the first Minister of the Soviet Ministry of General Machine Building; Alfred **Rosenberg**, German Nazi political figure, influential ideologue of the NSDAP, executed for war crimes; Sergey Pavlovich **Nepobedimiy**, Soviet designer of rocket weaponry; Nikolay Antonovich **Dollezhal**, Soviet mechanical engineer, a key figure in the Soviet atomic bomb project and chief designer of nuclear reactors; Vladimir Pavlovich **Barmin**, Soviet scientist, designer of the rocket launch complexes; Alexander Leonovich **Kemurdzhian**, Soviet engineer, first planetary rovers chassis designer; Konstantin Petrovich **Feoktistov**, Soviet cosmonaut and space engineer; Oleg Ivanovich **Skripochka**, Soviet cosmonaut and space engineer; Oleg Grigoryevich **Makarov**, Soviet cosmonaut; Gennadi Mikhailovich **Strekalov**, Soviet engineer and cosmonaut; Vassili **Nesterenko**, Soviet nuclear energy physicist; Maksim Zakharovich **Saburov**, Soviet engineer, economist and politician; **Zou** Jiahua, former Vice Premier of China; Georgy **Trefilov**, founder and the former co-owner of the "MARTA" holding; Dmitry **Grishin**, Russian businessman, investor and internet entrepreneur; Co-founder, Chairman and CEO of Mail.ru Group; Dmitry **Skylarov**, Russian programmer, hacker and cryptographer; Igor **Sysoev**, creator and developer of the Nginx web server and founder of NGINX, Inc.; and Alexander **Volkov** (fighter), professional Mixed Martial Artist, current UFC Heavyweight contender.

# The Russian Space Agency

**The Roscosmos State Corporation for Space Activities**, commonly known as Roscosmos, is a state corporation of the Russian Federation; responsible for space flights, cosmonautics programs, and aerospace research. It was established on February 25, 1992, by a decree of President Boris Yeltsin and has undergone restructuring in 1999 and 2004. The preceeding agency was the Soviet space program (1955-91). The 1990s saw serious financial problems due to the decreased cash flow, which led to improvisation and the agency taking a leading role in commercial satelite launches and "space tourism". A seat to the ISS costs from USD 20 to 81 million!

As of 2009, up to 50% of Russia's space budget was spent on the crewed space program, due to involvement in the ISS. Wages in the space industry are low, the average age of employees is high (46 years in 2007), and much of the equipment is obsolete. Many companies in the sector have been able to profit from contract and partnerships with foreign companies. In August 2013, the **United Rocket and Space Corporation** was formed as a joint-stock corporation to consolidate the Russian space sector. Under the 2013 plan, Roscosmos was to "act as a federal executive body and contracting authority for programs to be implemented by the industry". In 2016, the state agency was dissolved and the Roscosmos brand moved to the state corporation created in 2013.

Roscosmos coordinates with the Defense Ministry of the Russian Federation. On February 22, 2019, Roscosmos announced the construction of its new headquarters in the district of Fili, Moscow, the National Space Center. In March 2021, Roscosmos signed a Memorandum of cooperative construction of a lunar base called the **International Lunar Research Station** with the China National Space Administration. In April 2021, Roscosmos announced that it would depart the ISS program [but is still involved], saying that a new Space Station, the **Russian Orbital Service Station** will begin construction in 2025.

The Russian Space Force is the military counterpart of the Roscosmos, with similar mission objectives as the United States Space Force. It was formed after merging the space components of the Russian Air Force and the Aerospace Defense Forces (VKO) in 2015.

The Space Force controls Russia's Plesetsk Cosmodrome launch facility. Roscosmos operates a number of programs for Earth science, communication, and scientific research. Future projects include the Soyuz capsule successor, the Prospective Piloted Transport System, scientific robotic missions to one of the moons of Mars, as well as an increase in Lunar orbit research satellites. Currently, rocket development encompasses both a new rocket system, named *Angara*, as well as enhancements of the *Soyuz* rocket: *Soyuz-2* and *Soyuz-3*. Two modifications have been successfully tested, enhancng the launch capacity to 8.5 tons payload to low Earth orbit.

The newest spaceport is named Vostochny, in Tsiolkovsky, Amur Oblast. President Vladimir Putin directed the government to expedite the paperwork so construction could start on schedule. The first launch was *Soyuz-2.1a* on April 28, 2016, with ten successes out of eleven launch attempts. The facility was due to be completed in 2018. Vostochny (which means "Eastern" in Russian) is located above the 51st parallel north in the Amur Oblast, in the Russian Far East, and is intended to reduce Russia's dependency on the Baikonur Cosmodrome, located in the Independent Republic of Kazakstan (a separate country since the USSR dissolved in December 1991). Vostochny is 600 to 800 km from the Pacific Ocean, depending on launch azimuth, and spent first stages can be dropped into the water. The site is planned to be developed to a total area of about 30 km in diameter, so will thus be over 551 square kilometres. The nearby train station is Ledyanaya and the nearest city is Tsiolkovsky. Vostochny is in proximity to the Baikal-Amur Mainline and the Chita-Khabarovsk Highway. There is abundant local electricity production (for the hydrogen rocket fuel electrolysis and liquefaction); and the infrastructure supporting the former prototype Svobodny Cosmodrome.

# RSC Energia

**PAO S.P. Korolev Rocket and Space Corporation Energia** (Romanized: Raketno-kosmicheskaya korporatsiya *"Energiya")* also known as **RKK Energia**, is a Russian manufacturer of spacecraft and space station components. The company is the prime developer and contractor of the Russian crewed spaceflight program; it also owns a majority of Sea Launch. Its' name is derived from Sergei Korolev, and the Russian word for energy. **Energia** is the largest company of the Russian space industry and one of its' key players. It is responsible for all operations involving human spaceflight, is the lead developer of the Soyuz and Progress spacecraft, and the lead developer of the Russian segment of the International Space Station (ISS). In the mid-2000s, the company employed well over twenty-two thousand people. The enterprise has been awarded four Orders of Lenin; the Order of the October Revolution; and the Russian Federation President's Message of Thanks. Additionally, fourteen cosmonauts employed by **Energia** have been awarded the title "Hero of the Russian Federation".

The company was founded on August 26, 1946, and has been known successively as: Special Design Bureau No. 1 of RD Institute No. 88 (**OKB-1** of NII-88); TsKBEM (Central Design Bureau of Experimental Machine Building); **NPO Energia**; and S.P. Korolev **RSC Energia**. It is named after the first chief of its' Design Bureau, Sergei Korolev (in office 1946-66). His successors were: Vasily Mishin (1966-74); Valentin Glushko (1974-89); Yuriy Semenov (1989-2005); and Nikolai Sevastianov (2005-07). Until August 01, 2014, its' President and Chief Designer was Vitaly Lopata. In a *"long-awaited personnel reform in the [Russian] space industry"* facing severe budget constraints, Lopota was offered the position of Vice President for technological development in the **United Rocket and Space Corporation**.

Beginning with the first artificial satellite, *Sputnik 1*, and the first crewed spaceflight of *Vostok 1*, Korolev's Design Bureau was responsible for a major part of the Soviet space program. It was the main rival of **OKB-52** (later known as TsKBM, then the design bureau of Vladimir Chelomei) during the Soviet crewed lunar program and the Soviet space station programs. **OKB-1** was (among others) responsible for development of the crewed *Soyuz* spacecraft and its *Soyuz* rocket; the *N-1* "Moon Shot" rocket; large parts of the *Salyut* space station program; the uncrewed *Progress* resupply craft and designed the *Energia* rocket for the *Buran* space shuttle program. Since the early beginnings of the *Luna* program, it designed many space probes, among others of the *Venera, Zond* and Mars program.

**RSC Energia** continues to dominate a large part of the global space program, with its' *Soyuz* spacecraft becoming the only crewed spacecraft conducting regular flights and being the exclusive crew transport vehicle after the end of the US Space Shuttle program in July 2011 and until the maiden flight of SpaceX's Crew Dragon *Endeavour* capsule on May 30, 2020. The Chinese *Shenzhou* program is the only other program in the world with planned semi-regular crewed spaceflights.

# Ongoing projects:

**RSC Energia** now builds Russia's *Soyuz MS* spacecraft for three-person human spaceflight missions, and *Progress M* robotic spacecraft for cargo missions. Through its' Russian Orbital Segment (ROS) of the ISS, **Energia** provides its own cosmonauts for ISS expeditions, and has responsibility for all Russian scientific experiments. Through participation in the Sea Launch project, **Energia** produces the 'Block DM-SL' as the upper stage for the Ukrainian launch vehicle *Zenit-3SL*. Via the Universal Spacecraft Configuration (USC), **Energia** develops communications satellites, remote sensing satellites, navigation satellites, and uncrewed orbital servicing satellites. USC was a basis for *Yamal 100* and *Yamal 200* satellites.

## Future projects:

Further assembly of the ISS Russian segment: development [together with Krunichev Space Center] of the Multipurpose Laboratory Module (MPLM), and *Oka* space production modules (not permanently attached to the ISS).

Development of new spacecraft with three stages: 1. Modernization of *Soyuz TMA* spacecraft for human circumlunar missions (pending commercial orders for space tourism).

2. Development of the *Parom* space tug, to replace the *Progress M* cargo spacecraft.

3. Development of multi-aimed *Orel* spacecraft with capacity for six persons.

Development of the crewed lunar program, with a projected Moon landing in 2025, and creation of a lunar base by 2030 to extract helium-3. Development of the human Mars mission, with a landing beyond 2035. Development of the *Yamal 300* and *Yamal 400* communication satellites for **Gazprom Corporation**. Development of the *Smotr* remote sensing satellites. Development of a pod designed for clearing near-Earth space environs of satellite debris. This new device was planned to be assembled by 2020, and tested by 2023. The concept is to use a nuclear power source so that the craft could remain on active for up to fifteen years, primarily working in the geosynchronous orbit zone. Debris collected would be de-orbited to re-enter and burn up in Earth's atmosphere over oceans.

# The Soyuz spacecraft

Soyuz is the Russian word for Union. Soyuz is a series of spacecraft that have been in service since the 1960s, having made more than 140 flights to date. It was designed for the Soviet space program by the **Korolev Design Bureau**, and is now manufactured by **RSC Energia**. *Soyuz* succeeded the *Voskhod* spacecraft and was originally built as part of the Soviet crewed lunar programs. The first uncrewed *Soyuz* flight was on November 28, 1966. *Soyuz 3*, launched on October 26, 1968, was the program's first successful crewed mission. The *Soyuz* spacecraft are launched on *Soyuz* rockets from the **Baikonur Cosmodrome** in Kazakhstan. *Soyuz* was operated by the Soviet space program from 1967 to December 1991 when the USSR dissolved, then by **Roscosmos**, the Russian Space Agency from 1992 to the present day.

The Soviet Union's space program had experienced great success in its early years, but by the mid-1960s the pace of success had slowed. While the *Voskhod* program achieved the first multi-crewed spaceflight with Commander Vladimir Komarov, Konstantin Feoktistov and Boris Yegorov (October 12, 1964), and the first extravehicular activity (EVA) with Alexei Leonov on March 18, 1965, problems encountered led to its termination after only two flights, allowing the United States to surpass the Soviet achievements with Project Gemini. The *Soyuz* program was intended to rejuvenate the Soviet program by developing space rendezvous and docking capability, and practice extravehicular activity without tiring the cosmonaut. These capabilities would be required for the *Salyut* space station program.

The seven-metre-long (23 foot) seven-metric-ton *Soyuz* vehicle comprises three modules joined in line: a central, small, aerodynamic bell-shaped descent module of 7.5 cubic metres (260 cubic feet) with contoured couches for as many as three persons during ascent, descent and landing; a cylindrical service module mounted at the rear that provides propulsion engines, life support for about thirty person-days, instrumentation, and electrical power from a solar array; and a spheroidal orbital module in front that carries the docking system and contains living facilities and cargo for the orbital phase of the mission. The three modules remain together throughout the mission until the spacecraft is de-orbited. Only the descent module returns to Earth intact.

The orbital (or habitation) module houses all the equipment that will not be needed for re-entry; such as experiments, cameras or cargo. It also contains a toilet, docking avionics and the communications gear. On the latest Soyuz versions since TM, a small window was introduced, providing the crew with a forward view. The descent module (or re-entry capsule) is half covered by a heat-resistant surface to protect it during re-entry. It is initially slowed by the atmosphere, then by a braking parachute, followed by the main parachute. At one meter above the ground, solid-fuel braking engines mounted behind the heat shield are fired to give a "soft" landing. The internal volume of *Soyuz SA* is four cubic metres (140 cubic feet), of which 2.5 cubic metres is usable for crew (living space). At the rear of the vehicle is the Service Module. It has a pressurized container shaped like a bulging can; the instrument compartment. This contains systems for temperature control, electric power supply, long-range radio communications, radio telemetry, and instruments for orientation and control.

A non-pressurized part contains the main engine and a liquid-fueled propulsion system, using N2O4 and UDMH fuels, for maneuvering in orbit and initiating the descent back to Earth. The ship also has a system of low-thrust engines for orientation, attached to the intermediate compartment. The failure of several explosive bolts, causing incomplete separation between the service and re-entry modules led to emergency situations during the *Soyuz 5, Soyuz TMA-10,* and *Soyuz TMA-11* missions, which led to an incorrect re-entry orientation. Re-entry firing is usually done on the "dawn" side of the Earth, so that the spacecraft can be seen by recovery helicopters as it descends in the evening twilight, illuminated by the sun when it is above the shadow of the Earth. The *Soyuz* craft is designed to come down on land, usually somewhere in the deserts of Kazakhstan in Central Asia.

*Soyuz 1* had been launched on April 12, 1967, with the goal of docking with the crewed *Soyuz 2* craft, but even before the second craft was launched, serious problems with *Soyuz 1* made it clear that the *Soyuz 2* flight had to be canceled before the landing of *Soyuz 1.* This saved the lives of the three-man *Soyuz 2* crew: *Soyuz 1* ended with the death of cosmonaut Vladimir Komarov on April 13, 1967, due to a faulty parachute system. *Soyuz 2* would otherwise have flown with the same defective system as *Soyuz 1.* As a result, the *Soyuz* spacecraft was revised for the *Soyuz 2* mission (October 25-26, 1968) and the *Soyuz 3* flight of three days and 22 hours launched on October 26, 1968.

The *Soyuz 7K* spacecraft was developed in 1963, and the first crewed launch of a *Soyuz 7K-OK* took place on April 12, 1967. Its single test pilot, Vladimir Komarov was killed when the descent module's parachute failed to unfurl after re-entry and the module crashed at high speed. This was the first human death during a spaceflight. Despite early incidents, the *Soyuz* is widely considered the world's safest, most cost-effective human spaceflight vehicle, established by its unparalled length of operational history. The vehicle is protected during launch by a payload fairing, which is jettisoned along with the Spacecraft Abort System (SAS) at 2.5 minutes into launch. It has an automatic docking system. The craft can be operated automatically, or by a pilot independently of ground control.

After the United States of America won the race to put men on the Moon in July 1969, the Soviet Union adapted the *Soyuz 7K-L3* to ferry crews to space stations. *Soyuz 11*, version 7K-OKS with a docking port to allow internal crew transfer between spacecraft carried the inaugural three-man crew to the *Salyut 1* station in June 1971, but the three cosmonauts died when their descent module accidentally depressurized while returning to Earth. In redesigning the vehicle to forestall another such incident, one couch was removed to accommodate an independent life-support system for individual pressure suits. A modified *Soyuz 7K-TM* version flew in July 1975 for the *Apollo-Soyuz Test Project*, the first US-Soviet joint space venture. This *Soyuz* version was a technological bridge to the third-generation spacecraft. During the 1970s, an automated derivative of *Soyuz*, known as *Progress*, was developed as a space station resupply vehicle. Cargo and refueling modules replaced the orbital and descent modules in the *Soyuz* design. Its' operational use began in 1978, with a mission to the *Salyut 6* Soviet space station.

The first major redesign of *Soyuz* was introduced in 1979. Called the *Soyuz T "transportnyi"* and used until 1986, it had advanced equipment and capabilities, and restored the third crew seat, with the cosmonauts now wearing spacesuits. The *Soyuz TM "modifitsirovannyi"* fourth version (1986-2002), an upgrade featuring a variety of new systems, made its first crewed flight in 1987, when it carried *Mir*'s second crew to the then-embryonic space station. The *Soyuz TMA* (2003-12) *"antropometricheskii"* version debuted in 2002 with a crewed flight to the International Space Station (ISS). It featured digital control technology, improved parachute systems, and the design incorporated changes to meet certain National Aeronautics and Space Administration (NASA) requirements as an ISS "lifeboat", including eased height and weight restrictions for crew members. An upgraded version of *Progress* was also used to ferry cargo to the ISS. After the in-flight explosion of the US Space Shuttle orbiter *Columbia* in February 2003 and the consequent grounding of the shuttle fleet, *Soyuz TMA* spacecraft for a time provided the only means for ISS crew exchanges until US Space Shuttle flights resumed in July 2005.

A new *Soyuz* version, *TMA-M*, first launched on October 07, 2010, was used until 2016. It used a new computer, digital interior displays, updated docking equipment, and had 70 kg less mass. After the Space Shuttle program ended in July 2011, *Soyuz* again became the only spacecraft that could take astronauts to the ISS. An upgraded version, *Soyuz MS*, with more efficient solar arrays, a new TsVM-101 computer, a new *Kurs* NA approach and docking system (which weighs half as much and consumes a third of the power of the previous system), modified docking and attitude control engine positions providing redundancy, and extra shielding for micrometeorites debuted in July 2016. At least one *Soyuz* spacecraft is docked to the ISS at all times, for use as an escape craft in the event of an emergency.

Soyuz is intended to be replaced by the six-person *Orel* spacecraft. There have been three failed launches of a crewed Soyuz vehicle: *Soyuz 18a* on April 05, 1975; *Soyuz T-10a* on September 26, 1983; and *Soyuz MS-10* on October 11, 2018. The 1975 failure was aborted at T+295 seconds (four minutes and 55 seconds), at an altitude of 145 km. In 1983, the SAS successfully rescued the cosmonauts from an on-pad fire and explosion of the launch vehicle. In 2018, the SAS sub-system in the payload shroud successfully rescued the cosmonauts from a rocket failure two minutes and 45 seconds after liftoff, a few seconds after first stage separation and after the escape tower had already been jettisoned. While not being direct derivatives of *Soyuz*, the Chinese *Shenzhou* spacecraft uses *Soyuz TM* technology sold in 1984. The Indian Orbital Vehicle follows the same general layout as that pioneered by *Soyuz*.

# Facilities for the Buran spaceplane program

Maintenance, launches and landings of the *Buran*-class orbiters were to take place at the Baikonur Cosmodrome in the Kazakh SSR, Soviet Union. Several facilities at Baikonur were adapted or newly-built for these purposes.

**Site 110:** This was used for the launch of the *Buran*-class orbiters. Like the assembly and processing hall at Site 112, the launch complex was originally constructed for the Soviet lunar landing program, and later converted for the Energia-Buran program.

**Site 112:** This was used for orbiter maintenance and to mate the orbiters to their *Energia* launchers [thus fulfilling a role similar to the Vehicle Assembly Building at Kennedy Space Center, (KSC) USA]. The main hangar at the site, called *MIK RN* or *MIK 112*, was originally built for the assembly of the *N-1* Moon rocket. After cancellation of the N-1 program in 1974, the facilities at Site 112 were converted for the Energia-Buran program. It was here that Orbiter 1K was stored after the end of the *Buran* program. It was destroyed on May 12, 2002, when the hangar roof collapsed, due to lack of maintenance and recent heavy rainfall, killing eight workers whose roof equipment may have contributed to the collapse.

**Site 251:** This was used as the *Buran* landing facility, also known as *Yubileyniy [Jubilee] Airfield* (and fulfilling a role similar to the Spacecraft Landing Facility at KSC). It features one runway, called 06/24, which is 4,500 metres (4,900 yards) long and 84 metres wide, paved with "Grade 600" high-quality reinforced concrete. At the edge of the runway was a special mating-demating device, designed to lift an orbiter off its Antonov An-225 *Mriya* carrier aircraft and load it on a transporter, which would carry the orbiter to the processing building at Site 254. A purpose-built orbiter landing control facility, housed in a large multi-storey office building, was located near the runway. Yubileyniy Airfield was also used to receive heavy transport planes carrying elements of the *Energia-Buran* system. After the end of that program, Site 251 was abandoned but later re-opened as a commercial cargo airport. Besides serving Baikonur, Kazakh authorities now also use it for passenger and charter flights from Russia.

**Site 254:** This was built to service the *Buran*-class orbiters between flights (thus fulfilling a role similar to OPE at KSC). Constructed in the 1980s as a special four-bay building, it featured a large processing area flanked by several floors of test rooms. After cancellation of the *Buran* program, it was adapted for pre-launch operations of the *Soyuz* and *Progress* spacecraft.

**Atmospheric test flights:** An aerodynamic testbed, OK-GLI, was constructed in 1984 to test the in-flight properties of the *Buran* design. Unlike the American prototype Space Shuttle *Enterprise* (essentially a high-altitude glider), OK-GLI had four AL-31 turbofan engines fitted, so it could fly under its own power.

**Orbital flight of Buran in 1988:** Following a series of atmospheric test flights using the jet-powered OK-GLI prototype, the first operational spacecraft (*Buran*, Orbiter 1K) flew one uncrewed test mission. At 03:00 on November 15, 1988, *Buran* and the *Energia* carrier rocket lifted off from Pad 110/37 in Baikonur. The life support system was not installed for the flight, and no data was displayed on the CRT display screens in the Command Compartment.

The shuttle orbited the Earth twice, traveling 83,707 km (52,013 miles) in three hours and twenty-five minutes. On its return, it performed an automated landing in strong cross-winds on the shuttle runway (Site 251) at Baikonur Cosmodrome.

**Planned flights:** Uncrewed flights were planned for 1991-93. Orbiter 2K was to make an uncrewed flight of 7-8 days in December 1991. The first crewed space test flight, using Orbiter 3K, was scheduled to be with two cosmonauts for 24 hours in 1994.

**Cancellation of the Buran program:** After the first flight of the *Buran* shuttle, the project was suspended due to lack of funds and the deteriorating political situation in the Soviet Union in the early 1990s. Orbiters 2K (95% built) and 3K were never completed. The project was officially terminated on June 30, 1993, by Russian President Boris Yeltsin. To that point, 20 billion rubles had been spent on the *Buran* program. *"We had no civilian tasks for Buran, and the military ones were no longer needed"*. The program was designed to boost national pride, carry out research, and meet technological objectives similar to the US Space Shuttle program; including resupply of the *Mir* space station, which was launched in component stages from February 19, 1986, and remained in service until 2001. When *Mir* was finally visited by a spaceplane, it was a US Space Shuttle orbiter, not a *Buran*-class orbiter. The cost of a *Buran* launch carrying a twenty-ton payload was estimated at 270 million rubles, vs 5.5 million rubles on the *Proton* rocket. Therefore, the concept was hopelessly uneconomic.

# Salyut space stations

The *Salyut* program was managed by Kerim Kerimov, chairman of the State Commission for Soyuz missions. *Salyut 1* originated as a modification of the Soviet military's *Almaz* space station program then in development. After the July 1969 lunar landing of *Apollo 11*, the Soviets began shifting the focus of their crewed space program to orbiting space stations. Premier Leonid Brezhnev canceled the Soviet manned lunar program in 1974 after four disastrous *N-1* rocket launch failures. Another motivation for the Soviet orbital space station program was to one-up the US *Skylab* program then in development. The basic structure of *Salyut* would carry through to the *Mir* space station, with the last module of the program, *Zarya* (DOS-8), forming the core of the Russian segment of the International Space Station.

The world's first-ever space station, the Soviet *Salyut 1* (also called DOS-1, for Durable Orbital Station) was launched on April 19, 1971. *Soyuz 11* was the only mission to successfully dock with it, and the three-man crew stayed aboard for 22 days in June 1971. In that era, there were competing "civilian" and pure military Soviet space programs: *Salyut 1*, for example, was developed as a "civilian" program. Other civilian stations, called Long-term Orbital Stations (DOS), were considered successors of *Salyut 1*. These were the unsuccessful DOS-2 in 1972; DOS-3 in 1974; and later the successful *Salyut 4*, *Salyut 6* and *Salyut 7* stations. The space stations funded and developed by the military, known as *Almaz* stations, were roughly similar in size and shape to the civilian DOS stations. The details of their design, which is attributed to Vladimir Chelomey, are considered to be significantly different from the DOS stations.

Construction of *Salyut 1* began in early 1970, and nearly one year later it was shipped to Baikonur Cosmodrome. Final assembly was completed at the launch center. The launch, planned to coincide with the tenth anniversary of Yuri Gagarin's historic flight on April 12, 1961, was delayed for one week by technical problems. The *Salyut* was twenty metres long, four metres in diameter, and had 99 cubic metres (3,500 cubic feet) of interior space. The on-orbit dry mass was 18,425 kg (40,620 lbs). Three compartments, with a total volume of 100 cubic metres were pressurized, and two could be entered by the crew. The interior design used various colors (light and dark gray, apple green, and light yellow) for supporting the cosmonaut's orientation in weightlessness. The third pressurized compartment contained the control and communications equipment, the power supply, life support system and other auxillary equipment.

The fourth, unpressurized service compartment was about two metres in diameter, housing the engine installations and associated control equipment. The *Salyut* had buffer chemical batteries, reserve supplies of oxygen and water, and regeneration systems. Externally mounted were: two double sets of solar cell panels that extended like wings from the smaller compartments at either end, the heat regulation system's radiators, and orientation and control devices. *Salyut 1* was modified from one of the *Almaz* airframes. The unpressurized Service Module was the modified service module of a *Soyuz* spacecraft. The astrophysical 'Orion 1' Space Observatory obtained ultraviolet spectrograms of stars using a mirror telescope of the Mersenne system and a spectrograph of the Wadsworth system, using film sensitive to the far ultraviolet. Slitless spectrograms were obtained of the stars *Vega* and *Beta Centauri* between 2000 and 3800 A (200 and 380 nm). Viktor Patsayev aboard *Soyuz 11* became the first man to operate a telescope outside of the Earth's atmosphere.

*Soyuz 11* arrived at *Salyut 1* on June 07, 1971, and took three hours and 19 minutes to complete docking. The mission objectives of G. Dobrovolsky, V. Patsayev and V. Volkov were to: check the design, units, onboard systems, and equipment of the orbital piloted station; test the station's manual and autonomous procedures for orientation and navigation, as well as the maneuvering systems; study Earth's surface geology, geography, meteorology, and snow and ice cover; study physical characteristics, processes, and phenomena in the atmosphere and outer space in various regions of the electromagnetic spectrum; conduct medico-biological studies to determine the feasability of having cosmonauts perform various tasks, and study the influence of space flight on the human organism.

On June 29, after 23 days and flying 362 Earth orbits, the mission was cut short due to problems aboard the Salyut, including an electrical fire. *Salyut 1* was moved to a higher orbit in July-August 1971 to ensure that it would not be prematurely destroyed through orbital decay. Meanwhile the Soyuz capsules were being substantially redesigned to allow pressure suits to be worn by cosmonauts during launch, docking maneuvers, and re-entry. By September 1971, *Salyut 1* was running out of fuel, so it was decided to de-orbit the station on October 11. After 175 days in orbit, the pioneering space station burned up over the Pacific Ocean. *Pravda* (October 26, 1971) reported that 75% of *Salyut 1*'s studies were carried out by optical means and 20% by radio-technical means, while the remainder involved magnetometrical, gravitational, or other measurements. Synoptic readings were taken in both the visible and invisible parts of the electromagnetic spectrum.

The first *Almaz* station was *Salyut 2*, launched on April 03, 1973, atop a three-stage *Proton-K* rocket. Three days after the launch, the *Proton's* spent third stage exploded (due to pressure changes within the fuel tanks), resulting in a cloud of debris, some of which followed a similar trajectory to the station. Ten days later this debris struck *Salyut 2*, damaging the hull and causing depressurization. The station had also lost altitude control, leaving it unusable. Both solar panels were torn free, removing the power-generating capacity and to control the station's attitude. The two solar arrays were to generate a total of 3,120 watts of electricity. *Salyut 2* was equipped with thirty-two attitude control thrusters, as well as two RD-0225 engines, each capable of generating 3.9 kN of thrust for orbital maneuvers. The station's orbit quickly decayed, and it re-entered the atmosphere on May 28, 1973.

*Salyut 3*, (English: Salute 3), also known as OPS-2, or *Almaz 2*, was a Soviet space station. It was included in the Salyut program to disguise its true military nature. Launched by a three-stage *Proton* booster on June 25, 1974, it was the first space station to maintain its constant orientation relative to the Earth's surface. Its initial orbit was 219 by 270 km above mean sea level, which was considered to be low. The suspicions of Western observers were raised by the low altitude, combined with the choice of a crew from the Soviet Air Force, and the use of radio frequencies normally designated for military use.

The *Soyuz 14* mission brought the only one of three crews able to board and operate the station for 15 days: *Soyuz 15* came within forty metres, but the *Igla* docking system failed. The Soviet Union was reluctant to release details, but several sources report that *Salyut 3* contained several Earth-observation cameras, as well as a "self-defence" on-orbit cannon! Some accounts claim it was possibly a Nudelman-Rikhter "Vulkan" gun, a variant of the 23 mm Nudelman aircraft cannon. Some sources say three test firings took place during the *Soyuz 14* mission with Commander Pavel Popovich and Yury Artyuhkin.

*Salyut 3* consisted of an airlock chamber, a large-diameter work compartment, and a small-diameter living compartment, giving a total habitable volume of ninety cubic metres. It had two solar arrays, one docking port, and two main engines, each of which could produce 3.9 kN of thrust. Its launch mass was 18,900 kg. The station was equipped with a shower, a standing sleeping station, as well as a foldaway bed. The floor was covered with

Velcro to assist cosmonauts moving around. Some on-board entertainment included a magnetic chess set, a small library, and a cassette deck with some audio cassette tapes. Exercise equipment included a treadmill and Pingvin exercise suit. The first water-recycling facilities were tested on the station: the system was called Priboy.

The work compartment was dominated by the 'Agat-1' Earth-observation telescope, which likely had a resolution better than one metre. It was used in conjunction with a wide-film camera, and was primarily for military reconnaissance. Secondary objectives included the study of water pollution, agricultural land, possible ore-bearing landforms, and oceanic ice formation. The cosmonauts were able to develop film while aboard, with important images printed, then scanned by a TV imaging system for broadcast to Earth. Less important images were packed into a small Earth-return capsule, which could be ejected from the station. Pavel Popovich, the Commander of *Soyuz 14*, recalled that the station was equipped with fourteen cameras; including a topographical camera, a star camera, and a Volga infrared camera.

After the docking failure of *Soyuz 15* the next planned mission to *Salyut 3* was canceled. On September 03, 1974, the Earth-return film capsule was release, de-orbited by small engines, and parachuted down from 8.4 km altitude. The main parachute did not open, and the canister deformed upon landing, but all the film was recoverable. Apparently, the on-board station gun was then fired to depletion by remote control. *Salyut 3* was de-orbited on January 24, 1975, over the Pacific Ocean.

*Salyut 4* was a civilian space station launched on December 26, 1974, and was a complete success. Basically similar to *Salyut 1*, it had three solar panels mounted on the forward module, and an interior floor area of 34.8 square metres. The electric system produced an average of 2.00 kW of power, and the *Salyut* had 2,000 kg of scientific equipment. This station was equipped with the Delta Navigation System, that could autonomously calculate orbital elements without ground assistance. Instruments installed were: OST-1 (Orbiting Solar Telescope); the X-ray *Filin* telescope; lower body negative pressure gear; bicycle ergometer integrated physical trainer; penguin suits and alternate athletic suit; sensors for temperature and characteristics of upper atmosphere; ITS-K infrared telescope spectrometer and ultraviolet spectrometer; multispectral Earth resources camera; cosmic ray detector; embryological studies; new engineering instruments tested for orientation of the station by celestial objects; a teletypewriter and swivel chair (for vestibular function tests). The station was visited by the crews of *Soyuz 17* for 29 days, *Soyuz 18* for 62 days, and by the uncrewed *Soyuz 20* craft from November 17, 1975, to February 16, 1976. *Salyut 4* was de-orbited on February 02, 1977.

The next Soviet military space station was *Salyut 5*, which was the final *Almaz* space station.

*Salyut 6* was launched on September 29, 1977, by a *Proton 8K82K* rocket. It was the first space station to receive large numbers of crewed and uncrewed spacecraft for human habitation, crew transfer, international participation and resupply. It was the first "second generation" space station. In addition to a new propulsion system and its primary scientific instrument – the BST-1M multispectral telescope – the station had two docking ports, allowing two craft to visit simultaneously. Six long-term resident crews were supported by ten short-term visiting crews who typically arrived in newer *Soyuz* craft and departed in older craft, leaving the newer craft available to the resident crew as a return vehicle, and thereby extending the resident crew's stay past the design life of the *Soyuz*. Short-tem visiting crews routinely included international cosmonauts from the Intercosmos program: i.e., the Warsaw Pact countries allied to the Soviet Union. During periods of crew vacancy, the *Salyut* was visited by *Soyuz T-1* and *Kosmos 1267*. Following the launch of its successor, *Salyut 7*, it was de-orbited on July 29, 1982, almost five years after its own launch.

*Salyut 7* (aka DOS-6, short for Durable Orbital Station), was the backup vehicle to *Salyut 6*. With delays to the *Mir* program it was decided to launch the backup as *Salyut 7* on April 19, 1982, and it spent 3,215 days in Earth orbit to February 1991. It was the most advanced and comfortable space station of the *Salyut* series. There were about twenty windows with shades, and the color scheme was improved. *Salyut 7* carried electric stoves, a refrigerator, constant hot water, and redesigned seats at the command console. Two portholes were designed to allow ultraviolet light in, to help kill infections. The medical, biological and exercise sections were improved, to allow long stays in the station. The BST-1M telescope used in *Salyut 6* was replaced by an X-ray detection system.

*Salyut 7* was first crewed in May 1982 with two crew via *Soyuz T-5*, and was last visited in June 1986 by *Soyuz T-15*; occupied for a total of 816 days. In orbit, the station suffered technical failures, but it benefitted from the improved payload capacity of the visiting *Progress* and *Soyuz* craft, and the experience of its crews who improvised many solutions. Orbiting for eight years and ten months (a record not broken until the advent of *Mir*), it was visited by six main expeditions and four secondary flights. Aside from the many experiments and observations made, the station also tested the docking and use of large modules with an orbiting space station. The heavy Kosmos Modules were variants of the TKS spacecraft intended for the canceled *Almaz* military space station. They helped Soviet engineers develop the technology necessary to build *Mir* (the modular, expandable, third-generation space station).

# The Apollo-Soyuz Test Project

Although Soviet leader Nikita Krushchev made the Soviet Union's policy of *detente* official in his 1956 doctrine of "peaceful coexistence" at the Twentieth Congress of the Communist Party of the Soviet Union, the nation seemed to be in perpetual conflict with the USA. After John H. Glenn's February 20, 1962, orbital spaceflight, exchanges of letters between President John F. Kennedy and Kruschev led to the Dryden-Blagonravov agreement, formalized in October 1962. It called for cooperation on the exchange of data from weather satellites, a study of the Earth's magnetic field, and joint tracking of the NASA 'Echo II' balloon satellite. Kennedy interested Kruschchev in a joint crewed Moon landing, but after his assasination in November 1963, and Kruschchev's removal from office in October 1964, the competition between the two nations' crewed space programs heated up, and talk of cooperation became less common, due to tense diplomatic relations and over-riding military implications.

The Soviet Union and the United States, Cold War rivals, had signed several Arms Control Treaties in the 1960s and 1970s, and by the early 1970s had entered into a period of *detente*. On April 19, 1971, the USSR launched the first piloted orbital space station, *Salyut 1*. Meanwhile, the US had launched the *Apollo 14* mission on January 31, the third mission to land on the Moon (February 05, 1971), but each side gave little media coverage of their achievements. Both sides had severe criticisms of the other's engineering. Soviet spacecraft were designed with automation in mind: each *Soyuz* had been designed to minimize risk due to human error by having minimal manual controls. In contrast, the US Apollo spacecraft were designed to be flown by high-trained astronauts. The USSR criticised the *Apollo* spacecraft as being ***"extremely complex and dangerous"***. Christopher C. Kraft, director of the Johnson Space Center said, *"We in NASA rely on redundant components – if an instrument fails during flight, our crews switch to another in an attempt to continue the mission. Each Soyuz component, however, is designed for a specific function; if one fails, the cosmonauts land as soon as possible. The Apollo vehicle also relied on astronaut piloting to a much greater extent than did the Soyuz machine"*. Unquote.

With the United States ending its direct involvement in the Vietnam War with the signing of the Paris Peace Accords in January 1973, relations between the USSR and the US began to improve. The *Apollo-Soyuz Test Project* (ASTP) was made possible by this thaw in relations. According to Soviet leader Leonid Brezhnev, *"The Soviet and American spacemen will go up into outer space for the first major scientific experiment in the history of mankind. They know that from outer space our planet looks even more beautiful. It* ***is big enough for us to live peacefully on it, but it is too small to be threatened by nuclear war****"*. Thus, both sides recognized the ASTP as a political act of peace.

In April 1972, a Treaty was signed to participate in a joint crewed space flight as a symbol of this *detente*. It was an **Agreement Concerning Cooperation in the Exploration and Use of Outer Space for Peaceful Purposes**. Early concepts for a joint flight included the docking of a *Soyuz* craft to the American *Skylab* space station, or an *Apollo* vehicle docking with a *Salyut* space station. Once the Americans abandoned their *Skylab* station in 1974, the *Apollo-Salyut* concept seemed to be the logical choice, but since the Soviets had started to develop a universal docking adaptor for the mission and feared having to reveal details of their military-focused *Salyut* missions, the two powers opted to link a *Soyuz* spacecraft to an *Apollo* spacecraft.

Three test flights of an uncrewed version of the ASTP spacecraft were flown: *Kosmos 638*, launched April 03, 1974; *Kosmos 652*, launched May 15, 1974; and *Kosmos 672*, launched August 12, 1974. These three flights, and *Soyuz 16*, were all launched with an improved version of a *Soyuz* booster rocket. In an unprecedented move, Soviet planners offered to inform their NASA counterparts of the time of the launch of *Soyuz 16*, as long as they did not reveal that time to the Press. NASA officials refused to agree to that condition, and, accordingly, were informed of the launch an hour after it occurred, on December 02, 1974. The ASTP of 1975 was particularly significant for the USSR's policy of keeping the details of their space program secret from the Soviet people and the world at large, especially from Americans. *Soyuz 19* was the first Soviet spacecraft to which a foreign flight crew had access before flight: the *Apollo* crew were permitted to inspect it, and also the launch and crew training site. This was an unprecedented sharing of information with Americans about any Soviet space program.

Millions of people around the world watched on television as a US *Apollo* spacecraft docked with a Soviet *Soyuz 7K-TM* capsule on July 17, 1975.The two Commanders, Stafford and Leonov, exchanged the first international handshake in space throught the open hatch of the *Soyuz*; while the crews were high over the city of Metz in France. They exchanged flags and gifts (including tree seeds which were later planted in the two countries), listened to each other's music, signed certificates, visited each other's spacecraft, ate together, and conversed in each other's languages. ***Eksperimentalniy polyot Soyuz-Apollon (EPAS)***, lit. "Experimental flight Soyuz-Apollo", was officially designated as *Soyuz 19*. The unnumbered American vehicle was left over from the canceled Apollo missions, and was the last Apollo module to fly in space.

The three American astronauts, Commander Thomas P. Stafford, Command Module Pilot Vance D. Brand, and Docking Module Pilot D.K. "Deke" Slayton, and two Soviet cosmonauts, Commander Alexei Leonov and Flight Engineer Valery Kubasov, performed both joint and separate scientific experiments. After forty-four hours docked together, the craft separated to use the *Apollo* module to create an arranged eclipse of the Sun, to allow instruments on the *Soyuz* to take photographs of the solar corona. Another brief docking was made, then the Soviets remained in space for anther two days, and the Americans for five days. While *Soyuz 19* touched down, *Apollo* was still in orbit, with the astronauts performing space-science and Earth-observing experiments. The mission was considered a great success, both technically and as a public-relations exercise for both nations. The only serious problem was during re-entry and splashdown of the *Apollo* capsule, when the crew were accidentally exposed to fumes from hypergolic propellant venting from the spacecraft and re-entering via a cabin air intake. The three astronauts were hospitalized for two weeks in Honolulu, Hawaii.

The historic ASTP meetup showed that two different space vehicles could dock in orbit, but it had far-reaching diplomatic consequences as well. The ASTP mission became symbolic of each country's goals of scientific cooperation, while their news reports downplayed the technical prowess of the other. The Soviet press implied that it was leading the US in spaceflight, tying it to the Marxist-Leninist ideology, while the United States reported that the *Soyuz* craft was technically primitive. The two mission Commanders became lifelong friends, with Alexei Leonov being the godfather of Tom Stafford's younger children. An asteroid, 2228 Soyuz-Apollo, discovered in 1977, was named after the mission. The pre-flight work provided useful engineering experience for later joint American-Russian space flights, such as the *Shuttle-Mir* program and the International Space Station (ISS). The ASTP was the last crewed United States spaceflight for nearly six years until the first launch of the United States Space Shuttle on April 12, 1981, (the twentieth anniversary of Yuri Gagarin's historic mission), and the last crewed US spaceflight in a space capsule until SpaceX's *Crew Dragon* 'Demo-2' flight to the ISS on May 30, 2020.

# Intercosmos cosmonauts

These brave men were from countries politically aligned with the former Soviet Union. They trained in Russia at Soviet expense for their missions, usually only one mission. The program was formed in Moscow in April 1967, and ran until the dissolution of the USSR in December 1991; then was continued from 1992-94 by Russia. All members of the program from the USSR were given the Hero of the Soviet Union medal; or the 'Order of Lenin', the highest civilian award. Intercosmos included the allied east-European states of the Warsaw Pact, Eastern Bloc, CoMEcon, and other socialist states such as Afghanistan, Cuba, Mongolia, and Vietnam. In addition, pro-Soviet non-aligned states such as India and Syria took part. There were also astronauts from the United Kingdom (via Project Juno); France (Jean-Loup Cretien; 1982, 1988); and Austria (Franz Viehbock, on *Soyuz TM-13*, October 02, 1991).

Beginning in April 1967 with unpiloted research satellite missions, the first crewed Interkosmos mission was *Soyuz 28* in February 1978. So-called joint crewed spaceflights enabled fourteen non-Soviet cosmonauts to participate in *Soyuz* space flights from 1978-88. The program was responsible for sending into space the first citizen of a country other than the USA or USSR: Vladimir Remek of Czechoslovakia (aboard *Soyuz 28*, March 02, 1978); the first black and Hispanic person, Arnaldo Tamayo Mendez of Cuba (*Soyuz 38*, September 18, 1980); and the first Asian person in space, Pham Tuan of Vietnam (*Soyuz 37*, July 23, 1980). Of the countries involved, only Bulgaria flew two cosmonauts (Georgi Ivanov, *Soyuz 33*, April 10, 1979; and Aleksandr Aleksandrov, *Soyuz TM-5*, June 07, 1988); and the French spacionaut Jean-Loup Cretien flew on two separate flights: *Soyuz T-6*, June 24, 1982; and *Soyuz TM-7*, November 26, 1988.

The Soviet Union also made offers of joint human spaceflight on a commercial basis to the United Kingdom and Japan; resulting in the first Japanese cosmonaut (Toyohiro Akiyama, aboard *Soyuz TM-11*, December 02, 1990) and the first British astronaut (Dr. Helen Sharman, aboard *Soyuz TM-12*, for Project Juno, May 18, 1991). In the early 1980s, an offer was made to Finland, potentially with test pilot Jyriki Laukkanen. However, the pilots of the Test flight (Koelentue) refused on the grounds that participation would not benefit the Flight or test pilot activity in any way. The last flight was *Intercosmos 26* on March 02, 1994. All were launched from Baikonur, Kazakhstan SSR, USSR.

# The "lost cosmonauts"

There are numerous theories about early Soviet cosmonauts allegedly lost on missions. The prime tales are set out here: In 1959, pioneering space theoretician Hermann Oberth claimed that a pilot had been killed on a sub-orbital ballistic flight from Kasputin Yar in early 1958, but provided no source for his story. A 1959 edition of ***Ogoniok*** published an article and photos of three high-altitude parachutists: Colonel Pyotr Dolgov, Ivan Kachur, and Alexey Grachov. Officials state that Dolgov was killed on November 01, 1962, while carrying out a high-altitude parachute jump from a Volga balloon gondola from an altitude of 28,640 metres (93,960 feet). The helmet visor of Dolgov's *Sokol* space suit hit part of the gondola as he exited, depressurizing the suit and killing him. Alexey Grachov is also thought to have been involved, along with Dolgov and Ivan Kachur (known to have disappeared around this time), in testing the high-altitude equipment. Russian journalist Yaroslav Golovanov suggested that high-altitude testing was exaggerated into a story that those parachutists died on a space flight. In late 1959, ***Ogoniok*** carried pictures of a man identified as Gennady Zavadovsky testing high-altitude equipment; he would later appear on lists of "dead cosmonauts", but without a date of death or accident description.

Golovanov, who researched the "lost cosmonaut" claims in his book, *Cosmonaut #1,* found and interviewed the real Alexey T. Belokonov, a retired high-altitude parachutist, who revealed more about his colleagues Dolgov, Kachur, Mikhailov, Grachov, Zavadovsky and Ilyushin, and **confirmed they never flew to space**. In December 1959, an alleged high-ranking Czechoslovakian Communist leaked information about many purported unofficial space shots. Alexei Ledovsky was mentioned as being launched inside a converted *R-5A* rocket. Three more names of alleged cosmonauts claimed to have perished under similar circumstances were Andrei Mitkov, Sergei Shiborin, and Maria Gromova. Also in December 1959, the Italian news agency **Continentale** repeated the claims that a series of cosmonaut deaths on sub-orbital flights had been revealed by a high-ranking Czechoslovakian communist. **Continentale** identified the cosmonauts as Alexei Ledowsky, Serenty Schriborin, Andrei Mitkow, and Maria Gromova. No other evidence of Soviet sub-orbital crewed flights ever came to light.

While the 1964 US edition of the ***Guiness Book of World Records*** credits Yuri Gagarin's *Vostok 1* as the "earliest successful manned satellite", a footnote names nine putative lost cosmonauts: Alexei Ledovsky, Serenty Schiborin, Andrei Mitkov, Alexey Timofeyevich Belokonev, Ivan Kachur, Alexey Grachev, Pyotr Dolgov, and Vladimir S. Ilyushin; and Gennady Mikhailov (named by the Judica-Cordiglia brothers). Julius Epstein wrote several papers on "Soviet Failures in Space", including allegations of lost cosmonauts, which were read into the US **Congressional Record** in 1965 and 1971. The May 1987 issue #121 of *Dragon Magazine* features "Operation: Zodiac"; and the July 1987 issue #123 features the article "Operation: Zenith", both of which speculate about cosmonauts dying in space.

In 1960, science fiction author Robert A. Heinlein wrote in his article *Pravda means "Truth"* (reprinted in ***Expanded Universe***) that on May 15, 1960, while traveling in Vilnius, in the Soviet Lithuania, he was told by Red Army cadets that the Soviet Union had launched a human into orbit that day. Heinlein speculated that *Korabl-Sputnik 1* was an orbital launch (later officially denied and said to be uncrewed), and that the retro-rockets had fired in the wrong attitude, making recovery efforts unsuccessful. According to Yuri Gagarin's biography, these rumours were likely started as a result of two *Vostok* missions equipped with dummies (including a mannequin known as Ivan Ivanovich) and human voice tape recordings (to test if the radio worked) that were made just prior to Lt. Yuri Gagarin's historic flight of April 12, 1961.

In a US press conference on February 12, 1962, Colonel Barney Oldfield revealed that an uncrewed space capsule had indeed been orbiting the Earth since 1960, as it had become jammed into its booster rocket. According to the NASA *NSSDC Master Catalog, Korabl Sputnik 1*, designated at the time 1-KP or *Vostok 1P*, did launch on May 15, 1960, (eleven months before Yuri Gagarin). It was a prototype of the later *Zenit* and *Vostok* launch vehicles. The onboard Braking Engine Unit (TDU) had ordered the retro-rockets to fire to recover, but due to a malfunction of the attitude control system, the spacecraft was oriented upside-down, and the rocket firing put the craft into a higher orbit. The re-entry capsule lacked a heat shield as there were no plans to recover it. Engineers had planned to use the vessel's telemetry data to determine if the guidance system had functioned correctly, so recovery was unnecessary. So, another "lost cosmonaut" rumour is laid to rest!

The Judica-Cordiglia brothers are two Italian former amateur radio operators who made audio recordings which allegedly support the conspiracy theory that the Soviet space program covered up cosmonaut deaths in the 1960s. The pair claim to have recorded several failed "secret Soviet space missions". These recordings have been the center of public interest for more than fifty years, but ignore the fact that US reconnaissance satellites were keeping a very close watch on Soviet events. According to Mark Wade, editor of the space history web site ***Encyclopedia Astronautica***, *"The entire early history of the Soviet manned space program has been declassified and we have piles of memoirs of cosmonauts, engineers, etc., who participated.* ***We know who was in the original cosmonaut team, who flew, was dismissed, or was killed in ground tests.*** *Vladimir Ilushin (1927-2010) is not one of them".* Unquote.

Despite the unsuccessful first test launch of the new Soviet *N-1* rocket on January 20, 1969, it is alleged that a decision was made to send a crewed *Soyuz 7K-L3* craft to the Moon using an *N-1* booster. The attempt is alleged to have occurred on July 03, 1969, when it ended in an explosion, destroying the Baikonur launch pad and killing the cosmonauts on board. Note that the United States' *Apollo 11* crew launched on July 16, and successfully landed on the Moon on July 20, 1969. This possible Soviet launch may be more credible, since the prestige and propaganda prize of being the first nation to put men on the Moon was irresistible.

# The Cosmonauts A-Z

## From the Interkosmos program: 20 Eastern Bloc Cosmonauts

### Aleksandrov, Aleksandar P. Soyuz TM-5/4.

Lt. Col. Aleksandar Panayatov Aleksandrov was born in December 1951 in Omurtag, Bulgaria. He graduated from the Bulgarian Air Force Academy in 1974; and obtained a degree in Technical Sciences in 1983. Selected as a Research Cosmonaut on March 01, 1978, as part of the Soviet Union's Interkosmos program, he was the second Bulgarian to fly to space. Launched aboard *Soyuz TM-5* on June 07, 1988, for mission Mir EP-2, he became the first Bulgarian to reach a Soviet space station, *Salyut 6*. He logged nine days, 20 hours and ten minutes in space, returning to Earth on June 17 aboard *Soyuz TM-4*. Aleksandrov later became Deputy Director of the Institute of Space Research, Bulgarian Academy of Sciences. He received eight awards, including Military Pilot, First Class; Hero of the People's Republic of Bulgaria (1988); Hero of the Soviet Union (1988); the Order of Lenin (1988); the Order of Stara Planina (2003), First Class; and the Medal "For Merit in Space Exploration" (2011). He is married, with one child.

### Faris, Muhammed A. Soyuz TM-3/2.

Muhammed Ahmed Faris was born in May 1951 in Aleppo, Syria. He is a retired pilot and former cosmonaut. After graduating from Military Pilot School at the Syrian Air Force Academy in 1973, he joined the Air Force and eventually attained the rank of Colonel. He also served as an aviation instructor and a navigation specialist. Selected in the 1985 Intercosmos Group, which allowed cosmonauts from countries allied with the Soviet Union to participate in space missions, he flew on *Mir EP-1*, becoming the first Syrian and the second Arab in space.

Launched on July 22, 1987, aboard *Soyuz TM-3* as part of the first visiting crew to the Russian *Mir* space station, he conducted several experiments in space medicine and materials processing. He returned to Earth on July 30, 1987, aboard *Soyuz TM-2*, logging seven days, 23 hours and four minutes in space. Faris was awarded the title, Hero of the Soviet Union; and the Order of Lenin, the Soviet Union's highest civilian decoration.

### Farkas, Bertalan. Soyuz 36.

Brig. Gen. Bertelan "Bertsi" Farkas was born in August 1949 in Gyulahaza, Hungary. He is a Hungarian pilot and cosmonaut who became the first Hungarian citizen to travel into space. He graduated from the Gyorgy Kilian Aeronautical College in Szolnok, Hungary, in 1969 and then attended the Krasnodar Military Aviation Institute in Krasnodar, USSR (now Russia). He graduated in 1972, and qualified as a first-class military pilot in the Hungarian Air Force in 1976. Farkas was chosen as a Research Cosmonaut candidate in the fifth Interkosmos program in 1978, and trained at the Yury Gagarin Cosmonaut Training Center in Star City, Russia. He flew aboard *Soyuz 36*, launched on May 26, 1980, to the *Salyut 6* space station. He logged seven days, 20 hours and 45 minutes in space.

After landing on June 03, he was awarded Hero of the Soviet Union; received the Order of Lenin; and retired on June 03, 1980. He later worked as a scientist at the Technical University of Budapest (now known as the Budapest University of Technology and Economics), and received a degree from its Faculty of Transport Engineering in 1986. From 1996-97, he served as the Air Force attache at the Hungarian embassy in Washington, D.C.. He retired from the Hungarian Air Force in 1997 as a Brigadier General; then served as President of **Airlines Service and Trade Ltd.**, in Budapest and co-owned by Atlant-Hungary airlines. He is married, with three children.

## Gurragchaa, Jugderdemidiin. Soyuz 39.

Zhugderemidiyn Gurrachaa was born in December 1947 in Gurvan-Bulag, Bulgan, Mongolia. Gurragchaa is the given name, while Jugderdemidin is a patronymic. He is a Mongolian cosmonaut and military leader; the first Mongolian and second Asian to go into space. In 1966, he joined the Mongolian Air Force. He studied Aerospace Engineering at the Zhukovsky Air Force Engineering Academy in Ulaanbaatar and graduated in 1978. Gurragchaa was selected as part of the eighth Interkosmos program on March 01, 1978, and was in the rank of Major General at that time.

Launched aboard *Soyuz 39* on March 22, 1981, he docked with the *Salyut 6* space station. He and Vladimir Dzhanibekov carried out experiments on Earth science. Gurragchaa logged seven days, 20 hours and 42 minutes in space; and landed 170 km southeast of Dzhezkasgan. On March 30, 1981, he was awarded the title, Hero of the Soviet Union. He also received the Order of Lenin; Hero of the Mongolian People's Republic (1981); the Order of Honour (2011); the Order of Alexander Nevsky (2021); and the Order of Ghenghis Khan (2021). From 2000-04, he served as Mongolia's Defense Minister. From 2004-08, he was a member of the State Great Khural. He heads the fund for development of bandy in Mongolia. He is married, with two children.

## Habib, Munir H..

Lt. Col. Munir Habib Habib was born in September 1953 in Jablah, Syria. He is a former Syrian Air Force pilot and Research Cosmonaut. He graduated from Military Pilot School in Aleppo in 1973 and on September 30, 1985, was selected as part of the International Group as a backup for *Soyuz TM-3*. Habib was later again in military service. He retired from his position as a Research Cosmonaut on July 30, 1987. Habib is married, with two children.

## Jahn, Sigmund W.P. Soyuz 31.

Col. Sigmund Werner Paul Jahn (1937-2019) was born in February 1937 in the town of Morgenrothe Routenkranz, Vogtland region, Saxony, German Reich. He was a German pilot, cosmonaut, and achieved the rank of Major General in the National People's Army of the GDR. He enlisted into the predecessor of the East German Army in April 1955, and eventually worked his way into the East German Airforce. He finished basic training in 1956, and remained in the Jagdfliegergeschwader 8 (Fighter Aviation Squad 8) until 1960. From 1961-63, he worked as the Deputy Commander for Political Affairs in his squadron; then headed the Air Tactics/Air Combat division until 1965. At the same time, Jahn passed his Abitur and was sent to the Gagarin Air Force Academy in Morino, near Moscow. He achieved the rank of Lieutenant Colonel, and was selected for the first Interkosmos cosmonaut training program in November 1976. He and Eberhard Kollner were chosen: in addition to his flying experience and expertise with the Russian language, his blue collar background was of help.

He was the first German to fly into space as part of the Interkosmos program. Launched aboard *Soyuz 31* on August 26, 1978, Jahn logged seven days, 20 hours and 49 minutes in space. Experiments included the MKF-6 multispectral camera for remote sensing of the Earth's surface; material science involving crystallization, recrystallization, and the cultivation of a monocrystal; weightlessness; occupational psychological studies; biological experiments and others with organic polymers and inorganic substances. Following the flight, Jahn was promoted to Colonel.

Jahn was awarded Fliegerkosmonaut der Deutschen Demokratischen Republik; Hero of the German Democratic Republic (GDR, 1978); and Hero of the Soviet Union (1978). In 1985, he became a founding member of the Association of Space Explorers. After the dissolution of the GDR on October 02, 1990, he became a freelance consultant for the **German Aerospace Center** until 1993. Jahn then worked for the ESA for fifteen years, retiring in 2002. In a 1998 interview with **Superillu**, Jahn said about his launch, *"At first, it was as if thunder was thundering in the distance. The dull rumbling quickly got closer and closer. The rocket began to vibrate, as if trembling to get away from the crater of a volcano it was sitting on, as quickly as possible. I didn't see it from our capsule fifty metres above the Earth, but eyewitnesses later told me about this unique spectacle. It looked like a fire-breathing dragon, emitting a sea of flames and smoke. The rays from the five engines raged red, yellow, blue and violet. A fascinating sight. My heart rate was elevated. But this heart pounding wasn't fear, but rather stimulating. And what I saw then was total bliss: our Earth covered in bright blue. Simply fantastic".* Unquote.

Jahn received twenty-four awards. In 2001, the asteroid 1998 BF 14 was named after him (17737 Sigmundjahn). Two schools, a street (Dr. Sigmund Jahn Strasse), and a flight school are named in his honor. On the thirtieth anniversary of his flight, in a 2008 interview with **Der Spiegel**, Jahn commented, *"But the celebratory reports weren't music to my ears. I didn't want to be made into a folk hero [...] I found the spotlight more strenuous than traveling in space".* A group known as Mitteldeutscher Rundfunk organized a themed event for his eightieth birthday in 2017. It aired as part of a special TV program ***Sigmund Jahn and the Heroes of the Stars*** on February 12 and 13, 2017. Sigmund Jahn died on September 21, 2019, aged eighty-two years, in Strausberg, Brandenburg, Germany. Jahn was married, with two children.

## Kakalov, Georgi I. Soyuz 33.

Maj. Gen. Dr. Georgi Ivanov Kakalov (known as **Georgi Ivanov**) was born in July 1940 in Lovech, Kingdom of Bulgaria (now Bulgaria). He attended the Military Air Force School in Dolna Mitrpolia for five years, then served in the Bulgarian People's Army as a military pilot; later becoming an instructor and head of a Division. Selected in 1978 Intercosmos Group, he was launched aboard *Soyuz 33* on April 10, 1979, with Nikloai Rukavishnikov. Approaching the *Salyut 6* orbital space station on April 12, 1979, the *Soyuz 33* main engine had severe damage which caused a malfunction, and prevented docking to *Salyut 6*. Since the Soyuz had limited life support resources, the crew had to return to Earth immediately, completing thirty-one Earth orbits. After just one day, 23 hours and one minute in space, the landing was also problematic, with forces of more than nine G's; landing 320 km southeast of Dzhezkazgan, Kazakhstan. This flight remains the only example for manual landing, and is thus quoted in every astronautic textbook.

With his aborted *Soyuz 33* flight, Bulgaria became the fourth Interkosmos country (after Czechoslavakia, Poland and East Germany, in that order) and the sixth in the world to send a citizen into space. He was awarded the title, Hero of the Soviet Union on April 13, 1979; Hero of the People's Republic of Bulgaria (1979); the Order of Lenin; the Order of Stara Planina, First Class with Sword (2004); and the Medal "For Merit in Space Exploration" (Russian Federation). Ivanov earned a Ph.D. in Space Engineering; and retired as a cosmonaut on April 12, 1989. In 1990, he was a member of the National Assembly of Bulgaria, and took part in the creation of the new democratic Constitution of the Republic of Bulgaria. Since 1993, he has been the Executive Director of **Air Sofia**. He was married to Natalya Rousanova, and they have a daughter, but divorced in 1982. Currently married to Lidia, they have one son.

## Mendez, Arnaldo T. Soyuz 38.

Brigadier General Arnaldo Tamayo Mendez was born in January 1942 in Baracoa, Guantanamo province, Cuba. Orphaned as an infant, he was adopted at age one year; and began working as a shoeshine at age thirteen years. After the Cuban Revolution, he entered the Technical Institute "Rebel Army", completing a course for aviation technicians in December 1960, and went on to join the Cuban Revolutionary Armed Forces. He trained with MiG-15s, and was a certified combat pilot at age nineteen years. During the Cuban Missile Crisis of 1962, he flew twenty reconnaissance missions. In 1967, he joined the Communist Party of Cuba and spent the next two years serving with Cuban forces in the Vietnam War; returning in 1969 to study until 1971 at the Maximo Gomez Basic College of the Revolutionary Forces. He is a Cuban military officer, legislator, and former cosmonaut. Mendez was the first Caribbean, the first Latin American and first person of African heritage in space.

Selected on March 01, 1978, in the USSR's seventh Interkosmos Group, he flew on *Soyuz 38*, launched on September 18, 1980, and docked with the *Salyut 6* space station. He conducted experiments in an attempt to find what caused space adaptation syndrome (SAS), and on the crystallisation of sucrose in microgravity. Mendez logged seven days, 20 hours and 43 minutes in space. The risky night landing was 180 km (110 miles) from Dzhezkazgan. He was awarded the Hero of the Republic of Cuba medal (the first recipient), and the Order of Playa Giron. In Moscow he received the Order of Lenin, and was named a Hero of the Soviet Union. Since 1980, he has been a Deputy in the Cuban National Assembly, representing his home region of Guantanimo. He is married, with three children.

## Pham, Tuan. Soyuz 37/36.

Lt. Gen. Tuan Pham was born in February 1947 in Thai Binh province, northern Vietnam. He is a retired Vietnam Air Force aviator and cosmonaut. Tuan became the first Vietnamese citiizen and the first person from an Asian country to fly in space when he launched aboard the *Soyuz 37/36* mission as an Interkosmos Research Cosmonaut. He joined the Vietnam People's Air Force (VPAF) in 1965 initially as a radar mechanic student, then was selected for pilot training as a commissioned officer. He graduated at the Krasnodar Flight School in the Soviet Union as a MiG-17 pilot in 1967. From 1968-69, he participated in developing night-time interception techniques against US air raids. From 1970-73, he was with the 921st Fighter Regiment.

Pham recounted his successful engagement against a B-52D Stratofortress on the night of December 27, 1972, *"My plane was only about ten kilometres away from the B-52 bombers when I had detected them. Jettisoning my external [fuel] tank, I've immediately asked for an order to attack. Even though I was approaching the bombers beyond the sound barrier, I felt like time is moving very slowly. The enemy's escort group of F-4 fighters had not detected us. Just to be certain, I continued to shorten the distance, within the three km range and launched the missiles, and as I was escaping, I clearly saw the moments both [K-13 air-to-air] missiles exploding against the B-52, now going down in flames. ... the F-4s would certainly be after me now, but I've escaped"*. Unquote. This claim, which would be the only B-52 ever downed in air-to-air combat, is disputed by US records, which claim it was downed by a surface-to-air missile.

On April 01, 1979, Pham was selected as a member of the sixth international crew for the Interkosmos program. He launched from Baikonur Cosmodrome on July 23, 1980, as main pilot of *Soyuz 37* to the *Salyut 6* space station. There were some technical issues regarding the *Soyuz*' main engine. He performed experiments on the melting of mineral samples in microgravity; carried out plant experiments on azolla; and photographed Vietnam from orbit for mapping purposes. Pham logged seven days, 20 hours and 42 minutes in space; returning to Earth on July 31, 1980. In 1973, he was awarded Hero of the People's Armed Forces in Vietnam; Hero of the Soviet Union; Vietnam Labor Hero (1980); the Ho Chi Minh Order; and the Order of Lenin. He was given the position of Deputy Commander within the VPAF. In 1999, he was promoted to Lieutenant General. During 2000, Tuan was Director of the General Department of Defense Industry. He retired at the end of 2007. Pham is a non-elected member of the Vietnam National Assembly. He is married, and has two children.

## Prunariu, Dumitru. Soyuz 40.

Maj. Gen. Dr. Dumitru-Dorin Prunariu was born in September 1952 in Brasov (at that time, Orasul Stalin), Stalin Region, Romania. He is a Romanian cosmonaut who flew on *Soyuz 40* to the *Salyut 6* space laboratory. He graduated from the Physics and Mathematics High School in Brasov in 1971; and from Politehnica University of Bucharest, obtaining a degree in Aerospace Engineering in 1976. He then started working as a licensed engineer for **Industria Aero Nautica Romana** at an aircraft assembly facility. He enrolled in the Romanian Air Force Officers Training School the following year, 1977. Selected for spaceflight training in 1978, and having obtained the highest marks during three years' preparation, he was selected for a flight with Leonid Popov. Launched in May 1981 on *Soyuz 40* to the *Salyut 6* station, he logged seven days, 20 hours and 42 minutes in space. Prunariu conducted experiments in astrophysics, space radiation, space technology, space medicine and biology.

In 1985, Prunariu was a founding member of the **Association of Space Explorers** (ASE), which as of 2015 had over 400 members from thirty-six countries. From 1993-2004, he was the permanent representative of ASE at the United Nations Committee on the Peaceful Uses of Outer Space (UN COPUOS) sessions. Prunariu is a co-author of several books regarding space technology and space flight, and has presented or published numerous scientific papers. His Ph.D. thesis produced new developments in the field of space flight dynamics. He is the voice of the robot BURN-E in the Romanian-dubbed version of the animated film ***WALL-E*** made by Disney Pixar.

From 1992-93, Prunariu was a co-leader of the World Bank Project on reorganization of the higher education and research system in Romania. From 1998-2004, he was the President of the Romanian Space agency, and since 2000 an Associate Professor on Geopolitics, in Bucharest, Romania. From May 2004, he served as the itinerant Ambassador Extraordinary and Plenipotentiary of Romania to the Russian Federation. In 2007, he retired from the Romanian Air Force with the rank of Major General. At the end of 2015, he was promoted to Lt. General (retired). Currently he is working for the Romanian Space Agency as an expert within the Romanian Association for Space Technology and Industry (ROMSPACE). Dr. Prunariu has received nine honors, including Hero of the Socialist Republic of Romania; Hero of the Soviet Union; the Yuri Gagarin Medal of the International Astronautic Federation (1982); the Herman Oberth Gold Medal (1984); the Medal "for Merits in Space Exploration", Russian Federation (2011); and the inner main-belt asteroid 10707 Prunariu was named in his honor (April 13, 2017).

## Remek, Vladimir. Soyuz 28.

Vladimir "Volodya" Remek was born in September 1948 in Ceske Budejovice, Czechoslovakia. He is a Czech politician, diplomat and ambassador, as well as a former military pilot and cosmonaut. Remek graduated from aviation school in Kosoce in 1970, and was commissioned as a lieutenant in the Czechoslovak Air Force. He served as a fighter pilot, flying MiG-21s in the 1st Fighter Air Regiment. After studying from 1972, he graduated from Gagarin Air Force Academy in 1976; and was selected for Interkosmos-1 on December 06,

1976, as a Research Cosmonaut. As the first cosmonaut from a country other than the Soviet Union or the United States, and with the entry of the Czech Republic and Slovakia into the European Union, Remek is considered to be the first astronaut from the European Union.

He flew aboard *Soyuz 28* from March 02-10, 1978, becoming the first and only Czechoslovak in space. He logged seven days, 22 hours and 16 minutes in space, and retired on March 10, 1978. Remek was a member of the Communist Party of Czechoslovakia. Following the Velvet Revolution in 1989, he was relegated to a role as Director of the Museum for Aviation and Astronautics in Prague. After his retirement from the Czech Air Force in 1995, he represented Czech firm **CZ Strakonice** and joint venture **CZ-Turbo-GAZ** in Moscow. He was awarded Hero of the Czechoslovak Socialist Republic; and Hero of the Soviet Union (March 16, 1978). Asteroid 2552 Remek is named after him. During the 2004 European Parliament election, he was a successful Communist Party candidate for Bohemia and Moravia. He featured in a 2009 independent comedy film called ***Osadne***. Statues of him are located in Haje metro station, Prague and in Kosice, Slovakia. Remek is married, and has three children.

## Viehbock, Franz A. Soyuz TM-13/12.

Dr. Franz Artur Viehbock was born in August 1960 in Perchtoldsdorf, Vienna, Austria. He graduated from the Vienna University of Technology with a Master's degreee in Electrical and Electronic Engineering, and later earned a Doctorate in Electronic Engineering. As part of the Intercosmos program, he and Clemens Lothaller were selected on October 06, 1989, for the Soviet-Austrian space project 'Austromir 91'. He trained for two years at the Gagarin Academy, Russia. Launched aboard *Soyuz TM-13* on October 02, 1991, he visited the *Mir* space station, returning to Earth aboard *Soyuz TM-12* on October 10. He conducted 15 experiments in the fields of space medicine, physics, and space technology. Viehbock logged seven days, 22 hours and 12 minutes in space. After his mission, he gave numerous lectures about the experience. He then worked in the United States for **Rockwell International**, and later became the Director for International Business Development for **Boeing North American** in Vienna. In March 2000, he was appointed technology advisor to the governor of Lower Austria. In 2002, Viehbock became the president of **Berndorf Band GmbH**, and in January 2008 he was made a member of the board of the **Berndorf Corporation**. He is married, with three children.

**1986 January 02 - the fourth group of test pilots for project Buran**

**1987 January 05 - Shipka Group (Bulgaria)**

## Aleksandrov, Aleksandr P. Soyuz TM-5/4.

Lt. Col. Aleksandr "Sasha" Panayatov Aleksandrov was born in December 1951 in Omurtag, Bulgaria. He graduated from Moscow Bauman-High School in 1969; then from the Bulgarian Air Force Academy in 1974; and received a degree in Technical Sciences in 1983. With the Bulgarian Air Force he rose to the rank of lieutenant colonel. Selected among six semifinalists on March 01, 1978, in Interkosmos Group 2/Bulgaria as a Research Cosmonaut, he was backup to Georgi Ivanov on the *Soyuz 33* mission. He was then assigned to the prime crew of the *Soyuz TM-5* mission to the Russian *Mir* space station.

Aleksandrov flew on the 'Mir EP-2' mission (*Soyuz TM-5/Soyuz TM-4*) from June 06-17, 1988, becoming the first Bulgarian to reach a Soviet space station. He logged nine days, 20 hours and nine minutes in space. Aleksandrov later became Deputy Director of the Institute of Space Research, **Bulgarian Academy of Sciences**. He retired on June 17, 1988. Aleksandrov was awarded Pilot-cosmonaut of Bulgaria; Hero of the People's Republic of Bulgaria; Hero of the Soviet Union (1988); the Order of Georgi Dimitrov; the Order of Lenin (1988); and the Medal "For Merit in Space Exploration" (2011) - for outstanding contribution to the development of international cooperation in manned space flight. He is married, with one child.

## Stoyanov, Krasimir M..

Col. Krasimir Mihailov Stoyanov was born in January 1961 in Varna, Bulgaria. He graduated as a pilot from Georgi Benkovski Air Force Academy, Dolna Mitropolia, with a Master of Science degree in 1984. Selected on January 08, 1987, he went to the Senior Military Medical Institute in Sofia for several weeks of examinations in isolated conditions. The candidates who made it to the next round were Georgi Ivanov Kakalov, Aleksandr Aleksandrov, Georgi Yovchev and Ivan Nakov. Stoyanov was a Research Cosmonaut (Flight Engineer) backup for *Soyuz TM-5* mission. The first stage of training included theoretical studies, flight practice in jet aircraft, weightlessness simulation, splashdown training, physical exercise, and retrieval training in difficult terrain. The second phase was more specific, and concentrated on mastering the *Soyuz* spacecraft and the flight to the *Salyut* space station. Interkosmos member countries were largely relieved of financial costs, since the USSR financed almost all R & D activities, flights and technology sharing. In general, Interkosmos flights focused on five main areas of research: space physics, space meteorology, communications, space biology and medicine, and studies of the natural environment. Stoyanov retired as a cosmonaut on June 17, 1988. He is married, with two children.

**1987 March 26 - TsPK-8/NPOE-8 Cosmonaut Group**

**1988 February 12 - OS "Mir" Group (Afghanistan)**

## Masum, Mohammad Dauran-Ghulam.

Maj. Gen Mohammad Dauran-Ghulam Masum was born in January 1954 in Nidgrab District, Kapesa Province, Afghanistan. Sometimes his name is given as Darwan. "Dauran-Ghulam" is an additional name he gave himself. It means something like "servant of luck". As an Air Force pilot, he was selected for prime crew of *Soyuz TM-6*, but was replaced by Abdulahad Momand (probably because he is of Tadjik nationality), so was assigned as backup. He retired as a cosmonaut on September 07, 1988. Masum's most recent job was as Major General in the Afghanistan Air Force. He is married, with six children.

## Mohmand, Abdulahad. Soyuz TM-6/5.

Abdulahad "Abdulah" Mohmand was born in January 1959 in Sardah, Afghanistan. He is an Afghan-German and former Afghan Air Force aviator who became the first, and currently only, Afghan citizen to journey to space. Mohmand graduated from Kabul Polytechnic University in 1978; and from Gagarin Air Force Military Academy, Monino, in 1987. Selected on February 12, 1988, as a Research Cosmonaut, the nine-year Soviet-Afghan War was coming to a close, with the final troop withdrawal starting the in May 1988. It was decided by the Glavcosmos chairman that the Afghan's spaceflight would be moved earlier than planned, so that it would occur before the Soviet military completed their withdrawal from Afghanistan.

He flew on mission 'Mir EP-3' (*Soyuz TM-6/Soyuz TM-5*) from August 29 to September 07, 1988, as the Flight Engineer. Mohmand's main objective was to photograph and make observations of Afghanistan from space; to map the country's inaccessible mountainous regions, assess water and glacial run-off, as well as to find possible sources of oil and gas in lowland regions. He logged eight days, 20 hours and 26 minutes in space. Mohmand retired on September 07, 1988, and currently lives near Stuttgart, Germany; working in the printing service. He was a refugee, and now has German and Afghan citizenship. He now writes his surname as "Momand". He is married, with three children.

**1989 January 25 - IBMP-5/GKNII-3/NPOE-9/TsPK-10 Cosmonaut Group**

**1989 March 22 - The last group of test pilots for project Buran**

**1989 November 25 - Project Juno (UK-Soviet Union)**

## Mace, Timothy K.C.

Maj. Timothy Kristian Charles Mace (1955-2014) was born in November 1955 in Catterick, Yorkshire, UK. He was a British engineer and astronaut who graduated from the Military University in Shrivenham, Swindon, with a Bachelor of Science degree in Aeronautical Engineering in 1977. He served as a Major in the Royal Air Force and was a helicopter instructor pilot and parachutist. Mace was selected on November 25, 1989, as a Research Cosmonaut backup for *Soyuz TM-12* and was one of three candidates from the United Kingdom on the shortlist of twenty-five applicants in the 1992 ESA selection. However, he was not selected in 1992 or 1998; and retired in September 1998. Mace was married to the daughter of former Soviet cosmonaut V. Zolobov. Later, he lived in South Africa as a helicopter pilot for the former president of South Africa, Nelson Mandela. Mace died of cancer on September 14, 2014, in Tutzing, Germany. He was married, with two children.

## Sharman, Helen P. Soyuz TM-12.

Dr. Helen Patricia Sharman, (CMG, OBE, HonFRSC) was born in May 1963 in Sheffield, England. She is a British chemist and astronaut who became the first British person, first Western European woman, and first privately funded woman in space. She was also the first British citizen to go into space and she visited the *Mir* space station in May 1991. Sharman worked for **Mars Confectionery Ltd.** as a chemist before her astronaut career. She obtained a Bachelor of Science degree in chemistry at the University of Sheffield in 1984 and a Ph.D. degree from Birbeck, the University of London in 1987. In late 1989, Helen Sharman responded to a radio advertisement for astronauts. Her space mission was part of a cooperative Soviet-British mission called Project Juno. The aim was to enhance the relationship between the two nations in the twilight years of the Cold War by sending a British astronaut to the Soviet *Mir* space station. Sharman was selected in a process that gave weight to scientific, educational and aerospace backgrounds, as well as the ability to learn a foreign language.

She spent eighteen months in intensive flight training in Star City, Moscow, before flying to space on May 18, 1991, aboard *Soyuz TM-12*, with cosmonauts Commander Anatoly Artsebarsky and Sergei K. Krikalev. Due to financial constraints, some British experiments were limited. Sharman conducted medical and agricultural tests. She also communicated with British schoolchildren via radio. Their mission lasted seven days, 21 hours and 13 minutes. [Between 1986 and 2000, a total of 104 individuals visited *Mir*; including eleven women, forty-four *Mir* crew members, one tourist, eighteen double and four triple-flights, one quadruple and one quintuple flight]. In 2015, Sharman joined the faculty at Imperial College, London. Her books include the autobiography *Seize the Moment* (1993; written with Christopher Priest). Her biography is available in thirty-four different languages on Wikipedia. In 1992, Dr. Helen Sharman was made an Officer of the Order of the British Empire (OBE).

**1990 May 11 - TsPK-11 Cosmonaut Group (Soviet Union)**

Musabayev, Talgat A.

Severin, Vladimir G.

Sharipov, Salizhan S.

Vozovikov, Sergey Y.

Zalyotin, Sergei V.

**1992 March 03 - NPOE-10 Cosmonaut Group (Russia)**

Lazutkin, Aleksandr I.

Treshchov, Sergei Y.

Vinogradov, Pavel V.

**1994 April 01 - NPOE-11 Cosmonaut Group**

Kuzhelnaya, Nadezhda.

Tyurin, Mikhail V.

**1996 February 09 - MKS/RKKE-12 Cosmonaut Group (Russia)**

**1996 March 26 - MKS Supplemental Cosmonaut group**

Kononenko, Oleg D.

**1996 November - Shuttle-97 Group (Ukraine)**

Kadeniuk, Leonid K.

Pustovyi, Yaroslav.

**1997 July 28 - TsPk-12/RKKE-13 Cosmonaut Group**

**1998 February 24 - RKKE-14 Cosmonaut Group**

Korniyenko, Mikhail.

**1998 March 02 - OS "Mir" Steganik Group (Slovakia)**

## Bella, Ivan. Soyuz TM-29/28.

Brigadier General Ivan Bella was born in May 1964 in Brezno, Czechoslavakia. He is a Slovak Air Force officer who became the first Slovak citizen to fly in space. He graduated from Kosice Military Academy in 1987. Bella rose to the rank of Lt. Col. in the Slovak Republic Army as a pilot at the 33rd Air Force Base in Malacky, Slovakia. Selected in the March 25, 1998 Cosmonaut Group, he completed training as a research cosmonaut in August at the Gagarin Cosmonaut Training Center. Ivan Bella participated in an eight-day joint Russian-French-Slovak mission *Mir-Stefanik*, launched aboard *Soyuz TM-29* on February 20, 1999, docking with *Mir* on February 22, and returning to Earth on *Soyuz TM-28* on February 28, 1999. Bella logged seven days, 21 hours and 56 minutes in space. In 2004, Bella was appointed to serve as a military attache for Slovakia in Moscow, Russia.

## Fulier, Michal.

Col. Michal Fulier was born in February 1955 in Ceski Tesin, Moravia, Czechoslovakia. He received a Diploma at the Military Flying Institute at Kosice in 1974. He gained a higher leader position at the Slovakian Military Academy. He traveled to Star City, Russia, for Research Cosmonaut training from March 02, 1998. He did not fly, because Ivan Bella was chosen. Fulier was a backup for the *Soyuz TM-29* discovery spaceflight. He left the base on February 28, 1999. Fulier is married, with two children.

**2003 May 23 - TsPk-13/RKKE-15/IBMP-6 Cosmonaut Group**

**2003 Kazakhstan - Group 1**

## Aymakhanov, Mukhtar R.

**2006 September 04 - Angkasawan Group (Malaysia)**

## Kamaludin, Mohammed F..

Capt. Mohammed Faiz Kamaludin was born in January 1972 in Kuala Lumpur, Malaysia. A pilot, he was selected on January 09, 2006, as a Spaceflight Participant (UKP) for the Program Angkasawan Negara by the Malaysian National Space Agency (ANGKASA); one of four finalists for training in Russia. Kamaludin retired from the program on September 04, 2006. He is married, with children.

## Khaleed, Faiz. Soyuz TMA-11.

Maj. Faiz bin Khaleed was born in September 1980 in Kuala Lumpur, Malaysia. He is a military dentist with the Malaysian Armed Forces. Selected on September 04, 2006, for the Malaysian Angkasawan program, he was one of two candidates to undergo training in Star City, Russia. In 2007, he was selected to fly to the ISS aboard *Soyuz TMA-11* for 'Expedition 16'. On October 22, 2007, he was promoted to Major in the Malaysian Armed Forces; and retired from there in late 2013. He is currently single.

## Shukor, Sheik MusZaphar. Soyuz TMA-11.

Sheik MusZaphar Shukor Al Masrie bin Sheik Mustapha was born in July 1972 in Kuala Lumpur, Malaysia. He is of Malay, Minangkabau, and Arab descent. He received a Bachelor of Medicine and Surgery degree from Kasturba Medical College, affiliated to Manipal University, India. He previously worked as an orthopaedic medical officer, but never completed his specialization. Selected on September 04, 2006, for the Malaysian Angkasawan program, he was one of two prime candidates to undergo training in Star City, Russia. He was selected to fly to the International Space Station (ISS) aboard *Soyuz TMA-11* for 'Expedition 16', on October 10, 2007. He works part-time as a model.

## Vanajah, Siva.

Dr. Siva Vanajah was born in March 1971 in Kulim, Malaysia. She is an ethnic Indian and Malaysian engineer, and one of the four finalists from 11,000 applicants that applied for astronaut selection in 2003 to the Malaysian Angkasawan program, intended to send a Malaysian to the International Space Station. The finalists went to Moscow for training by the Russian Space Agency, but she lost her place to Sheik Muszaphar Shukor and Faiz Khaleed.

**2006 October 11 - TsPk-14/RKKE-16 Cosmonaut Group**

**2010 October 12 - TsPk-15/RKKE-17 Cosmonaut Group**

**2011 January-February - Enrolled in a United squad of Roscosmos cosmonauts (Russia) x13**

**2012 February - enrolled in a United squad of Roscosmos cosmonauts**

## Yurchikhin, Fyodor.

**2012 October 30 - TsPk-Addition Group x7**

**2014 August 14 - Individual set into a United detachment of Roscosmos cosmonauts**

## Aymakhanov, Mukhtar R.

**2018 August 10 - 17th Cosmonaut Group**

**2021 January 27 - 18th Cosmonaut Group**

## Aleksandrov, Aleksandr.

# Cosmonauts of the USSR and Russia, A-Z

## Afanasyev, Viktor M. Soyuz TM-11, TM-18, TM-29, TM-23.

Victor Mikhailovich Afanasyev was born in December 1948 in Bryansk, Russian SFSR, Soviet Union. He is a retired pilot, Colonel in the Russian Air Force and former cosmonaut. He graduated from Kachnskoye Military Pilot School in 1970; and from the Ordzhonikidze Aviation Institute, Moscow, in 1980. From 1970-76, he served in the Soviet Air Force, up to the position of aircraft flight commander. He attended the Test Pilot Training Center fom 1976-77, and from 1977-78 served at the State Research/Test Institute named after Valery Chkalov. He has a Class 1 military test pilot certification, and has logged over 2,000 flying hours in over forty aircraft types. Selected in 1985 Cosmonaut Group, Afanasyev trained part-time from 1985-87 at the Yuri Gagarin Cosmonaut Training Center, and proceeded to advanced training in 1988. From February 1989, he trained for a flight to the *Mir* orbital space station as the *Mir-7* mission backup crew commander. His first flight of 175 days was launched aboard *Soyuz TM-11* to the 'Mir EO-8' mission as Crew Commander (December 02, 1990 to May 26, 1991). He performed four EVAs totaling 20 hours and 55 minutes. From January 08 to July 09, 1994, he was aboard *Soyuz TM-18* for the 'Mir EO-15' mission as Crew Commander. From October 1996 to January 1998, he trained for the 'Mir 25' mission as backup crew.

From March 1998, he trained as the 'Mir EO-27' mission primary Crew Commander. Launched aboard *Soyuz TM-29* for 'Mir EO-27' (February 20 to August 28, 1999), he spent another 189 days in space; performing three EVAs. Launched aboard *Soyuz TM-23* to the ISS on October 21, 2001, for his fourth spaceflight, he returned to Earth aboard *Soyuz TM-23* on October 31 after eight days aboard the space station; landing near the township of Arkalyk, Kazakhstan, at 46 degrees, 44 minutes and 58" North, 69 degrees, 42 minutes, and 58" East. In 2001, he was assigned to the ISS Taxi-1 backup crew. Afanasyev logged a total of 555 days, 18 hours and 33 minutes in space; including seven EVAs totaling 38 hours and four minutes. He has a Class 1 cosmonaut certification; and retired on April 17, 2006. Afanasyev has received eight awards, including Hero of the Soviet Union (1991); the Order of Lenin (1991); the Order for Personal Courage (1994); the Order of Merit for the Fatherland, second class (2002), third class (1999); the Medal "For Merit in Space Exploration" (2011); and Grand Officer of the Legion of Honour (France). He is married, with two children.

## Aimbetov, Aydyn. Soyuz TMA-18/TMA-16M.

Aydyn Aimbetov was born in July 1972 in the Kazakh SSR, Soviet Union. He graduated from Kutakhov Armavir Higher Military Aviation School and became a military pilot. He served at Lugovaya and Taldykorgan, flying MiG-27 and Su-27 fighters. Aimbetov was selected to the first Kazakh cosmonaut class on September 09, 2002, from a pool of 2,000 candidates. He trained at Star City, Russia, from 2003-09. He had been scheduled to fly in Autumn 2009, but due to the world financial crisis, this fell through. Returning to Kazakhstan, Aimbetov stayed in the JSC (National Center of Space Research and Technologies) of the National Space Agency (NSA) of Kazakhstan. He founded the Institute for Space Development to promote the space industry in Kazakhstan.

Aimbetov started the Young Cosmonauts School in the Astana Pupil's Palace in 2012. In June 2015, he was selected to fly on the *Soyuz TMA-18* mission in place of singer Sarah Brightman who had withdrawn in May 2015, and her backup, Satoshi Takamatsu, who had also withdrawn. On September 02, 2015, Aimbetov became the third Kazakh-born cosmonaut to fly since Kazakhstan's independence in 1991, and the first to fly under the Kazakh flag, and as part of KazCosmos. He returned from his ten-day mission to the ISS on September 11, 2015, and touched down on the Kazakh Steppe aboard *Soyuz TMA-16M*.

## Amyakhanov, Mukhtar R.

Mukhtar Rabatuly Aymakhanov was born in January 1967 in Dzhusaly (Zhosaly), Kazakh SSR, USSR. He is a Russian (originally a Kazakh) cosmonaut. He graduated from Lenin Komsomol Higher Military Aviation School in Chernigov, Ukraine in 1988, and served as a pilot until 1993 in the Soviet and Kazakh Air Force, respectively. From 1993-98, he worked with the firms AO Central-Asian Depotitsystem Altyn-Kasna, YOO "Terminal" and TOO "MAAGBOOL" on the Kazakhstan stock exchange. In 1999, he founded his own firm, the TOO "Altyn Consalt", Almaty, a cooperative consulting society, which advises and rates firms and research projects on the field of law. He is also the managing director of this firm. After graduating from the Academy of Law "Adiljot", Astana, in 2000 he retired from active military duty and became Major of the Air Force Reserve, OKP.

Aymakhanov and Aidyn Aimbetov were selected as the first class of Kazakh Research Cosmonauts on September 09, 2002, for the Kazakhstan National Space Agency, KazCosmos. From 2003-09 they trained at Star City, Russia; until the world financial crisis indefinitely postponed the prospective Kazakhstan mission. On January 27, 2014, Amyakhanov was selected by MVK (former GMVK) and from August 14, 2014, he was a Test Cosmonaut (TC) member of the TsPK cosmonaut group. In September 2015, the mission finally took place aboard *Soyuz TMA-18*, when Aimbetov flew with Sergey Prokopyev as his backup. By that time, Aymakhanov had left KazCosmos and joined Roscosmos. On December 28, 2022, he left the position of a test cosmonaut in the Roscosmos squad after the decision of the Main Medical Commission; and retired on December 30. He will continue to work at the Gagarin Cosmonaut Training Center as a leading specialist in cosmonaut training. He is married, with one child.

## Anikeyev, Ivan N.

Ivan Nikolayevich Anikeyev (1933-92) was born in February 1933 in Liski, Voronezh Oblast, Russian SFSR, Soviet Union. Senior Lieutenant Anikeyev, aged 27 years, was selected on March 07, 1960, as one of the original twenty Test Cosmonaut candidates in TsPk-1, Air Force Group 1. Arrested by the militsia along with fellow candidates Valentin Filatyev and Grigory Nelyubov at Chkalovsky station for drunk and disorderly conduct, the officers of the security patrol were willing to ignore the whole incident if the cosmonauts apologised. Anikeyev and Filatyev agreed, but Nelyubov refused, and the matter was reported to the authorities. Because there had been previous incidents, all three were dismissed from the Soviet cosmonaut program for disciplinary reasons related to alcohol problems. He never flew in space, and "retired from the program" on April 17, 1963. To protect the image of the Soviet space program, images of the three trainees were airbrushed out of cosmonaut photos. This led to speculation about "lost cosmonauts", presumably to undisclosed accidents, but the actual reasons were often very mundane.

Later, Anikeyev was a Soviet Air Force pilot and navigation office on ground control. Cosmonaut veteran recounted that years afterward, while Anikeyev was at an evening party, someone stole his keys from his pocket, then stole his car. The thief hit and killed a pedestrian, and to avoid responsibility, returned the car keys to Anikeyev's pocket. The latter was sent to prison for a year before authorities realized he was innocent and released him, but Anikeyev was never given the chance to fly. His hobbies were music and theatre. Ivan Anikeyev died on August 08, 1992, aged fifty-nine years, in Bezhetsk, Tver Oblast, Russian Federation.

## Artemyev, Oleg G. Soyuz TMA-12M/Expn 39/40, MS-08/Expn 55/56, MS-21/Expn 66/67.

Oleg Germanovich Artemyev was born in December 1970 in Riga, Latvian SSR, (present-day Latvia) and grew up in Leninsk (now Baikonur), Kazakhstan. He graduated from the Tallinn Polytechnical School in 1990; and from the Bauman University, Moscow, in 1998, with a degree in Low Temperature Technology and Physics. He served in the Soviet Army in Vilnius, Lithuania, until 1991, and has worked at RKKE since 1998. Artemyev was a member of the pre-launch processing team of the *Zvezda* Service Module working on the EVA and teleoperation control system. Artemyev was selected as part of the May 29, 2003, RKKE-15 Cosmonaut Group. In 2008, he was part of a testing campaign for the *Orlan-MK* space suit. He graduated from the Russian Academy of State Service under the President of the Russian Federation in 2009, specializing in Personnel Management. As part of the Soyuz Processing team, he worked at the Baikonur Cosmodrome in 2010-11. He processed the *Soyuz TMA-21* craft that launched in 2011. Launched aboard *Soyuz TMA-12M* on March 25, 2014, he was a Flight Engineer member of the ISS 'Expedition 39/40' long-duration crew. He returned to Earth on September 11, 2014.

Launched on March 21, 2018, as Commander of *Soyuz MS-08*, he flew to the ISS to join 'Expedition 55/56'; returning to Earth on October 05, 2018. Launched on March 18, 2022, he commanded *Soyuz MS-21*, joining ISS 'Expedition 66', and became part of 'Expedition 67, later assuming command of the ISS on May 05, 2022. He spent 195 days in space before returning to earth on September 29, 2022. He will launch for the fourth time aboard *Soyuz MS-29*. He is an active Russian cosmonaut, with 529 days spent in space to date. Artemyev is married to Anna Malikhova.

## Artsebarsky, Anatoly P. Soyuz TM-12.

Anatoly Pavlovich Artsebarsky was born in September 1956 in Prosyana, Ukrainian SSR, Soviet Union. He is a retired test pilot, Russian Air Force officer and former cosmonaut. Selected in 1985, he launched aboard *Soyuz TM-12* atop a *Soyuz-U2* booster on May 18, 1991, to space station *Mir* for mission 'Mir EO-9'. Artsebarsky and Flight Engineer Sergei Krikalev stayed aboard *Mir*, while other crew (including Dr. Helen Sharman of Britain, the first woman to visit *Mir*) returned to Earth eight days later. The cosmonauts were almost stranded on *Mir*, since they were in orbit during the Soviet coup d'etat attempt against Mikhail Gorbachev of August 1991. For several days, the political situation seriously jeopardised their position. Artesbarsky logged 144 days, 15 hours and 21 minutes in space; including six EVAs totaling over 33 hours. Anatoly Artesbarsky was awarded Hero of the Soviet Union; Pilot-Cosmonaut of the USSR; the Order of Lenin; and the Medal "For Merit in Space Exploration". He is mentioned in the 2013 movie ***Gravity***, because fictional astronaut Matt Kowalski hopes to break Artsebarsky's spacewalking record. On October 07, 2019, he visited Teheran's Sharif University, together with cosmonaut Talgat Musabayev.

## Artyuhkin, Yuri P. Soyuz 14.

Col. Dr. Yuri Petrovich Artyuhkin (1930-98) was born in June 1930 in Pershutino, Moscow Oblast, Russian SFSR, USSR. He was an engineer, pilot in the Soviet Air Force, and former cosmonaut. Artyuhkin graduated from the Soviet Air Force Institute with a Doctorate in Engineering, specializing in military communication systems. Selected in 1963 Air Force Group 2, he would have flown on *Voskhod 3* if the mission had not been canceled. Launched on July 03, 1974, aboard *Soyuz 14* to dock with the orbital *Salyut 3*, the crew tested the *Almaz* station systems, such as the solar arrays. Increased solar activity raised human safety concerns, but it was decided radiation levels were within safe limits. Analysts presumed that much time was taken up with unreported military activities, since the mission purpose was to test *Salyut 3* for suitability as a crewed military reconnaissance platform. Some experiments described by the Soviets included studies of the heart and circulatory systems in orbit, studies of intercranial pressure, monitoring of blood composition, measuring of lung capacity and inhalation and exhalation rates, and the testing of a water purification system (which condensed moisture from the station's atmosphere).

The cosmonauts each exercised for two hours per day, and because of this were able to climb from their *Soyuz* descent module unaided when their mission ended on July 19, 1974. Artyuhkin logged 15 days, 17 hours and 30 minutes in space; landing some 140 km southeast of Dzhezkazgan, Kazakhstan. Artyuhkin left the Soviet space program in 1982 and held various positions in space-related fields. Most notably, he was involved in the development of the Soviet space shuttle, *Buran*, and in cosmonaut training. Yuri Artyuhkin was awarded six honors, including Hero of the Soviet Union; the Order of Lenin; the Order of the Red Star; and the Medal "For Distinction in Guarding the State Border of the USSR". He died of cancer on August 04, 1998, aged 68 years, in Zvyozdny Gorodok, Russia.

## Askyonov, Vladimir V. Soyuz 22, Soyuz T-2.

Vladimir Viktorovich Askyonov was born in February 1935 in Giblitsy, USSR. He is a retired pilot and cosmonaut. Selected in 1973 Civilian Specialist Group 5, he initially trained in the *Soyuz 7K-S* craft with Leonid Kizim, and helped train a group of fellow civilian Flight Engineers. Launched on September 15, 1976, he flew as Flight Engineer aboard *Soyuz 22*. Using the East German-made MKF-6 camera, with Commander Valery Bykovsky he took 2,400 photographs; including parts of East Germany, the Soviet Union and of the Moon. Their mission lasted seven days, 21 hours and 52 minutes; landing on September 23, 1976. He was awarded the title Hero of the Soviet Union (1976). From 1978-80, Askyonov trained as Flight Engineer on the *Soyuz T* spacecraft. Launched on June 05, 1980, aboard *Soyuz T-2*, with Commander Yuri Malyshev, they docked with the *Salyut 6* orbital space station on June 06, after testing new solar cells. After three days, 22 hours and 19 minutes in space, they landed on June 09, 1980. He was awarded a second title of Hero of the Soviet Union on July 17, 1980. Askyonov logged a total of 11 days, 20 hours and 11 minutes in space. From 1983-92, he was Deputy Chairman of the Board of the Soviet Peace Fund. In 1988, he became the Director of the State Research Center for the Study of Natural Resources of the State Hydrometeorological Service.

From 1990-92, Askyonov worked as General Director of the **NPO Planeta**. In 1992, he became Deputy Chairman of the Executive Committee of the Cosmos Association. From 1990-96, he was deputy chairman of the Board of Directors of **Mosbiznesbanka**. In 1996, he became Chairman of the Presidium of "Spiritual Movement of Russia". In 2001, he became the president of the Institute for Security and Sustainable Development Science Foundation. Vladimir Askyonov has received nine awards, including Order of Karl Marx (1976); two Order of Lenin (1976, 1980); Pilot-Cosmonaut of the USSR (1976); the Medal "Veteran of Labour"; the Gold Medal "For Services to Science and Humanity" (Czechoslovakia); Honoured Master of Sports; and Honorary Citizen of Ryazan (1976).

## Atkov, Oleg Y. Soyuz T-10V/11.

Dr. Oleg Yur'yevich Atkov was born in May 1949 in Khvorostyanka, USSR. He is a Russian cardiologist and former Soviet cosmonaut. With a doctorate from the Russian Academy of Medical Science, he was chosen to be the health specialist aboard two *Soyuz* flights; monitoring the health of the other cosmonauts and researching the effects of zero gravity on the human body. Atkov graduated from the I.M. Sechenov First Moscow State Medical University in 1973. He received his Doctorate in Cardiology from the Russian Academy of Medical Sciences, then started as a research fellow at the Myasnikov Institute of Clinical Cardiology of the Academy of Sciences, and soon became the senior research fellow. He gained eminence in his field when he discovered a method to diagnose cardiovascular diseases using ultrasound.

Selected in 1977, he joined the cosmonaut training group. On September 03, 1983, he was chosen to be part of the AMN selection group with Ural Sultanov and Magomed Tolboyev. Launched on February 08, 1984, aboard *Soyuz T-10V* to *Salyut 7* as Cosmonaut Researcher, he brought his own invention: a portable ultrasound cardiograph. The team objective was to study the acute stage of adaptation to weightlessness, phenomenology, and mechanisms of changes in hemodynamics, metabolism, and other functions. Dr. Atkov and his crew logged 236 days, 22 hours and 49 minutes in space; a new space endurance record (more than a month longer than *Soyuz T-5*). He returned to Earth aboard *Soyuz T-11*, and ended his cosmonaut career by retiring on October 01, 1984. After his extended stay in space, Dr. Atkov returned to work at the Myasnikov Institute of Clinical Cardiology to continue his research on the adaptation of weightlessness and cardiology. He is a professor of Medicine at the Russian National Research Medical University. Dr. Atkov has over 150 published articles, fifteen inventions, and holds several Patents. In 2000, he became the President of the **Russian Telemedicine Association**.

In 2002, he became the Head of the Health Department of the Ministry of Railways of the Russian Federation (soon reformed as the Russian Railways JSC). In August 2005, he was promoted to Vice-President **Russian Railways JSC**, while keeping his responsibility and focus in healthcare. In a 2012, interview with **3 Eyes Observer**, Dr. Atkov was questioned about the future of space exploration. He answered, *"We need to focus on returning to the Moon ... to establish an international base on the surface of the Moon in order to create* ***an outpost for Universe observation and solar system observation.*** *... There should be a focus on reaching Mars as well, but that those two goals should be prepared in safe conditions because of the known hazards that will be found"*. Unquote. Dr. Oleg Atkov has received eight awards: Hero of the Soviet Union; the Order of Lenin; Pilot-Cosmonaut of the USSR; the Medal "For Merit in Space Exploration" (Russian Federation); the State Prize of the USSR; the Lenin Komsomol Prize; Honoured Scientist of the Russian Federatio; and Kirti Chakra (India).

## Aubakirov, Toktar O. Soyuz TM-13/12.

Maj. Gen. Toktar Ongarbayuly Aubakirov was born in July 1946 in Karkaraly District, Karaganda Region, Kazakh SSR, Soviet Union. He is a retired test pilot, parachutist, Kazakh Air Force Officer and former cosmonaut, the first person from Kazakhstan to go into space. In 1965, he joined the Armavir Military Aviation Institute of the Anti-Air Defense Pilots. He served as a fighter pilot in the Soviet Air Force on the Far East borders of the USSR, until his acceptance into the Fedotov Test Pilot School in 1975. From 1976-91, he served as a test pilot at the Mikoyan Experimental Design Bureau (MiG aircraft), testing over 50 aircraft types. He was the first in the Soviet Union to make a non-stop flight crossing the North Pole, with two in-flight refuelings; and the first to take off from the aircraft-carrier *Tbilisi* aboard a MiG-29K jet aircraft.

Selected in 1991 Cosmonaut Group, he began training at the Gagarin Academy. Launched on October 02, 1991, aboard *Soyuz TM-13* with Franz Viehbock, he spent seven days, 22 hours and 12 minutes in space. He landed aboard *Soyuz TM-12* on October 10, 1991; some sixty-one km southwest of Arkalyk, Kazakstan. This mission was the last launched by the Soviet Union, which dissolved in December 1991, with Aubakirov becomng a citizen of the independent Republic of Kazakhstan, CIS. Aubakirov received eleven awards, including Hero of the Soviet Union (1988); the Order of Lenin (1988); Honoured Test Pilot of the USSR (1990); the Order of the October Revolution (1991); People's Hero of Kazakhstan (1995); and the Medal "For Merit in Space Exploration (2011). Since 1993, he has been the General Director of the National Aerospace Agency of the Republic of Kazakhstan. He was a member of the parliament of Kazakhstan, and is now a consultant. Aubakirov is married, with two sons.

## Avdeyev, Sergei V. Soyuz TM-15, TM-22, TM-28/29.

Sergey Vasilyevich Avdeyev was born in January 1956 in Chapayevsk, Samara Oblast, Russian SFSR, Soviet Union. He graduated from the Moscow Engineering Physics Institute in 1979 as an Engineer-Physicist. Selected on March 26, 1987, as part of the Energia Engineer Group 9, he trained from December 1987 to July 1989. Launched on July 27, 1992, aboard *Soyuz TM-15* (Mir EO-12), the mission lasted 188 days, 21 hours and 41 minutes; returning to Earth on February 01, 1993. Launched on September 03, 1995, aboard *Soyuz TM-22* (Mir EO-20), the mission lasted 179 days, one hour and 41 minutes. Avdeyev returned to Earth on February 29, 1996. Launched on August 13, 1998, aboard *Soyuz TM-28* (Mir EO-26/27), the mission lasted 379 days, 14 hours and 51 minutes. He returned to Earth aboard *Soyuz TM-29* on August 28, 1999.

Avdeyev has orbited the Earth 11,968 times, traveling about 515 million kilometres. He logged a total of 747 days, 14 hours and 14 minutes in space; including ten EVAs totaling 42 hours and two minutes. For a long time, Avdeyev held the record for ***time dilation*** experienced by a human being. In his 747 days aboard *Mir*, he went approximately 27,360 km/hr and thus aged roughly 0.02 seconds (20 milliseconds) slower from an earthbound person's perspective. This is due to the *special relativistic effect* of time dilation, and is not properly thought of as "time traveling" as described by mainstream culture. A common misconception is that the *Apollo* program astronauts hold the time dilation record – they did travel at a higher speed than Avdeyev (*Apollo 8* re-entered at 25,000 mph, or 40,000 km/hr), but *Apollo* missions were in space for only about six to eleven days. The book *Let's Do the Time Warp Again* (2002, by Scott Mowbray) says, *"... put Sergei Avdeyev one fiftieth of a second into the future ...* ***he's the greatest time traveler we have so far****"*. In August 2005, his record was taken by Sergei K. Krikalev; and the current record of 879 days was set by Gennady Padalka in 2015. It was broken again in late February 2024 by cosmonaut Oleg Kononenko on the ISS surpassing a career 1,000 days in orbit.

Avdeyev was awarded Hero of the Russian Federation (February 05, 1993) – for the successful implementation of spaceflight on the orbital scientific research complex *Mir* and displaying courage and heroism; the Order of Merit for the Fatherland, second class (1999), third class (1996); the Medal "for Merit in Space Exploration" (2011); Chevalier of the Legion of Honour (France, 1999); and was made an Honorary Citizen of Chapayevsk, Samara Oblast (2004). He is married, with two children.

## Aymakhanov, Mukhtar R. Soyuz TMA-18M.

Mukhtar Rabatuly Aymakhanov was born in January 1967 in Zhosaly, Kazakh SSR, USSR (now Kazakhstan, CIS). He graduated from the Chernihiv Higher Military Aviation School for Pilots. He and Aidyn Aimbetov were selected on November 09, 2002, as the first class of Kazakh cosmonauts for the Kazakhstan National Space Agency, KazCosmos. They trained at the Gagarin Academy, Star City, Moscow, from 2003-09, until the world financial crisis (Global Financial Crisis) of 2008 indefinitely potponed the prospective Kazakhstan mission. This finally took place in September 2015 aboard *Soyuz TMA-18M*, after prospective "space tourist" Sarah Brightman pulled out from the flight. In 2012, to pursue his cosmonaut dreams, Aymakhanov gave up his citizenship of Kazakhstan to become a Russian citizen. He joined Roscosmos, but on December 28, 2022, left the cosmonaut squad after the decision of the Main Medical Commission. Aymakhanov will continue to work at the Gagarin Academy as a leading specialist in cosmonaut training. He is married, with one child.

## Balandin, Aleksandr N. Soyuz TM-9.

Aleksandr Nikolayevich Balandin was born in July 1953 in Fryazino, Moscow Oblast, Russian SFSR, Soviet Union. He is a retired Flight Engineer and cosmonaut. In 1976, he graduated from the Bauman Moscow State Technical University, specializing in Flight Dynamics and Control. He worked as an engineer in the Moscow region at the Scientific Production Association (SPA) "Energy". Selected in the December 01, 1978 Cosmonaut Group (the fifth recruitment of civilian specialists from **SPA Energia**), he was prepared for flying the *Buran* reusable spacecraft (the program ran from 1979-84), the *Soyuz-TM* spacecraft and *Mir* orbital station. Balandin flew aboard *Soyuz TM-9*; launched atop a *Soyuz-U2* booster on February 11, 1990.

During the docking with *Mir*, cosmonauts aboard the space station noticed that three of the eight thermal blankets on the Descent Module of the approaching *Soyuz* had come loose from their attachments near the heat shield. The main concern was that the capsule might cool down, permitting condensation to form inside, and short-out its electrical systems. There was worry that the blankets might block the infrared vertical sensor, which oriented the module for re-entry. Other concerns were that the explosive bolts binding the Service Module to the Descent Module might fail to work after direct exposure to space; that the heat shield might be compromised by direct space exposure; and that an EVA to repair the blankets might cause additional damage. During an EVA, two blankets were folded back, and the third was ignored.

During re-entry, the crew ejected both the Orbital Module and the Service Module simultaneously in an effort to minimize the chances that a blanket could snag. The Descent Module suffered no damage as a result of its prolonged exposure to space conditions. The re-entry to Earth's atmosphere was normal. Balandin landed on August 09, some seventy kilometres northeast of Arkalyk, Kazakhstan. He logged 179 days, one hour and 17 minutes in space. Balandin retired on October 17, 1994. He was awarded five honors, including Hero of the Soviet Union; the Order of Lenin; the Medal "For Merit in Space Exploration"; and the Order of Bernardo O'Higgins.

## Baturin, Yuri M. Soyuz TM-28/27, TM-32/31.

Yuri Mikhailovich Baturin was born in June 1949 in Moscow, Soviet Union (modern-day Russia). He is a former Russian cosmonaut and politician, with federal state civilian service rank of First Class Active State Councillor of the Russian Federation. He graduated from the Moscow Institute of Physics and Technology in 1973. Selected as a cosmonaut in 1997, he launched on August 13, 1998, as a Research Cosmonaut aboard *Soyuz TM-28*, docking with the *Mir* space station for mission 'Mir EP-4'. Commander Gennady Padalka and Flight Engineer Sergei Avdeyev became the twenty-sixth resident crew of *Mir*. The first Russian politician to fly in space, Baturin returned to Earth aboard *Soyuz TM-27*, on August 25, 1998; landing at 47.96944 degrees North, 69.63056 degrees East in Kazakhstan.

From 1996-98, Baturin was Assistant to the President on National Security and Secretary of the Defense Council. Yuri Baturin is also an author in constitutional law. Launched on April 28, 2001, as Flight Engineer aboard *Soyuz TM-32* for ISS mission EP-1, he returned to Earth aboard *Soyuz TM-31* (the first *Soyuz* spacecraft to dock with the ISS), after seven days, 22 hours and four minutes in space; landing on May 06, 2001, some ninety km northeast of Arkalyk, Kazakhstan. This mission was notable for carrying the first "space tourist", Dennis Tito. Baturin logged a total of 19 days, 17 hours and 44 minutes in space. He is married, with a daughter, born in 1982; a student at the Moscow State Academy of Law.

## Belyayev, Pavel I. Voskhod 2.

Pavel Ivanovich Belyayev (1925-70) was born in June 1925 in Chelishchevo, Russia, USSR. He was a fighter pilot with extensive experience in different aircraft types. Belyayev began training as a fighter pilot in 1943, and saw action against the Japanese late in World War II. While attending the Soviet Air Force Academy in late 1959, he was selected as one of the original group of Soviet cosmonauts. He was the first Commander of the Cosmonaut Corps. Launched on March 18, 1965, he was Commander of *Voskhod 2*, when his co-pilot, Alexei Leonov became the first person to do a spacewalk, using a flexible extendable airlock. Before re-entry, the orbital module did not separate on time from the retrorocket module, so Belyayev had to perform a manual re-entry. Their craft landed hundreds of kilometres off course in inaccessible forest, with Belyayev and Leonov having to spend two days there after landing and before their recovery. Belyayev logged one day, two hours and two minutes in space. He was awarded Hero of the Soviet Union. Pavel Belyayev died on January 10, 1970 in Moscow, Soviet Union, after an operation on his stomach.

## Beregovoi, Georgi T. Soyuz 3.

Lt. Gen. Georgi Timofeyevich Beregovoi was born in April 1921 in Fedorivka, Poltava Oblast, Ukrainian SSR, Soviet Union. From 1938-87, he was a pilot in the Soviet Air Forces. Selected in 1962, he was Commander for *Soyuz 3* in 1968, and completed 81 Earth orbits over almost four days. Then aged forty-seven years, he was the oldest person to go into orbit. The mission achieved the first Russian space rendezvous with the uncrewed *Soyuz 2* (launched the previous day). His *Soyuz 7K-OK-P* No. 10 spacecraft came within 200 metres, but failed to achieve the planned docking of the two craft. He failed to notice that *Soyuz 2* was inverted (in relation to his craft), and he used up too much of his maneuvering fuel in the docking attempt. A second docking attempt the next day also failed, but Beregovoi continued to orbit, making topographical and meteorological observations for two more days. He also treated television viewers to a "live" tour of the spacecraft interior. The Soviets published a photo of *Soyuz 3*'s launch vehicle on the pad at Baikonur, marking the first time that the *R-7* booster was shown to the outside world.

Returning to Earth on October 30, 1968, Beregovoi logged three days, 22 hours and 50 minutes in space. He landed near Karaganda, Kazakhstan, cushioned by a blizzard's snowfall. The mission was proclaimed a "complete success", he was promoted to Major General and from 1972-87, he headed the **Yuri Gagarin Cosmonaut Training Center**, until retiring in 1987. Beregovoi was awarded two Hero of the Soviet Union; the first on October 26, 1944, for his military flying service in WWII, and the second for his spaceflight. The USSR concealed the fact that docking attempts had been unsuccessful, but the facts were disclosed after the December 1991 breakup of the Soviet Union, allowing historians to reassess the presumed "success" of Beregovoi's mission. He died on June 30, 1995, aged seventy-four years, in Moscow, Russia.

## Berezovoy, Anatoly N. Soyuz T-5/T-7.

Col. Anatoly Nikolayevich Berezovoy (1942-2014) was born in April 1942 in Enem, Adyghe Autonomous Oblast, Russian SFSR, USSR. He had Ukrainian heritage, and graduated from the Air Force Academy. Selected on April 27, 1970, in Air Force Group 5, he flew as Commander on *Soyuz T-5*, launched on May 13, 1982, with Flight Engineer Valentin Lebedev on the first mission to the *Salyut 7* space station. On May 17, the crew ejected a 28-kg amateur radio satellite from the *Salyut* trash airlock, and the Soviets called this the first launch of a communications satellite from a crewed space vehicle. In July, Lebedev reported that the *Arabidopsis* plants, chosen for their short 40-day lifecycle, had become the first plants to flower and produce seeds in the zero gravity of space.

Berezovoy returned to Earth aboard *Soyuz T-7* on December 10, 1982, after logging 211 days, nine hours and four minutes in space; including one EVA of two hours and 33 minutes. The spacecraft landed 70 km northeast of Arkalyk, Kazakhstan. He retired on October 31, 1992, and up to 1999 was a Deputy President of the **Russian Space Federation**. Anatoly Berezovoy received eight awards, including Hero of the Soviet Union; Order of Lenin; Order for Service to the Homeland in the Armed Forces of the USSR, third class; the Medal "For Merit in Space Exploration" (Russian Federation); the Kirti Chakra (India); and Order "The Sun of Freedom" (Afghanistan). He died on September 20, 2014, aged 72 years, in Moscow, Russia. He was married, with two children.

## Bondarenko, Valentin V.

Senior Lieutenant Valentin Vasilyevich Bondarenko (1937-61) was born in February 1937 in Kharkiv, Ukrainian SSR, Soviet Union. He was a Soviet fighter pilot who served in the Soviet Air Force's PribVO (the former Baltic Military District), and selected in 1960 for training as a cosmonaut. He was the youngest member of the original group of cosmonaut trainees. According to his fellow cosmonauts, Bondarenko was a mild-mannered person with a pleasant disposition. He had a good singing voice and played table tennis well. During the tenth day of a fifteen-day series of endurance tests in a pressurized 50% pure oxygen-filled chamber, he died from shock as the result of burns, after accidentally starting a fire. Having completed his duties for the day, he removed monitoring biosensors from his body and washed his skin with an alcohol-soaked cotton ball, which he discarded. The cotton wool landed on an electric hotplate, which he was using to brew tea, and a fierce blaze erupted in the oxygen-rich environment. Bondarenko tried to smother the flames with the sleeve of his woolen coveralls, which caught fire, and he suffered third-degree burns over most of his body. Because of the pressure difference, it took an onlooking doctor nearly thirty minutes to open the door of the experiment chamber.

Yuri Gagarin spent several hours at his colleagues' bedside as "deathwatch officer". Valentin Bondarenko died sixteen hours after the incident on March 23, 1961, aged 24 years, in Moscow, Soviet Union. This was less than three weeks before Gagarin's historic flight aboard *Vostok 1*. On June 17, 1961, the Presidium of the Supreme Soviet posthumously awarded Bondarenko the Order of the Red Star. The manned orbital flight director, Nikolai Kamanin blamed the death on the Institute's poor organization and control of the experiment. Until 1980, this tragic death was concealed by the Soviet government, along with Bondarenko's membership in the cosmonaut corps. It was one incident that led to speculation about "lost cosmonauts', but in fact there were mundane reasons for several exits from the Soviet space program. He was married, with a child.

## Borisenko, Andrey I. Soyuz TMA-21/Expedition 28, MS-02/Expedition 49.

Andrey Ivanovich Borisenko was born in April 1964 in Leningrad, Russian SFSR, Soviet Union. In 1981, he graduated from the Leningrad Physics and Mathematics School No. 30, then entered the Leningrad Military Mechanical Institute; graduating in 1987 with the qualification "Flight and Control Dynamics". From 1987-89, he worked for a military unit, then began working at **RSC Energia** in 1989 where he was responsible for the *Mir* motion control system and took part in the Mission Control Center – Moscow (MCC-M) on board systems operation analysis board. In 1999, he was a shift Flight Director at the MCC-M, first for the *Mir* station and then for the ISS.

Selected as a cosmonaut from **RSC Energia** on May 29, 2003, in 2003 RKKE Group, he trained to June 2005, then on July 05, 2005, he received his Test Cosmonaut qualification from the Interdepartmental Qualification Commission. From July 2005 to August 2008, he participated in advanced space flight training. Initially assigned as backup Flight Engineer and Soyuz Commander for 'Expedition 24/25', he was moved forward to backup for ISS 'Expedition 23/24', also training as Commander and Soyuz Flight Engineer. Launched on April 04, 2011, aboard *Soyuz TMA-21* to the ISS, he became Commander on May 22 for 'Expedition 28'. He undocked on September 16, landing with Alexander Samokutyaev and NASA's Ronald "Ron" Garan about 148 km (93 miles) southeast of the city of Dhezkazgan, Kazakhstan. After a checkup at a makeshift inflatable hospital, and attending the traditional greeting ceremony at the Karaganda airport, Borisov boarded a plane to return to Star City, Russia.

Launched on October 19, 2016, aboard *Soyuz MS-02* as Flight Engineer, with Sergey Ryzhikov and NASA's Shane Kimbrough, he docked with the ISS on October 21, joining 'Expedition 49', which ended on October 30. Kimbrough then took command of 'Expedition 50', during which the *Progress MS-04* robotic resupply spacecraft was due to arrive, but suffered a third-stage launch early-shutdown problem. Borisenko returned to Earth after an almost five-hour descent flight on April 10, 2017; landing on the Kazakh steppe after 173 days in space. In total, Borisenko has logged 337 days in space. In 2016, Borisenko and Mikhail Kornieko appeared as themselves in the comedy ***Yolki 5***. In 2018, in an exposition on space at the Lycee International de Saint-Germain-en-Laye organised by the Russian Section, he loaned his spacesuit to participate in the exposition. Borisenko is married, with a son and daughter.

## Budarin, Nikolai M. STS-71/Soyuz TM-21, STS-113/Soyuz TMA-1.

Nikolai Mikhailovich Budarin was born in April 1953 in Kirya, Chuvashia, Russian SFSR, Soviet Union. He is a retired mechanical engineer and former Russian cosmonaut; a veteran of three extended space missions aboard *Mir* and the International Space Station. From 1971-73, Budarin served the Soviet Army in Czechoslovakia. He then occupied engineering positions at **NPO Energia** where he was involved in experimental investigations and testing of space technology. From 1976-88, he was employed as an electrician, electrical foreman, test engineer, and head of a group until joining the cosmonaut corps as head of a group at the **Energia** check-out and testing facility. In 1979, he graduated from the night-time education department of Ordzhonikidze Moscow Aviation Institute with a Mechanical Engineering diploma. Selected in 1989 NPOE Cosmonaut Group, from September 1989 to January 1991, Budarin underwent basic cosmonaut training at the Gagarin Academy; and from February 1991 to December 1993 he took an advanced training course for the *Soyuz-TM* crew transport vehicle and the *Mir* station flight. From March 27 to early May 1995, he trained for a mission under the 'Mir EO-19' program as the Flight Engineer; and from May 15 to June 26, he trained for the STS-71 mission in the United States.

Launched aboard shuttle *Atlantis* on June 27, 1995, for STS-71 (Mir EO-19), he spent 75 days, 11 hours and 20 minutes in space, and returned to Earth aboard *Soyuz TM-21* on September 11, 1995; landing 108 km northeast of Arkalyk, Kazakhstan. Budarin launched aboard shuttle *Endeavour* on November 24, 2002, for STS-113 to the ISS for 'Expedition 6'. On January 15, 2003, Budarin was supposed to join Kenneth Bowersox in a six-hour EVA, however, he had failed to meet US standards in tests on a stationary bicycle in December 2002, so NASA barred him from the EVA on medical grounds. The Russian authorities said that the peculiarities of his cardiovascular system were known to them, and pronounced him to be healthy. He returned to Earth aboard *Soyuz TMA-1* on May 04, 2003, but the descent module followed a ballistic trajectory. A technical commission was established to find out why that had occurred, and revealed that the cause of switching to the ballistic descent mode was an inadequate reaction of the descent control unit (within the descent control system) to the signals from the gyroscope and the angular rate meter.

Since 2007, Budarin has been a member of the Russian parliament, the State Duma, representing the pro-Kremlin United Russia party. He has received six awards: Hero of the Russian Federation (1995); the Order Otan (Kazakhstan, 1998); the Order of Merit for the Fatherland, second class (2004), third class (1998); the Medal "For Merit in Space Exploration" (2011); and three NASA Space Flight Medals. He is married, with two sons.

## Bykovsky, Valery F. Vostok 5, Soyuz 22, Soyuz 31/29.

Maj. Gen. Valery Fyodorovich Bykovsky (1934-2019) was born in August 1934 in Pavlovsky Posad, Russia, USSR. He started flying lessons at age sixteen years, joined the Soviet Army in 1952, and became a jet fighter pilot in 1959. He began his cosmonaut training at the Zhukovsky Military Engineering Academy in 1960. Due to an injury sustained by Valentin Varlamov, Bykovsky was added to the original "Vanguard Six" pilots assigned to the *Vostok* missions. Launched on June 14, 1963, *Vostok 5* carried Bykovsky into a four-day, 23 hour and seven-minute mission, setting a new space endurance record, which remains a record for a solo spaceflight. He orbited the Earth eighty-one times. Valentina Tereshkova flew aboard *Vostok 6*, the final *Vostok* flight, which was launched on June 16, two days after Bykovsky's *Vostok 5*. The two craft spent three days in orbital planes 30 degrees apart, and, during Tereshkova's first orbit, approached each other to within five kilometres, when they were able to communicate by radio. Cameras placed inside both spacecraft transmitted live footage that was broadcast on Soviet state television. During his mission, Bykovsky was made a member of the Communist Party, and on April 30, 1966, he was promoted to Colonel.

As planned in all *Vostok* missions, Bykovsky ejected from the capsule at about seven kilometres above Earth and made a parachute landing in Kazakhstan at about 11.30 a.m. UTC on June 19, about three hours after Tereshkova. On June 22, Soviet Premier Nikita Kruschev attended a reception in Red Square, and awarded them each the title Hero of the Soviet Union. In the evening, a reception was held in the Kremlin in which both Bykovsky and Tereshkova were awarded the Order of Lenin. Bykovsky wrote, "*Service in the Air Force made us strong, both physically and morally. All of us cosmonauts took up sports and PT (physical training) seriously when we served in the Air Force. I know that Yuri Gagarin was fond of ice hockey. He liked to play goalkeeper. Gherman Titov was a gymnastics enthusiast, Andriyan Nikolayev liked skiing, Pavel Popovich went in for weightlifting. I don't think I am wrong when I say that sports became a fixture in the life of the cosmonauts*". Unquote.

Only five days after their return from space, Moscow hosted the International Women's Congress on June 24, where Tereshkova and Bykovsky were greeted by about 2,000 women from 119 countries. He became Chief of Cosmonaut Training for the *Apollo-Soyuz Test Project*, carried out in July 1975. Valery Bykovsky later flew as Command Pilot on the

*Soyuz 22*, a 190-hour mission launched on September 15, 1976. He commanded the *Soyuz 31* mission, which launched on August 26, 1978, with East German cosmonaut Sigmund Jahn to the *Salyut 6* space station. They returned to Earth on September 03, 1978, aboard *Soyuz 29*. He logged a total of 20 days, 17 hours and 48 minutes in space.

Much of his later career was devoted to promoting the Intercosmos program among the world's socialist nations. After his *Soyuz 31* mission, Bykovsky became the director of the House of Soviet Science and Culture in East Berlin, until 1990. Bykovsky established the **Russian Federation of Cosmonautics** (RFC) in 1998; with the main goals of increasing public interest in space exploration, providing educational opportunities for young people, and supporting research and development initiatives in the space sector. Bykovsky received nine Soviet and eight foreign awards, including Order of the Red Star (1961); the Medal "For Development of the Virgin Lands" (1963); two Hero of the Soviet Union (1963, 1976); three Order of Lenin (1963, 1976, 1978); Hero of Socialist Labour (Bulgaria, 1963); Hero of Socialist Labour (Vietnam, 1963); the Cross of Grunwald, First Class (Poland, 1963); the Order of the Red Banner of Labour (1976); the Gold Medal, Tsiokovsky Academy of Sciences; and Hero of the German Democratic Republic (East Germany, 1978). Col. Valery Bykovsky died on March 27, 2019, aged 84 years. He was married, with two sons.

## Chub, Nikolai A. Soyuz MS-24/Expedition 70/Soyuz MS-25.

Nikolai Aleksandrovich Chub was born in June 1984 in Novocherkassk, Russian SFSR, USSR (now Russia). He obtained a degree in Management and Infomatics from the South-Russian State Technical University (Novocherkassk Polytechnic Institute) in 2006 as the top member of his class. Selected in TsPk-16 Cosmonaut Group by Roscosmos in 2012, he is an active cosmonaut. He was selected as a crewmember for the *Soyuz MS-17* mission, but was replaced by NASA's Kathleen Rubins. He was assigned as a backup crewmember for *Soyuz MS-22*, and flew on *Soyuz MS-24* from September 15, 2023, for a long-duration mission with 'Expedition 70'. As of February 11, 2024, he was currently in space, and scheduled to return to Earth on *Soyuz MS-25* about April 02, 2024. He has made one EVA of seven hours and 41 minutes.

## Dezhurov, Vladimir N. Soyuz TM-21, STS-105/Expedition 3.

Lt. Col. Vladimir Nikolayevich Dezhurov was born in July 1962 in Yavas, Zubovo-Polyansky District, Mordovia, Soviet Union. He is a retired engineer, Lieutenant Colonel in the Russian Air Force, and former cosmonaut. He graduated from the S.I. Gritsevits Kharkov Higher Military Aviation School in 1983 with a Pilot Engineer's diploma, then served as a pilot and senior pilot in the Soviet Air Force. Selected in 1987 Cosmonaut Group, he trained at the Gagarin Academy from December 1987 to June 1989. From 1991, he was a correspondence student at the Yuri A. Gagarin Air Force Academy. Launched on March 14, 1995, aboard *Soyuz TM-21* to be Commander for the 'Mir-18' mission, after two days' independent flight the *Soyuz* docked with *Mir*. There were several technical problems during the 115-day mission, and the crew performed life science experiments. He landed aboard Space Shuttle *Atlantis* at KSC on July 07, 1995.

For his second mission, Dezhurov was launched aboard shuttle *Discovery* on STS-105 to the ISS on August 10, 2001. He spent about four months on the space station as a Flight Engineer with 'Expedition 3'. The crew enjoyed a unique view of the 2001 Leonid meteor storm. Dezhurov returned to Earth aboard shuttle *Endeavour* on December 17, 2001. He is twenty-fifth in the list of astronauts who have performed the most EVA time. Vladimir Dezhurov logged 244 days, five hours and 28 minutes in space; including nine EVAs totaling 37 hours and two minutes. He retired on July 12, 2004. Dezhurov has received three Air Force medals; the Hero of the Russian Federation medal; and Pilot-Cosmonaut title by Decree of the President of the Russian Federation. He is married, with two daughters.

## Dobrovolsky, Georgy T. Soyuz 11.

Georgy Timofeyevich Dobrovolsky (1928-71) was born in June 1928 in Odessa, Ukrainian SSR, Soviet Union (now Ukraine). He held the rank of Podpolkovnik in the Soviet Air Force. Selected in Air Force Group 2, he was launched on June 06, 1971, as Commander of *Soyuz 11*, which visited *Salyut 1* for twenty-two days. Launched two months earlier, it was the world's first space station, and the crew created the first manned orbital scientific station. Meteorological and plant growing experiments were carried out. This mission set a new empical record for time spent in space. Dobrovolsky logged 23 days, 18 hours and 21 minutes in space. Georgy Dobrovolsky died on June 30, 1971, in outer space, aged forty-three years, when the air pressure equalization valve on his *Soyuz 11* spacecraft opened just before re-entry, and the cabin atmosphere was lost. After a normal automated landing, the capsule was opened and the three-man crew was found dead. They are the only crew to date to have died at an altitude above the hypothetical Karman line (100 km, or 54 nautical miles; equal to 330,000 feet above mean sea level).

Cosmonauts in the confined three-person Soviet capsule were not yet equipped with pressure suits or independent life-support equipment, and the crew of Dobrovolsky, Vladislav Volkov and Viktor Patsayev died of asphyxiation in less than two minutes. They were the world's second crew to die during a space flight, and following this catastrophe the *Soyuz* craft was extensively redesigned. The crew were given a State funeral, and Georgy Dobrovolsky's ashes were buried in the Kremlin Wall Necropolis. He was posthumously awarded the title of Hero of the Soviet Union; the Order of Lenin; and the title of Pilot-Cosmonaut of the USSR.

## Dubrov, Pyotr V. Soyuz MS-18/Expedition 64/Soyuz MS-19.

Pyotr Valerievich Dubrov was born in January 1978 in Khabarovsk, Russian SFSR, Soviet Union. He is a Russian software engineer and active cosmonaut, selected by Roscosmos to begin training on October 29, 2012. He graduated from cosmonaut training on July 15, 2014. In 2020, he was assigned to the backup crew of *Soyuz MS-17* as Flight Engineer for 'Expedition 64' and traveled to the Johnson Space Center, Texas, to train on the ISS USOS. Launched on April 09, 2021, as Flight Engineer 1 aboard *Soyuz MS-18* (named *Yuri A. Gagarin*) atop a *Soyuz-2.1a* booster to the ISS, he returned to Earth on March 30, 2022, aboard *Soyuz MS-19*.

On October 15, 2021, during preparation of the ship's propulsion system for landing, the *Soyuz MS-18* inadvertently fired its thrusters for too long, which was the result of a procedural error in instructions sent from Ground Control. It had the result of changing the orientation of the ISS by as much as fifty-seven degrees. The space station's attitude control system then counteracted that motion by activating thrusters on the Russian Segment. The propellant reserves aboard the spacecraft required for landing were preserved by a time limit pre-set for engine testing. The crew were not in danger, and it was the second occurrence of an engine anomaly at the ISS. Dubrov logged 355 days, three hours and 45 minutes in space; including four EVAs totaling 29 hours and 49 minutes.

## Dyomin, Lev S. Soyuz 15.

Col. Dr. Lev Stepanovich Dyomin (1926-98) was born in January 1926 in Moscow, Russian SFSR, Soviet Union. He gained a doctoral degree in Engineering from the Soviet Air Force Engineering Academy, and the rank of Colonel in the Soviet Air Force. Dyomin flew aboard *Soyuz 15* (August 26-28, 1974). The mission was intended to dock with the *Salyut 3* space station (the second *Almaz* military station, launched on June 25, 1974), but the attempt failed. The *Igla* rendezvous system on the *Soyuz* failed, and they were also unable to dock manually. At that time he was the oldest cosmonaut at age forty-eight years, as well as the first grandfather to go into space. Lev Dyomin logged two days, zero hours and 12 minutes in space. He remained in the space program until 1982, when he left to pursue deep-sea research. He received seven awards, including Hero of the Soviet Union; the Order of Lenin; the Medal "For Battle Merit"; the Medal "For the Development of the Virgin Lands"; and the Order of the Banner of People's Republic of Bulgaria. Dyomin died on December 18, 1998, of cancer, aged seventy-two years, in Zvozdny Gorodok, Russia.

## Dzhanibekov, Vladimir A. Soyuz 27/26, Soyuz 29, Soyuz T-6, T-12, T-13.

Vladimir Aleksandrovich Dzhanibekov (born in May 1942 as Vladimir Krysin) is a retired Soviet Air Force Major General and a cosmonaut veteran of five orbital missions. He was from the remote area of Iskandar in what was then Bostanliq District, South Kazakhstan. Soon after his birth his family moved to Tashkent. In 1964 he married Liliya Munirovna Dzhanibekova, who was a descendant of Janibeg, the medieval ruler of the Golden Horde. As her father had no sons, Dzhanibekov took his wife's family name to honor her ancestry and continue her line of descent, an unusual step for a husband in the Soviet Union. In 1960, Vladimir entered Leningrad University to study physics, where he became involved in flying. In 1961, he enrolled in the V.M. Komarov Higher Military Flying School at Yeisk. Four years later Dzhanibekov graduated from **Rostov State University** and became a flying instructor in the Soviet Air Forces, serving in Taganrog from 1968-70. During the 1970 visit of cosmonaut Gherman Titov to the Air Forces training unit he was selected for the cosmonaut training pool, Air Force Group 5. Vladimir Dzhanibekov joined the Communist Party the same year.

Dzhanibekov served on five space missions: *Soyuz 27/26* return flight; *Soyuz 39; Soyuz T-6; Soyuz T-12;* and *Soyuz T-13*. His total time in space was 145 days, 15 hours and 56 minutes (about five months). He performed two EVAs with a total time of eight hours and 35 minutes. In 1985 he was promoted to the rank of Major General. After retiring from the cosmonaut program in 1986, he became involved with politics. He was Deputy to the **Supreme Soviet of Uzbek SSR** from 1985 until 1990. His photography and paintings, predominantly of space thematics, are owned by museums and private collectors. Starting in 1990, Dzhanibekov made four unsuccessful attempts to circumnavigate the globe by high-altitude balloon. Launched on September 08, 1990, his first flight lasted for thirty-one hours, spanning two nights, before landing at Omak, Washington. His last attempt in December 1994 reached 29,000 feet altitude (8,800 metres) when the balloon ballast failed.

Dzhanibekov received two Hero of the Soviet Union awards, five Orders of Lenin, Pilot-Cosmonaut of the USSR; the Order of the Red Star; the Order of Friendship (from the Russian Federation), and seven other Russian awards (plus four foreign awards). The minor planet 3170 Dzhanibekov, discovered in 1979, is named after him. The Kosmonavtlar metro station in Tashkent commemorates Uzbekistan's contribution to the Soviet space program, including that of Vladimir Dzhanibekov.

## Fedyaev, Andrey V. SpaceX Crew 6/Expedition 68/69.

Maj. Andrey Valerievich Fedyaev was born in February 1981 in Serov, Sverdlovsk Oblast, Russian SFSR, Soviet Union. He is an active Test Cosmonaut, who began training at the Gagarin Academy in 2012. In 2004, he received an Engineering degree in Air Transport and Air Traffic Control from the Balashov Military Aviation School. After graduation, he joined the Russian Air Force in the Three Hundred Seventeenth mixed aviation segment. Before his retirement with the rank of Major in 2013, he logged over 500 hours in Russian aircraft. Selected in TsPK-16 Cosmonaut Group, he was assigned on July 15, 2022, to the SpaceX

'Crew 6' mission to the ISS. He launched as a Mission Specialist on March 02, 2023, from KSC aboard SpaceX's reusable *Crew Dragon* spacecraft atop a *Falcon 9* booster. Joining 'Expedition 68/69', he logged 185 days, 22 hours and 43 minutes in space.

## Feoktistov, Konstantin P. Voskhod 1.

Dr. Konstantin Petrovich Feoktistov (1926-2009) was born in February 1926 in Voronezh, Russian SFSR, Soviet Union. He was a Russian engineer and cosmonaut in the former Soviet space program. During the German occupation of Voronezh, at the age of 16 years, Feoktistov fought with the Soviet Army, carrying out reconnaissance missions for the Voronezh Front. After being captured by a Waffen-SS Army patrol, he was shot in the neck, but was able to make his way to Soviet lines. After WWII was over, he enrolled in the Bauman Moscow Higher Technical School as an engineering student, graduating in 1949. He later earned a doctorate in Physics. He joined Mikhail Tikhonravov's OKB (design bureau), and in 1955, Feoktistov formed part of the team that went on to design the *Sputnik* satellites, the *Vostok* and *Voskhod* space capsules, and the *Soyuz* space capsule under the leadership of the Soviet Chief Designer, Sergey Korolev. During this time, he also worked on the design of an ion-propelled spacecraft to be capable of taking humans to Mars.

In 1964, Feoktistov was selected in Civilian Specialist Group 1 as part of a group of engineers for space training, and he was hastily assigned to the multi-disciplinary *Voskhod 1* crew as Engineer. He was the first Russian civilian to make a space flight, and the only cosmonaut in the Soviet Union who was not a member of the Communist Party of the Soviet Union. Feoktistov flew on *Voskhod 1* on October 12, 1964, the first spacecraft to carry three crew members. He logged one day, zero hours and 17 minutes in space. After his flight, training for any future space missions was discontinued for medical reasons.

He continued his space engineering work, and later became head of the Soviet space design bureau that designed the *Salyut* and *Mir* space stations. In October 1969, Feoktistov and Georgi Beregovoi traveled as guests of NASA throughout the USA. They were joined on the trip by astronauts including Gene Cernan and Neil Armstrong. When accompanied by Cernan, if a band was present the song *"Fly Me to the Moon"* was played. They enjoyed the Disneyland ride *Trip to the Moon*, then joked with the US astronauts that they went to Disneyland and not the Moon. Feoktistov resigned from his engineering position with **Energia**, then returned to the Bauman School in 1990 as a Professor. He wrote several books on space technology and exploration. He received eleven awards, including Hero of the Soviet Union; Order of Lenin; two Orders of the Patriotic War, First Class; two Orders of the Red Banner of Labour; the Order of the Badge of Honour; the Medal "For Victory over Germany in the Great Patriotic War 1941-1945"; the State Prize of the USSR; and Hero of Socialist Labour (Vietnam). He died on November 21, 2009, aged eighty-three years, in Moscow, Russia. The Feoktistov crater on the far side of the Moon is named in his honor.

## Filatyev, Valentin I..

Maj. Valentin Ignatyevich Filatyev (1930-90) was born in January 1930 in Malinovka, Ishimsky District, Tyumen Oblast, Russian SFSR, Soviet Union. As a senior lieutenant, he was selected on March 07, 1960, as one of the original twenty cosmonauts. He attended the Stalingrad Higher Air Force Pilots School, graduating in 1955, and served as a fighter pilot in the Air Defense Force. Filatyev was a Soviet cosmonaut who was dismissed from the Soviet space program for disciplinary reasons. On March 27, 1963, he, Grigory Nelyubov and Ivan Anikeyev were arrested for drunk and disorderly conduct by the militia. He agreed to apologise, but because there were previous incidents, all three were dismissed from the cosmonaut corps on April 17, 1963. His image was airbrushed out of cosmonaut photos, and efforts were made to conceal the reason for his dismissal. This airbrushing led to speculation about "lost cosmonauts" even though the actual reasons were often mundane. Filatyev never flew in space, and later became a teacher. He died on September 15 1990 in Oryol, USSR, aged sixty years. He was married, with one child.

## Filipchenko, Anatoly V. Soyuz 7, Soyuz 16.

Maj. Gen. Anatoly Vasilyevich Filipchenko (1928-2022) was born in February 1928 in Davydovka, Voronezh Governate, Russian SFSR, Soviet Union. He was of Ukrainian descent, and became a Major General in the Soviet Air Force. Filipchenko was Commander of *Soyuz 7* (October 12-17, 1969), part of a joint mission with *Soyuz 6* and '8', designed to test the ability to coordinate the simultaneous flight of three spacecraft. The plan was to dock with *Soyuz 8* and transfer crew, a technique in preparation for future lunar flights. The version of *Soyuz 7K-OK* spacecraft used for the joint mission carried a torus-shaped docking electronics equipment housing surrounding the motor assembly on the back of the Service Module. Mission objectives included: to checkout the modified structure of the *Soyuz* craft; further improvement of the control, orientation, and orbital stabilization systems and navigation aids; debugging the piloting systems by orbital maneuvering of the three spaceships; scientific observations and photographing of geological-geographical subjects, and exploration of Earth's atmosphere; and conducting experiments of engineering research and biomedical importance. The spacecraft failed to dock, and the *Soyuz 7* landed four days 22 hours and 40 minutes after launch; 155 km northwest of Karaganda.

*Soyuz 16* (December 02-08, 1974) was a crewed test flight for a joint Soviet-United States space flight which culminated in the *Apollo-Soyuz Test Project* (ASTP) mission in July 1975. Anatoly Filipchenko and Nikolai Rukavishnikov tested an androgynous docking ring and other systems to be used in the joint flight. These were modified environmental systems; new solar panels; and a new radar docking system. Air pressure was reduced from 760 to 540 mm, and cabin oxygen raised from 20% to 40% to test reducing the planned transfer time to the Apollo craft from two to one hour. On December 07, the docking ring was jettisoned with explosive bolts, to test emergency measures if the capture latches got stuck during the ASTP flight. The mission lasted five days, 22 hours and 23 minutes; landing thirty km northeast of Arkalyk, Kazakhstan.

He logged a total of ten days, 21 hours and three minutes in space. After leaving the space program in 1982, Filipchenko became the deputy director of the OKB in Kharkiv. He received ten awards, including two Hero of the Soviet Union; the Order of Lenin; the Medal "For Merit in Space Exploration"; the State Prize of the USSR; the Order of the Flag of the People's Republic of Hungary; the Order of the Banner of the Bulgarian People's Republic; the Medal "For the Strengthening Military Cooperation" (Czechoslovakia); and the Medal "Brotherhood in Arms" (German Democratic Republic). Anatoly Filipchenko died on August 07, 2022, aged ninety-four years.

## Gagarin, Yuri A. Vostok 1.

Major Yuri Alekseyevich Gagarin was born in March 1934 in Klushino, Russia, Soviet Union. He will forever be remembered as the first person to fly in space. He made history on April 12, 1961, when he made one orbit of the Earth aboard *Vostok 1*. Gagarin was awarded the title 'Hero of the Soviet Union' and received many other medals and titles. He became an international celebrity who promoted the Soviet space program and then worked on designing reusable spacecraft. Yuri was the third of four children and spent his childhood on a collective farm where his father worked as a carpenter and bricklayer, and his mother as a milkmaid. He was just seven years old when the German Nazis invaded the Soviet Union in 1941, and the Gagarin family were kicked out of their home.

At school, Yuri loved both mathematics and physics. While attending an industrial school he joined a flying club; making his first solo flight in 1955. After joining the Soviet Air Force, Gagarin attended Orenburg Aviation School, where he learned to fly Soviet MiG aircraft. Gagarin was one of 3,000 applicants to be the first Soviet cosmonaut. In 1960, he was one of the twenty candidates chosen. During the extensive physical and psychological testing of trainees, Gagarin excelled at the tests while maintaining a calm demeanor as well as his sense of humor. During his 106-minute spaceflight, Gagarin was not given manual control of *Vostok 1* because many scientists feared he would "go mad", but in case of an emergency there were override codes (1-2-5) left in an envelope! The rocket's nose cone separated at 154 seconds after launch and light poured into the cabin. Gagarin yelled into his microphone, *"I can see the Earth! I can see rivers, folds in the Earth. It's all easy to make out. Visibility is good. Everything is wonderfully visible!"*

At the House of Radio in Moscow, Yuri Levitan made a broadcast at 10:02 a.m. to more than 200 million Soviets: *"Govorit Moskva! Govorit Moskva! This is Moscow calling. This is Moscow calling. We are broadcasting a TASS statement announcing the world's first manned space flight. The pilot-cosmonaut of the spaceship 'Vostok' is a citizen of the Union of Soviet Socialist Republics, Major Yuri Alekseyevich Gagarin. The launch of the multi-stage rocket has gone well ... The spaceship has begun its flight around the orbit of the Earth. According to preliminary data, its maximum height is 302 kilometres. Two-way radio contact has been established and is continuing with cosmonaut Comrade Gagarin. Comrade Gagarin ... feels well at present. The systems providing all necessary life functions inside the cabin are functioning normally. The flight of the spaceship Vostok with pilot-cosmonaut Gagarin in orbit continues".*

A recording and film of the telephone call to Premier Nikita Krushchev has survived. Gagarin stood up smartly, while Krushchev showered him with congratulations. ***"You have made yourself immortal!"*** he said. *"With your feat you have glorified our Motherland!"*. He asked a few questions about what Gagarin had seen from space - *"I saw the Earth from a great height"*, said the cosmonaut helpfully. Krushchev said, *"I will be delighted to meet you in Moscow. Together with you, together with all our people, we will solemnly celebrate this great feat in space exploration. Let the whole world look and see what our country is capable of, what our great people and our Soviet science can do!". "Let other countries try to catch up with us!"* answered Gagarin, equally enthusiastically.

On April 14, Gagarin made it to the podium at Moscow's Vnukovo Airport soon after 1 p.m., where he saluted and made his official report: *"I am happy to report that the mission assigned to me by the Central Committee of the Communist Party and the Soviet government has been accomplished. The first flight in history has been made aboard a Soviet spaceship. I feel very well and ready to accomplish any other mission assigned to me by our Party and the government"*. He was word perfect, everybody applauded, and Kreshchev wiped a tear from his eye before giving Gagarin a bearhug and kissing him repeatedly on both cheeks. Later that day in Red Square, Moscow, Gagarin stood on Krushchev's right flanked by the nation's leading political figures in their hats and heavy overcoats. The national anthem blasted from loudspeakers and then, to a huge eruption of applause, Gagarin stepped up to the microphone: *"I am boundlessly happy that my beloved Motherland has been the first to achieve this flight, the first in the world to penetrate into outer space ... The first Sputnik, the first space ship, and the first space flight – such are the stages on my Motherland's great road to penetrating the secrets of nature ... Long live our Soviet Motherland! Long live our great and powerful Soviet people! Long live our Communist Party!"* Perhaps, like his speech at the airport, this one had also been practised back at the dacha.

The Soviet government showered Yuri Gagarin and his family with rewards. A secret decree from the Ministry of Defense, only published fifty years later, offers a revealing glimpse into the State's power and reach over all of their lives. Gagarin's parents would get an entire new house – 'with television, radio receiver, furniture' – while his wife, Valentina would, among many items, be provided with '3 dresses, a black suit, 2 hats, 2 women's handbags, and 3 pairs of shoes'. There were toys, dolls, and a pram for the two children, gifts of money for the three siblings, and, for Yuri Gagarin himself, a lengthy inventory of goods including '6 sets of silk underwear, a light summer coat, and an electric razor'. Gagarin's Chkalovsky apartment, now provided by the State with luxuries like a refrigerator and a washing machine, was doubled in size. Later, in 1966, the family was moved to an elite brand-new apartment block in the complex that shortly afterwards became known as Star City, outside Moscow.

After his successful mission, Gagarin helped train future cosmonauts, becoming the deputy training director for the Star City cosmonaut base. He rose quickly through the ranks of the Soviet Air Forces. The pad he launched from at Baikonur is now called "Gagarin's Start". Gagarin worked on his aerospace engineering studies and wrote a thesis about future space planes. Since 1962, there has always been a celebration in Russia called "Cosmonautics Day", to commemorate Gagarin's flight to space. "Yuri's Night" began in 2001 to celebrate his achievements and those of other astronauts. Yuri Gagarin was killed during a routine training flight at age thirty-four years in an accident while test-piloting a MiG-15 jet fighter on March 24, 1968, near Novosyolovo. It is possible an air vent failure occurred, or turbulence from another jet passing too close caused a loss of control at low altitude, but some believed that Soviet leader Leonid Brezhnev wanted Gagarin dead because he was jealous of the cosmonaut's fame. A notable quote from Gagarin is, *"To be the first to enter the cosmos, to engage single-handed in an unprecedented duel with nature – could anyone dream of anything greater than that?"*

## Gidzenko, Yuri P. Soyuz TM-22, TM-31/STS-102, Soyuz TM-34/33.

Yuri Pavlovich Gidzenko was born in March 1962 in Yelanets, Mykolaiv Oblast, Ukrainian SSR, Soviet Union. He graduated from the Higher Military Pilot school in Kharkiv in 1983; and from the Moscow State University of Geodesy and Cartography (MIIGAIK) in 1994 specializing in geodesy and cartography. A third class military pilot, he served as a senior pilot in the Air Force units of the Odessa military district. He logged 850 hours flying three aircraft types and has made 170 parachute jumps. Gidzenko attended basic space training from December 1987 to June 1989 and then advanced training as a Test Cosmonaut candidate.

Launched aboard *Soyuz TM-22*, he served as Commander of the long-duration 'Mir EO-20' *Euromir 95* expedition from September 03, 1995, to February 29, 1996; logging 179 days in space (including an extension of 44 days). Mission objectives included studying the effects of microgravity on the human body, experimenting with new materials, capturing cosmic dust and man-made particles in low Earth orbit, and to test new space equipment. He landed 105 km northeast of Arkalyk, Kazakhstan. Launched on October 31, 2000 as Commander aboard *Soyuz TM-31*, Gidzenko was part of the first 'Expedition 1' resident crew as Flight Engineer on the International Space Station; along with Sergei Krikalev and NASA's William Shepherd. They activated life support systems, unpacked station components, clothing, laptop computers, office equipment, cables and electrical gear left for them by Space Shuttle logistic supply flights of the previous two years. Over four months later, Gidzenko returned to Earth on *Endeavour* at the completion of STS-102; landing at KSC on March 21 at 02:33:06 EST. He logged 140 days in space.

Launched on April 25, 2002, aboard *Soyuz TM-34* along with ESA astronaut Roberto Vittori and South African "space tourist" Mark Shuttleworth, Gidzenko served as *Soyuz* Commander. Docking with the nadir port of the ISS *Zarya* module, they conducted joint operations with the resident 'Expedition 4' crew. He returned to Earth aboard *Soyuz TM-33* on May 05, 2002; landing 26 km southeast of Arkalyk after nine days in space. Gidzenko logged a total of 329 days, 22 hours and 44 minutes in space; including two EVAs totaling three hours and 43 minutes. He has received six awards, including Hero of the Russian Federation (April 01, 1996); the Order of Military Merit (March 02, 2000); the Order of Merit for the Fatherland, third class (2004), fourth class (2002); the NASA Distinguished Public Service Medal (2003); and the Medal "For Merit in Space Exploration (April 12, 2011). Gidzenko is married, with two sons.

## Glazkov, Yuri N. Soyuz 24.

Maj. Gen. Yuri Nikolayeveich Glazkov (1939-2008) was born in October 1939 in Moscow, USSR. He graduated from the Kharkov Military Engineering High School in 1962, receiving the Candidiate of Technical Sciences degree. He held the rank of Major General in the Soviet Air Force. He served as a flight engineer before being selected on October 23, 1965, in Air Force Group 3. Glazkov flew as Flight Engineer aboard *Soyuz 24* (February 07-25, 1977) to *Salyut 5*; the third and final mission to that station. He and Gorbatko entered the station wearing breathing apparatus, and made numerous tests of the atmosphere, before finding it safe. They re-activated the station after toxic fumes had apparently terminated the previous mission of *Soyuz 21*. They performed biological and materials experiments, and probably photographic reconnaisssance.

Glazkov logged 17 days, 17 hours and 25 minutes in space. In 1974, he was awarded a doctorate in Technical Sciences. Glazkov retired from the cosmonaut corps on January 26, 1982. In 1989, he was became the first Deputy Chief of the **Gagarin Cosmonaut Training Center**, a position from which he retired in May 2000. Yuri Glazkov wrote several books, including a 1977 technical guide to spacewalking, *Outside Orbiting Spacecraft* and a book about space exploration, *The world Around Us* in 1986. He also authored several science fiction novels; one of which, *The Black Silence* was illustrated by fellow cosmonaut Vladimir Dzhanibekov, and published in 1987. Glazkov received seventeen awards, including Hero of the Soviet Union (1977); the Order of Lenin; Order of the Red Star (1985); the USSR State Prize (1987); Order Dostyk (Kazakhstan, 1998); State Prize of the Russian Federation (1999); Order of Merit for the Fatherland, third class (2000), fourth class (1996); and the Tsiolkovsky Gold Medal (Academy of Sciences of the USSR). Yuri Glazkov died on December 09, 2008 in Moscow, Russia, aged 69 years. He was married, with two children.

## Gorbatko, Viktor V. Soyuz 7, Soyuz 24, Soyuz 37/36.

Maj. Gen. Viktor Vasilyevich Gorbatko was born in December 1934 in Ventsy-Zarya, Gulkevich district, Azov-Black Sea Krai, Soviet Union. A trained pilot, he was selected in 1960 Air Force Group 1 (the first group of cosmonauts), and was Research Engineer aboard *Soyuz 7* (October 12-17, 1969), a joint mission with *Soyuz 6* and '8'. Following an orbital correction during the night, *Soyuz 7* and '8' were expected to be less than one kilometre from each other when communications were regained at 9 a.m. The main mission objective was the testing of Soviet lunar equipment; a system for control of the simultaneous flight of three spacecraft. Docking did not occur, and the craft landed on October 17, five days after launch; 155 km northwest of Karaganda, Kazakhstan. *Soyuz 24* was launched on February 07, 1977, and docked with the *Salyut 5* space station the next day. The crew of Commander Gorbatko and Flight Engineer Yuri Glazkov continued the work started by the *Soyuz 21* crew. Being part of the *Almaz* military-based *Salyut* program, undisclosed projects were likely carried out. The mission lasted 17 days, 17 hours and 26 minutes; landing near Arkalyk on February 25. The *Soyuz* landed in a blowing snowstorm which jammed the locator beacon shut from snow: Gorbatko had to free it manually, so they could be found and recovered.

*Soyuz 37* was a flight launched on July 23, 1980, to the *Salyut 6* station. It was the thirteenth mission to and the eleventh successful docking at the orbiting facility. Commander Gorbatko was crewed with Flight Engineer Pham Tuan, the first Asian and first Vietnamese cosmonaut. They swapped craft with the long-duration crew and returned to Earth in *Soyuz 36* about seven days and 20 hours later. Viktor Gorbatko logged a total of 30 days, 12 hours and 47 minutes in space. Leaving the space program in 1982, he taught at the **Air Force Engineering Academy** in Moscow. Gorbatko's honors included eight Soviet and Russian Federation awards, and fourteen foreign awards: two Hero of the Soviet Union; three Order of Lenin; Hero of the Mongolian People's Republic; Hero of Socialist Labour (Vietnam); six decade Medals; and the Medal "Brotherhood in Arms" (East Germany). He died on May 17, 2017, in Moscow, Russia, aged 82 years. He was married, with two children. Viktor Gorbatko was buried in the Federal Military Memorial Cemetery.

## Grechko, Georgy M. Soyuz 17, Soyuz 26, Soyuz T-14.

Dr. Georgy Mikhaylovich Grechko (1931-2017) was born in May 1931 in Leningrad, USSR. He graduated from the Leningrad Institute of Mechanics with a Doctorate in Mathematics. Grechko was a member of the Communist Party of the Soviet Union. He went on to work at Sergei Korolev's design bureau and from there was selected for cosmonaut training for the Soviet Moon program. When that program was canceled, he went to work on the *Salyut* space stations. He flew to space three times, each bound for rendezvous with a different *Salyut* space station. The missions were *Soyuz 17*, (1974) to *Salyut 4; Soyuz 26*, December 1977 to *Salyut 6*; and *Soyuz T-14* (1985) to *Salyut 7*. On his 1974 mission, Grechko spent up to two hours per day exercising on a bicycle and treadmill, as well as experimenting with wearing negative pressure suits. Cosmonauts Grechko and Aleksei Gubarev spent thirty days in orbit, which set the Soviet record at that time. The 1977 mission of 96 days eclipsed the 84-day record set in 1974 by US *Skylab* astronauts Gerald Carr, Edward Gibson and William Pogue. Despite intensive exercise routines, both Grechko and Romanenko struggled to complete easy tasks shortly following their return, such as turning a radio dial or lifting a cup of tea. Grechko's total time in space was 134 days, 20 hours and 32 minutes.

Grechko was twice awarded Hero of the Soviet Union, and three Orders of Lenin, and six other awards. He was a Distinguished Member**, Association of Space Explorers**. He had a brief cameo role in Richard Viktorov's 1981 film ***Per Aspera Ad Astra***, and as a result attained pop culture status in his home city of Leningrad. He also appeared in the 1979 film ***Under the Constellation Gemini***. Grechko, along with Alexei Leonov, Vitaly Sevastyanov, and Rusty Schweickart (of the United States) established the **Association of Space Explorers** in 1984. He resigned from the Russian space program in 1992 to lecture on atmospheric physics at the Russian Academy of Sciences, and wrote his memoirs as *Cosmonaut No. 34 From Splinter to Aliens*, (2023). Grechko died at age eighty-five years because of several chronic illnesses. A minor planet, 3148 Grechko (discovered in 1979) is named after him.

## Gubarev, Aleksei A. Soyuz 17, Soyuz 28.

Maj. Gen. Aleksei Aleksandrovich Gubarev (1931-2015) was born in March 1931 in Gvardeitsi, Samara Oblast, USSR. He graduated from the Soviet Naval Aviation School in 1952 and served with the Soviet Air Force. Before being selected in Air Force Group 2, he undertook studies at the Gagarin Air Force Academy. Gubarev was originally trained for the Soviet lunar program and for military *Soyuz* flights, before training for *Salyut* missions.

*Soyuz 17* (January 10 to February 09, 1975) was the first of two long-duration missions to the *Salyut 4* space station, which was launched on December 26, 1974. Launched aboard a *Soyuz 7K-T No. 5* spacecraft, he spent 29 days, 13 hours and 19 minutes in space. Upon entering the new space station, he and Grechko found a note from its builders which said, *"Wipe your feet"!* The crew worked 15-20 hours per day, including their 2.5 hour exercise period. Activities included testing communication equipment for tracking ships and contacting mission control via a *Molniya* satellite. A major component of their mission was astrophysics, with the station's solar telescope activated on January 16, 1975. A new teleprinter was used for communications from ground crew, freeing the *Salyut* crew from constant interruptions during their work. They returned to Earth on February 09; landing 110 km northeast of Tselinograd, Kazakhstan, in a snowstorm with winds of 72 km/h, and wore gravity suits to ease the effects of re-adaptation.

Gubarev was Commander of *Soyuz 28* (March 02-10, 1978), the first Intercosmos flight EP-2, with Czech Research Cosmonaut Vladimir Remek. Military pilots from Soviet bloc nations were flown on missions of about eight days to a Soviet space station. This mission lasted seven days, 22 hours and 16 minutes. He logged a total of 37 days, 11 hours and 36 minutes in space. Aleksei Gubarev received five Soviet and five foreign awards, including two Hero of the Soviet Union; two Order of Lenin; two Gagarin Medals; the Medal "For Merit in Space Exploration" (Russian Federation); Hero of the Czechoslovak Socialist Republic; the Medal of Sino-Soviet Friendship (People's Republic of China); and the Medal "Brotherhood in Arms" (German Democratic Republic). Gubarev resigned as a cosmonaut in 1981 and took up an administrative position at the **Gagarin Cosmonaut Training Center**. In the 1980s, he worked at the Thirtieth Central Scientific Research Institute, Ministry of Defence (Russia). In 1982, Gubarev published a book, *The Attraction of Weightlessness*. He died on February 21, 2015, in Moscow, Russia, aged eighty-three years.

## Hermaszewski, Miroslaw. Soyuz 30.

Miroslaw "Mirek" Hermaszewski (1941-2022) was born in September 1941 in Lipniki, Reichskommisariat, Ukraine, USSR. He was a fighter pilot, Polish Air force officer and cosmonaut. He was a survivor of the Volhynian slaughter during which the Ukranian Insurgent Army murdered nineteen members of his family in March 1943. He graduated from the General Staff Academy, Warsaw, in 1971; graduated from Voroshilov General Staff Academy, Moscow, in 1982, and later became Brigadier General and commander of Deblin Air Force School. He later served as an advisor for the Polish Secretary of Defense. Selected on November 25, 1976, as a Research Cosmonaut, in 1978 he became the first Polish national in space when he flew aboard the *Soyuz 30* spacecraft. He logged seven days, 22 hours and two minutes in space; and retired on July 05, 1978. Hermaszewski died in Warsaw, Poland in December 2022. He was married, with two children.

## Ivanchenkov, Aleksandr S. Soyuz 29/31, Soyuz T-6.

Aleksandr Sergeyevich Ivanchenkov was born in September 1940 in Ivanteyevka, USSR. He is a retired engineer and former cosmonaut. Selected in Civilian Specialist Group 5, he flew on *Soyuz 29* (7K-T No. 46) to the *Salyut 7* space station for mission EO-2 on June 15, 1978. The station had been vacant for three months since the *Soyuz 26* mission ended after a record 96 days. During a spacewalk on July 29, a meteor passed below them, which briefly blinded them. A lunar eclipse was observed on October 07, extensive medical tests were carried out between October 17-18, and more material experiments were conducted in the Kristall and Splav furnaces on October 22. With Commander Kovalyonok, he established a new space-endurance record of 139 days; and returned aboard *Soyuz 31* on November 02, landing 180 km southeast of Jezkazgan, Kazakhstan. Their physical condition was slightly better than that of the previous long-duration crews: attributed to the hard exercise program carried out in the final mission days.

Ivanchenkov was the Flight Engineer aboard *Soyuz T-6*, which launched on June 24, 1982, to the *Salyut 7* space station, along with Commander Dzhanibekov and a Frenchman, Jean-Loup Cretien. A manual docking was needed, after problems arose with the craft's onboard automatic system. Cretien was quoted saying, *Salyut 7 "is simple, doesn't look impressive, but is reliable"*. During this mission, Ivanchenkov spent seven days, 21 hours and almost 51 minutes in space. He logged a career total of 147 days, 12 hours and 37 minutes in space; including one EVA of two hours and five minutes. Ivanchenkov received eight awards, including two Hero of the Soviet Union; two Order of Lenin; the Medal "For Merit in Space Exploration (Russian Federation); Hero of the German Democratic Republic; the Order of Karl Marx; and Commander of the Legion of Honour (France). He is married, with one child.

## Ivanishin, Anatoli A. Soyuz TMA-22/Expn 29/30, MS-01/Expn 48/49, MS-16/Expn 62/63.

Anatoli Alekseyevich Ivanishin was born in January 1969 in Irkutsk, Siberia, Soviet Union. In 1986 he unsuccesfully attempted to enter the Chernigov Higher Military Aviation School for Pilots; then entered the Irkutsk Polytechnic Institute, completing his first year in 1987. He again applied to the Chernigov School, was successful, and graduated in 1991 with a Gold Medal, then served in the Borisoglebsk military unit, Voronezh region, flying MiG-29 jet fighters. As a senior fighter pilot, he flew Su-27 fighter aircraft. He logged 507 hours flying time, and made 180 parachute jumps. In 2003, Ivanishin graduated from the Moscow State University in Economics, Statistics and Information Theory. From 1991, he served in combat units of the Russian Air Force. Selected on May 29, 2003, for Cosmonaut Group (TsPk-13), he passed the State exams with a rating of "excellent" on June 28, 2005. Ivanishin launched on November 11, 2011, aboard *Soyuz TMA-22* for ISS 'Expedition 29/30'. The landing of *Soyuz TMA-22* was delayed from March to late April 2012, due to a delay with the launch of the 'Expedition 31/32' crew. After 165 days in space, he returned to Earth on April 27, 2012.

Launched on July 07, 2016, aboard *Soyuz MS-01* for 'Expedition 48', Ivanishin was Commander for ISS 'Expedition 49'. Continuing delays with the introduction of the *Soyuz MS* spacecraft forced the launch of *Soyuz MS-02* to be delayed for technical reasons. He returned to Earth on October 29, 2016, after another 115 days in space. Launched on April 09, 2020, Ivanishin aboard *Soyuz MS-16* for 'Expedition 62/63' docked with the ISS six hours after launch. The 'Expedition 63' crew welcomed the *Crew Dragon* 'Demo-2' mission aboard the ISS on May 31, 2020; the first crewed test flight of the **SpaceX** Crew Dragon spacecraft. NASA's Doug Hurley and Robert Behnken stayed aboard the ISS for two months. Ivanishin retired on October 15, 2021, to devote most of his time to "scientific activities". He is married, with one son, Vladislav Ivanishin, born in 1993.

## Kadeniuk, Leonid K. STS-87.

Maj. Gen. Leonid Kostyantynovych Kadeniuk (1951-2018) was born in January 1951 in the village of Klishkivtsi, Chernivtsi Oblast, Ukrainian SSR, Soviet Union. He was a test pilot, Major General in the Ukrainian Air Force, and former cosmonaut. After graduating from high school with a silver medal, he entered the Chernihiv Higher Military Aviation School of Pilots, graduating in 1971 with a Pilot-Engineer diploma specializing in "Piloting and Operation of Aircraft". Kadeniuk became a Soviet Air Forces military pilot, flying over fifty aircraft types;

mainly on fighters, as well as on the American Northrop T-38 training aircraft. He was selected in 1976 Air Force Group 6, on a team of test pilots for the Soviet space shuttle, *Buran*. In 1977, he graduated as a test pilot, and from 1977-79 underwent general space training. From 1984-88, he was a test pilot of the State Research Institute of the USSR Air Force. In 1989, he graduated from the Moscow Aviation Institute – Faculty of Aircraft. From 1988-90, he trained as Commander of the Buran reusable spacecraft; but was removed from the team of test pilots due to marital issues. Kadeniuk was allowed to rejoin the Cosmonaut Detachment several years later. He took part in working out the glide path of descent for landing the *Buran* spaceplane.

From 1990-92, he trained as the Commander for *Soyuz TM* spacecraft. With the December 1991 breakup of the Soviet Union, Kadeniuk remained in the Russian Space Forces and adopted Ukrainian citizenship. Trained by NASA, he launched on November 19, 1997, as a Payload Specialist aboard Space Shuttle *Columbia* on STS-87 for a fifteen day, 16 hour and 35 minute mission. Later, he continued in the Ukrainian State Space Agency program. He was Deputy of Ukraine of the Fourth Convocation (State parliament), People's Ambassador of Ukraine; was awarded Hero of Ukraine; was President of the Aerospace Society of Ukraine, Advisor to the Prime Minister of Ukraine, Chairman of the State Space Agency of Ukraine, and made an honorary Doctor of Chernivtsi National University. Leonid Kadeniuk died on January 31, 2018, aged sixty-seven years in Kyiv, Ukraine.

## Kaleri, Aleksandr Y. Soyuz TM-14, TM-24, TM-30, TMA-3, TMA-01M/ Expedition 25/26.

Aleksandr "Sacha" Yuriyevich Kaleri was born in May 1956 in Jurmala, Latvian SSR (now Latvia). He is a former Russian cosmonaut and veteran of extended stays on the *Mir* space station and the International Space Station (ISS). In 1979, he graduated from the Moscow Institute of Physics and Technology, Dolgoprudny, Moscow region, as a specialist in Aircraft Flight Dynamics and control. In 1983, he completed post-graduate studies there as a specialist in the field of Mechanics of Fluids and Plasma. As Flight Engineer on *Soyuz TM-14* to station *Mir* for 145 days, 14 hours and ten minutes, the mission lasted from March 17 to August 10, 1992. Landing 136 km southeast of Dzhezkazgan, the landing system malfunctioned, causing the descent module to turn over. The occupants were trapped inside, upside down, until it could be righted. Kaleri was Flight Engineer on *Soyuz TM-24* to *Mir* for 196 days, 17 hours and 26 minutes (August 17, 1996 to March 02, 1997) as the 'Mir-22' Flight Engineer. The crew of Korzun, Kaleri and Ms Claudie Andre-Deshays joined American astronaut Mrs. Shannon Lucid and 'Mir 21' crewmates Yuri Onufriyenko and Yuri Usachev. Kaleri's landing after 3,160 orbits was at 47.82 degrees North, 69.40 degrees East.

For his third spaceflight, Kaleri launched aboard *Soyuz TM-30* on April 04, 2000, to *Mir* as the 'Mir EO-28' Flight Engineer. The crew were sent by **MirCorp**, a privately funded company, to reactivate and repair the station. The crew also resupplied *Mir* and used the engines of the *Progress M1-1* and 'M1-2' spacecraft to boost *Mir* to a higher orbit of 360 to 378 kilometres. This mission delayed the de-orbit of *Mir*, which ultimately occurred in March 2001. An EVA on May 12 was to repair damage to *Mir*'s exterior components, and record panoramas of the station's hull for experts on Earth to analyze the effecs of space on the ageing station itself. The cosmonauts undocked on June 15, 2000, after 70 days and 15 hours. The landing on June 16 was about 45 km southeast of Arkalyk, Kazakhstan. Launched aboard *Soyuz TMA-3* on October 18 as 'Flight Engineer 1', from October 20, 2003, to April 29, 2004, he was on the ISS for 'Expedition 8'; landing in *Soyuz TMA-3* some sixty km northeast of Arkalyk, Kazakhstan. His mission lasted 192 days, 13 hours and 36 minutes at the ISS.

Launched aboard *Soyuz TMA-01M* as Commander on October 07, 2010, Kaleri was a Flight Engineer for ISS 'Expedition 25/26'. Until his mission ended in March 2011, Kaleri participated in a wide range of research including fundamental physics, biometric experiments and investigations of crystal growth in space, as well as education outreach. Sacha Kaleri has logged a total of 769 days, six hours and 46 minutes in space, the fifth-longest time in space of any peron, and the longest time in space of any person not born in what is now Russia. He made five EVAs, totaling 23 hours; the longest being of six hours and 36 minutes on December 09, 1996, when providing more electrical power to the *Mir* station. Kaleri received eight awards, including Hero of the Russian Federation (1992); Chevalier of the Legion of Honour (France, 1997); the Order of Merit for the Fatherland, second class (2000), third class (1997), fourth class (2011); the Order of Friendship (2005); the Medal "For Merit in Space Exploration"; and the NASA Distinguished Public Service Medal. He is married, with a son.

## Khrunov, Yevgeny V. Soyuz 5/4.

Col. Yevgeny Vasilyevich Khrunov (1933-2000) was born in September 1933 in Prudy, Tula Oblast, USSR. He was an engineer and Colonel in the Soviet Air Force; drafted into the Soviet Army in 1952, later applying for pilot school. Khrunov graduated from the Air Force Military Engineering Academy, Monino, in 1968; and received a Candidate of Technical Sciences degree in 1971. He continued school in the military at Pavlograd in Ukraine, and later transferred to the Serov Higher Air Force School in Rostov Oblast, southwestern Russia. On August 06, 1958, he was promoted to senior lieutenant. In 1959, he and Gorbatko were interviewed in regards to becoming a cosmonaut, although they were not expressly told that was what they were being interviewed for.

Launched on January 15, 1969, he flew as Research Engineer on *Soyuz 5*, which docked with *Soyuz 4* in orbit. This mission was the first docking of two crewed spacecraft, and saw the first transfer of crew from one space vehicle to another. The transfer was accomplished with a spacewalk, two months before the United States' *Apollo 9* mission performed the first internal crew transfer. This mission, flown by Boris Volynov, Aleksei Yeliseyev and Khrunov, was memorable for its dramatic re-entry. The craft's service module did not separate, so it entered the atmosphere nose first. As the craft aerobraked, the descent module righted itself before the escape hatch burned through; but the parachute lines tangled and the landing rockets failed, resulting in a very hard landing which broke Volynov's teeth when he returned to Earth alone. Volynov logged three days, zero hours and 54 minutes in space aboard *Soyuz 5*; while Khrunov logged one day, 23 hours and 45 minutes; including one EVA of 37 minutes when he transferred to *Soyuz 4*.

After leaving the space program in 1980, he worked at the Thirtieth Central Scientific Research Institute, Ministry of Defence (Russia), and later was appointed to the Chief State Commmittee for foreign economic relations, until his retirement in 1989. Khrunov received twelve awards, including Hero of the Soviet Union; Order of Lenin; the Order of the Red Star; six different Jubilee Medals; Medals "For Impeccable Service", first, second and third classes; and the Medal "For Strengthening Military Cooperation". Khrunov died of a heart attack on May 20, 2000, aged sixty-six years, in Moscow, Russia. He was married, with two children.

## Kikina, Anna Y. SpaceX Crew 5 to ISS.

Anna Yuryevna Kikina was born in August 1984 in Novosibirsk, Russian SFSR, Soviet Union. She graduated with honors from the Novosibirsk State Academy of Water Transportation Engineering; and also earned a degree in Economics and Management. She worked as a tour guide in the Altai region, as well as being a swimming and paratroop instructor. Kikina was also a host for Radio Siberia. She was selected in 2012 TsPk-16 Group, and is an active cosmonaut, the only female currently in service at Roscosmos. In Spring 2021, toy manufacturer **Mattel** released a Barbie astronaut doll in Kikina's image. In September 2021, ***RIA Novosti*** reported that Kikina had beeen assigned to the *Soyuz MS-22* mission to the International Space Station, set to launch on September 21, 2022, for a 188-day mission. In December 2021, Roscosmos Director General Dmitry Rogozin announced she would fly an *"American commercial spacecraft"* in September 2022, making her the first Russian cosmonaut to fly aboard a *Crew Dragon*, and the first Roscosmos cosmonaut to fly aboard a US spacecraft since 2002. She launched in October 2022, aboard SpaceX *Crew Dragon*; returning to Earth with 'Crew 5' after 157 days, ten hours and one minute in space, with the capsule landing in the Gulf of Mexico. Kikina is married to Alexander Serdyuk, a physical training instructor at the Gagarin Cosmonaut Training Center.

## Kizim, Leonid D. Soyuz T-3, T-10, T-15.

Col. Gen. Leonid Denisovich Kizim (1941-2010) was born in August 1941 in Krasnyi Lyman, Donetsk Oblast, Soviet Union (now Lyman, Ukraine). He graduated from Higher Air Force School in 1975, and served as a test pilot in the Soviet Air Force. Selected as a cosmonaut on October 23, 1965, he flew as Commander on *Soyuz T-3, Soyuz T-10* and *Soyuz T-15*, and also served as backup Commander for *Soyuz T-2*. Launched on November 27, 1980, the *Soyuz T-3* mission to *Salyut 6* was the first *Soyuz* spacecraft to carry three cosmonauts (now equipped with pressure suits) since the fatal *Soyuz 11* flight ended on June 30, 1971. While their main objective was to refurbish the *Salyut*, part of the mission involved testing their newly improved spacecraft. They performed experiments using the *Splav* and *Kristall* units, and studied biological material carried aboard the *Soyuz* using the *Svetoblok* and *Oazis* units. A majority of their time was devoted to station maintenance, guided by the 'Principal Expedition 4' crew in TsUP ground control. *Soyuz 3* returned to Earth on December 10, 1980, after 12 days, 19 hours and seven minutes in space; landing 130 km (81 miles) west of Dzhezkazgan, Kazakhstan.

Launched on February 08, 1984, *Soyuz T-10* was the fifth expedition to the *Salyut 7* space station. The crew of Kizim, Flight Engineer Vladimir Solovyov and Research Cosmonaut Oleg Atkov entered a darkened and empty station using their flashlights, since *Soyuz 10a* had suffered a launch mishap. They commented on the burnt-metal odor of the drogue docking unit. By February 17 the station was fully reactivated. The crew made three EVAs to repair a ruptured fuel line, setting a record for spacewalk hours. The men returned to Earth aboard *Soyuz T-11* on October 02, 1984, landing 46 km east of Arkalyk, Kazakhstan. Launched on March 13, 1986, aboard *Soyuz T-15* with Flight Engineer Vladimir Solovyov, he was part of the only crew to visit two space stations during one spaceflight (*Mir*, its' first visit, for almost 55 days and *Salyut 7*, the final visit). *Soyuz T-15* was the final flight of the *Soyuz-T* marque, which had been in service from 1979-86. Due to the political pressure of launching *Mir* (meaning "Peace") in time for the Twenty-seventh Communist Party Congress, Soviet mission planners were left without the newer *Soyuz-TM* spacecraft or any of the planned modules to launch to *Mir*.

It was decided to launch an older *Soyuz-T* craft as *Soyuz T-15* on the dual-visit mission. The *Soyuz-T* was not equipped with the *Kurs* approach system used on *Mir*'s front docking port, so the crew had to approach *Mir*'s aft port using the *Igla* approach system, shut it down at 200 meters range, then manually maneuver around the station to dock manually at the station's front port. On May 04, *Mir*'s orbit was lowered by thirteen kilometres to expedite the approach to *Salyut 7* and conserve the *Soyuz*'s limited fuel supply for the transfer. After a crossing of 29 hours, *Soyuz T-15* docked with *Salyut 7* on May 06, 1986. The crew conducted two EVAs and collected experiment results, experimental apparatuses, and samples of materials to finish the work of the previous *Salyut 7* 'EO-4' crew, who had to leave early when Vladimir Vasyutin became ill. Kizim and Solovyov removed twenty instruments with a total mass of 350-400 kg from *Salyut 7* before returning to *Mir* on June 26. On July 06, 1986, Kizim became the first human to have spent a full year in space. US officials accused the Soviet Union of performing military experiments on their supposedly civilian space stations. After this flight, in an official Press conference, Kizim stated that *Mir* was not being used for any military purposes and that *"The US is accusing us of this sort of action in order to justify their own plans to extend the arms race into space"*. The Reagan Administration did not repeat these claims, so as not to negatively impact the planned 1987 Summit Meeting between President Ronald Reagan and Soviet leader Mikhail Gorbachev.

Leonid Kizim logged a total of 374 days, 17 hours and 56 minutes in space. He later served as Deputy Director Satellite Control-Center of the Russian Ministry of Defense. After May 1995 he was Director of the Military Engineering Academy of Aeronautics and Astronautics in St. Petersburg. Kizim left the cosmonaut program on June 13, 1987, but remained in the Soviet Air Force, and was later in the Russian Air Force. In 1993, he was placed in charge of the A.F. Mozhaysky Military-Space Academy, holding that position until retiring from the Russian Air Force on September 10, 2001, at the rank of Colonel General. He received six Soviet and six foreign awards, including two Hero of the Soviet Union (1980, 1984); the Order of Honour; Order of Friendship; three Order of Lenin; Order of Sukhbaatar (Mongolia); Knight of the Legion of Honour (France); and Order of Merit (Ukraine). Leonid Kizim died on June 14, 2010, leaving a wife and two children.

## Klimuk, Pyotr I. Soyuz 13, Soyuz 18, Soyuz 30.

Col. Gen. Pyotr Ilyich Klimuk was born in July 1942 in Kamarouka, Belarussian SSR, Soviet Union. He is a former cosmonaut and the first Belarussian to perform space travel. He entered the Soviet Air Force in 1964, and was selected in 1965 Cosmonaut Group. Klimuk flew on *Soyuz 13* (December 18-26, 1973) with Valentin Lebedev, the second flight of the re-designed *Soyuz 7K-T* spacecraft; landing 200 km southwest of Karaganda, Kazakhstan. The *Soyuz* was modified to carry the Orion-2 Space Observatory, the Soviet Union's first dedicated science mission; and was controlled from the new Kaliningrad Mission Control Center.

Klimuk launched on *Soyuz 18* (May 24 to July 26, 1975) with Vitaly Sevastyanov to the *Salyut 4* space station. Studies of the stars, planets, Earth and its atmosphere were made on June 02 and 03. Some 2,000 photographs of the Earth and 600 of the Sun were taken. On June 23, when they passed the 29-day endurance record set by *Soyuz 17* the cosmonauts made extensive medical experiments. By July 1975, the living conditions aboard the *Salyut* were starting to degrade, with the environmental control system failing, windows fogged over and green mould growing on the station walls. The crew donned exercise suits and increased their exercise period to over two hours per day. Controlled from the old Crimean Control Center, they set a new Soviet space endurance record of 62 days, 23 hours and 20 minutes; landing 56 km east of Arkalyk, Kazakhstan. It was ten days before Klimuk could take a ten-minute walk, and a week before he made a full recovery.

Klimuk flew on *Soyuz 30* (June 27 to July 05, 1978), Intercosmos flight EP-3, with Miroslaw Hermaszewski to the *Salyut 6* space station. They filmed the Aurora Borealis, and Hermaszewski photographed his homeland of Poland and conducted crystallization experiments. They were the first crew to visit the long-duration *Soyuz 29* resident crew of Kovalyonok and Ivanchenkov. The *Soyuz 30* mission lasted seven days, 22 hours and three minutes; landing in a Rostov state farm field 300 km west of Tselinograd. Klimuk logged a total of 78 days, 18 hours and 17 minutes in space. He resigned from the cosmonaut team in 1978 to take up a position as the Assistant to the Chief of the Gagarin Center. From 1991 to 2003, Kilmuk headed the Yuri Gagarin Cosmonaut Training Center. Klimuk is the author of two books on human spaceflight: *Beside the Stars*, and *Attack on Weightlessness*. He received twenty-one awards, including two Hero of the Soviet Union (1973, 1975); three Order of Lenin (1973, 1975, 1978); the Order "For Merit for the Fatherland", third class (2000), fourth class (1996); Order of Parasat (Kazakhstan, 1995); Order of the Legion of Honour, (France, 2004); Order of Friendship of Peoples (Belarus, 2007); Medal "One Hundredth Anniversary of the fall of the Ottoman yoke (NRB); the Tsiolkovsky Gold Medal; and Gagarin Gold Medal.

## Komarov, Vladimir M. Voskhod 1, Soyuz 1.

Vladimir Mikhaylovich Komarov (1927-67) was a Soviet test pilot, aerospace engineer, and cosmonaut. He commanded *Voskhod 1*, the first spaceflight to carry more than one crew member. Komarov was also the first man known to have died during a space mission. He was one of the oldest members of the inaugural cosmonaut selection in March 1960, and became the first cosmonaut to fly twice. Komarov began his formal education in 1935, and soon showed a natural aptitude for mathematics. In 1941, he left school because of the German invasion in WWII, and became a laborer on a collective farm. He showed an early interest in aeronautics, making model aircraft and his own propellor. Komarov joined the First Moscow Special Air Force School at the age of fifteen years, and graduated in 1945 with honors, becoming a pilot in 1949. In 1959, he graduated from the Zhukovsky Air Force Engineering Academy, Moscow. On October 12, 1964, Komarov and engineer Konstantin Feoktistov with medical doctor Boris Yegorov were launched aboard *Voskhod 1*. The *Voskhod* craft was 950 kg (2,100 lb) heavier than *Vostok* and featured many technical improvements, but there was no room for spacesuits to be worn. *Voskhod* crew remained aboard when making a hard-surface landing, using drogue parachutes and retrorockets.

*Soyuz 1* launched on April 23, 1967. The flight was plagued with technical issues. Back-up pilot, Yuri Gagarin was aware of many of the 203 design problems and had attempted to "bump" Komarov from the mission, knowing that the Soviet leadership would not risk a National Hero on the flight. At the same time, Komarov refused to pass on the mission, even though he believed it to be doomed. A solar panel would not deploy (reducing power), and, more seriously, problems with the orientation detectors complicated navigation and maneuvering the craft. The spacecraft span out of control for twenty-six hours, the mission was cut short, and during the eighteenth orbit, Komarov attempted a landing. An NSA listening station near Istanbul reported that Komarov was heard screaming, and he apparently cursed the Soviet bureaucracy as his doomed craft plunged into the atmosphere.

The drogue parachute deployed, but the spacecraft became entangled in its main parachute at an altitude of several miles and fell to Earth at high speed; around 140 km/hr (87 mph) to a spot near Karabutak, about 275 km (171 miles) east-southeast of Orenburg in Kazakhstan. The capsule was in flames with streams of molten metal dripping from the exterior. The entire base of the capsule had burned through during re-entry. Komarov's heat-shrunken body underwent a brief autopsy, then was cremated, and his ashes interred in the Kremlin Wall Necropolis. According to interviews with Venyamin Russayev, a former KGB agent, he said that prior to launch of *Soyuz 1* engineers had reported 203 design faults to Communist Party leaders. Their concerns *"were overruled by political pressures for a series of space feats to mark the anniversary of Lenin's birthday"*. Unquote. Named in Vladimir's honor, Kosmonaut Vladimir Komarov was a satellite tracking ship of the Soviet Union. Komarov is commemorated in two memorials left on the lunar surface: one left at Tranquility Base by the *Apollo 11* astronauts, and the Fallen Astronaut statue and plaque left by the *Apollo 15* crew.

## Kondakova, Yelena V. Soyuz TM-20, STS-84.

Yelena Vladimirovna Kondakova was born in March 1957 in Mytishchi, Russian SFSR, Soviet Union. She was the third Soviet or Russian female cosmonaut to travel to space, and the first woman to make a long-duration spaceflight. In 1980, Kondakova graduated from the Bauman University, where she specialized in production of aircraft. She then worked as an engineer in the One Hundred Thirteenth department of **NPO Energia** (RSC Energia since 1994). She was engaged in the long-term planning of spaceflights and training personnel to work in emergencies. In 1982-85, she was a member of the flight control group of the *Salyut 7* space station. Launched on October 04, 1994, as a Flight Engineer aboard *Soyuz TM-20*, she returned to Earth on March 22, 1995, after a 169 day, five hour and 21 minute stay (five months) at the *Mir* station.

In August 1996, Kondakova trained at JSC in Houston, Texas, then flew as a Mission Specialist on STS-84, launched aboard space shuttle *Atlantis* on May 15, 1997. The shuttle made the sixth docking with the *Mir* orbital station. Returning to Earth after nine days, five hours and 19 minutes on May 24, she had logged a total of 178 days, ten hours and 41 minutes in space. Kondakova retired from the cosmonaut corps on December 30, 1999, due to her election as a deputy to the State Duma as a member of the United Russia party. From 2003-11, she was Deputy of the State Duma of the Russian Federation of the Fourth and Fifth Convocations. In 2006, Kondakova graduated from the Diplomatic Academy of the Ministry of Foreign Affairs of the Russian Federation. On August 16, 2011, she withdrew from the United Russia party due to her disagreement with the results of the August intra-party elections, where it was obvious there were more ballots counted than were people on the site! At a Press conference that day, she said, ***"United Russia's primaries are an economically inexpedient game of democracy****. I realize that my announcement means the end of my political career"*. Unquote. From March 27, 2012 to 2018, Kondakova was a Trade Representative of the Russian Federation in Switzerland.

## Kondratiyev, Dmitri Y. Soyuz TMA-20/Expedition 26/27.

Col. Dmitri Yuryevich Kondratiyev was born in May 1969 in Irkutsk, Russian SFSR, Soviet Union. He completed Yak-52 flight training at the Alma-Aty Aviation Club in 1986; and graduated from the Kacha Air Force Pilot School in 1990 as a pilot-engineer. Also in 1990, Kondratiyev graduated from the Higher Military Air Force Pilot School in Volgograd; and from the Moscow State University for Economy, Statistics and Computer Science in 2000. In 2004, he graduated from the Yuri A. Gagarin Air Force Academy. He flew MiG-21, MiG-29 and Su-27 Russian fighter aircraft, and is a Class 1 Air Force pilot. Selected in December 1997 TsPK-12 Cosmonaut Group, he completed basic training in March 2000. From May 02, 2006, to April 09, 2007, he served as Director of Operations, Russian Space Agency, at the Johnson Space Center, USA.

Kondratiyev flew as Flight Engineer aboard *Soyuz TMA-20* on December 15, 2010, for ISS 'Expedition 26/27', becoming Commander on the ISS. His first EVA with Oleg Skripochka on January 21, 2011, was to complete installation of a new 100 MB/s high-speed data transmission system. In his second EVA on February 16, he and Skripochka installed a radio antenna, deployed a nano satellite, and installed two experiments then retrieved two exposure panels. The experiments they installed are the Molniya-Gamma experiment, which measures gamma splashes and optical radiation during terrestrial lightning and thunder conditions, and a high-speed data transmission system experiment that uses radio technology. He returned to Earth on May 24, 2011, and has logged 159 days, seven hours and 17 minutes in space; including two EVAs totaling ten hours and 13 minutes. He retired on July 25, 2012. Kondratiyev is married, with two children.

## Kononenko, Oleg D. Soyuz TMA-12, TMA-03M, TMA-17M, MS-11, MS-24/25/ Expn 70/71.

Oleg Dmitriyevich Kononenko was born in June 1964 in Chardzhou, Turkmen SSR (now Turkmenabat, Turkmenistan. In 1988, he graduated from the N.E. Zhukovsky Kharkiv Aviation Institute as a Mechanical Engineer. He then worked in the Russian Space Agency's Central Specialised Design Bureau TsSKB-Progress in Kuybishev; starting as an engineer and working his way up to the leading Design Engineer. His responsibilities included system design, analysis, and development of spacecraft electrical power sysytems. Selected in 1996 RKA Group, he has flown to the ISS five times: as a Flight Engineer aboard *Soyuz TMA-12* (April 08 to October 24, 2008) for 'Expedition 17'; aboard *Soyuz TMA-03M* (December 21, 2011 to July 01, 2012) to join 'Expedition 30' and become Commander of 'Expedition 31'; aboard *Soyuz TMA-17M* (July 22 to December 11, 2015) for 'Expedition 44/45'; aboard *Soyuz MS-11* (December 03, 2018 to June 24, 2019) to be Commander of 'Expedition 59', returning after 203 days, 15 hours and 16 minutes in space; and aboard *Soyuz MS-24/MS-25* (launched September 15, 2023) for 'Expedition 70/71', with the mission ongoing, and by March 2024 taking his total accumulated time to past 1,000 days in space.

As of mid-February 2024, Kononeko had logged over 881 days in space; breaking Gennady Padalka's world record for the most time spent in space on February 04, 2024, at 07:30:08 UTC. Kononenko has received the title, Hero of the Russian Federation; and Hero of Turkmenistan. He is married, with a son and a daughter.

## Korniyenko, Mikhail B. Soyuz TMA-18.

Mikhail Borisovich Korniyenko was born in April 1960 in Syzran, Kuibyshev (Samara) Region, Russian SFSR, USSR. He graduated from Secondary School No. 15, Chelyabinsk, Russia, in 1977. In May 1978, he was called to serve in the Russian Army. After his time in the Airborne Forces (VDV) in Kirovabad, Azerbaijan, he worked for the Moscow Municipal Militsiya from 1980-86. From 1981-87, he studied at the Moscow Aviation Institute, named after S. Ordzhonikdze. From 1986-91, Korniyenko worked in the Baikonur Launch Facility as a launch equipment specialist. In October 1995, he began working at the **Energia Rocket Space Corporation** (RSC Energia) as an engineer. He directly participated in testing the **RSC Energia** production on the testing ground.

Selected as a Test Cosmonaut candidate on February 24, 1998, Korniyenko completed basic training at the Gagarin Academy, and since 1999 trained in the International Space Station group. In 2005, he was assigned to *Soyuz TMA-10* and 'Expedition 15' backup as Flight Engineer 1. Launched on April 02, 2010, aboard *Soyuz TMA-18* with Aleksandr Skvortsov and NASA's Tracy Caldwell-Dyson, he became part of ISS 'Expedition 24'. He landed near Arkalyk, Kazakhstan on September 24, 2010. Sharing his experiences with an audience, Korniyenko said, *"The thing you miss there [in space] most of all is the Earth itself, I missed smells. I missed trees. I even dreamt of them. I even hallucinated. I thought I smelled a real fire and something being barbequed on it! I ended up putting pictures of trees on the walls to cheer up. You do miss the Earth there"*. Unquote.

In November 2012, NASA, the Russian Federal Space Agency (Roscosmos), and their international partners selected two veteran spacefarers for a one-year mission aboard the ISS in 2015. NASA's Scott Kelly and Roscosmos' Mikhail Korniyenko were chosen; launching in March 2015 and landing on March 01, 2016. Their mission goal was to better understand how the human body reacts and adapts to the harsh environment of space. Data obtained will help inform current assessments of crew performance and health, and determine better and validate countermeasures to reduce risks associated with future exploration as NASA plans for missions around the Moon, an asteroid and ultimately to planet Mars. In 2016, Kornieko and Andrey Borisenko appeared as themselves in the comedy film ***Yolki 5***. He is married, with a daughter.

## Korsakov, Sergey V. Soyuz MS-21/Expedition 67.

Sergey Vladimirovich Korsakov was born in September 1984 in Frunze, Kirghiz SSR, Soviet Union. He graduated (with honours) from the Bauman Moscow State Technical University in 2006 with a degree in rocketry. Selected by Roscosmos in TsPk-16 Cosmonmaut Group in 2012, he was scheduled to fly in April 2021 aboard *Soyuz MS-18*, but was replaced in March 2021, and then assigned to *Soyuz MS-21* which flew him as 'Flight Engineer 2' on March 18, 2022. It was the first mission to the ISS with three Roscosmos cosmonauts, including Commander Oleg Artemyev and Denis Matveev.

SpaceX Crew-3 launched to the ISS on November 10, 2021, with NASA astronauts Raja Chari, Thomas Marshburn and Kayla Barron and ESA astronaut Matthias Maurer for 'Expedition 66/67'. Korsakov joined 'Expedition 67' on November 11 for the second part of this four-part mission. Returning to Earth on September 29, 2022, Korsakov has logged a total of six months and 11 days (194 days) in space, as of August 07, 2023. He is married, with two children.

## Korzun, Valery G. Soyuz TM-24, STS-111/Expedition 5/STS-113.

Col. Valery Grigoryevich Korzun was born in March 1953 in Krasny Sulin, Rostov Oblast, Russian SFSR, Soviet Union. He graduated from the Kachinsk Military Aviation Pilot School in 1974; and from the Gagarin Military Aviation Academy in 1987. Korzun is a First Class military pilot with 1,473 flight hours in four main aircraft types. In 1987, after a successful tour as Commander of the Gagarin Military Air Force Academy, he was selected in TsPk-8 Cosmonaut Group; and was certified as a Test Cosmonaut in June 1989. From October 1992 to March 1994, he underwent training as Commander of the Soyuz-TM rescue spacecraft. From March 1994 to June 1995, he trained for flight aboard the orbital complex *Mir*. Launched with Alexander Kaleri and the first French woman in space, Claudi Andre-Deshays, he was Commander of *Soyuz TM-24* for the 'Mir 22' mission (August 17, 1996 to March 02, 1997), logging 196 days, 17 hours and 26 minutes in space. On December 02, 1996, Korzun's first EVA was to complete connections of the solar array to provide more power to the *Mir* station. His second EVA was to complete the MCSA cable installation, and attach a *Kurs* docking antenna. He landed 128 km east of Jezkazgan in Kazakhstan.

Launched on STS-111 aboard shuttle *Endeavour* (June 05 to December 07, 2002), Korzun joined 'Expedition 5' crew as Commander, and participated in ecological experiments and monitored the Earth's surface from space. Earth observation has many useful applications: including informing the ground of disaster situations such as fires and floods as quickly as they occur. Korzun and NASA's Peggy Whitsun made an EVA on August 16, 2002, to fortify the ISS against damaging space debris impacts by installing six protective panels onto the *Zvezda* module. During his fourth EVA, Korzun installed two ham radio antennas on *Zvezda*, to improve future amateur radio operations. He returned aboard *Endeavour* on STS-113, after 184 days, 22 hours and 14 minutes in space. Korzun logged a total of 381 days, 15 hours and 40 minutes in space; including four EVAs totaling 22 hours and 19 minutes. He was awarded Hero of the Russian Federation; Pilot-Cosmonaut of the Russian Federation; and Cavalier of Legion of Honour (France). He is married, with a son.

## Kotov, Oleg V. Soyuz TMA-10, TMA-17/Expn 22/23, TMA-10M/Expn 37/38.

Oleg Valeriyevich Kotov was born in October 1965 in Simferopol, Crimean Oblast, Ukrainian SSR. He graduated from the S.M. Kirov Military Medical Academy in 1988. After a career as a physician assigned to the Soviet space program, he was selected as a Test Cosmonaut candidate, and from June 1996 to March 1998, he completed basic cosmonaut training. He has flown three long-duration missions on the ISS; logging over 526 days in space. Kotov gained experience in practical training and medical support of spacewalks on *Mir*, and was a crew surgeon and instructor for biomedical training and science program training. He served as a Representative of the Gagarin Training Center at the Johnson Space Center from February to October 1999. In 2004, he became Chief of the CapCom Branch in the Cosmonaut Office. Launched aboard *Soyuz TMA-10* on April 07, 2007, with Fyodor Yurchikhin and "space tourist" Charles Simonyi, he spent 196 days and 17 hours in space; returning on October 21 in a ballistic re-entry and landing west of Arkalyk, about 340 km northwest of the intended landing site in Kazakhstan.

He commanded *Soyuz TMA-17*, and was assigned to 'Expedition 22' as a Flight Engineer, and to 'Expedition 23' as Commander aboard the ISS. Kotov returned to Earth on June 02, 2010, touching down on the central steppes of Kazakhstan. Launched on September 25, 2013, aboard *Soyuz TMA-10M* with Sergey Ryazansky and NASA's Michael Hopkins, he joined 'Expedition 37' as Flight Engineer, and became ISS Commander. 'Expedition 37/38' ran from September 2013 to March 2014. Kotov was awarded Hero of the Russian Federation; the Combat Heroism Russian Federation Armed Forces Medal of I and II degrees; and the Service Medal of I, II and III degrees. He is married, with two children.

## Kovalyonok, Vladimir V. Soyuz 25, 29, 31, Soyuz T-4.

Col. Gen. Vladimir Vasiliyevich Kovalyonok was born in March 1942 in Bielaye, Byelorussian SSR, Soviet Union. Selected in Air Force Group 4, he entered the Soviet space program on July 05, 1967, and was Commander of three missions. Kovalyonok's first mission was aboard *Soyuz 25* (October 09-11, 1977), the first flight to the *Salyut 6* space station, launched ten days earlier. The mission was aborted when five attempts to engage the station's docking latches failed. This failure, later traced to a faulty mechanism on the *Soyuz*, led to a new rule (enforced until 1994) whereby every crew had to have at least one person aboard who had previously flown in space [and had practical experience]. With battery power in their capsule for only about two days, Kovalyonok and Ryumin returned to Earth; landing 185 km northwest of Tselinograd, Kazakhstan, on October 11. The failure badly dented Soviet prestige, since their media had given the flight a high profile, noting that it came from the same launchpad as *Sputnik 1* and Gagarin's *Vostok 1*, and coincided with the sixtieth anniversary of the October Revolution in 1917.

Launched on June 15, 1978, aboard *Soyuz 29*, Kovalyonok and Aleksandr Ivanchenkov flew the long-endurance mission EO-2, which set a new record of 139 days, 14 hours and 47 minutes in space. They docked with *Salyut 6* on June 17; and exercised hard for three hours per day during their final month in orbit. Valery Bykovsky and Sigmund Jahn launched on August 26 aboard *Soyuz 31* to *Salyut 6*, and returned to Earth on September 03 aboard *Soyuz 29*. Kovalyonok returned to Earth aboard *Soyuz 31* on November 02 after 79 days, 15 hours and 23 minutes in space; landing 180 km southeast of Jezkazgan. The landing event was covered live by Soviet television. Kovalyonok and Ivanchenkov were the first crew to have difficulty talking after returning to Earth. Within five days the men had mostly recovered, and twenty-five days after their return they were fully recovered.

Launched on March 12, 1981, by a *Soyuz-U* booster, *Soyuz T-4* carried the crew of 'Salyut 6 EO-6', the sixth and final long-duration crew to this Soviet space station. Kovalyonok and Viktor Savinykh spent 74 days, 17 hours and 37 minutes in space. They undocked on May 26, and landed three hours later 125 km east of Dzhezkazgan, in Kazakh SSR. Kovalyonok logged a total of 216 days, nine hours and eight minutes in space, and retired from the cosmonaut team on June 23, 1984. From 1990-92, he was a Director of the Thirtieth Central Scientific Research Institute, Ministry of Defence, (Russia). Kovalyonok has seven national awards, including two Hero of the Soviet Union (1978, 1981); three Orders of Lenin (1977, 1978, 1981); the Order of Merit for the Fatherland, third class (1996); the Order of Military Merit (2000); the Medal "For Merit in Space Exploration"; and five foreign awards.

## Krikalev, Sergei K. Soyuz TM-7, TM-12/13, STS-88, Soyuz TM-31/Expn 11/ STS-102, Soyuz TMA-6.

Sergei Konstantinovich Krikalev was born in August 1958 in Leningrad, Russia, USSR (now St. Petersburg, Russia). He is a Russian mechanical engineer, former cosmonaut and former head of the Yuri Gagarin Cosmonaut Training Center from 2009 to 2014. Krikalev enjoys swimming, skiing, cycling aerobatic flying, and amateur radio operations, particularly from space (callsign U5MIR). Krikalev graduated from high school in 1975. In 1981, he received a degree in Mechanical Engineering from the Leningrad Mechanical Institute, now called the Baltic State Technical University. After graduation, he joined **NPO Energia**, the Russian industrial organization responsible for crewed space flight activities. He tested space flight equipment, developed space operations methods, and participated in ground control operations. When the *Salyut 7* space station failed in 1985, he worked on the rescue mission team, developing procedures for docking with the uncontrolled station and repairing the onboard system.

Krikalev was selected as a cosmonaut in 1985, completed his basic training in 1986, and was assigned to the *Buran Shuttle* program. In early 1988, he began training for his first long-duration flight aboard the *Mir* space station. The training included preparations for at least six EVAs, installation of a new module, the first test of the new *Manned Maneuvering Unit* (MMU), and the second joint Soviet/French space mission. *Soyuz TM-7* was launched on November 26, 1988, with Krikalev as Flight engineer, Commander Aleksandr Volkov and French astronaut Jean-Loup Cretien. Throughout his various missions aboard *Mir*, Krikalev regularly communicated with various amateur radio operators across the globe. He was stranded aboard *Mir* during the dissolution of the Soviet Union in late 1991. As the country that had sent him into space no longer existed, his return was delayed and Krikalev stayed in space for 311 consecutive days, twice as long as the mission originally called for. He is sometimes referred to as the last Soviet citizen. These events are documented and contextualized in Romanian filmmaker Andrei Ujica's 1995 documentary ***Out of the Present***. Krikalev's story inspired the 2017 film, ***Sergio and Sergei,*** directed by Ernest Daranas.

Krikalev has held the world record for being the man with the most time in space, at 803 days, nine hours and 39 minutes (now in third place). He flew on missions Mir EO-4 (*Soyuz TM-7*), launched November 26, 1988; Mir LD-3 (*Soyuz TM-12, Soyuz TM-13*), launched May 19, 1991; STS-60, launched February 03, 1994; STS-88, launched December 04, 1998; ISS 'Expedition 1' (*Soyuz TM-31*, STS-102), launched October 31, 2000; and was Commander of ISS 'Expedition 11' (*Soyuz TMA-6*), launched April 14, 2005. This was his third trip to the ISS, which ended on October 10, 2005. Krikalev made eight EVAs, with a total time of forty-one hours and eight minutes. In October 1992, NASA had announced that an experienced cosmonaut would fly aboard a future Space Shuttle mission. Named by the Russian Space Agency,

Roscosmos for training with the crew of STS-60, in April 1993 he was assigned as prime Mission Specialist. During the flight, Krikalev conducted significant portions of the Remote Manipulator System (RMS) operations. In December 1998, Krikalev and Robert Cabana became the first people to enter the ISS, when they turned on the lights in the *Unity* module for mission STS-88.

Sergei Krikalev was awarded the title of Hero of the Soviet Union, and he was the very first person to be awarded the title of **Hero of the Russian Federation** for his contributions to space exploration. He has received the Medal "For Merit in Space Exploration" and the Medal "In Commemoration of the 300th Anniversary of Saint Petersburg". The Soviet Union awarded him the Order of Lenin, and he has three foreign awards. He retired as a cosmonaut in 2007. On February 15, 2007, Krikalev was appointed Vice president of the S.P. Korolev Rocket and Space Corporation ***Energia***, in charge of crewed space flights. The Chilean rock band Fulano composed a song after him, named *Krikalev*, included in their 1997 album "Trabajos Inutiles" (Useless Works). From 1999 to 2007, Krikalev was President of the Russian Gliding Federation. In April 2014, he was appointed **Plenipotentiary Representative of the Governor of Sevastopol in Moscow and St. Petersburg**. In September 2016, Krikalev became a confidant of the United Russia political party in the elections to the State Duma of the 7th Convocation. Since 2017, he has been the President of the **Aircraft Sports Federation of Russia**. Krikalev is a member of the Expert Council of the National Prize "Crystal Compass" and President of the **International Environmental Foundation** "Clean Seas" since 2009.

## Kubasov, Valeri N. Soyuz 6, Soyuz 19, Soyuz 36.

Valeri Nikolayevich Kubasov (1935-2014) was born in January 1935 in Vyazniki, Ivanovo Industrial Oblast, Russian SFSR, Soviet Union (now Vladimir Oblast, Russia). He flew on two Soyuz missions as Flight Engineer. He graduated from the Moscow Aviation Institute in 1958 as an Aerospace Engineer, and then worked at the Korolev Design Bureau. Initially focusing on ballistic studies, he worked on the design of the Voskhod capsule. Kubasov authored several studies on the calculation of spaceship trajectories, and acquired a Master of Science degree in Engineering. In 1966, he was selected in Civilian Specialist Group 2, and accepted into the newly-established civilian cosmonaut corps. His first mission aboard *Soyuz 6* in October 1969 was unsuccessful due to technical issues, as the three spacecraft never met up. Along with Georgy Shonin, he performed the first welding experiments on samples of stainless steel and titanium in space. Launched aboard *Soyuz 19* in mid-July 1975, (the *Apollo-Soyuz Test Project* mission), he told US President Gerald Ford in a TV linkup that they got *"good space food ... some juice, some coffee, and a lot of water"*. On July 21, 1975, the *Soyuz 7K-TM* module No. 065A used for the ASTP landed in Kazakhstan.

His last mission was as Commander on *Soyuz 36* in 1990 during the Intercosmos program. Kubasov was involved in the development of the *Mir* space station. He retired from the Russian space program in November 1993, and was later Deputy Director of **RKK Energia**. Valeri Kubasov twice evaded death during his space career. He was part of the crew scheduled to fly *Soyuz 2*, which was found to have the same faulty parachute sensor that resulted in Vladimir Komarov's death aboard *Soyuz 1*. Later, he was grounded for medical reasons (a swelling on his right lung) before the *Soyuz 11* flight, which killed the three-man crew when the capsule was accidentally depressurised by a faulty air pressure equalization valve. Kubasov logged a total of 18 days, 17 hours and 59 minutes in space. He died on February 19, 2014, in Moscow, Russia, aged seventy-nine years. Kubasov was married, with a son and daughter.

## Kud-Sverchov, Sergey V. Soyuz MS-17/Expedition 63/64.

Sergey Vladimirovich Kud-Sverchov was born in August 1983 in Baikonur, Kazakh SSR, Soviet Union. He graduated (with honors) from Moscow State Technical University with a degree in Rocket Engineering. Kud-Sverchov worked as an engineer at **RSC Energia** from August 2006 until selected in 2010 RSA Group by Roscosmos, and he flew as a Flight Engineer for ISS 'Expedition 63/64'. With Commander Sergey Ryzhikov and NASA astronaut Kathleen Rubins, he launched aboard *Soyuz MS-17* on October 14, 2020. After a two-orbit rendezvous and docking procedure, the Soyuz docked with the *Rassvet* module of the ISS just three hours and three minutes after liftoff. This was the first time a crew reached the ISS in less than four orbits, and otherwise in approximately six hours. They returned to Earth on April 17, 2012, after Kud-Sverchov had logged 184 days, 23 hours and ten minutes in space; including one EVA of six hours and 48 minutes. In May 2023, he was assigned to the crew of *Soyuz MS-28*; due to launch in September 2025, and return in early 2026.

## Kuzhelnaya, Nadezhda.

Nadezhda Kuzhelnaya was born in November 1962 in Alexeyevskoye, Republic of Tatarstan, Russia. She was a child who enjoyed science fiction films and books on space. She joined a flying club in 1981, aspired to become an architect; and graduated from the Dnepropetrovsk Engineering and Building Institute in 1984. Inspired by Svetlana Savitskaya's August 19, 1982, flight aboard *Soyuz T-7*, she enrolled in the Moscow Aviation Institute. Upon graduating in 1988, she began working as an engineer for **RKK Energia**, designing equipment for spaceflights. In 1994, Kuzhelnaya applied to become a cosmonaut, and was accepted into the two-year course at the Gagarin Center. She married her flight instructor, Vladimir Morov, and they had a daughter; looked after by Morozov while Kuzhelnaya continued her training as a Flight Engineer, whose role would be to fly the Soyuz to dock with the International Space Station (ISS).

She was assigned to *Soyuz TM-32*, due to fly in 2001, but was cut from the mission to accommodate paying "space tourist" Dennis Tito. She was then allocated as backup to French astronaut Claudie Haignere for a mission later that year. From 1999-2004, Kuzhelnaya was the only female cosmonaut. In 2001, she was backup for Italian Roberto Vittori for the *Soyuz TM-34* mission. In May 2004, after an announcement that female cosmonauts would not be flying for several years, she retired to become a commercial pilot for Russian carrier **Aeroflot**. As of 2007, Kuzhelnaya was one of two female pilots (compared to around a thousand males) flying the Tupolev Tu-134, then Airbus A320 and Airbus A330 aircraft.

## Laveykin, Aleksandr I. Soyuz TM-2.

Aleksandr Ivanovich Laveykin was born in April 1951 in Moscow, USSR. Selected as a Test Cosmonaut on December 01, 1978, in Group NPOE-4, he flew with Commander Yuri Romaneko from February 05, 1987, as Flight Engineer on *Soyuz TM-2* for the long-duration expedition 'Mir EO-2' (also called Mir Principal Expedition 2). The core module of *Mir* had been launched into orbit on February 19, 1986. Between March and July 1986, it was visited twice by the crew of *Soyuz T-15*, who transferred equipment taken from the previous orbital station, *Salyut 7*. Before the arrival of mission EO-2, *Mir* was also visited by three *Progress* spacecraft, numbered 25, 26 and 27, as well as the uncrewed *Soyuz TM-1* spacecraft. From July 1986 to the arrival of EO-2 in February 1987, *Mir* remained uncrewed. During this time an associated relay satellite ceased operation, and computers on *Mir* were malfunctioning.

Aleksandr Laveykin logged 174 days, three hours and 25 minutes in space. His spaceflight was initially planned to last until December 1987, but due to minor heart irregularities detected by doctors on the ground, he was replaced by Aleksandr P. Aleksandrov in July 1987. Laveykin retired on March 28, 1994. He was awarded Hero of the Soviet Union; Pilot-cosmonaut of the USSR; the Order of Lenin; and the Medal "For Merit in Space Exploration". He is married, with one child.

## Lazarev, Vasili G. Soyuz 12, Soyuz 18a.

Col. Dr. Vasili Grigoryevich Lazarev (1928-90) was born in February 1928 in Poroshino, Altai Krai, Russian SFSR, Soviet Union. He was a Medical Doctor and Colonel in the Soviet Air Force. Selected in Voskhod Group (Medical Group 1), he launched as Commander aboard *Soyuz 12* on September 27, 1973, for a test flight of the newly redesigned *Soyuz 7K-T* spacecraft [built by Experimental Design Bureau (OKB-1)], and landed on September 29 some 400 km southwest of Karaganda, Kazakhstan. *Soyuz 18a* had an abortive launch on April 05, 1975, landing southwest of Gorno-Altaysk in the Altai Mountains, Kazakhstan (north of the Chinese border). Due to failure of the *Soyuz* launch vehicle, the mission was unable to dock with the *Salyut 4* space station for the planned sixty-day mission. The second and third stages of the booster failed to properly separate, when only three of the six locks released. When the safety system initiated separation, the re-entry capsule was already pointed toward Earth.

*Soyuz 18a* was recognized as a sub-orbital flight (lasting less than 22 minutes), but Lazarev had to petition the Soviet leader to receive his flight pay! The accident was partly disclosed by the usually secretive Soviets, since it occurred during preparations for their joint *Apollo-Soyuz Test Project* which flew three months later. Lazarev logged one day, 23 hours and 36 minutes in space. He remained in the Soviet space program until failing a physical exam in 1981. He never fully recovered from the injuries sustained by the astoundingly high 21.3 G-s (209 metres per second squared) acceleration of the abort and landing of *Soyuz 18a*. He was awarded Hero of the Soviet Union; Pilot-cosmonaut of the USSR; and the Order of Lenin. Dr. Vasili Lazarev died in Moscow on December 31, 1990, aged sixty-two years.

## Lazutkin, Aleksandr I. Soyuz TM-25.

Aleksandr Ivanovich Lazutkin was born in October 1957 in Moscow, Soviet Union. He attended the Moscow Aviation Institute and received a degree in Mechanical Engineering. Selected as a cosmonaut on March 03, 1992, he flew aboard *Soyuz TM-25* as Flight Engineer. Launched on February 10, 1997, he was on space station *Mir* during a 162-day mission when a collision occurred with the uncrewed *Progress M-23* supply craft, causing the cabin to decompress. This serious event knocked out the *Spektr* module's solar panels and took *Mir* out of its alignment with the Sun, also causing it to lose power. Quick action by the crewmen averted immediate disaster. Lazutkin and Michael Foale severed the connecting cables with the module, (using a tiny dinner knife) and sealed off the hatches. A few days later, the *Mir* station was eventually safely secured. Lazutkin landed on August 14, 1997, 170 km southeast of Dzhezkazgan. Aleksandr Lazutkin logged a total of 184 days, 22 hours and seven minutes in space.

## Lebedev, Valentin V. Soyuz 13, Soyuz T-5.

Dr. Valentin Vitalyevich Lebedev was born in April 1942 in Moscow, USSR. He is a former cosmonaut, whose stay of 211 days aboard the *Salyut 7* space station with Anatoly Berezovoy in 1982 was recorded in the ***Guiness Book of Records***. After graduating from the Moscow Aviation Institute in 1966, he worked for twenty-three years at the Central Design Bureau

"Energy" of the **Soviet Scientific Production Union** as an engineer, senior research Fellow, and a methodology instructor in the cosmonaut's detachment. In 1968, Lebedev led the specialists in Bombay (now Mumbai, India) supporting *Zond 5*, which circumnavigated the Moon. He continued leading technical detachments in the flying, design, testing and control of the spaceships *Soyuz* and *Soyuz T*; and the orbiting space stations *Salyut 4, Salyut 5* and *Salyut 6*. From 1971-73, he trained as a cosmonaut, and flew with Peter Klimuk aboard *Soyuz 13* (December 18-26, 1973), after which he was decorated and promoted to Pilot Cosmonaut.

In 1975, he defended his Ph.D. Thesis on *"Methods of formation of the dynamic test bench for the base service of spaceship and crew training"*. Launched aboard *Soyuz T-5*, he and Berezovoy were in space from May 13 to December 10, 1982, mostly aboard *Salyut 7*, where he was crew Engineer. Lebedev conducted over three hundred scientific experiments and studies. His Arabidopsis plant flowered in July, gaining a **Guiness Book of Records** entry under "First species of plant to flower in space". In 1985, Lebedev defended his doctoral thesis on *"Methods of carrying on astrophysical explorations aboard of orbital stations"*.

In the years following his second spaceflight, Lebedev became renowned as a specialist in the field of cosmonautics and geoinformation. From 1989-91, he worked as Deputy Director of Science at the **Institute of Geography AS USSR**, completely dedicated to scientific work. Under his leadership, a unified technology was developed for the creation of maps showing the dynamics of ecosystems, showing changes over time. One of his latest scientific projects, executed within a program called "Science to Moscow", is a work to remotely monitor Moscow transport traffic flow. His data became the basis for modifying street traffic, and development of an automated traffic management system called "Start". Lebedev retired from the Soviet space program in 1993. In 2000, he became the first cosmonaut ever elected to the **Russian Academy of Science**.

He is the author of 157 scientific works; author and co-author of nine monographs and two school textbooks for the Moscow Aviation Institute; twenty-seven articles in *Science and Life* magazine; and a large number of publications in domestic and foreign press. Lebedev was awarded twenty-six author's certificates for inventions introduced in the Soyuz and Salyut missions. He is the author of *Diary of a Cosmonaut: 211 Days in Space*, written entirely while he was serving aboard the space complex *TS Soyuz-5, Salyut-7, Soyuz T-7*. The book was published in the United States (1988, 1990), under the title *Angle of My Judgment* (1994). Lebedev has received eleven awards, including two Hero of the Soviet Union (1973, 1982); two Order of Lenin (1973, 1982); the Order "For Merit to the Fatherland", fourth class "for important contributions to science development", (2007); the Medal "For Construction of the Baikal-Amur Railway"; Honoured Scientist of the Russian Federation; and the Gold Medal of K. Tsiolkovsky by the AN USSR, for outstanding achievements in the field of space (1984). He is an honorary citizen of sixteen Russian cities, several CIS countries, the state of Texas, USA; and of the City of Fort Worth. 10015 Valenlebedev, a main-belt asteroid is named after him.

Lebedev is a member of the Communist Party of the Soviet Union, 1971 to present; Komsomol Central Committee member, 1974-78; President of USSR Acrobatics Federation, 1975-91; Member of the National Olympics Committee, 1976-91; Member, United RAS Scientific Committee on Geoinformation; Member, Editorial Council, *Herald of Computer and Information Technologies* magazine. He is married, with a son. He lives and works in Moscow, Russia.

## Leonov, Alexei A. Voskhod 2, Soyuz 19.

Maj. Gen. Alexei Arkhipovich Leonov (1934-2019) was born in May 1934 in Listvyanka, West Siberian Krai, Russian SFSR. He was a Soviet and Russian cosmonaut, Air Force Major General, writer and artist. He graduated from Secondary School No. 21 in 1953, and applied to the Academy of Arts in Riga, Latvia, but decided not to attend due to the high tuition costs.

He joined a Ukrainian preparatory flying school in Kremenchug, and made his first solo flight in May 1955. While studying art part-time in Riga, Leonov started an advanced two-year course to become a fighter pilot in the Chuguev Higher Air Force Pilots School in the Ukrainian SSR. Selected on March 18, 1965, in a group of 20 candidates as Air Force Group 1, he became the first person to conduct a spacewalk, exiting the capsule during the *Voskhod 2* mission for twelve minutes and nine seconds while connected to the craft by a 4.8-metre tether. His spacesuit had inflated in the vacuum of space to the point where he could not enter the rudimentary airlock. Leonov opened a valve to bleed off some of the pressure, and was barely able to reach the safety of the capsule. While on the mission, he drew a small sketch of an orbital sunrise, the first work of art made in outer space. In 1968, Leonov was selected to be Commander of a circumlunar *Soyuz 7K-L1* flight, but this was canceled because of delays. The *Apollo 8* mission of December 1968 had already made that step in the Space Race.

Leonov was also selected to be the first Soviet person to land on the Moon, aboard the LOK/N1 spacecraft, but this was also canceled. He was to have been Commander of the 1971 *Soyuz 11* mission to *Salyut 1*, but his crew was replaced with the backups, because Valery Kubasov was suspected to have contracted tuberculosis. Leonov was to have commanded the next mission to *Salyut 1*, but this was scrapped after the three cosmonauts died on the *Soyuz 11* return flight. By the time *Salyut 4* reached orbit, he had been switched to a more prestigious project. His second spaceflight was as Commander of *Soyuz 19*, part of the 1975 *Apollo-Soyuz Test Project* (ASTP). He became lasting friends with the US commander Thomas P. Stafford, with Leonov being the godfather of Stafford's younger children. He logged a total of seven days, zero hours and 33 minutes in space. From 1976-82, Leonov was the commander of the cosmonaut team, as "Chief Cosmonaut"; and deputy director of the Yuri Gagarin Cosmonaut Training Center, where he oversaw crew training. Leonov also edited the cosmonaut newsletter *Neptune.*

Alexei Leonov was an accomplished artist whose published books include albums of his artistic works and works he did in collaboration with his friend Andrei Sokolov. He drew portraits of the *Apollo* astronauts who flew with him during the *Apollo-Soyuz Test Project*. Arthur C. Clarke dedicated ***2010: Odyssey Two*** to Leonov and Soviet physicist Andrei Sakharov. The fictional spaceship in the book is named *Cosmonaut Alexei Leonov*. Together with Valentin Selivanov, Leonov wrote the script for the 1980 science fiction film ***The Orion Loop***. He was head of the **Banner of Peace in Space** project from 1990 until his death in late 2019. Leonov retired in 1991 and lived in Moscow. He was a member of the United Russia Party since December 2002, and a member of the Party's Supreme Council.

In 2004, Leonov and former American astronaut David Scott began work on a dual memoir covering the history of the Space Race between the United States and the Soviet Union. Titled *Two Sides of the Moon: Our story of the Cold War Space Race*, it was published in 2006. He received nineteen Russian awards, including Order of the Red Star (1961); two Hero of the Soviet Union (1965, 1975); two Order of Lenin (1965, 1975); the Order for Service to the Homeland in the Armed Forces of the USSR, third class (1975); five Jubilee Medals; Medals "For Impeccable Service", first, second and third classes; the Lenin Komsomol Prize (1980); the USSR State Prize (1981); the Order for Merit to the Fatherland, fourth class (2000); the Order of Friendship (2011); the Order "For Merit to the Fatherland, first class (2019); and eight foreign awards: Hero of Socialist Labour (People's Republic of Bulgaria, 1965); the Order of Georgi Dimitrov (People's Republic of Bulgaria, 1965); the Artur Becker Medal (GDR, 1965); the Order of Karl Marx (GDR, 1965); the Order of the Flag of the Republic of Hungary (1965); Hero of Labor (Democratic Republic of Vietnam, 1966); the Order of Civil Merit, first class (Syria, 1966); and the Order of Merit, third class (Ukraine, 2011). Alexei Leonov died on October 11, 2019, aged eighty-five years, after a long illness in Moscow, Russia. He was married, with two daughters and two grandchildren. His NASA friend, Tom Stafford gave a eulogy in Russian language at Leonov's funeral.

## Levchenko, Anatoly S. Soyuz TM-4/3.

Capt. Anatoly Semyonovich Levchenko (1941-88) was born in May 1941 in Krasnokutsk, Ukrainian SSR, Soviet Union. He became a test pilot and Captain in the Soviet Air Force. Selected on July 12, 1980, in the Cosmonaut Group, he was scheduled to be the backup Commander of the first *Buran* Soviet Space Shuttle flight. He made test-flights with Buran's counterpart, the prototype OK-GLI aircraft. In March 1987, he began extensive training for a *Soyuz* flight, the 'Mir LII-1' mission, intended to give him some experience in space. Launched on December 21, 1987, he occupied the third seat aboard *Soyuz TM-4*, which carried the first two crew members (Vladimir Titov and Musa Manarov) for the third long-duration expedition, 'Mir EO-3', and he returned with the crew of 'Mir EO-2', Romaneko and Alexandrov on *Soyuz TM-3* over a week later. Levchenko logged seven days, 21 hours and 58 minutes in space. He was awarded Hero of the Soviet Union; the Order of Lenin; and Pilot-Cosmonaut of the USSR. Anatoly Levchenko died of a brain tumor on August 06, 1988, in Moscow, Russian SFSR, aged forty-seven years. He was married, with one child. There is a memorial plate with his image installed on the wall of house 2 at Chkalova Street, where Levchenko once lived in Zhukovsky.

## Lonchakov, Yuri V. STS-100, Soyuz TMA-1/TM-34, Soyuz TMA-13/ Expedition 18.

Yuri Valentinovich Lonchakov was born in March 1965 in Balkhash, Dzhezkazkansk Region, Kazakh SSR (now Kazakhstan, CIS). He entered the Orenburg Air Force Pilot school, graduating (with honors) in 1986 as Pilot-Engineer. Next, he entered the Zhukovski Air Force Academy; graduating (with honors) in 1998 as pilot-engineer-researcher. He is a Class 1 Air Force Pilot, a paratroop training instructor with 526 jumps, and has logged over 1,400 hours flying time. Selected as a Test-Cosmonaut in December 1997, he flew aboard Space Shuttle *Endeavour* on STS-100 (April 19-31, 2001), the mission that installed the robotic Canadarm-2 to the ISS. This flight lasted 11 days, 21 hours and 31 minutes; landing at Edwards Air Force Base, California because of adverse weather at KSC. Launched on October 30, 2002, as Commander aboard *Soyuz TMA-1*, he spent ten days, 20 hours and 53 minutes in space. He returned on *Soyuz TM-34*; landing 100 km northeast of Arkalyk on November 10, 2002.

Launched on October 12, 2008, aboard *Soyuz TMA-13*, he docked with the Earth-facing docking port of the ISS's *Zarya* module on October 14. He was Flight Engineer for the 'Expedition 18' crew, and with "space tourist" Richard Garriott took part (along with NASA's Michael Fincke and Gregory Chamitoff) in filming a science-fiction movie ***Apogee of Fear***. On March 12, 2009, the crew prepared to evacuate the ISS by closing hatches between modules, and boarding the *Soyuz* craft always docked to provide emergency escape. Debris from the upper stage of a *Delta II* rocket passed close to the ISS, and five minutes later the crew were cleared to resume operations. As *Soyuz* Commander, Lonchakov returned to Earth aboard *Soyuz TM-34* with Michael Fincke and "spaceflight participant" Charles Simonyi; landing in Kazakhstan on April 08, 2009.

Lonchakov logged a total of 200 days, 18 hours and 39 minutes in space; performing two EVAs totaling ten hours and 27 minutes. In 2013, Lonchakov submitted his resignation, effective September 14. Although scheduled to fly on ISS 'Expedition 44', his replacement would be Gennady Padalka. Yury Lonchakov later took up a position at **Gazprom**, and on March 31, 2014, was appointed acting head of **GCTC**. Lonchakov was awarded Hero of the Russian Federation; and the honorary title Pilot-Cosmonaut of the Russian Federation. He is an "Honoured Man of the Aktubinsk Region". He received the Order "For Merit to the Fatherland", fourth class; the Medal "For Merit in Space Exploration"; and the NASA Space Flight Medal. He is married, with a son, Kirill, born in 1990.

## Lyakhov, Vladimir A. Soyuz 32/34, Soyuz T-9, Soyuz TM-6/5.

Col. Vladimir Afanasyevich Lyakhov (1941-2018) was born in July 1941 in Antratsyt, Ukrainian SSR, Soviet Union. He logged almost one year living aboard three different space stations. A Colonel in the Soviet Air Force, he was selected as a cosmonaut on May 05, 1967. Launched on February 25, 1979, atop a *Soyuz-U* rocket, he was Commander on *Soyuz 32*, flying with Valeri Ryumin to *Salyut 6*. The space station had been vacant since November 1978, and when the hatch was opened the cosmonauts smelled *"burnt steel"*, a scent Ryumin called *"the odor of space"*. The crew started routine activities and two types of medical exams. The first included daily checks of mood, their accomplishments and food intake. The second checked their psychological condition via observing their frequent communication sessions.

Their cardiovascular systems were checked every eight to ten days, and their body masses measured. By the second week, they exercised 2.5 hours each day, and averaged 3,100 calories of daily food intake, with 2.5 litres of water. The Soviets did not report that the *Salyut*'s propulsion system had problems, not revealed until March 16, when they said a *Salyut* fuel tank was leaking fuel into the nitrogen bellows which pressurized the fuel. Valves and regulators in the pressurization system were at potential risk. The tank was shut down, purged with nitrogen, and closed off. Experiments included hatching quail eggs; but the chicks grew much slower than on Earth, and lacked heads! On March 24, they installed a television monitor, allowing two-way TV links with ground control. Seeing family, instead of just hearing them, was considered to have great psychological importance, especially since longer flights were intended. They returned to Earth aboard *Soyuz 34* on August 21, 1979; landing 170 km southeast of Dzhezkazgan, and setting a space endurance record of 175 days.

Launched on June 27, 1983, Lyakhov was Commander aboard *Soyuz T-9*, the fourth expedition to *Salyut 7* following the failed docking of *Soyuz T-8*. Almost immediately after docking at the station's aft port, the crew entered *Kosmos 1443* and began transferring the 3.5 tons of cargo lining its walls. On July 27, a small object struck a *Salyut 7* viewport, blasting out a 4-mm crater. It may have been part of the Delta Aquariid meteor shower, or a small piece of orbital debris. The crew filmed scenes for the movie ***Return from Orbit***; and returned to Earth on November 23, after 145 days; landing 160 km east of Dzhezkazgan, Kazakhstan.

For his third mission, Lyakhov launched on August 29, 1988, aboard *Soyuz TM-6* for 'Mir EP-3' to space station *Mir*, taking Valeri Polyakov for the third long-duration mission 'Mir EO-3'/'Mir EO-4', while Lyakhov returned to Earth aboard *Soyuz TM-5* after five days. During descent, *Soyuz TM-5* suffered a computer software problem, combined with a sensor problem. The de-orbit engine did not behave as expected, shutting off prematurely. The Flight Director decided they would have to remain in orbit an extra day. After the computer was reprogrammed the next de-orbit burn was successful, and the crew landed safely on September 07, 1988. Vladimir Lyakhov logged a total of 333 days, seven hours and 47 minutes in space. He made three EVAs with a total time of seven hours and eight minutes; one outside of *Salyut 6*, and two outside *Salyut 7*. He was deputy director for cosmonaut training and deputy commander of the cosmonaut corps at the **Gagarin Cosmonaut Training Center**; and retired on September 07, 1994. Lyakhov was named a Hero of the Democratic Republic of Afghanistan; awarded the Order of the Saur Revolution; and the Order of The Sun of Freedom. Lyakhov died on April 19, 2019, in Astrakhan, Russia, aged seventy-six years. He was married, with two children.

## Makarov, Oleg. Soyuz 12, Soyuz 18a, Soyuz 27, Soyuz T-3.

Oleg Grigoryevich Makarov (1933-2003), born in Udomlya, Russian SFSR, was an engineer and former Soviet cosmonaut. He graduated from Bauman Moscow Higher Technical School in 1957 and started working at the Special Design Bureau Number One (which is now ***RSC Energia***) as an engineer, working on the *Vostok* spacecraft. In 1966 he was selected for cosmonaut training. He was originally part of the Soviet lunar program and was training with Alexei Leonov for the first human circumlunar flight. After the American success of *Apollo 8*, however, the Soviet mission was canceled.

His first spaceflight was *Soyuz 12* in September 1973, a test flight after the *Soyuz 11* fatalities of June 1971. His second mission was the abortive *Soyuz 18a* flight that made an emergency landing in the Altay Mountains, twenty-one minutes after launch. With his third launch in January 1978 on *Soyuz 27* he flew to space station *Salyut 6*, and returned to Earth five days later aboard *Soyuz 26*. His fourth and last mission was *Soyuz T-3* in 1982, during which several repairs were done on *Salyut 6*. Altogether Makarov spent twenty days, 17 hours and 44 minutes in space. After his final spaceflight he continued to work for **Energia**, both in the *Mir* space station program as well as the *Energia-Buran* spaceplane development. Makarov was made Hero of the Soviet Union (twice) and received the Order of Lenin (four times). He was made an Honorary Citizen of Baikonur, Dzhezkazgan, Kaluga, Karaganda, Rivne and Yakutsk. Oleg Makarov died in Moscow, Russia, in 2003 from a heart attack, aged seventy years.

## Malenchenko, Yuri I. Soyuz TM-19, STS-106, Soyuz TMA-2/Expn 7, TMA-11/Expn 16, TMA-19M/Expn 46.

Col. Yuri Ivanovich Malenchenko was born in December 1961 in Krushchev, Ukrainian SSR, USSR. He graduated from the Kharkiv Military Aviation School in 1983; then attended the Zhukovsky Air Force Engineering Academy, graduating in 1993. He became a fighter pilot and Colonel in the Russian Air Force. Selected in 1987 TsPK Cosmonaut Group, he trained from December 1987 to June 1989. Launched on July 01, 1994, he was Commander aboard *Soyuz TM-19* with Talgat Musabayev. They docked with *Mir* on July 03, and became part of the sixteenth resident crew. There were many problems with the mission, which finally ended with the first successful manual docking of a *Progress* supply ship at *Mir* by Malenchenko. After 125 days, 22 hours and 53 minutes in space, he landed on November 04, 1994, some eighty-eight km northeast of Arkalyk, Kazakhstan. Launched on September 08, 2000, aboard shuttle *Atlantis* on STS-106, Malenchenko was a Mission Specialist. The prime objectives were to unload supplies from a double SPACEHAB cargo module in the shuttle cargo bay, and from *Progress M1-3* which was docked to the aft port of the *Zvezda* module. He landed at KSC after 11 days, 19 hours and 12 minutes in space.

As Commander, Malenchenko and NASA astronaut Edward Lu launched from Baikonur on April 26, 2003, aboard *Soyuz TMA-2*, and docked with the ISS on April 28. He became Commander of 'Expedition 7', and on August 10, 2003, he became the first person to marry in space. The ceremony was conducted while he was 390 km (240 miles) above New Zealand while aboard the International Space Station (ISS), and his bride, Ekaterina Dmitrieva was in Texas, USA. He returned to Earth on October 28; landing near Arkalyk, with his mission lasting 184 days, 22 hours and 46 minutes. Launched on October 10, 2007, as Commander aboard *Soyuz TMA-11*, he joined ISS 'Expedition 16' as 'Flight Engineer 1'. Traveling up was "spaceflight participant" Sheikh Muszaphar Shukor of Malaysia, and returning to Earth with Malenchenko on April 19, 2008, were NASA's Dr. Peggy Whitson and "spaceflight participant"

Yi So-Yeon of South Korea. Due to a malfunction, *Soyuz TMA-11* performed a ballistic re-entry mush steeper than normal, and landed 475 km from the intended target, north of Arkalyk. Although the crew were recovered with no injuries, the spacecraft's hatch and antenna suffered burn damage during the unusual re-entry. This flight lasted 191 days, 19 hours and eight minutes.

Launched on December 15, 2015, Malenchenko arrived at the ISS aboard *Soyuz TMA-19M* to join the 'Expedition 46' crew. Due to a malfunction, he performed a manual docking. *Progress MS-1* was then scheduled for launch to the ISS, atop a *Soyuz 2.1a* booster, on December 21. The SpaceX *CRS-8* mission provided the ISS with the Bigelow Expandable Activity Module (BEAM). After 186 days in space, Malencheko returned to Earth on *Soyuz TMA-19M*, landing on June 18, 2016. Yuri Malenchenko logged a total of 827 days, nine hours and 20 minutes in space; including six EVAs totaling 34 hours and 52 minutes. His third spacewalk was on September 11, 2000; the sixth in support of the ISS assembly, and the fiftieth EVA in space shuttle history. As of December 2023, Malenchenko ranked third for accumulated career time in space, due to his time on both *Mir* and the ISS, of which he is a former Commander. He has received seven awards, including Hero of the Russian Federation; the National Hero of Kazakhstan medal; the Military Award of Excellence; the Jubilee Medal "70 Years of the Armed Forces of the USSR"; and Meritorious Service Medals; first, second and third class.

## Malyshev, Yuri V. Soyuz T-2, Soyuz T-11/10.

Col. Yuri Vasilyevich Malyshev (1941-99) was born in August 1941 in Nikolayevsk, Volgograd Oblast, USSR. He became a pilot and Colonel in the Soviet Air Force. He graduated from the Higher Air Force Academy at Chuguyev, Kharkov Oblast, in 1963; and from the Gagarin Military Academy in Monino in 1977. Selected in Air Force Group 4 on May 07, 1967 in TsPK-4, his basic cosmonaut training was from May 1967 to August 18, 1969. He flew as Commander on *Soyuz T-2* (June 05-09, 1980) to the *Salyut 6* space station. It was the twelvth mission to the station, and the tenth successful docking. The primary mission objective was to perform a crewed test of the new *Soyuz-T* spacecraft, which had new engine systems and a new *Argon* computer to control docking and re-entry procedures. Docking computer failures during the automatic approach to the *Salyut* were a recurring problem in future *Soyuz-T* missions. The flight lasted three days, 22 hours and 19 minutes; landing 200 km southeast of Dzhezkazgan.

*Soyuz T-11/10* (April 03-11, 1984) launched atop a *Soyuz-U* rocket was the sixth expedition to the *Salyut 7* space station. The crew were Commander Malyshev, Flight Engineer Gannady Strekalov, and Research Cosmonaut Rakesh Sharma of India, who were aboard the *Salyut* for seven days, 21 hours and 40 minutes. After *Soyuz T-11* undocked, *Salyut 7* was uncrewed until June 1985, and developed problems while unattended. The next crew, aboard *Soyuz T-13* had to perform a manual docking and make repairs to the station. Malyshev was the ninety-fifth human in earth orbit, logging a total of eleven days, 19 hours and 59 minutes (284 hours) in space. After retiring as a cosmonaut in 1993, he served as Deputy Director of the political department at the Gagarin Cosmonaut Training Center. He was awarded two Hero of the Soviet Union (1980, 1984); two Order of Lenin (1980, 1984); Pilot-Cosmonaut of the USSR (1980); and the Ashoka Chakra (India, 1984). Yuri Malyshev died on November 08, 1999, in Zvyozdny Gorodok, Moscow Oblast, Russian Federation, aged fifty-eight years. He was married, with two children.

## Manakov, Oleg G. Soyuz 12, Soyuz 18a, Soyuz 27, Soyuz T-3.

Oleg Grigoryevich Manakov (1933-2003) was born in January 1933 in Udomlya, Tver Oblast, USSR. He graduated from Bauman Moscow Higher Technical School in 1957 and started working at the Special Design Bureau Number One (now **RSC Energia**) as an engineer working on the *Vostok* spacecraft. Selected for cosmonaut training in 1966, he was originally part of the Soviet lunar program, but after the historic circumlunar flight of *Apollo 8* his mission was canceled. Manakov flew on *Soyuz 12* (Soyuz 7K-T No. 1) for one day, 23 hours and 15 minutes in September 1973 with Commander Vasily Lazarev; a test flight to prove the design changes made to Soyuz. The crew was reduced from three to two, making room for pressure suits and life support gear. His second brief flight was aboard *Soyuz 18a* that had a booster malfunction, and made an emergency landing in the Altay Mountains, twenty-one minutes after launch. Launched on January 10, 1978, aboard *Soyuz 27*, he flew as Flight Engineer with Commander Vladimir Dzhanibekov to *Salyut 6* and landed five days later aboard *Soyuz 26*. The descent module is displayed at the Sergei Pavlovich Korolyov Museum of Cosmonautics in Zhytomyr, Ukraine.

Manakov launched as Flight Engineer on November 27, 1980, aboard *Soyuz T-3*, a mission during which several repairs were made on *Salyut 6*. The landing was on December 10, 1980, 130 km from Dzhezkazgan. He also served on the backup crews for *Soyuz 17* and *Soyuz T-2*. Manakov logged a total of 20 days, 17 hours and 44 minutes in space. After his fourth and final spaceflight, he continued to work for **Energia**, both in the *Mir* space station program, as well as the *Energia-Buran* spaceplane development. Manakov was awarded two Hero of the Soviet Union (1973, 1978); four Order of Lenin (1973, 1975, 1978, 1980); the Order "Blue Nile" of Ethiopia (1981); and was made an Honorary Citizen of Baikonur, Dzhezkazgan, Karaganda, Rivne, and Yakutsk. Manakov died of a heart attack in Moscow on May 28, 2003, aged 70 years.

## Manarov, Musa K. Soyuz TM-4/6, TM-11.

Col. Musa Khiramanovich Manarov was born in March 1951 in Baku, Azerbaijan SSR, Soviet Union. An ethnic Lak, he is the first Azerbijani cosmonaut (now retired), and spent 541 days and 28 minutes in space. He graduated from the Moscow Aviation Institute in 1974 with an engineering qualification. He has flown ten aircraft types, and logged over 1,400 flying hours; holding the qualifications of Military Pilot, First Class; and Test Pilot, Third Class. Selected as a cosmonaut on December 01, 1978, he flew as a Flight Engineer for 'Mir EO-3' (*Soyuz TM-4/ Soyuz TM-6*). The mission lasted 365 days, 22 hours and 38 minutes, with Manarov spending December 21, 1987, to December 21, 1988, in space. His second mission was 'Mir EO-8' (*Soyuz TM-11*) from December 02, 1990 to May 26, 1991, (175 days, one hour and 50 minutes), a world record spaceflight duration at that time. Manarov performed EVAs totaling over twenty hours. He was awarded Hero of the Soviet Union; the Order of Lenin; the Order of the October Revolution; and the Medal "For Merit in Space Exploration". Representing the United Russia faction, Manarov was a member of the State Duma of the Russian Federation of the Fifth Convocation (2007-11). He is married, with two children.

## Matveev, Denis V. Soyuz MS-21.

Denis Vladimorovich Matveev was born in April 1983 in Leningrad, RSFSR, USSR (now St. Petersburg, Russia). He is an active Test Cosmonaut. Matveev received a degree in Computing and Networks from the Bauman Moscow State Technical University in 2006, and began working at the Yuri Gagarin Cosmonaut Training Centre in Star City, Russia, as a junior research assistant. He was promoted to Engineer in 2009, and selected in 2020 in TsPK-15 Cosmonaut Group. His training was from October 12, 2010, to July 31, 2012, when he was named a Test Cosmonaut. He was assigned as 'Flight Engineer 1' to *Soyuz MS-21*, (March 18 to September 29, 2022), which launched to the ISS atop a *Soyuz-2. 1a* rocket; the first mission to the ISS with three Roscosmos cosmonauts. The Soyuz docked at the *Prichal* nadir port of the ISS.

The arriving cosmonauts (Oleg Artemyev, Denis Matveev and Sergey Korsakov) gained international attention after entering the ISS in bright yellow suits with blue elements. International commentators saw in these colors the national colors of Ukraine, and interpreted the personal choice of suits by these cosmonauts as a sign of their sympathy for Ukraine in light of the ongoing Russian invasion of Ukraine from February 24, 2022; which strained space cooperation (such as the ISS program) after international sanctions were imposed on Russia. Roscosmos released a statement that *"the colors are to be read as the identifying colors of the Bauman Moscow State Technical University"*, from which all three cosmonauts graduated. After 194 days in space; including four EVAs totaling 26 hours and seven minutes, Matveev landed on the Kazakhstan steppe.

## Misurkin, Alexander A. Soyuz TMA-08M, Soyuz MS-06/Expn 54, Soyuz MS-20.

Maj. Alexander Alexanderovich Misurkin was born in September 1977 in Yershichi, Smolensk Oblast, Russian SFSR, Soviet Union. He is a Major in the Russian Air Force and a retired cosmonaut. He entered the Kacha High Air Force Pilot School, graduating in September 1998. He continued pilot training at the Armavir Military Aviation Institute, and graduated in October 1999 with a gold medal as a pilot-engineer. Selected in 2006 TsPK Cosmonaut Group, Misurkin completed his basic training in June 2009. From August 2009 to February 2011, he took advanced training specializing in the International Space Station (ISS) program.

Launched on March 28, 2013, as 'Flight Engineer 1' aboard *Soyuz TMA-08M* atop a *Soyuz-FG* rocket, this was the first crewed flight to use the "fast rendezvous" approach to the ISS, arriving in less than six hours after just four Earth orbits. Previous *Soyuz* launches had taken two days to dock with the ISS. The mission lasted 166 days, six hours and 15 minutes. Shortly after landing southeast of the town of Dzhezkazgan on the Kazakhstan steppe, he and astronaut Chris Cassidy participated in a number of exercises to assess their physiological state. The tests will be used to determine how crew members would feel after landing on Mars, to asses what type of activities they could perform soon after landing.

Launched on September 12, 2017, aboard *Soyuz MS-06*, he was Commander of the ISS for 'Expedition 54'. On February 02, 2018, he and Anton Schaplerov made an eight-hour 13-minute EVA to replace an old electronics box for a high-gain communications antenna; setting a new record for the longest Russian spacewalk. The mission lasted 168 days, five hours and 14 minutes. Launched on December 08, 2021, Misurkin commanded the *Soyuz MS-20* flight to the ISS. He carried two space tourists, Yusaku Maezawa and Yozo Hirano to space for 11 days, 19 hours and 35 minutes. All three landed safely on December 20, 2021. Misurkin logged a total of 346 days, seven hours and four minutes in space; including four EVAs totaling 28 hours and 14 minutes. He is married, with two children.

## Morukov, Boris V. STS-106.

Dr. Boris Vladimirovich Morukov (1950-2015) was born in October 1950 in Moscow, Soviet Russia, USSR. He received his Doctor of Medicine from the Second Moscow Medical Institute (now Russian State Medical University) in 1973; and gained his Ph.D. in 1979. He was a Russian physician at the State Research Center RF-Institute for Biomedical Problems.

He was first selected for medical-cosmonaut service in 1976. Based on his qualifications, he was continually offered selection, and was chosen as a Cosmonaut-Researcher in 1989. Completing his basic training in February 1992, he spent two decades providing medical operations support for crewed space flights. From 1979-80, he provided support for long-duration missions on *Salyut 6*; and from 1982-87 he coordinated scientific projects dedicated to the development of countermeasures to the negative metabolic changes that occur during hypokinesia and microgravity. Morukov participated in a joint American-Russian medical experiment on STS-60, Mir 18/STS-71 and all other *Mir-NASA* Project Missions. From 1995-98, he served as the Human Life-Sciences Experiments Coordinator for the *Mir-NASA* scientific program.

Morukov published over 100 scientific papers and has Patents for four inventions. From August 1998 to January 1999, he attended the Flight-Surgeon Training Course at Johnson Space Center, Texas, USA. Selected in 1989, he flew as 'Mission Specialist 5' aboard Space Shuttle *Atlantis* on STS-106 (September 08-20, 2000), and helped prepare the ISS for the arrival of the first permanent crew. Five astronauts and two cosmonauts delivered over 6,600 pounds of supplies, and installed batteries, power converters, a toilet and a treadmill on the Space Station. He logged 11 days, 19 hours and 12 minutes in space. In 1995, Morukov took an advanced course in emergency medical care; and in 1996 completed a training course in endocrinology and hematology. He also led a simulated 520-day Mars mission. Boris Morukov died on January 01, 2015, aged sixty-four years. He was married, with two children.

## Musabayev, Talgat A. Soyuz TM-19, TM-27, TM-32/31.

Talgat Armangeldyevich Musabayev was born in January 1951 in Kargaly, Dzhampul Oblast, Kazakh SSR. He is a Kazakh test pilot and former cosmonaut. His first two spaceflights were long-duration stays aboard the Russian space station *Mir*. Selected on May 11, 1990 as a Test Cosmonaut, he was Flight Engineer on *Soyuz TM-19/Mir-16* from July 01 to November 04, 1994 (125 days, 22 hours and 53 minutes). He was *Soyuz* commander and *Mir* Commander on *Soyuz TM-27/Mir-25* from January 29 to August 25, 1998 (207 days, 12 hours and 49 minutes). He was Commander on *Soyuz TM-32/31* from April 28 to May 06, 2001 (seven days, 22 hours and four minutes). Musabayev logged a total of 341 days, nine hours and 46 minutes in space; and retired on November 27, 2003. He is married, with two children.

## Nelyubov, Grigori G..

Grigori Grigoriyevich Nelyubov (1934-66) was born in March 1934 in Porfiryevka, Crimean ASSR, USSR. He was a pilot and Captain in the Soviet Air Force, selected in the original group of cosmonauts on March 07, 1960. In 1961, he was among six candidates evaluated for assignment on *Vostok* flight crews. Yuri Gagarin, Gherman Titov and Nelyubov were considered the top three candidates. For the dual launches of *Vostok 3* and '4', he was chosen as a backup for Andrian Nikolayev and Pavel Popovich. On March 27, 1963, he was arrested by the military security patrol at Chkalovskaya station, and following his repeated refusal to apologise was dismissed on April 17 from the space program for drunk and disorderly conduct. Nelyubov went back to flying interceptors in the Far East, but fell to drinking and depression. Nelyubov's image was airbrushed out of the "Sochi Six" photo which showed the top members of the original class of Soviet cosmonauts. He killed himself by stepping in front of a train while drunk on February 18, 1966, and his death at age thirty-one years was officially ruled a suicide. To protect the image of the Soviet space program, efforts were made to cover up the reason for his dismissal and death. His existence in the program was kept secret until the advent of Soviet glasnost in the late 1980s.

## Nikolayev, Andriyan G. Vostok 3, Soyuz 9.

Andriyan Grigoryevich Nikolayev (1929-2004) was born in September 1929 in Shorshely, a village in the Volga River valley, Russian SFSR, Soviet Union. He grew up on a collective farm, and entered Medical School before joining the Soviet Army. He was an ethnic Chuvash, and is considered as the first Turkic person who flew in space. During his training he was able to maintain a calm state in stressful situations, making him a good candidate as a cosmonaut. His future colleagues Yuri Gagarin, Gherman Titov and seventeen others joined him in March 1960.

The name *Vostok 3* is literally "Orient 3" or "East 3" in the Russian language. The *Vostok 3/4* flights were originally scheduled for November 1961, which would have had the propaganda value of the Soviet program having flown four men in space (along with a dual mission in the same calendar year) when the United States had yet to orbit a single astronaut, but Sergey Korolev's plans were frustrated by the *Zenit* photo reconnaissance satellite program which needed to use the *R-7* launch pads at Baikonur Cosmodrome. *Zenit* launches in December 1961, and June 1962 had failed, but another delay was forced when the US carried out a high-altitude nuclear test on July 09 (known as *Starfish Prime*), releasing high levels of radiation into the upper atmosphere and space, knocking out several satellites and making any crewed space launch unsafe for at least a month. Since the *Zenit* satellite was essentially a modified *Vostok*, the next flight doubled as an engineering test of certain *Vostok* components. After a successful test on July 28, preparations at LC-1 began for *Vostok 3*.

In August 1962, Nikolyaev was the first person to make a television broadcast from space. Launched from "Gagarin's Start" at Baikonur aboard *Vostok 3* atop a *Vostok 8K72K* rocket from August 11-15, 1962, with the callsign *Falcon*, Nikolayev orbited the Earth 64 times in nearly four days. This feat was not matched by NASA until the Gemini program (1965-66). *Vostok 3* and '4' were the first dual space flights. Two capsules were launched on precise trajectories that brought them, at their closest, within 6.5 km (4.0 miles) of each other, and visual contact was made. Pavel Popovich is quoted as sayng, *"I saw it at once"*. He said, *Vostok 3 "looked like a very small Moon in the distance"*. Unquote. Nikolayev and Popovich used VHF radio links, the first ship-to-ship communications in space. Pavel Popovich aboard *Vostok 4* used the callsign *Golden Eagle*. The two craft landed seven minutes and about 200 km apart, south of Karaganda.

Nikolayev was Commander aboard *Soyuz 9* with Flight Engineer Vitaly Sevastyanov. They launched on June 01, 1970, a mission lasting 17 days, 16 hours and 59 minutes. They used two-way TV links with their families, watched matches in the 1970 FIFA World Cup, played chess with ground control (the game was a draw), and voted in a Soviet election. In order to conserve attitude control gas, *Soyuz 9* was placed in a spin-stabilisation mode that made the cosmonauts dizzy and space sick. The mission paved the way for the *Salyut* space station missions, investigating the effects of long-term weightlessness on crew, and evaluating the work they could do in orbit. After landing in Kazakhstan, adjusting to gravity of Earth presented a minor problem. They had skipped their planned exercises in space, required help exiting the descent module, and were unable to walk for a few days. This proved the importance of providing crews with exercise equipment during missions. They spent two weeks in a quarantine unit originally designed for cosmonauts returning from Moon landings. On both missions Nikolayev set new endurance records for the longest time a human being had remained in orbit. Andriyan Nikolayev logged a total of 21 days, 15 hours and 20 minutes in space; and retired from the cosmonaut corps in 1982.

On November 03, 1963, he married Valentina Tereshkova, veteran of *Vostok 6*, but the marriage was effectively over in 1977 and they divorced in 1982. They had one daughter, Elena Andriyanova (now a doctor of medicine), before their marriage collapsed. On January

22, 1969, Nikolyev survived an assasination attempt on Leonid Brezhnev. He received twenty-one awards, including Order of the Red Star (1961); Honoured Master of Sports of the USSR (1962); the Order of Lenin (1962); two Hero of the Soviet Union (1962, 1970); the Star of the Republic of Indonesia, second class (1963); Hero of Labour of Mongolia (1965); the Order of the Red Banner of Labour (1976); was made an Honorary Citizen of the Chuvash Republic, Petrozavodsk (1980), Karaganda and Smolensk; awarded the USSR State Prize (1981); and the Order "For Service to the Homeland in the Armed Forces of the USSR, third class (1988). Nikolayev died of a heart attack on July 03, 2004, aged seventy-four years, in Cheboksary, the capital of Chuvasia in Russia. His daughter, Elena, who lives in Moscow, wanted him to be buried in Star City, but the President of Chuvasia demanded a farewell ceremony in Cheboksary, and Nikolayev was buried in his native village of Shorshely.

## Novitski, Oleg V. Soyuz TMA-06M/Expn 33/34, Soyuz MS-03/Expn 50/51, MS-18, MS-25/24.

Lt. Col. Oleg Viktorovich Novitski was born in October 1971 in Chervyen, Minsk Oblast, Byelorussian SSR, USSR (now Belarus). He is a Lieutenant Colonel in the Russian Air Force and an active Roscosmos cosmonaut. He entered the Borisoglebsk Military Pilot School in 1988; and graduated from the Kachinskoye Military Pilot School in 1994, specializing in command tactical aviation. From September to December 1995, he served as a pilot-instructor in the fighter aviation regiment of the V. Chkalov Training Center. Until June 2004, he served in a variety of Army positions. In 2006, Novitski graduated from the Gagarin Air Force Academy, specializing in military units/Air Force detachment management. He was then commander of an attack air squadron, logged 700 hours flying time and was awarded for bravery. He is a Class 2 military pilot and a qualified paratroop instructor and military diver.

Selected in 2006 TsPk-14 Cosmonaut Group, he launched on October 23, 2012, aboard *Soyuz TMA-06M* for 'Expedition 33/34'; returning to Earth on March 16, 2013, after 143 days, 16 hours and 15 minutes in space. Launched on November 17, 2016, aboard *Soyuz MS-03*, for ISS 'Expedition 50/51', he returned to Earth after 196 days in space on June 02, 2017. Launched on April 09, 2021, aboard *Soyuz MS-18*, with Pyotr Dubrov and NASA's Mark Vande Hei for ISS 'Expedition 64', he landed on October 17, 2021 with "space participants" Russian actress Yulia Peresild and film director Klim Shipenko, who shot scenes for the movie ***Vyzov***, aka *The Challenge*. Novitsky was elected Commander of *Soyuz MS-25*, to fly with Belarussian cosmonaut Marina Wasilewska and NASA's Tracy Caldwell Dyson for twelve days on the ISS; returning on *Soyuz MS-24* with NASA's Loral O'Hara. To date, Novitski has logged a total of 531 days, six hours and 58 minutes in space; including three EVAs. He has been awarded Hero of the Russian Federation; the Medal "For Participation in Military Operations"; the Medal "For Distinguished Military Service", I, II and III degree; the Medal "For Service in the Air Force"; and Veteran of Military Operations.

## Onufrienko, Yuri I. Soyuz TM-23, STS-108/Expedition 4.

Col. Yuri Ivanovich Onufrienko was born in February 1961 in Ryasne, Kharkov Oblast, Ukrainian SSR, Soviet Union. He graduated from the V.M. Komarov Eisk Higher Military Aviation School for Pilots in 1982 with a Pilot-Engineer's diploma; and earned a degree in Cartography from Moscow State University in 1994. He served in the Soviet (later Russian) Air Force, logging over 800 flying hours on the L-29, Sukhoi Su-7, Su-17 (M1-4), and L-39 aircraft. Selected as a cosmonaut candidate on January 25, 1989, in TsPk-10, he underwent a course of general space training from September 1989 to January 1991.

He was Commander of *Soyuz TM-23*, launched with Yury Usachov as Commander for the long-duration 'Mir EO-21' mission (February 21 to September 02, 1996); landing 107 km (66 miles) southwest of Akmola, Kazakhstan. During his stay, the seventh and final module for *Mir*, *Pirodia* arrived on April 26, 1996. Onufrienko performed his sixth spacewalk, for five hours and 42 minutes on June 13, 1996. He flew on STS-108 aboard space shuttle *Endeavour* (December 05, 2001, to June 91, 2002) to be Commander for ISS 'Expedition 4' for 196 days; performing two EVAs and returning after 6.5 months aboard *Endeavour*, ending mission STS-111 to land at Edwards Air Force Base, California. Onufrienko logged a total of 389 days, 19 hours and 38 minutes in space; including eight EVAs totaling 42 hours and 33 minutes. He received the Hero of Russia medal; the title of Pilot-Cosmonaut of the Russian Federation; the Gold Star Medal of the Hero of the Russian Federation; the NASA Space Flight and Public Service Medals; and was named a Chevalier of the Legion of Honour (France, 1997). He is married, with two sons and a daughter.

## Ovchinin, Aleksey N. Soyuz MS-10, Soyuz MS-12/Expn 59/60.

Maj. Aleksey Nikolaevich Ovchinin was born in September 1971 in Rybinsk, Yaroslavl, Russia. He is a retired Colonel in the Russian Air Force and former cosmonaut. Selected in 2006 TsPK-14 Cosmonaut Group, he was initially scheduled to fly in 2016 aboard *Soyuz TMA-20M* and join ISS 'Expedition 47/48'. Ovchinin launched on October 11, 2018, as Commander of *Soyuz MS-10* with astronaut Nick Hague, along with 5,600 pounds of supplies and six months' worth of research materials for the ISS. At first stage separation, one of the four 65-foot-long strap-on boosters did not separate from the core rocket via small propulsive gas jets. A nozzle lid was supposed to open, but did not because a tiny sensor mounting pin of the booster's ball joint was bent during asembly by six degrees. The still-attached booster angled into the rocket's core stage, tearing it open.

With the capsule traveling at 7,000 km/h (4,000 mph), and fifty km (thirty miles) high, an orange Master Caution lamp glowed on the control panel, the only visual clue there was a problem with the rocket. The *Soyuz*'s backup automated escape system kicked in, designed to jettison the capsule away from the exploding core stage and out of the path of the failing rocket. It's a stop-gap system, only for emergencies that occur between the loss of the main SAS at T+115 seconds and fairing release at T+157 seconds. An asymetric firing sequence using two pairs of small rocket engines freed the capsule from the dying rocket. Ovchinin reported a "rocket emergency" via radio, then commanded the capsule to go into ballistic re-entry mode. This is a steep and short landing profile, creating the atmospheric drag to slow the capsule. It also put the capsule into a controlled spin – one revolution every thirteen seconds – so when *Soyuz MS-10* deploys its parachutes, they won't tangle. The capsule's dashboard computer confirmed the exact time of the emergency at T plus 118 seconds. The now unpowered capsule rose to a peak at fifty-five miles above Earth, and the two men were on the edge of space for just three minutes. When the capsule went into free fall, the G forces peaked at 6.7 G-s, enough to impair their vision and force signifiant blood away from their brains. Now falling at 822 km/h (514 mph), a pair of pilot parachutes deployed, pulling a drogue chute, which slowed the *Soyuz* to 286 km/h (179 mph). The main parachute then deployed, slowing the capsule's descent to just 25.6 km/h (16 mph).

This deployment pattern was designed to swing the capsule, first at a 30-degree angle to expel heat, then to a straight orientation suitable for landing. At just one meter above the ground, six landing engines ignited, slowing the Soyuz to 3.5 mph for impact. The capsule bounced, tumbled and skidded to a stop. The mission had lasted 19 minutes and 42 seconds; landing near Dzhezkazgan, about 400 km (248 miles) northeast from the Baikonur Cosmodrome launchpad. Ovichinin deadpanned, *"That was a shorter trip than expected"*. Within ninety minutes, rescuers were banging on the side of their capsule. Search and Rescue personnel rolled it over 120 degrees and opened the hatch. NASA's Nick Hague called his wife, saying *"Hey, we're down. We're okay.* ***That was one wild ride****"*. Unquote.

Commander Ochinin, Flight Engineer 1 Nick Hague and Christina Koch launched on March 14, 2019 aboard *Soyuz MS-12,* taking five hours and 47 minutes to reach the ISS and join 'Expedition 59/60'. Haig said, *"It wasn't until we got to the Station and opened that hatch and I was able to give my crewmates on the other side a hug, that it just sank in. I thought, 'I'm here now. I have 204 days left. Let's get to work'."* Unquote. The mission lasted 202 days, 15 hours and 46 minutes, ending on October 03, 2019, and completing 3,248 Earth orbits. Hazza Al Mansuri of MBRSC, a "spaceflight participant" from the United Arab Emirates was on the landing flight. Ovchinin launched for his third mission on September 11, 2023, aboard *Soyuz MS-26* with crew of Flight Engineer Ivan V. Vagner and NASA's Donald Pettit to deliver fresh crew to the ISS for 'Expedition 71/72', a mission planned for about 180 days. To date, Aleksey Ovchinin has logged a total of over 374 days, 19 hours and 31 minutes in space.

## Padalka, Gennady I. Soyuz TM-28, TMA-4, TMA-14/Expn 19/20, TMA-04M/ Expn 31/32, TMA-16M/Expn 43/44.

Col. Gennady Ivanovich Padalka was born in June 1958 in Krasnodar, Soviet Union. He is a retired Russian Air Force officer and Roscosmos cosmonaut. He held the record for the most time spent in space, at 878 days, until Oleg Kononenko broke this record on February 04, 2024. Padalka graduated from the Yeysk Military Aviation College in 1979, then served as a pilot and senior pilot in the Russian Air Force. He logged 1,500 hours in six aircraft types as a First Class Pilot. He has performed over 300 parachute jumps as an Instructor of General Parachute Training. Until 1994, Padalka worked as an engineer-ecologist at the UNESCO International Center of Instruction Systems. He is an investigator for the Advanced Diagnostic Ultrasound in Microgravity Project, a US Government funded study investigating strategies for applying diagnostic telemedicine to space applications. Selected as a cosmonaut, he trained from June 1989 to January 1991 to qualify as a Test Cosmonaut. Launched on August 13, 1998 aboard *Soyuz TM-28,* he joined 'Mir Expedition 26', whose primary mission was to repair life support systems and prepare the station for de-orbit after 'Expedition 27'. He returned to Earth aboard *Soyuz TM-28* on February 28, 1999; landing in Kazakhstan. This mission lasted 198 days and 16 hours. From August 2000 to November 2001, he trained as the ISS 'Expedition 4' backup crew member.

In March 2002, Padalka was assigned as Commander of ISS 'Expedition 9' crew, and flew aboard *Soyuz TMA-4,* which docked with the ISS on April 21, 2004. Following a week of joint operations and handover briefings, they replaced the 'Expedition 8' crew for six months of science operations, maintenance of Station systems, and performed four EVAs, before landing in Kazakhstan on October 23, 2004. Padalka logged another 187 days, zero hours and 21 minutes in space; including 15 hours and 45 minutes of EVA time. Launched on March 26, 2009, aboard *Soyuz TMA-14,* he was Commander of 'Expedition 19' and the first six-person 'Expedition 20' crew; returning to Earth on October 11, 2009. He launched on May 15, 2012, aboard *Soyuz TMA-04M* as Flight Engineer for 'Expedition 31', then became Commander of 'Expedition 32'; returning to Earth on September 17, 2012. Launched on March 27, 2015, aboard *Soyuz TMA-16M* for ISS 'Expedition 43/44'; his ninth and final space walk was on August 20 for five hours and 51 minutes, and included deploying a 21-inch spherical satellite. He landed aboard *Soyuz TMA-16M* on September 12, 2015.

Padalka announced his retirement in April 2017, having logged a total of 878 days, 11 hours and 37 minutes in space, and is currently in second place for the most accumulated space time. Padalka said, *"I had to resign. I am tired of doing nothing. There are no prospects that I will fly again[to the International Space station]"* he told TASS, in explanation. Padalka was awarded Hero of the Russian Federation; the title of Russian Federation Test-Cosmonaut; the Fatherland Service Medal, fourth class; the Russian Federation prize for Science and Technology; Medals of the Russian Federation; and the Medal of the International Fund of Cosmonautics support for Service to Cosmonautics. He is married, with three daughters.

## Patsayev, Viktor I. Soyuz 11.

Victor Ivanovich Patsayev (1933-71) was born in June 1933 in Aktyubinsk, Kazakh SSR, Soviet Union (now Aqtobe, Kazakhstan). He was an engineer, was selected in 1968 USSR Civilian Specialist Group 3, and logged 23 days, 18 hours and 21 minutes in space. Launched on June 06, 1971, he served as design engineer on *Soyuz 11*. On June 07, the three-man crew arrived at *Salyut 1* space station (launched two months earlier) and he operated the Orion 1 Space Observatory, becoming the first man to operate a telescope outside the Earth's atmosphere. Meteorological and plant growing experiments were carried out. There was drama when an overheating instrument flooded part of the *Salyut 1* space station with smoke.

Patsayev became the first man to celebrate a birthday in space. His crewmates treated him to a surprise meal of veal, cookies, and blackberry juice. Late on June 29, 1971, *Soyuz 11* undocked from *Salyut 1*. Three hours later, the spacecraft's engines were fired to return to Earth. Volkov jokingly asked flight controllers to ensure that a supply of cognac – a traditional welcome-home gift – was waiting for them at their landing site. At an altitude of around 160 km (100 miles) explosive charges fired to separate the orbital (crewed) and instrument modules, but the air pressure inside the crew capsule then fell dramatically. Health trackers revealed the cosmonaut's heart rates soared as they searched for the leak; possibly a hatch improperly closed, but later investigation found it likely an air pressure equalization valve opened prematurely.

Within fifty seconds, Patsayev's pulse dropped to levels consistent with oxygen starvation; within 110 seconds all three men were dead. At this point, Mission control were unaware of the gravity of the situation, but attempts to contact the crew over VHF radio were met with silence. When helicopter crews spotted the undamaged Soyuz swinging gently beneath the parachute, there was still no word from the crew. The capsule was opened, and the crew were found dead. The three fatalities were the second Soviet mission to end with cosmonaut death, after Vladimir M. Komarov perished in April 1967. Viktor Patseyev died in outer space on June 30, 1971, and was buried in the Kremlin Wall Necropolis.

## Petelin, Dmitry A. Soyuz MS-22/Expedition 67/68/69/Soyuz MS-23.

Dmitry Aleksandrovich Petelin was born in July 1983 in Kustanai, Kazakh SSR, Soviet Union. He is an active Test Cosmonaut and aeronautical engineer. He graduated from the South Ural State University in 2006 with a degree in Aircraft and Helicopter Engineering. Petelin then worked for **NIK, LLC.** Selected in TsPK-16 Cosmonaut Group on October 26, 2012, he completed training at the Yuri Gagarin Academy on July 15, 2014. Petelin launched atop a *Soyuz-2.1a* booster as 'Flight Engineer 1' on September 21, 2022, aboard *Soyuz MS-22* for ISS long-duration 'Expedition 67/68/69'. 'Expedition 68' began upon the departure of *Soyuz MS-21* on September 29, 2022, and ended upon the departure of the uncrewed *Soyuz MS-22* on March 28, 2023; then 'Expedition 69' began.

The *Soyuz MS-22* was named in honor of Russian scientist Konstantin Tsiolkovsky, whose one hundred and sixty-fifth birthday fell on September 17, a few days before the launch of *MS-22*. On December 15, 2022, a micrometeorite impact made a 0.8 mm diameter hole in the ammonia cooling radiator of the docked *Soyuz*'s Service Module, and a *"visible stream of flakes"* was observed, concurrent with a loss of pressure in the external radiator cooling loop, so the outer surface of the *Soyuz* was examined using the cameras of the European Robotic Arm and the Canadarm-2. Since *Soyuz MS-22* was now unable to return with a crew, it was launched uncrewed back to Earth on March 28, 2023, to function as a replacement. Petelin returned to Earth on September 27, 2023, but until the uncrewed replacement *Soyuz MS-23* docked to the ISS on February 26, 2023, SpaceX 'Crew-5' was considered among the options to return the *Soyuz MS-22* crew, in case of emergency. Dmityy Petelin logged a total of 370 days, 21 hours and 23 minutes (one year and five days) in space; including six EVAs. He is married, with a daughter.

## Poleshchuk, Aleksandr F. Soyuz TM-16.

Capt. Aleksandr Fyodorovich Poleschuk was born in October 1953 in Cheremkhovo, Irkutsk Oblast, Russian SFSR, USSR. He graduated from the Moscow Aviation Institute in 1977 with a Mechanical Engineering diploma. He joined **RSC Energia** as a test engineer, where he was occupied with perfecting repair and assembly techniques performed during space flights. Selected as a Test Cosmonaut in February 1989 in NPOE-9 Group 14 Civil Specialists, he trained from September 1989 to January 1991. From then until March 1992, he had advanced training for the *Soyuz TM* craft and a flight to the *Mir* space station. In 1992, he was selected as backup for the *Soyuz TM-15* joint Russian-French mission; then was nominated as Flight Engineer of the prime crew for *Soyuz TM-16*. Launched with Gennady Manakov aboard *Soyuz TM-16* as Flight Engineer to the 'Mir-13' mission (January 24 to July 22, 1993), he logged 179 days, zero hours and 43 minutes in space; including two EVAs totaling nine hours and 58 minutes. Poleshchuk tested the androgynos peripheral docking sub-assembly of the *Kristall* module. From October 1994 to March 1995, he trained as backup Flight Engineer for the *Soyuz TM-21* flight and the 'Mir-18' expedition. He is married, with one daughter.

## Polyakov, Valeri V. Soyuz TM-6/7, Soyuz TM-18/20.

Dr. Valeri Vladimirovich Polyakov (1942-2022) was born Valeri Ivanovich Korshunov in April 1942 in Tula, USSR. His name changed after being legally adopted by his stepfather in 1957. He enrolled in the I.M. Sechenov First Moscow Medical Institute, and graduated with a Medical Doctoral degree. Next, he enrolled in the Institute of Medical and Biological Problems, Ministry of Public Health, Mocow, where he specialized in astronautics medicine. A qualified Physician, he was selected in Medical Group 3 on March 22, 1972. Launched on August 29, 1988, as Research Doctor/Flight Engineer aboard *Soyuz TM-6* to *Mir* for 240 days, he returned to Earth aboard *Soyuz TM-7* on April 27, 1989; landing some 140 km northeast of Dzhezkazgan. Upon landing, his first words were, *"We can fly to Mars"*.

Launched on January 08, 1994, aboard *Soyuz TM-18*, he spent about 437 days aboard *Mir*, completing over 7,000 Earth orbits; landing aboard *Soyuz TM-20* on March 22, 1995, setting the record for the longest time continuously spent in space by an individual. Upon landing, he insisted on walking to a nearby lawn chair, wishing to prove that humans could be physically capable of working on the surface of Mars after a long-duration transit voyage. He underwent medical assessments before, during, and after this flight; plus two follow-up examinations six months after coming home. Researchers found there was no impairment of cognitive functions, but he had experienced a clear decline in mood during the first few weeks of spaceflight. After returning to Earth, Polyakov did not suffer from any prolonged performance impairments. He logged a total of 679 days, 16 hours and 33 minutes in space; and retired from the Russian space program in June 1995.

Polyakov was then the Deputy Director of the **Ministry of Public Health** in Moscow. He was a member of the Russian Chief Medical Commission, participating in the qualification and selection of cosmonauts. He was a member of the International Space Researchers' Association and the International Academy of Astronautics. Polyakov received ten awards, including Hero of the Soviet Union; Hero of the Russian Federation; the Order of Lenin; Pilot-

Cosmonaut of the USSR; Officer of the Legion of Honour (France); Hero of the Democratic Republic of Afghanistan; the Prince of Asturias Award (Spain); the Order of Parasat (Kazakhstan); and the Order of the Sun of Freedom (Afghanistan). Valeri Polyakov died on September 07, 2022, aged eighty years, in Moscow, Russia. He was married, with one child.

## Popov, Leonid I. Soyuz 35/37, Soyuz 40, Soyuz T-7/T-5.

Maj. Gen. Leonid Ivanovich Popov was born in August 1945 in Oleksandriia, Kirovohrad Oblast, Ukrainian SSR, USSR. In 1968, he graduated from Chernigov Higher Air Force School with a degree in Electrical Engineering, and was a senior pilot of 627 Guards Fighter Aviation Regiment of 15 Air Defense Corps of Baku Air Defense District. Selected on April 27, 1970, in Air Force Group 5, he flew as Commander aboard *Soyuz 35* (April 09 to October 11, 1980) to *Salyut 6*, and returned to Earth aboard *Soyuz 37* after 184 days, 20 hours and 11 minutes in space. Launched atop a *Soyuz-U* booster aboard *Soyuz 40* (May 14-22, 1981) with Romanian Research Cosmonaut Dumitru Prunariu, this was the sixteenth expedition to *Salyut 6*, and the ninth international crew; landing 225 km southeast of Dzhezkazgan; and logging seven days, 20 hours and 41 minutes in space. *Soyuz T-7*, code name *Dnieper* (August 10-27, 1982), was the third mission to the *Salyut 7* space station; notable for carrying Svetlana Savitskaya, the first woman in space for almost 20 years. Popov returned to Earth aboard *Soyuz T-5*; landing after eight days on August 27 some 225 km east of Dzhezkazgan, Kazakhstan.

Leonid Popov logged a total of 200 days, 14 hours and 45 minutes in space, and retired on June 13, 1987. He graduated from the K.E. Voroshilov Military Academy of the General Staff of the Soviet Armed Forces in 1989. Leonid Popov received five Soviet and four foreign awards, including two hero of the Soviet Union; three Order of Lenin; the Medal "For Merit in Space Exploration" (Russian Federation); Hero of the Socialist Republic of Romania; Hero of the Republic of Cuba; and Hero of Socialist Labour (Vietnam). He is an Honorary Citizen of Arkalyk, Oleksandriia, and Butach. Popov is married, with two children.

## Popovich, Pavel. Vostok 4, Soyuz 14.

Pavel Popovich (1930-2009) was born in October 1930 in Uzyn, Kyiv Oblast, Ukrainian SSR, Soviet Union. He was the fourth cosmonaut in space, the sixth person in orbit, the eighth person and first Ukrainian in space. In 1951, he graduated as a construction engineer from a technical school in Magnitogorsk; and also received a pilot's degree in 1952, graduating from the Stalingrad Military Aviation School near Novosibirsk. Popovich then trained at the Military Officers of the Air Force Aviation Training School in Grozny until 1954, when he joined the Soviet Air Force. He flew Yak-11, Yak-18, Lavochkin La-9, and MiG-15 aircraft.

In 1960, Popovich was selected as one of the original group of twenty Air Force pilots to train for the Soviet space program. He passed his final exams in Cosmonaut Basic Training on January 18, 1961. Considered a strong candidate for the first manned spaceflight, he served as the CapCom for Yuri Gagarin's historic *Vostok 1* mission of April 12, 1961. From May to August 1961, he trained for *Vostok 2*, then *Vostok 3*. From November 1961 to May 1962, he trained for *Vostok 4* (vehicle 3A 3KA No. 6), which he commanded from August 12-15, 1962, with the callsign *Golden Eagle*. This mission lasted two days, 22 hours and 57 minutes, and was the first mission where two craft were in orbit simultaneously. Popovich made rendezvous with *Vostok 3*, coming within visual range of 6.5 km from the other craft. His life support system malfunctioned, with the cabin temperature dropping to ten degrees Celsius, with 35% humidity. He conducted experiments, and tested spacecraft systems including communication, guidance and landing. Popovich touched down 193 km southwest of Karaganda, Kazakhstan.

In January 1964, he became a cosmonaut instructor, and deputy commander to the second group of cosmonauts. He joined the Young Communist League in 1964, and married Marina Popovich. They had two daughters, but later divorced, and his second wife was Alevtina Oshegova. He was selected to command on of the planned Soviet Moon landings, and trained for this from 1966 until 1968, when the program was scrapped. In 1969, he was a senior cosmonaut instructor, and became Chief of Cosmonaut Training in 1972. Launched on July 03, 1974, Popovich commanded *Soyuz 14*, a mission to the *Salyut 3* space station. The crew tested the *Salyut* as a crewed military reconnaissance platform. After 15 days, 17 hours and 30 minutes in space, he landed on July 19 some 140 km southeast of Dzhezkazgan, Kazakhstan. In 1977, he received a post-graduate degree in Technical Sciences. From 1978, he was the deputy chief of the **Gagarin Cosmonaut Training Center**, responsible for research and testing work. In January 1982, he was removed from the list of active cosmonauts, so he could serve as Deputy Chief for Scientific Testing and Research at the Gagarin Center.

In 1991, Popovich became Director of the **Institute Rossiyskogo** for Monitoring of Land and Ecosystems; then worked as Chairman of the Board of Directors of the All-Russia Institute of Aero-Photo-Geodesic Studies (VISKhAGI), dealing with the compilation of a land inventory of Russia using images from space. He retired as a Soviet Air Force colonel and test pilot by order of the Defense Ministry on August 20, 1993, and became a member of the Supreme Soviet of the Ukrainian Soviet Socialist Republic's Sixth to Eleventh Convocations. He has been the Chairman of the Ukrainian diaspora organization in Russia. He received a Gold Tsiolkovsky Medal (Academy of Sciences of the USSR); and a De la Vaulx Medal from the Federation Aeronautique Internationale. In 2010, Unkraine celebrated the eightieth anniversary of Popovich's birth.

Pavel Popovich received eighteen awards, including Order of the Red Star (1961); Hero of the Soviet Union (1962, 1974); the Order of Lenin (1962, 1974); the Order of Friendship of Peoples (1982); the Order of Merit for the Fatherland, fourth grade (2000); and the Order of Prince Yaroslav the Wise, fourth class (Ukraine, 2005). Popovich also received Honorary Citizenship of eight cities. A bronze sculpture was established in his home area of Uzyn; his name was given to a mountain ridge in Antarctica; and to the Mars-crossing asteroid 8444 Popovich. He died on September 29, 2009, aged seventy-eight years, in Gurzuf, Crimea.

## Prokopyev, Sergey V. Soyuz MS-09/Expn 56/57, Soyuz MS-22/Expn 68/ Soyuz MS-23.

Sergey Valeryevich Prokopyev was born in February 1975 in Sverdlovsk, Russian SFSR, Soviet Union. He is an active cosmonaut and pilot, with the rank of Commander. Selected in TsPk-15 RKKE-17 Cosmonaut Group, he launched on June 06, 2018, aboard *Soyuz MS-09* for ISS 'Expedition 56/57'. On August 29, an air leak was discovered, caused by a hole aboard *Soyuz MS-09*, requiring an EVA on December 11 to apply a thermal blanket to the orbital module section. He returned to Earth on December 20, 2018, logging 196 days in space. Launched on September 21, 2022, aboard *Soyuz MS-22*, he became ISS Commander for 'Expedition 68'. Damage to the spacecraft extended the mission, and he returned to Earth aboard *Soyuz MS-23* on September 27, 2023 after a year and six days in space. He logged a total of 567 days, 16 hours and 13 minutes in space; including eight EVAs totaling 55 hours and 15 minutes. Prokopyev has received six awards: Hero of the Russian Federation (2019); Pilot-Cosmonaut of the Russian Federation (2019); the Medal of Y.A. Gagarin (Roscosmos); the Order of Military Merit; the Medal "For Distinction in Military service, second and third classes; and the Medal "For Participation in a Military Parade on Victory Day" (Ministry of Defense). He is married, with a daughter and son.

## Pustovyi, Yaroslav.

First Lieutenant Dr. Yaroslav "Yarko" Pustovyi was born in December 1970 in Kostroma, Russia, and grew up in Kyiv, Ukraine. He is a former electromagnetism researcher, Ukrainian astronaut and Canadian businessman. He graduated from the A.F. Mozhaksky Military-Space Academy in 1993 with a Master of Science degree. While working for the National Academy of Sciences for Ukraine as an electromagnetism researcher, he attended university at Kharkiv; earning a Doctor of Philosophy (Ph.D.) from the National University of Kharkiv in 1997, specializing in Radio Science. He was one of the first astronauts selected by the State Space Agency of Ukraine in 1996. He was backup Payload Specialist for mission STS-87, but did not fly in space. In 1998, he was awarded the Order of Merit for Ukraine and competed for the **Ansari X Prize** in 2003 on the Canadian Arrow team. The Ansari X Prize was a space competition in which the **X Foundatio**n offered a USD ten million prize for the first non-government organization to launch a reusable crewed spacecraft into space twice. It was aimed to spur development of low-cost spaceflight. It was won on October 04, 2004, by the Tier One project designed by Burt Rutan, financed by Paul Allen, and using the experimental spaceplane *SpaceShipOne*.

In January 1998, he was given the Ukrainian Order of Merit, third degree. Pustovyi moved with his family to Barrie, Ontario, Canada, in 2007 and worked in security. In 2009, he became a co-founder of **Space 1 Systems**. In October 2011, the Pustovyi family received Canadian permanent residency and settled in Simcoe County. In 2014, Pustovyi joined the **Canadian Space Commerce Association** as a board member, and became the organization's President later that year. He is married, with three children. Pustovyi is fluent in Ukrainian, Russian, English and French languages.

## Revin, Sergei N. Soyuz TMA-04M/Expedition 31/32.

Lt. Col. Sergei Nikolayevich Revin was born in January 1966 in Moscow, Soviet Union (now Russia). He graduated from the Moscow Institute of Electronic Technology in 1989, qualified as an Engineer-Physicist; then worked from 1989-93 as an engineer for the **NPO IT**, a scientific production organization of measuring equipment in Kaliningrad. From August 30, 1993, to April 1996, he worked as an engineer at the **NPO-Energia**. Selected in April 1996 RKKE Group, he completed his basic cosmonaut training in 1998. From July 1998 to January 2011, he was a Test Cosmonaut of the **Energia Rocket and Space Corporation**, then a Test Cosmonaut of the Gagarin Training Center. From October 1998 to April 2011, he trained as the 'Expedition 29/30' crew member to be a Flight Engineer aboard the ISS. He launched on May 15, 2012, aboard *Soyuz TMA-04M* with Gennady Padalka and NASA's Joe Acaba for ISS 'Expedition 31/32'. They returned to Earth on September 17, 2012; landing in Kazakhstan. Revin logged a total of 124 days, 23 hours and 52 minutes in space. In April 2017, he was stood down from cosmonaut duties for medical reasons. He is married, with a son.

## Romanchenko, Roman Y. Soyuz TMA-15/Expedition 20/21, TMA-07M/ Expedition 34/35.

Maj. Roman Yuriyevich Romanchenko was born in August 1971 in Shchyolkovo, Soviet Union. After graduating from the Star City High School in 1986, he entered the Leningrad Suvorov military school, graduating in 1988. He graduated from the Chernigov High Air Force School for Pilots in 1992 as a Pilot-Engineer, then served as a second commander in the Air Force. A Class 3 Air Force pilot, Romanchenko flew L-39 and Tu-134 aircraft, logging over 500 hours flying time. Selected in December 1997 TsPk Cosmonaut Group, his basic training was from January 1998 to November 1999. He served as backup to Oleg Kotov as Commander of *Soyuz TMA-10* and to Fyodor Yurchikhin as Commander of 'Expedition 15'.

He launched on May 27, 2009, as Commander aboard *Soyuz TMA-15* to be Flight Engineer for ISS 'Expedition 20/21'. After Sergey Volkov and Richard Garriott, he was the third second-generation space traveler. Romancheko returned to Earth on December 01, 2009, after 187 days and 20 hours in space. Launched on December 19, 2012, aboard *Soyuz TMA-07M* as Flight Engineer for ISS 'Expedition 34/35', he made an EVA of six hours and 38 minutes on April 19; and returned to Earth on May 14, 2013, after logging another 145 days and 14 hours in space. Roman Romanchenko logged a total of 333 days, 11 hours and one minute in space. On May 06, 2010, he was awarded Hero of the Russian Federation; and Pilot-Cosmonaut of the Russian Federation. He was elected to the State Duma in 2021, representing the Chertanovo constituency. Romanchenko was sanctioned by the UK on March 11, 2022, in relation to Russia's invasion of Ukraine. He is married, with a son and daughter.

## Romanenko, Yuri V. Soyuz 26/27, Soyuz 38, Soyuz TM-2/3.

Col. Yuri Viktorovich Romanchenko was born in August 1944 in Orenburg Oblast, Russian SFSR, USSR. In 1962, he enrolled at the Chernigov Higher Air Force School, Ukraine; graduating with honours in 1966. He remained with the School, training the students and practicing as a cosmonaut candidate. He is a professional pilot with experience in Yak-18, L-29, MiG-15, MiG-17 and MiG-21 aircraft, and has made thirty-nine parachute jumps. Selected in Air Force Group 5, in 1970, among sixteen other candidates he was cleared for space flights. Launched on December 10, 1977, he was Commander aboard *Soyuz 26* to *Salyut 6* mission 'EO-1'. He made one EVA of 90 minutes, and in preparation for a space walk forgot to attach his safety cord. He was grabbed by Grechko, and avoided a potentially fatal accident. After ninety-six days in space, he returned to Earth aboard *Soyuz 27*.

Launched on September 18, 1980, with the first Cuban cosmonaut, Arnaldo Tamayo, he flew aboard *Soyuz 38* for the seven-day mission *Salyut 6* 'EP-8'. Launched on February 06, 1987 with Aleksandr Laveykin aboard *Soyuz TM-2* for the 326-day 'Mir EO-2' mission, he returned to Earth on December 29, 1987 aboard *Soyuz TM-3*. He made three more EVAs, totaling eight hours and 48 minutes. He was discharged from military service on October 02, 1995, having reached the age limit. Romanenko logged a total of 430 days, 20 hours and 21 minutes in space; including four EVAs totaling 10 hours and 18 minutes. He received ten awards, including the Order of the Red Star (1976); two Hero of the Soviet Union (1978, 1980); three Order of Lenin (1978, 1980, 1987); Hero of the Republic of Cuba; the Order "Madara Horseman" (Bulgarian Socialist Republic); the Order of Klement Gottwald (Czechoslovak Socialist Republic); and the Medal "For Merit in Space Exploration" (Russian Federation). He is married, with two sons.

## Rozhdestvensky, Valery I. Soyuz 23.

Valery Ilyich Rozhdestvensky (1939-2011) was born in February 1939 in Leningrad, Soviet Union. He graduated from the Higher Military Engineering School of the Soviet Navy in Puskin, qualified in Engineering. From 1961-65, he was commander of a deep-sea diving unit in the Baltic Sea War Fleet. Selected in 1965 TsPK-3, he flew as Flight Engineer aboard *Soyuz 23* (October 14-16, 1976), the second visit to the *Salyut 5* space station. An equipment malfunction prevented docking, and the mission was aborted. The descent module landed on partially frozen Lake Tengiz, and sank under the surface. Recovery took nine hours, owing to fog and adverse weather. He logged two days, zero hours and six minutes in space. He then worked at the Gagarin Academy with the Soviet space program, and retired on June 24, 1986, to work with **Metropolis Industries**. Rozhdestvensky received five awards: Hero of the Soviet Union; Pilot-Cosmonaut of the USSR; the Order of Lenin; the Order for Service to the Homeland in the Armed Forces of the USSR, third class; and the Medal "For Merit in Space Exploration". He died on August 31, 2011, aged seventy-two years, in Moscow, Russia. He was married, with one child.

## Rukavishnikov, Nikolai N. Soyuz 10, Soyuz 16, Soyuz 33.

Nikolay Nikolayevich Rukavishnikov (1932-2002) was born in September 1932 in Tomsk, USSR. He studied at the Moscow Engineering and Physics Institute, then worked for Sergey Korolev's design bureau. Selected for cosmonaut training in 1967 Civilian Specialist Group 2, he became the fiftieth human to fly in space as Test Engineer aboard *Soyuz 10* (April 22-24, 1971), and logged one day, 23 hours and 45 minutes in space. With Vladimir Shatalov and Aleksei Yeliseyev, he was unable to dock with the *Salyut 1* space station. Launched with Anatoly Filipchenko, he was Flight Engineer aboard *Soyuz 16* (December 02-08, 1974), and the mission was a successful test of the *Soyuz 7K-TM* hardware to be used in the July 1975 *Apollo-Soyuz Test Project*. He logged another five days, 22 hours and 23 minutes in space.

Rukavishnikov was Commander of *Soyuz 33*, an Intercosmos mission to *Salyut 6*. Launched on April 10, 1979, the flight was scheduled to dock on April 12 (Cosmonautics Day). Upon approaching the *Salyut*, however, the final engine firing lasted only three seconds instead of six, and the *Igla* docking system switched off. The *Soyuz'* main engine had malfunctioned, and docking maneuvers were now impossible. Since *Soyuz 33* had limited life support resources, the crew had to return to Earth immediately. Choosing a steep-descent re-entry, which would subject the two-man crew to high 10-G loads, Rukavishnikov switched off all automatic landing programs, and skilfully landed the craft surprisingly close to the initially scheduled landing point. This mission logged one day, 23 hours and one minute in space. Rukavishnikov logged a total of nine days, 21 hours and nine minutes in space. He resigned from the space program in 1987, and went to work with **Energia**. Nikolau Rukavishnikov received seven awards, including two Hero of the Soviet Union; four Order of Lenin; Hero of the Mongolian People's Republic; Hero of the People's Republic of Bulgaria; the Order of Sukhbaatar (Mongolia); and the Order of Georgi Dimitrov (Bulgaria). He died of a heart attack on October 19, 2002, aged seventy years, in Moscow, Russia.

## Ryazanski, Sergei N. Soyuz TMA-10M/Expn 37/38, Soyuz MS-05/ Expn 52/53.

Sergei Nikolayevich Ryazanski was born in November 1974 in Moscow, Soviet Union. He is a biochemist and retired cosmonaut; the grandson of Mikhail Ryazansky, chief Soviet guidance designer of the *Sputnik* era. He attended Moscow State University and graduated as a biochemist in 1996, then began working as a researcher at the Institute of Biomedical Problems. He was selected as a Commander of the IMBP-6 Cosmomaut Group in 2003, but later transferred to the TsPK Cosmonaut Group. He completed his training in 2005, and in 2009 participated in a 105-day mission as part of phase two of the Mars 500 program.

He launched aboard *Soyuz TMA-10M* for ISS 'Expedition 37/38' from September 25, 2013, until returning to Earth on March 11, 2014. He performed three EVAs with Oleg Kotov, for a total duration of 20 hours and five minutes. Launched on July 28, 2017, Ryazansky flew as Commander of *Soyuz MS-05*, and served as Flight Engineer for ISS 'Expedition 52/53'. His fourth EVA of seven hours and 34 minutes tested a new version of the *Orlan* space suit, deployed five nano-satellites, and installed external experiments. He returned to Earth on December 14, 2017; after a mission lasting 138 days, 16 hours and 56 minutes. Ryazansky logged a total of 304 days, 22 hours and 21 minutes in space; including four EVAs totaling 27 hours and 39 minutes. He is married, with two sons and two daughters.

## Ryumin, Valery V. Soyuz 25, Soyuz 32/34, Soyuz 35/37.

Valery Victorovich Ryumin (1939-2022) was born in August 1939 in Komsomolsk-on-Amur, Khabarovsk Krai, Russian SFSR. In 1958, Ryumin graduated from the Kaliningrad Mechanical Engineering Technical College with the specialty "Cold Working of Metal". From 1958-61, Ryumin served in the Red Army as a tank commander. In 1966, he graduated from the Department of Electronics and Computing Technology of the Moscow Forestry Engineering Institute with the specialty "Spacecraft Control Systems". He was then employed at the Rocket Space Corporation **RSC Energia**, holding the positions of Ground Electrical Test Engineer, Deputy Lead Designer for Orbital Stations, Department Head, and Deputy General Designer for Testing. Ryumin helped develop and prepare all Soviet orbital space stations, beginning with *Salyut 1.*

In 1973, Ryumin joined the RSC Energia cosmonaut corps, and was selected in 1973 as part of Civilian Specialist Group 5. In 1977, he spent two days aboard *Soyuz 25*. In 1979, he spent 175 days aboard the *Salyut 6* space station, flying up on *Soyuz 32* and down on *Soyuz 34*. In 1980, Ryumin spent another 185 days aboard *Salyut 6,* flying up on *Soyuz 35* and returning aboard *Soyuz 37.* From 1981-89, he was Flight Director for the *Salyut 7* space station and the *Mir* space station. Since 1992, he was the Director of the Russian portion of the *Shuttle-Mir* and *NASA-Mir* programs. In January 1998, NASA announced Ryumin's selection to the crew of STS-91, and he served aboard the ninth and final *Shuttle-Mir* docking mission of June 02-12 1998. He accumulated 371 days, 17 hours and 24 minutes in space.

Valery Ryumin was married to fellow cosmonaut Yelena Kondakova, and they had two daughters and a son. His hobbies included tennis, angling, hunting, walking through forests, and travel. Ryumin died on June 06, 2022, and wqs buried in the Federal Military Memorial Cemetery. In the 2017 Russian film ***Salyut 7,*** a character (Valery Shudin) was based on Ryumin. He was featured on the USSR 1981 15-kopec postage stamp. Valery Ryumin was twice decorated as Hero of the Soviet Union; also Pilot-Cosmonaut of the USSR; the Order "For Merit to the Fatherland", Fourth Class; three Orders of Lenin, the Medal "For Merit in Space Exploration"; Hero of the Republic of Cuba; Hero of Socialist Labour (Vietnam); and the Order of Ho Chi Minh (Vietnam).

## Ryzhikov, Sergei N. Soyuz MS-02/Expn 49/50, Soyuz MS-17/Expn 63/64.

Lt. Col. Sergei Nikolayevich Ryzhikov was born in August 1974 in Bugulma, Tatar Autonomous SSR, Soviet Union. Selected in 2006, he is a veteran of two long-duration missions to the ISS. Ryzhikov graduated from secondary school in Nizhnevartovsk, Russia, in 1991 where he was part of the Young Aviators club. He graduated from the Kachinsky Higher Military Aviation School of Pilots in 1996 with a degree in command tactical fighter aviation. He then trained as a pilot in the Russian Air Force. In 1997, he was assigned as a pilot in the Seventy-Sixth Air Army, based out of Andreapol. Ryzhikov has logged 700 flying hours in L-39 Albatros and MiG-29 aircraft; as well as holding a parachute instructor qualification with over 350 jumps.

Selected by Roscosmos in October 2006, he began training at Star City in February 2007. In 2011, he served as a "cavenaut" into the ESA CAVES training in Sardinia; along with astronauts Thomas Pesquet, Tim Peake, Norishige Kanai and Randolph Bresnik. He launched on September 23, 2016 aboard *Soyuz MS-02* for 'Expedition 49'. Crew made four EVAs during 'Expedition 50' and Ryzhikov observed the arrival of three uncrewed resupply spacecraft. After 173 days in space, he landed on April 10, 2017. Launched on October 14, 2020, aboard *Soyuz MS-17* for 'Expedition 63' he became Commander of 'Expedition 64'. On November 18, he and Kud-Sverchkov made an EVA of six hours and 48 minutes. He returned to Earth on April 17, 2021, after almost 185 days on orbit. In April 2025, Ryzhikov will fly aboard *Soyuz MS-27* with Sergey Mikayev and Kirill Peskov; scheduled to return in September 2025. To date, he has logged a total of 358 days, two hours and 25 minutes in space; including one EVA of six hours and 48 minutes.

## Samokutyaev, Aleksandr M. Soyuz TMA-18, TMA-21/Expn 27/28.

Aleksandr Mikhailovich Samokutyaev was born in March 1970 in Penza, Russian SFSR, Soviet Union. He is a Russian politician and former cosmonaut. In 1987-88, he studied at Penza Polytechnical Institute; then went to Chernigov and graduated from Chernigov Higher Military Pilot School in 1992. In 2000, he graduated from the Gagarin Air Force Academy as a Pilot-Engineer. Samokutyayev has logged 680 hours flying time and performed 250 parachute jumps. Since 2000, he served at the Gagarin Academy as the Planning Division head. He qualified as a Test Cosmonaut on July 05, 2005. In July 2008, he was assigned to the backup crew for ISS 'Expedition 25'. During the launch on April 02, 2010, of *Soyuz TMA-18*, he served as backup Commander. On October 07, 2009, he was confirmed as a prime crewmember for 'Expedition 27'. He flew as Commander on April 04, 2011, aboard *Soyuz TMA-21* as Flight Engineer for ISS 'Expedition 27/28', spending 164 days aboard the ISS. He landed on September 16, 2011, some 149 km (93 miles) southeast of Dzhezkazgan, Kazakhstan. He performed two career EVAs, and in April 2017 Roscosmos decided to release Samokutyaev from his post on medical indicators. He has received various Russian Armed Force medals. He is married, with a daughter.

## Sarafanov, Gennady V. Soyuz 15.

Gennady Vasilyevich Sarafanov (1942-2005) was born in January 1942 in Sinenkiye, Saratov Oblast, USSR. He graduated from Balashov Military Pilot School in 1964; and from Gagarin Military Air Force Academy, Monino, in 1978. He was a Colonel in the Russian Air Force, selected as a Test Cosmonaut in TsPk-3 (Air Force Group 3) on October 28, 1965. Launched as Commander aboard *Soyuz 15* (August 26-28, 1974), the mission was to dock with *Salyut 3*, but failed to do so after electronics in the *Igla* docking mechanism malfunctioned. Without sufficient fuel for prolonged attempts at manual docking, the mission had to be abandoned. Observers concluded a flight of 19 to 29 days had been planned. The failed mission exposed a number of serious design flaws with the *Soyuz 7K-Y* spacecraft, namely the lack of reserve propellant and electrical power for repeated docking attempts. The *Igla* docking system needed major improvement. The backup spacecraft for this mission was placed in storage, but later flown as *Soyuz 20* despite being past its intended shelf life.

Sarafanov logged two days, zero hours and 12 minutes in space, resigned from the space program in 1986 and then lectured on technology. Gennady Sarafanov was awarded Hero of the Soviet Union; Pilot-Cosmonaut of the USSR; the Jubilee Medal "In Commemoration of the One Hundredth Anniversary since the Birth of Vladimir Il'ich Lenin"; the Medal "For Distinction in Guarding the State Border of the USSR; and the Medal "For the Development of Virgin Lands". He died on September 29, 2005, aged sixty-three years, in Moscow, Russia. He was married, with two children.

## Savinykh, Viktor P. Soyuz T-4, T-13, TM-5.

Viktor Petrovich Savinykh was born in March 1940 in Beryozkiny, Orichevsky District, Kirov Oblast, USSR. He is a former Soviet cosmonaut, scientist and organizer of personnel training in the higher education system. He was selected as a cosmonaut in December 1978 and flew as Flight Engineer on *Soyuz T-4, Soyuz T-13,* and *Soyuz TM-5* in 1988. In 1981, Savinykh spent seventy-four days (about two and a half months) aboard the *Salyut 6* space station. During his second mission on *Salyut 7* in 1985 Savinykh made several spacewalks to restore power to the station. In 1988 he spent ten days aboard the *Mir* space station. Savinykh logged a total of 252 days, 17 hours and 38 minutes in space (about eight and a half months). He retired on February 09, 1989. From 1989 to 2007 he was the rector, and since

2007 the president of the Moscow State University of Geodesy and Cartography. Savinykh is the author of the book *Notes from a Dead Station,* ISBN 5-901135-01-6. In March 2011, he was elected to the **Legislative Assembly of Kirov Oblast** of the fifth convocation in a single constituency from the United Russia Party.

Viktor Savinykh was twice awarded Hero of the Soviet Union; three Orders of Lenin; the State Prize of the Russian Federation for science and technology (1996); the Award of the President of the Russian Federation in the field of education for 2002; a Doctor of Technical Sciences, Professor, Academician of the Russian Academy of Science (2019); Full Cavalier of the Order For Merit to the Fatherland, First Class (2020); and has sixteen other honors. The minor planet 6890 was named after Viktor Savinykh in 2005.

## Savitskaya, Svetlana Y. Soyuz T-7, Soyuz T-12.

Svetlana Yevgenyevna Savitskaya (1948-) is a Russian former aviator and Soviet cosmonaut who flew aboard *Soyuz T-7* in August 1982; the second woman to travel to space. She became the first woman to fly in space twice, and the first woman to perform a spacewalk during her July 1984 *Soyuz T-12* mission. Savitskaya was born into a privileged family; her father, Yevgeny Savitsky, was a highly decorated fighter pilot during WWII, and later became the Deputy Commander-in-Chief of Soviet Air Defense. Her mother was a Moscow Communist party leader. Svetlana Savitskaya began parachuting at the age of sixteen years without her parents' knowledge and had already completed 450 jumps by her seventeenth birthday!

After graduating in 1966, she enrolled in the Moscow Aviation Institute (MAI), where she also took flight lessons. In 1971 she was licensed as a flight instructor. In May 1978, Svetlana went to work for the aircraft manufacturer **Yakolev**, as a test pilot. She became the first woman to reach 2,683 km/hr (about 1,667 miles/hr) in a MiG-25 aircraft. An experienced and highly educated female in the Soviet Space program, Savitskaya was reportedly an extremely serious, unbending and steely woman. While she and Valentina Tereshkova were both chosen for missions into space due to Soviet propaganda purposes, Savitskaya was much more trained and experienced in aeronautics, whereas Ms Valentina Tereshkova was chosen as a political stunt.

In 1979, Savitskaya participated in the selection process for the second group of female cosmonauts; she was admitted on June 30, 1980. Of the nine women selected, she was the only test pilot. *Soyuz T-7* launched on August 19, 1982. This made her the second woman in space, nineteen years after Valentina Tereshkova. Arriving at the *Salyut 7* space station the next day, this was the first time the station had a mixed gender crew. Her mission lasted seven days, 21 hours and 52 minutes. In a 1995 interview, she recalled encountering some sexism from her male crewmates. Svetlana stated that *"I was quickly able to establish a working, professional relationship with them"*. For her second mission, Savitskaya launched aboard *Soyuz T-12* on July 17, 1984, callsign *Pamir*. On July 25, she became the first woman to spacewalk, with a three-hour and 35-minute EVA outside *Salyut 7*. Along with cosmonaut Vladimir Dzhanibekov, she was the first to cut and weld metals in space. She performed six cuts of titanium and stainless steel, two coatings of anodized aluminum, six tests of tin and lead solder, and test cuts of a 0.5 mm titanium sample. This mission lasted eleven days, 19 hours and 14 minutes.

A third mission was proposed with an all-female crew in commemoration of International Women's Day, but due to several factors it was postponed. Eventually, the plan was not pursued due to Savitskaya's pregnancy and birth of her son, Konstantin, in October 1986. In February 1986, she graduated from the Bauman Moscow Higher Technical School. A committed communist, she was selected as a People's Deputy of the USSR from 1989, and a People's Deputy of Russia in 1990, a position she held until 1992. Savitskaya retired in 1993 from the Russian Air Force with the rank of Major. In 1994/95 she worked as an assistant professor in Economics and Investment at the Moscow State Aviation Institute. In 1996, she was elected as a **Deputy of the State Duma** representing the **Communist Party of the Russian Federation** and has been re-elected four times since then. She presently serves as Deputy Chair of the Committee on Defence, and is also a member of the Coordination Council Presidium of the **National Patriotic Union**

## Serebrov, Aleksandr A. Soyuz T-7/5, T-8, TM-8, TM-17.

Aleksandr Aleksndrovich Serebrov (1944-2013) was born in February 1944 in Moscow, Russian SFSR, Soviet Union. He graduated from the Moscow Institute of Physics and Technology in 1976; joined the design bureau **NPO Energia**, and was selected as a cosmonaut on December 01, 1978, in NPOE-4. He was a Flight Engineer on all four missions; flying twice to *Salyut 7* and twice to space station *Mir. Soyuz T-7* (August 19-28, 1982) was the first mission to include men and a woman, Svetlana Savitskaya, on the same crew. He returned to Earth on *Soyuz T-5* after seven days, 21 hours and 52 minutes in space. *Soyuz T-8* (April 20-22, 1983) was a mission to *Salyut 7*. The shroud protecting the *Soyuz* separated as planned, but tore with it a critical radar antenna boom needed for an autonomous docking. The crew attempted to fly a manual approach to the *Salyut*, but the mission was aborted to avoid a crash; returning to Earth after two days and 17 minutes in space. He launched aboard *Soyuz TM-8* for the 'Mir-5' mission (September 05, 1989 to February 19, 1990); returning to Earth after 166 days, six hours and 58 minutes; including five EVAs. He demonstrated the *Ikar* ("Ikarus"), that allowed the cosmonaut to fly around the station, similar to the NASA *Manned Maneuvering Unit* (MMU).

On *Soyuz TM-17* to *Mir* (July 01, 1993, to January 14, 1994) he performed five more EVAs; returning to Earth after 196 days, 17 hours and 45 minutes. His Commander, Vasili Tsibiyev, lost control of the *Soyuz* after undocking on January 14, colliding twice with the *Mir* space station, but causing minimal damage. Serebrov logged a total of 372 days, 22 hours and 52 minutes in space, and retired on May 10, 1995, to work as an advisor to then-President Boris Yeltsin on issues related to spaceflight. He once held the **Guiness World Record** (for four years) for the most spacewalks, and was the first to test drive a "space motorcycle", which was more like a floating armchair. He was awarded Hero of the Soviet Union; two Order of Lenin; and the Order of Friendship. Serebrov died suddenly at home, aged sixty-nine years on November 12, 2013, in Moscow, Russia. He was married, with one child.

## Serova, Yelena O. Soyuz TMA-14M/Expedition 41/42.

Yelena Olegovna Serova (nee Kuznetsova) was born in April 1976 in Vozdvizhenka village, Ussuriysk District, Primorsk Krai, in the far east of the Russian SFSR, Soviet Union. She is an engineer, politician and former cosmonaut. In March 2001, she graduated from the Aerospace Faculty of the Moscow Aviation Institute as an Engineer. In 2003, she graduated from the Moscow State Academy of Instrument Engineering and Information qualified as an Economist. Serova worked as an engineer of the second category for **RSC Energia**, and in the Mission Control Center. Selected in 2006 RKKE-16 Group at age thirty years, she completed basic cosmonaut training at Star City in 2009, then flew as Flight Engineer on *Soyuz TMA-14M*, launched atop a *Soyuz-FG* booster for ISS 'Expedition 41/42' (September 25, 2014 to March 12, 2015).

After Tereshkova (1963), Savitskaya (1982 and 1984), and Kondakova (1994 and 1997), Serova is the fourth female cosmonaut to travel in space. She returned to Earth after 167 days, five hours and 42 minutes in space. Serova was one of five cosmonauts selected to raise the Russian flag at the 2014 Winter Olympics opening ceremony in Sochi, Russia.

Between 2016-21, she served as a Deputy in the State Duma of the Russian Federation. In July 2017, Serova was a vice chairperson at a meeting to discuss a Bill on the protection of animals. She retired from the space program on September 23, 2016. She is married to cosmonaut Mark Serov, with one daughter.

## Sevastyanov, Vitali I. Soyuz 9, Soyuz 18.

Vitali Ivanovich Sevastyanov (1935-2010) was born in July 1935 in Krasnouralsk, Russian SFSR, Soviet Union. He was an engineer and Soviet cosmonaut, selected in Civilian Specialist Group 3. He trained as an engineer at the Moscow Aviation Institute and after graduation in 1959, joined Sergey Korolev's design bureau, where he worked on the design of the *Vostok* spacecraft. He also lectured at the (later Gagarin) Cosmonaut Training Center, teaching the physics of spaceflight. His cosmonaut training began in 1967.

Commander Andrian Nikolayev and Flight Engineer Vitali Sevastyanov flew aboard *Soyuz 9* (June 01-19, 1970). They conducted observations of celestial bodies and practised astronavigation, made physiological and biomedical experiments upon themselves, and investigated the social implications of prolonged spaceflight. They used two-way TV links with their families, and watched some of the 1970 FIFA World Cup matches. To conserve attitude control gas, their *Soyuz* was placed in a spin-stabilized mode that made them both dizzy and space sick. During their mission, the crew played the first documented chess game while in orbit. Their ground-based opponents were Nikolai Kananin (Head of Cosmonauts) and Viktor Gorbatko. Upon landing on the Kazakh steppe, they required help exiting the descent module and were struggling to walk for two days. Sevastyanov's second spaceflight was aboard *Soyuz 18* (May 24-July 26, 1975) for a two-month stay aboard *Salyut 4*; the second and final mission to that space station. Commander Pyotr Klimuk and Sevastyanov set a new Soviet space endurance record of 63 days, 23 hours and 20 minutes.

They were the backup crew for the failed *Soyuz 18a* mission, carried out that mission's objectives and continued the work of the previous crew of *Soyuz 17*. The Oasis space garden was started, and attempts were made to grow plants, including onions. By July, the environmental control system was failing, and they increased their exercise period to over two hours per day. After 61 days docked, their *Soyuz* was activated on July 24, and they landed two days later some 56 km east of Arkalyk, Kazakhstan. Soyuz 18 rock in Antarctica is named after the mission. In 1971, he was backup Flight Engineer for *Soyuz 11*. Vitali Sevastyanov was pulled from active flight status in 1976. Sevastyanov worked in Ground Control for *Salyut 6* before returning to spacecraft design in the 1980s to work on the *Buran* project.

Sevastyanov was president of the Soviet Chess Federation from 1977-86 and from 1988-89. During the 1980s, he was the host of a popular television program on space exploration entitled ***Man, Earth, Universe***. In 1993, Sevastyanov left the space program and was elected to the State Duma in 1994. Sevastyanov, along with Aleksey Leonov, Rusty Schweickart and Georgy Grechko established the **Association of Space Explorers** in 1984. Memberhip is open to anyone who has flown in space. Vitaly Sevasyanov logged a total of 80 days, 16 hours and 18 minutes in space. He died in Moscow, Russia, on April 05, 2010, aged seventy-four years.

## Severin, Vladimir G..

Vladimir Gayevich Severin was born in November 1956 in Zhukovskiy, Moscow Oblast, Russia. He graduated from the Moscow Aviation Institute with an Engineering diploma in 1982, and worked at **NPO Zvezda**. Selected as a cosmonaut by GMVK in MAP Group on May 11, 1990, his training was from October 01, 1990 to March 06, 1992. He retired from the space program in 1995, because the RKKE didn't want to take him over as a cosmonaut. He is married, with one child.

## Shargin, Yuri G. Soyuz TMA-5/4.

Lt. Col. Yuri Georgitevich Shargin was born in March 1960 in Engels, Saratov Oblast, Russian SFSR, USSR, to a Jewish family. He graduated from the Military Engineering Academy for Aeronautics and Astronautics, Leningrad, in 1982. He is a Lieutenant Colonel in the Russian Space Forces. Selected on February 09, 1996, in RKKA Group, he flew on October 14, 2004, aboard *Soyuz TMA-5* as Flight Engineer to the ISS. After nearly ten days, he returned aboard *Soyuz TMA-4*, logging nine days, 21 hours and 30 minutes in space. Due to his late addition to the crew, and lack of information about his activities in orbit, some questioned the motives of his flight. Chief Flight Director, Vladimir Solovyov assured, *"We, on the ISS, are not involved in military matters"*. Shargin retired from the cosmonaut corps in August 2008. He is divorced, with two children.

## Sharipov, Salizhan S. STS-89, Soyuz TMA-5/Expn 10.

Salizhan Shakirovich Sharipov was born in August 1964 in Uzgen, Osh Oblast (Andijan region), Uzbek SSR, Soviet Union (now Uzbekistan). He is a Kyrgyz cosmonaut of Uzbek descent; and ethnic Tajik, Uzbek and Russian. He graduated from the Soviet Air Force Pilot School in 1987, then worked as a pilot-instructor, and logged over 950 hours flying time; including on MiG-21 and L-39 aircraft. Sharipov became a cosmonaut-candidate on May 11, 1990, in TsPk-11 and completed general space training, becoming a Test Cosmonaut on March 11, 1992. In 1994, he graduated from Moscow State University with a degree in Cartography. From August 1997 to January 1998, he passed training as a member of the STS-89 crew at the Johnson Space Centre, Texas, USA. He is a co-author and investigator for the Advanced Diagnostic Ultrasound in Microgravity project. Salizhan Sharipov was launched from the US aboard Space Shuttle *Endeavour* and served as 'Mission Specialist' on STS-89 (January 22-31, 1998), during which the crew transferred over 8,000 pounds of scientific equipment, logistical hardware and water from *Endeavour* to orbital station *Mir*.

Next, he was 'Flight Engineer' on 'Expedition 10' to the ISS from October 16, 2004 to April 21, 2005. Launched from Baikonur aboard *Soyuz TMA-5* on October 14, the *Soyuz* docked with the ISS two days later. Sharipov conducted several science and human life sciences experiments for the Russian Space Program. SRS was a low-temperature synthesis in space hoped to develop new materials. He also used stem cells for possible development of a vaccination for HIV/AIDS treatment. Sharipov made two EVAs totaling ten hours and 34 minutes. The first, on January 26, 2005, included the installation of a new work platform and test robot outside the *Zvezda* module. The second walk on March 28 fitted the final three antennas on the *Zvezda* Service Module for the resupply spacecraft ATV, and a Global Positioning System (GPS) for the ATV. After 192 days in orbit, he returned to Earth on April 24, 2005; landing 90 km north of the town of Arkalyk, Kazakhstan. Sharipov logged a total of 201 days, 14 hours and 49 minutes (over 950 hours) in space, and conducted two EVAs. He retired as a cosmonaut on July 18, 2008. Sharipov has received five awards, including Hero of the Kyrgyz Republic (1998); the Gold Star Medal of the Hero of the Kyrgyz Republic (1998); Russian Federation Air Force Medals; and a NASA Space Flight Medal (1998). Sharipov is married, with two children.

## Shatalov, Vladimir A. Soyuz 4, Soyuz 8, Soyuz 10.

Lt. Gen. Vladimir Aleksandrovich Shatalov (1927-2021) was born in December 1927 in Petropavlovsk, Akmolinsk Governate, Kazak ASSR, RFSFR, Soviet Union (now Petropavl, North Kazakhstan Region, Kazakhstan). In 1941 during WWII, he took part in the defense of Leningrad for about forty-five days, and helped to build the "Road of Life" across the frozen Lake Ladoga. In 1956, he graduated from the Command Faculty of the Air Force Academy. From May 1960, he was deputy commander of an aviation regiment in combat units of the

Soviet Air Force. When he enrolled in the cosmonaut corps he had logged more than 2,500 flight hours. Shatalov trained from January 1963 to January 23, 1965, when he was appointed Cosmonaut of the Second Detachment (Military Space Programs). Selected in 1963 Air Force Group 2, he qualified as a cosmonaut on January 23, 1965, and flew on *Soyuz 4*, launched January 14, 1969, the first to carry out manual rendezvous and docking with *Soyuz-5;* on *Soyuz 8*, launched October 13, 1969, a joint mission with *Soyuz 6* and '7' when rendezvous failed; and on *Soyuz 10*, launched April 22, 1971, as the world's first mission to the world's first space station, *Salyut 1*. However, the docking was unsuccessful, and the crew returned to Earth without entering the *Salyut*. This was the first of numerous docking failures in the Soviet space program. Vladimir Shatalov logged a total of nine days, 21 hours and 55 minutes in space.

From 1971-91, Shatalov was a member of the State Commission on Manned Space Flights. On April 28, 1972, he defended his dissertation at the Gagarin Academy, and received the degree of Candidate of Technical Sciences. In 1980, he was a consultant for the science fiction film ***Per Aspera Ad Astra***. From 1987-91, as Commander he headed the Yuri Gagarin Cosmonaut Training Center. He was received thirty awards, including Medal for Battle Merit (1952); the Medal "For Distinction in Guarding the State Border of the USSR"; the Jubilee Medal "Twenty Years of Victory in the Great Patriotic War 1941-45"; two Hero of the Soviet Union (January and October, 1969); three Order of Lenin (1969, 1971, 1976); the Order of Karl Marx (GDR, 1977); Hero of Labour (Vietnam, 1980); the USSR State Prize (1981); the Order of the October Revolution (1982); and the Order of the Polar Star (Mongolia, 1983). A crater on the Moon was named after him. He was an Honorary Citizen of the cities of Kaluga, Kurgan (1969); Nalchik (Russia); Karaganda, Petropavl (Kazakhstan); Prague (Czech Republic); and Houston, Texas (USA). Shatalov died on June 15, 2021, in Moscow, Russia, aged ninety-three years. He was married, with two children.

## Shipenko, Klim A. Soyuz MS-19/Expn 65/66/Soyuz MS-18.

Klim Alekseevich Shipenko was born in June 1983 in Moscow, Russia. In 2002, he entered the California State University at Northridge (Film Production Department). He worked as an intern on the set of the film ***Something's Gotta Give***. Shienko studied at the Sal Dano Professional actors workshop. He launched on October 05, 2021, aboard *Soyuz MS-19* for 'Expedition 65/66', which began in October 2021 and ended on March 30, 2022. His fellow launch cosmonauts were cosmonaut Anton Shkaplerov and fellow "spaceflight participant" Russian actress Yulia Peresild, who spent almost twelve days on the ISS while they filmed about 35-40 minutes of scenes for a feature movie ***Vyzov*** (The Challenger). The film drama was a joint project of Roscosmos, Channel One, and the Yellow, Black and White Studio. Scenes were filmed between the launch of *Soyuz MS-19* and the return of *Soyuz MS-18*. Cosmonaut Pyotr Dubrov and astronaut Mark Vande Hei helped with filming. Shipenko logged 11 days, 16 hours and 13 minutes in space.

The movie film received opposition from the scientific and aerospace committees, as to the fact that they remove trained cosmonauts from their flights, ***a misuse of public money***. Igor Krasnov, Procurator General of Russian went so far as **to open an investigation into whether the use of space station resources is illegal**! After the landing of *Soyuz MS-18*, Roscosmos head Dmitry Rogozin revealed that Konstantin Ernst (CEO of Channel One) paid for Shipenko and Peresild's seats to fly. Awards Shipenko has received include the Sixteenth Golden Eagle Award, for the best feature film ***Salyut 7*** (2018); the Eighteenth Golden Eagle Award for the best feature film ***Text*** (2020); and the State Prize of the Russian Federation in the field of literature and art in 2022 (2023) – for the creation of the feature film ***The Challenger***. The scenes filmed on the ISS were in a race with Tom Cruise and Doug Liman to shoot the first narrative *feature film* in space. The short film ***Apogee of Fear***, shot in 2008, was the first narrative film filmed fully in space.

## Shkaplerov, Anton N. Soyuz TMA-22, TMA-15M/Expn 42/43, MS-06/ Expn 54/55, MS-07, MS-19/Expn 65/66.

Col. Anton Nikolaevich Shkaplerov was born in February 1972 in Sevastopol, Ukrainian SSR. He is a retired Colonel in the Russian Air Force, and a cosmonaut with four long-duration missions to the ISS totaling over 709 days in space. After graduating from high school in Sevastopol in 1989, he enrolled at Chernigov Higher Military Aviation School, before transferring to Kachinsk Higher Military Pilot School in 1992. He graduated (with honors) in 1994; then studied at Joukowski Military Air Engineering Academy from 1994-97, graduating as a Research Engineer Pilot. From 1997, he served in the Russian Air Force combat units in the Kaluzhkoy region, then became a pilot instructor. He also served as an aviation regiment Squadron Commander in the Russian Air Force, flying the L-39 and MiG-29 aircraft.

He was selected as a cosmonaut candidate in March 2003, and had two years of basic spaceflight training, passing with excellent grades in 2005. He worked at the Johnson Space Center as the Director of Operations for the Russian Space Agency. He was assigned as backup crew for ISS 'Expedition 22' in mid-2008, and was transferred to the backup crew of 'Expedition 27' in 2009. He launched aboard *Soyuz TMA-22* atop a *Soyuz-FG* rocket on November 14, 2011. On February 16, 2012, he performed an EVA of six hours and 15 minutes. Shkaplerov landed in Kazakhstan on April 27, 2012, after a mission of 165 days, seven hours and 31 minutes. He was commander of 'Expedition 42/43', launching on November 23, 2014, reaching low Earth orbit in nine minutes, and making a "fast docking" to the ISS less than six hours after the launch. The *Soyuz TMA-15M* mission was extended by a month; lasting for 199 days, 16 hours and 42 minutes; and landing on June 11, 2015, just over three hours after departing the ISS. The *Soyuz* accidentally fired its engines one day before undocking, and the one minute burn slightly shifted the ISS. 'Expedition 54/55' saw *Soyuz MS-06* launched on September 13, 2017, and landed on February 28, 2018, after 168 days in orbit. 'Expedition 54' began with the departure of *Soyuz MS-05* on December 14, 2017, and concluded upon the departure of *Soyuz MS-06* on February 27, 2018.

Shkaplerov's third flight saw him launched on December 17, 2017 as Flight Engineer aboard *Soyuz MS-07* with NASA's Scott Tingle and Norishige Kanai of JAXA. He became the ISS Commander, along with crew Aleksandr Misurkin, Mark T. Vande Hei, and Joseph M. Acaba from 'Expedition 54'; the other 'Expedition 55' crew being Tingle and Kanai. Launched on October 05, 2021, as Commander of *Soyuz MS-19*, his most recent fourth mission was 'Expedition 65/66', which began in October 2021 and ended on March 30, 2022. His fellow launch cosmonauts were "spaceflight participants" Russian film director Klim Shipenko and Russian actress Yulia Peresild, who spent about twelve days on the ISS while filming about 35-40 minutes for a movie ***Vyzov*** (The Challenger). The film drama was a joint project of Roscosmos, Channel One, and the Yellow, Black and White Studio. According to Dmitry Rogozin, head of Roscosmos, the film is an *"experiment to see if Roscosmos can prepare two ordinary people to fly in about three or four months"*. Shkaplerov logged a total of 709 days, eight hours and four minutes in space; including three EVAs totaling 21 hours and 39 minutes. He is married, with two daughters.

## Shonin, Georgi S. Soyuz 6.

Georgi Stepanovich Shonin (1935-97) was born in August 1935 in Rovenky, Luhansk Oblast (now in Ukraine), former Soviet Union. During World War II, Shonin's family hid a Jewish family from the Nazis. He graduated from Yeisk Military Pilot School in February 1957;

and from Zhukovsky Military Engineering Academy, Monino, in 1968. He was posted to the Baltic Fleet, and in 1958 transferred to the Northern Fleet, based in the Murmansk region; befriending another young flying officer named Yuri Gagarin.

Selected in Air Force Group TsPK-1, the original group of cosmonauts, he flew as Commander on *Soyuz 6* from October 10-16, 1969; logging four days, 22 hours and 42 minutes in space. This was part of a joint mission with *Soyuz 7* and '8'. The crew of Shonin and Valeri Kubasov were scheduled to take high-quality movie photography of the *Soyuz 7* & '8' docking, but the rendezvous systems on all three spacecraft failed! The cause is often quoted as being a helium pressurization integrity test. Due to unstable temperature, disparity occurred between the frequencies of the transmitters and receivers, which were stabilized by special quartz resonators. The piezo crystals were supposed to be in thermostats at a strictly constant temperature. Shonin and Kubasov carried out experiments in space welding, testing three methods: using an electron beam; a low-pressure plasma arc; and a consumable electrode. Kubasov almost accidentally burned through the hull of their crafts' living compartment, which in the absence of spacesuits would have proved catastrophic. They landed on October 16, 1969, one hundred and eighty kilometres northwest of Karaganda, Kazakhstan.

Shonin retired on April 28, 1979, for medical reasons. He was promoted to Major General, and later worked as Director of the Thirtieth Central Scientific Research Institute, Ministry of Defense (Russia), where he had management responsibilities for the development of the *Buran* Soviet space shuttle. He received ten honors, including Hero of the Soviet Union; the Order of Lenin; the Order of the October Revolution; the Order of the Red Banner of Labour; ten commemorative medals; three medals from the Mongolian People's Republic; and five medals from the Czechoslovak Socialist Republic. Shonin died of a heart attack in Zvyozdny Gorodok, Russian Federation, on April 07, 1997, aged sixty-one years. He was married, with four children.

## Skripochka, Oleg I. Soyuz TMA-01M/Expn 25/26, TMA-20M/ Expn 48, Soyuz MS-15/Expn 61/62.

Oleg Ivanovich Skripochka was born in December 1969 in Nevinnomyssk, Stavropol Krai, Russian SFSR, Soviet Union. He is a Russian engineer and cosmonaut. After graduating from the Bauman Moscow State Technical University in 1993 with a Diploma of Mechanical Engineer in rocket construction, he worked from August 1993 to August 1997 as an engineer in the **RSC Energia** project bureau on the development of transport and cargo vehicles.

Launched on October 07, 2010, he was 'Flight Engineer 1' aboard *Soyuz TMA-01M* for ISS 'Expedition 25/26'. He made three EVAs: on November 15, 2010, to remove Kontur and Expose-R experiments; to install a portable multipurpose workstation on the *Zvezda* service module, and handrail extensions between the *Poisk* module and both *Zvezda* and *Zarya* modules; on January 21, 2011 (to complete installation of a new high-speed data transmission system, and remove the plasma pulse generator); and on February 16, 2011 (to install a radio antenna, deploy a nano satellite, install two experiments and retrieve two exposure panels on a third experiment). After spending 159 days in space, he returned to Earth on March 16, 2011, with the capsule landing on its side near the township of Arkalyk in north central Kazakhstan. The weather was harsh, with high winds and viciously cold temperatures. At one point in the live NASA landing commentary, a spokesman said, *"You would think that was a scene out of the North Pole"*, to explain the extreme climatic conditions.

Launched on March 19, 2016, aboard *Soyuz TMA-20M*, he was 'Flight Engineer 4' for ISS 'Expedition 48', which ran from March 02 to June 18, 2016. Launched on April 08, the SpaceX *CRS-8* mission carried the Bigelow Expandable Activity Module (BEAM) to the ISS for two years of in-orbit habitat qualification. Launched on September 25, 2019, aboard *Soyuz MS-15* for ISS 'Expedition 61/62', he was Commander of the ISS from February 06 to April 09, 2020, and returned to Earth on April 17, 2020. Skripochka logged a total of 536 days, three hours and 49 minutes in space; including three EVAs during 'Expedition 26' totaling 16 hours and 40 minutes. On April 12, 2011, he was awarded Hero of the Russian Federation; and Pilot-Cosmonaut of the Russian Federation "for courage and heroism in the implementation of long-duration space flight on the International Space Station". He retired on December 01, 2021.

## Skvortsov, Aleksandr A. Soyuz TMA-18/Expn 23/24, TMA-12M/ Expn 39/40, MS-13/Expn 60/61.

Aleksandr Aleksandrovich Skvortsov was born in May 1966 in Schelkovo, Moscow Oblast, Russian SFSR, Soviet Union. He graduated from the Stavropol Air Force Pilot and Navigator school as pilot-engineer in 1987; and from the Military Red Banner Air Defense Academy in 1997. Svortsov flew L-39, MiG-23 and Su-27 aircraft, and as a Class 1 Air Force pilot has logged around 1,000 hours flight time. Selected in 1997 TsPK cosmonaut group, he qualified as a Test cosmonaut in November 1999. He is a veteran of three long-duration spaceflights aboard the ISS. Currently he is working towards a Law degree at the Russian Academy of Civil Service. Launched on April 02, 2010, aboard *Soyuz TMA-18*, Skvortsov served as a Flight Engineer for the ISS 'Expedition 23' to September 24 2010; celebrating his forty-fourth birthday in space on May 06. He was Commander for 'Expedition 24'. He landed on September 25 near Arkalyk, Kazakhstan. Sharing his experiences with an audience, Skvortsov said, *"I love spicy food. I make really hot chili sauce myself. I missed it badly [while in space]. I finished off all the spicy ketchup stock on the ISS and shocked everyone by making and eating wasabi sandwiches. That's my fuel, as good as a rocket's"*. Unquote.

He flew as a Flight Engineer aboard *Soyuz TMA-12M*, along with Oleg Artemyev and Steven Swanson on the 'Expedition 39' crew. They launched on March 25 and landed on September 11, 2014. For 'Expedition 60/61' he launched aboard *Soyuz MS-13* on July 20, 2019, arriving at the ISS the same day. The 100-day 'Expedition 60' mission ended on October 03, 2019, with the departure of *Soyuz MS-12*. Skvortsov's 'Expedition 61' was from July 20 to October 03, 2019. He logged a total of 545 days, 23 hours and eight minutes in space. Skvortsov has received six honors, including Hero of the Russian Federation; Pilot-Cosmonaut of the Russian Federation; the Jubilee Medal "70 Years of the Armed Forces of the USSR"; Russian Armed Forces Meritorious Service Medal of the first, second and third degree; and the Military Superior Service Medal, second degree.

## Solovyev, Anatoly Y. Soyuz TM-5/4, TM-9, TM-15, STS-71/Soyuz TM-21, Soyuz TM-26.

Anatoly Yakovlevich Solovyev (or Solovyov) was born in January 1948 in Riga, Latvia (at that time a Republic of the Soviet Union). He holds the world record for the number of spacewalks performed: sixteen. Solovyev completed two years' of evening school and in 1967 he enrolled at the Faculty of Physics and Mathematics of the State University of Latvia. He enrolled as a cadet in the Chernigov Higher Military School of Pilots in 1968. He joined the CPSU in 1971; and graduated from the Lenin Komsomol Chernigov Higher Military Aviation School in 1972, having flown the L-29, MiG-15 UTI and MiG-21. In 1975, he was awarded "Military Pilot, First Class".

Selected for cosmonaut training on August 23, 1976, as a member of TsPK-6, this group was uniquely chosen for the *Buran* "Soviet Space Shuttle" program, and as such had to also qualify as test pilots. In January 1979, he was appointed as a Test Cosmonaut on OK CPC (aerospace vehicles) working on *Buran*. Delays with *Buran* and increased need for space station crews resulted in his transferring to training for the Interkosmos program in January 1982. In 1985, he was assigned as reserve crew commander of *Soyuz T-15*. Next year, Solovyev was assigned as backup Commander of the *Soyuz TM-3* mission to *Mir* with Viktor Savinykh and Munir Habib (of Syria). He was Commander for the nine-day 'Mir EP-2' visiting crew launched aboard *Soyuz TM-5* in June 1998, and returned aboard *Soyuz TM-4*. He became the sixty-fifth Soviet cosmonaut, with the personal callsign *Spring*.

Solovyev was Commander for the 179-day *Soyuz TM-9* 'Mir EO-6' long duration mission of February to August 1990. This included an EVA to repair thermal blankets on their *Soyuz*, and a second EVA when repairs were unsuccessfully attempted on the *Kvant-2* module hatch. He was Commander for the 188-day *Soyuz TM-15* 'Mir EO-12' mission with Sergei Avdeyev; from July 1992 until February 1993. Solovyev made four more EVAs. He was Commander of 'Mir EO-19' with Nikolai Budarin. They launched aboard Space shuttle *Atlantis* on STS-71 in June 1995. This was the first Orbiter docking with *Mir*; and Solovyev performed three EVAs. After 75 days, he returned to Earth aboard *Soyuz TM-21* in September 1995. His final mission was as Commander of *Soyuz TM-26* for 'Mir EO-24' for 197 days from August 1997 until February 1998 with Pavel Vinogradov. Solovyev performed the final docking manually when he was unable to verify the accuracy of the *Kurs* automatic targeting. His EVA on August 22 was to connect power and survey damage to the depressurized *Spektr* module, holed in the collision with *Progress M-34* on June 25, 1997.

He performed six more EVAs to make a world record of sixteen; with a world record combined EVA duration of 82 hours and 21 minutes. Solovyev commented, *"It just came about. I never referred to my numbers as a record, nor did I ever think about beating someone else's accomplishments before going out on a [space] walk"*. Solovyev declined to serve under a less-experienced American commander on the first mission to occupy the ISS. He left the cosmonaut detachment in 1999, having reached the compulsory retirement age, and became the president of "For the Good of the Fatherland", a national organization recognizing the work of Russians devoted to cultural and social development. Solovyev received seven Soviet and five foreign awards, including Hero of the Soviet Union; the Order For Merit to the Fatherland, second and third classes; the Order of Lenin; the Order of the October Revolution (USSR); the Order of Friendship of Peoples (USSR); Hero of the People's Republic of Bulgaria; Officer of the Legion of Honour (France); and the Order of Stara Platina (Republic of Bulgaria). He is married, with two sons.

## Solovyov, Vladimir A. Soyuz T-10/11, Soyuz T-15.

Vladimir Alekseyevich Solovyov was born in November 1946 in Moscow, Russian SFSR, USSR. He is a retired former Soviet and Russian cosmonaut. Selected as a cosmonaut on December 01, 1978, he flew as Flight Engineer on two Soyuz missions. Launched on February 08, 1984, *Soyuz T-10* joined the *Salyut 7* space station, and the crew spent ten months (nearly 327 days) performing numerous medical and space manufacturing experiments. He returned aboard *Soyuz T-11* on October 02, 1984. Launched aboard *Soyuz T-15* on March 13, 1986, he returned to Earth 125 days later on July 16. During this misssion, the crew transferred equipment from *Salyut 7* to the new *Mir* space station. They were the last cosmonauts aboard the *Salyut*, and the first aboard *Mir*. He logged a total of 361 days, 22 hours and 49 minutes in space. For the next several years, Solovyov became the station *Mir* Flight Director (Russian Mission Control). He received eight awards, including two Hero of the Soviet Union;

the Order of Honour (Russian Federation); two Orders of Lenin; the Medal "For Merit in Space Exploration" (Russian Federation); Knight of the Legion of Honour (France); and Kirti Chakra (India). He retired on February 18, 1994, but came back to head the Russian segment of the International Space Station (ISS). He is married, with two children.

## Strekalov, Gennadi M. Soyuz T-3, T-8, T-11/10, TM-10, TM-21.

Gennady Mikhailovich Strekalov (1940-2004) was born in October 1940 in Mytishchi, USSR. His father was killed in 1945 while fighting for the Red Army in Poland. After leaving school, Strekalov began work as an apprentice coppersmith at OKB-1, Sergei Korolev's experimental design bureau, where he helped assemble *Sputnik 1*. Gennadi graduated from N.E. Bauman Moscow Higher Technical School in 1965 with an engineering diploma. In January 1974, Strekalov began training as a crew member for a mission as a Flight Engineer, and, in 1976, was part of the backup crew for the *Soyuz 22* mission. From October 1978 he trained as a Flight Engineer for *Soyuz* expeditions to *Salyut*-series space stations. His first flight was in 1980, as Research Engineer on the *Soyuz T-3* mission to the *Salyut 6* station.

His next flight was to be to the new *Salyut* 7 station. Launched in April 1983 on *Soyuz T-8*, part of the *Igla* rendezvous radar system was damaged. The crew unsuccessfully attempted a manual docking but used up too much propellant and had to abort the mission. Strekalov and Vladimir Titov were re-scheduled to fly to *Salyut 7* on September 26, 1983. Their *Soyuz* booster developed a serious fuel leak in the minutes before launching. The launch escape system finally operated only twenty seconds before the rocket exploded, devastating Baikonur's Site 1 launchpad. Their capsule was pulled away to safety, landing four kilometers from the pad. They were subjected to heavy acceleration of more than 10-Gs, but narrowly escaped with bruises. The catastrophic explosion led to him being one of only four people to ever use a launch escape system. Strekalov's next (his fourth) spaceflight was aboard *Soyuz T-11* in April 1984 with Yuri Malyshev and Indian cosmonaut Rakesh Sharma. They docked with *Salyut 7* and later returned in the re-entry module of *Soyuz T-10*. From August 01 to December 10, 1990, Strekalov was the Flight Engineer on *Soyuz TM-10*'s flight to *Mir*, with Gennady Manakov and Japanese TV reporter-cosmonaut Toyohiro Akiyama. After this, Strekalov formally retired and became head of the civilian section of the cosmonaut department.

He returned to flight status for the *Shuttle-Mir* program and on March 14, 1995, he flew aboard *Soyuz TM-21* to the *Mir* space station with Vladimir Dezhurov and American astronaut Norman E. Thagard. The crew returned on the Space Shuttle *Atlantis*, which had brought up their relief crew. The mission lasted 110 days, but Strekalov was fined approximately USD 10,000 in pension and benefit entitlements because he had refused to do an unplanned spacewalk to repair a stuck solar array. He believed the proposed EVA to be too dangerous and had argued for several days with Russian ground controllers. Strekalov challenged ***Energia***, and the fine was later overturned. Strekalov lived aboard the *Salyut 6, Salyut 7* and *Mir* space stations, spending a total of 268 days, 22 hours and 22 minutes in space (about nine months). After his spaceflights he returned to Russian aerospace firm OKB-1 (which was later renamed **RSC Energia**) as an administrator and worked there for the rest of his life. Gennady Strekalov was twice decorated as a Hero of the Soviet Union; received three Orders of Lenin; the Ashoka Chakra, Class 1 (India); and seven other awards. He died from cancer in Moscow, Russia, in December 2004. He married Lydia Telezhina and they had two daughters.

## Surayev, Maksim V. Soyuz TMA-16/Expn 22, TMA-13M/Expn 40/41.

Maksim Viktorovich Surayev was born in May 1972 in Chelyabinsk, in the Urals, USSR. He graduated (with honors) from the Kacha Air Force Pilot School as a pilot-engineer in 1994; and graduated (with honors) from the Zhukovski Air Force Academy in 1997 as pilot-engineer-researcher. A Class 3 Air Force pilot, he is qualified to fly L-39 and Su-27 aircraft, and has logged around 700 hours flight time. From December 1997 to November 1999, he completed basic Cosmonaut training. From March 2006 to April 2008, he was assigned as a backup to ISS 'Expedition 17' crew. From April 2008 until March 2009, he was a member of the ISS 'Expedition 19' backup crew.

Surayev flew as Commander aboard *Soyuz TMA-16* on September 30, 2009, to join ISS 'Expedition 22' as a Flight Engineer. After 167 days, he and Jeffrey Williams landed on the steppes of Kazakhstan on March 18, 2010. Surayev served as a Flight Engineer on ISS 'Expedition 40/41' crew. Launched on May 28, 2014, *Soyuz TMA-13M* docked with the ISS's *Rassvet* module about six hours after launch. He conducted scientific work, while the ISS was visited by the Russian *Progress* freighter, the American *Dragon* and *Cygnus* spacecraft, and the final mission of the European ATV – Georges Lemaitre ATV. He made EVAs on January 14, 2010, and October 22, 2014; totaling nine hours and 22 minutes. Surayev served as ISS Commander of 'Expedition 41' from September 10 to November 10, 2014, when he returned to Earth that day. In 2007, he received a Law degree from the Russian Academy of Civil Service. On December 30, 2010, Surayev received the Gold Star Medal of Hero of Russia, presented by Russian president Dmitry Medvedev. He served in the State Duma with the United Russia Party from 2016-21. He is married, with two daughters.

## Tarelkin, Evgeni I. Soyuz TMA-06M/Expn 33/34.

Capt. Evgeni Igorevich Tarelkin was born in December 1974 in Chita Oblast, Russian SFSR. He graduated from the Yeysk Air Force School in 1996; and the Gagarin Air Force Academy in 1998, before rising to the rank of Captain in the Russian Air Force. He served in the Air Force until selected as part of the TsPK-13 group in 2003. Launched on October 23, 2012, as Flight Engineer 1 aboard *Soyuz TMA-06M* atop a *Soyuz-FG* rocket, he spent six months aboard the ISS as part of 'Expedition 33/34'. This was the one hundred and fifteenth flight of a *Soyuz* spacecraft, and this mission was the first crewed flight from the remote Site 31 (the Tereshkova pad) since July 1984. On the docking day, the Automated Rendezvous Sequence was initated, and the *Soyuz* crew activated the *Kurs* Navigation System that provides accurate range and velocity data for the spacecraft's on-board computers. At 300 meters distance, Mission Control in Koroloev, Russia, gave a GO for the Flyaround. Once that was completed, station-keeping operations began at a distance of 190 meters from the ISS. This period was used to verify all systems and alignment with the space station.

*Soyuz TMA-06M* docked at the MRM-2 *Poisk* module, as the ISS flew above southern Ukraine. The standard one-hour leak and system check operation was done before the hatch opening procedures began. The *Soyuz* carried 32 Medaka fish used to test how conditions in space impact on living organisms. A fish tank inside the Japanese *Kibo* Lab Module has the Aquatic Habitat (AQH). The *Soyuz* undocked from the ISS on March 15, 2013, at 23:43 GMT, and landed at 03:06:30 GMT the following day; returning to Earth on March 16, 2013, and landing north of the town of Arkalyk. Tarelkin logged a total of 143 days, 16 hours and 15 minutes in space. In 2019, Tarelkin was Crew Commander in the SIRIUS-19 ground-based experiment.

## Tereshkova, Valentina V. Vostok 6.

Maj. Gen. Valentina Vladimorovna Tereshkova was born in March 1937 in the Bolshoye Maslennikovo, a village on the Volga River in Yaroslavi Oblast, 270 km (170 miles) northeast of Moscow, USSR. She is a Russian engineer, former Soviet cosmonaut, and current member of the **State Duma**. She came from a poor family and was first enrolled in school at the age of ten years and graduated at the age of seventeen years. Valentina began working at a factory and later in a textile mill. She continued her education by taking various correspondence courses and graduated from the Light Industry Technical School in 1960. Although she had no pilot training before joining the cosmonaut program, she was an avid parachutist and that skill helped her to be selected from four other women candidates to become the first woman in space.

Launched on June 16, 1963, Tereshkova spent three days in orbit aboard *Vostok 6*, completing forty-eight revolutions around the Earth in seventy-one hours. She received many honors and awards, including the Hero of the Soviet Union, and the Order of Lenin. She became a spokesperson for the Soviet Union and toured the world as a goodwill ambassador, promoting the equality of the sexes in the USSR. On November 03, 1963, she married Andriyan G. Nikolayev, another cosmonaut, but they divorced in 1982. Tereshkova went on to graduate from the **Zhuykosky Air Force Engineering Academy** in 1969 and earned a degree in Technical Science in 1976. From 1966 to 1991 Tereshkova was an active member in the **USSR Supreme Soviet**. She directed the Soviet Women's Committee in 1968 and promoted women's rights and gender equality in the Soviet Union. The committee was responsible for organizing women's conferences, publishing literature on women's issues, and advocating for women's rights in the government and society.

From 1974 to 1991 she served as a member of the **Supreme Soviet Presidium**. Tereshkova retired from the Russian Air force in 1997 with the rank of Major General. In 2008 she became the deputy chair of the parliament of Yaroslavl oblast as a member of the United Russia Party. In 2011, she was elected to the **State Duma**, and as of 2023 she still serves in that body. In 2020, she became the subject of political controversy when she introduced in the Duma the constitutional amendment that removed term limits for President Vladimir Putin. In 2022, Tereshkova faced international sanctions because of voting for the February 24, 2022, Russian invasion of Ukraine. Tereshkova's pioneering career, along with the feats of Sally Ride and Mae Jemison (the first African American woman to become an astronaut) helped pave the way for future ventures in space exploration.

## Titov, Gherman S. Vostok 2.

Gherman Stepanovich Titov (1935-2000) was born in September 1935 in Verkhneye Zhilono, near Barnaul, Russia. On August 06, 1961, the youngest space traveler of the era, 25-year and 11-month-old Gherman S. Titov made seventeen Earth orbits aboard *Vostok 2* for one day, one hour and eighteen minutes in space. Titov's call sign was *Eagle* (meaning "Open" in Russian), and he was the first to suffer from space sickness. He vomited, which would have been a very unpleasant experience in a weightless environment! Titov was the second human to orbit the Earth, and the first person to sleep in space. He slept for roughly one orbit and was surprised to wake with his arms floating in the air because of the absence of normal gravity. He stated, *"Once you have your arms and legs arranged properly, space sleep is fine ... I slept like a baby"*. Titov made the first manual photographs from orbit. He filmed Earth using a professional quality Konvas-Avtomat movie camera for ten minutes. Gherman Titov celebrated upon landing. His behaviour was described as "a fit of euphoria", and on his return flight from Kubishev for debriefing, he alarmed the medical staff by opening and downing a beer, in complete violation of the rules. He then indulged in some "fast living" and was tested to see if his behavior was the result of space sickness!

During a widely publicized 1962 visit to Seattle to visit the Century 21 World's Fair, Titov was asked by reporters how his space flight affected his philosophy of life. He replied: "***Sometimes people are saying that God is out there. I was looking around attentively all day, but I didn't find anybody there. I saw neither angels nor God***". This utterance was used in Soviet anti-religion propaganda, but it was mis-attributed to fellow cosmonaut Yuri Gagarin. Cosmonaut G.S. Titov was awarded the title, Hero of the Soviet Union on August 09, 1961. He also received two Orders of Lenin (on June 17 and August 09, 1961), and numerous medals and foreign Orders. He was a fine sportsman, and keen on gymnastics. Titov later played a major role in the Spiral Project, where he trained to become the first pilot of an orbital spaceplane. After the death of Yuri Gagarin, the Soviet government decided it could not afford to lose its second cosmonaut, so Titov's career as test pilot ended. On August 15, 1991, Titov was made an Honored Specialist of the Armed Forces of the USSR – for outstanding service to the Soviet state in the country's defense capacity and professional skill. In 1995, Titov was elected to the State Duma as a member of the **Communist Party of the Russian Federation**. On September 20, 2000, he died of cardiac arrest in his sauna at the age of sixty-five years in Moscow, Russia.

## Titov, Vladimir G. Soyuz T-8, T-10a, TM-4/Mir EO-3, STS-63, STS-86.

Col. Vladimir Georgiyevich Titov was born in January 1947 in Stretensk, Zabaykalsky Krai, Russia. He is a retired Russian Air Force Colonel and veteran of four spaceflights. The catastrophic explosion of a Soyuz rocket in 1983 led to him being one of only four people to use a launch escape system. He graduated from the Higher Air Force College in Chernihiv, Ukraine, in 1970; and from the Gagarin Air Force Academy in 1987. He flew ten aircraft types, and logged over 1,400 hours flying time. He is a Military Pilot, First Class, and a Test Pilot, Third Class. Selected as a cosmonaut in 1976, in September 1981 he was paired with Gennady Strekalov as backup crew for *Soyuz T-5* in 1982 and *Soyuz T-9* in 1983. Titov served as Commander on *Soyuz T-8* launched on April 20, 1983; and on *Soyuz T-10a* (experiencing the launch abort, twelve seconds after a large fire started at the base of the launch vehicle) on September 27, 1983. The *Soyuz* descent module was pulled clear by the launch escape system. After being subjected to a massive 15 to 17 G-s, the crew landed safely some four kilometres from their launch vehicle on the pad, which had exploded mere seconds after the *Soyuz* had separated. Their brief "flight" lasted five minutes and 30 seconds.

Titov was Commander on *Soyuz TM-4* launched on December 21, 1987, for mission 'Mir EO-3'. With cosmonauts Musa Manarov and Anatoli Levchenko, he linked up with the *Mir 1* space station and her crew. On October 28, 1992, NASA announced that an experienced cosmonaut would fly aboard the STS-60 mission. Titov was one of two candidates named by the Russian Space Agency for Mission Specialist training at the Johnson Space Center, Texas, USA. In April 1993, he was named as backup for Sergei Krikalev, who flew on STS-60. In September 1993, Titov was selected to fly on STS-63 aboard *Discovery* (February 02-11, 1995), the first flight of the new joint Russian-American Space Program. He became the second Russian to fly on an American Space Shuttle. Mission highlights included the rendezvous with *Mir*, operation of 'Spacehab', and the deployment and retrieval of the *SPARTAN-204* satellite.

He flew on STS-86 aboard *Atlantis* (September 25-October 06, 1997), NASA's seventh mission to rendezvous and dock with *Mir*. Two crew were exchanged; 10,400 pounds of science and logistics cargo were transferred to *Mir*; and experiment hardware and results were returned to Earth. Vladimir Titov logged a total of 387 days, zero hours and 52 minutes in space; including 18 hours and 48 minutes of EVA. He received 11 awards, including Hero of the Soviet Union; two Orders of Lenin (1983, 1988); the Order of the Red Star; Commander of the Legion of Honour (France); the Order of Stara Planina (People's Republic of Bulgaria); and the Order "The Sun of Liberty" (Afghanistan). In 1990, Titov and Musa Manarov were awarded the US Harmon Prize – the first Soviet citizens to win the award – in recognition of their world space endurance record. Titov retired from the Air Force and the Russian Space Agency in 1998, then became Director of Space and Communications for **The Boeing Company** in Moscow, Russia, and for the Commonwealth of Independent States (CIS), established after the fall of the Soviet Union. He is currently serving as General Director, Russian Operations of **Excalibur Almaz**. He is married, with two children.

## Tokarev, Valeri I. STS-96, Soyuz TMA-7/Expn 12.

Col. Valeri Ivanovich Tokarev was born in October 1952 in Kapustin Yar, Astrakan Oblast, Soviet Union. He is a Russian Air Force Colonel, and veteran of two space flights. Tokarev is a First Class Test Pilot, with flight experience in forty-four types of airplanes and helicopters; logging over 3,000 hours. In 1973, he graduated from Stavropol Higher Military Aviation School for Pilots and Navigators. In 1982, he completed studies at the Test Pilot Training Center (TPTC) in Akhtubinsk (with honors). In 1997, he received a Master's degree in State Administration. In 1987, Tokarev was selected to join the Cosmonaut Corps to test and fly the *Buran* spacecraft. With the ending of the *Buran* program in 1997, he was assigned as a Test Cosmonaut at the Yuri Gagarin Training Center. Launched on May 27, 1999, aboard *Discovery* for STS-96, this was the first Space Shuttle to dock with the International Space Station (ISS). As a Mission Specialist, he helped transfer four tons of logistics and supplies to the station. *Discovery* landed at KSC on June 06, after a ten-day mission.

Launched on October 01, 2005, along with NASA's William McArthur and US "spaceflight participant" Gregory Olsen, *Soyuz TMA-7* docked with the ISS on October 03, 2005, and Tokarev was a Flight Engineer for 'Expedition 12', performing the Russian part of over thirty science experiments. His physics research included investigating the upper layers of the atmosphere. Tokarev, McArthur and "spaceflight participant" Marcos Pontes left the ISS on April 06, 2006, and returned to earth aboard *Soyuz TMA-7* on April 08; landing near the town of Arkalyk in Kazakhstan. He logged a total of 199 days, 15 hours and six minutes in space; performing two EVAs totaling 11 hours and five minutes. He retired in June 2008, and was awarded Hero of the Russian Federation. On March 26, 2009, Tokarev joined the United Russia Party. On September 08, 2013, he was elected head of Zvyozdny Gorodok for five years, receiving 67.9% of the vote.

## Treshkov, Sergei Y. STS-111/Expedition 5/STS-113.

Sergei Yevgenyevich Treshkov was born in August 1958 in Volynsky District, in the Lipetsk Region of Russia, USSR. In 1976, he graduated from technical school as an electric welder. In 1982, he graduated from the Moscow Power Institute specializing in Engineering Teacher of Electrical Power disciplines. From 1982-84, he served as a group leader in an Air Force regiment. His responsibilities included the analysis and planning of cosmonaut duties aboard a space station, and their inflight technical training. In 1992, he enrolled in the **RSC Energia** cosmonaut detachment, and during 1992-94 he completed the basic Cosmonaut training. Treshkov spent 1994-96 in advanced Test Cosmonaut training. From June 1997 to February 1998, he trained as a Flight Engineer for the *Mir* space station backup 'Expedition 25' crew.

From June 1999 to July 2000, he trained as a Flight Engineer for the *Soyuz TM* backup ISS contingency crew; initially for the ISS 'Expedition 3' crew. Launched aboard Space Shuttle *Endeavour* on June 05, 2002, on Mission STS-111 he joined ISS 'Expedition 5'. The crew studied cold plasma, crystal growth, radiation effects in space and the human body. There were also medical and psychological experiments. On August 26 he made an EVA of five hours and 21 minutes. Treshkov returned to Earth on December 07, 2002, aboard *Endeavour,* ending mission STS-113. He logged a total of 184 days, 22 hours and 15 minutes in space.

## Tsibliyev, Vasili V. Soyuz TM-17, Soyuz TM-25.

Lt. Gen. Vasili Vasiliyevich Tsibliyev was born in February 1954 in Orekhovka, Crimea, USSR. He is a retired Russian cosmonaut and former head of the Yuri Gagarin Cosmonaut Training Center. Selected as a cosmonaut on March 26, 1987, he flew as Commander on *Soyuz TM-17* from July 01, 1993 to January 14, 1994; and on *Soyuz TM-25* from February 02 to August 14, 1997. He was Commander of *Mir* when it was hit by a *Progress* spacecraft in 1997. Having logged 381 days, 15 hours and 52 minutes in space, Tsibliyev retired as a cosmonaut on June 19, 1998. From 2003-09, he headed the **Gagarin Cosmonaut Training Center**. He received nine awards, including Pilot-Cosmonaut of the Russian Federation (January 14, 1994); Hero of the Russian Federation (1994); the Order of Merit for the Fatherland, third class (1998); the Medal "For Merit in Space Exploration" (2011); the Medal "For Distinction in Military Service", second class; the Medal for Battle Merit; and the NASA Exceptional Public Service Medal. He is married, with two children.

## Tyurin, Mikhail V. STS-105/108, Soyuz TMA-9/Expn 14, Soyuz TMA-11M.

Mikhail Vladislavovich Tyurin was born in March 1960 in Kolomna, Russia, USSR. He graduated from the Moscow Aviation Institute in 1984 with a degree in Engineering, and a specialization in creating mathematical models related to mechanical flight. He then worked at the **Energia Corporation** as an engineer. His personal scientific research is connected with the psychological aspects of cosmonaut's training for the manual control of spacecraft motion. Selected in 1994 for cosmonaut training, in 1998 he started training as a Flight Engineer for the 'Expedition 3' crew. Launched on August 10, 2001, on STS-105 aboard Space Shuttle *Discovery*, he docked with the ISS on August 12; spent about four months aboard the space station; and conducted three EVAs. During the long-duration mission, the crew had a unique view of the 2001 Leonid meteor storm. For his third EVA of two hours and 46 minutes on December 03, 2001, Tyurin and Dezhurov used a cutting tool to remove a rubber seal (left from the previous cargo ship) that had prevented a *Progress* resupply craft from firmly docking with the ISS. Tyurin returned to Earth aboard the shuttle *Endeavour* (ending mission STS-108) and landing on December 17, 2001.

With NASA astronaut Michael Lopez-Alegria and "spacefight participant" aka commercial tourist Ms Anousheh Ansari, he launched aboard *Soyuz TMA-9* on September 18, 2006, for the ISS. He served as Soyuz Commander, and joined 'Expedition 14'. As a Flight Engineer, Tyurin spent 215 days aboard the ISS. He returned to Earth aboard *Soyuz TMA-9* with Lopez-Alegria and "spaceflight participant" Charles Simonyi on April 21, 2007; landing southwest of Karaganda, Kazakhstan. He was Commander of *Soyuz TMA-11M*, which docked with the ISS on November 07, 2013; six hours after launch from Baikonur Cosmodrome. During this mission, Tyurin was involved in studying the cardiovascular and neurological systems as a part of understanding how the human body adapts to zero gravity. After 188 days in space, he returned to Earth on May 14, 2014; landing southeast of Dzhezkazgan in Kazakhstan. As of June 2010, Tyurin had performed four EVAs, the last of five hours and 38 minutes being on November 22, 2006; for a total EVA time of 17 hours and 14 minutes.

## Usachov, Yuri V. Soyuz TM-18, TM-23, STS-101, STS-102/Expn 2.

Yury Vladimirovich Usachov was born in October 1957 in Donetsk, Rostov Oblast, Russia. He is a former cosmonaut who resides in Star City, Moscow. Usachov is a veteran of four spaceflights, including two long-duration missions. Usachov graduated from the Donetsk Public School in 1975. In 1985, he graduated from the Moscow Aviation Institute with an engineering diploma, then went to work for **RSC Energia**. He participated in groups working with EVA training, future construction in space, public relations, and ergonomics. In 1989, he was appointed to the position of cosmonaut candidate at the Yuri Gagarin Cosmonaut Training Center, and underwent a three-year course of general space training.

Usachov was a member of the backup crew for the *Mir-13, Mir-14,* and *Mir-19* missions. From January 08 to July 09 1994, he served as Board Engineer on the Space Station *Mir. Soyuz TM-18* docked with *Mir* on January 10, and Usachov, along with cosmonauts Viktor Afanasyev and Valeri Polyakov became the 15th resident crew. On January 14, there was alarm when the departing *Soyuz TM-17* craft struck the *Kristall* module two glancing blows during the customary inspection fly-around prior to the de-orbit burn. The 'EO-15' crew checked carefully, and found no damage. Three uncrewed *Progress M-21, M-22,* and *M-23* spacecraft supplied *Mir* during his stay. On this first mission, Usachov was aboard *Mir* for 182 days.

From February 21 to September 02, 1996, Usachov served as a Board Engineer on the 'Mir EO-21' expedition. He launched on *Soyuz TM-23* on February 21 with Yuri Onufriyenko, and docked with *Mir* two days later. One month later, NASA's Mrs Shannon Lucid joined them for 188 days, until replaced by French astronaut Claudie Andre-Deshays. The seventh and final *Mir* module, *Priroda*, arrived on April 26. Usachov next flew on mission STS-101, which launched on May 09, 2000. This was the third space shuttle mission devoted to ISS construction, and he was a Mission Specialist. Usachov and NASA astronaut Susan Helms were the first crew to enter the ISS. On Flight Day 8, Usachov spoke with Russian media gathered at the Russian Mission Control Center. He returned to Earth on shuttle *Atlantis*, touching down at Kennedy Space Center on May 29, 2000.

Yury Usachov made his fourth spaceflight aboard the shuttle *Discovery* on mission STS-102, which launched on May 08, 2001, and docked with the ISS the next day. The 'Expedition 2' crew of Usachov (as commander) and NASA astronauts Susan Helms and James Voss lived and worked on the ISS for the next 167 days. They prepared the *Destiny* module for upcoming experiments. A major focus of the mission was to gain a better understanding of how to protect crew members from radiation while living in space. During 'Expedition 2', there were four Space Shuttle and one *Soyuz TM-32* craft visiting the ISS. Usachov has performed seven career spacewalks. Six were during his stay aboard *Mir*, the longest being of five hours and 51 minutes. Usachov is married to Vera Sergeevna Usachove (nee Nazarova), and they have one daughter. In August 1994, he was awarded Hero of the Russian Federation; two Orders of Merit for the Fatherland, (1996 and 2002); the Medal "For Merit in Space Exploration", (2011); Pilot-Cosmonaut of the Russian Federation, (1994); Chevalier of the Legion of Honour (France, 1997); the NASA Space Flight Medal; and the NASA Distinguished Public Service Medal.

## Vasyutin, Vladimir V. Soyuz T-14.

Vladimir Vladimirovich Vasyutin (1952-2002) was born in March 1952 in Kharkiv, Kharkov Oblast, Ukrainian SSR, USSR. He graduated from Higher Air Force School and from Test Pilot School, both in Kharkov. Vasyutin was a Lieutenant General in the Soviet Air Forces, and took cosmonaut basic training in August 1976. Selected on December 01, 1978 (TsPK-6), he was assigned to the TKS program for a new generation of manned military spacecraft that would be docked to the existing *Salyut* space stations.

Vasyutin flew as Commander aboard *Soyuz T-14* to the *Salyut 7* station, for part of the long-duration mission *Salyut 7* 'EO-4'. However, he fell ill soon after arriving at the station and was unable to perform his duties. Originally scheduled to have a six-month stay, his illness (which he concealed from doctors before the launch) forced the crew to make an emergency return to Earth. He most likely had a prostate infection, which manifested as inflammation and a fever. He logged 64 days, 21 hours and 52 minutes in space; and retired on February 25, 1986. He later became Deputy Faculty Chief, VVA – **Gagarin Air Force Academy**, Monino. Vladimir Vasyutin received seven awards, including Hero of the Soviet Union; the Order of Honour (Russian Federation); the Order of Lenin; the Medal "Brotherhood in Arms" (Polish People's Republic); and the Medal "For Strengthening Military Cooperation" (USSR). Vasyutin died of cancer on July 19, 2002, aged fifty years. He was married, with two children.

## Vagner, Ivan V. Soyuz MS-16/Expn 62/63.

Ivan Viktorovich Vagner was born in July 1985 in Severonezhsk, Soviet Union. He is a Russian engineer and active cosmonaut, selected in October 2010. He graduated from the Baltic State Technical University with a Master's degree in Aeronautical Engineering in 2008, before working as an engineer for **RKK Energia**. Vagner completed basic spaceflight training in 2012. After being removed from the crew of 'Expedition 53/54' in September 2016 due to delays with the launch of the *Nauka* laboratory module, in February 2020 he was moved up to the prime crew for *Soyuz MS-16*. He flew to space on April 09, 2020, as a Flight Engineer on *Soyuz MS-16* for 'Expedition 62/63' on the ISS. They welcomed the first crewed test flight of **SpaceX** *Crew Dragon* from May 31, 2020. This was the first crewed orbital spacecraft to launch from US soil since the final Space Shuttle flight STS-135 in July 2011. Ivan Vagner logged a total of 195 days, 18 hours and 49 minutes in space. On April 05, 2022, Vagner was awarded the title, Hero of the Russian Federation.

## Viktorenko, Aleksandr S. Soyuz TM-3/2, TM-8, TM-14, TM-20.

Col. Aleksandr Stepanovich Viktorenko (1947-2023) was born in March 1947 in Olginka, Oktyabriskiy District, North Kazakhstan region, Kazakhstan SSR, USSR. He was a Colonel in the Russian Air Force, and a Soviet and Russian cosmonaut. Viktorenko had a brother, Vladimir, who was also a cosmonaut. Selected on March 23, 1978, Intercosmos Group, he launched on July 22, 1987, as Commander of *Soyuz TM-3/2* (Mir EP-1). He logged 160 days, seven hours and 26 minutes in space; landing on December 29, 1987, some 140 km northeast of Arkalyk. Launched on September 05, 1989, he commanded *Soyuz TM-8* to mission 'Mir EO-5'. The *Soyuz-U2* rocket was painted with slogans. During final approach to *Mir*, the *Kurs* rendezvous and docking system malfunction when four meters away, so Viktorenko took manual control and was successful. He landed on February 19, 1990 some fifty-five km northeast of Arkalyk. Launched on March 17, 1992, he commanded *Soyuz TM-14* for mission *Mir* 'EO-11'. He logged another 145 days, 14 hours and ten minutes in space; landing on August 10, 1992 about 136 km southeast of Dzhezkagan. The capsule suffered a landing system malfunction, causing the descent module to turn over. It came to rest upside down, trapping the crew inside until it could be righted.

*Soyuz TM-20* (Mir EO-17) launched on October 03, 1994, and docked with *Mir* two days later; spending a month on the space station. Viktorenko and Yelena Kondakova (the second Russian woman in space) returned to Earth on March 22, 1995. He was the first to test a "space motorcycle" and the first cosmonaut to launch after the fall of the Soviet Union. The Orbital Motorcycle was a concept proposed by Russian scientists as a way to commute between spacecraft and space stations. It would use thrusters and gyroscopes to maneuver in zero gravity, and could also be used for spacewalks and repairs. He retired on May 30, 1997, having logged a total of 489 days, one hour and 33 minutes in space; including 19 hours and 33 minutes of EVA. Aleksandr Viktorenko received eight awards, including Hero of the Soviet Union; Officer (1988), then Commander of the Legion of Honour (1999); the Order of the October Revolution (1990); the Order of Friendship of Peoples (1992); the Order of Merit for the Fatherland, third class (1995); and the Medal "For Merit in Space Exploration" (2011). Viktorenko died on August 10, 2023, aged seventy-six years. He was married, with two children.

## Vinogradov, Pavel V. Soyuz TM-26, Soyuz TMA-8/Expn 13, Soyuz TMA-08M/ Expn 35/36.

Pavel Vladimorovich Vinogradov was born in August 1953 in Magadan, USSR. He is a retired engineer, and former cosmonaut and commander of the ISS. Vinogradov graduated from Moscow Aviation Institute in 1977. From 1977-83, he specialized in software development for automated interactive designing systems of recoverable vehicles, development of aerodynamics and aerodyne arrangement design models, and computer graphics. In 1983, he started working for the Head Design Bureau of **RSC Energia**; on flight procedure verification for the *Soyuz TM* and *Buran* spacecraft (and their launch preparation), and development of automated crew training systems. As of January 2023, he had flown into space three times; aboard both *Mir* and the International Space Station. Selected on March 03, 1992, in NPOE Cosmonaut Group, he trained from October 1992 to February 1994; then from May 1994 to February 1995 he completed advanced Test-Cosmonaut training. He was assigned to the *Soyuz TM-24* flight, but the crew was grounded when Commander Gennadi Manakov experienced heart problems.

Launched on August 05, 1997, aboard *Soyuz TM-26*, he flew as Flight Engineer to the *Mir* space station. The crew repaired the power cable and harness/connectors in the severely damaged *Spektr* module, and restored much of the lost power. They also repaired and replaced *Mir*'s oxygen generators. On this mission, Vinogradov spent 197 days, 17 hours and 34 minutes in space; with five EVAs. In January 2004, he began training as the ISS 'Expedition 13' crew commander. He was launched aboard *Soyuz TMA-8* on March 29, 2006, docking with the ISS for 'Expedition 13' on March 31. He returned on a *Soyuz* on September 28, 2006, landing on the steppes of Kazakhstan near Arkalyk at 21:13 EDT.

Vinogradov was launched aboard *Soyuz TMA-08M* on March 28, 2013, for ISS 'Expeditions 35/36' from March to September 2013. On May 12, he assumed command of the ISS. 'Expedition 36' was from May 13 to September 10, 2013, with six crew who studied how space affects the eyes and brain; to better understand Earth-bound health conditions. They also promoted augmented-reality Earth observations from the space station and on the ground using new software.

On June 16, 2013, the fiftieth anniversary of Valentina Tereshkova's flight in *Vostok 6* (the first spaceshot by a woman), there were two women in space: NASA's Karen L. Nyberg (aboard the ISS), and China's Wang Yaping (aboard the *Tiangong-1* station on the *Shenzhou 10* mission). Vinogradov logged a total of 546 days, 22 hours and 32 minutes in space; including seven EVAs totaling 38 hours and 25 minutes. After his third flight, his accumulated space time puts him among the top ten astronauts. He holds the record for the oldest person to perform an EVA. He was awarded the medal, Hero of the Russian Federation. Vinogradov is married, with one child.

## Volk, Igor P. Soyuz T-12.

Ogor Petrovich Volk (1937-2017) was born in April 1937 in Zmiiv, Kharkiv Oblast, Ukrainian SSR, USSR. He was a Soviet test pilot and cosmonaut. He studied at the Kirovograd Military Aviation School of Pilots; graduating in 1956 to become a pilot in the Soviet Air Forces. After graduating from the Fedotov Test Pilot School in 1965, he joined the Gromov Flight Research Institute. He logged over 7,000 hours in over eighty aircraft types. Volk showed outstanding abilities in complex tests of various airplanes at critical angles of attack, stall, and spin. He did notable work on the MiG-21I testbed maiden flight. Selected in Air Force Group 5 on July 12, 1977, he was assigned to the *Buran* program. As part of his preparations for a space shuttle flight, he made test flights with *Buran*'s counterpart OK-GL1 aircraft. In July 1984, Volk flew aboard *Soyuz T-12* (aka Salyut 7 EP-4) to give him some experience in space. He was a Research Cosmonaut on the seventh expedition to *Salyut 7*, with one goal being to evaluate the effects of long-duration spaceflight on a pilot's skills and ability to fly and land the spaceplane safely after an extended stay in space. The mission was from July 17-29, 1984; landing 140 km (87 miles) southeast of Dzezkazgan.

Svetlana Savitskaya also flew on *Soyuz T-12* as Flight Engineer, and made the first-ever female EVA on July 25, 1984, lasting for three hours and 30 minutes. This achievement (to beat NASA's Kathryn Sullivan on mission STS-41-G) was considered more important than gaining space experience for a potential Buran crew member. At the time of this mission, the *Buran* Soviet space shuttle program was still a State secret. **The British Interplanetary Society** magazine *Spaceflight* queried why a test pilot was occupying a *Soyuz* test seat usually reserved for researchers of foreign cosmonauts. Cosmonauts Vladimir Dzhanibekov and Ms Savitskaya tested the URI multipurpose tool; cut, welded, soldered, and coated metal samples; conducted Rezonans tests and collected station air samples. After this orbital mission, Volk then served as the head of the pilot-cosmonaut training department for the *Buran* program, and, later (after the project's cancellation) he was employed by Gromov Flight Research Institute (1965-2002) as a Flight Tests Deputy Chief.

Volk logged eleven days, 19 hours and 14 minutes in space. As recognition for his contributions as a test pilot and cosmonaut, he was awarded the title, Hero of the Soviet Union on July 29, 1984. Volk was a Member of the Executive committee of the Green Movement (since 1989); an active supporter of the environmental movement in the USSR and Russia. From 2003-05, he was the first Vice President of the Federation of Cosmonautics of Russia (FKR), then Chairman of the Executive Committee of the FKR. In May 2016, Volk supported the program of environmentalists in the elections and primaries of United Russia in the Moscow Oblast. He actively supported the environmental projects of the ***EkoGrad*** magazine. Volk received nine awards, including Honored Test Pilot of the USSR; the Order "For Merit to the Fatherland", fourth class; the Order of Lenin; the Order of the Red Banner of Labor; the Medal "For Merit in Space Exploration"; and the Order of Friendship of Peoples. He was an inventor, was married, and had two children. Igor Volk died on January 03, 2017 while on holiday in Plovdiv, Bulgaria, aged seventy-nine years.

## Volkov, Aleksandr A. Soyuz T-14, Soyuz TM-7, Soyuz TM-13.

Col. Aleksandr Aleksandrovich Volkov was born in May 1948 in Gorlovka (Horlivka), Donetsk Oblast, Ukraine, USSR. He is a retired test pilot and cosmonaut, a veteran of three space flights; including twice to the *Mir* space station. He graduated from the Chugayev Higher Air Force School in Kharkov province, Ukraine, USSR, in 1970, then served as a test pilot in the Soviet Air Force, and reached the rank of Colonel. He received cosmonaut training from 1976-79.

Launched on September 17, 1985, aboard *Soyuz T-14* Alexander Volkov was a Research Cosmonaut for the ninth expedition to the *Salyut 7* space station. The three-man crew conducted scientific and technical experiments with military applications. Commander Vladimir Vasyutin became unwell in late October, and the planned six-month mission was terminated early. Russian psychologists cited unexpected "mood and performance issues", but Vasyutin had a UTI, inflammation and a fever. The mission lasted 64 days, and 21 hours; completing 1,021 Earth orbits. Volkov returned to Earth on November 21, 1985. Launched on November 26, 1988, aboard *Soyuz TM-7* for 'Mir EO-4' (aka Principal Expedition 4), Volkov was Commander on the fourth long-duration mission to Russian space station *Mir* (151 days and 11 hours from November 26, 1988, to April 26, 1989). Their research focused on medical and technology experiments in support of the Hermes shuttle project of the European Space Agency (ESA).

He was Commander of *Soyuz TM-13* (Mir EO-10) from October 12, 1991, to March 1992 (175 days). Volkov controlled the docking procedures. The mission included cosmonaut Franz Viehbock from Austria and Toktar Aubakirov from the soon-to-be-independent region of Kazakhstan, as the Soviet Union collapsed in December 1991. Sergei Krikalev and Volkov were "the last citizens of the USSR"; having gone into orbit as Soviet citizens, they returned to Earth 176 days later as Russian citizens. Volkov then worked as Commander of the Cosmonaut Team at the Cosmonauts Training Center from January 1991 to August 1998. He prepared Russian and foreign cosmonauts for future flights to *Mir* and the ISS. Volkov logged a total of 391 days, 11 hours and 52 minutes in space. Aleksandr Volkov received eight awards: Hero of the Soviet Union (1985); the Order of Lenin (1985); the Order of the October Revolution (1989); the Order of Friendship of Peoples (1992); the Medal of the "Order of Merit for the Fatherland" (1996), second class; the Order of the Legion of Honor (France); the Medal "For Merit in Space Exploration" (2011); and the Order of Merit, third class, (Ukraine, 2011). His son, Sergei A. Volkov is the first second-generation cosmonaut.

## Volkov, Sergei A. Soyuz TMA-12/Expn 17, TMA-02M/Expn 28/29, TMA-18M/ Expn 45/46.

Col. Sergei Aleksandrovich Volkov was born in April 1973 in Chuhuiv, Ukrainian SSR, Soviet Union. He is a retired engineer, fighter pilot and cosmonaut; a member of three missions to the ISS. Sergei Volkov is the first second-generation cosmonaut, the son of Aleksandr Volkov (see above). He graduated from Star City High School in 1990, and entered the Tambov Air Force Academy for Pilots; graduating in 1995 with a degree of Pilot/Engineer. He is a Class 3 military pilot. Selected in 1997 TsPK Cosmonaut Group, he underwent general cosmonaut training from December 1997 to November 1999. From August 2001 to February 2003, he trained as part of the 'Expedition 7' backup crew as a *Soyuz TMA* Commander and ISS Flight Engineer. From March 2003 to December 2004, he trained as part of the 'Expedition 11' primary crew for launch on the Space Shuttle. From January 2005 to February 2006, he trained as part of a group of Test Cosmonauts for missions to the ISS. In February 2006, he was appointed as a member of the 'Expedition 13' backup crew. In June 2006, he became a member of the 'Expedition 17' prime crew as *Soyuz TMA* Commander and ISS Commander. *Soyuz TMA-12* was launched on April 08, 2008, and he was the youngest ISS Commander to date. This mission returned to Earth on October 24, 2008, aboard *Soyuz TMA-12*.

Launched aboard *Soyuz TMA-02M* on June 07, 2011, Volkov joined the crew of 'Expedition 28'. The crews worked with 111 experiments involving about 200 researchers across the fields of human life sciences, physical sciences and Earth observation, and conducted technology demonstrations. NASA sponsored seventy-three of the experiments. On July 10, 2008, Volkov made an EVA of six hours and 23 minutes. On July 15, a second EVA in Russian *Orlan* suits lasted five hours and five minutes. Volkov's third EVA of six hours and 23 minutes on August 03 was to move a small crane; install a laser communications terminal; and remove and inspect antennas. Launched aboard *Soyuz TMA-18M* on September 02, 2015, Volkov was ISS Commander for 'Expedition 45/46'. His fourth EVA was during 'Expedition 46', and lasted four hours and 45 minutes. He returned to Earth on March 02, 2016. Sergei Volkov logged a total of 547 days, 22 hours and 21 minutes in space; including four EVAs totaling 23 hours and 20 minutes. He retired from the Cosmonaut group in February 2017. Volkov was awarded Hero of the Russian Federation; and Russian Federation Armed Forces medals. He is married, with a son.

## Volkov, Vladislav N. Soyuz 7, Soyuz 11.

Vladislav Nikolayevich Volkov (1935-71) was born on November 23, 1935, in Moscow, Russian SFSR, Soviet Union. A talented athlete and boxer, he graduated from the Moscow Aviation Institute in 1959. As an aviation engineer at Korolyov Design Bureau, he was involved in the development of the *Vostok* and *Voskhod* spacecraft. Launched on October 12, 1969, aboard *Soyuz 7*, Volkov was the Flight Engineer, part of a joint mission with *Soyuz 6* and '8'. Following an orbital correction during the night, *Soyuz 7* and '8' were expected to be less than one kilometre from each other when communications were regained at 9 a.m. The plan was to dock with *Soyuz 8* and transfer crew, but this was not achieved due to equipment failures. Other objectives were to test a system for control of the simultaneous flight of three spacecraft, and debug the piloting systems by orbital maneuvering of the spaceships in relation to one another. The *Soyuz 7* mission lasted four days, 22 hours and 40 minutes; landing on October 17, 1969, at 09:25:05 GMT at 155km northwest of Karaganda, Kazakhstan.

In 1970, Volkov became the first chairman of the Soviet Society of Sportsacrobatics. From 1973 to 2015, the Pilot Cosmonaut-Volkov-Award (later the Volkov Cup, then Zolotov Cup from 2016) was given for the best sportsacrobatics. He was assigned to *Soyuz 11* along with Georgy Dobrovolsky and Viktor Patseyev. They launched on June 06 and from June 07, 1971, they spent 22 days on *Salyut 1*, the world's first space station. The three crew of *Soyuz 11* were asphyxiated upon re-entry on June 30, 1971, the second Soviet space flight to end fatally. After a normal re-entry, the capsule was opened and the corpses of the three cosmonauts were found inside. They lay dead in their seats, with dark blue splotches on their faces and blood trickling from their noses and ears. It was later discovered that an air pressure equalization valve had opened just prior to leaving orbit, (during the separation of the orbital and instrument modules) allowing the capsule's atmosphere to vent away into space, causing fatal hypoxia within 110 seconds. Volkov had died in outer space, aged thirty-five years. His ashes were interred in the Kremlin Wall Necropolis, on Red Square in Moscow. Future crews wore pressure suits for both launches and landings.

Vladislav Volkov was twice decorated as Hero of the Soviet Union (first on October 22, 1969, and then posthumously on June 30, 1971). He was awarded two Orders of Lenin and the title of Pilot-Cosmonaut of the USSR. A street in Moscow; the lunar crater Volkov and the minor planet 1790 Volkov are named in his honor. A tomato variety from Ukraine was named Cosmonaut Volkov by his space scientist friend and gardener, Mikhailovich Maslov. An account of Volkov's life and space career appears in the 2003 book, *Fallen Astronauts: Heroes Who died Reaching for the Moon* by Colin Burgess. In *Russia in Space*, Brian Harvey described a fleet of "comships" used to track Soviet space missions. The fleet included the *Vladislav Volkov*, built during the 1970s, and used until dissolution of the Soviet Union in late 1991.

## Volynov, Boris V. Soyuz 5, Soyuz 21.

Col. Dr. Boris Valentinovich Volynov was born in December 1934 in Irkutsk, Siberia, Soviet Union. He is an electrical engineer, and following the death of Alexei Leonov in October 2019, he is the last surviving member of the original group of cosmonauts. Volynov is also considered to be the first Jew in space. In 1953, he completed basic pilot training in Pavlodar, Kazakhstan; and graduated from an aviation school in Novosibirsk in 1955. From September 1961 to January 1968, he studied at the faculty of engineering of the Zhukovsky Air Force Engineering Academy; graduating with a Diploma of Pilot-Engineer-Cosmonaut. Volynov received his Ph.D there in 1980. Selected in Air Force Group 1, he was assigned as one of two possible commanders for *Voskhod 1*, but his prime crew were bumped from the mission just three days before the scheduled launch date. Volynov then spent a year training for *Voskhod 3*, but following the death of Sergey P. Korolev (Korolyov) on January 14, 1966, the flight was canceled ten days before launch. He was then transferred to the *Soyuz* group under Vasily Mishin, and assigned as backup for *Soyuz 3*.

Volynov was Commander for *Soyuz 5*, which launched on January 15, 1969, and transferred two crew to *Soyuz 4*. He returned to Earth alone in *Soyuz 5*, ending a mission which was memorable for its dramatic re-entry. The craft's service module did not separate, so it entered the atmosphere nose-first. Gaskets sealing the light metal entry hatch began to burn, filling the compartment with dangerous fumes and smoke. The descent module righted itself before the escape hatch burned through, but the parachute lines partially tangled and the soft-landing rockets failed. The resulting hard landing on January 18 at 07:59:12 GMT almost wrecked the module, and broke most of Volynov's upper teeth. He landed far short of the target landing site, abandoned the capsule, and walked several kilometres to find shelter at a local peasant's house. It was seven years before Volynov flew to space again.

Launched aboard *Soyuz 21* on July 06, 1976, Volynov and Fight Engineer Vitaliy Zholobov were scheduled to spend from 54 to 66 days aboard the *Salyut 5* space station. Zholobov became unwell, apparently because of nitric acid fumes leaking from the *Salyut*'s propellant tanks (or perhaps because of lack of sleep), and the decision was made to terminate their mission early. As Volynov tried to undock from *Salyut*, the latch failed to release properly, then the mechanism jammed. Instructions for two emergency procedures were received, and one orbit later the latches finally disengaged. Since *Soyuz 21* was returning early, it was outside the normal recovery window, it encountered strong winds, and the retrorockets fired unevenly. The capsule made a hard landing near midnight, some 200km southwest of Kokchetav, Kazakhstan. Volynov was awarded the Order of the Red Star (1961); two Hero of the Soviet Union (1969, 1976); Order "For Merit to the Fatherland", fourth class; two Orders of Lenin (1969, 1976); the Medal "For Distinction in Guarding the State Border of the USSR"; the Order of the Banner of the People's Republic of Bulgaria; the Order of Friendship (2011); and was made an Honorary Citizen of Prokopyevsk. He logged a total of 52 days, seven hours and 17 minutes in space. He is married, with two children.

## Vozovikov, Sergey V..

Maj. Sergey Yuriyevich Vozovikov (1958-93) was born in April 1958 in Alma-Ata, Kazakh SSR. He graduated from Higher Military Pilot School, Armavir, in 1979. Selected on May 11, 1990, he was a member of the Soviet Air Force Cosmonaut Training Group TsPk-11. His training was to take place from October 01, 1991 to March 06, 1992, but was cut short when he drowned on July 11, 1993, while on water recovery training in the Black Sea, near Anapa Rayon, Russia. Vozovikov was an experienced diver, but became tangled in a fishing net and could not free himself. His fishing mate, Aleksandr Lazutkin was unable to help him. He was married, with two children.

## Yegorov, Boris Y. Voskhod 1.

Dr. Boris Bosisovich Yegorov (1937-94) was born in November 1937 in Moscow, Soviet Union. He was a Soviet physician and cosmonaut who became the first medical doctor to fly in space. He graduated from the First Moscow State Medical University in 1961; and earned his doctorate in Medicine, specializing in disorders of the sense of balance. He then began working as a researcher at the Institute of Biomedical Problems and the Institute of Medical and Biological Problems. His work focused on space medicine, and he soon became an expert in the field.

Selected in Civilian Specialist Group 1 in 1964, he flew on *Voskhod 1*, which launched on October 12. This was the first mission to carry more than one crew member, the first to use a multi-person spacecraft and Yegorov conducted experiments to study the effect of spaceflight on the human body; and monitored the health of his fellow crew members. Yegorov logged one day, zero hours and 17 minutes in space. He was awarded Hero of the Soviet Union (1964), then became a professor at the Moscow Medical Institute. He was involved in the development of medical equipment for use in space. Yegorov received the Order of Lenin; the Order of the Red Banner of Labour; the Medal "For the Development of Virgin Lands"; Hero of Socialist Labour (Vietnam); and the Banner Order of the Hungarian People's Republic. He died on September 12, 1994, in Moscow, aged fifty-six years.

## Yeliseyev, Aleksei S. Soyuz 5, Soyuz 8, Soyuz 10.

Aleksei Stanislavovich Yeliseyev was born in July 1934 in Zhizdra, Kaluga Oblast, Russian SFSR, Soviet Union. He is a retired Soviet engineer and cosmonaut who flew on three missions in the *Soyuz* program as a Flight Engineer. He graduated from the Bauman Higher Technical School in 1957; then had postgraduate studies at Moscow Institute of Physics and Technology (1962). Yeliseyev worked in Sergey Korolev's design bureau before being selected for cosmonaut training in Civilian Specialist Group 2. The *Soyuz 5* mission used the *Soyuz 7K-OK* spacecraft launched on January 15, 1969, which docked with *Soyuz 4* in orbit for four hours and 35 minutes. *Soyuz 4* had launched first, and was the active vehicle in the docking. While over the Soviet Union, Yeliseyev transferred via a spacewalk to *Soyuz 4*, joining Commander Vladimir Shatalov for the landing. *Soyuz 5* landed in the Ural Mountains, 200 km southeast of Kostanay, near Orenburg, Kazakhstan on January 16, 1969 at 07:59:12 GMT.

Launched on October 13, 1969, *Soyuz 8* was a joint mission with *Soyuz 6* and '7' that saw three *Soyuz* craft in orbit together, carrying a total of seven cosmonauts. The mission plan was for *Soyuz 8* (Soyuz 7K-OK NO. 16) to dock with *Soyuz 7* and transfer crew, while *Soyuz 6* was to film the operation from nearby. Due to equipment failures this objective was not achieved. This was the last time that the Soviet-crewed Moon landing hardware was tested in orbit, and the failure seems to have been one of the deciding factors to end their program. The mission lasted four days, 22 hours and 51 minutes to October 18 09:09:58 GMT. Launched on April 22, 1971, aboard *Soyuz 10*, cosmonauts Shatalov, Yeliseyev and Rukavishnikov were able to navigate their craft to the *Salyut 1* space station, but the automatic control system failed during approach, owing to a serious design oversight. They made a "soft dock" (contact without a full link) on April 24, 1971, and large quantities of propellant were used in maneuvering. Then the probe would not come out of the station's docking cone until the power supply was shut down. This was the first of numerous docking failures in the Soviet space station program. The capsule returned to Earth the same day, landing 120km to the northeast of Karaganda, Kazakhstan; and the *Soyuz* automatic control system was then redesigned.

Yeliseyev logged a total of eight days, 22 hours and 20 minutes in space; including one EVA of 37 minutes. Following his retirement from the space program in 1985, he held an administrative position at the **Bauman School** for several years before retiring fully.

Yeliseyev received eight awards, including two Hero of the Soviet Union (1969); four Orders of Lenin (January and October 1969, 1971, 1976); the State Prize of the USSR (1980); the Jubilee Medal "In Commemoration of the One Hundredth Anniversary since the Birth of Vladimir Il'ich Lenin"; the Order of Georgi Dimitrov (Bulgaria); and Hero of the People's Republic of Bulgaria. He is married, with one child.

## Yurchikhin, Fyodor N. STS-112, Soyuz TMA-10/Expn 15, Soyuz TMA-19/Expn 24/25, Soyuz TMA-09M/Expn 36/37, Soyuz MS-04/Expn 51/52.

Dr. Fyodor Nikolayevich Yurchikhin was born in January 1959 in Batumi, Adjarian ASSR, Georgian SSR. He is a retired Russian cosmonaut of Greek descent, a mechanical engineer and RSC Energia test pilot. He entered the Moscow Aviation Institute, and qualified in 1983 as a Mechanical Engineer, specializing in airspace vehicles. From September 1983 to August 1997, he worked at the Russian Space Corporation ***Energia***. He eventually became a lead engineer for *Shuttle-Mir* and *NASA-Mir* programs. In 2001, he graduated from the Russian Presidential Academy of National Economy and Public Administration (RANEPA) with a Ph.D. in Economics.

Selected in 1997 RKKE Cosmonaut Group, he launched on mission STS-112 aboard Space Shuttle *Atlantis* on October 07, 2002, for a ten-day and 20 hours ISS Assembly Flight 9A. He launched on April 07, 2007, aboard *Soyuz TMA-10* as Flight Engineer for the ISS long-duration 'Expedition 15' mission; returning to Earth on October 21. He launched on June 16, 2010, aboard *Soyuz TMA-19* for Expedition 24/25' on the ISS; and landed (after 163 days) on November 26, 2010. He participated in the all-Russian census from space, answering questions during a linkup with Mission Control on October 18, 2010. Launched on May 28, 2013, Yurchikhin served as Commander aboard *Soyuz TMA-09M* as Flight Engineer for 'Expedition 36', and was ISS Commander for 'Expedition 37'. He returned to Earth on November 11, 2013. On April 20, 2017, he launched on *Soyuz MS-04* on his fifth flight, a six-month mission to the ISS as part of 'Expedition 51/52' for which he was Commander. Returning to Earth on September 03, 2017, Yurchikhin logged a total of 672 days, 20 hours and 38 minutes in space; including nine EVAs totaling 59 hours and 28 minutes.

## Zaikin, Dmitri A..

Col. Dmitri Alekssevich Zaikin (1932-2013) was born in April 1932 in Yekaterinovka village, Rostov Oblast, southeastern Russia. He graduated from Military Fighter Pilot School, Armavir (Krasnodr Krai) and Frunze (now Bishek) in 1955. While serving as chief pilot with the Ninety Fifth IAD (fighter aviation division) in Belorussia, he was elected for cosmonaut training in 1960; one of twenty Air Force pilots in the first cosmonaut group. He graduated from the Zhukovsky Military Engineering Academy of Monino in 1968. Zaikin left the space service on October 1969 for medical reasons (a stomach ulcer) while training for *Soyuz* missions, then became an instructor and lead engineer at the Yuri Gagarin Cosmonaut Training Center until 1982. The training center's statement read, *"Characterized by humility, [a] high sense of responsibility for his work, hard work, diligence and dedication ... he was respected and [had a] well-deserved reputation among his colleagues"*. Zaikin retired from active military duty in 1987 at age sixty-five years, and died of natural causes on October 20, 2013, in Moscow, aged eighty-one years. He was married, with two sons.

## Zalyotin, Sergei V. Soyuz TM-30, Soyuz TMA-1/34.

Lt. Col. Sergei Viktorovich Zalyotin was born in April 1962 in Tula Oblast, Russian SFSR, Soviet Union. He attended the Borisoglebsk Higher Military School, graduating in 1983, and became a fighter pilot in the Russian Air Force. He gained a qualification of

Engineer Ecologist by doing correspondence study; and a Master's degree in Ecological Management from the International Center of Training Systems. Zalyotin was selected as a cosmonaut candidate in TsPK-11 on May 11, 1990, and trained until March 1992 to become a Test Cosmonaut. Launched aboard *Soyuz TM-30* as Commander, he was then Commander of the final resident crew aboard *Mir* mission 'EO-28' from April 04 to June 16, 2000, a mission lasting for 72 days, 19 hours and 41 minutes. Zalyotin made one EVA of five hours and three minutes.

He briefly visited the International Space Station for ten days, 20 hours and 53 minutes from October 30 to November 10, 2002, as Commander aboard *Soyuz TMA-1/Soyuz TM-34*. He logged a total of 83 days, 16 hours and 34 minutes in space. Zalyotin was awarded the title, Hero of the Russian Federation; and Pilot-Cosmonaut of the Russian Federation. He retired from the cosmonaut corps to become chief of the laboratory **RGNII TsPk**. On October 03, 2004, he was elected as Deputy of the Tula provincial Duma; returned to the TsPk cosmonaut group on January 11, 2011, and later retired again for medical reasons. He is married, with one child.

## Zholobov, Vitali M. Soyuz 21.

Vitali Mikhaylovich Zholobov was born in June 1937. He is a retired Soviet Ukranian cosmonaut who flew on *Soyuz 21* as the flight engineer, and a former head of Kherson Regional Council in Ukraine. Selected in Air Force Group 2, he spent two months on the *Salyut 5* space station. The flight was scheduled to last for sixty days, but because a noxious odor was detected on board it was reduced to forty-nine days. He reported to the Mission Control Center that the smell was similar to that of a propellant which was known to be toxic. The mission was aborted, since the research and technology programs were already finished. He logged 49 days, six hours and 23 minutes in space; from June 06 to August 24, 1976. Zholobov stayed in the space program until 1981, when he resigned to become Director of a geological science research group. In 1994, he was elected in the Kherson region as the Chairman of the Kherson Regional Council, and held the position until 1996. From June 12, 1996 to February 04, 1997, he held the position of deputy General Director of the **National Space Agency** of Ukraine. From 1999, he was chairman of the board of the Kyiv organization "Slava", and since 2001 has been the president of the All-Ukranian association **Slava**. From April 11 2002 to 2015, Zholobov was the president of the **Aerospace Society of Ukraine**.

## Zudov, Vyacheslav D. Soyuz 23.

Vyacheslav Dmitriyevich Zudov was born in January 1942 in Bor, Gorki Oblast, Russia. He is a retired Colonel, pilot, parachutist and Soviet cosmonaut. He graduated from the Higher Military Pilot School, Balashov, in 1963; and graduated from Gagarin Air Force Academy in 1979. Selected in TsPK-3 as a Test Cosmonaut on October 23, 1965, Zudov took cosmonaut basic training from November 1965 to December 1967. He was assigned for a second *Soyuz 4/5* mission (later canceled); and later was Head, Political Affairs, **TsPK**. He then worked as a commercial scientist and for customs management. Zudov flew as Commander on *Soyuz 23* on October 14-16, 1976, logging two days, zero hours and six minutes in space, and retired on May 14, 1987. Zudov has been awarded the titles, Hero of the Soviet Union, and Pilot-Cosmonaut of the USSR; the Jubilee Medal "Twenty Years of Victory in the Great Patriotic War 1941-45"; the Jubilee Medal "In Commemoration of the 100th Aniversary since the Birth of Vladimir Il'ich Lenin"; the Jubilee Medal "50 Years of the Armed Forces of the USSR"; and the Medal "For Merit in Space Exploration" (Russian Federation). Zudov is married and has two children.

## Foale, C. Michael. STS-45, STS-56, STS-63, STS-84/Mir EO-23/STS-86, STS-103, Soyuz TMA-3/Expn 8.

Dr. Colin "Michael" Foale (CBE) was born in January 1957 in Louth, Lincolnshire, England. He earned a first-class honors BA degree in Natural Sciences in 1978; and a Ph.D. in Laboratory Astrophysics in 1982. To pursue a career in the US Space Program, he moved to Houston, Texas, to work on Space Shuttle navigation problems for **McDonnell Douglas**. Selected in NASA Group 12, Foale flew as a Mission Specialist on STS-45 (launched March 24, 1992), STS-56 (April 08, 1993), and STS-63 (February 03, 1995) when he performed four EVAs. He was the second Briton in space, and the first to perform a spacewalk. Foale launched on May 15, 1997, aboard *Atlantis* on STS-84 for mission 'Mir-23/24' of 144 days, 3 hours and 47 minutes; and on STS-103 aboard *Discovery* on December 20, 1999, for seven days and 23 hours. Foale's final and longest-duration mission launched aboard *Soyuz TMA-3* to the ISS 'Expedition 8' on October 18, 2003, for 194 days, 18 hours and 33 minutes. Up to April 17, 2008, he held the record for the most time spent in space by a U.S. Citizen: 374 days, 11 hours and 19 minutes. As of 2024, he also holds the cumulative time-in-space record for a British citizen. Foale is married to Rhonda Butler of Kentucky, with two children.

## McAuliffe, S. Christa. STS-51-L.

Sharon Christa McAuliffe (nee Corrigan, 1948-86) was born in September 1948 in Boston, Massachusetts, USA. Inspired by the Mercury and Apollo programs, she wrote on her NASA application form: *I watched the Space Age being born, and I would like to participate*. She received a Bachelor's degree in Education and History in 1970; and a Master of Arts degree in Education, Supervision, and Administration in 1978. After being selected in 1985 from the final ten of over 11,000 applicants to be the first Teacher in Space, she had an immediate rapport with the media. On January 28, 1986, McAuliffe boarded the Space Shuttle *Challenger* in Cape Canaveral, Florida. The vehicle exploded just 73 seconds after liftoff, at an altitude of 48,000 feet (14.63 km) killing all seven astronauts aboard. At the time of her death, McAuliffe had a son, Scott, aged nine years, and a daughter, Caroline, aged six years. Several American schools have been named in honor of Christa McAuliffe, and forty world-wide; as are the crater McAuliffe on the Moon, a crater on the planet Venus, and the asteroid 3352 McAuliffe. She was portrayed by Karen Allen in the 1990 TV movie, ***Challenger***. In 2017, McAuliffe was inducted into the International Air & Space Hall of Fame at San Diego, California.

## Ramon, Ilan. STS-107.

Col. Ilan Ramon was born in June 1954 in Ramat Gan, Israel. He earned a Bachelor of Science degree in Electronics and Computer Engineering from Tel Aviv University in 1987. When he joined the Israeli Air Force, he Hebraized his surname from Wolfferam to Ramon, and became a fighter pilot. He participated in A-4 Skyhawk Basic Training and Operations from 1974-76. Ramon logged over 3,000 flight hours on the A-4, Mirage IIIC, and F-4, and over 1,000 flight hours on the F-16 jet. Selected in the 1996 NASA group, he launched aboard space shuttle *Columbia* on January 16, 2003, for mission STS-107, which conducted eighty experiments. Personally non-religious, Ramon performed traditional observances while in orbit: *I feel I am representing all Jews and all Israelis*. He was the first astronaut to request kosher food, and he marked the Sabbath. On February 01, 3003, just sixteen minutes before scheduled touchdown, *Columbia* disintegrated over east Texas upon re-entry into Earth's atmosphere, resulting in the deaths of all seven crew. He logged 15 days, 22 hours and 20 minutes in space. Among the recovered 40% from the contents of the shuttle were 37 pages from Ramon's diary. Awarded posthumously, Ramon is the only foreign recipient of the United States Congressional Space Medal of Honor. He received three military decorations, and Israel named fourteen places in his honor, including seven schools.

PART FIVE

# Epilogue

How were astronauts affected by their journeys to space? Some first-hand accounts are detailed here. Lt. Yuri Gagarin's first view of Earth came 154 seconds after launch on April 12, 1961, when the rocket's nose cone separated. He yelled into his microphone, *"I can see the Earth! I can see rivers, folds in the Earth. It's all easy to make out. Visibility is good. Everything is wonderfully visible!"* Major Gagarin later said, *"To be the first to enter the cosmos, to engage single-handed in an unprecedented duel with nature – could anyone dream of anything greater than that?"* John H. Glenn's first spaceflight was on February 20, 1962, orbiting the Earth three times. He later said, *"As I hurtled through space, one thought kept crossing my mind – every part of this rocket was supplied by the lowest bidder"*. On the *Apollo 8* astronauts post-flight world tour, Frank Borman said that we *"are first and foremost not Germans or Russians or Americans, but Earthmen"*. He said that *"raging nationalistic interests, famines, wars, pestiliences don't show from that distance.* ***From out there it really is "one world""***. A year to the day before the Genesis reading from space and the taking of the December 24, 1968, Earthrise photograph by *Apollo 8* astronaut Bill Anders, Dr. Martin Luther King had said, *"If we are to have peace on Earth ... we must develop a world perspective. As nations and individuals, we are interdependent. ... We are made to live together because of the inter-related structure of reality"*.

The Earthrise photograph, said the ***New York Times*** was a *"humbling reminder of the world's insignificance"*, while the ***LA Times*** commented that it *"made introverts of us all"*. The poet, Archibald MacLeish, wondered if, in a matter of hours, human beings had shifted perspective, away from a conception of themselves as *"God-directed actors at the center of a noble drama"*, to one in which *"Man may at last become himself"*. Ending his article in the ***New York Times***, MacLeish wrote, *"To see the Earth as it truly is, small and blue and beautiful in that eternal silence where it floats, is to see ourselves as riders on the Earth together, brothers on that bright loveliness in the eternal cold – brothers who know now they are truly brothers"*. Earthrise was picked by ***Life*** magazine as one of its one hundred photos of the twentieth century. Anne Morrow Lindbergh said that after Earthrise no one would ever look at the Earth the same way again: *"As Earthmen, we may have taken another step into adulthood"*.

According to Commander Frank Borman, "*Rookie Bill Anders was thirty-five, slightly built, a devout Roman Catholic, and very serious minded. ... Anders was one hell of a worker, a superb technician and all in all a great guy. Some of the guys regarded him as a younger version of Frank Borman in his single-minded concentration on work, his aversion to unnecessary conversation*". Unquote. Anders stated that the Earthrise photograph *"really undercut my religious beliefs. The idea that things rotate around the Pope, and up there is a big supercomputer wondering whether Billy was a good boy yesterday? It doesn't make any sense. I became a big buddy of Richard Dawkins"*. According to Anders: "***We came all this way to explore the Moon, and the most important thing is that we discovered the Earth***".

On conservation of the planet, he said, "*If you can imagine yourself in a darkened room with only one clearly visible object, a small blue-green sphere about the size of a Christmas-tree ornament, then you can begin to grasp what the Earth looks like from space. I think that all of us subconsciously think that the Earth is flat ... Let me assure you that, rather than a massive giant, it should be thought of as the fragile Christmas-tree ball which we should handle with considerable care*". Unquote. Bill Anders said that his faith faltered when he heard a broadcast of 'O Holy Night' disintegrate as they went around the dark side of the Moon: *"If music could not survive an encounter with the Universe, was there much hope for religious doctrine? And if the church could not claim to be Universal, could it claim the authority to declare transcendent laws?"*

Gherman Titov flew aboard *Vostok 2* on August 06, 1961. He was then questioned if he had seen God while orbiting the Earth. He replied, *"Some say God is living there [in space]. I was looking around very attentively, but I did not see anyone. I did not detect either angels or gods ... I don't believe in God. I believe in man – his strength, his possibilities, his reason"*. Jim McDivitt was asked to respond to Titov's comments. He said, *"I did not see God looking Into my space cabin window as I did not see God looking into my car's windshield on Earth, but I could recognize His work in the stars as well as when walking among the flowers in a garden. If you can be with God on Earth you can be with God in space as well"*. In June 1965, Ed White made the first spacewalk by an American, and took some photographs of Earth. After twenty minutes outside his craft, he was told to return. *"This is great! I don't want to come back inside,"* he replied. White said he had felt ecstatic during the twenty-minute spacewalk, and that getting back inside the capsule was the saddest moment of his life. Rusty Schweickart had a metaphysical experience during an EVA outside *Apollo 9* in 1969 while he was staring at the Earth and contemplating it's place in the Universe. Back on Earth, he then practised Trancendental Meditation.

After his *Apollo 14* flight of February, 1971, Dr. Ed Mitchell wrote, *"You develop an instant global consciousness, a people orientation, an intense dissatisfaction with the state of the world, and a compulsion to do something about it. From out there on the Moon, international politics look so petty. You want to grab a politician by the scruff of the neck and drag him a quarter of a million miles out and say, "**Look at that, you son of a bitch**"*. Unquote. On his way back to Earth, Mitchell had a powerful savikalpa samadhi experience, and he claimed to have conducted private ESP experiments with his friends on Earth. He retired from the US Navy with the rank of Captain, and founded **Edgar D. Mitchell & Associates** of Monterey, California, a "commercial organization promoting ecologically-pure products and services designed to alleviate planetary problems". In 1973, he became founding chairman of the **Institute of Noetic Sciences** (IONS) in Palo Alto, California, for the purpose of consciousness research and other "related phenomena". Mitchell publicly said he was *"ninety percent sure that many of the thousands of unidentified flying objects, or UFOs, recorded since the 1940s, belong to visitors from other planets"*.

Dr. Edgar Mitchell's 1996 book, *The Way of the Explorer*, discusses his journey into mysticism and space. He said, *"We all know that UFOs are real: now the question is where they come from"*. A NASA spokesman stated: *"NASA does not track UFOs. NASA is not involved in any sort of cover-up about alien life on this planet or anywhere in the Universe"*. Mikhail Korniyenko flew to the ISS from April to September 2010, and from March 2015 to March 01, 2016. He said, *"The thing you miss there [in space] most of all is the Earth itself, I missed smells. I missed trees. I even dreamt of them. I even hallucinated. I thought I smelled a real fire and something being barbequed on it! I ended up putting pictures of trees on the walls to cheer up. You do miss the Earth there"*. Unquote.

During a pre-flight interview, Heidemarie Stefanyshyn-Piper described her philosophy about human exploration of space: *"To me, **exploring space is just a natural progression of where humans are going.** As we become more advanced, we have more technology to go farther. Thousands of years ago, people would go just beyond the next hill ... then it* *was "OK, let's go into the skies". ... So looking into the skies and looking at the stars and at the planets and thinking, what's out there? ... We're curious. We, as humans always want to know what's out there. ... We're still developing the means to go out there farther. ...* ***what else can we learn, and how can we go there?*** *Maybe we can learn something that we can bring back here and help solve some of the problems we have on Earth".* Unquote.

The last word could go to former Soviet leader Leonid Brezhnev who said, *"The Soviet and American spacemen will go up into outer space for the first major scientific experiment in the history of mankind. They know that from outer space our planet looks even more beautiful. It* ***is big enough for us to live peacefully on it, but it is too small to be threatened by nuclear war****".* In April 1972, a Treaty was signed to participate in a joint crewed space flight as a symbol of this *detente*. It was an **Agreement Concerning Cooperation in the Exploration and Use of Outer Space for Peaceful Purposes**.

## Supporting astronauts on long-duration space flights with VR.

Dr. Jay Buckey flew on mission STS-90 in 1998, a life sciences mission called Neurolab. Since 2001, he has been a Professor of Medicine at Dartmouth College's Geisel School of Medicine, where he is also director of the Space Medicine Innovations Laboratory. One of his areas of study has been testing and developing self-directed, autonomous, interactive, media-based behavioral health tools and virtual reality for use in isolated and confined environments. He and his colleages are evaluating the use of this technology to help people in confinement deal with three main issues: conflict resolution, stress management, and mood improvement. Virtual reality forests, beaches and cities are being tested at Australia's Antarctic stations to see whether they provide support for expeditions there. This research will also inform the development of programs supporting astronauts on long-duration space flights, like a mission to Mars. In a 2018 interview, Dr. Buckey explained that VR, by allowing people to virtually immerse themselves in natural settings, has restorative effects similar to the effects of exposure to nature itself.

## Some useful applications of space technologies

As of late 2021, twenty-eight Earth-observation satellites were in orbit around our planet, measuring moisture, air quality, ozone, gravity, magnetic field, smoke, aerosols, methane, carbon dioxide, and nitrogen oxides, as well as providing high-resolution imagery of Earth. 'Sentinel 6 Michael Freilich' is one of fifteen such satellites planned to launch through 2026. All enable us to measure the invisible threats to our planetary life support system, and provide visible evidence of how our planet is changing over time. It was a NASA satellite sending back data that revealed the growing hole in the ozone layer over the South Pole. This information was acted upon and resulted in the **Montreal Protocol**, which was the first international agreement (and arguably the most successful one) to address a global environmental problem.

It could be possible to put big data centers in space, and so take the emissions that result from refrigerating our data storage literally off the planet. In 2016, it was reported that the world's data centers used more than Britain's total electricity consumption. Using three percent of the global electricity supply, and accounting for about two percent of total greenhouse gas emissions, data centers have the same carbon footprint as the aviation industry.

# Bibliography


Aldrin, Edwin E., Jr., & W. Warga. *Return to Earth*. Random House, 1973.

Aldrin, Edwin E., Jr., & M. McConnell. *Men from Earth*. Bantam Books, 1989.

Aldrin, Edwin E., Jr., & K. Abraham. *Magnificent Desolation: The Long Journey Home from the Moon*. Three Rivers, 2009.

Aldrin, Edwin E., Jr. *No Dream is Too High: Life Lessons From a Man who Walked on the Moon*. National Geographic, 2016.

Armstrong, Neil, Michael Collins, and Edwin E. Aldrin, Jr. *First on the Moon*. Little, Brown, 1970.

Borman, Frank & R.J. Serling. *Countdown: An Autobiography*. William Morrow, 1988.

Buckbee, Ed, and W. Schirra. *The Real Space Cowboys*. Apogee Books, 2005. ISBN 1-894959-21-3.

Burgess, Colin. *Fallen Astronauts: Heroes Who Died reaching for the Moon*, 2003.

Burgess, Colin. *Moon Bound: Choosing and Preparing NASA's Lunar Astronauts*. Springer-Praxis, 2013.

Carpenter, Scott, et al. *We Seven: By the Astronauts Themselves*. Simon and Schuster, 1962.

Cernan, Eugene, and D. Davis. *The Last Man on the Moon*. St. Martin's, 1999.

Clark, Phillip. *The Soviet Manned Space Program*. New York: Orion Books, 1988.

Collins, Michael. *Carrying the Fire*. Adventure Library, 1998.

Collins, Michael. *Flying to the Moon: An Astronaut's Story*. RR Donnelley & Sons Co., 2010.

Donovan, James. *Shoot for The Moon ... Voyage of Apollo 11*. Little, Brown and Company, 2019.

Doran, Jamie, and P. Bizony. *Starman: The Truth Behind the Legend of Yuri Gagarin*. New York: Walter, 2011.

Evans, Michelle. *The X-15 Rocket Plane*. University of Nebraska Press, 2013.

French, Francis & C. Burgess. *In the Shadow of the Moon*. University of Nebraska Press, 2007.

Glenn, John. *John Glenn: A Memoir*. Ballantine, 1999.

Hansen, James R. *First Man. The Life of Neil A. Armstrong*. Simon and Schuster, 2005.

Hunt, Linda. *Secret Agenda: The United States Government , Nazi Scientists, and Project Paperclip, 1945 to 1990*. Thomas Dunne Books, 1991.

Koppel, Lily. *The Astronaut Wives Club*. Grand Central, 2013.

Kranz, Gene. *Failure is Not an Option*. Simon and Schuster, 2000.

Lovell, Jim, & J. Kluger. *Lost Moon: The Perilous Journey of Apollo 13*. Houghton Miflin, 1994.

Ley, Willy. *Rockets, Missiles, and Outer Space*. New York: Signet, 1969.

Mullane, Richard. *Riding Rockets: The Outrageous Tales of a Space Shuttle Astronaut*. 2006.

Murray, Charles, & C. Bly Cox. *Apollo: The Race to the Moon*. Simon and Schuster, 1989.

NASA Public Affairs. *The Kennedy Space Center Story*. NASA Public Affairs Office, 1991.

Neufeld, Michael J. *Von Braun: Dreamer of Space, Engineer of War*. New York: Knopf. 2007.

Newkirk, Dennis. *Almanac of Soviet Manned Space Flight*. Houston: Gulf, 1990.

Petrov, G.I., ed. *Conquest of Outer Space in the USSR, 1967-70*. NASA, 1973.

Piszkiewicz, Dennis. *The Nazi Rocketeers*. Westport, CT: Praeger, 1995.

Potter, Christopher. *The Earth Gazers: On Seeing Ourselves*. Pegasus Books, 2018.

Scott, David, Alexei Leonov, & C. Toomey. *Two sides of the Moon: Our Story of the Cold War Space Race*. Thomas Dunne Books, 2004.

Schefter, James. *The Race: The Uncensored Story of How America Beat Russia to the Moon*. New York: Doubleday, 1999.

Shepard, Alan, and Deke Slayton. *Moon Shot: The Inside Story of America's Race to the Moon*. Turner Publishing, 1994.

Slayton, Donald K., and M. Cassutt. *Deke! : US Manned Space from Mercury to the Shuttle*. Forge, 1994.

Stott, Nicole. *Back to Earth: What Life in Space Taught Me About Our Home Planet and Our Mission to Protect it*, 2021. ISBN 978-154-16750-49

Treadwell, Terry C. *Stepping Stones to the Stars: The Story of Manned Spaceflight*. History Press, 2010.

Tyree, Graeme K. *Beyond the Sky – A History of Space Exploration*, 2023. ISBN 9800473-69538-5

von Braun, Wernher. *Space Frontier*. Holt, Rinehart and Winston, 1971.

Walker, Stephen. *Beyond: The Astonishing Story of the First Human to Leave Our Planet and Journey into Space*, 2021. ISBN 978-0-06-297815-8.

Ward, Bob. *Dr. Space: The Life of Wernher von Braun*. Naval Institute Press, 2005.

Wendt, Guenter, and R. Still. *The Unbroken Chain*. Apogee Books, 2001.

Wolfe, Tom. *The Right Stuff*. Bantam Books, 1980.

Worden, Al, and F. French. *Falling to Earth*. Smithsonian Books, 2011.

Young, John, and J. Hansen. *Forever Young: A Life of Adventure in Air and Space*. University Press of Florida, 2012.

Thanks to all the good people at Wikipedia. You rock!

# Index of Names, A-Z

Made in the USA
Monee, IL
08 November 2026

68033802R00267